ANNUAL REVIEW OF PLANT PHYSIOLOGY AND PLANT MOLECULAR BIOLOGY

EDITORIAL COMMITTEE (1988)

ANNUAL REVIEW OF PLANT PHYSIOLOGY AND PLANT MOLECULAR BIOLOGY

VOLUME 39, 1988

WINSLOW R. BRIGGS, *Editor*

Carnegie Institution of Washington, Stanford, California

RUSSELL L. JONES, *Associate Editor*

University of California, Berkeley

VIRGINIA WALBOT, *Associate Editor*

Stanford University

ANNUAL REVIEWS INC.   4139 EL CAMINO WAY   PO BOX 10139   PALO ALTO, CALIFORNIA 94303–0897 USA

℞ ANNUAL REVIEWS INC.
Palo Alto, California, USA

International Standard Serial Number: 0066–4294
International Standard Book Number: 0–8243–0639–2
Library of Congress Catalog Card Number: A-51-1660

Typesetting by Kachina Typesetting Inc., Tempe, Arizona; John Olson, President Typesetting coordinator, Janis Hoffman

PRINTED AND BOUND IN THE UNITED STATES OF AMERICA

PREFACE

Those of you who have received the 1988 Prospectus for Annual Reviews, Inc. may already have noticed an important change in the *Annual Review of Plant Physiology:* It has a new name—the *Annual Review of Plant Physiology and Plant Molecular Biology.* When the members of the Editorial Committee held their annual meeting in March of 1987 to draw up a list of possible topics and authors for Volume 40, to appear in 1989, they also spent part of a day looking at how plant physiology itself and plant biology in general have been evolving. As all of you know, the changes in the past decade have been immense. Foci have shifted, new techniques have appeared both in plant biology laboratories and elsewhere, and the powerful tools of molecular biology have been reflected by a steady increase in the number of articles appearing in the *Annual Review of Plant Physiology* that have a strong molecular focus. After looking both at the field and at the volume itself as it has appeared in recent years the Committee voted unanimously to change the name of the volume. The original suggestion to the Board of Directors of Annual Reviews, Inc. was "Annual Review of Plant Physiology, Biochemistry, and Molecular Biology". The Board reasonably concluded that this title was a bit cumbersome, and approved the present one as a good compromise.

Since the Editorial Committee is self-perpetuating, with one member rotating off each year, we had an opportunity to make one other change: replace a retiring nonmolecular biologist with a molecular biologist. As the total Committee consists of eight people, we feel that we can still represent important areas of Plant Physiology that do not have molecular foci and assume that they will continue to be reviewed in timely fashion. We hope that readers of the *Annual Review of Plant Physiology and Plant Molecular Biology* will agree with the wisdom of this decision, and recognize the value of the slight shift in emphasis.

WINSLOW R. BRIGGS, EDITOR

Annual Review of Plant Physiology
and Plant Molecular Biology

CONTENTS

PREFATORY

Growth and Development of a Botanist, *Ralph O. Erickson*   1-22

MOLECULES AND METABOLISM   *Refs*   *pp*

Plant Growth-Promoting Brassinosteroids, *N. Bhushan
   Mandava*   23-52   29

Fatty Acid Metabolism, *John L. Harwood*   101-38   37

Genetic Analysis of Legume Nodule Initiation, *Barry
   G. Rolfe and Peter M. Gresshoff*   13 3 297-319   22

Metabolism and Physiology of Abscisic Acid, *Jan A. D.
   Zeevaart and Robert A. Creelman*   439-73   34

Electron Transport in Photosystems I and II, *L.-E.
   Andréasson and T. Vänngård*   379-411   32

Enzymatic Regulation of Photosynthetic CO_2 Fixation
   in C3 Plants, *Ian E. Woodrow and Joseph A. Berry*   533-94   61

ORGANELLES AND CELLS

Coated Vesicles, *D. G. Robinson and Hans Depta*   53-99   46

Immunocytochemical Localization of Macromolecules
   with the Electron Microscope, *Eliot M. Herman*   84   139-55   16

Cell Wall Proteins, *Gladys I. Cassab and Joseph
   E. Varner*   321-53   32

Chloroplast Development and Gene Expression, *John
   E. Mullet*   475-502   27

Plant Mitochondrial Genomes: Organization, Expression,
   and Variation, *Kathleen J. Newton*   503-32   29

TISSUES, ORGANS, AND WHOLE PLANTS

Photocontrol of Development in Algae, *M. J. Dring*   117   157-74   17

The Control of Floral Evocation and Morphogenesis,
   Georges Bernier   175-219   44

Physiological Interactions Between Symbionts in
   Vesicular-Arbuscular Mycorrhizal Plants, *Sally E.
   Smith and Vivienne Gianinazzi-Pearson*   179   221-44   23

Water Transport in and to Roots, *J. B. Passioura*   *8 7* 245-65 (20).

The Control of Leaf Expansion, *J. E. Dale*   *188* 267-95 28

Metabolism and Compartmentation of Imported
  Sugars in Sink Organs in Relation to Sink
  Strength, *Lim C. Ho*   *146* 355-78 (23)

POPULATION AND ENVIRONMENT

The Chromosomal Basis of Somaclonal Variation,
  Michael Lee and Ronald L. Phillips   *143* 413-37 (24)

INDEXES

Author Index   595-620

Subject Index   621-29

Cumulative Index of Contributing Authors, Volumes 31–39   630-31

Cumulative Index of Chapter Titles, Volumes 31–39   632-37

ARTICLES OF INTEREST FROM OTHER *ANNUAL REVIEWS*

From the *Annual Review of Genetics, Volume 21 (1987)*

Arabidopsis thaliana, E. M. Meyerowitz
RNA 3' End Formation in the Control of Gene Expression, D. I. Freidman, M. J. Imperiale, and S. Adhya

From the *Annual Review of Phytopathology, Volume 25 (1987)*

Physiological Plant Pathology Comes of Age, R. K. S. Wood
The Impact of Molecular Genetics on Plant Pathology, A. Kerr
Salt Tolerance and Crop Production, D. Pasternak
Molecular Markers for Genetic Analysis of Phytopathogenic Fungi, R. W. Michelmore and S. H. Hulbert
Rhizobium—The Refined Parasite of Legumes, M. A. Djordjevic, D. W. Gabriel, and B. G. Rolfe
Fungal Endophytes of Grasses, M. R. Siegel, G. C. M. Latch, and M. C. Johnson

From the *Annual Review of Biophysics and Biophysical Chemistry, Volume 17 (1988)*

Fourier Transform Infrared Techniques for Probing Membrane Protein Structure, M. S. Braiman and K. J. Rothschild
Pulsed-Field Gel Electrophoresis of Very Large DNA Molecules, C. R. Cantor, C. L. Smith, and M. K. Mathew
Computer Methods for Analyzing Sequence Recognition of Nucleic Acids, G. D. Stormo
Cellular Mechanics as an Indicator of Cytoskeletal Structure and Function, E. L. Elson
RNA Structure Prediction, D. H. Turner, N. Sugimoto, and S. M. Freier
The Physical Basis for Induction of Protein-Reactive Antipeptide Antibodies, H. J. Dyson, R. A. Lerner, and P. E. Wright
Genetic Studies of Protein Stability and Mechanisms of Folding, D. P. Goldenberg
A Critical Evaluation of Methods for Prediction of Protein Secondary Structures, G. E. Schulz
Secondary Structure of Proteins Through Circular Dichroism Spectroscopy, W. C. Johnson, Jr.
Proton Circuits in Biological Energy Interconversions, R. J. P. Williams
The Submicroscopic Properties of Cytoplasm as a Determinant of Cellular Function, K. Luby-Phelps, F. Lanni, and D. L. Taylor
The Forces That Move Chromosomes in Mitosis, R. B. Nicklas

From the *Annual Review of Cell Biology, Volume 3 (1987)*

Ubiquitin-Mediated Pathways for Intracellular Proteolysis, *M. Rechsteiner*
Oligosaccharide Signalling in Plants, *C. A. Ryan*
Intracellular Transport Using Microtubule-Based Motors, *R. D. Vale*

From the *Annual Review of Microbiology, Volume 41 (1987)*

Genetics of Azotobacters: Applications to Nitrogen Fixation and Related Aspects of
Metabolism, *C. Kennedy and A. Toukdarian*
Enzymatic "Combustion": The Microbial Degradation of Lignin, *T. K. Kirk and R.
L. Farrell*
Genetic Research with Photosynthetic Bacteria, *P. A. Scolnik and B. L. Marrs*

From the *Annual Review of Biochemistry, Volume 57 (1988)*

Amino Acid Biosynthesis Inhibitors as Herbicides, *G. M. Kishore, D. M. Shah*
The Biology and Enzymology of Eukaryotic Protein Acylation, *D. A. Towler, J. I.
Gordon, S. P. Adams, and L. Glaser*
Cell-Surface Anchoring of Proteins Via Glycosyl-Phosphatidylinositol Structures,
M. A. J. Ferguson and A. F. Williams

For the convenience of readers, a detachable order form/envelope is bound into the back of this volume.

Ralph O. Erickson

Ann. Rev. Plant Physiol. Plant Mol. Biol. 1988. 39:1–22

GROWTH AND DEVELOPMENT OF A BOTANIST

Ralph O. Erickson

Department of Biology, University of Pennsylvania, Philadelphia, Pennsylvania 19104

FOREWORD

I suppose that my earliest interest in plants stems from my childhood in northern Minnesota and Michigan. Winters there are severe but the long summer days among trees and lakes are delightful. Our family was close to nature in many ways. We often spent days picking wild blueberries and raspberries which my mother canned for desserts through the year. Pin cherries and wild strawberries made excellent jelly. We used our pocket knives to make whistles of poplar twigs, and selected symmetrical maple crotches for sling shots. Later, when my commitment to botany was firm, I was drawn to studies of plant growth and development from a variety of educational and research experiences, but with the strong influence of Edgar Anderson and David R. Goddard. I have had little conventional training in plant physiology, and it may be interesting to try to trace the development of my teaching and research interests from such diverse fields as plant taxonomy, plant anatomy, cytology and cytogenetics, evolutionary theory and ecology, and an interest in statistics and numerical analysis. I can probably be accused of dilettantism.

Discussing this theme in a personal vein, with a bit of apprehension, has resulted in a sort of selective autobiography. One author (24) has written that "autobiography is a most peculiar genre or form....[It] presuppose[s] a particular kind of arrogance, a conviction that one's life is in some serious way exemplary...." However, I may hope that frequent use of the first person pronoun will not be taken as conceit, but as the candor that I intend.

0066-4294/88/0601-0001$02.00

ROOTS

My grandparents were emigrants from Sweden in the 1870s, a decade or so after the Sioux massacre of settlers in the Minnesota River valley in 1862. My mother's parents were from Småland and Västergötland and homesteaded near Bernadotte in southern Minnesota, where they raised ten children. My father's parents came from the Åland Islands and settled near St. Hilaire in northern Minnesota, also as homesteaders. My father, Charles, the second of four children, left the farm at age 19 or 20, against his father's wishes, to attend high school, went on to Gustavus Adolphus College, and graduated in 1913. My mother, Stella Sjostrom, and he were married in 1913, and after two years in Duluth, Minnesota, moved to Rock Island, Illinois, where my father attended Augustana Seminary, graduating in 1918 with a B. D. degree. He was then a Lutheran pastor in Clearbrook, in northern Minnesota. I was born in Duluth, 27 October 1914, the first of six children. When I was five, my mother suffered a severe "nervous breakdown," and the family, then of four children, were dispersed. I was sent to my grandparents Sjostrom, who had retired to St. Peter, Minnesota, and there I attended kindergarten. I have virtually no memory of these years, but I am told that they were unhappy.

My father left the church in Clearbrook for nearby Leonard, a town of 75 people about 15 miles from the Red Lake Indian Reservation (Chippewa). He cleared land beside a small lake and single-handedly built a four-room house for the family, which is still lived in. He was pastor of the church in Leonard, and principal of the three-room school, where I advanced through the first six grades in four years. In sixth grade I heard the seventh and eighth grade recitations, since they were in the same room. At the end of the year I was given, and passed, the State Board examinations for graduation from eighth grade and was then entitled to enter high school when I was not quite eleven.

SI QUAERIS PENINSULAM AMOENAM...

My father was called to a church in Iron River in the upper peninsula of Michigan, which in 1925 was a fairly prosperous iron mining town. The school authorities were dubious about my starting high school, and instead I entered eighth grade. Some of my classmates were the children of the immigrant Italian, Polish, and Finnish miners. It was a rough town. I suppose that my school experience was unusual, since I was two or three years younger than my classmates, and in one's early teens this is important. In high school it meant that I could not join in sports such as football, basketball, or hockey, nor could I join a gang. Also the puritanical character of midwestern Lutheranism at that time meant that, as preacher's kids, we were not allowed to do such "sinful" things as see movies or go to dances.

After classes, piano practice, band rehearsal, and delivering newspapers, I had time to read, and I read widely. In addition to pulps, Jules Verne, and H. G. Wells, I tackled the Bible, and such books as Dostoyevski's *The Brothers Karamazov*. My father's library included, in addition to homiletics and concordances, "Dr. Eliot's Five-foot Shelf of Books" (*The Harvard Classics*), and I believe that I read the entire collection, with or without comprehension. I recall reading Darwin's *Origin of Species* and *Voyage of the Beagle*. I also spent much time with the *Encyclopedia Americana*. At least I learned some words. On entering college, I scored above the median for college graduates on a standardized vocabulary test.

During the summers, we spent much of our time at nearby Fortune Lake where my father organized and built a summer Bible Camp. Having grown up on a farm, he knew carpentry. He also had a woodworking shop at home, which I could sometimes use under strict supervision. At the camp he, my brother and I, and occasionally others cleared brush, built roads and paths, erected buildings, built tables and benches, even rowboats, painted, and installed electric wiring—whatever was required. After a time, I realized that I could cut rafters or hang a door as accurately as a professional carpenter who worked at the camp for a while. We were not paid and so could spend a part of our time at tennis, swimming, diving, and boating. We longed for a canoe. My fond memories of Fortune Lake include recollections of loons, ducks, herons, woodchucks, and deer. I did not care to fish, but heard stories about the bass, trout, and pickerel. Fortune Lake was one of a chain of five or six, and occasionally I would row the whole length to check the beaver lodges or perhaps look for lady's slippers. I remember some irritation that I could not easily learn the trees and other plants. Names people used were contradictory, and I suppose I did not realize then that I needed a manual.

My father also taught us photography, since he had earned a part of his college expenses by photographing barns, livestock, and families and selling the picture postcards he made to the farmers.

GUSTAVUS ADOLPHUS COLLEGE

When the stock market crashed in 1929, the mines closed almost immediately. With widespread unemployment, the church was unable to pay my father's full salary and, when I graduated from high school in 1930, college seemed out of the question. During the next year I had a part-time job and took a course or two at the high school. The following year, my father announced that I could go to Gustavus Adolphus College, St. Peter, Minnesota. As I recall, tuition was about $75 per semester and I lived with relatives for two years, then in a dormitory as a proctor. I worked for meal tickets as an attendant in the library (more opportunity to read), and as a reader for a blind classmate.

At college I was still two years younger than my classmates and felt exceedingly shy. I was regularly permitted to take one or two courses beyond the required four per term. This entailed some scattering of effort, but I graduated magna cum laude. My major was biology, but I also took most of the math courses offered and not quite enough chemistry for a second major. In addition to zoology, comparative anatomy, human physiology, etc, there was one botany course, taught by a zoologist. One of the assignments was to turn in dried specimens of 10 plants. I did many more. My roommate for two years was a born naturalist, who kept plants and tropical fish in our room. The two of us spent many extra hours in the biology lab and in the field, collected material for use in the biology course, frog eggs in season, and many other things. He became a high school biology teacher in St. Peter and we continued our joint biology ventures for some time.

Music was important at Gustavus; the a capella choir was nearly as important as the football team. I took music seriously. I had voice and piano lessons one year (even played a recital), played clarinet in the band and bassoon in the orchestra, and sang in the choir. I also wanted some music theory. In high school the music teacher, remarkably, gave a course in harmony to a few of us, and in college I was the only person who wanted the course in advanced harmony and counterpoint. One term Dr. Nelson agreed to meet with me once a week at the piano, to play and correct exercises I had written. This was one of my most demanding and satisfying courses. Shortly before the spring choir tour my senior year, Dr. Nelson was ill for about three weeks, and it fell to me to direct the choir in rehearsals and the first concert, which included the Bach motet, *Singet dem Herrn*. This was nearly disastrous, since I felt I had to devote my full time to studying the music, instead of attending classes. In later years my performing skills have atrophied but music continues to be an important part of our family life.

At graduation from college, only two of my classmates had job prospects. I was qualified for certification as a high school teacher, but there were no opportunities. I spent the summer at Fortune Lake wondering what I might do. Late in August a letter arrived from the president of Gustavus offering me the position as assistant (really instructor) in biology, which had just become vacant. The salary was very low even by 1935 standards, but I immediately hitchhiked to St. Peter. The biology faculty consisted of Dr. J. A. Elson and me. I spent four years teaching there, in sole charge of the elementary biology labs and the botany course, 24 contact hours per week. In my fourth year I introduced a course in genetics, using *Drosophila* and segregating ears of corn in the lab.

Funds for lab material were limited; for instance, gophers caught in nearby fields were dissected in the zoology course, instead of specimens bought from a supply house. The microscope slide collection was inadequate, but the department had a microtome and with an improvised paraffin oven I prepared slides of stem, root, and leaf sections, shoot apices, slides for animal histology, even parasitic flat worms. I converted an ice-box into an incubator, prepared whole mounts of blastoderms and introduced lab exercises on chick embryology. I also made many 2 × 2 inch lantern slides. At times student volunteers helped with this work. I had read Cooper's article on embryo sac development in lilies (7) so I bought some Easter lily plants with flower buds of various ages at a local greenhouse and prepared sections of anthers and ovaries. Since I was into mass production, I could select choice slides for myself of the crucial stages of pollen and embryo sac development, which have continued to be useful in teaching for many years, and I recently learned that some of my slides are still in use at Gustavus.

Something more should be said about Gustavus. The announced mission of the college was, and is, training for Christian leadership. A daily chapel service was compulsory, there were evening prayer meetings, etc. In my third year of teaching, as I recall, a decision was made to ordain the Gustavus professors into the Lutheran ministry. I took little part in the religious life of the college because my interests were elsewhere. The faculty at Gustavus were dedicated teachers, but it occurred to me later that I knew of none who were engaged in scientific research or any other scholarly work. I graduated with only a slight understanding of academe in the wider sense.

SUMMER SCHOOL

After my first year of teaching, I attended a summer session at the Douglas Lake Biological Station of the University of Michigan, taking systematic botany and plant anatomy. The former course was devoted to the local flora and consisted of all-day field trips in which we filled our vasculums, then sat down in some pleasant place to key out our specimens with Gray's *Manual*. It pretty well satisfied my desire to be able to identify plants. C. D. LaRue's plant anatomy was less cut and dried. LaRue was a challenging and entertaining lecturer and he taught us the paraffin technique. I recall some dissatisfaction with the static descriptions in Eames & McDaniels, our text. When I asked how fast the onion root grew and how rapidly cells in the meristem divided, I found no answers. To say that I could not imagine how the root tip could grow so as always to look the same in sections, is perhaps invoking too much hindsight. It was interesting that George Avery shared LaRue's laboratory that summer, working up his sections of *Aesculus* shoots to explore the possible role of auxin in the initiation of cambial activity in the spring (6). This was my first view of research in progress.

The following two summers I had courses at the Lake Itasca Forestry and Biological Station and at the Minneapolis campus of the University of Minnesota. Among them was a field course on the ecology of Itasca Park, a course in genetics, an elementary plant physiology and a seminar course concerned with the structure of chlorophyll and the physiology of photosynthesis, for which I was not prepared. The structure of grana was not then known. This is the extent of my formal training in ecology, genetics, and plant physiology.

SHAW'S GARDEN

Summer experiences and extracurricular fooling around in the laboratory at Gustavus reinforced my desire for graduate study in biology. I had made unsuccessful inquiries about the possibility of doing full-time graduate work in the botany department at Minnesota, and during my fourth year of teaching, I resolved to make a more serious effort to get into graduate school, realizing by that time that my chances of being accepted were slim. My academic record at Gustavus was good, but I was sure that my grades and my recommendations would be discounted. With the advice of people in botany at Minnesota I applied to 12 schools, mainly Ivy League and State Universities, at which work in plant cytology was going on. I got rejections, or no word, from all but one.

Edgar Anderson wrote that he was impressed with my application, that there were no opportunities for support at the time, but that he would "by hook or by crook" see that I could come to Washington University. It is still a mystery to me what merit he could see in my application. Shortly before the start of classes in September 1939, I was awarded a University Fellowship. I hitchhiked to St. Louis, found a boarding house near the Missouri Botanical Garden, and became a graduate student in the Henry Shaw School of Botany. When I had paid tuition, room, and board, I had ten dollars per month. I took Jesse Greenman's course on the flowering plants, which was quite another thing than a local flora. Greenman had studied at Berlin with Adolph Engler and his course was a *grosses Praktikum*, intended to acquaint us with virtually all the plant families, through lectures, and dissection and drawing of boiled-up specimens of dried flowers, fruits, etc., filched from herbarium sheets. I enjoyed it and still value it greatly. During my three years at the Garden, I made a point of walking through the conservatories at least once a week. At the main campus of the University, a course in physical chemistry fulfilled my old intention to major in chemistry, and I had a cytology course, a seminar course in animal embryology, and others.

I chose to do a taxonomic problem for a master's degree, which was narrowed down to a revision of the *Viorna* section of *Clematis*, under Greenman's direction. In retrospect, I can perhaps see Edgar Anderson's hand in this choice. (By a curious coincidence, I was given a cordial welcome to the Academy of Natural Sciences in Philadelphia and the botany department of the University of Pennsylvania in 1942, when I made a bus trip to Eastern herbaria to study *Clematis* specimens.) My thesis was published (9) as my thickest paper to date, and it earned me several pages of testy criticism from M. L. Fernald in *Rhodora*. However, I am cited in Fernald's *Manual* as author of one variety of *Clematis*, so I can claim to have had a bit of taxonomic competence.

The most valuable part of my experience in St. Louis was association with Edgar Anderson. It was strenuous. From the day I arrived he subjected me to a continual barrage of discussion, wisecracks, pithy anecdotes about other biologists, genetic questions intended to stump me, a continuing contest of wits. I often went with him on field trips, and *Faltboot* trips on Ozark rivers; taught his wife, Dorothy, and him to play recorders; accepted many invitations to the barn at the Gray Summit Arboretum, which he had converted to a summer place; and listened to a certain amount of advice on how I should conduct my life. I also learned a great deal about species of *Iris*, *Tradescantia*, *Acer*, and other genera: their geographical distribution, morphological variation, cytology and, introgressive hybridization. Anderson paid me to help with his study of F2 segregation in a species cross between *Nicotiana alata* and *N. langsdorfii*, photographing and measuring flowers and leaves, and making preliminary extractions of tissue for auxin analysis in F. W. Went's laboratory. Anderson had published (2) on the hindrance to recombination imposed by linkage, and in these studies he wished to document a further hindrance apparently imposed by developmental constraints. I do not believe that this work was published. Anderson was at that time beginning his survey of the indigenous varieties of maize. One summer he paid me to plant, hoe, and self-pollinate a number of strains of maize from the Hopi and other southwest Indians, from Mexico and Guatemala. The latter grew to about 20 feet and pollinating them required a ladder when they tasseled in September, contrasting with an 18-inch Hopi strain. I learned a great deal about the diversity of *Zea*.

Anderson gave a great deal of thought to methods of analyzing and representing variation in natural populations. Although he was far from naive in statistics and mathematics, he preferred graphical methods, such as pictorialized scatter diagrams (3), rather than formal statistical analyses of his data. I was impressed with the outcome of his association with R. A. Fisher, whose discriminant function (23) was worked out using Anderson's data on three species of *Iris* as an example, and has become a useful technique in multivariate analysis. One might argue that Anderson's ideographs (1, Plate 23) visually demonstrate the relationships among the three species as well as Fisher's Figure 1 does. I was also taken with his graphs of internode length vs number (4) and made similar plots of growing *Clematis* vines. I joked that he had plotted the first derivative of the plants.

My doctoral problem grew out of these discussions, or perhaps it was tactfully assigned to me. The idea of making a thorough field study of the glade leather leaf (*C. fremontii* var. *riehlii*) was roughly formulated in the spring of my first year. I was able to buy a used Model A Ford and a sleeping bag that summer (from hoeing corn), and spent a large part of my time on the Ozark glades (dolomitic barrens with an attractive endemic flora), during all seasons for the next two years, and a lesser part for another two years. At Anderson's urging I took a microscope to the glades and made squash preparations of anthers to look for irregularities in meiosis, which might indicate introgressive hybridization with another *Clematis*. I found none and went on to a study of the ecology, reproduction, and natural variation of the population. This constituted a major in botany and a minor in trespassing. I had found that walking up to the door of a farmhouse to ask the farmer's wife for permission to look at a glade tended to frighten her. A farmer once found some plants with bags over them on his glade and called the State Police, thinking that someone was growing hashish The Police brought one of the bagged plants to the Gray Summit Arboretum and I was called in for an explanation. The specimen, which I had bagged to find out if it would self-pollinate, was then annotated and deposited in the herbarium at Shaw's Garden.

I wrote up my work on the glades as a dissertation and received my Ph.D. degree in 1944. This was well enough regarded to be reprinted (10). In later years, I have had intentions of continuing field work. I made a few collections of two species of *Uvularia* in western New York state, with the idea of studying their relationship, but did nothing with them. Later there were several trips with Robert B. Platt to shale barrens of Virginia and West Virginia, where I learned to know the *Clematis* species, closely related to *C. fremontii*, which are restricted to the barrens. The *Clematis* leaves that I collected have served as samples for the analysis of variance by many biometry classes, but nothing else has come of these efforts. On several occasions I have taken a break from other things and driven to Missouri to revisit the glade *Clematis*, often with one or two students.

WESTERN CARTRIDGE COMPANY

My going to St. Louis in September 1939 nearly coincided with the German invasion of Poland. A year later the Selective Service Act was passed. I was classified 1-A, appealed, and was granted deferment as a student. In the spring of 1942 student deferments were abolished, and through Anderson's acquaintance with the research director at a defense plant, who was an amateur botanist with a master's degree in botany, I was offered a job as a chemical microscopist. Western Cartridge Co., East Alton, Illinois (later a part of Olin Industries) manufactured small arms ammunition, including smokeless powder. The lab to which I was assigned was mainly concerned with problems of variability of the powder charge in cartridges, and with a polarizing microscope and the guidance of Chamot's & Mason's *Chemical Microscopy*, I was able to learn something about the composition of powder grains (nitrocellulose, nitroglycerin and a plasticizer) from thin sections. I boned up on the processing of wood pulp and on the chemistry of cellulose, nitrocellulose, and polymers generally, all with a crowd of chemical engineers. It seemed to a friend and me that some of the variability of the ballistic tests might arise in the blending of various batches of powder, and we made a statistical test. We had a barrel of marked powder poured through the blending tower with many unmarked barrels, then counted marked grains in samples of the output. Our statistics showed that the blending was very poor, but so far as I know nothing was done about it. After a year I was put in charge of a laboratory to study dry cells (flashlight batteries). So now the topic was electrochemistry, specifically of the Leclanché cell. With a punch press and other equipment, we did not succeed in a year's time in making experimental cells that equalled production cells in their service life. At least I learned some more chemistry, a bit of chemical engineering, and broadened my understanding of microscopy.

ROCHESTER

A position at the University of Rochester became available in the spring of 1944, and Anderson suggested my name to David R. Goddard. He planned to be in Terre Haute, Indiana, on a consulting job and asked if I could meet him there. I played hooky from my job, had the first of countless exhilarating discussions with Goddard and in effect was promised the job then. I now needed approval from the War Manpower Board to change jobs, but that turned out to be a breeze, since I was to be an instructor, teaching in the Navy V-12 program at Rochester. My superiors at Western Cartridge offered me a handsome raise and painted a rosy picture of my future there. When I pointed out that I expected a much lower salary at Rochester, that discussion ended, as did my career in industry.

The biology group at Rochester was largely assembled by Benjamin Willier several years before I came. It was a congenial and exciting group. Curt Stern was a great geneticist who had the collaboration of Ernst Caspari, Warren Spencer, and others on classified work for the Manhattan Project. There were many luncheon discussions of genetics and many other topics, as well as a journal club and "Festschrift." Sherman Bishop, A. W. Küchler, and I organized a memorable seminar on biogeography. I am indebted to Donald R.

Charles for patiently guiding me through an analysis of covariance and the solution of a discriminant function, as well as introducing me to *The Calculus of Observations* (39). I sat in on Dave Goddard's course in plant physiology, and for the first time heard critical lectures on thermodynamics and metabolic cellular physiology. He and I had many free-wheeling discussions of science. I particularly remember that we both felt that the nucleic acids richly deserved study, a few years before Watson & Crick.

I had told Goddard that I wanted to do research in experimental cytology, having only a vague idea of what that meant. Having been fascinated by Darlington's (8) speculations about the evolution of sexuality, and the contrast between mitosis and meiosis, I thought that it would be interesting to try to do something that might illuminate the difference between the latter two processes. This narrowed down to a plan to study the respiration of microsporocytes, microspores, and pollen. Thinking that it would be good to select a plant that had large anthers and was easy to manage, I made a little survey, and concluded that it would be hard to beat the Easter lily, *Lilium longiflorum*. I ordered some bulbs to set out in the greenhouse on a staggered schedule, and Goddard taught me the ins and outs of using the Fenn microrespirometer. This resulted in papers on the respiration of anthers (11), on growth of the flower bud and its parts using log bud length as a developmental index (12), and later, on nucleic acids in the anthers (33). Lilies have now been used by other workers, notably H. Stern (37), in a number of important studies of biosynthetic aspects of microsporogenesis, and Moens (32) for a study of the synaptinemal complex in meiosis.

My wife, Elinor Borgstedt, and I were married after my first year in Rochester. We have two daughters and two granddaughters. Elinor had been a secretary to the research director at Western Cartridge for a year, and was a music student at the University of Illinois at Urbana. She transferred to the Eastman School of Music at Rochester and earned her B. Mus. degree there. She has had a rewarding career as an organist and choir director, and now devotes her efforts to the piano. When long-playing records were announced in 1947, we bought six from the very first list, assembled an amplifier kit from war surplus parts, connected a turntable and speaker, and were both overjoyed with the music. We have been audio fans ever since.

PENN

During my second year at Rochester, Goddard accepted a professorship at the University of Pennsylvania and was succeeded at Rochester by F. C. Steward. With two colleagues at Penn, Goddard obtained research grants from the National Cancer Institute and the American Cancer Society for studies of cell division in plants. After three years at Rochester, I had been promoted to an assistant professorship, but as it turned out I did not serve in that capacity, since Goddard offered me a position as a research associate in his program. After some soul searching and negotiation, we moved to Philadelphia. Maurice Ogur, a biochemist with particular interests in nucleotides and nucleic acids, joined the group as a research associate, as well as three excellent technicians, Kathie Sax, Gloria Rosen, and Connie Holden.

Goddard and I had reasoned that root meristems, as well as anthers, were favorable for studying cell division, and we began experiments with the primary roots of corn seedlings. They were grown in the presence of certain alkaloids, which were candidates for cancer therapeutic agents. The control and poisoned roots were fixed, sectioned, stained, and examined for mitotic abnormalities. After many weeks of study of the slides, I was frustrated at trying to imagine what had happened, for instance, to nuclei that had surely been in metaphase and after treatment looked something like interphase nuclei. I proposed that we abandon this traditional approach and try first to learn something about how roots and their cells grow. I knew in some detail about the work which Richard H. Goodwin, my predecessor at Rochester, and William Stepka had done in describing the growth pattern of *Phleum roots* (25). (In a footnote they acknowledge the assistance of Don Charles.) Their microscopic method of studying the minute grass roots was not feasible, since we had chosen to work with the much larger roots of *Zea* in anticipation of getting biochemical and metabolic data. At a meeting of our group, I suggested the kinds of data we should try to get and what I had in mind for a growth analysis. I put together a special camera rig to automatically record the displacement of marks placed on the roots, we worked out methods of counting cells, etc. I also supposed that I could handle the math involved in coordinating and interpreting the data. This is the rashest statement I have ever made. It took about 18 months of study to arrive at the analysis presented in the first papers (18, 21, 22), and it is apparent from recent publications by several authors that much remained to be done. In addition to the root work, we made the study of nucleic acids in *Lilium* anthers referred to above (33). The collaboration with Ogur was a great education in biochemical principles and methods of analysis.

After two years Ogur left Penn and a year later I was appointed associate professor. I had participated with John Preer in a biometry course (mostly statistics) for two years, but I now had additional teaching and some administrative duties. Goddard had great talents as an administrator as well as a teacher and researcher, and it was by his efforts that the departments of botany, microbiology, and zoology were merged in 1954 into a greatly strengthened division of biology. In 1961 he became provost of the university and served Penn eminently throughout the turbulent 1960s. Unfortunately, however, the analytical (chemical) work on roots was published only in summary form (18). The work that had been done on respiratory metabolism of root segments would have yielded estimates of the energetic requirement for growth, but this could not be analyzed or interpreted without Goddard's participation.

With the assistance of Roman Maksymowych, the root studies continued at a reduced scale. We undertook to explore the effects of inhibitors of root growth and cell division, based on the growth analysis that had been worked out on a descriptive basis. However it seemed too laborious to work out elemental growth rates, so we photographed the growing roots with an automatic camera, and on the basis of hourly readings of total root length and measurements of mature cell lengths in sections of roots fixed at the end of each run, were able to calculate average rates of elongation and of cell production. A variety of substances were tried and three distinctive patterns of inhibition were found. Metabolic inhibitors such as cyanide and azide inhibit elongation in a manner suggestive of enzyme inhibition, with no effect on the rate of cell production. Substituted nitrogen bases are potent inhibitors of cell division with no effect on elongation for many hours. Several alkaloids depress both processes. Unfortunately, these results have not been adequately analyzed nor published, since there are certain points that I have not fully understood until recently. Another study was based on data on growth and cell division in *Phleum* roots, which Goodwin kindly provided. Reanalysis of the data for cell division rates showed that all the cells in the apical part of the meristem divide, whereas in the basal part, the proportion of cells that divide falls progressively to zero (13).

The studies of lily anthers and of the corn root were motivated by the idea of doing "experimental cytology." In both cases, however, my interest shifted from the cells per se, to questions of how the organs, the flower and anther, or the root, grow. I was impressed in both cases by the great regularity and coordination of cellular processes into a predictable morphogenesis. If there is a question of whether growth should be modeled as a stochastic process or a deterministic one, I would certainly argue for the latter. While there seemed to be sufficient opportunities to devote a career of research to either anthers or roots, I began to wonder whether the same regularity would be found in other developing systems, such as shoot apical meristems, and set out to obtain growth data.

Zygmunt Hejnowicz joined my lab in 1963–64 and collaborated in root studies. He worked out a method of recording growth using fluorescent marks illuminated with near-UV light, with the idea of applying it to a study of gravitropic curvature of roots, and made a study of the inhibition of root growth by auxin (27). I owe a great deal to him for many animated discussions of growth problems, particularly their mathematical and physical aspects, and we have had the pleasure of visiting him in Poland.

Hejnowicz suggested that we use celery, *Apium graveolens*, for studies of the shoot apex, since it has a large and relatively flat apex. After dissecting out a few young leaves from an otherwise intact potted plant it was possible to focus an Ultropak objective on the apex, and with an automatic camera to obtain photographs with cellular detail. After a great deal of effort we gave this up because of inadequacies in the Ultropak image, based as it is on shifting light reflections from the cell surfaces. We had also attempted to photograph shoot apices of *Xanthium*, chosen because we could assure vegetative growth by keeping these short-day plants on a non-inductive light schedule. This photography was similarly unsuccessful and I began to think of the possibility of an indirect approach, like using log bud length as an index of the development of lily flower buds. Recalling the plots of internode lengths of growing *Clematis* plants which I had made long before, I began daily measurements of internodes of *Xanthium* plants, and saw only that they were quite variable in length and apparently erratic in their growth. There had been some discussion of leaf growth with graduate students, and this led to daily measurements of leaf length in *Xanthium*. One day during a discussion with Mike Michelini of a semilog plot of this data, I found myself writing the formula for the plastochron index on the blackboard, as if by an inspiration (20). The plastochron index has now been used by many authors, including ourselves, in a variety of ways.

Richards & Kavanagh (34) published their analysis of Avery's (5) data on the growth of a tobacco leaf, marked with a grid of points, in 1943. As it happened, I read their paper when that issue of *The American Naturalist* appeared in the current literature box at Western Cartridge. I did not understand it fully then but was convinced that it was important, since it dealt with differentials of spatial dimensions as well as time. Later at Penn, I felt that their analysis could be repeated more satisfactorily with *Xanthium* leaves, and eventually I was able to complete it (14). This two-dimensional analysis of elemental growth rates was then applied to younger *Xanthium* leaves, to a fern prothallus (Mae Chen's Master's thesis), and to the thallus of *Marchantia*. The intention also was to use this analysis for data on shoot apical meristems, but as I indicated, this did not work out. This kind of analysis has scarcely been followed up by other workers but it has been important in our thoughts about the nature of plant growth, and growth analysis, including root growth analysis.

Of the many courses I have taught at Penn, the one I most enjoyed was developmental plant morphology, given occasionally to a class of graduate and undergraduate students, sometimes with the collaboration of a colleague or a visiting botanist. It was a mix of talks by members of the class and myself, with small projects in the lab. The first class, in fall 1952, was a remarkable group, some of whom are now professional botanists. Paul B.

Green took this course as an undergraduate and, in the next term, he enrolled for an independent study course, made a study of the growth of the *Nitella* internode cell, and published it. He went on to graduate school and, a few years later, returned to Penn as a member of the biology department. During his years at Penn, Green was a stimulating colleague. While our formal collaboration (in print) has been minimal, we shared discussions continuously. I trust that I was helpful to Green in some technical matters and his viewpoint has certainly had, and continues to have, a great influence on my thinking.

Wendy K. Silk came to Penn as a graduate student in 1969, with a degree in biomathematics, and for personal reasons stayed only a year. When she had completed her graduate work at the University of California, Berkeley, she came to my lab as a postdoctoral fellow. She had made a compartmental analysis of the uptake and release of gibberellic acid by excised hypocotyl segments of lettuce, *Lactuca*, and wished to analyze growth of the segments in detail. This analysis did not work out well and we began to talk about the curvature of the hypocotyl hook. She set up a time-lapse camera to photograph intact lettuce seedlings and made a thorough analysis of the kinematics of hook maintenance in the growing hypocotyl, which required rather deep study of continuum mechanics. I then suggested that there should be a general article on the kinematics of plant growth (36). In my view, this work has provided a sort of capstone to the growth studies, answering questions that I had only dimly perceived. It has been followed by a number of theoretical papers on plant growth by other authors.

PHYLLOTAXIS AND OTHER THINGS

A topic that I found confusing at first was phyllotaxis, as discussed by taxonomists, morphologists and plant anatomists, and at one point I decided to look into the copious classical literature. I felt that it must somehow be important to understanding plant morphogenesis, implying as it does a very close regulation of the process of leaf initiation at the shoot apex. Aristid Lindenmayer was then a member of the Penn botany department, and in many luncheon discussions, he was very helpful in my early puzzlement about phyllotaxis. (It may be that these discussions played some part in Lindenmayer's later formulation of the cellular automata known as L-systems.) I read Church's 1904 monograph, failing to understand his emphasis on orthogonal parastichies as implying some mysterious physical analogy. I also studied F. J. Richards's papers and quite a few others. Van Iterson's thesis of 1907 was far more difficult since it is in German and bristles with equations, but it impressed me as a much more comprehensive work than Church's. After some time, I came to feel that the supposed conflict between Church's and van Iterson's models was minor, and could be resolved by rather simple notational changes in the equations used. When I felt that I understood this, a Dutch friend suggested that I write to van Iterson. He responded, generously sending me a copy of his monograph, which I have had bound, and treasure. I have recently completed a review of phyllotaxis (17).

This concern with phyllotaxis has had some unexpected consequences. Maksymowych at Villanova University had described striking changes in the morphology and the growth pattern of vegetatively grown *Xanthium* shoots, as the result of a single application of gibberellic acid. Among other things he made transverse sections of the shoot apex and young leaves of control and treated plants. When I saw them, I exclaimed that the treatment had altered the phyllotaxis. I proposed that we make a careful analysis of the arrangement of leaf primordia at the apex in these plants. We found that the normal pattern had been changed to a stable higher-order pattern (31) and speculated that the effect had some similarity to changes that occur on photoperiodic floral induction. Roger Meicenheimer (19) then induced *Xanthium* plants to flower and found that indeed the shoot apex underwent an identical but transient change in its phyllotaxis. The gibberellic acid effect is one of the few instances of an experimental modification of leaf arrangement in plants.

A second development is at the molecular level. Arthur Veen, a student with Lindenmayer at Utrecht University, had written a computer program to simulate growth of a shoot with the initiation of leaves in phyllotactic patterns (38). Their model was developed on a cylindrical surface, and I was sufficiently interested to write a preliminary program to carry out the simulation in a plane. Veen came to my lab to work with me on it. At that time, Lewis Tilney had developed an elegant technique of high-resolution electron microscopy of negatively stained microtubules. Lewis Routledge, working with Bernard Gerber, was using the technique for studies of bacterial flagella. When Tilney and Routledge showed me their pictures and asked how to analyze the obviously helical arrangement of subunits, my immediate impression was that they resembled certain of van Iterson's models. I suggested measuring distances between the units and certain angles, and constructing cylindrical models. After further discussion, I proposed to do the analysis if they would provide micrographs, references, etc. Veen and I were then deeply involved with computer modeling of phyllotaxis, so that the mathematical work and computations went quickly. In little more than a month the manuscript on tubular packing of spheres was completed (15). I had not been aware of the closely related work on cylindrical crystals by William F. Harris until our manuscript was completed, but I then sent him a copy. This led to voluminous correspondence, a visit by him to my laboratory, a visit to his at the University of the Witwatersrand, and to a far more rigorous analysis of tubular packings, from a crystallographic point of view (26).

The *Science* article on tubular packings (15) attracted the attention of others than biologists. At Penn I was one of the founders of a discussion group of people with varied interests in "form." For about five years this "Form Forum" met monthly, with wine and cheese, for talks and discussions of a remarkably wide range of topics, usually with demonstrations of paintings, sculptures, architectural renderings, tilings, polyhedra, electron micrographs, computer simulations, music, and poetry. I have also participated in two mathematical conferences on polyhedra. It is my hope that this broad approach to the geometry of form may be valuable in biological problems, such as the analysis of cellular patterns.

CALIFORNIA

I have taken three sabbatical leaves from Penn, and in each case have chosen to go to a California institution. In 1954–1955, with the award of a Guggenheim Fellowship, I worked at Frits W. Went's phytotron at the California Institute of Technology where I was incarcerated daily with Lloyd T. Evans, Harry R. Highkin, William S. Hillman, Margaretta G. Mes, Paul E. Pilet, Roy Sachs, and Went, when he was not travelling. There was much discussion of plant physiology, with a slant toward problems of floral induction. My idea was to grow *Xanthium* and perhaps other plants in a range of environmental conditions, using the plastochron index to assess temperature and light effects on the development pattern of the plants. I shared Anton Lang's dissatisfaction with the Cal Tech phytotron (30). My complaint was that growing conditions were not under control! Because carts were moved twice-daily to meet the schedules all of us had requested for our plants, there was no way, short of being an outright stinker, of knowing on a given day whether one's plants would be next to a flat of oat seedlings, or under the shade of a coffee bush. It seemed to me that the temperature coefficients for leaf growth and for the rate of leaf initiation (inverse of the plastochron) were nearly three, implying that the ratio of relative elongation rate to initiation rate (that is, the relative plastochron rate) was nearly constant over a broad range of temperatures. I took this to be an evidence of temperature regulation of morphogenesis, as to leaf initiation and growth, such that plants grown at different temperatures look much the same. Horie et al (29) have since described similar findings with cucumber plants.

But I was unable to get respectable data to support these ideas, and did not publish them. Looking around for other things to do, I set up lights and a 16-mm movie camera, which I had brought from Philadelphia, in a machinery room, which it turned out was air-conditioned. Time-lapse equipment was not then easily available or affordable, but I had brought home-made timers and mechanisms to operate cameras. There was an excellent machinist at the phytotron who, understandably, did not welcome others to use his shop. But when I had satisfied him that I was not likely to abuse his machines and tools, he gave me nearly free rein, excellent instruction in shop practices, and good advice about getting my gadgets to work. Several striking scenes of *Xanthium* growing from seed to plastochron 15 or so, in continuous light or with a non-inductive dark period, resulted, and some footage on *Coleus*, before the year was out. I also rigged up an automated 35-mm camera, and made the photographs of *Xanthium* leaves, which were analyzed much later (14). Despite my dissatisfaction at the time, it was a profitable year.

Our second California trip was to La Jolla in 1966–1967, where Herbert Stern had welcomed me to his lab at the new campus of the University of California, San Diego. This might have been an opportunity to learn modern biochemical techniques at the bench. However, Stern and I got to talking about some published work on the inhibition of cell division in roots of the broad bean, *Vicia faba*, by a thymine analog, and I decided to resume inhibition experiments with *Zea* roots. I had again brought camera equipment with me, and a student who wanted to learn the paraffin technique assisted. We could find no mitotic figures in sections of roots grown in the presence of purine and pyrimidine analogs, although the overall rate of elongation was normal for as long as 18 hours, during which time the meristem appeared to be "used up." To the question of where these inhibitors were incorporated, Yasuo Hotta suggested the simple expedient of using radioactively labeled inhibitors, putting root segments into the cocktail of scintillation vials and counting them. To our slight surprise, the label appeared not only in the former meristematic region but also in cells quite some distance behind it; but time ran out before this result was reconciled with the growth data. I also set up a time-lapse movie camera to photograph growing thalli of *Marchantia* that a postdoctoral fellow provided, and later analyzed them for the pattern of growth in area. I was enamored of the CDC 6600 computer at UCSD, far more satisfactory than the IBM machine at Penn, and spent a part of my time at the computer center working on problems such as the numerical solution of differential equations.

At Stanford University, in the fall of 1978, we had the pleasure of renewing our long acquaintance with Paul and Margaret Green. Following some discussions with Paul, I began calculations of the effect of growth deformation on the multi-net pattern of cellulose microfibrils in cell walls. This resulted quickly in a geometrical and statistical model of changes in the microfibrillar pattern, which agrees satisfactorily with experimental data on the walls of growing *Nitella* internode cells (16).

In January, we moved to the University of California, Davis, where I had a visiting professorship for the spring term. I collaborated with Wendy Silk in a graduate seminar on plant growth analysis and gave a few other lectures. We set up an automatic camera to record the growth of marked *Avena* coleoptiles, with the intention of following changes in phototropic curvature (curvature in the mathematical sense.) These preliminary experiments were not entirely satisfactory, but we did get some promising photographs.

One can conclude from these experiences that experiments do not always work out in a new laboratory with a time limitation. By and large, however, the scholarly leave is a valuable institution. The new viewpoints, acquaintances, and intangible benefits that result are ample justification for the inconvenience of moving and the frustrations one encounters. I wish I had taken leaves more often.

COMPUTERS AND CALCULATORS

The history of the computer has been written more than once (28, 35) but there may be some interest in my personal experiences as a user. The situation before the computer revolution is well stated in the preface of Whittaker & Robinson (39), written in 1924 "Each student should have a copy of Barlow's tables of squares, etc...a stock of computing paper (i.e., paper ruled into squares...), and...computing forms for...Fourier analysis... With this modest apparatus nearly all the computations hereafter described may be performed, although time and labour may often be saved by the use of multiplying and adding machines when these are available." As a boy I was impressed by the facility at mental arithmetic of one of my uncles, a bookkeeper, but I was all for machines. In high school I bought a cheap slide rule, possibly the only one in the school, and later I had a simple adding machine with dials to be turned with a stylus, which I could laboriously multiply with. I wore them both out. At Washington University and later, I was usually able to find a mechanical desk calculator of some sort, but when I came to Penn in 1947 there was no calculator in the botany department, only an ancient Monroe in the zoology office. I immediately ordered a Marchant, and soon additional machines were obtained for the biometry course and research. When I could afford it, I bought a Curta hand-held calculator to use at home and was delighted with the watch-like precision of its construction. In the summer of 1951, when the analysis of our root growth data needed to be done, a few of us moved to the Morris Arboretum with Marchant calculators and, on a pleasant terrace, punched the machines every day for about three weeks.

In 1972 Hewlett-Packard announced their first pocket scientific calculator, the HP-35, and I of course bought it for myself and the lab. As improved models appeared, I acquired and used several, including the current HP-28C, which has memory exceeding that of the IBM mainframe computer at which I learned FORTRAN.

By a curious coincidence, at the first scientific meeting I attended, the AAAS Christmas meeting at Richmond in 1938, I saw an exhibit of computing equipment from the Moore School of Electrical Engineering at Penn. It undoubtedly had to do with Weygandt's differential analyzer, an analog computer (35). So I was not as surprised as I might otherwise have been when I learned of the first digital computer, the ENIAC built at the Moore School. Penn had a computer lab when I came, which at one time housed a Univac, and later other machines. Occasionally I visited this lab and found that the staff were friendly enough, but the computers were not. One needed to know a great deal about the machines to use them. In January 1964, Penn acquired an IBM 7040, which was one of the first computers with an operating system designed to accommodate ordinary users. A full-scale computer center was quickly organized and I immediately took the FORTRAN course. At the end of the course each student was to write a program and run it. The instructor suggested a payroll problem, but I wrote my own program to calculate Fibonacci numbers and plastochron ratios. It ran on the first submission. I then proceeded to program the computation of elemental rates of growth in area of the *Xanthium* leaf, and completed the analysis for the 1964 Edinburgh Congress (14). I have now had experience with three or four mainframe computers, as well as two desktop computers.

When microcomputers appeared on the market we were in Silicon Valley, and I spent some time learning about the first Apple and other micros, skeptical at first about their usefulness for serious work. I bought an AIM 65 and set it up in our bedroom in Davis. Soon I had wired up a small speaker and written a program to play tunes through one of the output ports: *Papa Haydn's Dead and Gone*, Brahms's *Lullaby*,... The AIM was perhaps the least expensive, and one of the most educational of the early machines, with provisions for expansion of the hardware and the monitor program. I brought it back to Penn and used it to good effect for about five years. The machine at which I am writing this text is a far more competent personal computer, an MTU-130, which in its turn is about five years old....

LOOKING FORWARD

The years I have written about have of course seen a great revolution in biology, with the development of molecular genetics and many other advances. Reflecting on the state of what might be called the biometry of growth, say in 1936, one realizes that this received very scanty treatment in plant physiology textbooks, and none at all in plant anatomy. One learned of the "grand period of growth," of auxanometers, and of Julius Sachs's root-marking experiments of the 1860s. Plant anatomists wrote of gliding growth, and used the word plastochron in a purely descriptive sense. In the 1870s Kreusler et al in Germany had studied the growth of *Zea* plants, but in the English-speaking world serious consideration of plant growth seems to have begun with defining of the efficiency index, or relative growth rate, by Blackman and by Briggs, Kidd & West in the early 1920s. Robertson's ideas about a master reaction in control of growth were widely quoted, and Huxley's allometric coefficient was fashionable. Biologists had a definite prejudice against mathematics and statistics and for many years I felt that my work was mostly quietly ignored. At the first presentation of our root work at a meeting, an older botanist took me aside and offered me the fatherly advice to soft-pedal the math.

In the intervening years the analysis of the growth of whole plants and plant organs has advanced considerably, particularly in connection with agricultural research. Statistics such as the absolute and relative growth rates, unit leaf rate, and leaf area ratio in the analysis of growth of individual plants, and related quantities for the growth of crops, appear to be firmly established, with general agreement about how they are to be estimated. Methods for fitting of growth curves, particularly the F. J. Richards function, have been highly developed. There is much activity in mathematical modeling of the growth of plants and crops.

The concept of elemental growth rates was introduced by Richards & Kavanagh in 1943 (34) and much of what I have discussed above, such as root growth analysis and analysis of growth of the *Xanthium* leaf, is related to this idea. This has led to the consideration of the kinematics of plant growth (36) in the context of continuum mechanics. The importance of distinguishing between material and spatial specifications has become clear. Kinematics deals only with motion without consideration of mass and force, but with kinematics as a basis we can look forward to the development of the dynamics of plant growth. Since a large part of plant physiology has to do with growing tissues, it will be important to deal effectively with the kinematics of growth, in order to make valid estimates of biosynthetic rates, for instance. Studies of the role of water and solute transport in tissue growth will have to take account of the kinematics of the growing tissue, as will considerations of the energetics of tissue growth. It is also true that morphogenesis, the development of the form of plant organs from meristematic tissue, is to a large extent a matter of tissue deformation, and it will be necessary to consider the forces that give rise to kinematic displacements and their origin. Work in these directions is under way, and we can look forward to further advances in the empirical analysis of plant growth processes, and in theoretical treatments of plant growth.

Literature Cited

1. Anderson, E. 1936. The species problem in *Iris*. *Ann. Mo. Bot. Gard.* 23:457–509
2. Anderson, E. 1939. The hindrance to gene recombination imposed by linkage: an estimate of its total magnitude. *Am. Nat.* 73:185–88
3. Anderson, E. 1967. *Plants, Man and Life.* Berkeley: Univ. Calif. Press
4. Anderson, E., Schregardus, D. 1944. A method for recording and analyzing variations of internode pattern. *Ann. Mo. Bot. Gard.* 31:241–47
5. Avery, G. S. Jr. 1933. Structure and development of the tobacco leaf. *Am. J. Bot.* 20:565–92
6. Avery, G. S. Jr., Burkholder, P. R., Creighton, H. B. 1937. Production and distribution of growth hormone in shoots of *Aesculus* and *Malus* and its probable role in stimulating cambial activity. *Am. J. Bot.* 24:51–58
7. Cooper, D. C. 1935. Macrosporogenesis and development of the embryo sac of *Lilium henryi*. *Bot. Gaz.* 97:346–55
8. Darlington, C. D. 1937. *Recent Advances in Cytology.* Philadelphia: Blakiston. 2nd ed.
9. Erickson, R. O. 1943. Taxonomy of *Clematis* section *Viorna*. *Ann. Mo. Bot. Gard.* 30:1–62
10. Erickson, R. O. 1945. The *Clematis fremontii* var. *riehlii* population in the Ozarks. *Ann. Mo. Bot. Gard.* 32:413–60. Reprinted in Ornduff, R., ed. 1967. *Papers on Plant Systematics.* Boston: Little, Brown. Pp. 319–66
11. Erickson, R. O. 1947. Respiration of developing anthers. *Nature* 159:275
12. Erickson, R. O. 1948. Cytological and growth correlations in the flower bud and anther of *Lilium longiflorum*. *Am. J. Bot.* 35:729–39
13. Erickson, R. O. 1961. Probability of division of cells in the epidermis of the *Phleum* root. *Am. J. Bot.* 48:268–74
14. Erickson, R. O. 1966. Relative elemental rates and anisotropy of growth in area: a computer programme. *J. Exp. Bot.* 17:390–403
15. Erickson, R. O. 1973. Tubular packing of spheres in biological fine structure. *Science* 181:705–16
16. Erickson, R. O. 1980. Microfibrillar structure of growing plant cell walls. *Lect. Notes Biomath.* 33:192–212
17. Erickson, R. O. 1983. The geometry of phyllotaxis. In *The Growth and Functioning of Leaves*, ed. J. E. Dale, F. L. Milthorpe, 3:53–88. Cambridge: Cambridge Univ. Press
18. Erickson, R. O., Goddard, D. R. 1951. An analysis of root growth in cellular and biochemical terms. *Growth, Symp.* 10:89–116
19. Erickson, R. O., Meicenheimer, R. D. 1977. Photoperiod induced change in phyllotaxis in *Xanthium*. *Am. J. Bot.* 64:981–88
20. Erickson, R. O., Michelini, F. J. 1957. The plastochron index. *Am. J. Bot.* 44:297–305; 47:350–51
21. Erickson, R. O., Sax, K. B. 1956. Elemental growth rate of the primary root of *Zea mays. Proc. Am. Philos. Soc.* 100:487–98
22. Erickson, R. O., Sax, K. B. 1956. Rates of cell division and cell elongation in the growth of the primary root of *Zea mays. Proc. Am. Philos. Soc.* 100:499–514
23. Fisher, R. A. 1936. The use of multiple measurements in taxonomic problems. *Ann. Eugen. London* 7:179–88
24. Gilman, R. 1987. Noted with pleasure. *N. Y. Times Book Rev.* 1 Feb.
25. Goodwin, R. H., Stepka, W. 1945. Growth and differentiation in the root tip of *Phleum pratense. Am. J. Bot.* 32:36–46
26. Harris, W. F., Erickson, R. O. 1980. Tubular arrays of spheres: geometry, continuous and discontinuous contraction, and the role of moving dislocations. *J. Theor. Biol.* 83:215–46
27. Hejnowicz, Z., Erickson, R. O. 1968. Growth inhibition and recovery in roots following temporary treatment with auxin. *Physiol. Plant.* 21:302–13
28. Hodges, A. 1983. *Alan Turing: the Enigma.* New York: Simon & Schuster
29. Horie,T., de Wit, C. T., Goudrian, J., Bensink, J. 1979. A formal template for the development of cucumber in its vegetative stage. *Proc. Kon. Nederl. Akad. Weten.* 82:443–79
30. Lang, A. 1980. Some recollections and reflections. *Ann. Rev. Plant Physiol.* 31:1–28
31. Maksymowych, R., Erickson, R. O. 1977. Phyllotactic change induced by gibberellic acid in *Xanthium* shoot apices. *Am. J. Bot.* 64:33–44
32. Moens, P. B. 1968. The structure and function of the synaptinemal complex in *Lilium longiflorum* sporocytes. *Chromosoma* 23:418–51
33. Ogur, M., Erickson, R. O., Rosen, G. U., Sax, K. B., Holden, C. 1951.

Nucleic acids in relation to cell division in *Lilium longiflorum*. *Exp. Cell Res.* 11:73–89

34. Richards, O. W., Kavanagh, A. J. 1943. The analysis of relative growth gradients and changing form of growing organisms: illustrated by the tobacco leaf. *Am. Nat.* 77:385–99

35. Shurkin, J. 1985. *Engines of the Mind, a History of the Computer*. New York: Washington Square Press

36. Silk, W. K., Erickson, R. O. 1979. Kinematics of plant growth. *J. Theor. Biol.* 76:481–501; 83:701–3

37. Stern, H., Hotta, Y. 1977. Biochemistry of meiosis. *Philos. Trans. R. Soc. London Ser. B* 277:277–94

38. Veen, A. H., Lindenmayer, A. 1977. Diffusion mechanism for phyllotaxis, theoretical, physico-chemical and computer study. *Plant Physiol.* 60:127–39

39. Whittaker, E., Robinson, G. 1944. *The Calculus of Observations. An Introduction to Numerical Analysis*. New York: Dover Publications. 4th ed., reprint

Ann. Rev. Plant Physiol. Plant Mol. Biol. 1988. 39:23–52

PLANT GROWTH-PROMOTING BRASSINOSTEROIDS

N. Bhushan Mandava*

Office of Pesticides and Toxic Substances, US Environmental Protection Agency, Washington, D.C. 20460

CONTENTS

INTRODUCTION 23
BACKGROUND 24
ISOLATION OF BRASSINOLIDE 25
TERMINOLOGY 26
CHEMISTRY OF BRASSINOSTEROIDS 28
SYNTHESIS OF BRASSINOSTEROIDS AND STRUCTURE-ACTIVITY
    RELATIONSHIP 32
BIOLOGICAL ACTIVITY AND PROPOSED MODE OF ACTION 33
EFFECTS ON NUCLEIC ACID AND PROTEIN METABOLISM AND
    ON OTHER ENZYMES 39
PRACTICAL APPLICATIONS IN AGRICULTURE 41
PROPOSED FUNCTIONS 42
PROPOSED BIOSYNTHESIS 43
PROBLEMS AND PROSPECTS 45

INTRODUCTION

In 1979 a novel plant growth-promoting steroidal lactone, termed brassino-lide, was found in the pollen of rape *(Brassica napus)*. Since then, chemists

*Present Address: Todhunter, Mandava, and Associates, 1625 K Street, N.W., Suite 975, Washington, D.C. 20036

0066-4294/88/0601-0023$02.00

and biologists have isolated brassinolide and identified its related compounds in various higher and lower plants.

Brassinolide and its analogues have been synthesized and made available to biologists for basic physiological and biochemical studies and field experiments. Synthesis has enabled a structure-activity evaluation of these compounds, and the tasks of elucidating their mechanism of action and of establishing their interactions with other endogenous growth-regulating hormones are now under way. The biosynthetic pathways for the brassinosteroids remain to be determined.

Since suitable bioassay systems can now detect steroidal compounds with biological activity similar to that of brassinolide, the role of the brassinosteroids in plant growth and development can be studied. Use of these substances in improving agricultural productivity through increases in crop yield and biomass is being investigated.

BACKGROUND

It has long been recognized that hormones play an important role in the reproduction of plants and that pollen is rich in hormones and other growth substances (10, 31, 103). In the 1930s and 1940s, United States Department of Agriculture (USDA) workers and others recognized that pollen extracts promoted growth. Using the bean first-internode bioassay, Mitchell and his coworkers (63) found that extracts from immature bean seeds elicited growth, an effect Japanese scientists later attributed to gibberellins. Several growth substances from reproductive plant parts, especially pollen, were screened by a bean second-internode bioassay. Over 60 kinds of pollen were screened, and a representative, effective group was further evaluated (see Table 1) (45). Mitchell and associates (62) reported that a few samples, notably the pollen from rape plant (*Brassica napus* L.) and alder tree (*Alnus glutinosa* L.), produced an unusual response that combined elongation (the typical gibberellin response) with swelling and curvature. These workers proposed that the pollen of rape plant contains a new group of lipoidal hormones, termed brassins. Subsequently, Mandava, Mitchell, and coworkers (46, 49) reported an active fraction predominantly containing the glucosyl esters of fatty acids. Later work (25) showed that, although they produced elongation, these glucosyl esters lacked the biological activity present in the purified brassin fraction. Milborrow & Pryce (59) asserted that this biological response could be explained as the effect of gibberellins rather than of any endogenous lipid(s) in brassins.

In 1968, Marumo and his coworkers (54) had reported that certain novel growth substances in *Distylium racemosum* caused lamina bending in a rice bioassay. They attributed this response to the presence of new auxins. Insufficient material and unsuitable analytical instrumentation prevented further pursuit of this work at that time.

ISOLATION OF BRASSINOLIDE

In 1972 Mitchell & Gregory (60) demonstrated that brassins could enhance crop yield, crop efficiency, and seed vigor. Consequently in 1975 the USDA mounted a major effort (a) to evaluate brassins for yield increases on major crops such as wheat, corn, soybeans, and potatoes at several locations in the United States and Puerto Rico; (b) to identify and synthesize the active constituent(s) in brassins; and (c) to evaluate the plant growth-enhancing properties of the active constituent(s) under greenhouse and field conditions.

In an effort to isolate the active constituent(s), a large amount (500 lb) of bee-collected rape pollen was extracted with 2-propanol in 50-lb batches (44). The extract was partitioned among carbon tetrachloride, methanol, and water. The methanol fraction containing the biological activity was chromatographed on a series of large silica gel columns, a process that reduced the amount of material to about 100 g. Further purification by column chromatography and HPLC gave 10 mg of crystalline material, termed brassinolide. Biological activity was monitored by a bean second-internode bioassay throughout the purification stages.

Table 1   Sources of pollen that induces brassin activity in the bean second-internode bioassay

Scientific name	Common name
Aescules hippocastanum L.	horse chestnut
Alnus glutinosa (L.) Gaertn.	alder, European
Sinapis arvensis L.	mustard
(=*Brassica kaber* L.)	
Brassica napus L.	rape
Carduus nutans L.	nodding thistle
Crataegus sp.	hawthorn
Echium vulgare L.	viper's bugloss
Pyrus communis L.	pear
Robinia pseudoacacia L.	black locust
Rhus sp.	sumac
Secale cereal L.	rye
Sisymbrium irio L.	London rocket
Thea sinensis L.	tea
(=*Camellia sinensis* (L.) Ktze.)	
Typha sp.	cattail
Ulmus sp.	elm
Zea mays	corn

Essentially, the lipid fraction of the brassin complex that contained glucosyl esters was refined to yield the crystalline brassinolide. Brassinolide is present at 200 parts-per-billion (ppb) in pollen and at about 10 parts-per-million (ppm) in the purified lipid fraction. Had the USDA not undertaken this large-scale extraction effort, the controversy over the role of glucosyl esters, auxins in *Distylium racemosum,* and gibberellins in immature bean seeds and pollen would have remained unresolved, and brassinolide would not have been identified.

Understanding of the physical, chemical, and biological properties of the parent compound made detection of brassinolide and its related steroidal compounds in other plants easier. The USDA approach of team research (involving chemists, chemical engineers, biochemists, plant physiologists, and plant pathologists) is a proven way to reach a projected goal.

The discovery of brassinolide (after more than 10 demanding years of research by USDA scientists and a cost to the US taxpayers of more than a $1 million) stimulated worldwide interest. Especially intrigued were the Japanese scientists who had been unable to isolate similar compounds from *Distylium racemosum* (S. Marumo, personal communication). (Gibberellin research, incidentally, followed a similar pattern. Imperial Chemical Industries in England and the USDA played a key role in determining the structure of gibberellic acid, which had first been detected in Japan. Later, Japanese scientists explored the field very extensively.) The major initial emphasis in brassinolide research was on the synthesis and isolation of the hormone and its related compounds, followed by biological evaluation. It is not surprising that of over 100 papers published on this subject, about half of them came from Japan during the 5 years after the discovery of brassinolide.

TERMINOLOGY

Brassins, after the genus *Brassica,* comprise a complex mixture of lipids purified from rape (*Brassica napus* L.) pollen. The characteristics biological activity—elongation, curvature, and splitting of the treated internode in the bean second-internode bioassay—is called *brassin activity*. Brassinolide and/or its related compounds are known collectively as *brassinosteroids*. The rice-lamina inclination bioassay, the mung bean epicotyl bioassay, and the bean first-internode bioassay have also been used to measure brassin activity.

Brassinosteroids (BRs) are a group of naturally occurring polyhydroxysteroids. A sequential numerical suffix designates the brassinosteroids occurring in nature. BR_1 denotes brassinolide, and others follow the sequence BR_2, BR_3, BR_4,. . . .BR_n (Table 2). Of the 16 natural BRs identified, 15 have been isolated from higher plants and one from a lower plant.

All brassinosteroids thus far characterized from plants are 5-α-cholestane

Table 2 Assignment of BR numbers to brassinosteriods

BR designation	Trivial name	Reference(s)
BR_1	Brassinolide	26, 47, 50, 76
BR_2	Brassinosterone[a]	46, 78, 105, 106
	(castasterone and brassinone)	
BR_3	Dolicholide	9, 108–110
BR_4	Dolichosterone	4, 9, 108–110
BR_5	6-Deoxobrassinosterone	2, 6, 33, 104, 107, 111
	(6-deoxocastasterone)	
BR_6	6-Deoxodolichosterone	6, 104, 108
BR_7	Typhasterone[a] (typhasterol and	2, 3, 75, 106
	2-deoxybrassinosterone)	
BR_8	Theasterone[a] (teasterone and	2, 3, 35, 68, 104, 112
	3-epi-2-deoxybrassinosterone)	
BR_9	24-Epibrassinosterone[a]	112
	(24-epicastasterone)	
BR_{10}	Homodolicholide	9, 108–110
BR_{11}	Homodolichosterone	9, 108–110
BR_{12}	Homobrassinosterone[a]	1–3, 34, 112
	[(24S)-24-ethylbrassinosterone]	
BR_{13}	6-Deoxohomodolicholide	111
BR_{14}	28-Norbrassinolide	1–3, 112
BR_{15}	28-Norbrassinosterone[a]	1, 2, 6, 78, 104, 112
	(brassinone)	
BR_{16}	25-Methyldolichosterone	113

[a] Names proposed or preferred in this review. Previously published names are shown in parentheses.

derivatives and can be classified as C_{27}, C_{28}, and C_{29} steroids. In this review I retain where practical the trivial names originally given to BRs. However, the ketone precursor to brassinolide was given two names—castasterone (105) and brassinone (47). To be consistent with the names given other steroid ketones (dolichosterone, homodolichosterone, etc), this ketone should be renamed brassinosterone. To avoid confusion, brassinone (BR_{15}), originally isolated by Abe et al (2), should be referred to as 28-norbrassinosterone, a precursor ketone to 28-norbrassinolide (BR_{14}). I further recommend that the BR_1, BR_2,. . . .BR_n designation be applied only to natural BRs, and not to others derived by synthesis. For example, brassinosteroid analogues should be named after the parent natural BRs (e.g. 24-β-brassinolide for brassinolide isomeric at C-24). Under the BR numbering system, BR_1–BR_9 are all C_{28} steroids (normal brassinosteroids), BR_{10}–BR_{13} and BR_{16} are C_{29} steroids (homo brassinosteroids), and BR_{14} and BR_{15} are C_{27} steroids (nor brassinosteroids).

The notation of the stereochemistry of the side chain in brassinosteroids is a difficult problem in a review of this nature. According to R,S notation [also known as the Sequence Rule (68a)], brassinolide is $2\alpha,3\alpha,22(R),23(R)$-

tetrahydroxy-24(S)-methyl-B-homo-7-oxa-5α-cholestan-6-one. This notation is undoubtedly precise, but the priorities sometimes change with differences in substitution. For example, both α-ecdysone and brassinolide are 22(R)-hydroxy steroids, though they have opposite configurations. To circumvent this problem, Nes (68a) has suggested the use of a modified Fieser notation (α,β instead of R,S notation) for naming the substituents in the steroid side chain. According to this notation, brassinolide is 2α,3α,22α,23α-tetrahydroxy-24α-methyl-B-homo-7-oxa-5α-cholestan-6-one. Because of its simplicity, the Nes convention (α,β notation) for naming the side-chain substituents in BRs is favored in this review.

CHEMISTRY OF BRASSINOSTEROIDS

Brassinolide, present at 200 ppb in rape (*Brassica napus* L.) pollen, was the first brassinosteroid isolated in a crystalline form and characterized by physical and spectroscopic analysis. The structure was shown by X-ray diffraction (26) to be 2α,3α,22(R),23(R)-tetrahydroxy-24(S)-methyl-B-homo-7-oxa-5α-cholestan-6-one (Figure 1, I a). According to α,β notation for representing the side-chain stereochemistry in steroids, the chemical name for brassinolide is 2α,3α,22α,23α-tetrahydroxy-24α-methyl-B-homo-7-oxa-5α-cholestan-6-one. Rape pollen was shown to contain another brassinosteroid, termed brassinone (47). The same precursor ketone to brassinolide, which Yokota et al named castasterone (BR_2), was found in the leaves and insect galls of chestnut (*Castanea* spp.), and its structure (Figure 1, II a) was determined by spectroscopic methods (106).

An analytical method consisting of boronation (with methyl boronic acid) of the vicinal glycol groups followed by identification of the derivative by gas chromatography-mass spectrometry (GC-MS) was reported to identify brassinosteroids from other plant sources (34–36). This method was based on the information that BR_1 and BR_2 contain two vicinal *cis*-glycol groups (one in the steroid nucleus and the other in the side chain). The derivatized boronates were analyzed by means of selective ion monitoring (SIM) in mass spectrometry (36). The SIM method was reportedly capable of detecting BRs at ppb levels in the plant tissue (34, 91).

Bioassay data (Table 1), the identification of several brassinosteroids (Table 3) in both higher and lower plants, and results from immunoassay (32) suggest that BRs, like gibberellins and auxins, are widely distributed in plants. Although they are likely present in all parts of the plant (32), pollen is still the richest source of BRs (Table 3). Pollen and seeds contain 1–1000 ng of BR per kg of tissue, shoots 1–100 ng/kg, fruit 1–10 ng/kg, and leaf 1–10 ng/kg.

Table 3 Distribution of brassinosteroids in plants

Common name	Plant species	Plant part	Brassinosteroids
Rape	*Brassica napus* L.	pollen	BR_1, BR_2
Chinese cabbage	*Brassica campestris*	fruit	BR_1, BR_2, BR_{12}, BR_{14}, BR_{15}
Japanese chestnut	*Castnea crenata*	gall, stem, leaf, flower	BR_1, BR_2, BR_5, BR_{15}
Tea	*Thea sinensis*	leaf	BR_1, BR_4, BR_7, BR_8, BR_{12}, BR_{15}
—	*Distylium racemosum*	gall, leaf	BR_1, BR_2, BR_4, BR_{15}
Kidney bean	*Phaseolus vulgaris* L.	seed	BR_1–BR_6, BR_{13}, BR_{16}
Japanese morning glory	*Pharbitis purpura* (*Ipomoea*)	fruit	BR_2, BR_{15}
Cattail	*Typha latifolla*	pollen	BR_7
Rice	*Oryza sativa*	shoot	BR_2, BR_4
Corn	*Zea mays*	pollen	BR_2, BR_7, BR_8
Japanese black pine	*Pinus thunbergil*	pollen	BR_2, BR_7
Spruce	*Picea sitchensis*	shoot	BR_2, BR_7
Algae	*Hydrodyction reticulatum*	colony	BR_9, BR_{12}

Among the sources investigated, brassinosterone (BR_2) occurs most frequently (Table 3): BR_2 has been found in 12 sources, BR_1 in 7, BR_4 and BR_7 in 5 sources, Br_5 in 4, BR_{12} in 3 sources (including blue algae), and other steroids (BR_3, BR_6, BR_8, and BR_{14}) in 2 sources each.

All brassinosteroids contain a steroid nucleus (some with the oxygen function in the B ring) with a side chain at C-17 similar to the side chain in cholesterol. Other common features for all brassinosteroids, in addition to β-oriented angular C-18 and C-19 methyl groups, are (*a*) α-orientation at C-5 (A/B ring junction; (*b*) α-oriented hydroxyl groups at C-22 and C-23 (side chain); and (*c*) α-oriented hydroxyl groups (*cis*-geometry) at C-2 and C-3 in ring A of the steroid nucleus (exceptions: typhasterol contains only one hydroxyl at C-3 in the α position and theasterol contains a β-hydroxyl at C-3 only; both compounds lack hydroxyl at C-2).

Brassinosteroids generally differ in functional groups at C-24 (steroid side chain): CH_3, brassinolide (BR_1) and brassinosterone (BR_2); $=CH_2$, dolicholide (BR_3) and dolichosterone (BR_4); $=CH-CH_3$, homodolicholide (BR_{10}) and homodolichosterone (BR_{11}); and C_2H_5, homobrassinosterone (BR_{12}). Brassinosteroids lacking the substituent at C-24 have been isolated (1–3, 6, 78, 104). They include 28-norbrassinolide (BR_{14}) and 28-norbrassinosterone (BR_{15}), which exhibited weak biological activity in the rice bioassay (1–3, 78) and no biological activity in the bean first- and second-internode bioassays (96, 97).

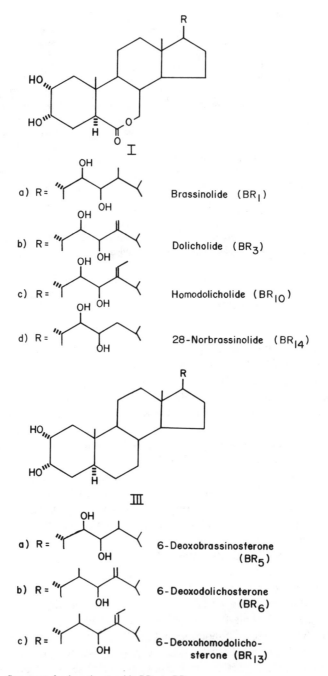

a) R =   Brassinolide (BR$_1$)

b) R =   Dolicholide (BR$_3$)

c) R =   Homodolicholide (BR$_{10}$)

d) R =   28-Norbrassinolide (BR$_{14}$)

a) R =   6-Deoxobrassinosterone
(BR$_5$)

b) R =   6-Deoxodolichosterone
(BR$_6$)

c) R =   6-Deoxohomodolicho-
sterone (BR$_{13}$)

Figure 1 Structures for brassinosteroids BR$_1$ to BR$_{15}$.

a) R = α-OH   Typhasterone (BR$_7$)

b) R = β-OH   Theasterone (BR$_8$)

a) R =   Brassinosterone (BR$_2$)

b) R =   Dolichosterone (BR$_4$)

c) R =   24-β-Methyl-28-Nor-
          brassinosterone (BR$_9$)

d) R =   Homodolichosterone (BR$_{11}$)

e) R =   Homobrassinosterone (BR$_{12}$)

f) R =   28-Norbrassinosterone (BR$_{15}$)

SYNTHESIS OF BRASSINOSTEROIDS AND STRUCTURE-ACTIVITY RELATIONSHIP

The unique structural features of brassinolide made its synthesis, and that of its isomers and analogues, especially interesting. Basically, experiments in synthesis aimed (a) to confirm the structure assigned to brassinolide (and other BRs) by physical and spectroscopic methods; (b) to show that the BRs alone, not trace compounds present in the extracted material, produce the observed biological activity; (c) to establish structure-activity relationships for brassinolide and structurally related compounds; and (d) to enable economical production, and distribution to investigators, of BRs or their analogues (since isolation of BRs from natural sources is prohibitively expensive and time consuming).

Chemists have long been fascinated by the isolation, synthesis, and biosynthesis of steroids. (Among the Nobel Prizes awarded to chemists, about 20 have been given for steroid-related contributions.) Steroids possess a unique, fused nucleus; various substituents occur in the nucleus, and various side chains attach to it (e.g. cholesterol); this reactivity and these stereochemical aspects give rise to several enantiomeric forms. Several schemes have been developed to design steroid molecules in a controlled (stereoselective and stereospecific) manner that avoids the formation of unnatural and unwanted isomers during synthesis.

The general synthetic scheme involves (a) modification of the side chain to introduce the desired substituents, and (b) introduction of a glycol group at C-2 and C-3 of ring A and the oxygen function(s) in ring B. The side chain of brassinolide and its analogues has been synthesized using two approaches, both involving readily available sterols as the starting materials: (a) The steroid chain has been built up starting from a steroidal C-22 aldehyde derived from a known sterol (14, 20, 37); and (b) a sterol with appropriate C-20 and C-24 alkyl substituents in the side chain has been used (5, 64–67, 94–97). Both the methods require the introduction of vicinal glycol group at C-22 and C-23 to define the stereochemistry. Once the desired side chain is built, both methods introduce the desired substituents in rings A and B. Using these approaches, researchers have synthesized all the natural BRs and several of their enantiomers and analogues. The second method is less expensive because the less costly sterols are abundant and fewer steps are required in the synthetic process.

In addition to the natural brassinosteroids (BR_1–BR_{16}), several synthetic analogues were prepared to establish the structure-activity relationship (13, 28, 64–67, 70, 77, 79–90, 99). From evaluations of the synthetic analogues (50, 94–97; M. J. Thompson, N. B. Mandava, unpublished results) in several test systems, the following structural requirements for eliciting brassin activity were determined:

1. *cis*-Vicinal glycol function at C-2 and C-3 in ring A. Compounds containing a single hydroxyl group at either C-2 or C-3 (see BR_7 and BR_8) in either α- or β-form, or *trans*-vicinal glycol function, show either weak or no biological acitivity.
2. *trans*-A/B ring junction.
3. Oxygen function at C-6 in the form of either a ketone or a lactone. Brassinosteroids (BR_5, BR_6, and BR_{13}) lacking a ketone group at C-6 are not active. Furthermore, synthetic analogues of BRs containing 6-oxa function (6-oxa-7-ketone) instead of the 7-oxa group (as present in natural BRs), or replacement of oxygen by hetero atoms such as nitrogen, are devoid of biological activity.
4. Cholesterol (steroid) side chain with vicinal glycol group at C-22 and C-23. All natural steroids have *cis* geometry for hydroxyls at C-22 and C-23. Compounds with *trans*-geometry for hydroxyls at these carbon atoms elicit biological activity but no significant variation in activity. All compounds with α-orientation at C-22, C-23, and C-24 were more active than the corresponding β-oriented compounds.
5. Substitution at C-24. Biological data on natural and synthetic BRs indicate that compounds with an α-orientation at C-24 show maximum activity. Among lactones and ketones, the degree of activity follows the order: Me > Et > H at C-24. Lactones and ketones containing methylene or E-ethylidene functions at C-24 were also reported to be active in the rice bioassay.

BIOLOGICAL ACTIVITY AND PROPOSED MODE OF ACTION

The bioassay first used to detect and isolate brassinolide (BR_1) and brassinosteroid-like substances from pollen (Table 1), and later to determine the structure-activity relationship of the synthetic BRs and their analogues, was the bean second-internode bioassay (50, 61). Although in this bioassay gibberellins cause only elongation of the treated and upper internodes, BRs characteristically evoke both cell elongation and cell division resulting in elongation, swelling, curvature, and splitting of the second internode. Such activity is called brassin activity. Morphological changes such as swelling and splitting of the treated internode are affected by the concentration of BRs (50). Auxins and cytokinins are not detected by this bioassay. A rice bioassay in which active agents evoke lamina-inclination has been reported as specific for BRs; auxins produce a response in this test, but only at much higher concentrations than those of BRs (53, 100–102). Gibberellins evoke a straight growth response without the bending characteristic for BRs (93). The bean second-internode and rice test systems (50, 100) are regarded as bioassays specific for BRs. With five other assays—namely, bean first-internode bioas-

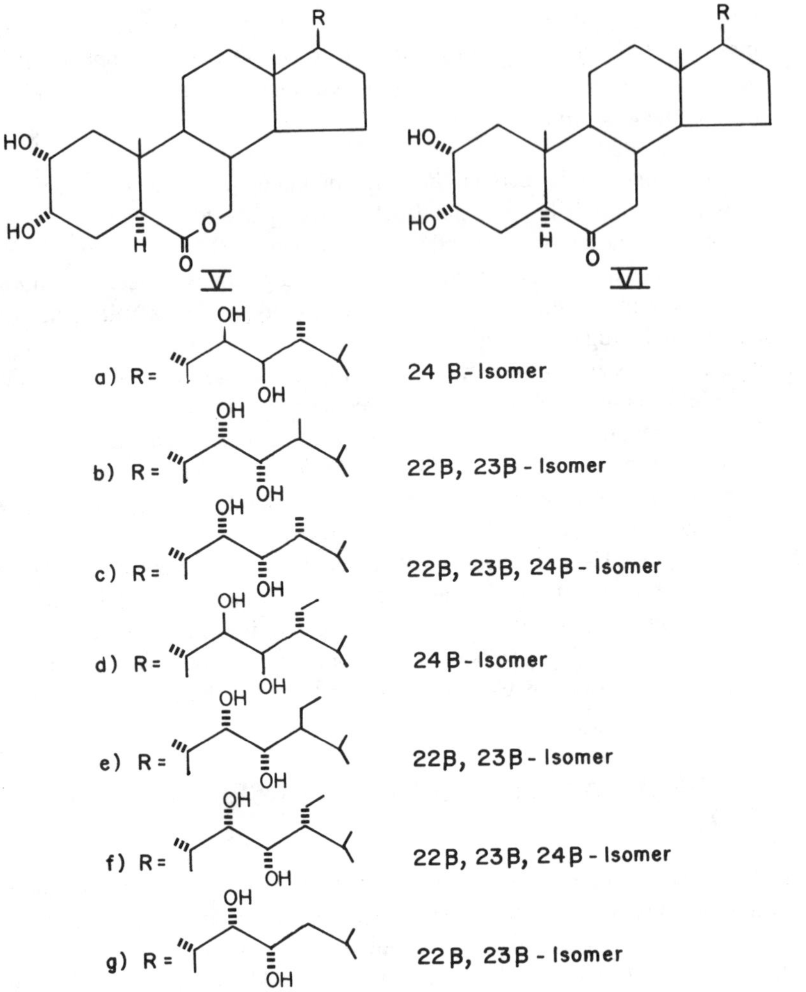

Figure 2   Structures for biologically active synthetic analogues of brassinosteroids.

say (55), mung bean epicotyl bioassay (24), auxin-induced ethylene production in etiolated mung bean tissue (7, 8), radish (*Raphanus sativus* L.) bioassay (88, 92), and wheat bioassay (98)—these were used to evaluate the structure-activity relationship in BRs and their congeners, establishing the following order of activity.

1. *Lactones:* These exhibit the highest activity. Of all known synthetic and natural BRs, brassinolide is the most active in any test system. The order of activity is as follows: brassinolide > its synthetic isomers (Figure 2 V a–c) > homobrassinolide > its 3 synthetic isomers (Figure 2, V d–f) >

dolicholide > homodolicholide > 28-norbrassinolide and its isomer (Figure 2, V g).

2. *Ketones:* Of several 6-keto compounds tested, brassinosterone (BR_2) is the most active BR; it equals homobrassinolide and other isomers of BR_1 in biological activity. Among the ketones, the activity follows the order: brassinosterone > its 3 synthetic isomers (Figure 2, VI a–c) > homobrassinosterone > its 3 synthetic isomers (Figure 2, VI d–f) > dolichosterone > homodolichosterone > 28-norbrassinosterone and its isomer (Figure 2, VI g).

3. *6-Deoxo Compounds:* These are the least active among the BRs, suggesting that the lack of oxygen function in the form of either a ketone or lactone is essential for biological activity.

Brassinosteroids were evluated (48, 114, 115) in several bioassays. In the bean second-internode bioassay (which also responds to gibberellins, but not to auxins and cytokinins), BRs evoke characteristic brassin activity. In all the bioassays, BRs produce activity at concentrations much lower (nM to pM range) than those effective for gibberellins (usually μM range). BR also induces elongation of normal and dwarf pea epicotyls, dwarf bean apical segments, mung bean epicotyls, cucumber hypocotyls, Azuki bean epicotyls, and sunflower (*Helianthus annuus* L.) hypocotyls. Additionally, BR inhibits betacyanin formation in *Amaranthus* and retards senescence in leaf discs of *Rumex* (48). Katsumi (39) reported that cucumber (*Cucumis sativus* L.) hypocotyls are very sensitive to BR. The effects of BR and GA in these test systems are additive. Information supporting this additivity was obtained from pretreatment studies with known inhibitors (24, 39, 115). For example, the growth retardant ancymidol inhibits the growth response induced by GA, and not that induced by BR (24, 39, 115). Dicyclohexyl carbodiimide, a membrane-bound ATPase inhibitor, affected only BR-induced elongation in cucumber hypocotyls but did not inhibit the GA-mediated hypocotyl growth in the same test (39). This suggests that ATPase is involved in eliciting the response to BR alone. In the radish bioassay (92), the effects of BR and GA are additive.

Takeno & Pharis (93) and also J. F. Worley and N. B. Mandava (unpublished results) found that BR does not affect the growth of dwarf rice (*Oryza sativa* var. Tan-ginbozu and Waito-C) in a bioassay known to be very sensitive to GAs. Abscisic acid interacts very strongly with the BR-induced elongation in beans (*P. vulgaris* L.) and mung beans (*P. aureus*)—effects that are reversible, as was deduced from tissue pretreatment studies (50, 116; N. B. Mandava, J. H. Yopp, unpublished results).

Braun & Wild (11, 12) found that BR stimulates leaf elongation of wheat (*Triticum aestivum*) plants. Their results are comparable to those of Gregory (23) in brassin-treated barley seeds. Also the increases in fresh and dry weights of leaves and shoots in wheat and mustard *(Sinapis alba)* are comparable to the results obtained in the bean second-internode test (41–43). Apparently, auxins and cytokinins are ineffective in wheat and mustard plants.

Younger tissues are more responsive than older tissues to BR (23, 24, 60). This could be due to either higher endogenous auxin levels in the tissue or higher sensitivity of young tissue to BR. For instance, the response to BR of 7-day-old mung bean seedlings (N. B. Mandava, D. W. Spaulding, unpublished results) was 2–3-fold greater than that of the 14-day-old plants used by Henry et al (30). As for the independent mechanisms of action of BR and GA, the differential effect of DCCP on BR- and GA-induced elongation also supports the idea that the effects of BR and GA are simply additive. Thus BR appears to have no interactive relationship to GA (24, 39, 48, 116).

Whether BR is applied to the basal part of the mung bean cutting in aqueous solutions or to the meristem or cotyledonary node of the intact mung bean seedling as a lanolin paste, it always affects the growth of the epicotyls, not the hypocotyls. Several morphological changes occur, such as drooping of the primary leaves, abnormal swelling of the epicotyls, and epinasty, apparently owing to cell division and perhaps involving ethylene production. GA elicits a weaker growth response than BR in epicotyl, a response easily arrested with ancymidol (which has no effect on the BR-mediated response). Other morphological responses due to BR treatment are not seen with GA at any concentration (N. B. Mandava, M. J. Thompson, J. H. Yopp, unpublished results).

When aqueous solutions of BR were applied to mung bean cuttings (with or without meristem and/or the primary leaves or decapitated tissues), and also when BR in lanolin paste was applied to meristem or cotyledonary node, a several-fold increase in BR-induced elongation of epicotyls, together with fresh weight increases over controls, were noticed (N. B. Mandava, D. W. Spaulding, unpublished results). These results strongly suggest that the primary effect of BR is on the growing region of the plant. In mung beans this is reflected in the growth of the epicotyls and in pinto beans in the second internodes. It also appears that application of BR to intact plants such as lettuce, cucumber, mustard, and wheat grown under hydroponic conditions contributes to the growth of the whole plant, including the roots (11, 12, 23; L. E. Gregory, N. B. Mandava, unpublished results). However, in mung bean cuttings, root growth is inhibited but not the formation of root initials (24).

Yopp et al (115) showed the similarity between the effects of BR and auxin on the elongation of apical hook segments of dwarf pea, maize mesocotyls, and Azuki bean epicotyls; the retardation of bean hypocotyl hook opening; the retardation of bean hypocotyl hook opening; the promotion of lateral root formation in mung bean hypocotyls; and the increase in fresh weight of slices of Jerusalem artichoke tissue. However, BR failed to show any activity in auxin bioassays involving suppression of lateral bud growth and inhibition of elongation of intact cress roots. BR was less active in some auxin bioassays but much more active than auxin in others (114, 115).

Yopp et al (115, 116) also demonstrated strong synergistic interaction between auxin and BR. This synergism, however, occurs only when the tissue is pretreated with BR and then exposed to auxin (IAA). When the order is reversed, there is no effect (39, 55). A similar synergistic response was observed in the promotion of second leaf sheath elongation (93)—second leaf lamina bending in intact dwarf rice seedlings, and lamina bending in rice tissue sections (98–102). Inhibitors such as 2,3,5-triiodobenzoic acid (TIBA) and p-chlorophenoxyisobutyric acid (PCIB) nullified both BR-induced bending and BR-auxin synergized bending (39, 116). Katsumi's work on cucumber hypocotyl elongation with BR supports the synergism, which depends on the IAA concentration (39). This work showed that once PCIB and kinetin have inhibited the BR-induced elongation, increasing BR concentration does not produce recovery (39).

Further evidence for BR-auxin synergism comes from the data on auxin-induced ethylene production in etiolated mung bean hypocotyl segments (7, 8, 74). BR stimulates this response in a synergistic manner. However, in light-grown mung bean sections, BR does not increase ethylene production (J. H. Yopp, unpublished results). Cohen & Meudt (19) emphasized that the action of BR is not related to IAA metabolism. They demonstrated that BR affects neither IAA synthesis nor its breakdown in the first-internode sections of Phaseolus vulgaris. These investigators examined only the IAA metabolism and not the fate of BR in the BR-auxin synergistic response.

Although BR does not inhibit the initiation of adventitious root meristems in the mung bean epicotyl bioassay, it inhibits their outgrowth to form the visible adventitious roots, apparently interfering with the action of endogenous auxin (L. E. Gregory, unpublished results). These results suggest that, at least in root systems, the action of BR is either independent of auxin or interferes with endogenous auxin. In most tests, BR action is mediated through endogenous auxin, a fact established by the exogenous application of auxin, resulting in a synergism.

Tissue sensitivity of hormones has been the subject of considerable interest in recent years. Losing support is the old concept of hormonal control in which the concentration of a specific substance at its receptor in the cellular compartment determines the magnitude of some cell-regulatory process(es). Trewavas (97a) recently proposed that growth substance sensitivity is the prime controlling variable in plant development and that this sensitivity varies with number of receptors in the responsive cells. Meudt et al (55) and Katsumi (39) noted that both tissue sensitivity and site specificity to BR are responsible for curvature of bean first internodes and cucumber hypocotyl sections. These workers found that pretreatment with BR resulted in sensitization of the tissue, which on further treatment with auxin exhibited a several-fold growth response (synergism). On the other hand, when the tissue was pretreated with exogenous auxin before exposing it to BR, no increase in auxin response (no synergism) was observed. Sections pretreated with auxin (in the absence of BR) did not show any sensitivity to auxin. In other words, BR serves as a potentiator (or modulator) in enhancing the auxin response. However, several questions remains to be answered: How sensitive is the tissue to BR (e.g. what are the rate-limiting steps, saturation limits, etc)? Does the tissue sensitized by pretreatment with BR respond to auxin in other test systems? Sasse (73) examined the place of BR in the sequential response to plant hormones in wheat coleoptile and dwarf pea segments, concluding that the peak of sensitivity due to BR lies between those of gibberellin (cytokinin) and auxin.

Marre, Lado, and their coworkers (10, 15–18, 52) reported that growth and acid secretion in Azuki bean epicotyls are affected by BR and auxin. They interpreted this to mean that the growth responses to BR and auxin are additive (perhaps "synergistic"). However, in maize root segments (69), BR significantly stimulates root growth. This root growth in maize is associated with an increase of acid secretion, but auxin inhibits such action. L. E. Gregory (unpublished results) found that BR induces both root and leaf growth in lettuce. The Italian group (15–18, 58, 69) concluded that, in roots, the actions of BR and auxin are elicited through different pathways. In the absence of sequential-treatment and/or tissue-sensitivity data, however, Katsumi (39) doubts this interpretation.

In test systems that require darkness, BR generally does not cause any growth. For example, Avena coleoptiles are insensitive to BR in a standard bioassay system normally carried out in the dark (112; M. J. Growchowsha, N. B. Mandava, unpublished results); but in the light, coleoptiles respond to BR just as they do to auxin (M. J. Growhowska, N. B. Mandava, unpublished results). The importance of radiant energy and the spectral quality of light was emphasized for BR-induced growth in beans, which correlated with chlorophyll content and assimilation of photosynthate (41–43). In soybean and mung bean tissues, the growth-promoting effect of BR occurs in light but not in dark (24, 112). The most important spectral region corresponds to red light (660 nm, 2.6 W/cm^2) (112).

Brassinosteroids were screened in several cytokinin bioassays (48, 92, 98). BR increased (48) the expansion of dark-grown cucumber cotyledons, but the response was only 50% of that induced by an equimolar amount of kinetin. Cytokinins, but not BR, produced lateral leaf expansion of etiolated dwarf pea hooks. BR was only weakly active at 0.1 μM in the stimulation of lateral expansion of pea apexes, but it stimulated elongation. BR and kinetin produced opposite effects in bioassays involving the retardation of senescence of dark-grown *Xanthium* leaf discs, wherein BR promoted senescence. In dark-grown *Amaranthus* seedlings, kinetin promoted betacyanin formation, but BR was inactive in this test. While BR did not promote division in cultured cells of carrot *(Daucus carota)*, it did produce cell enlargement (71). On the other hand, Takematsu et al (92) found that BR and auxin in combination stimulated growth of callus tissues of a number of plants more effectively than did auxin and benzyladenine. Wada et al (98) found that BR is highly active in the wheat leaf unrolling bioassay; cytokinins are weakly active, and auxins inhibit the response.

The effect of BR was compared with that of such known growth regulators as fusicoccin, malformin, podolactones, dihydroconiferyl alcohol, and prostaglandins (J. M. Sasse, J. H. Yopp, N. B. Mandava, unpublished results) in bioassays involving pea segments and the greening of barley leaves and radish cotyledons. Some of these and other regulators—namely, auxin synergists (chlorogenic acid and diethyl dithiocarbamate), colchicine (a microtubule disorganizer), and AVG (aminoethyl vinyl glycine, a rhizobitoxine analogue)—were also tested against BR-induced responses in several test systems (116). BR was the most effective of all the growth regulators in eliciting the elongation of the tissue.

To summarize, in many test systems BR interacts strongly with auxins, perhaps synergistically. The response of BRs and GAs appear to be both independent and additive. In systems designed to assay for cytokinins, BR effects vary. ABA interacts strongly with BR and prevents the effects BR induces. BR alone and in combination with auxin induces the synthesis of ethylene, perhaps between SAM and ACC.

The biological activity of the BRs distinguishes them from the known groups of plant hormones. As pointed out by Geuns (27), since BR is active in various bioassays thought to be specific for different groups of plant hormones, the specificity and usefulness of such assays must be treated with caution.

EFFECTS ON NUCLEIC ACID AND PROTEIN METABOLISM AND ON OTHER ENZYMES

Cell elongation and cell division resulting in tissue growth require the synthesis of nucleic acids and proteins. Plant hormones such as the auxins, gibberellins, and cytokinins regulate nucleic acid metabolism in plants (40). Mandava and his coworkers attempted to determine whether BR affects nucleic acid and protein metabolism. First, putative inhibitors of RNA and protein synthesis were tested for their effects on the BR-induced response in mung bean epicotyle cuttings. RNA synthesis inhibitors (e.g. actinomycin D, 6-methyl purine, 2,6-diaminopurine, 8-azaguanine, 2-azauracil, 2-thiouracil, and 5-fluorouracil) and protein synthesis inhibitors (including cycloheximide, puromycin, and chloramphenicol) were examined for their effects on BR-induced epicotyl growth. All interfered with epicotyl growth, actinomycin D and cycloheximide to the greatest extent (47, 51). The effects caused by these inhibitors appears to be overcome by BR when the inhibitor-treated tissue was washed with water and then exposed to BR. This procedure overcame the inhibitory response and produced an additional growth-promoting effect. If the BR-pretreated tissue is further treated with the inhibitors, the BR-induced promotion is completely arrested. These effects are reversible and concentration dependent. The inhibitors affect cell elongation and cell division, a result obtained from sequential response studies with BR and the inhibitors. Cerana et al (15–18) reported that inhibitors such as actinomycin D and cordycepin completely suppress the BR-induced growth and proton extrusion in Azuki bean epicotyls. These inhibitors also affect the BR-induced response in pea segments (72). These studies clearly indicate that the growth effects induced by BR as a result of cell elongation and cell division, like those induced by auxins and gibberellins, depend upon the synthesis of nucleic acids and cellular proteins.

In another study, Kalinich et al (38) found that BR treatment significantly increased RNA and DNA polymerase activities, and the synthesis of RNA, DNA, and protein in beans (*P. vulgaris* L.) and mung beans (*P. aureus Roxb*). This finding suggests the involvement of BRs in transcription and replication during tissue growth. Under the influence of BR, changes in enzyme activities apparently affect nucleic acid metabolism such that the levels of accumulated RNA, DNA, and proteins in the tissue increase during growth. Although BR causes an increase in cellular protein, it does not substantially affect peroxidase and polyphenol oxidase activities in mung beans. This suggests an induction of the synthesis of specific proteins by BR rather than an indiscriminate increase in overall protein synthesis (J. F. Kalinich, N. B. Mandava, unpublished results).

BR treatment increases ATPase activity in Azuki bean epicotyls and maize roots (15–18). This activity is correlated with strong acid secretion in those plants. Katsumi (39) reported that a membrane-bound ATPase inhibitor, dicyclohexyl carbodiimide, affected BR-induced elongation in cucumber hypocotyls. These studies suggest the involvement of ATPase in BR-induced growth in plants. BR is also implicated in increasing the rate of dark CO_2 fixation, a reliable indication of the rate of cytoplasmic malate synthesis via phosphoenolpyruvate (PEP) carboxylase. Additionally, BR-induced leaf growth in wheat corresponds with increased carboxylase activity, which results in increased accumulation of Fraction 1 protein (11, 12). The overall BR stimulation of in vitro ribulose bisphosphate carboxylase (RubPC-ase) activity corresponds with the enhanced in vivo CO_2-fixation rate in mustard and wheat plants. By enhancing the activation state of these enzymes, BR appears to increase the levels of soluble proteins and reduce those of sugars in wheat and mustard plants (11).

PRACTICAL APPLICATIONS IN AGRICULTURE

Mitchell and his coworkers (60; J. W. Mitchell, J. F. Worley, L. E. Gregory, unpublished results) reported that brassins increased growth in beans and soybeans and in woody plants such as Siberian elm. After the isolation of brassinolide and subsequent synthesis of BRs and their analogues, the USDA conducted limited tests in greenhouses and small field plots on a few vegetable and root crops with BR analogues containing $22\beta,23\beta$-oriented hydroxyls and 24β-oriented methyl (Figure 2, V c), and $22\alpha,23\alpha$-oriented hydroxyls and 24β-oriented methyl (Figure 2, V a). Meudt and coworkers (56, 57) reported that BR enhanced maturation and increased crop yields of several varieties of vegetables, including lettuce, radishes, pepper, and bush beans, as well as barley and potatoes (L. E. Gregory, unpublished results). Yields also improved when BR was applied at $1–10$ μg per 10 mg of lanolin directly on axils of the first pair of leaves or on the eyes of potatoes in greenhouse experiments, and at $0.1–0.001$ mg/liter of spray containing 0.1% Tween 20 on the leaves of barley, lettuce, and other crops in the field experiments. BR significantly increased the fresh weights and yield of these crops (6–40%), regardless of the mode of application. BR also increased the growth of wheat and mustard plants (11, 12). Growth responses induced by BR occur mainly in the slow-growing tissues; therefore the phenotypic variability of the seedlings is reduced (23, 60).

In Japan, BRs (especially brassinolide and homobrassinolide) have been evaluated for use in increasing crop yield, stress tolerance, and disease resistance (22, 29, 92, 112). BR increased the ear weights of both tillers and main stem as well as the grain-setting and grain weight in wheat when applied to the crop at the beginning of anthesis. BR sprayed on the ear and silk of corn plants significantly increased yield (18–33%). BR accelerated the growth of rice and tobacco plants, especially under low-temperature conditions, when seeds were coated with the hormone before germination. BR increased the yield (112) of potatoes (normal and sweet varieties) when applied to the planting stage. The yield of the fruit crops, corn, wheat, and such vegetables as tomato, cucumber, and eggplant increased when the plants were treated with BR at their flowering stages. BR application to crops also helped to overcome environmental stress. Crops such as eggplant, cucumber, and rice were more resistant to cold when they were treated with BR, perhaps owing to recovery of endogenous auxins by exogenous application of BR. BR helped to prevent rice sheath blight, cucumber gray mold, and tomato late blight, but not rice blast or cucumber powdery mildew. BR diminishes herbicidal injury to rice by simetrin, butachlor, and pretilachlor, and to wheat by simazine, perhaps by reducing transpiration and herbicidal absorption and by counteracting the photosynthetic inhibition caused by herbicides. BR also appears to overcome harmful salt tolerance in water-cultivated rice plants (29).

Although these findings are encouraging, they result from preliminary experiments in greenhouses and limited field plots. Several parameters must be investigated further (e.g. formulation, time, and method of application) before the full potential of BRs for increasing the biomass and yield of crops, and for controlling diseases and environmental stress, can be realized.

PROPOSED FUNCTIONS

The role of endogenous BRs in plant growth and development has not been fully established. Evidence of biological activity in various test systems suggests that BRs have a role in cell elongation; data from a few test systems suggest a role in cell division. Available information on the numerous effects of BRs in a wide range of test systems may be summarized as follows:

1. When BR is applied to growing tissues, it usually promotes growth (e.g. cell enlargement). This growth is sometimes less apparent when BR is applied to intact plants than when isolated or excised parts are used (24, 76, 115). BR promotes growth in young growing tissues, particularly the meristems (23, 24, 48, 60, 115).
2. BR has little or no effect on nongrowing mature tissue (24). Tissue response to BR is affected by the spectral quality and intensity of the light source used to grow the plants (40–43, 112).
3. BR (a) elicits proton extrusion (acid secretion) and (b) inhibits adventitious root emergence (but not formation of root initials) from hypocotyl and epicotyl cuttings (15–18, 24, 50, 52).

Such effects of exogenous application do not necessarily imply that endogenous BR functions as a growth regulator, but a growth-regulatory role for BR is inferred from the following observations:

1. In many test systems BR is active at concentrations ($\leqslant$ nM) similar to (or lower than) those associated with the optimal activities of other plant hormones (24, 48, 55, 88, 93, 98, 115). The concentrations of endogenous BRs are in parts-per-billion to parts-per-trillion levels.
2. BR frequently induces responses similar to those induced by the main classes of growth-promoting hormones (auxins, gibberellins, and cytokinins) in various plant bioassay systems (7, 8, 24, 40–43, 48, 55, 88, 93, 98, 115). In addition, BR elicits growth responses previously obtained with either auxin or gibberellin but not both (112, 116).

3. BR appears to be translocated, since it affects the growth of the whole plant when used to pretreat the seed (23, 60). Furthermore, as shown in the mung bean, treating hypocotyls (24, 50, 51) affects epicotyl growth, and treating the meristem results in root growth.
4. The growth-promoting effects of BR can sometimes be arrested by abscisic acid (N. B. Mandava, J. H. Yopp, unpublished results) and other growth inhibitors (116).
5. BRs appear to be widely distributed in higher plants and may be present in lower plants (50, 104, 112).

BRs may thus be regarded as a new group of plant hormones (22, 29, 50, 104, 112) with a regulatory function in cell elongation and cell division; they may also have a role in the control of RNA and/or protein synthesis (38). They interact with plant hormones and other growth substances. Such interactions are not unusual. For example, both auxin and gibberellin control cell elongation in shoots; both auxin and cytokinin (and sometimes gibberellin) control cell division; gibberellin and abscisic acid control seed germination; and ethylene, whose formation is stimulated by auxin, modifies lateral transport of auxin (93). Hence only through interactions like those reported above can BR play a regulatory role in plant growth and development; however, the nature of such interactions is obscure at this stage of brassinosteroid research.

PROPOSED BIOSYNTHESIS

Sterols such as sitosterol occur widely in plants, and the question naturally arises whether they have functions similar to those of cholesterol in animals (27). Androgens and estrogens are steroid hormones and are derived from cholesterol via oxidative reactions.

In the absence of information to the contrary, one can visualize the synthesis of BRs from plant sterols. The major sterols in rape pollen, for example, are 24-methylenecholesterol and brassicasterol (M. J. Thompson, N. B. Mandava, unpublished results). Either of these or 22-dehydrocampesterol could serve as an intermediate to BR via a stereospecifically controlled synthesis. Similar biosynthetic processes could produce BR in other plants.

Based on the proposed structures of BRs, one could develop the following arguments:

1. *Side chain hydroxylation:* It appears that the hydroxylation on the steroid side chain proceeds from the precursor sterols such as VII or VIII to give a vicinal glycol with substitution of C-22 and C-23. During introduction of the glycol group, a *cis* geometry is established. This conclusion is based on the natural occurrence of typhasterone and theasterone (2, 3, 35, 75,

104, 106, 112), which lack other functional groups in rings A and B but exhibit a weak brassin activity in the rice lamina-bending bioassay (112).

2. *Hydroxylation of ring A:* The next step in the biosynthesis of BRs could be the introduction of a vicinal glycol group of C-2 and C-3 of ring A of the steroid nucleus. This may involve a dehydration step from either a 3α- or a 3β-hydroxyl to give an unsaturation between C-2 and C-3, followed by hydroxylation. The biosynthetic sequence may follow a pathway similar to that of the ring A hydroxylation step in sapogenins, or it may involve enzymes such as dehydrogenase prior to hydroxylation. Compounds such as typhasterone and theasterone could undergo hydroxylation at C-2 and C-3, irrespective of their geometry at C-3 (possibly through a dehydration step).

3. *6-Ketone formation:* Oxidation of the steroid B ring is another step that might proceed independently. Because 6-deoxobrassinosterone (BR_5) and 6-deoxodolichosterone (BR_6) occur along with brassiosterone (BR_2) and dolichosterone (BR_4) in *P. vulgaris* seed, one could postulate that oxidation in ring B is elaborated after the introduction of the vicinal glycol group in ring A and the formation of *trans*-A/B ring fusion. These deoxo derivatives may be considered as the precursors of their corresponding ketones (brassinosterone and dolichosterone). These nonoxidized B-ring brassinosteroids show very weak activity in the bioassays, suggesting that they are not easily convertible into biologically active congeners (104, 112).

4. *Alkylation at C-24:* The classification of BRs into C27, C28, and C29 steroids is based on the functional groups at C-24 (C-28 substituent)—that is, no substituent (as in C27 steroids); a methyl or an exomethylene (as in C28 steroids); or an ethyl or an (E)-ethylidene groups (as in C29 steroids). The alkyl (methyl and ethyl) substituents at C-24 are all α-oriented in BRs from higher plants. But recently 24β-brassinosterone (BR_9), containing a 24β-methyl (24S-methyl) group, was isolated from green algae *(Hydrodictyon reticulatum)* (29, 112). Such differences in the C-24 substituents are well documented for sterols from higher plants (though this is not the case for lower plants) and could be attributed to their different biosynthetic pathways. Further work is needed to assess the biosynthetic relationship between BR and the major phytosterols with respect to C-24 alkylation.

5. *B-ring oxidation:* It has been suggested that the B-ring lactone could originate from a 6-ketosteroid precursor (e.g. brassinosterone and dolichosterone) via a Baeyer-Villiger-type reaction (26, 50). Recent work (47, 112) indicates that such C-6 precursor ketones and BRs containing lactones are present in pollen and other plants (Tables 1 and 3). Additionally, these naturally occurring ketones and their synthetic analogues were highly active in the bioassay test systems (7, 24, 88, 96, 97, 101).

In view of the extremely low concentrations (ppb levels) of brassinolide and the other lactones in pollen, fruit, seed, and other plant parts (Tables 1 and 3), determining the biosynthetic pathway may prove difficult. Perhaps the initial step is to determine which precursor sterol (see Figure 3) is enzymatically converted to a ketone (or to the intermediate steps postulated in Figure 3) and then to the corresponding lactone. Establishing the biosynthetic sequence from a known plant sterol to the corresponding brassinosteroid should then not prove difficult.

PROBLEMS AND PROSPECTS

This area of plant hormone research has not yet caught the attention of many plant scientists. The limiting factor at this moment is the availability of brassinolide and/or its analogues from commercial sources. Chemists have already contributed their share to BR research by isolating and identifying several brassinosteroids from various higher plants. They will continue by isolating BRs from both higher and lower plants. In the past, chemists devised appropriate analytical (chromatographic and colorimetric) methods and then biologists used them to identify plant hormones and other growth regulators. Considerable progress has been made using this approach with other plant hormones. However, because BRs occur in extremely low concentrations (ppb to ppt levels), their analysis requires an even more sophisticated analytical methodology (derivitization techniques for isolating BRs followed by characterization utilizing GC-mass spectrometry for selectively monitoring the characteristic fragment ions). Thus, unless biologists are trained in modern analytical chemical methods, any future efforts must be undertaken jointly by chemists and biologists. (The Japanese have already demonstrated the fruitfulness of this approach.) Immunoassays also show great promise in detecting BRs (32), and further work is needed to combine this technique with other analytical methods.

Chemists have already synthesized all 16 of the isolated BR compounds as well as several other analogues. The structure-activity relationship for brassinosteroids has been established. Three patents have been granted on the synthesis of brassinosteroids and several other analogues (64, 77, 96). Industry must now either follow the various methods outlined or modify them for commercial production of either BRs or their analogues. Several Japanese companies are exploring the commercial production of BRs, including the most active compound, brassinolide. The next major task confronting chemists and biochemists is to investigate the metabolism and biosynthesis of BRs. Using radiolabeled compounds (deuterium-labeled BRs have already been synthesized), researchers can establish the biosynthetic and metabolic pathways for BRs.

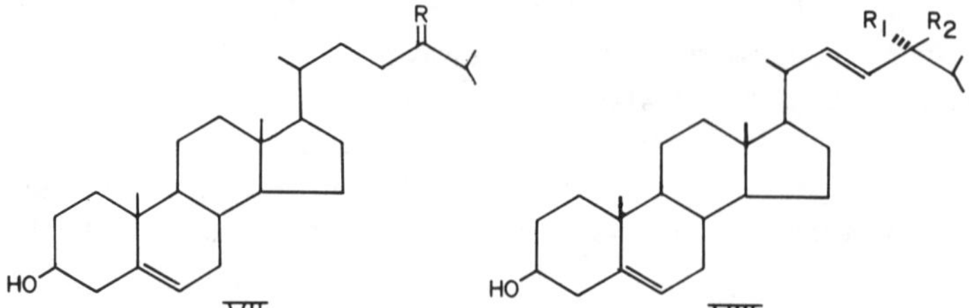

a) R=CH$_2$; 24-Methylenecholesterol   a) R$_1$=CH$_3$, R$_2$=H; Brassicasterol

b) R = trans-CH-CH$_3$; Isofucosterol   b) R$_1$=H, R$_2$=CH$_3$; Stigmasterol

Precursor Sterols

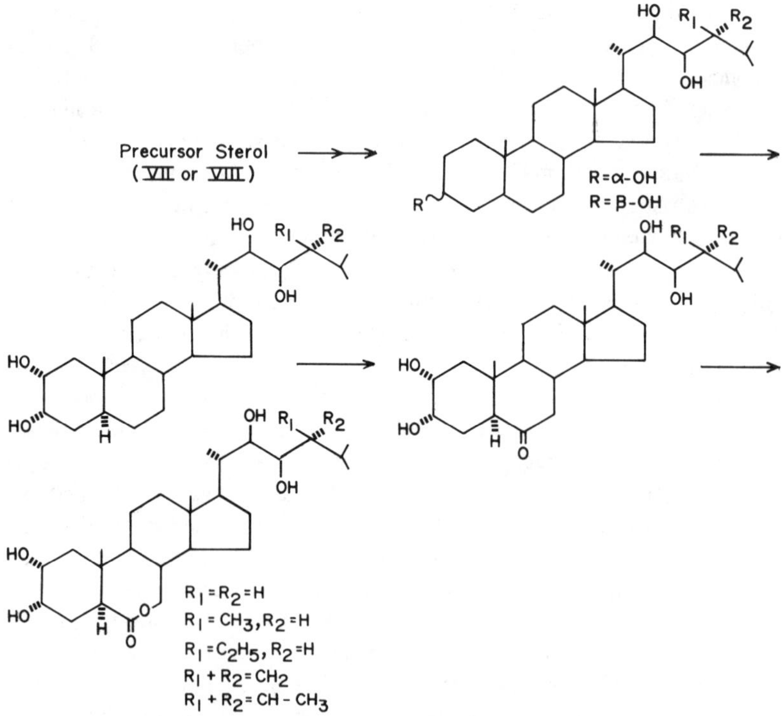

R$_1$=R$_2$=H
R$_1$ = CH$_3$, R$_2$=H
R$_1$=C$_2$H$_5$, R$_2$=H
R$_1$ + R$_2$= CH$_2$
R$_1$ + R$_2$= CH - CH$_3$

Figure 3   Proposed biosynthetic pathway for brassinosteroids.

Little has been accomplished toward understanding the physiological function(s) of the BRs. BRs appear to differ from other hormones, though owing to the design of current bioassays the effects of BRs and these other plant hormones appear to be synergistic or additive. A few workers have claimed that BR functions only in conjunction with auxin. In certain tests this may be the case, but other aspects require study, especially the growth response in the whole plant. The consensus is that BRs exhibit a mode of action different from that of other hormones. Although testing brassinosteroids in isolated plant parts is rapid and useful, extrapolating the data (including those on interactions with other hormones and on each tissue's capacity to respond) to establish a BR's mode of action in the whole plant can be dangerous.

The mechanism of action of the BRs is unknown. Such questions as the following remain to be answered through experimentation.

Where in plant tissue do BRs act—the cell wall, the cytoplasmic region, or the nucleus? How do BRs affect such enzymes as peroxidase, polyphenol oxidase, ATPase, and RNA and DNA polymerases? Despite their association with an increase in cellular proteins, BRs do not appear to alter the levels of polyphenol oxidase and peroxidase activity, at least in the hypocotyls and epicotyls of mung beans (N. B. Mandava, J. F. Kalinich, unpublished results). These results indicate that BRs induce the synthesis of specific proteins rather than an indiscriminate increase in overall protein synthesis. ATPase may be involved in the BR-induced H^+ extrusion in Azuki beans (52); further work is needed to establish its role conclusively.

It has been reported that BR application to beans results in increased RNA and DNA polymerase activities that contribute to increased cellular levels of RNA, DNA, and protein (38). Furthermore, putative inhibitors of RNA and protein synthesis have been found to have a profound effect on BR-induced growth in beans (N. B. Mandava, J. H. Yopp, unpublished results), but little is known about the effects of DNA-replication inhibitors. Additionally, the effects of BR on adenylate cyclase, cAMP, protein kinase, phosphorylase, and a host of other enzymes implicted in growth phenomena require detailed investigations. No evidence exists concerning BR effects on polyamines such as putrescine, spermidine, and spermine (22). They may have a role in plant stress, because preliminary reports (21, 30) indicate that BR (and its stimulation of rapid growth) has a key role in plant stress.

Ahead for investigators of the brassinosteroids lie: verification of any special protein/enzymes formed when plants are treated with BRs; determination of whether BRs affect lipid, carbohydrate, and organic acid levels during growth and development; study of in vivo synthesis of BRs with the use of labeled sterols; investigation of the metabolic fate of BRs during growth and development; determination of any toxic or other environmental effects in anticipation of commercial applications of BRs; and development of microbiological and recombinant DNA methods for evaluating the potential of BRs in modern agriculture.

Concerted efforts are already under way. In Japan, an industrial-academic consortium with the support of the national government is investigating the practical applications of BRs.

Literature Cited

1. Abe, H., Morishita, T., Uchiyama, M., Marumo, S., Munakata, K., et al. 1982. Identification of brassinolide-like substances in Chinese Cabbage. *Agric. Biol. Chem.* 46:2609–11
2. Abe, H., Morishita, T., Uchiyama, M., Takatsuto, S., Ikekawa, N., et al. 1983. Occurrence of three new brassinosteroids: brassinone, (24S)-24-ethylbrassinone and 28-norbrassinolide, in higher plants. *Experientia* 39:351–53
3. Abe, H., Morishita, T., Uchiyama, M., Takatsuto, S., Ikekawa, N. 1984. A new brassinolide-related steroid in the leaves of *Thea sinensis. Agric. Biol. Chem.* 48:2171–72
4. Abe, H., Nakamura, K., Morishita, T., Uchiyama, M., Takatsuto, S., Ikekawa, N. 1984. Endogenous brassinosteroids of the rice plant: Castasterone and dolichosterone. *Agric. Biol. Chem.* 48: 1103–4
5. Akhrem, A. A., Lakhvich, F. A. Khripach, V. A., Kovganko, N. 1983. New way to introduce functional groups of brassinosteroids into A and B rings of delta-5-sterols. *Dokl. Akad. Nauk. SSSR* 269:366–68
6. Arima, M., Yokota, T., Takahashi, N. 1984. Identification and quantification of brassinolide-related steroids in the insect gall and healthy tissues of the chestnut. *Phytochemistry* 23:1587–91
7. Arteca, R. N., Bachmaa, J. M., Yopp, J. H., Mandava, N. B. 1985. Relationship of steroidal structure to ethylene production by etiolated mung bean segments. *Physiol. Plant.* 64:13–16
8. Arteca, R. N., Tsai, D. S., Schlagnhaufer, C., Mandava, N. B. 1983. The effect of brassinolide on auxin-induced ethylene production by etiolated mung bean segments. *Physiol. Plant.* 59:539–44
9. Baba, J., Yokota, T., Takahashi, N. 1983. Brassinolide-related new bioactive steroids from *Dolichos lablab* seed. *Agric. Biol. Chem.* 47:659–61
10. Bonetti, A., Cerana, R., Lado, P., Marre, E., Marre, M. T., Romani, G. 1983. Mechanism of action of the pollen

hormone brassinolide. In *Pollen Biology and Implications for Plant Breeding,* ed. L. Mulcahy, E. Ottaviano, pp. 9–14. Amsterdam: Elsevier
11. Braun, P., Wild, A. 1984. The influence of brassinosteroid, a growth-promoting steroidal lactone, on development and carbon dioxide fixation capacity of intact wheat and mustard seedlings. In *Advances in Photosynthesis Research. Proc. 6th Congr. Photosynthesis,* ed. C. Sybesma, 3:461–64. The Hague: Nijhoff
12. Braun, P., Wild, A. 1984. The influence of brassinosteroid on growth and parameters of photosynthesis of wheat and mustard plants. *J. Plant Physiol.* 116:189–96
13. Brooks, C. J. W., Ekhato, I. V. 1982. Highly regioselective reduction of ring B seco-5 alpha-steroid anhydrides to afford the lactone grouping characteristic of brassinolide. *J. Chem. Soc. Chem. Commun.* 1982:943–44
14. Donaubauer, J. R., Greaves, A. W., McMorris, T. C. 1984. A novel synthesis of brassinolide. *J. Org. Chem.* 49:2833–34
15. Cerana, R., Bonetti, A., Marre, M. T., Romani, G., Lado, P., Marre, E. 1983. Effects of a brassinosteroid on growth and electrogenic proton extrusion in Azuki bean epicotyls. *Physiol. Plant.* 59:23–27
16. Cerana, R., Colombo, R., Lado, P. 1983. Changes in malate content and dark CO_2 fixation associated with brassinosteroid- and indolacetic acid-induced changes in proton pump activity of maize roots. *Physiol. Veg.* 21:875–81
17. Cerana, R., Lado, P., Anastasia, M., Ciuffreda, P., Allevi, P. 1984. Regulating effects of brassinosteroids and of sterols on growth and H^+ secretion in maize roots. *J. Plant Physiol.* 114:221–25
18. Cerana, R., Spelta, M., Bonetti, A., Lado, P. 1985. On the effects of cholesterol on H^+ extrusion and on growth in maize root segments: Comparison with brassinosteroid. *Plant Sci.* 38:99–105

19. Cohen, J. D., Meudt, W. J. 1983. Investigations on the mechanism of the brassinosteroid response. I. Indole-3-acetic acid metabolism and transport. *Plant Physiol.* 72:691–94

20. Fung, S., Sidall, J. B. 1980. Stereoselective synthesis of brassinolide: a plant growth promoting steroidal lactone. *J. Am. Chem. Soc.* 102:6580–81

21. Fujita, F. 1985. Prospect of application of brassinolide to the crop production of agriculture. *Chem. Biol.* 23:717–25 (In Japanese)

22. Galston, A. W. 1983. Polyamines as modulators of plant development. *BioScience* 33:382–83

23. Gregory, L. E. 1981. Acceleration of plant growth through seed treatment with brassins. *Am. J. Bot.* 68:586–88

24. Gregory, L. E., Mandava, N. B. 1982. The activity and interaction of brassinolide and gibberellic acid in mung bean epicotyls. *Physiol. Plant.* 54:239–43

25. Grove, M. D., Spencer, G. F., Pfeffer, P. E., Mandava, N. B., Warthen, J. D., Worley, J. F. 1978. 6-Beta-glucopyranosyl fatty acid esters from *Brassica napus* pollen. *Phytochemistry* 17:1187–92

26. Grove, M. D., Spencer, F. G., Rohwedder, W. K., Mandava, N. B., Worley, J. F., et al. 1979. A unique plant growth promoting steroid from *Brassica napus* pollen. *Nature* 281:216–17

27. Geuns, J. M. C. 1983. Plant-steroid hormones. *Biochem. Soc. Trans.* 11:543–48

28. Hayami, H., Sato, M., Kanemoto, S., Morizawa, Y., Oshima, K., Nozaki, K. 1983. Transition-metal catalyzed silylmetalation of acetylenes and its application to the stereoselective synthesis of steroidal side chain. *J. Am. Chem. Soc.* 105:4491–92

29. Hamada, K. 1986. Brassinolide: some effects for crop cultivations. *Conf. Proc. Int. Seminar Plant Growth Regul. Tokyo, Japan, Oct. 15, 1985*

30. Henry, E. W., Dungy, L. J., Bracciano, D. M. 1981. The effect of brassinolide on growth and enzyme activity in mung bean (*Phaseolus aureusc* Roxb). *8th Proc. Plant Regul. Soc. Am.*, pp. 110–26

31. Hewitt, F. R., Hough, T., O'Neill, P., Saase, J. M., Williams, E. G., Rowan, K. S. 1985. Effect of brassinolide and other growth regulators on germination and growth of pollen tubes of *Prunus avium* using a multiple hanging-drop assay. *Aust. J. Plant Physiol.* 12:201–11

32. Horgan, P. A., Nakagawa, C. K., Irvin, R. T. 1984. Production of monoclonal antibodies to a steroid plant growth regulator. *Can. J. Biochem. Cell Biol.* 62:715–21

33. Ikeda, M., Takatsuto, S., Sassa, T., Ikekawa, N., Nukina, M. 1983. Identification of brassinolide and its analogues in chestnut gall tissue. *Agric. Biol. Chem.* 47:655–57

34. Ikekawa, N., Takasuto, S. 1984. Microanalysis of brassinosteroids in plants by gas chromatography/mass spectrometry. *Mass Spectrom.* 32:55–70

35. Ikekawa, N., Takatsuto, S., Kitsuwa, T., Saito, H., Morishita, T., Abe, H. 1984. Analysis of natural brassinosteroids by gas chromatography and gas chromatography-mass spectrometry. *J. Chromatogr.* 290:289–302

36. Ikekawa, N., Takatsuto, S., Marumo, S., Abe, H., Morishita, T., et al. 1983. Identification of brassinolide and its 6-oxo analog in the plant kingdom by selected ion monitoring. *Proc. Jpn. Acad.* 59:9–12

37. Ishiguro, M., Takatsuto, S., Morisaki, M., Ikekawa, N. 1980. Synthesis of brassinolide, a steroidal lactone with plant-growth promoting activity. *J. Chem. Soc. Chem. Commun.* 1980:962–64

38. Kalinich, J. F., Mandava, N. B., Todhunter, J. A. 1985. Relationship of nucleic acid metabolism to brassinolide-induced responses in beans. *J. Plant Physiol.* 120:207–14

39. Katsumi, M. 1985. Interaction of a brassinosteroid with IAA and GA$_3$ in the elongation of cucumber hypocotyl sections. *Plant Cell Physiol.* 26:615–25

40. Key, J. L. 1969. Hormones and nucleic acid metabolism. *Ann. Rev. Plant Physiol.* 20:449–74

41. Krizek, D. T., Mandava, N. B. 1982. Influence of spectral quality on the growth response of intact bean plants to brassinosteroids, a growth promoting steroidal lactone. I. Stem elongation and morphogenesis. *Physiol. Plant.* 57:317–23

42. Krizek, D. T., Mandava, N. B. 1982. Influence of spectral quality on the growth response of intact bean plants to brassinosteroid, a growth promoting steroidal lactone. II. Assimilate partitioning and chlorophyll content. *Physiol. Plant.* 57:324–29

43. Krizek, D. T., Worley, J. F. 1973. The influence of high intensity on the internodal response of intact bean plants to brassins. *Bot. Gaz.* 13:147–50

44. Mandava, N. B., Kozempel, M., Worley, J. F., Matthees, D., Warthen, J. D.,

et al. 1978. Isolation of brassins by pilot plant extraction of rape *(Brassica napus)* pollen. *Ind. Eng. Chem.* 17:351–54

45. Mandava, N. B., Mitchell, J. W. 1971. New plant hormones: chemical and biological investigations. *Indian Agric.* 15:19–31

46. Mandava, N. B., Mitchell, J. W. 1972. Structural elucidation of brassins. *Chem. Ind.* 1972:930–32

47. Mandava, N. B., Rao, M. M., Thompson, M. J., Spaulding, D. W. 1982. Brassinolide and other plant growth promoting substances in pollen. *5th Int. Congr. Pesticide Chem. (IUPAC), Kyoto, Japan, Aug. 29–Sept. 4.* (Abstr. IIId-15)

48. Mandava, N. B., Sasse, J. M., Yopp, J. H. 1981. Brassinolide, a growth promoting steroidal lactone. II. Activity in selected gibberellin and cytokinin bioassays. *Physiol. Plant.* 53:453–61

49. Mandava, N. B., Sidwell, B. A., Mitchell, J. W., Worley, J. F. 1973. Production of brassins from rape pollen: a convenient preparatory method. *Ind. Eng. Chem. Prod. Res. Dev.* 12:138–39

50. Mandava, N. B., Thompson, M. J. 1983. Chemistry and functions of brassinolide. In *Proceedings of the Isopentenoid Symposium,* ed. W. D. Nes, G. Fuller, L.-S. Tsai, pp. 401–31. New York: Dekker

51. Mandava, N. B., Thompson, M. J., Yopp, J. H. 1987. Effects of selected putative inhibitors of RNA and protein synthesis on brassinosteroid-induced growth in mung bean epicotyls. *J. Plant Physiol.* 128:63–68

52. Marre, E. 1982. Hormonal regulation of transport: data and perspectives. In *Plant Growth Substances,* ed. P. F. Wareing, pp. 407–17. London: Academic

53. Marumo, S. 1983. A new plant growth regulator, brassinolide. In *Biological and Bioactive Natural Products.* (Agricultural and Biological Chemistry Series), pp. 102–34. Tokyo: Agric. Chem. Soc. Japan (In Japanese)

54. Marumo, S., Hattori, H., Nanoyama, Y., Munakata, K. 1968. The presence of novel plant growth regulators in leaves of *Distylium racemosum* Sieb et Zucc. *Agric. Biol. Chem.* 32:528–29

55. Meudt, W. J., Thompson, M. J. 1983. Investigations on the mechanism of the brassinosteroid response. II. A modulation of auxin action. *10th Proc. Plant Growth Regul. Soc. Am.,* pp. 306–11

56. Meudt, W. J., Thompson, M. J., Bennett, H. W. 1983. Investigations on the mechanism of the brassinosteroid response. III. Techniques for potential enhancement of crop production. *10th Proc. Plant Growth Regul. Soc. Am.,* pp. 312–18

57. Meudt, W. J., Thompson, M. J., Mandava, N. B., Worley, J. F. 1984. Method for promoting plant growth. *Can. Patent No. 1173659.* Assigned to USA. 11 pp.

58. Michelis, M. I., Lado, P. 1986. Effect of a brassinosteroid on growth and H^+ extrusion in isolated radish cotyledons: comparison with the effect of benzyl adenine. *Physiol. Plant.* 68:603–7

59. Milborrow, B. V., Pryce, R. J. 1973. The brassins. *Nature* 243:46

60. Mitchell, J. W., Gregory, L. E. 1972. Enhancement of overall growth, a new response to brassins. *Nature* 239:254

61. Mitchell, J. W., Livingston, G. A. 1968. Methods of studying plant hormones and growth regulating substances. US Dept. Agric. *Agric. Handb. No. 336.* Washington, DC: GPO

62. Mitchell, J. W., Mandava, N. B., Worley, J. F., Plimmer, J. R., Smith, M. V. 1970. Brassins: a new family of plant hormones from rape pollen. *Nature* 225:1065–66

63. Mitchell, J. W., Skaggs, D. P., Anderson, W. P. 1951. Plant growth stimulating hormones in immature bean seeds. *Science* 114:159–61

64. Mori, K. 1980. Synthesis of a brassinolide analog with high plant growth promoting activity. *Agric. Biol. Chem.* 44:1211–12

65. Mori, K. 1981. Homobrassinolide and its synthesis. *Eur. Patent No. 0040517A2.* Assigned to Sumitomo Chemical Co. Ltd., 24 pp.

66. Mori, K. 1985. Synthetic chemistry of brassinolide and related brassinosteroids, new plant growth promoting substances. *J. Synth. Org. Chem.* 43:849–61 (In Japanese)

67. Mori, K., Sakakibara, M., Ichikawa, Y., Ueda, H., Okada, K., et al. 1982. Synthesis of (22 S, 23 S)-homobrassinolide and brassinolide from stigmasterol. *Tetrahedron* 38:2099–109

68. Morishita, T., Abe, H., Uchiyama, M., Marumo, S., Takatsuto, S., Ikekawa, N. 1983. Evidence for plant growth promoting brassinosteroids in leaves of *Thea sinesis. Phytochemistry* 22:1051–53

68a. Nes, W. R., McKean, M. L. 1977. *Biochemistry of Steroids and Other Isopentenoids.* Baltimore: University Park Press. 690 pp.

69. Romani, G., Marre, M. T., Bonetti, A., Cerana, R., Lado, P., Marre, E. 1983. Effects of brassinosteroid on growth and

electrogenic proton extrusion in maize root segments. *Physiol. Plant.* 59:528–32

70. Sakakibara, M., Mori, K. 1983. Improved synthesis of brassinolide. *Agric. Biol. Chem.* 47:663–64

71. Sala, C., Sala, F. 1985. Effect of brassinosteroid on cell division and enlargement in cultured carrot (*Daucus carota* L.) cells. *Plant Cell Rep.* 4:144–47

72. Sasse, J. M. 1985. Some characteristics of brassinolide-induced elongation. *12th Int. Plant Growth Substances Conf., Heidelberg, Aug.* (Abstr. RO6-04)

73. Sasse, J. M. 1985. The place of brassinolide in the sequential response to plant growth regulators in elongating tissue. *Physiol Plant.* 63:303–8

74. Schlagnhaufer, C., Arteca, R. N., Yopp, J. H. 1984. A brassinosteroid-cytokinin interaction on ethylene production by etiolated mung bean segments. *Physiol. Plant.* 60:347–50

75. Schneider, J. A., Yoshihara, K., Nakanishi, K., Kato, N. 1983. Typhasterol (2-deoxycastasterone): a new plant growth regulator from cattail pollen. *Tetrahedron Lett.* 24:3859–60

76. Steffens, G. L., Buta, J. G., Gregory, L. E., Mandava, N. B., Meudt, W. J., Worley, J. F. 1979. New plant-growth regulators from higher plants. In *Advances in Pesticide Science*, ed. H. Geissbuhler, pp. 343–46. London: Academic

77. Suntry Ltd. 1983. (22 R, 23 R)-B-Homo-7-oxa-5 alpha-cholestan-6-one—2 alpha, 3 alpha, 22, 23 tetraol. 24-Desmethyl brassinolide or norbrassinolide synthesis. *Jpn. Patent 82163400A2.* (cf. CA 98:198606)

78. Suziki, Y., Yamaguchi, I., Takahashi, N. 1985. Identification of castasterone and brassinone from immature seeds of *Pharbitis purpurea. Agric. Biol. Chem.* 49:49–54

79. Takatsuto, S., Ikekawa, N. 1982. Synthesis of (22 R, 23 R)-28-homobrassinolide. *Chem. Pharmacol. Bull.* 30:4181–85

80. Takatsuto, S., Ikekawa, N. 1983. Stereoselective synthesis of plant growth-promoting steroids, dolicholide and 28-norbrassinolide. *Tetrahedron Lett.* 24:773–76

81. Takatsuto, S., Ikekawa, N. 1983. Remote substituent effect on the regioselectivity in the Baeyer-Villiger oxidation of 5 alpha-cholestan-6-one derivatives. *Tetrahedron Lett.* 24:917–20

82. Takatsuto, S., Ikekawa, N. 1983. Stereoselective synthesis of the plant-growth promoting steroids, dolicholide

and dolichosterone. *J. Chem. Soc. Perkin Trans. I* 1983:2133–37

83. Takatsuto, S., Ikekawa, N. 1984. Synthesis and activity of plant growth-promoting steroids, (22 R, 23 R, 24 S)-28-homobrassinosteroids, with modifications in rings A and B. *J. Chem. Soc. Perkin Trans. I* 1984:439–47

84. Takatsuto, S., Ikekawa, N. 1984. Short-step synthesis of plant growth-promoting brassinosteroids. *Chem. Pharm. Bull.* 32:2001–2004

85. Takatsuto, S., Ikekawa, N. 1986. Synthesis of [26, 28−^{2}H6] brassinolide, [26, 28−^{2}H6] castasterone, [26, 28−^{2}H6] typhasterol and [26, 28−^{2}H6] teasterone. *Chem. Pharmacol. Bull.* 34:1415–18

86. Takatsuto, S., Yazawa, N., Ikekawa, N. 1984. Synthesis and biological activity of brassinolide analogues, 26,27-bisnor-brassinolide and its 6-oxo analogues. *Phytochemistry* 23:525–28

87. Takatsuto, S., Yazawa, N., Ikekawa, N., Morishita, T., Abe, T. 1983. Synthesis of (24 R)-28-homobrassinolide analogues and structure-activity relationship of brassinosteroids in the rice-lamina inclination test. *Phytochemistry* 22:1393–97

88. Takatsuto, S., Yazawa, N., Ikekawa, N., Takematsu, T., Takeuchi, Y., Koguchi, M. 1983. Structure-activity relationship of brassinosteroids. *Phytochemistry* 22:2437–41

89. Takatsuto, S., Yazawa, N., Ishiguro, M., Morisaki, M., Ikekawa, N. 1984. Stereoselective synthesis of plant growth promoting steroids, brassinolide, castasterone, typhasterol and their nor analogues. *J. Chem. Soc. Perkin Trans. I* 1984:139–46

90. Takatsuto, S., Ying, B., Morisaki, M., Ikekawa, M. 1971. Synthesis of 28-norbrassinolide. *Chem. Pharmacol. Bull.* 29:903–5

91. Takatsuto, S., Ying, B., Morisaki, M., Ikekawa, N. 1982. Microanalysis of brassinolide and its analogues by gas chromatrography and gas chromatography-mass spectrometry. *J. Chromatogr.* 239:233–41

92. Takematsu, T., Takenchi, Y., Koguchi, M. 1983. New plant growth regulators. Brassinolide analogues: their biological effects and application to agriculture and biomass production. *Chem. Regul. Plants* 18:2–15 (In Japanese)

93. Takeno, K., Pharis, R. P. 1982. Brassinolide-induced bending of the lamina of dwarf rice seedlings: an auxin-mediated phenomenon. *Plant Cell Physiol.* 23:1275–81

94. Thompson, M. J., Mandava, N. B.,

Flippen-Anderson, J. L., Worley, J. F., Dutky, S. R., Robbins, W. E. 1979. Synthesis of brassinosteroids. New plant growth promoting steroids. *J. Org. Chem.* 44:5002–4

95. Thompson, M. J., Mandava, N. B., Meudt, W. J., Lusby, W. R., Spaulding, D. W. 1981. Synthesis and biological activity of brassinolide and its 22,23-isomer. Novel plant growth promoting steroids. *Steroids* 38:567–80

96. Thompson, M. J., Mandava, N. B., Worley, J. F., Dutky, S. R., Robbins, W. E., Flippen-Anderson, J. L. 1982. Plant growth promoting brassinosteroids. *US Patent No. 4,346,226*. Aug. 24. 12 pp.

97. Thompson, M. J., Meudt, W. J., Mandava, N. B., Dutky, S. R., Lusby, W. R., Spaulding, D. W. 1982. Synthesis of brassinosteroids and relationship of structure to plant growth-promoting effects. *Steroids* 39:89–105

97a. Trewavas, A. J. 1982. Growth substance sensitivity: the limiting factor in plant development. *Physiol. Plant.* 55:60–72

98. Wada, K., Kondo, H., Marumo, S. 1985. A simple bioassay for brassinosteroids: a wheat leaf-unrolling test. *Agric. Biol. Chem.* 49:2249–51

99. Wada, K., Marumo, S. 1981. Synthesis and plant growth-promoting activity of brassinolide analogues. *Agric. Biol. Chem.* 45:2579–85

100. Wada, K., Marumo, S., Abe, H., Morishita, T., Nakamura, K., et al. 1984. A rice lamina inclination test—a micro-quantitative bioassay for brassinosteroids. *Agric. Biol. Chem.* 48:719–26

101. Wada, K., Marumo, S., Ikekawa, N., Morisaki, M., Mori, K. 1981. Brassinolide and homobrassinolide promotion of lamina inclination of rice seedlings. *Plant Cell Physiol.* 22:323–35

102. Wada, K., Marumo, S., Mori, K., Takatsuto, S., Morisaki, M., Ikekawa, N. 1983. The rice lamina inclination-promoting activity of synthetic brassinolide analogues with a modified side chain. *Agric. Biol. Chem.* 47:1139–41

103. Worley, J. F., Mitchell, J. W. 1971. Growth responses induced by brassins (fatty plant hormones) in bean plants. *J. Am. Soc. Hort. Sci.* 96:270–73

104. Yokota, T. 1984. Brassinosteroids in higher plants. *Chem. Regul. Plants* 19:102–9 (In Japanese)

105. Yokota, T., Arima, M., Takahashi, N. 1982. Castasterone, a new phytosterol with plant-hormone activity from chestnut insect gall. *Tetrahedron Lett.* 23:1275–78

106. Yokota, T., Arima, M., Takahashi, N., Crozier, A. 1985. Steroidal plant growth regulators, castasterone and typhasterol (2-deoxycastasterone) from shoots of sitka spruce *(Picea sitchensis)*. *Phytochemistry* 24:1333–35

107. Yokota, T., Arima, M., Takahashi, N., Takatsumo, S., Ikekawa, N., Takematsu, T. 1983. 2-Deoxycastasterone, a new brassinolide-related bioactive steroid from *Pinus* pollen. *Agric. Biol. Chem.* 47:2419–20

108. Yokota, T., Baba, J., Koba, S., Takahashi, N. 1984. Purification and separation of eight steroidal plant growth regulators from *Dolichos lablab* seed. *Agric. Biol. Chem.* 48:2529–34

109. Yokota, T., Baba, J., Takahashi, N. 1982. A new steroidal lactone with plant growth-regulatory activity from *Dolichos lablab* seed. *Tetrahedron Lett.* 23:4965–66

110. Yokota, T., Baba, J., Takahashi, N. 1983. Brassinolide-related bioactive sterols in *Dolichos lablab:* brassinolide, castasterone and an analog, homodolicholide. *Agric. Biol. Chem.* 47:1409–11

111. Yokota, T., Morita, M., Takahashi, N. 1983. 6-Deoxocastasterone and 6-deoxodolicholide: putative precursors for brassinolide-related steroids from *Phaseolus vulgaris. Agric. Biol. Chem.* 47:2149–51

112. Yokota, T., Takahashi, N. 1986. Chemistry, physiology and agricultural application of brassinolide and related steroids. In *Plant Growth Substances 1985,* ed. M. Bopp, pp. 129–38. Berlin/ Heidelberg: Springer-Verlag

113. Yokota, T., Koba, S., Kim, S. K., Takatsuto, S., Ikekawa, N., Sukakibara, M., Okada, K., Mori, K., Takahashi, N. 1987. Diverse structural variations of the brassinosteroids in *Phaseolus vulgaris* seeds. *Agric. Biol. Chem.* 51:1625–31

114. Yopp, J. H., Colclasure, G. C., Mandava, N. B. Effect of brassin-complex on auxin and gibberellin mediated effects in the morphogenesis of the etiolated bean hypocotyl. *Physiol. Plant.* 46:247–54

115. Yopp, J. H., Mandava, N. B., Sasse, J. M. 1981. Brassinolide, a growth promoting steroidal lactone. I. Activity in selected auxin bioassays. *Physiol. Plant.* 53:445–52

116. Yopp, J. H., Mandava, N. B., Thompson, M. J., Sasse, J. M. 1981. Brassinosteroids in selected bioassays. *8th Proc. Plant. Regul. Soc. Am.,* pp. 110–26

Ann. Rev. Plant Physiol. Plant Mol. Biol. 1988. 39:53–99

COATED VESICLES

D. G. Robinson and Hans Depta

Abteilung Cytologie des Pflanzenphysiologischen Instituts, Universität Göttingen, D-3400 Göttingen, West Germany

CONTENTS

INTRODUCTION 54
COATED MEMBRANES IN THE CELL 54
    The Plasma Membrane (PM) 54
    The Golgi Apparatus 57
    Other Membranes 57
    Coated Vesicles (CVs) 58
ISOLATION OF COATED VESICLES 59
    Sources for Isolation 59
    Methods for Purification 59
    CV Isolation from Plants 60
    Sizes and Yields of CVs 61
THE COAT PROTEINS 61
    Individual Proteins and Their Properties 61
    Coat Architecture and Coat-Vesicle Interactions 68
    Dissociation and Reassembly of Coat Proteins in vitro 69
    The Dynamics of Coat Proteins in vivo 73
LIPIDS IN COATED VESICLES 75
ENZYME ACTIVITIES ASSOCIATED WITH COATED VESICLES 76
    Kinases 76
    ATPase 76
    Glucan Synthases 77
FUNCTIONS OF COATED VESICLES 79
    In Animal Cells 79
    In Yeasts 84
    In Plants 85
COATED VESICLES IN PLANTS: PERSPECTIVES FOR THE FUTURE ... 86

0066-4294/88/0601053$02.00

INTRODUCTION[1]

Membranes bearing a coat on their cytoplasmic surface first became recognized as a general feature of eukaryotic cells in the 1960s. Circular, vesicular profiles bearing such coats were reported in numerous cell types and given a variety of names in the early electron microscopy literature—e.g. spiny-coat vesicles, bristle-coat vesicles, alveolate vesicles (see 157, 262)—but only the most noncommittal of these, the "coated vesicle" (CV), has stood the test of time. Research on CVs remained of a descriptive nature until 1975 when a paper from Pearse (172) detailed a method for the isolation of CVs from porcine brain tissue. Proceeding parallel to the biochemical characterization of CVs, the last ten years has also seen a renaissance of studies dealing with a classic phenomenon of animal physiology: endocytosis. The demonstration that CVs play an integral role in this process has constituted one of the most significant advances in cell biology in recent times. The range and importance of the extracellular macromolecules taken up into animal cells via CVs (see, for example 74 and Table 1 in 234) have produced much interest in and support for CV research in the biomedical area. By contrast the role(s) of coated vesicles in plant cells has yet to be defined.

This is the first review on CVs to appear in this series. In surveying the literature (completed in June 1987) it became clear that in our report papers devoted to CVs in animal cells would outnumber those dealing with plants by at least ten to one. Despite this imbalance, we hope to generate critical interest among plant scientists.

COATED MEMBRANES IN THE CELL

The Plasma Membrane (PM)

Coats on the cytoplasmic surface of the PM have been demonstrated by thin sectioning, by negative staining, by freeze etching, and by cleaving, thus reducing the possibility that they are artefacts of preparation. In cross sections of the PM the coat usually takes the form of bristles or spikes (see, for

[1]*Abbreviations*: AChE—Acetylcholine esterase; AChR—Acetylcholine receptor; CHO—Chinese hamster ovary; CP—Coated pit; CV—Coated vesicle; DCCD—N, N'-dicyclohexylcarbodiimide; D_2O—Deuterium oxide; DTT—Dithiothreitol; EGF—Epidermal growth factor; EGTA—Ethylene glycol-bis (β-amino-ethyl ether) N,N,N',N'-tetra-acetic acid; EM—Electron microscopy; ER—Endoplasmic reticulum; FCCP—Carboxyl-p-trifluoromethoxyphenyl hydrozone; FITC—Fluorescein isothiocyanate; FPLC—Fast performance liquid chromatography; HEPES—N-2-Hydroxyethylpiperazine-N-2-ethanesulfonic acid; IgG—Immunoglobulin G; LDL—low density lipoprotein; MES—2- N-morpholino ethanosulfonic acid; NPA—Naphthylphthalamic acid; NBD-Cl—7-chloro-4-nitrobenz-2 oxa-1,3 diazole; NEM—N-ethyl-maleimide; PCR—Partially coated reticulum; PM—Plasma membrane; PMSF—Phenylmethylsulfonyl fluoride; RME—Receptor-mediated endocytosis; SDS—Sodium dodecylsulfate; VSV—Vesicular stomatitis virus

example, Figure 11 in 213 and Figure 5 in 260) and becomes increasingly recognizable the more that portion of the PM begins to protrude into the cell. This does not mean that elements of the coat become more defined during this invagination (or fusion!) process, it merely refers to the fact that one tends to register such coated pits (CPs) before one becomes aware of coated, non-indented, portions of the PM. This is particularly the case when the pits are filled with an electron-dense substance, be it of natural origin or added exogenously. The best examples of this are probably oocytes from a variety of animals (206) where the necessity to take up large quantities of proteins from the maternal vascular system is especially apparent.

Coats can easily be recognized on the cytoplasmic surface of the PM in conventionally fixed material, but their structure (in some animal cells at least) can be enhanced by fixing in the presence of tannic acid (137). The structure of the coating is, however, best appreciated when the PM is visualized in surface view. This is occasionally possible with tangential sections through the PM (e.g. Figure 10 in 213) but is more convincing when larger portions of the PM are made visible. The technique of ultrarapid freezing followed by deep etching and low-angle rotary shadowing, especially in the hands of Heuser and colleagues (87, 88, 91, 92), is particularly illuminating in this regard. As demonstrated in their papers the coat is seen to consist of rod-like units arranged into hexagons and pentagons that come together to make a tightly fitting lattice. Studies on fibroblasts (87) suggest that the PM is initially coated with a planar array of hexagons. Pentagons are then created at dislocations in the lattice and somehow move toward the periphery, presumably inducing curvature and thereby the formation of a CP. Heuser's technique has also been applied recently to plant cells (80), revealing similar polygonal structures at the PM.

Another surface replica technique that has proven useful in demonstrating patches of coating material and CPs at the PM entails adhering cells between two surfaces and then rapidly separating them. Portions of the PM remain attached to one of the surfaces and can then be replicated (see Figure 1b). This can be performed under aqueous ("wet-cleaving"—cleaving followed by chemical fixation, critical point drying, and then replication—e.g. 2, 238) or dehydrated ("dry-cleaving"—chemical fixation, critical point drying, cleavage, and then replication—e.g. 241) conditions. The latter procedure has been of particular value to plant cytologists and has allowed determinations of CP density in different cell types, thereby encouraging speculations about their function (55, 84, 191).

Negative staining has also been employed to visualize the cytoplasmic surface of the PM of plant protoplasts (50, 101, 251, 252). In this method the protoplasts are attached to polylysine-coated formvar grids and burst osmotically. After washing, the adhering pieces of PM are fixed briefly with

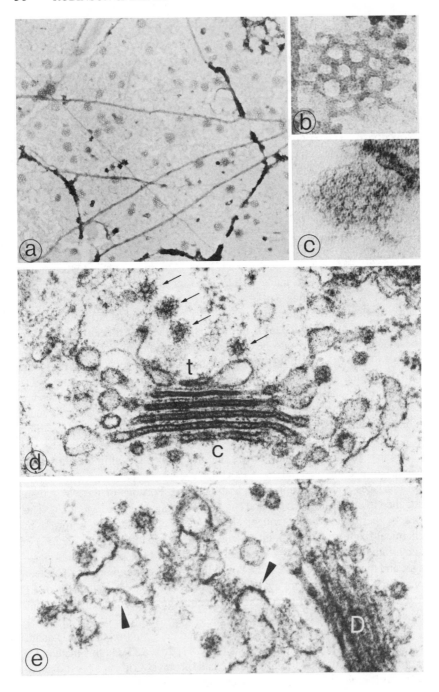

glutaraldehyde and then negatively stained. Extensive patches of coating material consisting exclusively of hexagons as well as CPs (see Figure 1a) are easily recognizable. The density of CPs can be high in such preparations (8 μm^{-2}), which means that for protoplasts with diameters around 40 μm almost 40,000 CPs must be present on the PM (251).

The Golgi Apparatus

Numerous papers dealing with both animal and plant organisms (59, 157, 199) attest to the fact that the Golgi apparatus is the other major membrane type with coated structures on its cytoplasmic surface. The most distinct kind of coating, antigenically similar to that occurring at the PM, occurs primarily on the peripheral extension of the *trans*-face cisternae. This is true for both animal (e.g. 2, 79, 144, 164, 165, 240) and plant (17, 98, 213) cells (see Figure 1d). In the latter case the coats are often seen on small vesicles that appear to bud-off from large vesicles (released cisternae?) located at the *trans* face of the dictyosome.

According to the recent publication of Orci et al (163), another kind of coating is present on budding vesicles at the medial and *cis* cisternae of the Golgi apparatus of Chinese hamster ovary (CHO) cells. This coating is antigenically different from that at the *trans* face or PM and is more amorphous. It is possible that it is related to the coating visible on budding transition vesicles at the endoplasmic reticulum (see, for example, 277 and references therein).

Other Membranes

A membrane structure bearing numerous coated, budding vesicles has been described in the alga *Chara* (181) and in a number of higher-plant cells and protoplasts (93, 98, 146, 182). Because of its branching, tubular appearance this organelle has been given the name "partially coated reticulum" (PCR; see

---

Figure 1   Some coated membranes in plant cells. (*a–c*) Polygonal lattice as visualized by three different methods of preparation. (*a*) Negatively stained plasma membrane from protoplast isolated from a suspension-cultured tobacco cell. The long fibrous elements are microtubules ($\times$ 25,000; micrograph courtesy of L. C. Fowke). (*b*) Coated pit in the plasma membrane of a tobacco leaf protoplast prepared by dry cleaving. Both hexagonal and pentagonal units in the lattice are clearly visible ($\times$ 182,000; micrograph courtesy of J. Derksen). (*c*) Tangential section through a coated membrane associated with the contractile vacuolar apparatus in *Euglena gracilis* ($\times$ 143,000; authors' micrograph). (*d,e*) Coated membranes in the vicinity of the Golgi apparatus in root cap cells of *Allium cepa*. (*d*) Coated vesicles (arrows) present at the *trans* (t) rather than *cis* (c) face of a dictyosome ($\times$ 83,000; authors' micrograph). (*e*) Partially coated reticulum (arrowheads) with attached, budding, coated vesicles near a dictyosome (D) ($\times$ 93,000; authors' micrograph).

Figure 1e). Although immunological data are not yet available, the coat of the PCR is likely to be the same as that present on the PM and Golgi apparatus; it is at least morphologically very similar.

A counterpart to the PCR in animal cells is not immediately apparent. Whereas Pesacreta & Lucas (182) have maintained that direct connections between the PCR and Golgi apparatus are not obvious and rarely seen, we feel that the PCR probably is a ramifying extension of the *trans* Golgi cisternae. As such it would be equivalent to a similar structure already described in some animal cells (194, 205).

Elements of the contractile vacuole in ciliates and flagellates are often seen to be coated (reviewed in 83, 157). Tangential sections, however, indicate that there are two types of coating material: One is the typical polygonal structure described above (see Figure 1c) and the other is a more punctate type (139, 258).

Coated Vesicles (CVs)

The demonstration of CPs at the PM and of similar coated profiles at the Golgi apparatus suggests that CVs should also exist in the cell. Putative CVs have been demonstrated in the cytoplasm of thin-sectioned plant and animal cells on countless occasions. It would appear that there are at least two populations of free CVs: Those found in the vicinity of the PM have a diameter of 100–150 nm (e.g. 42, 43, 66 for animal cells; 146, 157, 198, 252 for plants).

Doubts, however, have arisen as to whether CVs exist in situ. The fact that they can be isolated (see below) is no proof of their existence, since they might be artefactually formed during cell/tissue homogenization by cleavage of CPs from the PM or Golgi apparatus. Pastan, Willingham, and colleagues have been most resolute in defense of this argument (167, 263–265). These authors maintain that CPs are invariably attached to the PM via a long (up to 10 μm), smooth-surfaced neck or stalk. Instead of a CV being released, a larger, smooth-surfaced vesicle termed a "receptosome" is envisaged as being released from the stalk. One might expect that the matter could easily be resolved by serial sectioning, but this is not the case. Whereas some authors (57, 183) report that up to 50% of the putative CVs just beneath the PM are without a stalk, Goldenthal et al (72) have not been able to find any evidence for free CVs. It is difficult to adjudicate on this point, since many animal cells have such undulating surfaces that the presence of stalked CPs at the PM is not surprising. As far as plant cells are concerned, invaginations of this type are not seen. In particular there is no evidence whatsoever for stalked CPs in protoplasts which, because of the absence of a cell wall and the osmotic conditions, are perfectly spherical. CVs in protoplasts, at least, are definite structural entities.

ISOLATION OF COATED VESICLES

Sources for Isolation

The first attempts at CV isolation using chicken oocytes and guinea pig brains as experimental material were published in 1969 (103, 217). In the intervening years a number of different animal tissue and cell types have been employed, including adrenal cortex and medulla (41, 140), liver (29, 107, 111, 187), and placenta (40, 172, 174, 177). However, for the purposes of investigation into coat protein structure and function, mammalian brain tissue is by far the most frequently used source for CV isolation. As far as nonanimal systems are concerned, CVs have been successfully isolated from yeasts and a number of different plants, including photosynthetic and etiolated tissues, suspension-cultured cells, and protoplasts.

Methods for Purification

CVs are found in the postmicrosomal supernatant of cell homogenates. The first step in their isolation therefore usually involves the removal of endo- and plasma membranes by centrifuging the (filtered) homogenate at 20,000–40,000 g for 30–60 min. CVs can then be separated from other components in the supernatant fraction by a variety of procedures. The most frequently used method employs a combination of rate zonal and isopycnic centrifugations in sucrose density gradients (e.g. 108, 140, 172). Such a procedure is not only time consuming but also subjects the CVs to high concentrations (50% w : w) of sucrose that may destabilize CVs, causing a loss of vesicle contents (175) and dissociating the coat proteins (154). Indeed it has been remarked upon on several occasions (15, 103, 141, 173) that CV fractions prepared in this way contain appreciable amounts (up to a third; 143) of empty "cages" (for definition see below). In order to avoid this problem, centrifugations in other, more isoosmotic media—e.g. 8% sucrose in D_2O (154) or Ficoll-D_2O (176, 177)—are now to be preferred.

Gel permeation chromatography, using glass microbeads, was introduced by Pfeffer & Kelly (184) as an alternative to gradient centrifugations for purifying CVs. According to Altstiel & Branton (5), better resolution is obtained with Sephacryl-S 1000 and CV fractions obtained usually have less contamination than those prepared by other means. The method is now in regular use for CV isolation from brain postmicrosomal supernatants (e.g. 11, 81, 228). The purity of gradient-prepared CV fractions can also be increased by subjecting these fractions to a gel electrophoresis on agarose. CVs normally travel faster than the contaminants, but the mobilities of CVs from different sources are by no means identical (211).

An elegant method of CV isolation, introduced by Merisko et al (143), is that of immunoadsorption. For the solid phase, heat and formalin-fixed *Staphylococcus aureus* (Staph A) cells were employed and coated with polyclonal antibodies prepared against porcine brain clathrin (see below for definition). When antibody-coated Staph A cells and crude tissue homogenates are mixed, a monolayer of CVs is formed on the surface of the Staph A cells. Contaminants can then be washed away by centrifugation. Despite its obvious advantages, this method has not yet found frequent use (see, however, 185, 228).

The medium used for homogenizing, preparing gradients, equilibrating, and eluting columns is nearly always the same. It contains 0.1 M of a non-ionizing buffer (usually MES or HEPES) at a pH where CVs are stable (pH 6.5). In addition it has 1 mM EGTA and O.5 mM $MgCl_2$, and there are also a number of "protective" agents present—e.g. sodium azide (2–4 mM), DTT (0.1–0.3 mM), and phenylmethylsulfonyl fluoride (PMSF, 0.2–0.6 mM).

CV Isolation from Plants

Depending on whether the plant in question is nongreen or green there are one or two major sources of contamination in the postmicrosomal supernatant. The first of these are ribosomes, which occur in considerable quanitities in resuspended postmicrosomal pellets (47, 145) and can easily be removed by treatment with pancreatic RNase (47). This is a procedure that is only seldom (e.g. 177, 209) carried out on animal systems. The other source of contamination, which is never encountered in animal homogenates, is that of ribulose bisphosphate carboxylase, the major protein of photosynthetically active cells. As was shown by Depta et al (48) this enzyme complex becomes separated from CVs during the purification of the latter on Ficoll-D_2O gradients.

Because of their density and sedimentation characteristics these contaminants do not interfere with the isolation of CVs in the "one-step" method of Mersey et al (146). By directly applying a cleared homogenate (10 min precentrifugation at 1000 g) to a sucrose gradient and centrifuging in a vertical rotor these authors have obtained an enriched CV fraction within 3 hr. Additional purification was achieved through a subsequent rate zonal centrifugation on a linear 25–50% gradient.

In the isolation of CVs from plant tissues two factors must be stressed. The first is the presence of higher levels of endogenous proteases than in homogenates from animal cells. These can only be combated by including appropriate inhibitors (added fresh at each stage in the purification procedure)—e.g. PMSF or chymostatin (67). The other is the existence, in many plants, of considerable quantities of phenolic compounds. These necessitate

the inclusion of agents that prevent or counteract phenolic oxidase activity—
e.g. metabisulfite or thiourea (51).

Sizes and Yields of CVs

With the exception of the large (200 nm) CVs isolated from chicken oocytes
under isotonic conditions (176), most CVs isolated from animal tissues have a
diameter of the order of 100 nm (269). Although there is some variation in the
values given in the literature, it would appear that the average diameter for
brain CVs lies around 70 nm, whereas CVs from other tissues, including
plants, may have larger diameters. Sometimes this reflects the presence of
two or more populations of CVs in the fraction. This is certainly the case for
CVs from bovine adrenal medulla (41) and from rat brain and liver (235),
where difference in size is also correlated with differences in electrophoretic
mobility (107). An indication that this may also be true for plants has been
given recently by Depta et al (48), who have shown that the CVs in the highly
enriched CV fraction of a Ficoll-D_2O gradient of bean leaves had a different
diameter (<85 nm) from those in a relatively contaminated fraction (>85 nm)
from the same gradient.

The yield of CVs depends, of course, on the method used for isolation, and
will vary according to the relative degree of purity of the CV fraction
obtained. In terms of availability, ease of handling, and high yields, brain
tissue is by far the best source for CV isolation. Yields varying from 13–30
mg CV protein per kg fresh weight are typical for brain tissue (11, 47, 146,
173, 184). When working with plants, however, the concentration of protein
in the homogenate is much lower than that for animal tissues, owing to
dilution with vacuolar sap. This means that much larger volumes of homogenate have to be centrifuged (depending on the cell type, at least four-fold) in
order to obtain postmicrosomal pellets comparable in amount to those from
brain tissue. Recoveries are, in contrast to brain, much poorer, varying
between 0.4 mg per kg fresh weight for bean leaf tissue (48) and 2 mg per kg
zucchini hypocotyl tissue (H. Depta and D. G. Robinson, unpublished
observations) when CVs are isolated using Ficoll-D_2O gradients.

COAT PROTEINS

Individual Proteins and Their Properties

TERMINOLOGY   When Pearse published a method in 1975 dealing with the
isolation of a highly pure CV fraction from pig brain tissue (172), she also
introduced the word "clathrin" (from the Greek meaning cage or bar) as a
name for the major polypeptide from CVs that was visible in sodium dodecyl
sulfate–polyacrylamide gel electrophoresis (SDS-PAGE). Several years later

it became clear that a number of other distinctly different polypeptides were present in the coat of CVs (see Table 1). Of these, two bind tightly to clathrin. These were both originally given the name "light chains" by Kirchhausen & Harrison (115) and Ungewickell & Branton (248). Since then the terms "clathrin heavy chains(s)" and "clathrin light chains" have entered the literature. The reader should note that in this review we use these terms in their original form—i.e. we speak of clathrin rather than "clathrin heavy chain", and of light chains but not "clathrin light chains".

CLATHRIN   Almost 90% of the protein in animal CVs belongs to the coat (261). The major coat protein is clathrin. In SDS-PAGE of CVs isolated from a variety of animal cell types, the molecular mass of clathrin is 180 kDa (e.g. 173, 174, 187, 259). There is a close similarity in amino acid composition of clathrin isolated from various sources (259), with the principal residues being glutamic acid (15%), aspartic acid (10%), and leucine (10%). By contrast clathrin from yeast and higher plants (see Figure 2) has a molecular mass of 190 kDa (47, 48, 145, 146, 171). Immunoblotting has been carried out to determine whether antigenic similarities exist between the various clathrin sources. According to Payne & Schekman (171), polyclonal antibodies to yeast clathrin do not react with brain clathrin, nor do the polyclonal brain clathrin antibodies obtained by Louvard et al (128) cross-react with yeast clathrin. However a monoclonal antibody to brain clathrin may cross-react with yeast clathrin (see 171). Cross-reactivity between several polyclonal antibodies to bovine brain clathrin and carrot cell clathrin has now also been demonstrated (36a).

Under nondissociating conditions, animal clathrin has a molecular mass of around 630 kDa (190, 248) with a sedimentation coefficient variously reported between 8.1 S (190) and 9 S (26). When this complex is treated with

Table 1 The major coat proteins in coated vesicles from animal cells

Molecular mass of polypeptide in SDS-PAGE (kDa)	Name of polypeptide	Important references
180	clathrin	116, 172, 190, 248
180	assembly polypeptide	3
100–110	assembly polypeptides (at least 6, in two different subsets)	108, 180, 203, 275, 276
55	tubulin[a]	112, 186
50	assembly polypeptide	108, 180, 275, 276
30–38	α,β light chains	22, 94, 100, 115, 121, 174, 244, 248

[a] Principally in brain coated vesicles

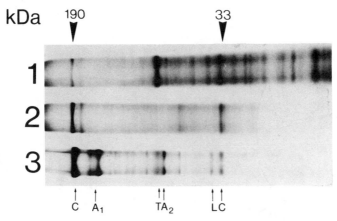

Figure 2 SDS-PAGE (silver staining) of coated vesicle fractions purified according to the method of Depta & Robinson (47). Lane 1—bean leaves (5 μg protein); lane 2—zucchini hypocotyls (5 μg protein); lane 3—bovine brain (3 μg protein). C—clathrin; A_1—100–110-kDa assembly polypeptides; T—tubulin; A_2—50-kDa assembly polypeptide; LC—light chains.

high concentrations of chaotropic reagents—e.g. 7 M urea (249), 6 M guanidine-HC1 (109) or 1 M potassium thiocyanate (267)—3 polypeptides of molecular mass between 30 and 36 kDa are released. These are the light chains (see below). This indicates that the soluble 9S species is hexameric, comprising three clathrin polypeptides and three light chain polypeptides, a fact confirmed by chemical cross-linking experiments (115) as well as immunoelectronmicroscopically (244).

In the electron microscope the 9-S hexameric complex has been shown to have the form of a manx cross or three-legged swastika. Because of its unusual shape this structure has been given the name "triskelion" (40, 248). When prepared for electron microscopy by spraying onto mica with glycerol followed by rotary shadowing, triskelion populations appear to have a uniform clockwise orientation (see Figure 3) However, more recent investigations by Kirchhausen et al (117) have shown that the orientation of the triskelions is a question of the method of preparation. According to these authors the triskelion in solution is not a planar structure; instead, the vertex is raised to produce a "pucker" form. Because of the air-drying involved in the glycerol-spray technique, this feature of triskelion structure becomes lost. Depending on their affinity for mica, triskelions can attach either "feet-first," and have a uniform clockwise orientation (occurs in the presence of Tris-buffers, which produce a low affinity) or "vertex-first," leading to a predominantly anticlockwise configuration (high-affinity conditions with non-Tris buffers).

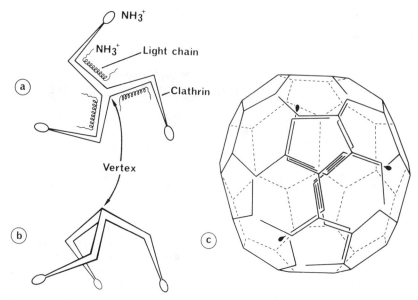

Figure 3   Schematic representation of the structure and organization of clathrin and light chains in animal coated vesicles. (*a*) Polar view of a hexameric triskelion with terminal (NH₃) domains and light-chain binding sites on proximal portion of the clathrin leg. (*b*) Side view showing the "pucker" form of the triskelion (overemphasized). (*c*) Association of triskelions to form a polyhedral structure consisting of hexagonal and pentagonal units (structures redrawn after figures published in 89, 116, 117).

The legs of a triskelion have a length between 44 and 50 nm, with the bend or joint occurring about 19 nm from the vertex, thus dividing the leg into proximal and distal regions. At the tip of the distal segment there is a knob-like terminal domain (116) that appears hinged at an angle to the plane of the distal segment (89, 254). The terminal domain accounts for approximately 30% of the clathrin molecule. This can be deduced from the sizes of the products of limited proteolysis of triskelions that have been reassembled into cages (see below). Subtilisin, thermolysin, elastase, and trypsin all release a primary fragment of molecular mass between 50 and 60 kDa (116, 223, 248), leaving behind cages comprised of triskelions with clathrin of molecular mass 120–125 kDa and a leg length of 30 nm. Such triskelions are, after disassembly (see below), still capable of reassembling into cages in vitro (223, 248). Estimates of the primary structure of clathrin are also in agreement with the existence of a globular terminal domain. Thus circular dichroism measurements (190, 246, 248) indicate that clathrin has about 50–55% of its amino acid residues in an α-helix conformation. However, only about 30% of the amino acids in the terminal domain are arranged in an α-helix compared to 80% in the case of the 120-kDa residual clathrin.

A triskelion configuration for clathrin from nonaminal systems has, up till now, been demonstrated only for yeast (152). The contour length of each leg was measured to be about 5 nm longer than for brain triskelions—i.e. about 49 nm, which is in keeping with the higher molecular mass of clathrin in this case.

LIGHT CHAINS   The presence of a doublet of polypeptides with molecular masses lying between 30 and 36 kDa in SDS-PAGE of coated-vesicle fractions was first noted by Pearse (174). Unlike clathrin, whose structure is more or less constant throughout the animal kingdom, these "light-chain" polypeptides show both species as well as tissue specificity. At least four distinct light chains have been detected in mammalian cells (23, 39, 94), each cell possessing only two, named α (or a or CAP_1—clathrin associated protein; 125) and β (or b or CAP_2). The light chains are present equimolar to clathrin in the triskelion, but the two forms (α, β) appear to be randomly distributed among individual triskelions (118). In SDS-PAGE of brain CVs the light chains appear much more prominent and have a higher M_r than those present in other cell types (113, 166). Most recently it has been shown that α and β light chains are homologous proteins coded for by different genes (100).

As already mentioned, light chains can be removed from hexameric triskelions by treatment with chaotropic agents. They can, however, be more readily purified by boiling CV preparations (126). The denatured proteins are then centrifuged-out, leaving the light chains in solution. This property of extreme heat stability suggests that the light chains are extended, rod-shaped molecules with a high proportion of their amino acids in an α-helix conformation. Another property common to all light chains is their ability to bind calmodulin (124) and calcium (149).

Antibodies against light-chain polypeptides have been used in a number of studies on the binding and orientation of the light chains on clathrin (22, 118, 121, 244, 267). Such antibodies fall into two classes: (a) those that recognize antigenic determinants in both free light chains and in hexameric triskelions or CVs, and (b) those that react only with free light chains. Thus a number of antigenic sites become cryptic as a result of the light chain–clathrin interaction. Through the preparation of enzymically and cyanogen bromide–derived light-chain peptides (24), it has been possible to locate all of the cryptic antigenic sites in a distinct region of the molecule (between residues 93 and 157). This region also shows the greatest degree of homology in amino acid sequence between the various light-chain forms (100, 119) and is characterized through a series of closely packed helical regions (119).

The direct visualization of biotinylated light chains on triskelions by means of avidin-ferritin (249) initially suggested that the light chains were situated near the vertex of the triskelion. A more precise location has been possible with indirect immunocytochemical methods (118, 244). These methods have shown that the light-chain binding site extends along much of the proximal segment of the triskelion leg. Monoclonal antibody CVC-6, whose antigenic determinant lies outside of the clathrin-binding region of the light chain at the COOH-terminal (22), has been shown to bind much closer to the vertex of the triskelion (118). This indicates that the COOH-terminal of the light chain is directed towards the triskelion vertex, and is therefore oriented like that of clathrin (116).

Using SDS-PAGE, investigators have found polypeptides that migrate electrophoretically in the region of the light chains for CV preparation from yeast (152, 171), for soybean protoplasts (146), for bean leaves (48), and for zucchini hypocotyls (see Figure 2). Using essentially the same method for CV isolation, the gels of Payne & Schekman (171) reveal only one light-chain band in yeast, whereas those of Mueller & Branton (152) clearly show two (at 36 and 33 kDa). This may be related to the fact that two different yeast strains were used in these studies.

Cole et al (36a) have claimed that the light chains in plant cells have molecular masses of 60 and 57 kDa, respectively. Being 75% larger than their counterparts in animal cells would certainly have consequences for their binding to clathrin and for the packing of triskelia into a polygonal lattice. Not only are there theoretical objections to the conclusions of Cole et al (36a), there are methodological problems with this paper as well. The authors have based their claim on the cross-reactivity of bands in SDS-PAGE of carrot cell CVs with polyclonal antibodies raised against coat proteins of bovine brain CVs. Of the four antibody types used, only one reacts strongly in immunoblots with one of the putative light-chain bands. Strangely, this antibody is classified as being specific for bovine CV light chains, yet the authors indicate that it also cross-reacts with both the light chains and clathrin of porcine brain CVs. Yet another antibody that is said to recognize both clathrin and the light chains of bovine brain CVs shows no cross-reactivity with the light chains in immunoblots of porcine brain CVs, although the authors claim that it does. Considering the dubious nature of their antibodies we feel that a strong cross-reaction with polypeptides at around 60 kDa more likely indicates their being proteolytic fragments of clathrin (see above).

THE ASSEMBLY POLYPEPTIDES   Keen et al (108) noted that among the proteins dissociated from the surface of brain CVs by Tris-HCl treatment (see below) there were two smaller groups of polypeptides with molecular masses around 100 and 50 kDa. These polypeptides could be separated from clathrin by gel filtration and were shown to promote the reassembly of clathrin in vitro (see below). Further purification of the assembly polypeptide mixture by ion-exchange chromatography (246, 275) and hydroxyapatite adsorption chromatography (180) gives rise to two fractions: one containing only 100-kDa polypeptides (designated HA-I) and the other consisting of a stoichiometric complex of 100- and 50-kDa polypeptides (designated HA-II). Both the HA-I and HA-II fractions from bovine brain CVs contain three 100-kDa polypeptides. Human placenta CVs have, in contrast, fewer 100-kDa polypeptides, and these do not coelectrophorese with those from brain tissue (180). Moreover, whereas bovine brain CVs appear to possess on the average only one 100-kDa polypeptide per triskelion (three is saturation), those from human placenta seem to contain two. As with the light chains this suggests that there may exist a tissue/species specificity for these minor CV components as well. Supporting this contention is a study by Robinson (203) involving monoclonal antibodies raised against three of the major 100-kDa family of polypeptides from bovine brain CVs. Only two of the three bands of 100-kDa polypeptides in CVs from other tissues (e.g. liver, adrenal gland) seem to cross-react with these antibodies.

Another assembly protein in brain CVs with a molecular mass of 180 kDa has recently been discovered (3). This polypeptide had not previously been detected because it was obscured by clathrin, which has the same molecular mass. However, gel filtration of solubilized coat proteins by FPLC with a Superose 6 column enables a clear separation of the 180-kDa polypeptide from both the 100/50-kDa assembly polypeptides and triskelions.

Bands at 100–110 kDa and 50–60 kDa are visible in SDS-PAGE of CVs from yeast (152) and soybean protoplasts (146). Faint bands corresponding to these molecular mass ranges are also visible in SDS-PAGE of carrot cell and bean leaf CVs (47, 48). It is possible that some of them may have become lost during the preparation procedure, although one notes that this is not the case for brain CVs prepared by identical means. Clearly, if the assembly of plant CVs follows the same principles as that of animal CVs (see below) one should expect the presence of assembly polypeptides. However, their molecular masses could well be different from their counterparts in animal cells, preventing easy recognition in SDS-PAGE of CV preparations.

TUBULIN   An inspection of the numerous published SDS-gels of brain CVs reveals a strong band around 55 kDa in addition to the major bands corresponding to clathrin, the light chains, and the assembly polypeptides. Several papers (112, 186, 260) provide strong evidence (identical molecular masses and isoelectric points, and common antigenic determinants) that this band represents the two subunits of tubulin (α, 54 kDa; β, 56 kDa). Because of its tight associaton with the 50-kDa assembly polypeptide (which also becomes phosphorylated, see below) it has been suggested (186) that the latter acts as a linker between the CV and microtubules, thereby facilitating CV-mediated intracellular transport. The observations of Wiedenmann et al (261) are, however, not in agreement with this supposition. These authors have shown that tubulin is tenaciously bound to the CV: It is not released by a high-salt (1M KCl) treatment that releases a significant proportion of the other coat proteins, and is only solubilized at pHs above 10 whilst the other coat proteins begin to be liberated at pH 8. Although tubulin is known to interact with phospholipids (122), removal of 90% of the lipid of brain CVs by Triton X-100 treatment fails to solubilize CV-associated tubulin (261).

Tubulin does not appear to be a significant component in CVs from other tissues—e.g. adrenal cortex (140). Nor does it appear to be a major polypeptide in CVs from fungi or plant sources, although a definite statement as to its presence can only be determined through immunoblotting with tubulin antibodies.

Coat Architecture and Coat-Vesicle Interactions

Negatively stained (e.g. 40, 41) and deep-etched rotary shadowed (87, 89) preparations of CVs show clearly that the coat proteins are arranged as a shell in the form of a polyhedral lattice. The various lattice types seen in reconstituted cages and CV preparations from different sources comply with Euler's theorem in that there are a variable number of hexagonal units together with the 12 pentagons necessary for icosahedral symmetry. The most frequently encountered forms in brain CV preparations have 4 or 8 (see Figure 3) hexagons, and represent the smallest vesicle (41, 89). Other CV types—e.g. in fibroblasts (87) or oocytes (206), as well as the larger forms of reassembled cages—may have 20, 30, or more hexagons.

Measurement of the edge length of pentagons or hexagons gives values ranging from 14 nm to 18 nm in negatively stained preparations (41, 89) and a value of 15.5 nm for deep-etch samples (89). Since this value closely corresponds to the distance between the vertex of a triskelion and the kink in the leg of the triskelion, it has been proposed (40, 116) that each vertex in the polyhedral lattice is occupied by a single triskelion vertex and that the edges are composed of the overlapping proteins of 4 triskelion legs (see Figure 3). In the original model of Crowther & Pearse (40) the kink in the triskelion leg is seen to occur before the vertex of a polygon. However, as a result of the more recent investigations of Heuser & Kirchhausen (89), a much more tightly packed arrangement of triskelions is envisaged, with the bend in the triskelion leg occurring directly under the vertex.

Heuser & Kirchhausen (89) also inferred from their deep-etch studies of CVs and cages that, for reasons of optimal packing, the terminal domain of each triskelion leg had to point into the center of the polyhedral lattice—i.e. toward the vesicle. This has been confirmed by Vigers et al (254), who have made 3-D reconstructions from a tilt series of micrographs of cages embedded in vitreous ice. According to these authors the inwardly protruding terminal domains create an internal shell to the clathrin lattice. Removal of the terminal domains through trypsin treatment leads to a loss of this inner shell. A further recent study from Pearse's laboratory (255), using the same methods, has added to this picture by showing that the 100- and 50-kDa assembly proteins are also located internally to the layer of terminal domains. Thus within the clathrin lattice there are two shells of density before the vesicle membrane is reached.

Similar high-resolution EM studies on the cage architecture of CVs from fungi or plants have not yet been published. However, from their appearance in negatively stained preparations (e.g. 47, 48) it would seem that the structure of the polyhedral lattice around plant CVs follows the same general rule as for animal CVs.

In order to determine how the coat complex interacts with the membrane of the CV there have been several investigations involving in vitro binding assays (81, 242, 249). Essentially the assay involves incubating radiolabeled (with either ^3H or ^{125}I) brain triskelions with CVs stripped of their coat proteins and then removing vesicles with bound triskelions from the mixture either by centrifugation or filtration. Triskelion binding is of high affinity ($K_d = 2 \times 10^{-9}M$), occurs at pHs between 6.2 and 6.8, does not require Ca^{2+} or Mg^{2+}, and is dependent on stripped vesicle concentration but not on temperature. These binding experiments, however, require stripped vesicles prepared by a 10 mM Tris-HCl pH 8.5 treatment that does not remove appreciable amounts of the 100- and 50-kDa assembly proteins (see Table 2). By contrast, 0.5 M Tris-HCl pH 7.0 stripping that removes significant amounts of the assembly proteins prevents triskelion binding (249). Triskelion binding to vesicles stripped with 10 mM Tris-HCl pH 8.5 is also eliminated by treating the vesicles with proteases that have been shown to digest the assembly proteins (242). It appears, therefore, that triskelions bind to the vesicle via the assembly proteins, a fact in accordance with the more recent ultrastructural results discussed above. On the other hand, clathrin has been shown to bind to phospholipid vesicles in vitro (231), so that a direct interaction of triskelions with the membrane cannot be ruled out.

Dissociation and Reassembly of Coat Proteins in vitro

The coat proteins of CVs isolated from animal tissue (usually brain) can be dissociated from the surface of the membrane vesicle in several ways (Table 2). Although the exact conditions may vary from group to group there are essentially three methods in common use for dissociating coat proteins: by treatment with low (2–3 M) concentrations of urea; with high (0.5–1 M) concentrations of protonated amines (usually Tris-HCl); or by raising the pH of the isolation medium (see above) to pH 8–8.5 with low (around 10 mM) concentrations of Tris-HCl. These treatments do not denature the coat pro-

Table 2 Methods for releasing coat proteins from animal coated vesicles[a]

Dissociating conditions[b]	Proteins released into solution[c]	References
1. 2 M Urea, 0.3 M sucrose, 10 mM HEPES, pH 7.0 (5 min, RT)	clathrin, 100 kDa[d], 50 kDa, LC	15
2. 2 M Urea, 40 mM Tris-maleate, pH 6.5 (ic)	clathrin	104
3. 2.8 M Urea, 2 mM DTT, 100 mM Tris-HCl, pH 8.0 (dialysis, 12 hr, 4°C)	clathrin, 100 kDa, 50 kDa, LC	54
4. 2 M Urea, 1 mM EDTA, 0.1% 2-mercaptoethanol, 200 mM Tris-HCl, pH 7.5 (dialysis, 12 hr, 4°C)	clathrin, LC	176, 225
5. 2 M Urea, 1 mM EDTA, 20 mM NaCl, 30 mM Tris-HCl, pH 7.5 (dialysis, 12 hr, 4°C)	clathrin, 110 kDa, 50 kDa LC	81, 248
6. 0.5 M Tris-HCl, pH 7.0, 1 mM EGTA, 0.5 mM MgCl$_2$, 0.1 M Na-MES (15 min, RT)	clathrin (80),100 kDa (most), 50 kDa (most) LC (most)	81, 108, 180, 275
7. 0.75 M Tris-HCl, pH 6.2, 0.25 mM EGTA, 0.1 mM MgCl$_2$, 25 mM Na-MES (30 min, RT)	clathrin, LC	115
8. 10 mM Tris-HCl, pH 8.5, 1 mM EGTA, 0.1 M MES (60 min, RT)	clathrin (most), 100 kDa (50), 50 kDa (50),LC (most)	153, 270
9. 10 mM Tris-HCl, pH 8.5, 1 mM EDTA, 0.2 mM DTT (60 min, RT)	clathrin (90),110 kDa (50), 50 kDa (50), LC (50)	81
10. 2 M Urea, 1 mM EDTA, 20 mM NaCl, 30 mM Tris-HCl, pH 7.5 (dialysis 12 hr, 4°C)	clathrin,110 kda, 50 kDa, LC	81, 248
11. 1.2 M Na-thiocyanate, 2 mM EDTA, 0.1% mercaptoethanol, 50 mM Tris-HCl, pH 8.0 (ic)	most (?) proteins	180
12. 0.25 M MgCl$_2$, pH 6.5, 1 mM EDTA, 0.1 MES (60 min, RT)	all (?) proteins all proteins	270
13. 25 mM Na$_2$CO$_3$, pH 10 (15 min, RT)	all proteins	108

[a] Abbreviations: ic—carried out immediately "in the centrifuge"; RT—room temperature; LC—light chains
[b] In addition to the substances given 1–3 mM NaN$_3$ and/or 0.2 mM PMSF are often added to prevent proteolysis, etc.
[c] Figures (or words) in parentheses indicate percent amounts of individual proteins released from the surface of a coated vesicle preparation. However, most authors have not presented quantitative data, and some have restricted their comments to clathrin alone. In these cases we have judged which additional proteins had been released by inspecting the published gels.
[d] Actually meant are the 100–110-kDa family of polypeptides.

teins. In the case of clathrin, denaturation is known to occur at higher urea concentrations (54). The dissociation is usually carried out by resuspending a CV pellet in the appropriate medium and allowing it to stand at room temperature for periods of up to 60 min before centrifuging for 1–2 hr at $100–150 \times 10^3$ g in order to separate the stripped vesicles from the coat proteins. Whether such long periods are really necessary in all cases is perhaps debatable, since it has been shown by light scattering (155) that dissociation is complete within 10 sec of making the pH change.

The amounts and types of coat protein released into solution by such treatments vary according to the conditions employed (see Table 2), but the production of a completely clathrin-free stripped vesicle population only appears possible under the extremely harsh conditions of pH 12 (261). As a rule, proportionately more clathrin is dissociated from CVs than the other coat proteins, and a stripping efficiency of over 75% for this protein is common. As far as the nonclathrin coat proteins are concerned, treatments with 0.5 M Tris-HCl at neutral pH, or with 10 mM Tris-HCl at elevated pH, seem to be more effective stripping procedures than 2 M urea. The one exception to this generalization is that of tubulin, which can only be removed by pH 12 treatment (261).

In all cases the clathrin released into solution remains in the form of a triskelion (248). However, in contrast to clathrin prepared by 0.5 M Tris-HCl treatment, the triskelions released by 2 M urea are without light chains (54, 188). There is also a difference between triskelions released with 0.5 M Tris-HCl and those released in 2 M urea in terms of their ability to reassemble into ordered structures (see below).

As far as fungi and plants are concerned clathrin, a 55-kDa polypeptide, and the 36-kDa light chain are released from CVs isolated from yeast by 2 M urea treatment at room temperature for 1 hr (171). These and some additional minor polypeptides are also released by 0.5 M Tris-HCl at pH 7.5 (123). Information on the dissociation of coat proteins from plant CVs has not yet been published, but recent experiments from our laboratory indicate that a number of proteins, including clathrin, can be released by urea or Tris-HCl treatments of plant CVs.

Upon removal of the dissociating factor (urea, Tris) by dialysing against the buffer used for isolating the coated vesicles, coat proteins from brain tissue spontaneously reassemble to build structures morphologically similar to the coat of a CV. These are variously termed "baskets" (103) or "cages" (224). The ability to reassociate depends in part on the way in which the coat proteins were dissociated and on the conditions for their reassembly. According to Keen et al (108), sedimentable cages from 0.5 M Tris-extracted coat proteins are only obtained when the dialysis is carried out with a buffer having a much lower (5 mM) ionic strength than that of the isolation medium. Moreover, Hanspal et al (81) have reported that 30–35% of the triskelions released from brain CVs by 0.5 M Tris pH 7.0 in the presence of isolation buffer are sedimentable at 100×10^3 g in the form of "structureless aggregates." In this regard unusual effects of urea on the coat proteins have not been reported. Indeed Woodward & Roth (270) have mentioned that cages from urea-treated preparations form spontaneously when the dissociation medium is diluted a mere 5–10 fold. By contrast, coat proteins released by elevating the pH of the CV suspension do not reassemble into cages simply by returning the pH to 6.5; this occurs only after dialysis overnight against isolation buffer.

The reassembly of clathrin triskelions from brain tissue is not absolutely dependent upon the presence of the light chains (188, 267). However, when the light chains are removed from the triskelions through limited proteolysis with elastase, irregular-shaped cages and cage aggregates are formed upon reassembly (115). Since the loss of the light chains from intact cages with elastase does not result in a change in cage morphology (115) one can infer that the light chains must play some role in regulating triskelion assembly (245).

Hexameric triskelions from brain can be induced to polymerize into cages, in the absence of other proteins, when mM Ca^{2+} is present in the assembly medium (108, 115, 275). The cages formed are heterogeneous in structure (300S; diameter 70–125 nm) and tend to separate into two major size classes upon centrifugation in 10–30% sucrose density gradients (275). When Ca^{2+} is absent, triskelions can only assemble into cages when the so-called assembly polypeptides (50 kDa, 100–110 kDa, 180 kDa) are present (3, 108, 188, 275). The cages are smaller (150S; diameter 70–80 nm) and more homogeneous. Pearse & Robinson (180) have now shown that clathrin polymerization in the presence of the HA-II group of 100–110-kDa polypeptides (see above) alone gives rise to a homogeneous population of small (diameter around 65 nm) cages. In contrast the HA-I group leads to the production of a mixed population of larger (diameter tending to be over 80 nm) cages. When the 100–110-kDa polypeptides are subjected to limited proteolysis with elastase the resulting polypeptides of 76 kDa and 65 kDa are not able to promote clathrin reassembly (276).

The kinetics of cage assembly have been investigated by van Jaarsveld et al (99) and found to be strictly pH dependent. Above pH 7 it is very slow, but it is very rapid at pHs approaching 6. The polymerization of clathrin appears to be a high-order reaction. It has not been possible to detect intermediates under normal conditions of assembly. However by reducing the strength of the dialyzing buffer to 2 mM a homogeneous population of 27S clathrin aggregates has been obtained (189). These apparently consist of six triskelions. This species can be converted into 150S cages by increasing the buffer concentration to 50 mM. An interesting anomaly in terms of clathrin assembly is the observation of Sorger et al (230), who have shown that dialysis against a buffer containing 12% v:v saturated ammonium sulfate results in the production of "puckered" cubes. The cubes are made with a triskelion at each corner. Since the cage length is 45 nm, this means that each edge is composed of two completely overlapping antiparallel legs.

While the reassembly in vitro of urea-extracted coat proteins from yeast CVs has been reported (123), there have as yet appeared no communications on the reassembly of coat proteins from plant CVs.

The Dynamics of Coat Proteins in vivo

The assembly and dissociation of the coat proteins in vivo is probably related to the needs of the cell and reflects the relative extent of exocytosis and endocytosis, as well as intracellular transport (see below). As such, CVs do not seem capable of fusing with membranes (5); for fusion to occur the coat must be partially or completely removed. On the other hand a progressive polymerization of clathrin and other coat proteins could well be interpreted as representing the mechanical side of CV formation and release. Observations of both abnormal cellular conditions, as well as of special developmental stages, tend to support the need for a dynamic equilibrium between "free" and "bound" clathrin. Thus blockage of intracellular transport by inhibitors of energy metabolism in pancreatic exocrine cells has been shown to cause the accumulation of cages and CVs, particularly at the *trans* pole of the Golgi apparatus (144). Moreover, there is good evidence for an increased synthesis and accumulation of free clathrin before the onset of endocytosis in developing *Xenopus* oocytes (142, 193). However, direct proof of the coexistance of soluble and bound forms of clathrin in cells actively engaged in transport events has not been easy to obtain.

Immunocytochemical studies involving antibodies against clathrin have been carried out to elucidate the dynamics of CV assembly in situ. In some (8, 106, 265) the general background staining in the cytoplasm is so low that the authors have concluded that the majority of the cellular clathrin is present in a bound form—i.e. as CPs or CVs. By contrast, Louvard et al (128), who have used a monoclonal antibody against clathrin, have demonstrated a diffuse staining of the cytoplasm. The ensuing issue of whether a soluble form or pool of clathrin triskelions in the cytoplasm exists was finally decided by Goud et al (75). These authors used an enzyme-linked immunoassay to determine the total amount of clathrin in cell homogenates prepared from a number of tissue. The same assay was applied to cytosolic and total membrane fractions in order to determine the relative proportions of soluble and bound clathrin. A great variation in total clathrin content was observed, with values ranging from 0.07% for rat liver to 0.75% for rat brain. However, it is clear from these studies that the relative physiological status of the cell (e.g. endocytically or exocytically active) could lead to considerable differences in the relative amounts of soluble to bound clathrin. Thus of the various cell/tissue types investigated by these authors, the two most active in secretory and intracellular transport (liver and brain) had the smallest ratios of soluble to bound clathrin. Whereas the detection of a soluble form of clathrin has now been confirmed by other workers on animal cells (e.g. 26, 142) there have been no comparable studies carried out as yet on plants.

Having established that both soluble and bound forms of clathrin exist in the cell, and knowing that the assembly of clathrin triskelions into a polyhedral lattice is modulated by the 100/50-kDa proteins, it is logical to ask what causes the coats of CVs to dissociate? A number of papers, particularly from Rothman's group, have been of great value in providing an answer to this question. The first clue was provided from experiments in which brain CVs or cages were incubated with the cytosol in the presence of ATP (168). In both cases the coat proteins were dissociated into a form incapable of reassembling. The factor in the cytosol responsible for the "uncoating" has now been isolated and is a 70-kDa protein (19, 219). The enzyme has a specific requirement for ATP, is active in both monomeric and dimeric forms, and is present in the cell in amounts similar to that of clathrin. It turns out that this uncoating ATPase has distinct similarities to at least two of the 70-kDa family of heat-stress proteins (32, 247).

The presence of light chains in the coat is a necessary prerequisite for the binding of the uncoating ATPase, since light chain–free cages do not dissociate. Readdition of light chains to such cages leads to a recovery of the uncoating capacity (224). Also necessary for the activity of the uncoating ATPase are the terminal domains of clathrin triskelions. Truncated triskelions whose legs have been shortened by limited proteolysis (see above) can assemble to cages that neither bind the uncoating ATPase nor are capable of eliciting ATP hydrolysis (220). Thus it would appear that the binding site for the uncoating ATPase lies at the vertex of the clathrin polyhedra, where both light chains and terminal domains from overlapping triskelia (see Figure 3) are close to one another. By subjecting the uncoating ATPase to chymotrypsin digestion Chappell et al (33) have been able to recognize functional domains in the molecule. Thus, while ATPase activity is restricted to a large 44-kDa fragment located at the NH_2-terminal, the clathrin-binding portion of the molecule is located in the COOH-terminal residue.

Evidence has been provided by Schmid et al (222) for a double function of ATP in the uncoating process. In addition to providing energy through hydrolysis, ATP apparently promotes the binding of the uncoating ATPase to the clathrin lattice. As a result, Rothman and coworkers (208, 221) have envisaged the uncoating process as a two-stage event. The first stage, which requires energy, involves the binding of the uncoating ATPase and a conformational change that forces the leg of the triskelion away from the vertex. The exposed terminal domain is then "captured" by the uncoating protein, preventing it from falling back into the vertex position. Presumably, after each leg is displaced in this manner, the triskelion is released, still bound to the uncoating ATPase. According to Schmid & Rothman (220), the uncoating ATPase rapidly leaves the free triskelion to bind again to a clathrin lattice, providing there are cages or CVs still available.

Unfortunately the latter types of investigation have as yet no parallel with plant cells, but one notes that heat-stress proteins, including some with a molecular mass around 70 kDa (114), have been detected in plants, suggesting that a similar uncoating mechanism may also work there. The 70-kDa heat-shock protein has been localized in the cytolasm of cultured tomato cells held at 25°C by immunocytochemical means (156). It is interesting that this protein tends to regroup and is found predominantly around the nucleus and in the nucleoli after increasing the temperature to 40°C.

LIPIDS IN COATED VESICLES

Not much work has been done on the lipids of CVs from animals, and no information is available as yet for CVs from fungi or plants. Both bovine and porcine CVs contain 20–25% lipid by weight, the most important of which are the phospholipids (representing about two thirds of the total: 35–43% phosphatidylcholine, 24–33% phosphatidylethanolamine, 14–30% phosphatidylinositol and phosphatidylserine), although sphingomyelin (6–20%) and cholesterol are also present (16, 68, 172, 173). Because the PM in animal cells is known to be relatively rich in cholesterol (233) whereas the endomembranes have a much lower ratio of cholesterol to phospholipid (36), it has been suggested (20, 179) that those CVs originating at the PM should also be low in cholesterol in order to conserve the general composition of cellular membranes. The recent work of Helmy et al (85) shows that this is not the case. Endocytotic CVs from rat liver cells have a cholesterol:phospholipid ratio of almost 1 and possess three times the cholesterol of exocytotic CVs.

Based on a size of 50–100 nm diameter Pearse (172) has estimated that the lipid bilayer of a CV must contain about 10,000 phospholipid molecules. However, the phospholipids do not appear to be uniformly distributed between the two bilayers. By treating bovine brain CVs with phospholipase A2, which cannot attack lipids located in the inner leaflet of a membrane (253), Altstiel & Branton (5) have shown that phosphatidylethanolamine and phosphatidylserine are preferentially located in the outer leaflet of the CV membrane. This is in agreement with the fact that phosphatidylcholine tends to be located predominantly in the extracellular-facing leaflet of the PM (162), which is topographically equivalent to the lumen-facing leaflet of a CV.

ENZYME ACTIVITIES ASSOCIATED WITH COATED VESICLES

Kinases

The existence of protein kinase activity in CVs was first demonstrated in 1982 by Kadota et al (102) and Pauloin et al (170). Since then it has become apparent that at least two kinases are present (11, 237). They have been characterized principally from brain CVs but have also been detected in liver CVs (11, 31, 186). One of the kinases is responsible for the phosphorylation of the 50-kDa assembly polypeptide. It is not dependent upon the presence of either Ca^{2+}, calmodulin, or cyclic nucleotides but is inhibited by N-ethylmaleimide (NEM) (31). The other kinase phosphorylates specifically the β light chain (11, 169, 250)

The two kinases differ from one another in several respects. The 50-kDa kinase does not phosphorylate foreign proteins, utilizes only ATP, and is not stimulated by polycations such as polylysine. In contrast the β-light-chain kinase can also phosphorylate casein and phosvitin, can use both ATP and GTP as substrates, and is stimulated by polylysine (11, 237, 250). Moreover the two kinases can be separated from one another by gel chromatography and also differ in terms of their extractability with 1 M NaCl from CV preparations (11). Both kinases appear to phosphorylate serine and threonine residues and both are components of the coat rather than being integral membrane proteins. Strong evidence now exists (31, 110) that the 50-kDa kinase substrate is the 50-kDa polypeptide itself—i.e. the protein autophosphorylates. Whether this is also true for the β-light-chain kinase is not clear.

ATPase

Brain and liver CVs have been shown to possess an ATPase activity (15, 151). They also possess an inwardly directed proton pump as judged from experiments designed to demonstrate the acidification of the CV lumen [fluorescence quenching with acridine orange (236, 272); luminal trapping of ^{14}C-methylamine (62, 64)]. A number of properties shared by both the ATPase and proton pumping activities suggest that the two are coupled. Thus both are sensitive to the sulfhydryl agent NEM, the alkylating agent 7-chloro-4-nitrobenz-2oxa-1,3diazole (NBD-Cl), and the carboxyl agent N,N^1-dicyclohexylcarbodiimide (DCCD), but are resistant to the Na^+,K^+-ATPase inhibitor ouabain (53, 62, 63, 64). The electrogenic nature of the proton pump is to be inferred from studies involving substances that dissipate the membrane potential (ψ)—e.g. the K^+-ionophore valinomycin—and those that abolish the pH gradient (ΔpH)—e.g. the H^+-ionophore carboxyl-p-

trifluoromethoxyphenyl hydrazone (FCCP). The former leads to an increase in the development of the pH gradient, and both give rise to higher ATPase activities (53, 64, 272).

Is the CV proton pump related to electrogenic ATPases in other organelles? Usually proton translocating ATPases in eukaryotic cells belong to one of three groups: the so-called E_1E_2 type characteristic of plasma membranes, the F_1F_2 type typical mitochondria, and the endomembrane type (134). The types differ in terms of inhibitor sensitivity and in some other properties. The proton pumping ATPase of CVs does not have a specific cation requirement but does show anion (Cl^-) sensitivity (53). It is insensitive to vanadate, ouabain, and azide but is inhibited by NEM and NBD-Cl (271). In addition, no stable phosphorylated intermediate is formed (62). These are all characteristic features of lysosomal, endosomal, Golgi, and ER proton ATPases of animal cells (71, 82, 161, 195, 274). On the other hand the mitochondrial F_1F_2 type of ATPase also does not form a phosphorylated intermediate. In common with this type of ATPase, the CV proton pump is inhibited by chlorpromazine and oligomycin (53; cf, however, 64). These effects are probably concentration dependent since other mitochondrial ATPase inhibitors, such as aureovertin and efropeptin, seem to be without effect on the CV ATPase (64, 271). Moreover there are clear differences in the molecular mass of the respective protein complexes: with 360 kDa for the F_1 portion (responsible for ATP hydrolysis) of the mitochondrial ATPase and 200–250 kDa (63) or 530 kDa [possibly a dimer form (271, 273)] for the CV proton pumping ATPase. On the basis of the foregoing the CV protein pumping ATPase appears to belong to the endomembrane type of H^+-ATPases.

Glucan Synthases

Whether kinase or ATPase activities are present in CVs from plants or fungi is not yet known. On the other hand plant CVs have been shown to possess an enzyme activity that is certainly not expected to occur in animal CVs, namely glucan synthase. Griffing et al (77) provided evidence for the presence of glucan synthase I activity in CVs isolated from soybean protoplasts. Since this enzyme is usually regarded as a marker for Golgi membranes (192) the authors concluded that, although not carrying exportable wall polysaccharides, the CVs represent instead a transport vehicle for precursor glucan synthase (synthase II), which is found in the PM.

In our laboratory we have confirmed that a glucan synthase activity is present in CVs isolated from zucchini hypocotyl tissue (201). However, the activity is that of glucan synthase II rather than I. Normally the product of glucan synthase II is principally a 1,3-β-glucan (46, 200) and is often interpreted as reflecting the synthesis of wound callose. The product formed by CVs is also 1,3-β-glucan and is in the form of short, needle-like microfibrils (see Figure 4). Incorporation of radioactive glucose from UDPG-^{14}C into an ethanol-insoluble product is much higher (about 20% of that originally present in the substrate) than that previously recorded for glucosyl transferase activities. It is also stimulated several fold by Ca^{2+} ions and spermine (49). Whether the presence of an enzyme normally regarded as being a PM-marker (192) is indicative of an endocytotic origin for the CVs used in these experiments is unclear. But the possibility that they transport a cryptic form of the enzyme to the PM also cannot be ruled out at present. In this regard one notes that there is a degree of similarity between CVs and chitosomes, which also can be induced to synthesize a crystalline polysaccharide in vitro (10, 212).

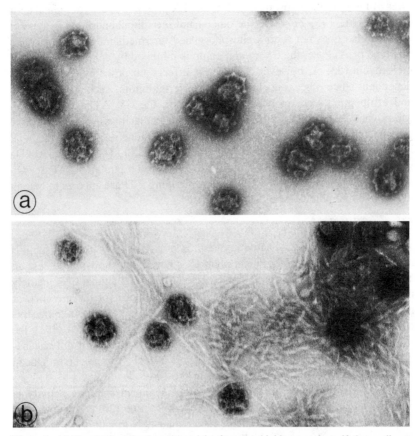

Figure 4   (*a*) Negatively stained coated vesicles from zucchini hypocotyls purifed according to the method of Depta & Robinson (47). (*b*) 1,3-β-glucan fibrils synthesized by zucchini coated vesicles. CVs incubated for 30 min at 30°C in the presence of 0.5 mM UDPG, 10 mM cellobiose, 10μM free Ca^{2+}, 800 μM spermine ($\times$ 100,000; authors' micrographs).

FUNCTIONS OF COATED VESICLES

In Animal Cells

CVs AS SPECIFIC TRANSPORT VESICLES   Work on animal cells over the last decade has provided strong evidence for the participation of CVs in a number of cellular transport events, including receptor-mediated endocytosis, transepithelial transport, the intracellular transport of secretory and lysosomal proteins, and the recycling of receptors and membrane. During this period numerous review articles have appeared that are devoted to one or more of these aspects of CV function (6, 21, 60, 73, 74, 179, 232). Superficially it would appear that CVs are universal agents of intracellular transport; however, evidence is accumulating that, within the total cellular complement of CVs, there are many subpopulations, each having particular roles. We have already seen that CVs from the same cell or tissue type can differ in size and surface charge (see above); these are clearly the physical manifestations of their multifunctional role in the cell.

CVs move along a number of different intracellular transport routes, but what controls the nature of the substances carried ("cargo") and what determines their destination ("target")? Pearse & Bretscher (179) have reasoned that, to allow for efficient targeting, some sort of recognition marker must exist on the cytoplasmic surface of the CV. Moreover, receptors specific for cargo molecules must also be present on the innermost surface of the CV (or at least in the CP during the formation of the CV). While it is conceivable that the cytoplasmic domain of a transmembrane receptor could act as a targeting marker, there may be receptors that do not completely traverse the membrane; indeed, the cargo itself could be integrated into the membrane, thus not requiring a receptor at all. For such cases Pearse & Bretscher (179) therefore envisaged the presence of so-called "adaptor" molecules in the membrane of a CV that carry both the necessary information for targeting and, via interaction with receptors, for cargo molecule recognition.

Why then do such vesicles need a clathrin lattice on their cytoplasmic surface? There are at least two reasons. First, the extent of clathrin polymerization controls the size of the vesicle, and the small size of CVs in general reduces the chances of random entrapment of unspecific cargo in vesicles. If, in addition, as Pearse & Bretscher (179) have suggested, adaptor molecules also interact with some component of the coat, an even greater selectivity is imposed on the transport process. That CVs of different sizes exist is probably a geometrical consequence of the relative concentrations of cargo, receptor, and adaptor molecules in a particular membrane type. Second, the presence of a coat around a vesicle can be taken as increasing targeting efficiency by preventing inadvertent, random fusion events.

Can the foregoing theoretical considerations be backed up by experimental facts? Specifically, can the following questions be answered: What evidence exists for the presence of specific cargo, receptor, and adaptor molecules in CVs? Is there any information pertaining to possible interactions between the coat proteins and receptor or adaptor molecules? What is known about the recognition properties of CVs; must the coat be released before recognition can occur?

Receptor, cargo, and adaptor molecules in CVs   Receptor molecules have been demonstrated in CVs by in vivo labeling with a radiolabeled ligand followed by CV isolation (e.g. 107), by performing ligand-binding assays on isolated CVs (e.g. 140, 228), as well as immunocytochemically in situ with gold-labeled receptor antibodies or ligands (e.g. 69). In keeping with the location of the ligand-binding site on the luminal side of the CV, CV-held receptors are regarded as being cryptic—i.e. ligand-binding in vitro is considerably stimulated by pretreatment with detergent. Among the most important receptors (together with their ligands) shown to be present in CVs in this way are the LDL-receptor [for cholesterol–low density lipoprotein complexes (9, 140)], the mannosyl-6-phosphate receptor [for lysosomal enzymes (29, 30, 70, 214)], the asialoglycoprotein receptor [for galactose-terminating glycoproteins (69, 107, 232)], and the transferrin receptor [for iron-containing protein (18, 96, 177)].

Specificity of cargo molecule transport by CVs is not restricted to extracytoplasmic ligands. It also includes integral membrane proteins in those cases where membrane transport is the principal function of the CV in question. A good example of this are brain CVs, which participate in the retrieval of membrane from the PM of the nerve cell terminal after the fusion of neurotransmitter-containing synaptic vesicles (90, 91). Two proteins characteristic for the synaptic membrane [95 + 65 kDa (27, 135)] have now been identified in brain CVs (185). In the same paper evidence was presented for the existence of adaptor molecules. Those brain CVs carrying the 95- and 65-kDa proteins also contained two other polypeptides (molecular masses 38 and 29 kDa) that were not present in other subpopulations of the brain CV fraction nor in CVs from liver cells.

Interactions between receptor molecules and coat proteins   The structure of a number of CV-associated receptor molecules is now known (reviewed in 74). All are transmembrane glycoproteins, but there is considerable variation in their cytoplasmic domains. These can be short (less than 40 amino acids—e.g. for the LDL- and asialoglycoprotein receptors) or long (up to 500 amino acids in the case of the epidermal growth factor-EGF-receptor); some—e.g. the LDL-receptor—have their COOH-terminal in the cytoplasm, while others—e.g. the asialoglycoprotein and transferrin receptors—have their NH$_2$-terminal in this orientation. The only possible feature in the cytoplasmic domain common to several receptors is the presence of clusters of negatively charged amino acids, which can be assumed to give rise to an α-helical conformation. Whether this is sufficient to induce the formation of a clathrin coat remains speculative. Direct proof that receptor molecules can interact with the coat proteins of CVs has been provided by Pearse (178). She has shown that rat liver mannosyl-6-phosphate receptors can associate in vitro with bovine brain 100/50-kDa assembly protein complexes to form spherical (30–100 nm diameter) aggregates. When the assembly is performed in the presence of triskelions, structures resembling CVs are produced. These have a central, membrane-like core of receptor molecules, an intermediate layer of assembly protein complexes, and an external clathrin lattice.

Recognition properties of CVs   Here little is known. Before the demonstration by Pfeffer & Kelly (185) that antibodies are capable of reacting with the cytoplasmic domain of integral membrane proteins in a CV, despite the presence of an intact cage around the vesicle, it had been widely assumed that the coat would prevent such a recognition. The statement that a CV sheds its coat immediately after its separation from the CP is met often in the literature, but it may have been overemphasized since there are clear cases where the coat is retained for some time. The best examples relate to the phenomenon of transcytosis in specialized epi- and endothelial cells—e.g. in the placenta. Here there is good evidence that CVs participate in both uptake and discharge processes in transcytosis (1, 204).

RECEPTOR-MEDIATED ENDOCYTOSIS (RME)   Since, as already mentioned, this subject has been dealt with in detail in a number of review articles, only a brief summary is necessary here. The first step in RME entails the formation of a CP and the collection of receptors in it. Some receptors—e.g. those for LDL, transferrin, and asialoglycoprotein—appear to enter cells constitutively—i.e. with or without attached ligand (4, 9, 95, 256). Others—e.g. the EGF receptor—can only be internalized when coupled to the ligand (52, 218). In the former case the majority of the receptors tend to be clustered in or around CPs at any given time (7, 14, 257), whereas in the latter case they remain randomly distributed on the cell surface until the ligand is bound (138, 266). Not all cell-surface receptor-ligand complexes get internalized via CPs—e.g. the H63 antigen in 3T3 fibroblasts (179), receptor-bound thrombin in mouse embryo cells (13), and luteinizing-hormone receptors in ovarian granulosa cells (202). On the other hand, CPs are not specific for a single type of receptor. While different receptors have been demonstrated in the same CP

(8, 34) there is also some evidence for competition between receptors for vacant sites within a CP (21).

Although the exact mechanism of CP formation is not yet fully understood there is mounting evidence that the assembly of clathrin triskelions on the cytoplasmic face of the PM is often triggered by the coupling of ligand to receptor. Thus a considerable increase in the number of CPs in the PM of lymphoid cells is recorded within minutes after the addition of IgGs (37). Another example in this context are the results obtained with macrophages during "frustrated phagocytosis." It is known that macrophages possess numerous receptors on their cell surface that bind to the Fc portion of IgGs (243). Thus antibody-coated particles—e.g. erythrocytes—can be specifically recognized and ingested by the macrophages (227). Macrophages can also be frustrated in their attempts at phagocytosis by placing them on antibody-coated cover slips (147). Clathrin lattices and CPs become visible within minutes after the first contact has been made between the PM and the antibody-coated surface (148, 238). This increase in PM-bound clathrin even takes place at 4°C and apparently occurs at the expense of Golgi-associated clathrin (238).

The fate of the receptor-ligand complexes that are internalized through the formation of a CV depends upon whether they are degraded in the lysosome or recycled back to the PM. There are three possibilities (reviewed in 74): Either the ligand is degraded and the receptors are returned to the PM (e.g. for LDL, asialoglycoproteins), or both receptor and ligand are degraded (e.g. for EGF), or both receptor and ligand are recycled to the PM (e.g. for transferrin). In all of these cases the receptor-ligand complex enters the endosomal compartment of the cell by fusion of the PM-derived CV. Except in the case of transferrin, where the iron is stripped from the apoprotein portion of the ligand and the apoprotein-receptor complex is returned to the PM, the acid environment of the endosome seems to cause the dissociation of the receptor-ligand complex.

COATED VESICLES AND EXOCYTOSIS   Secretory and PM-located (glyco) proteins are transported to the cell surface via at least three vesicle-transport steps: ER to the Golgi, within the Golgi stack, and from the Golgi to the PM. Morphological evidence for the participation of CVs in these events has existed for some time (reviewed in 60, 105). This is now supported by cytochemical and biochemical data, particularly with respect to the synthesis and intracellular transport of acetylcholine receptors (AChR) and acetylcholine esterase (AChE). Thus the elimination of endocytic AChRs by cell-surface labeling with α-bungarotoxin does not lead to any change in the relative amounts of AChRs detectable in CVs (28). The presence of newly synthesized AChE in CVs has been demonstrated cytochemically in cells from chicken myotubes recovering from a pretreatment with diisopropyl fluorophosphate, which effectively inactivates intra- and extracellular AChE (12). The reaction product in this method is a dense, iron-containing precipitate whose development also allows the identification of exocytotic CVs in vitro by virtue of their increased density (12, 85). CV fractions density-shifted in this manner not only contain newly synthesized AChE (whose synthesis can be prevented by cycloheximide) but also AChR.

Unfortunately, none of the above results indicates which of the intracellular transport steps is mediated by a CV. Rothman's group (207, 209) have addressed this problem using Chinese hamster ovary (CHO) cells infected with vesicular stomatitis virus (VSV). In response to infection the cells synthesize large amounts of the integral membrane glycoprotein, G-protein, which is transported to the PM via the Golgi apparatus. By performing pulse-chase experiments and isolating CVs at the appropriate times these investigations have provided evidence for the participation of CVs in both the ER-to-Golgi and Golgi-to-PM transport steps. These experiments have recently been subjected to alternative interpretations (163). It appears that VSV-G protein is constitutively endocytosed by CHO cells (79) so that with time it becomes increasingly difficult to differentiate between G-protein-containing exocytotic and endocytotic CVs. In addition, when intracellular transport in CHO cells is blocked by dropping the temperature to 20°C, G-protein accumulates in a tubular, vesiculating network at the *trans* face of the Golgi (136, 215). Unfortunately antibodies to G-protein did not react with those budding, coated, vesicles that reacted positively towards clathrin antibodies. It should also be noted that, as far as the ER-to-Golgi transport step is concerned, the coated structures visible within the Golgi stack and at the ER do not react with clathrin antibodies (79, 163, 165).

COATED VESICLES AND SORTING AT THE *TRANS* GOLGI COMPARTMENT   There is considerable evidence for the participation of CVs in the transport of lysosomal enzymes. Both the mannosyl-6-phosphate receptor (25, 29) and lysosomal enzymes (30) or their precursors (226) have been detected in CVs. Since from the foregoing it seems unlikely that clathrin CVs are responsible for the ER-to-Golgi step, they probably reflect the vesicle transfer step from Golgi to lysosome. Immunocytochemical evidence supports this, the mannosyl-6-phosphate receptor having been detected in budding vesicles and cisternae at the *trans* face of the Golgi apparatus (70); but there are also conflicting data (reviewed in 58) implicating the *cis* Golgi cisternae as the exit site for lysosomal enzymes.

Segregation of lysosomal and secretory proteins is apparently not the only sorting process that seems to occur at the *trans* face of the Golgi apparatus. A number of studies (reviewed in 58, 86) have shown that extracellular, electron-dense markers make their way to the *trans* Golgi compartment upon internalization. These investigations have been confirmed more recently in studies involving transferrin and its receptor. Thus transferrin receptors, tagged at the cell surface with labeled transferrin receptor antibodies, are later found in the *trans* Golgi compartment (268). In addition cell-surface transferrin receptors, which have been desialylated through neuraminidase treatment, have been shown to be returned to the *trans* Golgi compartment for resialylation (229). Finally, Fishman & Fine (61) have presented evidence for the transport of previously internalized transferrin from the *trans* Golgi compartment in exocytotic CVs. Their experiments involved pefusing rat liver with either ^{125}I-transferrin or asialotransferrin followed by isolation of CVs using the AChE-mediated density-shift method for differentiating exo- from endocytotic CVs, as described above. One to two hours after internalization both radioactive and resialylated transferrin could be detected in the exocytotic CV population. Although in the last two papers resialylation at the *trans* Golgi compartment was inferred rather than demonstrated, the enzyme responsible, sialyl transferase, has now been convincingly localized immunocytochemically at this site (205).

The *trans* Golgi compartment can therefore be regarded as being the crossroads of outgoing and ingoing vesicular traffic at which the sorting of at least three different types of cargo takes place. Several models that attempt to accommodate these features have recently been published. In one (61) it has been postulated that membrane or membrane-bound proteins are segregated into the ramifying tubular extensions of the *trans* Golgi compartment where they are incorporated into CVs for transport to the PM. Soluble, secretory proteins, in contrast, are thought to be released from the central, cisternal portion in smooth vesicles. Another model (78) envisages three types of vesicle exiting from the *trans* Golgi compartment: clathrin-coated vesicles destined for the lysosome, non-clathrin-coated vesicles representing the transport of membrane proteins, and large smooth vesicles containing secretory proteins. The first two types of vesicle are restricted to the tubular periphery of the *trans* Golgi compartment, while the latter is considered a product of the central portion. A number of authors have drawn attention to the fact that the large secretory vesicles, typical of cells with a regulated discharge, are initially partially coated (164, 240). Kelly (111) has suggested that these coated regions, which are lost during maturation of the secretory granule, represent the sites of CV release. Such CVs are then thought to recycle back to the Golgi apparatus carrying proteins inadvertently incorporated into the secretory vesicle.

In Yeasts

Endocytosis has been demonstrated in yeast using both cell-foreign (lucifer yellow, FITC-dextran, α-amylase, and VSV) as well as natural (α-factor) ligands (44, 131, 132, 196). Upon internalization these ligands invariably reach the vacuole, probably via a compartment equivalent to the endosome (133), although the identification of this organelle at the EM-level remains to be carried out. Most of these investigations have been performed on spheroplasts, although the removal of the cell wall does not appear to be a prerequisite for successful endocytosis. Thus the results of Makarow (131) show clearly that a relatively large molecule (bacterial α-amylase, molecular mass 55 kDa) can permeate the yeast cell wall and be internalized at the PM.

Yeasts have great potential for studies in intracellular transport owing to the availability of a large number of secretory mutants (160), some of which are also known to be defective in endocytosis (195). However, whether or not CVs play a role in these events is unclear. On the one hand neither CPs nor CVs have been demonstrated in situ in yeast cells. On the other hand when the gene for clathrin is deleted from the yeast genome the secretion of invertase, for example, still occurs in appreciable quantities although the cells grow slower than the wild type (171). While indicating that the presence of clathrin is not obligatory for secretion (and presumably endocytosis as well), these results do not rule out the participation of nonclathrin CVs or smooth-surfaced vesicles in transporting macromolecules to and from the PM.

In Plants

As with yeasts, there is little evidence for the participation of CVs in exocytosis in plant cells. The presence of large numbers of CPs in the PM of the developing cell plate (65) is often cited as an example of the exocytotic function of CVs. It is also possible that the glucan synthase activity measured in plant CVs (see above) represents an enzyme being transported from the Golgi apparatus to the PM. Both cases, however, are interpretations rather than factual demonstrations.

By contrast there is good evidence for the internalization of extracellular electron-dense markers via CPs and CVs in protoplasts prepared from bean leaves and from suspension-cultured soybean cells (93, 101, 239). Both cationic ferritin and colloidal gold-lectin conjugates were employed as unspecific cell surface–binding ligands. The presence of these markers in CPs, CVs, small smooth vesicles, multivesicular bodies, and partially coated reticulum, as well as in the peripheral elements of the Golgi apparatus (especially at the *trans* face), was observed after 30–60min incubation in ligand suspension. Whereas the sequence of entry could not be established with certainty in these studies, the initial internalization steps appear to involve organelles similar to those that participate in endocytosis in animal cells.

Plant cells usually possess a cell wall whose existence has been considered to have potential consequences for endocytosis. First, its semirigid nature allows the development of a turgor pressure, which could act negatively on the formation and release of vesicles at the PM (38). Second, the cell wall could be considered a barrier to the passage of molecules to the cell surface that are destined for internalization via CPs and CVs. The first objection may well apply to smooth-surfaced vesicles, but the gradual polymerization of clathrin at the cytoplasmic surface of the PM (see above) may well allow the plant cell to counteract the effects of outwardly directed turgor pressure. At any rate, as judged from experiments carried out by Hübner et al (98) on maize root cap cells, it seems that a CP/CV-mediated endocytosis also occurs in walled plant cells. These authors put to use the fact that heavy metal salts (e.g. lead, lanthanum, uranyl) can be passively transported along the apoplast up to the endodermis (e.g. 56, 197). Hübner et al (98) were thus able to demonstrate the presence of electron-dense precipitates in CPs, CVs, partially coated reticulum, and the Golgi apparatus of root cap cells after germinating maize seedlings had been placed in lead or lanthanum nitrate solutions for 1–2 hr at room temperature.

With respect to the second problem it has been claimed that the pore size of the plant cell wall lies around 4 nm (35). This value has been determined from experiments involving the penetration of polyethylene glycol molecules. Similar experiments on yeast cells (216) have also indicated that a molecule larger than 700 Daltons should not pass through the cell wall, and yet a large molecule such as α-amylase can cross the yeast cell wall in both directions (131, 210). The literature on higher plants also contains examples of the permeability of the cell wall to proteins. For example, antibodies against putative hormone receptors in the PM or to cell wall fragments, prevent growth when applied to tissue segments (97, 127); and centrifugation experiments (150) indicate that proteins can easily be removed from the cell walls of intact tissue. Proving that such molecules can cross the wall does not necessarily mean that they become endocytosed at the PM, let alone that CPs and CVs are involved in such a process. However there are indications from the older literature (reviewed in 158) that proteins can be taken up into root cells. In these investigations cell-foreign proteins—e.g. hemoglobin, lysosome—were given exogenously to seedlings and their localization determined either by autoradiography or fluorescence microscopy. Most of the protein remains within the cell wall, but there are indications that some might be taken up into the cytoplasm. This often entails the formation of large heterophagic vacuoles, as exemplified by the study of Nishizawa & Mori (159), but might also involve CPs and CVs. Clearly this area deserves suitable reinvestigation.

COATED VESICLES IN PLANTS: PERSPECTIVES FOR THE FUTURE

It will have become clear to the reader of this review that research into the biochemistry and function of plant CVs is just beginning to get under way.

Inevitably, owing to the huge lead enjoyed by cell biologists working on animal CVs, much of the work to be done on plant CVs will be of a repetitive nature. Nevertheless, posing the question "Do plant CVs function this way too?" and performing the appropriate experiments could well bring some unexpected and worthwhile answers. We would therefore like to take this opportunity to focus on some aspects of plant CVs that certainly deserve attention in the years to come.

The presence of CVs associated with the Golgi apparatus in plant cells suggests that they may also be involved in the segregation of secretory from vacuolar proteins, as is the case in animals. To prove this, experiments are necessary that demonstrate the presence of these molecules in different vesicle subpopulations from the same cell. Certain proteins occurring as isoenzymes in the cell wall and in the vacuole could be exploited in this regard. Peroxidase is such a case (129, 130), and cytochemical evidence is at hand for the presence of peroxidase in CVs in situ (76). It should therefore not be too difficult to determine which of the isoenzymes is present in CVs.

Endocytosis no longer appears to be a phenomenon restricted to animal cells. Preliminary data strongly implicate the involvement of CVs in this process in plants, but we know little more than this at the present. Are CVs in plants primarily responsible for retrieval of membrane, performing only a fluid-phase endocytosis (234) as they do so; or do they selectively internalize molecules from the extracellular (wall) space? With respect to the latter, one notes that none of the ligands (e.g. peptide hormones, insulin, LDL, etc.) taken up by animal cells via receptor-mediated endocytosis are physiologically relevant in plants. Suggesting candidates for cell surface–binding ligands in plant cells is easy, but identifying their binding sites as functional receptors is a much more difficult proposition. Falling into the category of potential ligands are plant growth regulators, phytoalexin elicitors (45), exoenzymes, and cell wall structural proteins. Indeed recent work (127) with antibodies against a putative plasma membrane–associated receptor for the plant growth regulator auxin suggests that it might be possible to perform experiments designed to show the internalization and recycling of physiologically relevent receptors in plant cells. Not only are these ligands chemically different from those internalized via CPs in animal cells, the corresponding receptors may differ as well. This is certainly the case with the soybean agglutinin receptor from wheat mesophyll cells, which is a galactolipid (120).

Highly enriched fractions of plant CVs are now available. Although they may contain vesicles of different origin and destination they constitute a membrane vesicle fraction with uniform sidedness and one that is exactly inverse to that of the PM. Providing ligands and receptors get concentrated in CPs as happens in animals cells, CV fractions from plants could then represent a prepurified source for the detection and characterization of receptor molecules. Thus the high levels of binding of radioactive 1-N-naphthyl phthalamic acid (NPA, a specific inhibitor of polar auxin transport) that we have obtained with CV fractions from zucchini hypocotyls (H. Depta, unpublished observations) could be seen under this aspect. Moreover, if the binding affinity is high enough it should be possible to detect receptor molecules through "ligand screening" of immunoblots (16, 141).

CV fractions also represent a subcellular fraction hitherto discarded or neglected by plant scientists. In this context it might well be worthwhile to reinvestigate those cases where a cytosolic location for a particular factor has previously been claimed. But there are also more specific questions about plant CVs that need to be answered. For instance: what is the functional significance of a 190-kDa as opposed to a 180-kDa clathrin molecule? Are there assembly polypeptides present in plant CVs? Are they interchangeable with those from animal CVs? What are the factors determining the uncoating of plant cells? The very fact that plant CVs have some unusual properties (e.g. CVs with high levels of glucan synthase activity) suggests that there might be unexpected answers in store for us in the years to come.

ACKNOWLEDGMENTS

Discussions with Russell L. Jones held during the course of the preparation of this review are greatly appreciated. We also acknowledge the help of Bettina v. Linde-Suden, Heike Freundt, and Bernd Raufeisen on the technical side. Our work cited here was made possible by generous support from the Deutsche Forschungsgemeinschaft.

Literature Cited

1. Abrahamson, D. R., Rodenwald, R. 1981. Evidence for the sorting of endocytic vesicle contents during the receptor-mediated transport of IgG across the newborn rat intestine. *J. Cell Biol.* 91:270–80

2. Aggeler, J., Takemura, R., Werb, Z. 1983. High-resolution three-dimensional views of membrane-associated clathrin and cytoskeleton in critical-point-dried macrophages. *J. Cell Biol.* 97:1452–58

3. Ahle, S., Ungewickell, E. 1986. Purification and properties of a new clathrin assembly proteins. *EMBO J.* 5:3143–49

4. Ajoika, R. S., Kaplan, J. 1986. Intracellular pools of transferrin receptors result from constitutive internalization of unoccupied receptors. *Proc. Natl. Acad. Sci. USA* 83:6445–49

5. Altstiel, L., Branton, D. 1983. Fusion of coated vesicles with lysosomes: measurement with a fluorescence assay. *Cell* 32:921–29

6. Anderson, R. G. W., Kaplan, J. 1983. Receptor mediated endocytosis. In *Modern Cell Biology,* ed. B. Satir, 1:1–52 New York: Alan R. Liss

7. Anderson, R. G. W., Brown, M. S., Goldstein, J. L. 1977. Role of the coated endocytic vesicle in the uptake of receptor-bound low density lipoprotein in human fibroblasts. *Cell* 10:351–64

8. Anderson, R. G. W., Vasile, E., Mello, R. J., Brown, M. S., Goldstein, J. L. 1978. Immunocytochemical visualization of coated pits and vesicles in human fibroblasts: relation to low density lipoprotein receptor distribution. *Cell* 15:919–33

9. Anderson, R. G. W., Brown, M. S., Beisiegel, U., Goldstein, J. L. 1982. Surface distribution and recycling of the LDL receptor antibodies. *J. Cell Biol.* 93:523–31

10. Bartnicki-Garcia, S., Bracker, C. E., Reys, E., Ruiz-Herrera, J. 1978. Isolation of chitosomes from taxonomically

diverse fungi and synthesis of chitin microfibrils in vitro. *Exp. Mycol.* 2:173–92

11. Bar-Zvi, D., Branton, D. 1986. Clathrin-coated vesicles contain two protein kinase activities. *J. Biol. Chem.* 261:9614–21

12. Benson, R. J., Porter-Jordan, K., Buoniconti, P., Fine, R. E. 1985. Biochemical and cytochemical evidence indicates that coated vesicles in chick embryo myotubes contain newly synthesized acetylcholinesterase. *J. Cell Biol.* 101:1930–40

13. Bergman, J. F., Carvey, D. H. 1982. Receptor-bound thrombin is not internalized through coated pits in mouse embryo cells. *J. Cell Biochem.* 20:247–58

14. Bliel, J., Bretscher, M. 1982. Transferrin receptor and its recycling in HeLa cells. *EMBO J.* 1:351–55

15. Blitz, A. L., Fine, R. E., Toselli, P. A. 1977. Evidence that coated vesicles isolated from brain are calcium-sequestering organelles resembling sarcoplasmic reticulum. *J. Cell Biol.* 75:135–47

16. Bomsel, M., Paillerets, C. de, Weintraub, H., Alfsen, A. 1986. Lipid bilayer dynamics in plasma and coated vesicle membranes from bovine adrenal cortex. Evidence for two types of coated vesicles involved in the LDL receptor traffic. *Biochem. Biophys. Acta* 859:15–25

17. Bonnett, H. T., Newcomb, E. H. 1966. Coated vesicles and other cytoplasmic components of growing root hairs of radish. *Protoplasma* 62:59–75

18. Booth, A. G., Wilson, M. J. 1981. Human placental coated vesicles contain receptor-bound transferrin. *Biochem. J.* 196:355–62

19. Braell, W. A., Schlossman, D. M., Schmid, S. L., Rothman, J. E. 1984. Dissociation of clathrin coats coupled to the hydrolysis of ATP: role of an uncoating ATPase. *J. Cell Biol.* 99:734–41

20. Bretscher, M. S. 1976. Directed lipid flow in cell membranes. *Nature* 260:21–23

21. Bretscher, M. S., Pearse, B. M. F. 1984. Coated pits in action. *Cell* 38:3–4

22. Brodsky, F. M. 1985. Clathrin structure characterized with monoclonal antibodies. I. Analysis of multiple antigenic sites. *J. Cell Biol.* 101:2047–54

23. Brodsky, F. M., Parham, P. 1983. Polymerisation in clathrin light chains from different tissues. *J. Mol. Biol.* 167:197–204

24. Brodsky, F. M., Galloway, C. T.,

Blank, G. S., Jackson, A. P., Seow, H-F., Drickamer, K., Parham, P. 1987. Localization of clathrin light-chain sequences mediating heavy-chain binding and coated vesicle diversity. *Nature* 326:203–5

25. Brown, W. J., Farquhar, M. G. 1984. Accumulation of coated vesicle bearing mannose 6-phosphate receptors for lysosomal enzymes in the Golgi region of I-cell fibroblasts. *Proc. Natl. Acad. Sci. USA* 81:5135–39

26. Bruder, G., Wiedenmann, B. 1986: Identification of a distinct 9S form of soluble clathrin in cultured cells and tissue. *Exp. Cell. Res.* 164: 449–62

27. Buckley, K., Kelly, R. B. 1985. Identification of a transmembrane glycoprotein specific for secretory vesicles of neural and endocrine cells. *J. Cell. Biol.* 100:1284–94

28. Bursztajn, S., Fischbach, G. 1984. Evidence that coated vesicles transport acetylcholine receptors to the surface membrane of chick myotubes. *J. Cell Biol.* 98:498–506

29. Campbell, C. H., Fine, R. E., Squicciarini, J., Rome, L. H. 1983. Coated vesicles from rat liver and calf brain contain cryptic mannose 6-phosphate receptors. *J. Biol. Chem.* 258:2628–33

30. Campbell, C. H., Rome, L. H. 1983. Coated vesicles from rat liver and calf brain contain lysosomal enzymes bound to mannose 6-phosphate receptors. *J. Biol. Chem.* 258:13347–52

31. Campbell, C., Squicciarini, J., Shia, M., Pilch, P. F., Fine, R. E. 1984. Identification of a protein kinase as an intrinsic component of rat liver coated vesicles. *Biochemistry* 23:4420–26

32. Chappell, T. G., Welch, W. J., Schlossman, D. M., Palter, K. B., Schlesinger, M. J., Rothman, J. E. 1986. Uncoating ATPase is a member of the 70 kilodalton family of stress proteins. *Cell* 45: 3–13

33. Chappell, T. G., Konforti, B. B., Schmid, S. L., Rothman, J. E. 1987. Surface distribution and recycling of the LDL receptor as visualized by anti-receptor antibodies. *J. Cell Biol.* 93: 523–31

34. Carpentier, J. L., Gorden, P., Anderson, R. G. W., Goldstein, J. L., Brown, M. S., Cohen, S., Orci, L. 1982. Colocalization of ^{125}I-epidermal growth factor and ferritin–low density lipoprotein in coated pits: a quantitative electron microscopic study in normal and mutant human fibroblasts. *J. Cell Biol.* 95:73–77

35. Carpita, N., Sabularse, D., Montezinos,

D., Delmer, D. P. 1979. Determination of the pore size of cell walls in living plant cells. *Science* 205:1144–47

36. Colbeau, A., Nachbaur, J., Bignais, P. M. 1971. Enzymatic characterization and lipid composition of rat liver subcellular membranes. *Biochem. Biophys. Acta* 249:462–92

36a. Cole, L., Coleman, J. O. D., Evans, D. E., Hawes, C. R., Horsley, D. 1987. Antibodies to brain clathrin recognise plant coated vesicles. *Plant Cell Rep.* 6:227–30

37. Connolly, J. L., Green, S. A., Greene, L. A. 1981. Comparison of changes in surface morphology of PC12 cells induced by NGF, EGF and insulin. *J. Cell Biol.* 91:597 (Abstr. 12001)

38. Cram, W. J. 1980. Pinocytosis in plants. *New Phytol.* 84:1–17

39. Creutz, C. E., Harrison, J. R. 1984. Clathrin light chains and secretory vesicle binding proteins are distinct. *Nature* 308:208–10

40. Crowther, R. A., Pearse, B. M. F. 1981. Assembly and packing of clathrin into coats. *J. Cell Biol.* 91:790–97

41. Crowther, R. A., Finch, J. T., Pearse, B. M. F. 1976. On the structure of coated vesicles. *J. Mol. Biol.* 103:785–98

42. Croze, E. M., Morré, D. M., Morré, D. J. 1983. Three classes of spiny (clathrin-)coated vesicles in rodent liver based on size distributions. *Protoplasma* 117:45–52

43. Croze, E. M., Morré, D. J., Morré, D. M., Kartenbeck, J., Franke, W. W. 1982. Distribution of clathrin and spiny-coated vesicles on membranes within mature Golgi apparatus elements of mouse liver. *Eur. J. Cell Biol.* 28:130–38

44. Chvatchko, Y., Howald, I., Riezman, H. 1986. Two yeast mutants defective in endocytosis are defective in pheromone response. *Cell* 46:355–64

45. Darvill, A. G., Albersheim, P. 1984. Phytoalexins and their elicitors—a defense against microbial infection in plants. *Ann. Rev. Plant Physiol.* 35:243–75

46. Delmer, D. P. 1977. The biosynthesis of cellulose and other plant cell wall polysaccharides. In *Recent Advances in Phytochemistry*, ed. F. Loewus, 11:45–77. New York: Academic

47. Depta, H., Robinson, D. G. 1986. The isolation and enrichment of coated vesicles from suspension-cultured carrot cells. *Protoplasma* 130:162–70

48. Depta, H., Freundt, H., Hartmann, D., Robinson, D. G. 1987. Preparation of a homogeneous coated vesicle fraction from bean leaves. *Protoplasma* 136:154–60

49. Depta, H., Andreae, M., Blaschek, W., Robinson, D. G. 1988. Glucan synthase II activity in a coated vesicle fraction from zucchini hypocotyls. *Eur. J. Cell Biol.* In press

50. Doohan, M. E., Palevitz, B. A. 1980. Microtubules and coated vesicles in guard cell protoplasts of *Allium cepa* L. *Planta* 149:389–401

51. Driessche, E. van, Beeckmans, E., Dejaegere, R., Kanarek, L. 1984. Thiourea: the antioxidant of choice for the purification of proteins from phenol-rich plant tissues. *Anal. Biochem.* 141:184–88

52. Dunn, W. A., Hubbard, A. C. 1984. Receptor-mediated endocytosis of epidermal growth factor by hepatocytes in the perfused rat liver: ligand and receptor dynamics. *J. Cell Biol.* 98:2148–59

53. Dyke, R. W. van, Scharschmidt, B. F., Steer, C. J. 1985. ATP-dependent proton transport by isolated brain clathrin-coated vesicles. Role of clathrin and other determinants of acidification. *Biochem. Biophys. Acta* 812:423–36

54. Edelhoch, H., Prasad, K., Lippoldt, R. E., Nandi, P. K. 1984. Stability and structure of clathrin. *Biochemistry* 23:2314–20

55. Emons, A. M. C., Traas, J. A. 1986. Coated pits and coated vesicles on the plasma membrane of plant cells. *Eur. J. Cell Biol.* 41:57–64

56. Evert, R. F., Botha, C. E. J., Mierzwa, R. J. 1985. Free-space marker studies on the leaf of *Zea mays* L. *Protoplasma* 126:62–73

57. Fan, J. Y., Carpentier, J.-L., Gorden, P., van Obberghen, E. V., Blackett, N. M., Grunfeld, C., Orci, L. 1982. Receptor-mediated endocytosis of insulin: role of microvilli, coated pits, and coated vesicles. *Proc. Natl. Acad. Sci. USA* 79:7788–91

58. Farquhar, M. G. 1985. Progress in unraveling pathways of Golgi traffic. *Ann. Rev. Cell Biol.* 1:447–88

59. Farquhar, M. G., Palade, G. E. 1981. The Golgi apparatus (complex) (1954–1981)-from artifact to center stage. *J. Cell Biol.* 91:775–1035

60. Fine, R. E., Ockleford, C. D. 1984. Supramolecular cytology of coated vesicles. *Int. Rev. Cytol.* 91:1–43

61. Fishman, J. B., Fine, R. E. 1987. A trans Golgi-derived exocytic coated vesicle can contain both newly synthesized cholinesterase and internalized transferrin. *Cell* 48:157–64

62. Forgac, M., Cantley, L. 1984. Characterization of the ATP-dependent proton pump of clathrin-coated vesicles. *J. Biol. Chem.* 259:8101–5

63. Forgac, M., Berne, M. 1986. Structural characterization of the ATP-hydrolyzing portion of the coated vesicle proton pump. *Biochemistry* 25:4275–80

64. Forgac, M., Cantley, L., Wiedenmann, B., Altstiel, L., Branton, D. 1983. Clathrin-coated vesicles contain an ATP-dependent proton pump. *Proc. Natl. Acad. Sci. USA* 80:1300–3

65. Franke, W. W., Herth, W. 1974. Morphological evidence for de-novo formation of plasma membrane from coated vesicles in exponentially growing cultivated plant cells. *Exp. Cell Res.* 89:447–51

66. Friend, D. S., Farquhar, M. G. 1967. Functions of coated vesicles during protein absorption in the rat vas deferens. *J. Cell Biol.* 35:357–76

67. Gallagher, S. R., Carroll, E. J. Jr., Leonard, R. T. 1986. A sensitive diffusion plate assay for screening inhibitors of protease activity in plant cell fractions. *Plant Physiol.* 81:869–74

68. Geisow, M. J., Childs, J., Burgoyne, R. D. 1985. Cholinergic stimulation of chromaffin cells induces rapid coating of the plasma membrane. *Eur. J. Cell Biol.* 38:51–56

69. Geuze, H. J., Slot, J. W., Strous, G. J. A. M., Lodish, H. F., Schwartz, A. L. 1983. Intracellular site of asialoglycoprotein receptor-ligand uncoupling: double-label immunoelectron microscopy during receptor mediated endocytosis. *Cell* 32:277–87

70. Geuze, H. J., Slot, J. W., Strous, G. J. A. M., Hasilik, A., Figura, K. von. 1984. Ultrastructural localization of the mannose 6-phosphate receptor in rat liver. *J. Cell Biol.* 98:2047–54

71. Glickman, J., Croen, K., Kelly, S., Al-Awqati, Q. 1983. Golgi membranes contain an electrogenic H$^+$ pump in parallel to a chloride conductance. *J. Cell Biol.* 97:1303–8

72. Goldenthal, K. L., Pastan, I., Willingham, M. C. 1985. Serial section analysis of clathrin-coated pits in rat liver sinusoidal endothelial cells. *Exp. Cell Res.* 161:342–52

73. Goldstein, J. L., Anderson, R. G. W., Brown, M. S. 1979. Coated pits, coated vesicles, and receptor-mediated endocytosis. *Nature* 279:679–85

74. Goldstein, J. L., Brown, M. S., Anderson, R. G. W., Russell, D. W., Schneider, W. T. 1985. Receptor-mediated endocytosis: concepts emerging from the LDL receptor system. *Ann. Rev. Cell Biol.* 1:1–39

75. Goud, B., Huet, C., Louvard, D. 1985. Assembled and unassembled pools of clathrin: a quantitative study using enzyme immunoassay. *J. Cell Biol.* 100:521–27

76. Griffing, L. R., Fowke, L. C. 1985. Cytochemical localization of peroxidase in soybean suspension culture cells and protoplasts: intracellular vacuole differentiation and presence of peroxidase in coated vesicles and multivesicular bodies. *Protoplasma* 128:22–30

77. Griffing, L. R., Mersey, B. G., Fowke, L. C. 1986. Cell-fractionation analysis of glucan synthase I and II distribution and polysaccharide secretion in soybean protoplasts. *Planta* 167:175–82

78. Griffiths, G., Simons, K. 1986. The trans Golgi network: sorting at the exit site of the Golgi complex. *Science* 234:438–42

79. Griffiths, G., Pfeiffer, S., Simons, K., Matlin, K. 1985. Exit of newly synthesized membrane proteins from the trans cisterna of the Golgi complex to the plasma membrane. *J. Cell Biol.* 101:949–64

80. Hawes, C. R., Martin, B. 1986. Deep etching of plant cells: cytoskeleton and coated pits. *Cell Biol. Int. Rep.* 10:985–91

81. Hanspal, M., Luna, E., Branton, E. 1984. The association of clathrin fragments with coated vesicle membranes. *J. Biol. Chem.* 259:11075–82

82. Harikumar, P., Reeves, J. P. 1983. The lysosomal proton pump is electrogenic. *J. Biol. Chem.* 258:10403–10

83. Hausmann, K., Patterson, D. J. 1984. Contractile vacuole complexes in algae. In *Compartments in Algal Cells and Their Interaction,* ed. W. Wiessner, D. G. Robinson, R. C. Starr, pp. 139–46. Berlin: Springer-Verlag

84. Hawes, C. R. 1985. Conventional and high voltage electron microscopy of the cytoskeleton and cytoplasmic matrix of carrot (*Daucus carota* L.) cells grown in suspension culture. *Eur. J. Cell Biol.* 38:201–10

85. Helmy, S., Porter-Jordan, K., Dawidowicz, E. A., Pilch, P., Schwartz, A. L., Fine, R. E. 1986. Separation of endocytic from exocytic coated vesicles using a novel cholinesterase mediated density shift technique. *Cell* 44:497–506

86. Herzog, V. 1981. Pathways of endocytosis in secretory cells. *Trends Biochem. Sci.* 6:319–22

87. Heuser, J. E. 1980. Three-dimensional

visualization of coated vesicle formation in fibroblasts. *J. Cell Biol.* 84:550–83

88. Heuser, J. E. 1983. Procedure for freeze-drying molecules absorbed to mica flakes. *J. Mol. Biol.* 169:155–95

89. Heuser, J., Kirchhausen, T. 1985. Deep-etch views of clathrin assemblies. *J. Ultrastruct. Res.* 92:1–27

90. Heuser, J. E., Reese, T. S. 1973. Evidence for the recycling of synaptic vesicle membrane during transmitter release at the frog neuromuscular junction. *J. Cell Biol.* 57:315–44

91. Heuser, J. E., Reese, T. S. 1981. Structural changes after transmitter release at the frog neuromuscular junction. *J. Cell Biol.* 88:564–80

92. Heuser, J. E., Reese, T. S., Dennis, M. J., Jan, Y., Jan, L., Evans, L. 1979. Synaptic vesicle exocytosis captured by quick-freezing and correlated with quantal transmitter release. *J. Cell Biol.* 81:275–300

93. Hillmer, S., Depta, H., Robinson, D. G. 1986. Confirmation of endocytosis in higher plant protoplasts using lectin-gold conjugates. *Eur. J. Cell Biol.* 42:142–49

94. Holmes, N., Biermann, S. J., Brodsky, F. M., Bharuca, B., Parham, P. 1984. Comparison of the primary structures of clathrin light chains from bovine brain and adrenal gland by peptide mapping. *EMBO J.* 3:1621–27

95. Hopkins, C. R. 1985. The appearance and internalization of transferrin receptors at the margins of spreading human tumor cells. *Cell* 40:199–208

96. Hopkins, C. R., Trowbridge, I. S. 1983. Internalization and processing of transferrin and the transferrin receptor in human carcinoma A 431 cells. *J. Cell Biol.* 97:508–21

97. Huber, D. J., Nevins, D. J. 1981. Wall-protein antibodies as inhibitors of growth and of autolytic reactions of isolated cell walls. *Physiol. Plant.* 53:533–39

98. Hübner, R., Depta, H., Robinson, D. G. 1985. Endocytosis in maize root cap cells. Evidence obtained using heavy metal salt solutions. *Protoplasma* 129:214–22

99. Jaarsveld, P. P. van, Nandi, P. K., Lippoldt, R. E., Saroff, H., Edelhoch, H. 1981. Polymerization of clathrin protomers into basket structures. *Biochemistry* 20:4129–35

100. Jackson, A. P., Seow, H.-F., Holmes, N., Drickamer, K., Parham, P. 1987. Clathrin light chains contain brain-specific insertion sequences and a region of homology with intermediate filaments. *Nature* 326:154–57

101. Joachim, S., Robinson, D. G. 1984. Endocytosis of cationic ferritin by bean leaf protoplasts. *Eur. J. Cell Biol.* 34:212–16

102. Kadota, K., Usami, M., Takahashi, A. 1982. A protein kinase and its substrate associated with the outer coat and the inner core of coated vesicles from bovine brain. *Biomed. Res.* 3:575–78

103. Kanaseki, T., Kadota, K. 1969. The "vesicle in a basket": a morphological study of the coated vesicle isolation from the nerve endings of the guinea pig brain, with special reference to the mechanism of membrane movements. *J. Cell Biol.* 42:202–20

104. Kartenbeck, J. 1978. Preparation of membrane-depleted polygonal coat structures from isolated vesicles. *Cell Biol. Int. Rep.* 2:457–64

105. Kartenbeck, J. 1980. Coated secretory vesicles. In *Coated Vesicles,* ed. C. D. Ockleford, A. Whyte, pp. 243–54. London: Cambridge Univ. Press

106. Kartenbeck, J., Schmid, E., Müller, H., Franke, W. W. 1981. Immunological identification and localization of clathrin and coated vesicles in cultured cells and in tissues. *Exp. Cell Res.* 133:191–211

107. Kedersha, N. L., Hill, D. F., Kronquist, K. E., Rome, L. H. 1986. Subpopulations of liver coated vesicles resolved by preparative agarose gel electrophoresis. *J. Cell Biol.* 103:287–97

108. Keen, J. H., Willingham, M. C., Pastan, I. H. 1979. Clathrin-coated vesicles: isolation, dissociation and factor-dependent reassociation of clathrin baskets. *Cell* 16:303–12

109. Keen, J. H., Willingham, M. C., Pastan, I. 1981. Clathrin and coated vesicle proteins. Immunological characterization. *J. Biol. Chem.* 256:2538–44

110. Keen, J. H., Chestnut, M. H., Beck, K. A. 1987. The clathrin coat assembly polypeptide complex. Autophosphorylation and assembly activities. *J. Biol. Chem.* 262:3864–71

111. Kelly, R. B. 1985. Pathways of protein secretion in eukaryotes. *Science* 230:25–32

112. Kelly, W. G., Passaniti, A., Woods, J. W., Daiss, J. L., Roth, T. F. 1983. Tubulin is a molecular component of coated vesicles. *J. Cell Biol.* 97:1191–99

113. Kilmartin, J. V., Wright, B., Milstein, C. 1982. Rat monoclonal antitubulin antibodies derived by using a new nonsecreting rat cell line. *J. Cell Biol.* 93:576–82

114. Kimpel, J. A., Key, J. L. 1985. Heat shocks in plants. *TIBS* 10:353–57

115. Kirchhausen, T., Harrison, S. C. 1981.

Protein organization in clathrin trimers. *Cell* 23:755–61

116. Kirchhausen, T., Harrison, S. C. 1984. Structural domains of heavy clathrin chains. *J. Cell Biol.* 99:1725–34

117. Kirchhausen, T., Harrison, S. C., Heuser, J. 1986. Configuration of clathrin trimers: evidence from electron microscopy. *J. Ultra. Mol. Struct. Res.* 94: 199–208

118. Kirchhausen, T., Harrison, S. C., Parham, P., Brodsky, F. M. 1983. Location and distribution of the light chains in clathrin trimers. *Proc. Natl. Acad. Sci. USA* 80:2481–85

119. Kirchhausen, T., Scarmats, P., Harrison, S. C., Monroe, J. J., Chow, E. P., Mattaliano, R. J., Ramachandran, K. L., Smart, J. E., Ahn, A., Brosius, J. 1987. Clathrin light chains LCA and LCB are similar, polymorphic, and share repeated heptad motifs. *Science* 236:320–24

120. Kogel, K. H., Ehrlich-Rogozinski, S., Reisener, H. J., Sharon, N. 1984. Surface galactolipids of wheat protoplasts as receptors for soybean agglutinin and their possible relevance to host-parasite interaction. *Plant Physiol.* 76:924–28

121. Kohtz, D. S., Georgieva-Hanson, V., Kohtz, J. D., Schook, W. I., Puszkin, S. 1987. Mapping two functional domains of clathrin light chains with monoclonal antibodies. *J. Cell Biol.* 104:897–903

122. Kumar, N., Blumenthal, R., Heukart, M., Weinstein, J. V., Klausner, R. D. 1982. Aggregation and calcium-induced fusion of phosphatidyl-choline vesicle-tubulin complexes. *J. Biol. Chem.* 257:15137–44

123. Lemmon, S. K., Lemmon, V. T., Jones, E. V. 1985. Purification of yeast clathrin and production of monoclonal antibodies to the 180 kD protein. *J. Cell Biol.* 101:47a (Abstr. 175)

124. Linden, C. D. 1982. Identification of the coated vesicle proteins that bind calmodulin. *Biochem. Biophys. Res. Commun.* 109:186–93

125. Lisanti, M. P., Schook, W., Moskowitz, N., Ores, C., Puszkin, S. 1981. Brain clathrin and clathrin-associated proteins. *Biochem. J.* 201:297–304

126. Lisanti, M. P., Shapiro, L. S., Moskowitz, N., Hua, E. L., Puszkin, S., Schook, W. 1982. Isolation and preliminary characterization of clathrin-associated proteins. *Eur. J. Biochem.* 125:463–70

127. Löbler, M., Klämbt, H. D. 1985. Auxin-binding protein from coleoptile membranes of corn (*Zea mays* L.). II.

Localization of a putative auxin receptor. *J. Biol. Chem.* 260:9854–59

128. Louvard, D., Morris, C., Warren, G., Stanley, K., Winkler, D., Reggio, H. 1983. A monoclonal antibody to the heavy chain of clathrin. *EMBO J.* 2:1655–64

129. Mäder, M., Walter, C. 1986. De-novo synthesis and release of peroxidases in cell suspension cultures of *Nicotiana tabacum* L. *Planta* 169:273–77

130. Mäder, M., Meyer, Y., Bopp, M. 1975. Lokalisation der Peroxidase-Isoenzyme in Protoplasten und Zellwänden von *Nicotiana tabacum* L. *Planta* 122:259–68

131. Makarow, M. 1985. Endocytosis in *Saccharomyces cerevisiae:* internalization of α-amylase and fluorescent dextran into cells. *EMBO J.* 4:1861–66

132. Makarow, M. 1985. Endocytosis in *Saccharomyces cerevisiae:* internalization of enveloped viruses into spheroplasts. *EMBO J.* 4:1855–60

133. Makarow, M., Nevalainen, L. T. 1987. Transport of a fluorescent macromolecule via endosomes to the vacuole in *Saccharomyces cerevisiae. J. Cell Biol.* 104:67–75

134. Manolson, M. F., Percy, J. M., Apps, D. K., Xie, X.-S., Stone, D. K., Poole, R. J. 1988. Endomembrane H⁺-ATPases from plants and animals show immunological crossreactivity. In *Membrane Proteins: Proceedings of the Membrane Protein Symposium,* ed. S. C. Goheen, L. Hjelmeland, M. McNamee, R. Gennis. Bio-Rad Publication

135. Mathew, W. D., Isavalev, L., Reichardt, L. F. 1981. Identification of a synaptic vesicle specific membrane protein with broad distribution in neuronal and neurosecretory tissue. *J. Cell Biol.* 91:257–69

136. Matlin, K., Simons, K. 1984. Sorting of an apical plasma membrane glycoprotein occurs before it reaches the cell surface in cultured epithelial cells. *J. Cell Biol.* 99:2131–39

137. Maupin, P., Pollard, T. D. 1983. Improved preservation and staining of HeLa cell actin filaments, clathrin-coated membranes, and other cytoplasmic structures by tannic acid-glutaraldehyde-saponin fixation. *J. Cell Biol.* 96:51–62

138. Maxfield, F. R., Schlessinger, J., Shecter, J., Pastan, I., Willingham, M. C. 1978. Collection of insulin, EGF, α₂-Macroglobulin in the same patches on the surface of cultured fibroblasts and common internalization. *Cell* 14:805–10

139. McKanna, J. A. 1974. Permeability modulating membrane coats. I. Fine structure of fluid segregation organelles of peritrich contractile vacuoles. *J. Cell Biol.* 63:317–22

140. Mello, R. J., Goldstein, M. S., Anderson, R. G. W. 1980. LDL receptors in coated vesicles isolated from bovine adrenal cortex: binding sites unmasked by detergent treatment. *Cell* 20:829–37

141. Merisko, E. M. 1985. Evidence for the interaction of α-actinin and calmodulin with the clathrin heavy chain. *Eur. J. Cell Biol.* 39:167–72

142. Merisko, E. M. 1986. The distribution of clathrin during amphibian oogenesis. *Eur. J. Cell Biol.* 42:118–25

143. Merisko, E. M., Farquhar, M. G., Palade, G. E. 1982. Coated vesicle isolation by immunoadsorption on *Staphylococcus aureus* cells. *J. Cell Biol.* 92:846–57

144. Merisko, E. M., Farquhar, M. G., Palade, G. E. 1986. Redistribution of clathrin heavy and light chains in anoxic pancreatic acinar cells. *Pancreas* 1:110–23

145. Mersey, B. G., Fowke, L. C., Constabel, F., Newcomb, E. H. 1982. Preparation of a coated vesicle-enriched fraction from plant cells. *Exp. Cell Res.* 141:459–63

146. Mersey, B. G., Griffing, L. R., Rennie, P. J., Fowke, L. C. 1985. The isolation of coated vesicles from protoplasts of soybean. *Planta* 163:317–27

147. Michl, J., Unkeless, J. C., Pieczonka, M. M., Silverstein, S. C. 1983. Modulation of Fc receptors of mononuclear phagocytes by immobilized antigen-antibody complexes. Quantitative analysis of the relationship between ligand number and Fc receptor response. *J. Exp. Med.* 157:1746–57

148. Montesano, R., Mossar, A., Vassalli, P., Orci, L. 1983. Specialization of the macrophage plasma membrane at sites of interaction with opsonized erythrocytes. *J. Cell Biol.* 96:1227–33

149. Mooibroek, M. J., Michiel, D. F., Wang, J. H. 1987. Clathrin light chains are calcium-binding proteins. *J. Biol. Chem.* 262:25–28

150. Morrow, D. L., Jones, R. L. 1986. Localization and partial characterization of the extracellular proteins centrifuged from pea internodes. *Physiol. Plant.* 67:397–407

151. Moskowitz, N., Schook, W., Lisanti, M., Hua, E., Puszkin, S. 1982. Calmodulin affinity for brain coated vesicle proteins. *J. Neurochem.* 38:1742–47

152. Mueller, S. C., Branton, D. 1984. Identification of coated vesicles in *Saccharomyces cerevisiae*. *J. Cell Biol.* 98:341–46

153. Nandi, P. K., Edelhoch, H. 1984. The effects of lysotropic (Hofmeister) salts on the stability of clathrin coat structure in coated vesicles and baskets. *J. Biol. Chem.* 259:11290–96

154. Nandi, P. K., Irace, G., Jaarsveld, P. P. van, Lippoldt, R. E., Edelhoch, H. 1982. Instability of coated vesicles in concentrated sucrose solutions. *Proc. Natl. Acad. Sci. USA* 79:5881–85

155. Nandi, P. K., Prasad, K., Lippoldt, R. E., Alfsen, A., Edelhoch, H. 1982. Reversibility of coated vesicle dissociation. *Biochemistry* 21:6434–40

156. Neumann, D., Nieden, U. zur, Manteuffel, R., Walter, G., Scharf, K-D., Nover, L. 1987. Intracellular localization of heat shock proteins in tomato cell cultures. *Eur. J. Cell Biol.* 43:71–81

157. Newcomb, E. H. 1980. Coated vesicles: their occurrence in different plant cell types. In *Coated Vesicles*, ed. C. D. Ockleford, A. Whyte, pp. 55–68. Cambridge: Cambridge Univ. Press

158. Nishizawa, N., Mori, S. 1977. Invagination of plasmalemma: its role in the absorption of macromolecules in rice roots. *Plant Cell Physiol.* 18:767–82

159. Nishizawa, N., Mori, S. 1978. Endocytosis (heterophagy) in plant cells: involvement of ER and ER-derived vesicles. *Plant Cell Physiol.* 19:717–30

160. Novick, P., Field, C., Schekman, R. 1980. Identification of 23 complementation groups required for post-translational events in the yeast secretory pathway. *Cell* 21:205–15

161. Okhuma, S., Moriyama, Y., Takano, T. 1982. Identification and characterization of a proton pump on lysosomes by fluorescein isothiocyanate-dextran fluorescence. *Proc. Natl. Acad. Sci. USA* 79:2758–62

162. Op den Kamp, J. A. F. 1979. Lipid asymmetry in membranes. *Ann. Rev. Biochem.* 48:47–71

163. Orci, L., Glick, B. S., Rothman, J. E. 1986. A new type of coated vesicular carrier that appears not to contain clathrin: its possible role in protein transport within the Golgi stack. *Cell* 46:171–84

164. Orci, L., Halban, P., Amherdt, M., Ravazzola, M., Vassalli, J.-D., Perrelet, A. 1984. A clathrin-coated, Golgi-related compartment of the insulin secreting cell accumulates proinsulin in the presence of monensin. *Cell* 39:39–47

165. Orci, L., Ravazzola, M., Amherdt, M., Louvard, D., Perrelet, A. 1985.

Clathrin-immunoreactive sites in the Golgi apparatus are concentrated at the trans pole in polypeptide hormone-secreting cells. *Proc. Natl. Acad. Sci. USA* 42:5385–89

166. Parham, P., Androlewicz, M. J., Brodsky, F. M., Holmes, N. J., Ways, J. P. 1982. Monoclonal antibodies: purification, fragmentation, and application to structural and functional studies of class I MHC antigens. *J. Immunol. Methods* 53:133–73

167. Pastan, I. H., Willingham, M. C. 1981. Receptor-mediated endocytosis of hormones in cultured cells. *Ann. Rev. Physiol.* 43:239–50

168. Patzer, E. J., Schlossman, D. M., Rothman, J. E. 1982. Release of clathrin from coated vesicles dependent upon a nucleoside triphosphate and a cytosol fraction. *J. Cell Biol.* 93:230–36

169. Pauloin, A., Jolles, P. 1984. Internal control of the coated vesicle pp50-specific kinase complex. *Nature* 311:265–67

170. Pauloin, A., Bernier, I., Jolles, P. 1982. Presence of cyclic nucleotide Ca^{2+} independent protein kinase in bovine coated vesicles. *Nature* 258:574–76

171. Payne, G. S., Schekman, R. 1985. A test of clathrin function in protein secretion and cell growth. *Science* 230:1009–14

172. Pearse, B. M. F. 1975. Coated vesicles from pig brain: purification and biochemical characterization. *J. Mol. Biol.* 97:93–98

173. Pearse, B. M. F. 1976. Clathrin: a unique protein associated with intracellular transfer of membrane by coated vesicles. *Proc. Natl. Acad. Sci. USA* 73:1255–59

174. Pearse, B. M. F. 1978. On the structural and functional components of coated vesicles. *J. Mol. Biol.* 126:803–12

175. Pearse, B. M. F. 1980. Coated vesicles. *TIBS* 5:131–34

176. Pearse, B. M. F. 1982. Structure of coated pits and vesicles. In *Membr. Recycl. Ciba Found. Symp.* 92:246–65

177. Pearse, B. M. F. 1982. Coated vesicles from human placenta carry ferritin, transferrin, and immunoglobulin G. *Proc. Natl. Acad. Sci. USA* 79:451–55

178. Pearse, B. M. F. 1985. Assembly of the mannose-6-phosphate receptor into reconstituted clathrin coats. *EMBO J.* 4:2457–60

179. Pearse, B. M. F., Bretscher, M. S. 1981. Membrane recycling by coated vesicles. *Ann. Rev. Biochem.* 50:85–101

180. Pearse, B. M. F., Robinson, M. S. 1984. Purification and properties of 100

kD proteins from coated vesicles and their reconstitution with clathrin. *EMBO J.* 3:1951–57

181. Pesacreta, T. C., Lucas, W. J. 1984. The plasma membrane coat and a coated vesicle-associated reticulum of membranes: their structures and possible interrelationship in *Chara corallina. J. Cell Biol.* 98:1537–45

182. Pesacreta, T. C., Lucas, W. J. 1985. Presence of a partially-coated reticulum and a plasma membrane coat in angiosperms. *Protoplasma* 125:173–84

183. Petersen, O. W., Deurs, B. van. 1983. Serial section analysis of coated pits and vesicles involved in adsorptive pinocytosis in cultured fibroblasts. *J. Cell Biol.* 96:277–81

184. Pfeffer, S. R., Kelly, R. B. 1981. Identification of minor components of coated vesicles by use of permeation chromatography. *J. Cell Biol.* 91:385–91

185. Pfeffer, S. R., Kelly, R. B. 1985. The subpopulation of brain coated vesicles that carries synaptic vesicle proteins contains two unique polypeptides. *Cell* 40:949–57

186. Pfeffer, S. R., Drubin, D. G., Kelly, R. B. 1983. Identification of three coated vesicle components as alpha and beta tubulin linked to a phosphorylated 50-kD polypeptide. *J. Cell Biol.* 97:40–47

187. Pilch, P., Shia, M., Benson, R. J., Fine, R. E. 1983. Coated vesicles participate in the receptor-mediated endocytosis of insulin. *J. Cell Biol.* 93:133–38

188. Prasad, K., Lippoldt, R. E., Edelhoch, H. 1985. Coat formation in coated vesicles. *Biochemistry* 24:6421–27

189. Prasad, K., Lippoldt, R. E., Edelhoch, H. 1986. An intermediate polymer in the assembly of clathrin baskets. *Biochemistry* 25:5214–19

190. Pretorius, H. T., Nandi, P. K., Lippoldt, R. E., Johnson, M. L., Keen, J. H., Pastan, I., Edelhoch, H. 1981. Molecular characterization of human clathrin. *Biochemistry* 20:2777–82

191. Quader, H., Deichgräber, G., Schnepf, E. 1986. The cytoskeleton of *Cobaea* seed hairs: patterning during cell-wall differentiation. *Planta* 168:1–10

192. Quail, P. H. 1979. Plant cell fractionation. *Ann. Rev. Plant Physiol.* 30:425–84

193. Raikhel, A. S. 1984. Accumulations of membrane-free clathrin-like lattices in the mosquito oocyte. *Eur. J. Cell Biol.* 35:279–83

194. Rambourg, A., Clermont, Y., Hermo, L. 1979. Three-dimensional architecture of the Golgi apparatus in Sertoli cells of the rat. *Am. J. Anat.* 154:455–76

195. Rees-Jones, R., Al-Awqati, Q. 1984. Proton-translocating adenosine triphosphatase in rough and smooth microsomes from rat liver. *Biochemistry* 23:2236–40

196. Riezman, H. 1985. Endocytosis in yeast: several of the yeast secretory mutants are defective in endocytosis. *Cell* 40:1001–9

197. Robards, A. W., Robb, M. E. 1974. The entry of ions and molecules into roots: an investigation using electron-opaque tracers. *Planta* 120:1–12

198. Robertson, J. G., Lyttleton, P. 1982. Coated and smooth vesicles in the biogenesis of cell walls, plasma membranes, infection threads and peribacteroid membranes in root hairs and nodules of white clover. *J. Cell Sci.* 58:63–78

199. Robinson, D. G. 1985. Plant membranes: endo- and plasma membranes of plant cells. In *Cell Biology Monographs*, Vol. 3, ed. E. E. Bittar. New York: John Wiley

200. Robinson, D. G., Quader, H. 1981. Towards cellulose synthesis *in vitro*. *J. Theor. Biol.* 92:483–95

201. Robinson, D. G., Andreae, M., Depta, H., Hartmann, D., Hillmer, S. 1987. Coated vesicles in plants: characterization and function. In *Plant Membranes. Structure, Function, Biogenesis UCLA Symp. Mol. Cell. Biol. (NS)* 63:341–58

202. Robinson, M., Rhodes, J., Albertini, D. 1983. Slow internalization of human chorionic gonadotropin by cultured granulosa cells. *J. Cell Physiol.* 117:43–50

203. Robinson, M. S. 1987. 100 kD coated vesicle proteins: molecular heterogeneity and intracellular distribution studied with monoclonal antibodies. *J. Cell Biol.* 204:887–95

204. Rodenwald, R. 1983. Intestinal transport of antibodies in the newborn rat. *J. Cell Biol.* 58:189–211

205. Roth, J., Taatjes, D. J., Lucocq, J. M., Weinstein, J., Paulson, J. C. 1985. Demonstration of an extensive transtubular network continuous with the Golgi apparatus that may function in glycosylation. *Cell* 43:287–95

206. Roth, T. F., Porter, K. R. 1964. Yolk protein uptake in the oocyte of the mosquito *Aedes aegypti* L. *J. Cell Biol.* 20:313–32

207. Rothman, J. E., Fine, R. E. 1980. Coated vesicles transport newly synthesized membrane glycoproteins from endoplasmic reticulum to plasma membrane in two successive stages. *Proc. Natl. Acad. Sci. USA* 77:780–84

208. Rothman, J. E., Schmid, S. L. 1986. Enzymatic recycling of clathrin from coated vesicles. *Cell* 46:5–9

209. Rothman, J. E., Brusztyn-Pettegrew, H., Fine, R. E. 1980. Transport of membrane glycoprotein of vesicular stomatitis virus to the cell surface in two stages by clathrin-coated vesicles. *J. Cell Biol.* 86:162–71

210. Rothstein, S. J., Lazarus, C. M., Smith, W. E., Baulcombe, D. C., Gatenby, A. A. 1984. Secretion of a wheat α-amylase expressed in yeast. *Nature* 308:662–65

211. Rubinstein, J. L. R., Fine, R. E., Luskey, B. D., Rothman, J. E. 1981. Purification of coated vesicles by agarose gel electrophresis. *J. Cell Biol.* 89:357–61

212. Ruiz-Herrera, J., Sing, V. O., Woude, W. Van Der, Bartnicki-Garcia, S. 1975. Microfibril assembly by granules of chitin synthetase. *Proc. Natl. Acad. Sci. USA* 72:2706–10

213. Ryser, U. 1979. Cotton fibre differentiation: occurrence and distribution of coated and smooth vesicle during primary and secondary wall formation. *Protoplasma* 98:223–39

214. Sahagian, G. G., Steer, C. J. 1985. Transmembrane orientation of the mannose-6-phosphate receptor in isolated clathrin-coated vesicles. *J. Biol. Chem.* 260:9838–42

215. Saraste, J., Kuismanen, E. 1984. Pre- and post-Golgi vacuoles operate in the transport of Semliki Forest virus membrane glycoproteins to the cell surface. *Cell* 38:535–49

216. Scherrer, R., Louden, L., Gerhardt, P. 1974. Porosity of the yeast cell wall and membrane. *J. Bacteriol.* 118:534–40

217. Schjeide, O. A., Lin, R. I. S., Grellert, E. A., Galey, F. R., Mead, J. F. 1969. Isolation and preliminary chemical analysis of coated vesicles from chicken oocytes. *Physiol. Chem. Phys.* 1:141–63

218. Schlessinger, J. 1980. The mechanism and role of hormone-induced clustering of membrane receptors. *TIBS* 5:210–14

219. Schlossman, D. M., Schmid, S. L., Braell, W. A., Rothman, J. E. 1984. An enzyme that removes clathrin coats: purification of an uncoating ATPase *J. Cell Biol.* 99:723–33

220. Schmid, S. L., Rothman, J. E. 1985. Two classes of binding sites for uncoating protein in clathrin triskelions. *J. Biol. Chem.* 260:10050–56

221. Schmid, S. L., Rothman, J. E. 1985. Enzymatic dissociation of clathrin cages

in a two-stage process. *J. Biol. Chem.* 260:10044–49

222. Schmid, S. L., Braell, W. A., Rothman, J. E. 1985. ATP catalyzes the sequestration of clathrin during enzymatic uncoating. *J. Biol. Chem.* 260:10057–62

223. Schmid, S. L., Matsumoto, A. K., Rothman, J. E. 1982. A domain of clathrin that forms coats. *Proc. Natl. Acad. Sci. USA* 79:91–95

224. Schmid, S. L., Braell, W. A., Schlossman, D. M., Rothman, J. E. 1984. A role for clathrin light chains in the recognition of clathrin cages by "uncoating ATPase". *Nature* 311:228–31

225. Schook, W., Puszkin, S., Bloom, W., Ores, C., Kochwa, S. 1979. Mechanochemical properties of brain clathrin interaction with actin and α-actinin and polymerization into basket-like structures or filaments. *Proc. Natl. Acad. Sci. USA* 76:116–20

226. Schulze-Lohoff, E., Hasilik, A., Figura, K. von. 1985. Cathepsin D precursors in clathrin-coated organelles from human fibroblasts. *J. Cell Biol.* 101:824–29

227. Show, D. R., Griffin, F. M. 1981. Phagocytosis requires repeated triggering of macrophage phagocytic receptors during particle ingestion. *Nature* 289:409–11

228. Silva, W. I., Andres, A., Schook, W., Puszkin, S. 1986. Evidence for the presence of muscarinic acetylcholine receptors in bovine brain coated vesicles. *J. Biol. Chem.* 261:14788–96

229. Snider, M. D., Rogers, D. C. 1985. Intracellular movement of cell surface receptors after endocytosis: resialylation of asialo-transferrin receptor in human erythroleukemia cells. *J. Cell Biol.* 100:826–34

230. Sorger, P. K., Crowther, R. A., Finch, J. T., Pearse, B. M. F. 1986. Clathrin cubes: an extreme variant of the normal cage. *J. Cell Biol.* 103:1213–19

231. Steer, C. J., Klausner, R. D., Blumenthal, R. 1982. Interaction of liver clathrin coat protein with lipid model membranes. *J. Biol. Chem.* 257: 8533–40

232. Steer, C. J., Wall, D. A., Ashwell, G. 1983. Evidence for the presence of the asialoglycoprotein receptor in coated vesicles isolated from rat liver. *Hepatology* 3:667–72

233. Steer, C. J., Bisher, M., Blumenthal, R., Steven, A. C. 1984. Detection of membrane cholesterol by filipin in isolated rat liver coated vesicles is dependent upon removal of the clathrin coat. *J. Cell Biol.* 99:315–19

234. Steinman, R. M., Mellman, I. S., Mueller, W. A., Cohn, Z. A. 1983. Endocytosis and the recycling of plasma membrane. *J. Cell Biol.* 96:1–27

235. Steven, A. C., Hainfield, J. F., Wall, J. S., Steer, C. J. 1983. Mass distributions of coated vesicles isolated from liver and brain: analysis by scanning transmission electron microscopy. *J. Cell Biol.* 97:1714–23

236. Stone, D. K., Xie, X.-S., Racker, E. 1983. An ATP-driven proton pump in clathrin-coated vesicles. *J. Biol. Chem.* 258:4059–62

237. Takahashi, A., Usami, M., Kadota, T., Kadota, K. 1985. Properties of protein kinases in brain coated vesicles. *J. Biochem.* 98:63–68

238. Takemura, R., Stenberg, P. E., Bainton, D. F., Werb, Z. 1986. Rapid redistribution of clathrin onto macrophage plasma membranes in response to Fc receptor-ligand interaction during frustrated phagocytosis. *J. Cell Biol.* 102:55–69

239. Tanchak, M. A., Griffing, L. R., Mersey, B. G., Fowke, L. C. 1984. Endocytosis of cationized ferritin by coated vesicles of soybean protoplasts. *Planta* 162:481–86

240. Tooze, J., Tooze, S. A. 1986. Clathrin-coated vesicular transport of secretory proteins during the formation of ACTH-containing secretory granules in AtT20 cells. *J. Cell Biol.* 103:839–50

241. Traas, J. A. 1984. Visualization of the membrane bound cytoskeleton and coated pits of plant cells by means of dry cleaving. *Protoplasma* 119:212–18

242. Unanue, E. R., Ungewickell, E., Branton, D. 1981. The binding of clathrin triskelions to membranes from coated vesicles. *Cell* 26:439–46

243. Unkeless, J. C. 1980. Fc receptors on mouse macrophages. In *Mononuclear Phagocytes*, ed. R. van Furth, 1:735–51. The Hague: Martinus Nijhoff

244. Ungewickell, E. 1983. Biochemical and immunological studies on clathrin light chains and their binding sites on clathrin triskelions. *EMBO J.* 2:1401–8

245. Ungewickell, E. 1984. First clue to biological role of clathrin light chains. *Nature* 311:213

246. Ungewickell, E. 1984. Characterization of clathrin and clathrin-associated proteins. *Biochem. Soc. Trans.* 12: 978–80

247. Ungewickell, E. 1985. The 70-kd mammalian heat shock proteins are structurally and functionally related to the uncoating proteins that release

clathrin triskelia from coated vesicles. *EMBO J.* 4:3385–91

248. Ungewickell, E., Branton, D. 1981. Assembly units of clathrin coats. *Nature* 289:420–22

249. Ungewickell, E., Unanue, E. R., Branton, D. 1982. Functional and structural studies on clathrin triskelions and baskets. *Cold Spring Harbor Symp. Quant. Biol.* 46:723–31

250. Usami, M., Takahashi, A., Kadota, K. 1984. Protein kinase and its endogenous substrates in vesicles. *Biochim. Biophys. Acta* 798:306–12

251. Valk, P. van der, Fowke, L. C. 1981. Ultrastructural aspects of coated vesicles in tobacco protoplasts. *Can. J. Bot.* 59:1307–13

252. Valk, P. van der, Rennie, P. J., Connolly, J. A., Fowke, L. C. 1980. Distribution of cortical microtubules in tobacco protoplasts. An immunofluorescence microscopic and ultrastructural study. *Protoplasma* 105:27–43

253. Verkleij, A. J., Zwaal, R. F. A., Roelfsen, B., Comfurius, P., Kastelija, D., van Deenen, L. L. M. 1973. The asymmetric distribution of phospholipids in the human red cell membrane. A combined study using phospholipases and freeze-etch microscopy. *Biochem. Biophys. Acta* 323:178–93

254. Vigers, G. P. A., Crowther, R. A., Pearse, B. M. F. 1986. Three-dimensional structure of clathrin cages in ice. *EMBO J.* 5:529–34

255. Vigers, G. P. A., Crowther, R. A., Pearse, B. M. F. 1986. Location of the 100 kD-50 kD accessory proteins in clathrin coats. *EMBO J.* 5:2079–85

256. Wall, D. A., Hubbard, A. L. 1981. Galactose-specific recognition system of mammalian liver: receptor distribution on the hepatocyte cell surface. *J. Cell Biol.* 90:687–96

257. Wall, D. A., Wilson, G., Hubbard, A. L. 1980. The galactose-specific recognition system of mammalian liver: the route of internalization in hepatocytes. *Cell* 21:79–93

258. Wessel, D., Robinson, D. G. 1979. Studies on the contractile vacuole of *Poterioochromonas malhamensis* Peterfi. I. The structure of the alveolate vesicles. *Eur. J. Cell Biol.* 19:60–66

259. Whyte, A., Ockleford, C. D. 1980. Structural aspects of coated vesicles at the molecular level. In *Coated Vesicles,* ed. C. D. Ockleford, A. Whyte, pp. 283–302. Cambridge: Cambridge Univ. Press

260. Wiedenmann, B., Mims, C. T. 1983. Tubulin is a major constituent of bovine brain coated vesicles. *Biochem. Biophys. Res. Commun.* 115:303–11

261. Wiedenmann, B., Lawley, K., Grund, C., Branton, D. 1985. Solubilization of proteins from bovine brain coated vesicles by protein perturbants and Triton X-100. *J. Cell Biol.* 101:12–18

262. Wild, A. E. 1980. Coated vesicles: morphologically distinct subclass of endocytic vesicles. In *Coated Vesicles,* ed. C. D. Ockleford, A. Whyte, pp. 1–24. Cambridge: Cambridge Univ. Press

263. Willingham, M. C., Pastan, I. 1980. The receptosome: an intermediate organelle of receptor-mediated endocytosis in cultured fibroblasts. *Cell* 21:67–77

264. Willingham, M. C., Pastan, I. 1985. Receptosomes, endosomes, CURL: different terms for the same organelle system. *TIBS* 10:190–91

265. Willingham, M. C., Keen, J. H., Pastan, I. 1981. Ultrastructural immunocytochemical localization of clathrin in cultured fibroblasts. *Exp. Cell Res.* 131:329–38

266. Willingham, M. C., Maxfield, F. R., Pastan, I. 1979. α_2- Macroglobulin binding to the plasma membrane of cultured fibroblasts. Diffuse binding followed by clustering in coated regions. *J. Cell Biol.* 82:614–25

267. Winkler, F. K., Stanley, K. K. 1983. Clathrin heavy chain, light chain interactions. *EMBO J.* 2:1393–1400

268. Woods, J. M., Doriaux, M., Farquhar, M. G. 1986. Transferrin receptors recyble to cis as well as trans Golgi cisternae in Ig-secreting myeloma cells. *J. Cell Biol.* 103:277–86

269. Woods, J. W., Woodward, M. P., Roth, T. F. 1978. Common features of coated vesicles from dissimilar tissues: composition and structure. *J. Cell Sci.* 30:87–97

270. Woodward, M. P., Roth, T. F. 1978. Coated vesicles: characterization selective dissociation and reassembly. *Proc. Natl. Acad. Sci. USA* 75:4394–98

271. Xie, X.-S., Stone, D. K. 1985. Isolation and reconstitution of the clathrin-coated vesicle proton translocation complex. *J. Biol. Chem.* 261:2492–95

272. Xie, X.-S., Stone, D. K., Racker, E. 1983. Determinations of clathrin-coated vesicle acidification. *J. Biol. Chem.* 258:14834–38

273. Xie, X.-S., Stone, D. K., Racker, E. 1984. Activation and partial purifica-

tion of the ATPase of clathrin-coated vesicles and reconstitution of the proton pump. *J. Biol. Chem.* 259:11676–78

274. Yamashiro, D. J., Maxfield, F. R. 1984. Acidification of endocytic compartments and the intracellular pathways of ligands and receptors. *J. Cell Biochem.* 26:231–46

275. Zaremba, S., Keen, J. H. 1983. Assembly polypeptides from coated vesicles mediate reassembly of unique clath-

rin coats. *J. Cell Biol.* 97:1339–47

276. Zaremba, S., Keen, J. H. 1985. Limited proteolytic digestion of coated vesicle assembly polypeptides abolishes reassembly activity. *J. Cell Biochem.* 28:47–58

277. Zhang, Y.-H., Robinson, D. G. 1986. The endomembranes of *Chlamydomonas reinhardii:* a comparison of the wildtype with the wall mutants CW2 and CW15. *Protoplasma* 133:186–94

Ann. Rev. Plant Physiol. Plant Mol. Biol. 1988. 39:101–38
Copyright © 1988 by Annual Reviews Inc. All rights reserved

FATTY ACID METABOLISM*

John L. Harwood

Department of Biochemistry, University College, Cardiff, CF1 1XL, United Kingdom

CONTENTS

INTRODUCTION 101
SOURCE OF ACETYL-CoA FOR FATTY ACID SYNTHESIS 102
ACETYL-CoA CARBOXYLASE IS A MULTIFUNCTIONAL PROTEIN.............. 103
PLANT FATTY ACID SYNTHETASE IS A TYPE II ENZYME 105
   Acyl Carrier Protein (ACP) 105
   Molecular Nature of the Fatty Acid Synthetase... 109
   Subcellular Sites of Fatty Acid Synthesis 112
   Chain Termination Mechanisms... 114
ELONGATION OF FATTY ACIDS OCCURS IN PARTICULATE FRACTIONS 116
DESATURATION REACTIONS 118
   Biosynthesis of Oleic Acid 118
   Membrane-Bound Enzymes Using Glycerolipid Substrates Form the
       Polyunsaturated Fatty Acids 120
   Other Desaturase Reactions 126
OXIDATION OF FATTY ACIDS... 126
   Formation of Ricinoleic Acid... 126
   β-Oxidation of Fatty Acids 127
   Lipoxygenase... 128
CONCLUSIONS AND FUTURE DIRECTIONS 129

Dedicated to the memory of David Bishop, whose untimely death has robbed us of a fine lipid biochemist and friend.

INTRODUCTION

It is many years since fatty acid metabolism was last covered in this series (but see 158 for some aspects). Because of the limitations of space and the large

*Abbreviations: ACP = acyl carrier protein; MGDG = monogalactosyldiacylglycerol; PC = phosphatidylcholine; VLCFA = very long chain (>C18) fatty acid

0066-4294/88/0601-0101$02.00

advances in our knowledge over the last few years, this account will deal largely with biosynthesis; a few remarks will be made about particular aspects of degradation. The reader is referred to reviews on biosynthesis (55, 189, 191), degradation (39, 40, 85), and recent specialized aspects (193).

SOURCE OF ACETYL-CoA FOR FATTY ACID SYNTHESIS

Although carbon dioxide fixed by photosynthesis is the ultimate source of carbon for fatty acid synthesis, acetyl-CoA can be regarded as a more direct precursor. Acetyl-CoA itself can be generated by the pyruvate dehydrogenase complex but, until recently, it was disputed whether there was sufficient activity of this enzyme in chloroplasts for the acetyl-CoA to be provided directly to the chloroplastic acetyl-CoA carboxylase. Previous work has been summarized well (189). It indicated the alternative of a mitochondrial pyruvate dehydrogenase complex coupled to acetyl-CoA hydrolase which generated free acetate. This acetate was free to move to the chloroplast, where it could be rapidly converted to acetyl-CoA in the stroma, thus accounting for the well-known ability of isolated chloroplasts to use [^{14}C]acetate very well (see 55).

The above indirect pathway for the generation of acetyl-CoA was proposed from experiments with spinach leaf fractions (123), and its application to other tissues has been questioned (46). Certainly, there have now been a large number of reports of pyruvate dehydrogenase activity in chloroplasts from a variety of species (e.g. 17, 96, 201). However, the necessary pyruvate would be provided by glycolysis and, even then, the uninterrupted pathway from 3-phosphoglycerate seems to be missing in many cases (187). This aspect has been reviewed (45, 97). Recently, it has been demonstrated that chloroplasts from mustard cotyledons are capable of supplying sufficient pyruvate from glycolysis to satisfy the plastidic pyruvate dehydrogenase. However, the authors were careful to point out that other sources of pyruvate, such as from the cytoplasm, were also possible (99). Thus, it seems that acetyl-CoA for fatty acid synthesis can be supplied by both pathways, their relative importance depending on the tissue studied (69), although plastidic pyruvate dehydrogenase would be favored logically (98). Interestingly, the pyruvate dehydrogenase from chloroplasts differs from the mitochondrial enzyme in a number of species (17).

In contrast to the controversy that has surrounded the sources of acetyl-CoA for fatty acid synthesis in chloroplasts, there seems to be general agreement that nongreen plastids contain not only an active fatty acid synthesizing system, but also the necessary adjunct glycolytic enzymes and pyruvate dehydrogenase. Evidence comes from a variety of plastids, including the leucoplasts of developing castor oilseed (107, 108) and of cauliflower (80), the chromoplasts of daffodil flowers (100), and, probably, plastids from safflower and linseed cotyledons (16).

ACETYL-CoA CARBOXYLASE IS A MULTIFUNCTIONAL PROTEIN

Acetyl-CoA carboxylase catalyzes the ATP-dependent formation of malonyl-CoA from acetyl-CoA and bicarbonate. This reaction is the first committed step for de novo fatty acid synthesis; in addition, the malonyl-CoA is used for elongation reactions. In mammals the enzyme consists of a single polypeptide chain containing three functional domains, while in bacteria the enzyme consists of three separable proteins—biotin carboxyl carrier protein (BCCP), biotin carboxylase, and BCCP: acetyl-CoA transcarboxylase. Early attempts to purify acetyl-CoA carboxylase from plants yielded data suggesting that the enzyme resembled that from bacteria or, alternatively, had an intermediate form (see 189). It is now clear that these results were influenced by the high activity of endogenous proteinases in many plant tissues; and evidence has gradually accumulated that the plant acetyl-CoA carboxylase, like that from mammals, is a multifunctional protein.

In leaves, acetyl-CoA carboxylase has been shown to be localized in chloroplasts (84, 110, 130, 199). Its activity, which is low in the dark, is increased significantly upon illumination (70). The enzyme from maize was shown to be sensitive to changes in pH, ATP, ADP, and Mg^{2+} that were likely to occur during illumination (129). The nucleotide and Mg^{2+} changes could each account for a two-fold increase, and the increase of pH for a three-fold elevation of activity (Table 1). The combined effects could bring about a 24-fold increase of acetyl-CoA carboxylase activity. Similar alterations are also known to affect acetate activation and fatty acid synthesis in spinach chloroplasts (70, 165). Moreover, the wheat germ carboxylase was also shown to be tightly controlled in vitro through its requirement for ATP and its inhibition by ADP and AMP (31). However, as discussed (60) it is unwise to generalize too much from these results since other plant acetyl-CoA carboxylases show different responses. For example, while ATP protected the soya bean enzyme from loss of activity during dilutions, preincubation of the purified enzyme with ATP had no effect on activity. However, it was noted that, in this case also, ADP and AMP were inhibitory (21).

Several workers have tested the plant acetyl-CoA carboxylase for stimulation by tricarboxylic acids (21, 36, 110), such as is observed for mammalian carboxylases. However, no significant stimulation has been observed nor is there any evidence for regulation by phosphorylation/dephosphorylation. Nevertheless, as in animals, the activity of plant acetyl-CoA carboxylase

Table 1 Possible regulatory factors for maize acetyl-CoA carboxylase[a]

Parameter	In vivo change on illumination		Activity change in vitro[b]		Additive fold change
	From	To	From	To	
pH	7.1	8.0	50	155	3.1 ×
Mg^{2+}	2	5 mM	70	165	7.3 ×
ATP	0.3–0.8	0.8–1.4 mM	80	135	12.3 ×
ADP	0.6–1.0	0.3–0.6 mM	65	125	23.6 ×

[a] Data taken from reference 129.
[b] Activity expressed as nmol/min/mg protein

is likely under many circumstances to be rate limiting for de novo fatty acid synthesis. For example, in ripening castor bean or rape seeds accumulation of lipid correlates well with the measured activity of acetyl-CoA carboxylase (175, 205).

Early attempts to purify acetyl-CoA carboxylases from plant tissues produced preparations that yielded 3–6 subunits on PAGE under denaturing conditions (189). These preparations are now thought to have been severely degraded by endogenous proteinases (60). The first clear advance in this area came when Egin-Buhler et al (33) used the proteinase inhibitor phenylmethylsulfonyl fluoride (PMSF). They found that acetyl-CoA carboxylase purified from wheat germ or parsley cells, in the presence of PMSF, was composed mainly of a high-molecular-mass polypeptide of 240 kDa and 210 kDa, respectively. Their purification procedure was simplified and improved by the use of an affinity chromatography step with avidin-monomer-Sepharose 4B. This gave a molecular mass of about 420 kDa for the native enzyme and about 220 kDa for the enzyme subunit (32). The purified enzyme showed 60% of the acetyl-CoA activity with propionyl-CoA but only 10% with butyryl-CoA. The kinetic constants were reported and the lack of stimulation by citrate confirmed (32).

Acetyl-CoA carboxylase from rape seed has also been purified by employing avidin-Sepharose (181). The final preparation was almost homogenous and contained a major polypeptide of 220 kDa. The kinetic constants were determined and, in general, were of comparative values to those for the parsley (32), castor bean (36), and maize enzymes (129).

Purification of acetyl-CoA carboxylase from maize leaves yielded a preparation that seemed to contain all of the functional activity of the carboxylase in a single peptide of 60 kDa. The native enzyme had a molecular mass of 500 kDa (129). In an alternative procedure for determining the molecular mass of acetyl-CoA carboxylase from plant leaves, Nikolau et al (131) examined biotin-containing enzymes by Western blotting following SDS-PAGE. Biotin was probed in such proteins by using (^{125}I)-Streptavidin. No high-molecular-mass bands were found, but all plants examined (two C3-plant leaves and two C4-plant leaves) contained bands of about 62 kDa that were suggested to correspond to acetyl-CoA carboxylase (131). In contrast, acetyl-CoA carboxylase has been purified from oilseed rape leaves, maize leaves (70) and soya bean leaves (21) with molecular masses in the range 220–240 kDa. Indeed, it has been noted that the use of proteinase inhibitors *and* a rapid purification protocol is essential for success (70), and it seems probable that plant acetyl-CoA carboxylases are all high-molecular-mass multifunctional proteins. In that regard they are similar to the enzyme from animals, although obviously differing in their control mechanism.

PLANT FATTY ACID SYNTHETASE IS A TYPE II ENZYME

The isolation of acyl carrier protein (ACP) from three plants in 1967 (176) implied that the plant fatty acid synthetase was, like that from *E. coli,* a Type II enzyme with individual partial reactions catalyzed by separate proteins. In fact, even prior to 1982 (when three laboratories simultaneously described the purification or partial purification of proteins catalyzing the individual reactions), it was customary and necessary to add exogenous ACP to assays for the fatty acid synthetase.

Acyl Carrier Protein (ACP)

Overath & Stumpf (145), studying fatty acid synthesis in avocado mesocarp, found that extracts could be separated into two components that were inactive alone but when recombined could catalyze the synthesis of palmitate and stearate. One of the fractions was stable to heat and acid treatment, was destroyed by proteinases, and could substitute for and be substituted by *E. coli* ACP in reconstituted fatty acid synthesizing systems. The purification of plant ACP from lettuce and avocado (176) showed that they had molecular masses of about 11 kDa and a single phosphopantetheine residue. Like the *E. coli* ACP, those from plants have a preponderance of acidic amino acyl residues, which accounts for their easy purification on anionic columns. A partial sequence of 17 amino acids around the prosthetic-group region of spinach ACP was then determined (106). Complete sequences have recently been obtained for ACPs from barley (73), spinach (88), and oilseed rape (180). The barley ACP was found to contain two sequences identical with those of *E. coli* ACP in the midregion of the molecule containing the 4'-phosphopantetheine attachment site. Nine extra residues were present at the N-terminal end of the barley protein, thus accounting for its higher molecular weight. This difference in primary structure may be particularly significant in view of the observation that the activity of *E. coli* ACP is rather sensitive to modifications of its N-terminus (cf 2). Modifications to that region alter ACP's activity not only with fatty acid synthetase but also with other competing enzymes such as glycerol-3-phosphate acyltransferase. Thus, it is not surprising that plant and *E. coli* ACPs are not entirely interchangeable in the other fatty acid synthesizing system. Moreover, it was noted many years ago (176) that spinach ACP was a poor substrate for the dehydrase components of *E. coli* fatty acid sythetase and its use led to the accumulation of significant amounts of intermediate chain-length hydroxy acids. In contrast, not only does *E. coli* ACP give the same products as plant ACP in barley (73) or spinach (176) fatty acid synthesizing systems, it actually gives higher total activity. This may imply that one of the plant fatty acid synthetase components has not evolved structurally to accommodate the changes in plant ACP (73). The possibility of modifications in ACP structure controlling the nature or fate of fatty acid synthesis products has important implications with the discovery of isoforms of ACP in plants (see below).

The sequence of the major isoform of spinach ACP (termed ACP-1) showed about a 70% homology with barley and about 40% with *E. coli* ACP (87). It is interesting that the homology with rabbit ACP was worse, in keeping with the type II organization of the plant fatty acid synthetase. This organization resembles *E. coli*'s in that regard and contrasts with other proteins such as cytochrome *c*, which in plants shows closer homology with animals than with bacteria (135). When antibodies were raised against spinach ACP it was found that they had partial cross-reactivity against various plant ACPs. In general, dicotyledonous species were found to compete more effectively than monocotyledonous species in radioimmunoassays where spinach [^{3}H]palmitoyl-ACP was the radiolabeled ligand (86). This suggested that ACPs from other dicotyledonous plants may show an even higher homology with spinach ACP than the 70% observed with barley ACP.

Kuo & Ohlrogge (87) compared the predicted secondary structure of spinach ACP-1 to that of *E. coli* ACP. They found that the three β-turns predicted in the *E. coli* model (155) corresponded well to the three regions in spinach with highest β-turn probability. Thus, the high degree of sequence conservation observed with the prosthetic-group region also extended to the secondary structure of most of the molecule. The same authors also noted, however, that the secondary structure of the barley ACP lacked a high prediction probability of a β-turn after the prosthetic group.

In early studies on the amino acid sequence of ACP, Matsumura & Stumpf (106) found that the amino acid composition of some peptide fragments showed heterogeneity, leading them to suggest that spinach ACP might exist in two forms. Such isoforms of ACP have been demonstrated in barley (74), spinach, castor oilseed, and soybean (138). The isoforms of spinach leaf ACP were separated by HPLC and found to be immunologically related. The N-terminal sequence of ACP-II was determined and found to be very similar to ACP-I, with only 6 differences in the first 20 residues (138) (Figure 1). Three isoforms were found in barley and ACP-I and ACP-II purified to homogeneity. Their amino terminal sequences are also shown in Figure 1, where it will be seen that ACP-I contains additional amino acyl residues at its N-terminus. In fact, amino acid composition data had shown that whereas barley ACP-II had 75 residues, ACP-I had 87 (74). As seen in Figure 1, seven of these extra residues are located at the N-terminus. Thus, in both barley and spinach leaves different isoforms of ACP are present, probably coded for by different genes.

Whether barley seeds were grown in the dark or in the light, both ACP isoforms were present. However, ACP-I tended to increase relative to ACP-II with more light exposure (74). In spinach seeds only ACP-II was present, while in leaves both forms were present with ACP-I predominating. Similarly, in castor oil seeds only one higher-molecular-mass form of ACP was present, while leaves from the same plant showed two forms. In castor bean the two ACP isoforms differed by an apparent kDa on SDS-PAGE (138). This implies a more substantial structural difference for ACP isoforms in this species. In general these observations show that there is a tissue-specific expression of ACP isoforms in plants. Experiments on the formation of ACP by in vitro translation systems have been carried out in two laboratories. mRNA isolated from greening barley leaves was translated with a rabbit

BARLEY I	NH_2-Ala-Ala-Met-Gly-Glu-Ala-Gln-Ala-Lys-Lys-Glu-Thr-Val-Asp-Lys-Val-Cys-Met-Ile-Val-
BARLEY II	NH_2-Ala-Lys-Lys-Glu-Thr-Val-Glu-Lys-Val- ? -Asp-Ile-Val-
SPINACH I	NH_2-Ala-Lys-Lys-Glu-Thr-Ile-Asp-Lys-Val-Ser-Asp-Ile-Val-
SPINACH II	NH_2-Ala-Ala-Lys-Pro-Glu-Met-Val-Thr-Lys-Val-Ser-Asp-Ile-Val-

BARLEY I	Lys-Lys-Gln-Leu-Ala-Val-Pro-Asp-Gly-Thr-Pro-
BARLEY II	Lys-Ser-Gln-Leu-Ala-Leu-Ser-Asp-Asp-Asp-Glu-
SPINACH I	Lys-Glu-Lys-Leu-Ala-Leu-Gly-
SPINACH II	Lys-Ser-Gln-Leu-Ala-Leu-

Figure 1   N-terminal amino acid sequences of isoforms of barley and spinach leaf acyl carrier proteins (data from 74, 138).

reticulocyte lysate. Two polypeptides were precipitated by antibarley ACP-I serum with apparent molecular masses of 18 and 19 kDa. These molecular masses were apparently at least 2kDa greater than ACP-I and ACP-II by SDS-PAGE (74). poly-A RNA was isolated from young spinach leaves and translated by a wheat germ system to yield two radioactive peptides precipitable by antiACP serum. These proteins had apparent molecular masses of 20.2 and 21.2 kDa which in this case was around 6 kDa larger than native ACP (137). Thus, both barley and spinach ACPs are synthesized as precursor forms containing additional amino acids consistent with a transit sequence necessary for transport into the plastid where most if not all of the leaf ACP is localized (136). Moreover, previous observations with the barley mutant "albostrians," which lack plastid 70S ribosomes and yet can still make fatty acids and polar lipids, suggested strongly that the proteins needed for these processes were nuclear encoded (28). The results of ACP in vitro translation experiments are fully consistent with this proposal.

The discovery of isoforms of ACP immediately posed the question of what role they play in plant lipid formation. The possibility that the different isoforms were present in different parts of the cell was discounted by experiments showing that the spinach chloroplast contained both isoforms and in the same ratio as whole leaf extracts (138). [Earlier experiments had shown that the only detectable ACP in spinach leaves was present in chloroplasts (136).] Moreover, barley ACP-I and ACP-II were both capable of supporting in vitro fatty acid synthesis, with identical products being accumulated (74). This excluded the possibility that only one form of ACP was used by fatty acid synthetase. However, these workers suggested that, in view of the known altered specificity of ACP seen with structural modifications (discussed above), the two isoforms might interact differently with different ACP-dependent enzymes. In particular they theorized that one form could be used for de novo synthesis of fatty acids while the other participated in glycerolipid formation (74).

The above suggestions were tested directly with the spinach ACP isoforms. In keeping with the results with barley, both isoforms of spinach ACP caused identical rates and patterns of products with a fatty acid synthesizing system (48). Moreover, detailed examination of the malonyl-CoA : ACP transacylase failed to reveal any difference in ACP isoform reactivity. In marked contrast, oleoyl-ACP thioesterase and glycerol-3-phosphate acyltransferase were both strongly affected by ACP isoforms (Table 2). These two enzymes determine the fate of oleate produced by the plastid. Thus, when oleoyl-ACP thioesterase activity is predominant then oleate is released for reesterification to oleoyl-CoA and export to the endoplasmic reticulum for possible further modification. In contrast, direct acylation of glycerol-3-phosphate keeps the acyl chain within the plastid and would be characteristic of the so-called "prokaryotic pathway" (cf 68, 159). The apportionment of acyl chains between these two pathways may be controlled by a number of factors including glycerol-3-phosphate concentration and the relative activity of thioesterase versus acyltransferase. The experiments with ACP isoforms suggest that the relative abundance of these may also help to control the "prokaryotic" and "eukaryotic" pathways (48). However, since triacylglycerol synthesis in seeds occurs predominantly outside the plastid, high acylthioesterase activity would be needed, and the occurrence of ACP-II as the major isoform in spinach seeds is not in keeping with such a requirement.

To date, therefore, we do not understand the reason for different isoforms of ACP nor for their tissue-specific distribution. However, the different activity of some plant ACP–requiring enzymes with different ACPs means that experiments with *E. coli* ACP must be treated with caution. So far, it has not been feasible to use plant ACP for in vitro experiments because of the difficulties in purifying large amounts (135). However, the recent synthesis, cloning, and expression of a spinach ACP-I gene in *E. coli* means that we may soon be able to obtain much larger amounts of a plant ACP (8) and so allow further study of the ACP-dependent reactions. Further cloning experiments will, of course, be necessary for other ACP isoforms in order to evaluate fully the different specificity of the 12 ACP-dependent enzymes (135, 191) participating in fatty acid metabolism.

Molecular Nature of the Fatty Acid Synthetase

The identification (145) and later isolation (176) of plant ACP provided good evidence that the plant fatty acid synthetase was a dissociable type II enzyme. However, it was not until 1982 that proteins catalyzing the individual partial reactions were purified. Partial purification of β-ketoacyl-ACP synthetase, β-ketoacyl-ACP reductase, acetyl-CoA:ACP transacylase, and malonyl-CoA:ACP transacylase was achieved from barley chloroplasts (72). From avocado fruit, Caughey & Kekwick (20) purified the β-ketoacyl-ACP reduc-

Table 2 Relative activities of *E.coli* and of spinach ACP isoforms in three spinach ACP-dependent reactions[a]

ACP form	Maximum recorded activity (nmol/min)		
	MT[b]	OT	GAT
E. coli ACP	0.40	1.8	3.9
Spinach ACP-I	0.39	0.9	1.7
Spinach ACP-II	0.47	0.2	2.9

[a] Data from reference 48
[b] Abbreviations: MT = malonyl-CoA:ACP transacylase; OT = oleoyl-ACP thioesterase; GAT = glycerol-3-P acyl transferase

tase and malonyl-CoA : ACP acyltransferase to homogeneity and also purified the enoyl-ACP reductase. The partial purification of β-ketoacyl-ACP synthetase was also reported from cell suspension cultures of parsley (166). However, the most thorough study was that by Shimakata & Stumpf, mainly with spinach leaves. They reported purifying β-ketoacyl-ACP reductase, β-hydroxyacyl-ACP dehydrase, and enoyl-ACP reductase to homogeneity (168) and partly purifying of acetyl-CoA : ACP transacylase (171), β-ketoacyl-ACP synthetase I (170), and β-ketoacyl-ACP synthetase II (169).

Acetyl-CoA : ACP transacylase may be rate limiting for the overall fatty acid synthetase (171). The other partial reaction that was also relatively low—β-ketoacyl-ACP synthetase I—did not seem to be rate limiting in spinach (171). In barley, acetyl-CoA transacylase had an approximate molecular mass of 82 kDa (72). The enzyme from spinach leaves was purified 175-fold and found to have a molecular mass of 48 kDa. Its pH optimum was 8.1 and it was relatively specific for acetyl-CoA with butyryl-CoA, hexanoyl-CoA, and octanoyl-CoA being transacylated at 35, 18, and 7% the rate of acetyl-CoA, respectively (171). It was inhibited by typical -SH reagents such as p-hydroxymercuribenzoate, by arsenite (171), and, possibly, by thiolactomycin (132). Two isoforms have been separated from *Brassica* (218). When purified acetyl-CoA : ACP transacylase was added back to a crude spinach extract, it resulted in a small shift in the chain length of fatty acyl products toward the C_{10}–C_{12} range. However, when the same experiment was repeated with a system reconstituted from purified enzymes, up to 88% of the products were laurate (171). This led to the suggestion that the activity of acetyl-CoA : ACP transacylase could be one of the factors influencing the chain length of fatty acid synthetase products in vivo. [Increasing the acetyl-CoA concentration with the barley system, which was also likely to raise the relative activity of the transacylase, also increased the proportion of medium chain products (72).]

Malonyl-CoA : ACP transacylases have been purified from a number of plant tissues including avocado (20), barley (72), spinach (186), soybean (47), and leek (95), as well as from the cyanobacterium *Anabaena variabilis* (186). These enzymes usually have molecular masses of around 40 kDa, and some general properties are listed in Table 3. The reaction mechanisms for the spinach and *Anabaena* enzymes were compared to, and found to be different from, the *E. coli* transacylase. A random sequence mechanism was proposed for the plant enzymes (186). Tissue-specific isoforms have been reported. Thus, in soybean two isoforms were found in leaf tissue but only one in seeds (47). In leek, the malonyl-CoA : ACP transacylases from epidermal and parenchymal cells had molecular masses of 38 kDa and 45 kDa, respectively (95).

Two forms of β-ketoacyl-ACP synthetase have been separated from spin-

Table 3 Comparative aspects of malonyl-CoA:acyl carrier protein transacylases (MCT) from different plant tissues

Source (Reference)	Molecular mass (kDa)	K_m (malonyl-CoA)	K_m (ACP)	Remarks
Avocado fruit (20)	40.5	3.3 μM	42 μM	
Barley chloroplasts (72)	41			
Spinach leaves (186)	31	0.5 mM	0.4 mM	
A. Variabilis (186)	36	0.3 mM	0.4 mM	
Soybean (47)	43	9.4 (MCT$_1$) μM		Both forms present in leaves, MCT$_1$ predominant in seeds
	43	15.0 (MCT$_2$)μM		
Leek leaves (95)	38	13.7 μM		Epidermal cells
	45	21.7 μM		Parenchymal cells

ach leaf extracts. These condensing enzymes differ markedly in their substrate specificity, with synthetase 1 utilizing acyl-ACPs in the range C_2–C_{14} effectively with poor activity towards palmitoyl-ACP (170). In contrast, β-ketoacyl-ACP synthetase 2 was only active with myristoyl-ACP and palmitoyl-ACP substrates and was, therefore, needed for stearate formation (169). These observations explained earlier results showing that soluble plant extracts could catalyze the formation of stearate by elongation as well as de novo (11) but that the ability to synthesize stearate was often lost during attempts at purification (65, 75). Moreover, β-ketoacyl-ACP synthetase 1 was insensitive to arsenite whereas β-ketoacyl-ACP synthetase 2 was inhibited severely (169, 170) thus explaining the selective action of arsenite in preventing stearate but not palmitate synthesis (64, 72, 76, 79). The effect of arsenite had led to the original suggestion that stearate and palmitate were formed by different enzyme systems (64). Moreover, β-ketoacyl-ACP synthetase 1 was much more sensitive to cerulenin than synthetase 2 (169, 170), again explaining previous results with crude systems (76, 79). The spinach enzymes have molecular masses of 56 kDa for synthetase 1 (170) and 57.5 kDa for synthetase 2 (169), which were lower than the values for the parsley (166) or barley (72) enzymes.

The β-ketoacyl-ACP reductase has been reported in two forms and, in the case of avocado, the NADPH form has been resolved from that using NADH (20). In other tissues, such as safflower or spinach, NADPH is a more effective substrate than NADH (167, 168). The NADPH-specific reductase has been purified from avocado fruit and rape seed where it had molecular masses of 29 kDa and a double band of 29–30 kDa, respectively (166).

β-Hydroxyacyl-ACP dehydrase has been purified from safflower (167) and to homogeneity from spinach (168). The latter enzyme is reported to be tetrameric with a subunit molecular mass of 19 kDa. A range of 2-enoyl-ACPs were tested for activity; those in the range of C_4–C_{16} were utilized, with maximal activity shown toward the C_8 substrate. The dehydrase was stereospecific for the D-β-hydroxyacyl-ACP substrate (168).

Two forms of the enoyl-ACP reductase have been detected. Type I utilizes crotonyl-ACP and is NADH-specific (168). Type II uses 2-decenyl-ACP as substrate and NADPH in preference to NADH. Both types are present in safflower, castor bean, and rape seeds (179), although only type I seems to be present in leaf tissue (168) or avocado (20). The type I enoyl-ACP reductase has been purified to homogeneity from spinach leaves (168) and rape seed (183). The spinach enzyme had a molecular mass of 115 kDa but a subunit molecular mass of 32.5 kDa, suggesting that it existed as a tetramer (168). The rape seed enzyme's molecular mass was 140 kDa, but careful SDS-PAGE showed that it had dissimilar subunits of 33.6 kDa and 34.8 kDa (183). The amino acid compositions of the two enzymes were remarkably similar. It is interesting that in rape seeds the type I enoyl-ACP reductase had no activity towards a C_{18} substrate, suggesting that the (NADPH) type II enzyme had to be used for stearate formation (179). Since the type I enzyme was the only one detected in spinach leaves, this enzyme must have been active with the C_{18} substrate, in contrast to the rape type I enzyme. However, the spinach enoyl-ACP reductase was only tested with substrates up to C_{16} (168).

Subcellular Sites of Fatty Acid Synthesis

In leaf tissues, it is clear that chloroplasts represent the major, perhaps exclusive, site of de novo fatty acid synthesis. The most convincing evidence comes from the localization of acyl carrier protein in chloroplasts derived from spinach leaf protoplasts (136), but other results support this conclusion (cf 191). In addition, plastids from nonleaf tissues have also been shown to be active, as originally suggested by Weiare & Kekwick (213). These observations have led to the formulation of the "ACP-track" hypothesis (Figure 2) where de novo synthesis in the plastid leads first to palmitate formation. The palmitoyl-ACP is then elongated to stearoyl-ACP, which can be desaturated to oleoyl-ACP. Hydrolysis of the latter acyl-thioester then gives rise to free oleate, which can then form oleoyl-CoA for esterifications and further modifications outside the chloroplast. This elegant and simple concept (190) forms a working basis for current thinking about the organization of fatty acid synthesis in the plant cell. However, it should be borne in mind that matters are seldom as simple as one would like, and data are available that cannot be explained easily by such a scheme. For example, nonchloroplast fractions are capable of de novo fatty acid synthesis in certain cases (10, 210), and further desaturation of oleate within the chloroplast seems to occur to a greater or

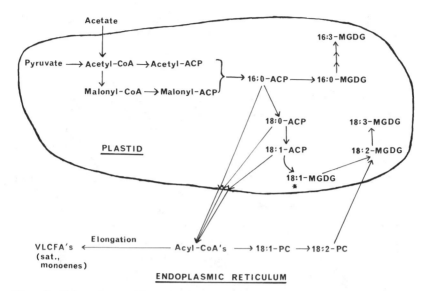

Figure 2   Major pathways of fatty acid synthesis in plant cells. Asterisk (*) indicates other lipids are probably involved in "16:3" plants

lesser degree in different plant types (133, 134). Thus, although it is likely that the plastid is predominantly responsible for oleate formation in most tissues, it would be dangerous to generalize too greatly.

Microsomal fractions from germinating peas have been used to study fatty acid synthesis ever since the observation that they contained a very active elongation system (103). In palmitate formed by such fractions from [^{14}C]malonyl-CoA, radiolabel was distributed in a manner indicating de novo synthesis in contrast to the very long chain fatty acids (10). Moreover, results from partial proteolysis experiments indicated that two populations of membranes were present, both of which contained fatty acid synthesizing ability (161). Because of the high activity of chloroplast preparations, including some thylakoid fractions (213), it was suggested that one of the active membranes could have been derived from plastids. However, further experiments with lettuce and pea leaf chloroplasts showed the soluble (stromal) nature of the fatty acid synthetase enzymes, although loose association with thylakoids was seen, especially in the case of lettuce chloroplasts (210). The fatty acid synthesis measured in microsomal fractions from such tissues could not, therefore, be due to the presence of chloroplastic membranes (210). Other experiments with the pea microsomal fraction showed that all fatty acid synthesizing activities were located on the outside of the vesicles [equivalent to the cytosolic face of the endoplasmic reticulum, which is the major membrane present (79)] and also that the particulate fatty acid synthetase had different properties compared to the soluble activity (161). It is also interesting, in view of the loose association of plastid fatty acid synthetase with membranes (210), that ACP is also localized there (180).

Chain Termination Mechanisms

As outlined above, the product of plant fatty acid synthetase (FAS) is usually palmitoyl-ACP, which can then be elongated by a specific elongation system. This system may only differ from the FAS by the use of β-ketoacyl-ACP synthetase 2. Thus, the products of de novo synthesis will be limited by the substrate specificity of the condensing enzymes to C_{16} and C_{18} fatty acids. Normally these will be, by far, the major products of de novo fatty acid synthesis in vivo and in vitro. For example, soluble fatty acid synthetase from potato tubers produced 25% C_{14} and 61% C_{16} products (75); that from lettuce chloroplasts (210) produced 15% C_{16} and 82% C_{18} products.

However, there are certain plant tissues where medium-chain-length products are formed in large amounts. Commercially important in this regard are coconut and palm, where the kernel oils contain up to 90% of C_8–C_{14} fatty acids (49). The mechanism of chain termination in such tissues is intriguing and unknown at present. The various possibilities have been summarized (148, 179, 191). In some instances they have been tested with experimental systems but often with negative results. Specific control mechanisms are:

1. A medium-chain acyl-ACP thioesterase. By analogy with the mammary gland system, where a specific thioesterase ensures medium-chain-length products, the presence of such an enzyme would prevent further chain-lengthening and provide an unesterified product, which would be available for conversion of acyl-CoA and transfer into triacylglycerol. Direct tests for such an enzyme have proved negative; moreover, the products of fatty acid synthesis in coconut are almost exclusively acyl-ACPs (143), whereas they would be free fatty acids if a thioesterase were operating.

2. A specific acyl-ACP acyltransferase. Here the required enzyme would transfer fatty acids, as they were being synthesized, to membrane-located acceptor lipids. There is no evidence to support this hypothesis in the literature. Indeed, when microsomes were added to a coconut fatty acid synthesizing system the resulting products had even longer chain lengths than before (179).

3. A tissue-specific fatty acid synthetase. There is no evidence at present that such tissues as coconut contain a special fatty acid synthetase independent of the conventional C_{16}–C_{18} type II enzyme.

4. The availability of malonyl-CoA. Several observations have suggested that the level of malonyl-CoA may influence the chain length of products. For example, in potato lowering the level of malonyl-CoA allowed the production of significant amounts of C_8–C_{14} fatty acids (75). Similar

results were obtained for an avocado system (65) but not with a barley chloroplast preparation (72). In a coconut endosperm system capable of making significant amounts of short- and medium-chain-length products, their proportion was increased from 45% of the total to 90% (with commensurate chain-shortening) as the concentration of malonyl-CoA was lowered (179). Thus, in several plant systems there is experimental evidence that malonyl-CoA concentrations may be important.

5. Nucleotide concentrations. Although the concentration of NADH and NADPH influences the chain length of products for the barley chloroplast system (72), it had insignificant action with the potato synthetase (191). Because of the involvement of pyridine nucleotides in many metabolic reactions, their use as controlling factors for fatty acid synthesis does not, in any case, seem likely.

6. Level of acyl carrier protein. Several laboratories have reported that increased ACP levels cause a build-up of medium-chain fatty acid products in vitro (72, 75, 177, 179, 182). These observations are explained by postulating that there is competition for intermediates between the condensation and chain-termination reactions. Transfer of products to free ACP will be favored by a high concentration of the latter since all other sites will be saturated.

7. Acetyl-CoA availability. While acetyl-CoA concentration had no effect on product chain length in potato (75), increased levels caused a decrease in chain length in barley (72) and caused a dramatic chain-length reduction in coconut (179). However, in the latter case total synthesis was also reduced markedly.

8. Rate-limiting nature of acetyl-CoA:ACP transacylase. This enzyme has been reported to be rate limiting for several plant fatty acid synthetase systems (cf 191). Moreover, in reconstitution experiments, Shimakata & Stumpf (171) showed that increased levels of the spinach transacylase promoted the formation of medium-chain products. This result may be related to the observations with acetyl-CoA concentrations documented under point 7 above. High activity of the acetyl-CoA:ACP transacylase would be expected to flood the system with primer substrates, thus increasing the chances that chain lengthening would be limited (191). However, correlation between the enzyme's level and the in vivo pattern of fatty acid products is not good (179).

9. A specific β-ketoacyl-ACP synthetase. The discovery in plants of specific condensing enzymes that control the chain length of products at the C_{16}–C_{18} level provides a precedent for this hypothesis. There is no experimental evidence in its favor, but it would be a relatively specific and simple way of ensuring a tissue-specific pattern of fatty acid products. In vitro tests have failed to demonstrate the existence of a

specific condensing enzyme, since such preparations make only long-chain acids in contrast to the tissue from which they were isolated (143). This result demonstrates one of the perennial difficulties in trying to define plant fatty acid synthesis with subcellular or more highly purified fractions—i.e. they lose many of the tissue-specific properties. This problem is especially acute for the examination of desaturase reactions (discussed below).

ELONGATION OF FATTY ACIDS OCCURS IN PARTICULATE FRACTIONS

Very-long-chain fatty acids ($>C_{18}$;VLCFA) are significant components or precursors of the constituents of the surface layers (wax, cutin, suberin) of plants. In a few cases they also accumulate in unusual seed oils. Furthermore, although C_{20} (and C_{22}) fatty acids are major components of algal and lower-plant membranes, their formation has been studied hardly at all.

Work up until 1979 has been reviewed (55). Since that time information has accumulated from four systems that have proven useful to biochemists examining elongation mechanisms. These are the germinating pea, aged potato slices, fractions from the epidermal cells of leek, and various maturing seeds that accumulate fats rich in VLCFAs.

That the VLCFAs are formed by malonyl-CoA-dependent elongation of preformed chains was first demonstrated by chemical degradation studies to elucidate the position of radiolabeling (63, 102). The microsomal fraction from germinating peas was found to be responsible for VLCFA formation. Malonyl-CoA was the source of the C_2 addition unit, and NADPH was used as reductant (64, 102). Although exogenous ACP gave some stimulation, stearoyl-ACP was not elongated in detectable amounts (10). The stimulation by ACP is puzzling, since it is by no means certain that there is any ACP outside the chloroplast (plastid) (cf 136). Indeed, in the pea system the nature of the primer acyl substrate is unknown. Not only was stearoyl-ACP inactive but stearoyl-CoA also showed no activity with microsomes from germinating pea (10) or palmitoyl-CoA with microsomes from pea leaves (86). However, exogenous lipids, including nonesterified fatty acids but especially various PCs, were capable of providing acyl chains for elongation (10, 79). Although ACP, as mentioned above, stimulated total synthesis of fatty acids, the increased acids accumulated as acyl-ACPs, a fraction which did not contain VLCFAs. In fact, the latter products were found exclusively in the acyl lipid fraction (160). The absence of VLCFAs in the acyl-ACP fraction was in agreement with (a) the generally accepted opinion that these molecules are not suitable substrates for the elongases responsible for the synthesis of VLCFAs (10, 18) and (b) the absence of such acids in the acyl thioesters isolated from several plant tissues (162). Further experiments showed that the elongases responsible were located on the outside of the microsomal vesicles— equivalent to the cytoplasmic face of the endoplasmic reticulum. Two populations of membranes seemed to contain the elongases (161), which are sensitive to thiocarbamate herbicides (cf 62).

In aged potato tubers, VLCFA synthesis can be induced at high rates (215). Synthesis of VLCFAs in potato, as in pea, was sensitive to thiocarbamate herbicides (9, 55). Recently, we obtained evidence indicating three separate chain-specific elongases, responsible for C_{20}, C_{22}, and C_{24} acid formation, respectively, in potato (211). This demonstration of several elongase enzymes agreed with work with leek (see below and 4, 90) and with genetic studies of barley (209).

The leek system has been studied principally by Cassagne, Lessire, and coworkers. Microsomes from leek are particularly active (19), and attempts have now been made to purify elongases from that source (4, 90). Like the potato (211) and pea systems (55), leek microsomes use malonyl-CoA as the C_2 donor and show no requirement for ACP (3). Although VLCFAs are transferred into glycerolipids, the products of the elongation reaction appear to be acyl-CoAs (94), the transfer being effected by an acyl-CoA transacylase (1). This contrasts with the pea system, where stearoyl-CoA was an ineffective substrate and VLCFAs were not detected in the acyl-CoA pool in vitro (160). Furthermore, in the leek system experiments in vivo with [^{14}C]acetate showed clearly that VLCFAs were present in the radiolabeled acyl-CoA pool (111). However, it should be borne in mind that regardless of whether an acyl-CoA or a complex lipid substrate is used by a particular system, the acyl chain will have to be transferred to the elongase because C_2 addition takes place at the carboxyl end. Experiments with the leek system also showed that stearoyl-CoA was used more effectively as a primer than palmitoyl-CoA (93). Solubilization of elongating activity in two laboratories has shown that it is possible to isolate a fraction that synthesizes only eicosanoic acid (4, 91). A second (C_{20})elongase was separated also (91). This confirmed the results with potato that separate chain-length-specific elongases are found in plants (211). The nature of the elongase is as yet unclear. It may be a high-molecular-weight multifunctional protein (4) or have several protein components (91). The transfer of newly synthesized VLCFAs from the endoplasmic reticulum to the plasma membrane (92, 112) has also been studied.

Various oilseeds that accumulate VLCFAs have also been used to study elongases. In these tissues, where the VLCFAs are unsaturated, the double bond is introduced at the C_{18} (oleoyl) level and retained during elongation. Acyl-CoA substrates seem to be used (147, 149, 150). The systems are similar to the *Crambe* and rape systems previously described (54).

DESATURATION REACTIONS

Biosynthesis of Oleic Acid

Most plant unsaturated fatty acids are three related C_{18} molecules: oleic, linoleic, and α-linolenic acids. These compounds make up 80% or more of the total fatty acids of most higher-plant tissues. Their *cis* double bonds are introduced by the aerobic mechanism, but there is some controversy about whether the mechanism is stepwise or concerted (50).

Stearoyl-ACP was identified as the substrate for oleate formation in *Chlorella* and spinach chloroplasts (126). This result was confirmed with a nonphotosynthetic system, developing safflower seeds (77). The plant enzymes (in contrast to those from other eukaryotes such as animals and yeasts, which are membranous and use stearoyl-CoA) are soluble to a large extent, being concentrated in proplastids or the chloroplast stroma. They require an input of electrons from NADPH or from water via photosystems I and II. Ferredoxin can act as the intermediate electron carrier, although its use in vivo remains to be demonstrated (191).

The safflower stearoyl-ACP desaturase has been purified about 200-fold by DEAE and affinity chromatography. In its native state it exists as a dimer with a molecular mass of 68 kDa (104). Its properties have been reviewed (191). Of particular interest is the substrate specificity. Stearoyl-CoA and palmitoyl-ACP were only 5% and 1% as active as stearoyl-ACP. Although the K_ms for palmitoyl-ACP and stearoyl-ACP were similar (0.51 μM and 0.38 μM, respectively), the V_{max} for the latter was more than 100 times that for palmitoyl-ACP. Thus, the properties of stearoyl-ACP desaturase explain very well the presence of palmitate as the major saturated fatty acid, the low level of stearate, and the dominance of oleate over palmitoleate in plant tissues.

One common way poikilotherms adapt to low temperature is by an increase in desaturation (58, 101). Several explanations of the mechanism of this adaptation have been proposed. Perhaps the one most supported by experimental evidence involves the fact that O_2 is required for desaturation and the solubility of O_2 is significantly higher at low temperatures. Thus, if O_2 is rate limiting for desaturation, then low temperatures will promote unsaturation by increasing substrate (O_2) availability (52, 153). Measurements of the K_m for oxygen with the purified safflower stearoyl-ACP desaturase, however, did not suggest that oxygen could be a major controlling factor for the synthesis of oleic acid (104). It may be pertinent to observe that safflower is somewhat exceptional in showing poor temperature adaptation of its lipids (191) and, moreover, that the conversions of oleate to linoleate and of linoleate to linolenate, in general, seem to be more susceptible to temperature change (58, 153).

As far as the synthesis of oleate, all the reactions of plant fatty acid synthesis involve acyl-ACP substrates. The systems of fatty acid synthetase, palmitate elongation, and stearate desaturation are tightly coupled in the chloroplast so that examination of oleate synthesized from [^{14}C]acetate by isolated organelles shows that it is made de novo. However, once oleoyl-ACP is produced it can undergo two reactions in the plastid. It is either made available for further modifications outside the chloroplast or it begins to be incorporated into glycerolipids. The two alternative reactions are, respectively, oleoyl-ACP hydrolysis and transfer of oleate to glycerol-3-phosphate. Different aspects of these reactions have been summarized (158, 189).

Acyl-ACP hydrolases first studied in detail by Shine et al (172) were purified 400-fold from avocado mesocarp (138a) and 700-fold from safflower (104). These enzymes were highly specific for oleoyl-ACP, thus again ensuring that de novo synthesis of fatty acids can proceed efficiently as far as oleate.

Two acyl transferases present in chloroplasts can utilize acyl-ACPs. A soluble enzyme uses glycerol-3-phosphate and acylates the sn-1 position. A particulate enzyme then utilizes the monoacylglycerol-3-phosphate and acylates the sn-2-position to yield phosphatidate. When mixtures of acyl-CoA and acyl-ACP substrates were provided for either acyl transferase, a 10–20-fold preference for acyl-ACP substrates was seen. Oleoyl-ACP was preferentially used for the first transfer while palmitoyl-ACP was preferred for monoacylglycerol-3-phosphate acylation (37). The substrate preferences of these two plastid acyltransferases ensure that the phosphatidate initially formed in chloroplasts has the "prokaryotic" pattern with C_{18} (oleate) at the sn-1 position and C_{16} (palmitate) at the sn-2 position (158). In contrast, enzymes that utilize acyl-CoAs and are located outside the plastid generate glycerolipids with saturated acids at the sn-1 position—i.e. the typical "eukaryotic" structure.

The proportion of oleoyl-ACP being used for acyl transfer or for hydrolysis determines the fate of the acid for further modification. If oleic acid is released by thioesterase action then the acyl chain is reesterified, this time to CoA in the plastid envelope (6, 82). Oleoyl-CoA can then serve to acylate glycerolipids in the endoplasmic reticulum and become further modified by the "eukaryotic pathway" (158). In contrast, acyl chains acylated to glycerol-3-phosphate remain characteristic of "prokaryotic" lipids and may never leave the plastid. Therefore, the relative activity of acyl-ACP transferase and acyl-ACP hydrolase in a given tissue can be expected to play a major role in controlling the proportion of acyl groups exported from the plastid. In this regard the recent difference in activity between acyl groups esterified to ACP isoforms is particularly interesting (48).

Membrane-Bound Enzymes Using Glycerolipid Substrates Form the Polyunsaturated Fatty Acids

By analogy with the aerobic desaturases of animal tissues, it was thought likely at first that desaturation of oleate to linoleic acid would utilize acyl-CoA substrates. Indeed, early experiments (summarized in 158, 189, 191) showed that oleoyl-CoA was an effective precursor in many tissue preparations. Almost invariably both the substrate oleate and the product linoleate were rapidly transferred to the *sn*-2 position of endogenous PC in the membrane preparations used. This raised the possibility that it was PC and not oleoyl-CoA that was the true substrate for the desaturase. Attempts to differentiate between these possibilities were made with safflower microsomes; unfortunately, the results were equivocal because the methodology used was shown later to give rise to artifacts (194).

The possible use of complex lipids as substrates was originally suggested following labeling experiments carried out by Nichols, James, and coworkers (127, 128). These experiments were extended using *Chorella* and subfractions therefrom to include the use of exogenous substrates (51). A careful follow-up of these experiments with a microsomal fraction from safflower suggested that β-oleoyl-PC and not oleoyl-CoA was the true substrate for desaturation (194). More recently other workers have studied the $\Delta12$-desaturase and there is general agreement that, in most systems studied, oleoyl-PC is the preferred substrate for this reaction (23, 125, 184, 197). Indeed, the substrate may be even better defined in some cases in vitro. Thus, microsomal fractions from pea leaves were shown to utilize PC substrate with the 1-sat.,2-oleoyl combination of acyl groups preferentially (125). By utilizing an HPLC separation method to isolate individual molecular species of PC it was possible to prelabel different species with [^{14}C]oleate and then to follow the formation of [^{14}C]linoleate under desaturating conditions in potato tuber microsomes (26). In this case, two major species were utilized (16:0/18:1 and 18:2/18:1), with desaturation actually highest with the 1-linoleoyl,2-oleoyl substrate (Table 4). Little desaturation was observed with PCs containing oleate at the *sn*-1 position (Table 4). This was in agreement with other workers, although some desaturation in both positions of PC was seen (125, 188).

Many of these experiments rely on the simultaneous presence in microsomal membranes of an acyltransferase and an oleoyl-PC desaturase. Stymne & Glad (195) published evidence for acyl exchange between acyl-CoA and PC. Their system has been developed further and is used to explain the so-called "linoleate enrichment cycle," where oleate and linoleate at the *sn*-2 position of PC constantly equilibrate with an acyl-CoA pool, thus ensuring that a rapid supply of new oleate substrate is available for the $\Delta12$-desaturase and, hence, a high proportion of linoleate at the *sn*-2 position of accumulating glyceroli-

Table 4 Incorporation of [^{14}C]oleate from [^{14}C]oleoyl-CoA into molecular species of phosphatidylcholine and further desaturation by potato microsomes[a]

Molecular species of radiolabeled PC	Distribution of radioactivity (% total)		With NADH
	Without NADH		
	5 min	10 min	10 min
18:3/[^{14}C]18:2	0	0	2
18:2/[^{14}C]18:2	0	0	13
18:3/[^{14}C]18:1	tr.[b]	4	5
16:0/[^{14}C]18:2	0	0	12
18:2/[^{14}C]18:1	24	41	36
16:0/[^{14}C]18:1	76	55	33

[a] Data taken from reference 26.
[b] tr. = trace

pids (195). In contrast, experiments with developing sunflower (154) and pea leaf microsomes (120), as well as with those from safflower (115), suggested that the main route of entry of oleate into PC was by the acylation of 1-lysoPC by oleoyl-CoA. The O-lysophospholipid acyltransferase responsible has been solubilized from developing safflower seeds (115) and pea leaves (124). The safflower enzyme could be solubilized by a variety of detergents, including octylglucoside, deoxycholate, or lysoPC. The solubilized acyltransferase was most active with oleoyl-CoA (K_m = 9.5 μM) but also showed good reaction rates with various saturated acyl-CoAs and linolenyl-CoA. LysoPC was, by far, the best monoacylphospholipid acceptor (115).

In the case of the pea leaf microsomal system, evidence was published to show that the lysoPC acyltransferase was tightly associated with the oleoyl-PC desaturase (121). For the desaturation to proceed, a supply of reduced equivalents is necessary, and NADH is usually thought to be involved. Unlike the acyltransferase, the NADH:cytochrome b_5 oxidoreductase from pea microsomes was not easily solubilized by cholate (124). However, this oxidoreductase (assayed using the NADH:ferricyanide oxidoreductase reaction) was successfully solubilized from potato microsomes, using the zwitterionic detergent CHAPS, and purified to homogeneity. The enzyme was found to have molecular mass of 44 kDa and to show a visible absorption spectrum typical of a flavoprotein (38). Cytochrome b_5 has also been purified from the same tissue (12).

The effect of catalytic hydrogenation of microsomal lipids on enzyme activities involved in oleate desaturation in potato tubers has been studied. Maximum hydrogenation involved a decrease of linolenate and linoleate (45→8% and 27→5%) and increases in stearate and oleate (7→34% and 5→24%, respectively). At these levels it was found that NADH:ferricyanide oxidoreductase was stimulated (200–250% control) whereas NADH:cytochrome c reductase, lysoPC acyltransferase, and the desaturase were inhibited (40%, 100%, and 100%, respectively) (27). These results contrast with other data that more closely resemble the naturally occurring homeoviscous adaptation where an increase in membrane saturation is accompanied by enhanced desaturase activity (152, 178, 208). In addition to the regulation of desaturase activity by membrane fluidity, other factors have been shown to control the enzyme's level or activity. Diurnal fluctuations in the levels of oleate and linoleate have been reported, especially in PC (15). Water stress reduced oleate desaturation (146) while the enzyme system can be induced to very high levels in aging potato or artichoke tubers (13, 215).

Although oleate desaturation has been discussed above in terms of oleoyl-PC as the main substrate in plants, it cannot be emphasized too strongly that this is another case where it is extremely dangerous to generalize too much with different plant systems. Thus, data have been reported indicating that other phospholipids, such as phosphatidylethanolamine, may act as substrates in plants (42, 163, 202). Moreover, it is particularly salutary that Kates and coworkers, who were among the first to demonstrate convincingly that complex lipid substrates were used for oleate desaturation (151), have recently described the use of oleoyl-CoA as substrate in soybean cell suspension cultures (34, 35). Thus, it seems likely that different tissues will show their own specific characteristics with regard to desaturation, and that results obtained with one preparation should only be applied to a different system with extreme caution.

The most prevalent fatty acid in higher plants is α-linolenic acid, which makes up about 65% of the total leaf fatty acids (49, 56). In spite of its importance, we know little about its synthesis, mainly because it is difficult to obtain active subcellular fractions in which to study its formation. Time-course labeling studies with leaves clearly showed that α-linolenate is formed by a $\Delta15$-desaturase acting on linoleate (53). Other possible pathways have now been discounted (191). The stereochemistry of hydrogen removal was studied and found to involve the concerted removal of hydrogens (116), and labeling studies with organisms such as *Chorella* indicated the possible involvement of glycerolipids as substrates for the $\Delta15$-desaturase (128). By the use of very young, rapidly expanding leaves it was possible to obtain systems where rapid synthesis of linolenate could be observed. When such preparations were used for [^{14}C]acetate or ^{14}CO incorporation experiments, [^{14}C]linolenate was observed first in MGDG (67, 173). In contrast to its putative role in linoleate formation, PC was labeled slower than the galactolipid when linolenate synthesis was measured. This led to the idea that MGDG could be the substrate for the $\Delta15$-desaturase (81, 214). Experiments with isolated chloroplasts incubated with [^{14}C]acetate also gave results in keeping with a direct involvement of MGDG (157).

Although it seemed likely that the chloroplast was the site of linoleate desaturation, isolated plastids proved poor at synthesizing linolenate and were unable to utilize exogenous linoleate. In order to test these organelles directly for Δ15-desaturase activity with linoleoyl-MGDG, a new approach was needed. Because intact chloroplasts seemed to be needed for polyunsaturated fatty acid synthesis [the importance of using intact chloroplasts was confirmed recently (5)], the potential substrate could not be added with a detergent. However, the addition of a crude fraction containing lipid-exchange enzymes to isolated chloroplasts allowed good rates of desaturation to be demonstrated (Table 5). This showed directly that MGDG could serve as a substrate for the Δ15-desaturase (78). Linoleoyl-PC was also desaturated but at slower rates, either because this glycerolipid was also a substrate or, more likely, because linoleate was transferred to MGDG before desaturation (139, 140). In contrast to results on oleate desaturation in *Chorella* or soybean (51, 105), the amount of desaturation of exogenously supplied substrate (78) compared well with that for endogenously labeled MGDG (157).

Further experiments have generally confirmed the role of MGDG in linolenate formation in photosynthetic tissues (66, 139, 140). Of particular interest are experiments with the substituted pyridazinone herbicide San 9785, which has a selective action in specifically inhibiting Δ15-desaturation in susceptible plants (89, 119). When this compound was used with [14C]acetate labeling experiments, the block in Δ15-desaturation seemed to be associated with linoleate attached to MGDG. No change in the labeling pattern of PC was seen (Table 6; 25, 118, 216). Furthermore, experiments with [14C]linoleate showed that [14C]linolenate appeared in MGDG, that this desaturation was blocked by San 9785, and that this particular location was not due to specific acyltransferases (118).

Table 5 Ability of isolated chloroplasts to desaturate exogenously supplied linoleoyl-glycerolipids[a]

14C-lipid substrate	Incubation	Distribution of radioactivity (% total 14C-fatty acids)			
		16:0	18:1	18:2	18:3
MGDG	control	tr.[c]	n.d.	61	39
	+ chloroplasts	tr.	n.d.	54	46
	+ chloroplasts + s.p.[b]	n.d.	n.d.	46	54
PC	control	n.d.	n.d.	100	n.d.
	+ chloroplasts + s.p.	3	tr.	97	n.d.
	+ chloroplasts + s.p. + UDP-gal+G-3-P	tr.	tr.	77	23

[a] Data from reference 78
[b] Abbreviations: s.p. = lipid exchange protein fraction; G-3-P = glycerol-3-phosphate
[c] tr. = trace

Table 6 Inhibition of linoleate desaturation in leaves by San 9785[a]

System	Lipid	San 9785 (mM)	Distribution of radiolabel (% total fatty acids)			
			16:0	18:1	18:2	18:3
Spinach[b]	MGDG	0	35	15	21	30
(^{14}C-acetate)		0.2	41	15	28	17
	PC	0	26	64	7	3
		0.2	24	66	7	3
Barley	MGDG	0	—	24	49	27
(^{14}C-oleate)		0.1	—	31	64	5
	PC	0	—	41	59	tr.
		0.1	—	45	55	n.d.

[a] Data taken from references 216 (spinach) and 118 (barley).
[b] Fatty acids from spinach were separated on the basis of double bonds but those compounds indicated are the major components of the saturated and various unsaturated fractions.

Light has been shown in some tissues to stimulate fatty acid desaturation (71). Greening cucumber cotyledons were shown to desaturate oleate and linoleate actively, whereas etiolated tissues did not (122). In contrast, oleate desaturation in PC was independent of light, whereas linoleate desaturation in MGDG was enhanced by light in greening oat leaves (141). Furthermore, in maize seedlings previous exposure to light had no effect on oleate or linoleate desaturation. Moreover, experiments following the desaturation of prelabeled MGDG in isolated spinach chloroplasts confirmed the absence of a light requirement for linoleate desaturation (5). Whether or not light affected linolenate formation, these studies all confirmed the role of MGDG in the acid's synthesis.

The de novo synthesis of fatty acids was discussed earlier as being to a large extent (if not exclusively) confined to plastids. In contrast, oleate is desaturated while esterified to PC, the major extra-chloroplastic lipid (56). The final desaturation involving MGDG (which is a plastidic lipid) then takes place in the plastid. Thus, for the sequential desaturation of stearate to linolenate the first and third reactions will be predominantly located in the plastid, while the $\Delta12$-desaturase is to a large extent found in the endoplasmic reticulum (see Figure 2). A cooperation between plastids and endoplasmic reticulum is, therefore, necessary for the formation of leaf fatty acids. This was first proposed by Tremolieres & Mazliak (204). Further experiments have added to its validity (174, 203, 214), and the overall details have been discussed (29, 57). However, a clear difference is seen between the so-called "16:3-plants" and the "18:3-plants." "16:3-Plants" are those that contain hexadecatrienoic acid at the sn-2 position of MGDG, whereas "18:3-plants" contain linolenate there. It is the latter group of organisms that utilize organellar cooperation to a large extent; in contrast, it is believed that the "16:3-

plants" operate a "prokaryotic" type of metabolism with sequential desaturation of C_{16} and C_{18} acids attached to MGDG within the chloroplast (30, 68). Space does not permit a full discussion of these pathways here but the reader is referred to a recent review for a comprehensive discourse (156).

There are several consequences of the presence of two possible metabolic pathways for trienoate fatty acid formation. Two aspects bear on the substrate specificity of the $\Delta12$- and $\Delta15$-desaturases: first, the possibilities of extra-chloroplastic lipids being involved in linolenate synthesis, and second, the ability of chloroplast lipids to serve as substrates for complete desaturation sequence from palmitate to hexadecatrienoate and from oleate to linolenate. Regarding the first point, there have been several reports in the literature that linoleate may be desaturated on substrates other than MGDG. Thus, for example, some linolenate may be derived from desaturation at the sn-2 position of PC (134, 217). This is especially true in nongreen tissues such as developing oilseeds (158). However, a recent test of linoleoyl-CoA desaturation failed to demonstrate any activity in soybean cells, so that thioesters do not seem to be substrates for $\Delta15$-desaturation in plants (34, 35).

The idea that chloroplast lipids, such as MGDG, can be substrates for a whole sequence of desaturation is not only of relevence to the topic of "prokaryotic" and "eukaryotic" pathways in higher plants but also has parallels in lower organisms. A series of experiments by Murata and colleagues using the cyanobacterium Anabaena variabilis showed that all (chloroplastic) lipids were capable of serving as substrates for successive desaturations at the C_{18} level (117). MGDG was, quantitatively, the most important in this regard and furthermore served as a substrate for palmitate desaturation to hexadecadienoate (61, 117). Similar experiments have been performed with the eukaryotic green alga Dunaliella salina, an organism particularly well characterized by Thompson's group (22). These ideas are extended to chloroplasts operating "prokaryotic" pathways, and a direct comparison can be made between recent experiments on palmitate desaturation to hexadecatrienoate in spinach (198) and the same reactions on the MGDG of A. variabilis (164). Observations with Arabidopsis mutants low in linolenate and hexadecatrienoate also suggest that several glycerolipids may serve as substrates for a $\Delta15$-desaturase (14). Mutants (of soybean) have also been used in an attempt to identify the $\Delta15$-desaturase protein (212).

Finally, it is important to note that the mechanism of lipid movement throughout the plant cell has not been clarified. The lipid exchange proteins that have been purified from several plant tissues including leaves (83) may play a role here. These exchange proteins cause the movement not only of phospholipids but also of glycolipids between membranes. The sequence of a nonspecific exchange protein has recently been reported (220).

Other Desaturase Reactions

Although oleate, linoleate, and linolenate are, by far, the most prevalent unsaturated fatty acids in plants, a large number of other unsaturated fatty acids exist (49) and, in some cases, their synthesis has been examined. The formation of the unique *trans*-Δ3-hexadecenoic acid seems to take place at the *sn*-2 position of phosphatidylglycerol (55, 185). Unlike other desaturations, this reaction seems to require light (55, 141). In *Borago officinalis*, γ-linolenate and octadecatetraenoate are made, and these acids are thought to be formed by a Δ6-desaturase acting on linoleate at position-2 of PC and by the Δ15-desaturase acting on γ-linolenate at position-2 of MGDG, respectively (196). In marked contrast, the Δ5-desaturase that makes Δ5-*cis*-eicosanoic acid in *Limnanthes alba* seeds utilizes eicosanoyl-CoA substrate (113).

In algae and lower plants, other fatty acids such as arachidonate and eicosapentaenoate are major components (61, 71). Their synthesis has been little studied and is an obviously important area for future work.

OXIDATION OF FATTY ACIDS

Oxidative reactions involving fatty acids in plants include α-, β- and ω-oxidations; hydroxylations; and other more specialized processes, such as lipoxygenation. Significant progress has been made in several of these areas recently, and these are highlighted below. For an excellent overall review of our knowledge prior to 1980 of oxidative reactions involving fatty acids see references 39 and 40.

Formation of Ricinoleic Acid

Ricinoleic acid [D(+)-12-hydroxy-*cis*-9-octadecenoic acid] is the major component of the storage triacylglycerols of castor bean seeds, representing about 90% of the total fatty acids (49). The in vitro synthesis of this acid was partly characterized in 1966 using a microsomal fraction from developing seeds and oleoyl-CoA as substrate (41). The enzyme responsible was shown to be a mixed-function oxidase requiring O_2 and NADH. However, as discussed above, it is appreciated now that the efficacy of acyl-CoA substrates for fatty acid modification reactions often lies in the transfer of the acyl group from the thioester to the true substrate. Accordingly, the synthesis of ricinoleic acid was reexamined with experiments designed to elucidate the immediate substrate for oleate hydroxylation in castor bean (114). Several interesting observations were made. First, the data suggested strongly that oleoyl-CoA was not the immediate substrate for hydroxylation. Although oleoyl-PC was rapidly formed and could have been the substrate, the lack of appreciable levels of ricinoleoyl-PC and the appearance of an unidentified ricinoleoyl-nonpolar lipid product mean that oleoyl-PC might not be the true substrate.

Second, a lack of inhibition by CO (99%) or cyanide (1 mM) suggested that the hydroxylation reaction did not involve either a cytochrome P450 or a cyanide-sensitive iron porphyrin system. Third, the hydroxylase activity in microsomes was unstable, whereas that in the 12,000-g supernatant was not. Some activity could be restored to the microsomes by addition of catalase and also by soluble heat-sensitive components of the 105,000-g supernatant. Parallels were drawn between the hydroxylase system and the oleate desaturase of other plants, such as in their CO insensitivity and NADH requirement (114). It would be interesting in view of the above results to attempt further to define the substrate requirement and other characteristics of this hydroxylation system.

β-Oxidation of Fatty Acids

Although it is well known that the fatty acyl-CoA oxidizing system of germinating fatty seedlings is located in specialized organelles called glyoxysomes, until recently the situation in nonfatty plant tissues was less clear. In fact, it was generally believed that, as in other eukaryotes, the mitochondria were a major (or exclusive) site for such β-oxidation. Our knowledge in 1980 has been summarized well (7, 39). However, when mitochondria from nonfatty plant tissues were specifically examined for activity, little evidence for the location of β-oxidation there could be found (59).

The glyoxysomal fatty acyl oxidizing system operates by a series of reactions similar to the well-characterized β-oxidation pathway of mammalian mitochondria. However, there are important differences (39). In particular, the initial step is catalyzed by an acyl-CoA oxidase that transfers electrons directly to O_2, thus producing H_2O_2 (24).

Direct evidence that peroxisomes from nonfatty plant tissues contained all the enzymes needed for β-oxidation was first obtained with spinach leaves (43). Further reports have now been published for Jerusalem artichoke tubers (103), mung bean hypocotyls, and potato tubers (44). By the use of step gradients, it was possible to obtain peroxisomal fractions devoid of detectable isocitrate lyase or malate synthetase activity. They also contained minimal activities of usual mitochondrial markers such as fumarase or cytochrome oxidase (Table 7).

When the peroxisomal fractions were used to catalyze palmitoyl-CoA oxidation, a stoichiometric relationship was demonstrated between O_2 uptake and H_2O_2 formation in agreement with the operation of acyl-CoA oxidase (24). Experiments with KCN (an inhibitor of the cyanide-sensitive electron-transport chain) or salicylhydroxamic acid (an inhibitor of the cyanide-resistant plant mitochondrial electron-transport chain) showed that neither the cytochrome pathway nor the alternative pathway of respiration was involved in the oxidation of palmitoyl-CoA (44). When kinetic parameters were com-

Table 7   Activity of marker enzymes and those of β-oxidation in mitochondrial and per-oxisomal fractions from mung-bean hypocotyls or potato tubers[a]

| | Mung bean hypocotyls | | Potato tubers | |
Enzyme	Peroxisomal fraction	Mitochondrial fraction	Peroxisomal fraction	Mitochondrial fraction
Enoyl-CoA hydratase	8,143[b]	329	6,124	123
3-Hydroxyacyl-CoA DH	5,101	241	14,741	284
Thiolase	1,330	82	615	156
Catalase	2,125,726	107,652	11,618,265	220,263
Fumarase	3,189	10,202	n.d.	3,413
Cytochrome oxidase	1,758	5,975	1304	2,550

[a] Data taken from reference 44.
[b] Activities expressed as pkat/mg protein.

pared, the initial oxidation of palmitoyl-CoA by peroxisomes was found to be similar to those of the glyoxysomal acyl-CoA oxidase and rat liver (144). By adding the remaining enzymes needed for β-oxidation to form a reconstituted system, the postulated product of the acyl-CoA oxidase, namely *trans*-2-enoyl-CoA, was implicated, in that a stoichiometry was shown between oxygen consumption and NADH production.

Careful comparison of the distribution of marker enzymes and the partial reactions of β-oxidation on discontinuous sucrose density gradients (43, 44) excluded a significant role for mitochondria in such fatty acid oxidation. Although data have been obtained recently with pea mitochondrial fractions (219), other experiments with highly purified organelles from both nonfatty (43, 44, 103) and fatty (24, 109) tissues show that microbodies are the only site for β-oxidation.

Obviously, β-oxidation of fatty acids is of prime importance during the germination of fatty seedlings. Therefore, it was unexpected that peroxisomes from nonfatty tissues contained rather similar activities of two of the enzymes compared to glyoxysomes from sunflowers (44). However, bearing in mind the much greater numbers of glyoxysomes in fatty tissues, the overall capacity for β-oxidation must be considerably greater there. At present, the fate of the acetyl-CoA produced by peroxisomes of nonfatty tissues is not known.

Lipoxygenase

Plant lipoxygenases are extremely active enzymes that catalyze the degradation of fatty acids containing a *cis, cis,* 1,4-pentadiene structure. The biochemical function is not at present understood, although the (harmful) effects of their activity are well known in the food industry. Helpful reviews of their biochemistry are available (40, 206, 207). One recent aspect that deserves a mention is the role of lipoxygenase in the synthesis of 12-oxo-phytodienoic acid and jasmonic acid (206). This pathway has analogies with eicosanoid production in animals. Moreover, jasmonic acid may act as a previously unrecognized plant hormone (206). Several other current lines of research on lipoxygenase biochemistry are described in reference 193.

CONCLUSIONS AND FUTURE DIRECTIONS

It will be obvious from this review that although considerable advances have recently been made in the enzymology of fatty acid synthetase, several other aspects of fatty acid metabolism are still unclear. However, progress has been made in defining elongation and desaturation systems and, with the purification of active preparations from membranes, we can look forward to exciting new information. These seem to me to be areas where attention should be focussed.

Many aspects of the de novo system should also be followed up. In particular, improved protein isolations will allow us to gain knowledge of acetyl-CoA carboxylase and enzymes of the synthetase complex, of which the condensing enzymes are ill defined at the moment. The presence of isoenzymes for many of these proteins is an unexpected and intriguing finding. As information becomes available, control mechanisms may be better understood and the complicated interaction of different parts of the plant cell in ensuring the correct balance of end-products should become clearer. Above all, molecular biology can be expected to contribute greatly in the next few years.

I end with a cautionary note. It will be clear from a number of the examples I have given that simple conclusions drawn from a single set of experiments with one plant tissue often cannot be applied generally. Indeed, as J. B. S. Haldane once said, "My own suspicion is that the universe is not only queerer than we suppose, but queerer than we can suppose."

It is to be hoped that biochemists researching plant fatty acids in the future will not be too dogmatic and will always extrapolate prudently from their results.

Literature Cited

1. Abdul-Karim, T., Lessire, R., Cassagne, C. 1982. Involvement of an acyl-coenzyme A transacylase in the insertion of C20–C30 fatty acids in the microsomal lipids from leek epidermal cells. *Physiol. Veg.* 20:679–89
2. Abita, J. P., Lazdunski, M., Ailhaud, G. P. 1971. Structure function relationships of the acyl carrier protein of *Escherichia coli. Eur. J. Biochem.* 23:412–20
3. Agrawal, V. P., Lessire, R., Stumpf, P. K. 1984. Biosynthesis of very long

chain fatty acids in microsomes from epidermal cells of *Allium porrum. Arch. Biochem. Biophys.* 230:580–89
4. Agrawal, V. P., Stumpf, P. K. 1985. Characterisation and solubilisation of an acyl chain elongation system in microsomes of leek epidermal cells. *Arch. Biochem. Biophys.* 240:154–65
5. Andrews, J., Heinz, E. 1987. Desaturation of newly synthesized monogalactosyldiacylglycerol in spinach chloroplasts. *J. Plant. Physiol.* 131:75–90
6. Andrews, J., Keegstra, K. 1983. Acyl-

CoA synthetase is located in the outer membrane and acyl-CoA thioesterase in the inner membrane of pea chloroplast envelopes. *Plant Physiol.* 72:735–40

7. ap Rees, T. 1980. Assessment of the contribution of metabolic pathways to plant respiration. In *The Biochemistry of Plants, Vol 2.*, ed. P. K. Stumpf, E. E. Conn, pp. 1–29. New York: Academic

8. Bereman, P. D., Hannapel, D. J., Guerra, D. J., Kuhn, D. N., Ohlrogge, J. B. 1987. Synthesis, cloning and expression in *Escherichia coli* of a spinach acyl carrier protein-1 gene. *Arch. Biochem. Biophys.* 256:90–100

9. Bolton, P., Harwood, J. L. 1976. Effect of thiocarbamate herbicides on fatty acid synthesis in potato. *Phytochemistry* 15:1507–9

10. Bolton, P., Harwood, J. L. 1977. Fatty acid synthesis by a particulate preparation from germinating pea. *Biochem. J.* 168:261–69

11. Bolton, P., Harwood, J. L. 1977. Some characteristics of soluble fatty acid synthesis in germinating pea seeds. *Biochim. Biophys. Acta* 489:15–24

12. Bonnerot, C., Galle, A. M., Jolliot, A., Kader, J. C. 1985. Purification and properties of plant cytochrome b_5. *Biochem. J.* 226:331–34

13. Bonnerot, C., Mazliak, P. 1984. Induction of the oleoyl-phosphatidylcholine desaturase activity during the storage of plant organs. A comparison between potato and Jerusalem artichoke tubers. *Plant Sci. Lett.* 35:5–10

14. Browse, J., McCourt, P., Somerville, C. 1986. A mutant of *Arabidopsis* deficient in $C_{18:3}$ and $C_{16:3}$ leaf lipids. *Plant Physiol.* 81:859–64

15. Browse, J., Roughan, P. G., Slack, C. R. 1981. Light control of fatty acid synthesis and diurnal fluctuations of fatty acid composition in leaves. *Biochem. J.* 196:347–54

16. Browse, J., Slack, C. R. 1985. Fatty acid synthesis in plastids from maturing safflower and linseed cotyledons. *Planta* 166:74–80

17. Camp, P. J., Randall, D. D. 1985. Purification and characterisation of the pea chloroplast pyruvate dehydrogenase complex. *Plant Physiol.* 77:571–77

18. Cassagne, C., Lessire, R. 1978. Biosynthesis of saturated very long chain fatty acids by purified membrane fractions from leek epidermal cells. *Arch. Biochem. Biophys.* 191:146–52

19. Cassagne, C., Lessire, R., Bessoule, J. J., Moreau, P. 1987. Plant elongases. In *The Metabolism, Structure and Function of Plant Lipids*, ed. P. K. Stumpf, J. B.

Mudd, W. D. Nes, pp. 481–88. New York: Plenum

20. Caughey, I., Kekwick, R. G. O. 1982. The characteristics of some components of the fatty acid synthetase system in the plastids from the mesocarp of avocado (*Persea americana*) fruit. *Eur. J. Biochem.* 123:553–61

21. Charles, D. J., Cherry, J. H. 1986. Purification and characterisation of acetyl-CoA carboxylase from developing soybean seeds. *Phytochemistry* 25:1067–71

22. Cho, S. H., Thompson, G. A. 1987. Metabolism of galactolipids in *Dunaliella salina*. See Ref. 19, pp. 623–29

23. Citharel, B., Oursel, A., Mazliak, P. 1983. Desaturation of oleoyl and linoleoyl residues linked to phospholipids in growing roots of yellow lupin. *FEBS Lett.* 161:251–56

24. Cooper, T. G., Beevers, H. 1969. β-Oxidation in glyoxysomes from castor bean endosperm. *J. Biol. Chem.* 244:3514–20

25. Davies, A. O., Harwood, J. L. 1983. Effect of substituted pyridazinones on chloroplast structure and lipid metabolism in greening barley leaves. *J. Exp. Bot.* 34:1089–1100

26. Demandre, C., Trémolières, A., Justin, A. M., Mazliak, P. 1986. Oleate desaturation in six phosphatidylcholine molecular species from potato tuber microsomes. *Biochim. Biophys. Acta* 877:380–86

27. Demandre, C., Vigh, L., Justin, A. M., Jolliot, A., Wolf, C., Mazliak, P. 1986. Effects of the catalytic hydrogenation of microsomal lipids upon four enzyme activities involved in oleic acid desaturation. *Plant Sci.* 44:13–21

28. Dorne, A. J., Corde, J. P., Joyard, J., Borner, T., Douce, R. 1982. Polar lipid composition of a plastid ribosome-deficient barley mutant. *Plant Physiol.* 69:1467–70

29. Drapier, D., Dubacq, J.-P., Trémolières, A., Mazliak, P. 1982. Cooperative pathway in young pea leaves: oleate exportation from chloroplasts and subsequent integration into complex lipids of added microsomes. *Plant Cell Physiol.* 23:125–35

30. Dubacq, J.-P., Drapier, D., Trémolières, A. 1983. Polyunsaturated fatty acid synthesis by a mixture of chloroplasts and microsomes from spinach leaves: evidence for two distinct pathways for the synthesis of trienoic acids. *Plant Cell Physiol.* 24:1–9

31. Eastwell, K. C., Stumpf, P. K. 1983. Regulation of plant acetyl-CoA carboxy-

lase by adenylate nucleotides. *Plant Physiol.* 72:50–55
32. Egin-Buhler, B., Ebel, J. 1983. Improved purification and further characterisation of acetyl-CoA carboxylase from culture cells of parsley *(Petroselinum hortense)*. *Eur. J. Biochem.* 133:335–39
33. Egin-Buhler, B., Loyal, R., Ebel, J. 1980. Comparison of acetyl-CoA carboxylase from parsley cell cultures and wheat germ. *Arch. Biochem. Biophys.* 203:90–100
34. Ferrante, G., Kates, M. 1986. Identification of oxygenated and related products of oleoyl-CoA and linoleoyl-CoA metabolism by cell fractions of soybean suspension cultures. *Biochim. Biophys. Acta* 876:417–28
35. Ferrante, G., Kates, M. 1986. Characteristics of the oleoyl- and linoleoyl-CoA desaturase and hydroxylase systems in cell fractions from soybean cell suspension cultures. *Biochim. Biophys. Acta* 876:429–37
36. Finlayson, S. A., Dennis, D. T. 1983. Acetyl-CoA carboxylase from the developing endosperm of *Ricinus communis*. Isolation and characterisation. *Arch. Biochem. Biophys.* 225:576–85
37. Frentzen, M., Heinz, E., McKeon, T. M., Stumpf, P. K. 1983. Specificities and selectivities of glycerol-3-phosphate acyl transferase and monoglycerol-3-phosphate acyltransferase from pea and spinach chloroplasts. *Eur. J. Biochem.* 129:629–36
38. Galle, A.-M., Bonnerot, C., Jolliot, A., Kader, J.-C. 1984. Purification of a NADH-ferricyanide reductase from plant microsomal membranes with a zwitterionic detergent. *Biochem. Biophys. Res. Commun.* 122:1201–5
39. Galliard, T. 1980. Degradation of acyl lipids: hydrolytic and oxidative enzymes. In *The Biochemistry of Plants*, ed. P. K. Stumpf, E. E. Conn, 4:85–116. New York: Academic
40. Galliard, T., Chan, H. W.-S. 1980. Lipoxygenases. See Ref. 39, pp. 132–62
41. Galliard, T., Stumpf, P. K. 1966. Enzymatic synthesis of ricinoleic acid by a microsomal preparation from developing *Ricinus communis* seeds. *J. Biol. Chem.* 241:5806–12
42. Gennity, J. M., Stumpf, P. K. 1985. Studies on the Δ12 desaturase of *Carthamus tinctorius* L. *Arch. Biochem. Biophys.* 239:444–54
43. Gerhardt, B. 1981. Enzyme activities of the β-oxidation pathway in spinach leaf peroxisomes. *FEBS Lett.* 126:71–73
44. Gerhardt, B. 1983. Localisation of β-

oxidation enzymes in peroxisomes isolated from non-fatty plant tissues. *Planta* 159:238–46
45. Givan, C. V. 1983. The source of acetyl-CoA in chloroplasts of higher plants. *Plant Physiol.* 57:311–16
46. Givan, C. V., Hodgson, J. M. 1983. The formation of coenzyme A from acetyl-coenzyme A in pea leaf mitochondria: a requirement for oxaloacetate and the absence of hydrolysis. *Plant Sci. Lett.* 32:233–42
47. Guerra, D. J., Ohlrogge, J. B. 1986. Partial purification and characterisation of two forms of malonyl-coenzyme A: acyl carrier protein transacylase from soybean leaf tissue. *Arch. Biochem. Biophys.* 246:274–85
48. Guerra, D. J., Ohlrogge, J. B., Frentzen, M. 1986. Activity of acyl carrier protein isoforms in reactions of plant fatty acid metabolism. *Plant Physiol.* 82:448–53
49. Gunstone, F. D., Harwood, J. L., Padley, F. B., eds. 1986. *The Lipid Handbook*. London: Chapman and Hall
50. Gurr, M. I. 1974. The biosynthesis of unsaturated fatty acids. In *MTP International Review of Science*, ed. T. W. Goodwin, 4:181–236. London: Butterworths
51. Gurr, M. I., Robinson, P., James, A. T. 1969. The mechanism of formation of polyunsaturated fatty acids by photosynthetic tissue. The tight coupling of oleate desaturation with phospholipid synthesis in *Chlorella vulgaris*. *Eur. J. Biochem.* 9:70–78
52. Harris, P., James, A. T. 1968. The effect of low temperatures on fatty acid biosynthesis in plants. *Biochem. J.* 112:325–30
53. Harris, R. V., James, A. T. 1965. Linoleic and α-linolenic acid biosynthesis in plant leaves and a green alga. *Biochim. Biophys. Acta* 106:456–64
54. Harwood, J. L. 1975. Fatty acid biosynthesis. In *Recent Advances in the Chemistry and Biochemistry of Plant Lipids*, ed. T. Galliard, E. I. Mercer, pp. 43–93. London: Academic
55. Harwood, J. L. 1979. The synthesis of acyl lipids in plant tissues. *Prog. Lipid. Res.* 18:55–86
56. Harwood, J. L. 1980. Plant acyl lipids: structure, distribution and analysis. See Ref. 39, pp. 1–55
57. Harwood, J. L. 1980. Fatty acid synthesis. In *Biogenesis and Function of Plant Lipids*, ed. P. Mazliak, P, Benveniste, C. Costes, R. Douce, pp. 143–52. Amsterdam: Elsevier
58. Harwood, J. L. 1983. Adaptive changes

in the lipids of higher-plant membranes. *Biochem. Soc. Trans.* 11:343–46

59. Harwood, J. L. 1985. Plant mitochondrial lipids: structure, function and biosynthesis. In *Higher Plant Cell Respiration*, ed. R. Douce, D. A. Day, pp. 37–71. Berlin: Springer-Verlag

60. Harwood, J. L. 1987. Medium- and long-chain fatty acid synthesis. See Ref. 19, pp. 465–72

61. Harwood, J. L., Pettitt, T. P., Jones, A. L. 1987. Lipid metabolism. In *The Biochemistry of the Algae and Cyanobacteria*, ed. L. J. Rogers, J. R. Gallon. Oxford: Oxford Univ. Press. In press

62. Harwood, J. L., Ridley, S. M., Walker, K. A. 1988. Herbicides which inhibit lipid metabolism. In *Herbicides and Plant Metabolism*, ed. A. D. Dodge. Cambridge: Cambridge Univ. Press. In press

63. Harwood, J. L., Stumpf, P. K. 1970. Synthesis of fatty acids in the initial stage of seed germination. *Plant Physiol.* 46:500–8

64. Harwood, J. L., Stumpf, P. K. 1971. Control of fatty acid synthesis in germinating seeds. *Arch. Biochem. Biophys.* 142:281–91

65. Harwood, J. L., Stumpf, P. K. 1972. Palmitic and stearic synthesis by an avocado supernatant system. *Arch. Biochem. Biophys.* 148:282–90

66. Hawke, J. C., Stumpf, P. K. 1980. The incorporation of oleic and linoleic acids and their desaturation products into the glycerolipids of maize leaves. *Arch. Biochem. Biophys.* 203:296–306

67. Heinz, E., Harwood, J. L. 1977. Incorporation of carbon dioxide, acetate and sulphate into the glycerolipids of *Vicia faba* leaves. *Hoppe-Seyler's Z. Physiol. Chem.* 38:897–908

68. Heinz, E., Roughan, P. G. 1983. Similarities and differences in lipid metabolism of chloroplasts isolated from 18:3 and 16:3 plants. *Plant Physiol.* 72:273–79

69. Heinz, K.-P., Teede, H.-J. 1987. Regulation of acetyl-CoA synthesis in chloroplasts. See Ref. 19, pp. 505–7

70. Hellyer, A., Bambridge, H. E., Slabas, A. R. 1986. Plant acetyl-CoA carboxylase. *Biochem. Soc. Trans.* 14:565–68

71. Hitchcock, C., Nichols, B. W. 1971. *Plant Lipid Biochemistry*. London: Academic

72. Hoj, P. B., Mikkelsen, J. D. 1982. Partial separation of individual enzyme activities of an ACP-dependent fatty acid synthetase from barley chloroplasts. *Carlsberg Res. Commun.* 47:119–41

73. Hoj, P. B., Svendsen, I. B. 1983. Barley acyl carrier protein: its amino acid sequence and assay using purified malonyl-CoA : ACP transacylase. *Carlsberg Res. Commun.* 48:285–305

74. Hoj, P. B., Svendsen, I. B. 1984. Barley chloroplasts contain two acyl carrier proteins coded for by different genes. *Carlsberg Res. Commun.* 49:483–92

75. Huang, K. P., Stumpf, P. K. 1971. Fatty acid synthesis by a soluble fatty acid synthetase from *Solanum tuberosum*. *Arch. Biochem. Biophys.* 143:412–27

76. Jaworski, J. G., Goldschmidt, E. E., Stumpf, P. K. 1974. Properties of the palmityl acyl carrier protein: stearyl acyl carrier protein elongation system in maturing safflower seed extracts. *Arch. Biochem. Biophys.* 163:769–76

77. Jaworski, J. G., Stumpf, P. K. 1974. Properties of a soluble stearyl-acyl carrier protein desaturase from maturing *Carthamus tinctorius*. *Arch. Biochem. Biophys.* 162:158–65

78. Jones, A. V. M., Harwood, J. L. 1980. Desaturation of linoleic acid from exogenous lipids by isolated chloroplasts. *Biochem. J.* 190:851–54

79. Jordan, B. R., Harwood, J. L. 1980. Fatty acid elongation by a particulate fraction from germinating pea. *Biochem. J.* 191:791–97

80. Journet, E.-P., Douce, R. 1985. Enzymic capacities of purified cauliflower bud plastids for lipid synthesis and carbohydrate metabolism. *Plant Physiol.* 79:458–67

81. Joyard, J., Chuzel, M., Douce, R. 1979. Is the chloroplast envelope a site of galactolipid synthesis? Yes. In *Advances in the Biochemistry and Physiology of Plant Lipids*, ed. L.-A., Appelqvist, C. Liljenberg, pp. 181–86. Amsterdam: Elsevier

82. Joyard, J., Stumpf, P. K. 1981. Synthesis of long-chain acyl-CoA in chloroplast envelope membranes. *Plant Physiol.* 67:250–56

83. Kader, J. C., Julienne, M., Vergnalle, C. 1984. Purification and characterisation of a spinach-leaf protein capable of transferring phospholipids from liposomes to mitochondria or chloroplasts. *Eur. J. Biochem.* 139:411–16

84. Kannangara, C. G., Jensen, C. J. 1975. Biotin carboxyl carrier protein in barley chloroplast membranes. *Eur. J. Biochem.* 54:25–30

85. Kindl, H. 1984. Lipid degradation in higher plants. In *Fatty Acid Metabolism and Its Regulation*, ed. S. Numa, pp. 181–204. Amsterdam: Elsevier

86. Kolattukudy, P. E., Buckner, J. S. 1972. Chain elongation of fatty acids by

cell-free extracts of epidermis of pea leaves. *Biochem. Biophys. Res. Commun.* 46:801–7

87. Kuo, T. M., Ohlrogge, J. B. 1984. A novel general radioimmunoassay for acyl carrier proteins. *Anal. Biochem.* 136: 497–502

88. Kuo, T. M., Ohlrogge, J. B. 1984. The primary structure of spinach acyl carrier protein. *Arch. Biochem. Biophys.* 234: 290–96

89. Lem, N. W., Williams, J. P. 1981. Desaturation of fatty acids associated with monogalactosyldiacylglycerol: the effects of San 6706 and San 9785. *Plant Physiol.* 68:944–49

90. Lessire, R., Bessoule, J.-J., Cassagne, C. 1985. Solubilization of C_{18}–CoA and C_{20}–CoA elongases from *Allium porrum* L. epidermal cell microsomes. *FEBS Lett.* 187:314–20

91. Lessire, R., Bessoule, J.-J., Cassagne, C. 1987. Acyl-CoA elongation systems in *Allium porrum* microsomes. See Ref. 19, pp. 525–27

92. Lessire, R., Hartmann-Bouillon, M.-A., Cassagne, C. 1982. Very long chain fatty acids: occurrence and biosynthesis in membrane fractions from etiolated maize coleoptiles. *Phytochemistry* 21:55–59

93. Lessire, R., Juguelin, H., Moreau, P., Cassagne, C. 1985. Elongation of acyl-CoAs by microsomes from etiolated leek seedlings. *Phytochemistry* 24: 1187–92

94. Lessire, R., Juguelin, H., Moreau, P., Cassagne, C. 1985. Nature of the reaction product of [1-^{14}C]stearoyl-CoA elongation by etiolated leek seedling microsomes. *Arch. Biochem. Biophys.* 239:260–69

95. Lessire, R., Stumpf, P. K. 1983. Nature of the fatty acid synthetase systems in parenchymal and epidermal cells of *Allium porrum* L. leaves. *Plant Physiol.* 73:614–18

96. Liedvogel, B. 1985. Acetate concentration and chloroplast pyruvate dehydrogenase complex in *Spinacea oleracea* leaf cells. *Z. Naturforsch.* 40C: 182–88

97. Liedvogel, B. 1986. Acetyl coenzyme A and isopentenyl-pyrophosphate as lipid precursors in plant cells—biosynthesis and compartmentation. *J. Plant Physiol.* 124:211–22

98. Liedvogel, B. 1987. Lipid precursors in plant cells: the problem of acetyl-CoA generation for plastid fatty acid synthesis. See Ref. 19, pp. 509–12

99. Liedvogel, B., Bauerle, R. 1986. Fatty acid synthesis in chloroplasts from mustard (*Sinapsis alba* L.) cotyledons: formation of acetyl-CoA by intraplastid glycolytic enzymes and a pyruvate dehydrogenase complex. *Planta* 169:481–89

100. Liedvogel, B., Kleinig, H. 1980. Fatty acid synthesis in isolated chromoplasts and chromoplast homogenates. ACP stimulation, substrate utilisation and cerulenin inhibition. In *Biochemistry and Function of Plant Lipids*, ed. P. Mazliak, P. Benveniste, C. Costes, R. Douce, pp. 107–10. Amsterdam: Elsevier

101. Lynch, D. V., Thompson, G. A. 1984. Retailored lipid molecular species: a tactical mechanism for modulating membrane properties. *Trends Biochem. Sci.* 9:442–45

102. Macey, M. J. K., Stumpf, P. K. 1968. Long chain fatty acid synthesis in germinating peas. *Plant Physiol.* 43: 1637–47

103. Macey, M. J. K., Stumpf, P. K. 1982. β-Oxidation enzymes in microbodies from tubers of *Helianthus tuberosus*. *Plant Sci. Lett.* 28:207–12

104. McKeon, T. M., Stumpf, P. K. 1982. Purification and characterisation of the stearoyl-acyl carrier protein desaturase and the acyl-acyl carrier protein thioesterase from maturing seeds of safflower. *J. Biol. Chem.* 257:12141–47

105. Martin, B. A., Rinne, R. W. 1986. A comparison of oleic acid metabolism in the soybean genotypes Williams and A5, a mutant with decreased linoleic acid in the seed. *Plant Physiol.* 81:41–44

106. Matsumura, S., Stumpf, P. K. 1968. Partial primary structure of spinach acyl carrier protein. *Arch. Biochem. Biophys.* 125:932–41

107. Miernyk, J. A., Dennis, D. T. 1982. Isozymes of the glycolytic enzymes in endosperm from developing castor oilseeds. *Plant Physiol.* 69:825–28

108. Miernyk, J. A., Dennis, D. T. 1983. The incorporation of glycolytic intermediates into lipids from the developing endosperm of castor oil seeds. *J. Exp. Bot.* 34:712–18

109. Miernyk, J. A., Trelease, R. N. 1981. Control of enzyme activities in cotton cotyledons during maturation and germination. IV. β-Oxidation. *Plant Physiol.* 67:341–46

110. Mohan, S. B., Kekwick, R. G. O. 1980. Acetyl-CoA carboxylase from avocado plastids and spinach chloroplasts. *Biochem. J.* 187:667–76

111. Moreau, P., Juguelin, H., Lessire, R., Cassagne, C. 1984. *In vivo* incorporation of acetate into the acyl moieties of

polar lipids from etiolated leek seedlings. *Phytochemistry* 23:67–71

112. Moreau, P., Juguelin, H., Lessire, R., Cassagne, C. 1986. Intermembrane transfer of long chain fatty acids synthesised by etiolated leek seedlings. *Phytochemistry* 25:387–91

113. Moreau, R. A., Pollard, M. R., Stumpf, P. K. 1981. Properties of a Δ5-fatty acyl-CoA desaturase in the cotyledons of developing *Limnanthes alba*. *Arch. Biochem. Biophys.* 209:376–84

114. Moreau, R. A., Stumpf, P. K. 1981. Recent studies of the enzymic synthesis of ricinoleic acid by developing castor beans. *Plant Physiol.* 67:672–76

115. Moreau, R. A., Stumpf, P. K. 1982. Solubilisation and characterisation of an acyl-coenzyme A: O-lysophospholipid acyltransferase from the microsomes of developing safflower seeds. *Plant Physiol.* 69:1293–97

116. Morris, L. J., Harris, R. V., Kelly, W., James, A. T. 1968. The stereochemistry of desaturations of long-chain fatty acids in *Chlorella vulgaris*. *Biochem. J.* 109:673–78

117. Murata, N., Nishida, I. 1987. Lipids of blue-green algae (Cyanobacteria). In *Biochemistry of Plants*, ed. P. K. Stumpf, E. E. Conn, 9:315–47. New York: Academic

118. Murphy, D. J., Harwood, J. L., Lee, K. A., Roberto, F., Stumpf, P. K., St. John, J. B. 1985. Differential responses of a range of photosynthetic tissues to a substituted pyridazinone, Sandoz 9785. Specific effects on fatty acid desaturation. *Phytochemistry* 24:1923–29

119. Murphy, D. J., Harwood, J. L., St. John, J. B., Stumpf, P. K. 1980. Effect of a substituted pyridazinone compound BASF 13-338 on membrane lipid synthesis in photosynthetic tissues. *Biochem. Soc. Trans.* 8:119–20

120. Murphy, D. J., Mukherjee, K. D., Latzko, E. 1984. Oleate metabolism in microsomes from developing leaves of *Pisum sativum* L. *Planta* 161:249–54

121. Murphy, D. J., Mukherjee, K. D., Woodrow, I. E. 1984. Functional association of a monoacylglycerophosphocholine acyltransferase and the oleoylglycerophosphocholine desaturase in microsomes from developing leaves. *Eur. J. Biochem.* 139:373–79

122. Murphy, D. J., Stumpf, P. K. 1979. Light-dependent induction of polyunsaturated fatty acid biosynthesis in greening cucumber cotyledons. *Plant Physiol.* 63:328–35

123. Murphy, D. J., Stumpf, P. K. 1981. The origin of chloroplastic acetyl-CoA. *Arch. Biochem. Biophys.* 212:730–39

124. Murphy, D. J., Woodrow, I. E., Latzko, E., Mukherjee, K. D. 1983. Solubilisation of oleoyl-CoA thioesterase, oleoyl-CoA: phosphatidylcholine acyltransferase and oleoyl phosphatidylcholine desaturase. *FEBS Lett.* 162:442–46

125. Murphy, D. J., Woodrow, I. E., Mukherjee, K. D. 1985. Substrate specificities of the enzymes of the oleate desaturase system from photosynthetic tissue. *Biochem. J.* 225:267–70

126. Nagai, J., Bloch, K. 1968. Enzymatic desaturation of steryl acyl carrier protein. *J. Biol. Chem.* 243:4626–33

127. Nichols, B. W. 1968. Fatty acid metabolism in the chloroplast lipids of green and blue-green algae. *Lipids* 3:354–60

128. Nichols, B. W., James, A. T., Breuer, J. 1967. Interrelationships between fatty acid biosynthesis and acyl-lipid synthesis in *Chlorella vulgaris*. *Biochem. J.* 104:486–96

129. Nikolau, B. J., Hawke, J. C. 1984. Purification and characterisation of maize leaf acetyl-coenzyme A carboxylase. *Arch. Biochem. Biophys.* 228:86–96

130. Nikolau, B. J., Hawke, J. C., Slack, C. R. 1981. Acetyl-CoA carboxylase in maize leaves. *Arch. Biochem. Biophys.* 211:605–12

131. Nikolau, B. J., Wurtels, E. S., Stumpf, P. K. 1984. Tissue distribution of acetyl-CoA carboxylase in leaves. *Plant Physiol.* 75:895–901

132. Nishida, I., Kawaguchi, A., Yamada, M. 1984. Selective inhibition of type II fatty acid synthetase by the antibiotic thiolactomycin. *Plant Cell Physiol.* 25:265–68

133. Norman, H. A., St. John, J. B. 1986. Metabolism of unsaturated monogalactosyldiacylglycerol molecular species in *Arabidopsis thaliana* reveals different sites and substrates for linolenic acid synthesis. *Plant Physiol.* 81:731–36

134. Norman, H. A., Smith, L. A., Lynch, D. V., Thompson, G. A. 1985. Low temperature-induced changes in intracellular fatty acid fluxes in *Dunaliella salina*. *Arch. Biochem. Biophys.* 242:157–67

135. Ohlrogge, J. B. 1987. Biochemistry of plant acyl carrier proteins. See Ref. 117, pp. 137–57

136. Ohlrogge, J. B., Kuhn, D. N., Stumpf, P. K. 1979. Subcellular localisation of acyl carrier protein in leaf protoplasts of

Spinacia oleracea. Proc. Natl. Acad. Sci. USA 76:1194–98

137. Ohlrogge, J. B., Kuo, T. M. 1984. Spinach acyl carrier protein: primary structure, mRNA translation and immunoelectrophoretic analysis. In *Structure, Function and Metabolism of Plant Lipids*, ed. P.-A. Seigenthaler, W. Eichenberger, pp. 63–66. Amsterdam: Elsevier

138. Ohlrogge, J. B., Kuo, T. M. 1985. Plants have isoforms of acyl carrier protein that are expressed differently in different tissues. *J. Biol. Chem.* 260:8032–37

138a. Ohlrogge, J. B., Shine, W. E., Stumpf, P. K. 1978. Purification and characterisation of plant acyl-ACP and acyl-CoA hydrolases. *Arch. Biochem. Biophys.* 189:382–91

139. Ohnishi, J.-I., Yamada, M. 1980. Glycerolipid synthesis in *Avena* leaves during greening of etiolated seedlings. II. α-Linolenic acid synthesis. *Plant Cell Physiol.* 21:1607–18

140. Ohnishi, J.-I., Yamada, M. 1982. Glycerolipid synthesis in *Avena* leaves during greening of etiolated seedlings. III. Synthesis of α-linolenoyl-monogalactosyldiacylglycerol from liposomal linoleoyl-phosphatidylcholine by *Avena* plastids in the presence of phosphatidylcholine-exchange protein. *Plant Cell Physiol.* 23:767–73

141. Ohnishi, J.-I., Yamada, M. 1983. Glycerolipid synthesis in *Avena* leaves during greening of etiolated seedlings. IV. Effect of light on fatty acid desaturation. *Plant Cell Physiol.* 24:1553–57

142. Deleted in proof

143. Oo, K. C., Stumpf, P. K. 1979. Fatty acid synthesis in the developing endosperm of *Cocus nucifera. Lipids* 14:132–43

144. Osumi, T., Hashimoto, T., Ui, N. 1980. Purification and properties of acyl-CoA oxidase from rat liver. *J. Biochem. (Tokyo)* 87:1735–46

145. Overath, P., Stumpf, P. K. 1964. Properties of a soluble fatty acid synthetase from avocado mesocarp. *J. Biol. Chem.* 239:4103–4220

146. Pham Thi, A. T., Borrel-Flood, C., de Silva, J. V., Justin, A. M., Mazliak, P. 1987. Effects of drought on [1-^{14}C]-oleic and [1-^{14}C]-linoleic acid desaturation in cotton leaves. *Physiol. Plant.* 69:147–50

147. Pollard, M. R., McKeon, T. M., Gupta, L. M., Stumpf, P. K. 1979. Studies on biosynthesis of waxes by developing jojoba seed. II. Demonstration of wax biosynthesis by cell-free homogenates. *Lipids* 14:651–62

148. Pollard, M. R., Singh, S. S. 1987. Fatty acid synthesis in developing oil seeds. See Ref. 19, pp. 455–63

149. Pollard, M. R., Stumpf, P. K. 1980. Long chain (C$_{20}$ and C$_{22}$) fatty acid biosynthesis in developing seeds of *Tropaeolum majus. Plant Physiol.* 66:641–48

150. Pollard, M. R., Stumpf, P. K. 1980. Biosynthesis of C$_{20}$ and C$_{22}$ fatty acids by developing seeds of *Limnanthes alba. Plant Physiol.* 66:649–55

151. Pugh, E. L., Kates, M. 1973. Desaturation of phosphatidylcholine and phosphatidylethanolamine by a microsomal enzyme system from *Candida lipolytica. Biochem. Biophys. Acta* 316:305–16

152. Pugh, E. L., Kates, M., Szabo, A. G. 1980. Fluorescence polarization studies of rat liver microsomes with altered phospholipid desaturase activities. *Can. J. Biochem.* 58:952–58

153. Rebeille, F., Bligny, R., Douce, R. 1980. Role de l'oxygène et de la température sur la composition en acides gras des cellules isolées d'érable (*Acer pseudoplantanus* L.). *Biochim. Biophys. Acta* 620:1–9

154. Rochester, C. P., Bishop, D. G. 1984. The role of lysophosphatidylcholine in lipid synthesis by developing sunflower (*Helianthus annuus* L.) seed microsomes. *Arch. Biochem. Biophys.* 232:249–58

155. Rock, C. O., Cronan, J. E. 1979. Reevaluation of the solution structure of acyl carrier protein. *J. Biol. Chem.* 254:9778–85

156. Roughan, P. G. 1987. On the control of fatty acid compositions of plant glycerolipids. See Ref. 19, pp. 247–54

157. Roughan, P. G., Mudd, J. B., McManus, T. T., Slack, C. R. 1979. Linoleate and α-linolenate synthesis by isolated spinach chloroplasts. *Biochem. J.* 18:571–74

158. Roughan, P. G., Slack, C. R. 1982. Cellular organisation of glycerolipid metabolism. *Ann. Rev. Plant Physiol.* 33:97–132

159. Roughan, P. G., Slack, C. R. 1984. Glycerolipid synthesis in leaves. *Trends Biochem. Sci.* 9:383–86

160. Sanchez, J., Harwood, J. L. 1981. Products of fatty acid synthesis by a particulate fraction from germinating pea. *Biochem. J.* 199:221–26

161. Sanchez, J., Jordan, B. R., Kay, J., Harwood, J. L. 1982. Lipase-induced alterations of fatty acid synthesis by sub-

cellular fractions from germinating peas. *Biochem. J.* 204:463–70

162. Sanchez, J., Mancha, M. 1980. Separation and analysis of acylthioesters from higher plants. *Phytochemistry* 19:817–20

163. Sanchez, J., Stumpf, P. K. 1984. The effect of the hypolipidemic drugs WY14643 and DH990 and lysophospholipids on the metabolism of oleate in plants. *Arch. Biochem. Biophys.* 228:185–96

164. Sato, N., Seyama, Y., Murata, N. 1986. Lipid-linked desaturation of palmitic acid in monogalactosyldiacylglycerol in the blue-green alga *Anabaena variabilis* studied *in vivo*. *Plant Cell Physiol.* 27:819–35

165. Sauer, A., Heise, K. 1983. On the light dependence of fatty acid synthesis in spinach chloroplasts. *Plant Physiol.* 73:11–15

166. Schuz, R., Ebel, J., Hahlbrock, K. 1982. Partial purification of β-ketoacyl-ACP synthase from a higher plant. *FEBS Lett.* 140:207–9

167. Shimakata, T., Stumpf, P. K. 1982. The procaryotic nature of the fatty acid synthetase of developing *Carthamus tinctorius* L. (safflower) seeds. *Arch. Biochem. Biophys.* 217:144–54

168. Shimakata, T., Stumpf, P. K. 1982. Purification and characterisation of β-ketoacyl-[acyl-carrier-protein]reductase, β-hydroxyacyl-[acyl-carrier-protein]dehydrase and enoyl-[acyl-carrier-protein]reductase from *Spinacia oleracea* leaves. *Arch. Biochem. Biophys.* 218:77–91

169. Shimakata, T., Stumpf, P, K. 1982. Isolation and function of spinach leaf β-ketoacyl-[acyl-carrier-protein]synthetases. *Proc. Natl. Acad. Sci. USA* 79:5808–12

170. Shimakata, T., Stumpf, P. K. 1983. Purification and characterisation of β-ketoacyl-ACP synthetase 1 from *Spinacea oleracea* leaves. *Arch. Biochem. Biophys.* 220:39–45

171. Shimakata, T., Stumpf, P, K. 1983. The purification and function of acetyl coenzyme A : acyl carrier protein transacylase. *J. Biol. Chem.* 258:3592–98

172. Shine, W. E., Mancha, M., Stumpf, P. K. 1976. The function of acyl thioesterases in the metabolism of acyl-CoA's and acyl-acyl carrier proteins. *Arch. Biochem. Biophys.* 172:110–16

173. Siebertz, H. P., Heinz, E. 1977. Labelling experiments on the origin of hexa- and octa-decatrienoic acids in galactolipids of leaves. *Z. Naturforsch.* 32C:193–205

174. Siebertz, H. P., Heinz, E., Joyard, J., Douce, R. 1980. Labelling *in vivo* and *in vitro* of molecular species of lipids from chloroplast envelopes and thylakoids. *Eur. J. Biochem.* 108:177–85

175. Simcox, P. D., Garland, W., De Luca, V., Canvin, D. T., Dennis, D. T. 1979. Respiratory pathways and fat synthesis in the developing castor oil seed. *Can. J. Bot.* 57:1008–14

176. Simoni, R. D., Criddle, R. S., Stumpf, P. K. 1967. Purification and properties of plant and bacterial acyl carrier proteins. *J. Biol. Chem.* 242:573–81

177. Singh, S. S., Nee, T., Pollard, M. R. 1984. Neutral lipid synthesis in developing *Cuphea* seeds. See Ref. 137, pp. 161–65

178. Skriver, L., Thompson, G. A. 1979. Temperature-induced changes in fatty acid unsaturation of *Tetrahymena* membranes do not require induced fatty acid desaturase synthesis. *Biochim. Biophys. Acta* 572:376–81

179. Slabas, A. R., Harding, J., Hellyer, A., Sidebottom, C., Gwynne, H., Kessell, R., Tombs, M. P. 1984. Enzymology of plant fatty acid biosynthesis. *Dev. Plant Biol.* 9:3–10

180. Slabas, A. R., et al. 1987. Oil seed rape acyl carrier protein and gene structure. See Ref. 19, pp. 697–700

181. Slabas, A. R., Hellyer, A. 1985. Rapid purification of a high molecular weight subunit polypeptide form of rape seed acetyl-CoA carboxylase. *Plant Sci.* 39:177–82

182. Slabas, A., Roberts, P., Osmesher, J. 1982. Characterisation of fatty acid synthesis in a cell free system from developing oil seed rape. In *Biochemistry and Metabolism of Plant Lipids*, ed. J. F. G. M. Wintermans, P. J. C. Kuiper, pp. 251–56. Amsterdam: Elsevier

183. Slabas, A. R., Sidebottom, C. M., Hellyer, A., Kessel, R. M. J., Tombs, M. P. 1986. Induction, purification and characterisation of NADH-specific enoyl acyl carrier protein reductase from developing seeds of oil seed rape *(Brassica napus)*. *Biochim. Biophys. Acta* 877:271–80

184. Slack, C. R., Roughan, P. G., Browse, J. R. 1979. Evidence for an oleoyl phosphatidylcholine desaturase in microsomal preparations from cotyledons of safflower *(Carthamus tinctorius)* seed. *Biochem. J.* 179:649–56

185. Sparace, S. A., Mudd, J. B. 1982. Phosphatidylglycerol synthesis in spinach chloroplasts: characterisation of the newly synthesized molecule. *Plant Physiol.* 70:1260–64

186. Stapleton, S. R., Jaworski, J. G. 1984. Characterisation and purification of malonyl-coenzyme A : [acyl-carrier-protein]transacylases from spinach and *Anabaena variabilis*. *Biochim. Biophys. Acta* 794:240–48

187. Stitt, M., ap Rees, T. 1979. Capacities of pea chloroplasts to catalyse the oxidative pentose pathway and glycolysis. *Phytochemistry* 18:1905–11

188. Stobart, A. K., Stymne, S. 1985. The regulation of the fatty acid composition of the triacylglycerols in microsomal preparations frm avocado mesocarp and the developing cotyledons of safflower. *Planta* 163:119–25

189. Stumpf, P. K. 1980. Biosynthesis of saturated and unsaturated fatty acids. See Ref. 39, pp. 177–203

190. Stumpf, P. K. 1981. Plants, fatty acids, compartments. *Trends Biochem. Sci.* 6:173–76

191. Stumpf, P. K. 1984. Fatty acid biosynthesis in higher plants. See Ref. 85, pp. 155–99

192. Stumpf, P. K., Conn, E. E., eds. 1980. *The Biochemistry of Plants,* Vol. 4. New York: Academic

193. Stumpf, P. K., Mudd, J. B., Nes, W. D., eds. 1987. *Metabolism, Structure and Function of Plant Lipids*. New York: Plenum

194. Stymne, S., Appleqvist, L.-A. 1978. The biosynthesis of linoleate from oleoyl-CoA via oleoyl-phosphatidylcholine in microsomes of developing safflower seeds. *Eur. J. Biochem.* 90:223–29

195. Stymne, S., Glad, G. 1981. Acyl exchange between oleoyl-CoA and phosphatidylcholine in microsomes of developing soya bean cotyledons and its role in fatty acid desaturation. *Lipids* 16:298–305

196. Stymne, S., Stobart, A. K. 1986. Biosynthesis of γ-linolenic acid in cotyledons and microsomal preparations of the developing seeds of common borage. (*Borago officinalis*). *Biochem. J.* 240:385–93

197. Stymne, S., Stobart, A. K., Glad, G. 1983. The role of the acyl-CoA pool in the synthesis of polyunsaturated 18-carbon fatty acids and triacylglycerol production in the microsomes of developing safflower seeds. *Biochim. Biophys. Acta* 752:198–208

198. Thompson, G. A., Roughan, P. G., Browse, J. A., Slack, C. R., Gardiner, S. E. 1986. Spinach leaves desaturate exogenous [14C]palmitate to hexadecatrienoate. *Plant Physiol.* 82:357–62

199. Thomson, L. W., Zalik, S. 1981. Acetyl coenzyme A carboxylase activity in developing seedlings and chloroplasts of barley and its virescens mutant. *Plant Physiol.* 67:655–61

200. Deleted in proof

201. Treede, H.-J., Heise, K.-P. 1985. The regulation of acetyl coenzyme A synthesis in chloroplasts. *Z. Naturforsch.* 40C:496–502

202. Trémolières, A., Drapier, D., Dubacq, J. P., Mazliak, P. 1980. Oleoylcoenzyme A metabolization by subcellular fractions from growing pea leaves. *Plant Sci. Lett.* 8:157–269

203. Trémolières, A., Dubacq, J. P., Drapier, D., Muller, M., Mazliak, P. 1980. *In vitro* cooperation between plastids and microsomes in the synthesis of leaf lipids. *FEBS Lett.* 114:135–38

204. Trémolières, A., Mazliak, P. 1974. Biosynthetic pathway of α-linolenic acid in developing pea leaves. *In vivo* and *in vitro* study. *Plant Sci. Lett.* 2:193–201

205. Turnham, E., Northcote, D. N. 1983. Changes in the activity of acetyl-CoA carboxylase during rapeseed formation. *Biochem. J.* 212:223–29

206. Vick, B. A., Zimmerman, D. C. 1987. The lipoxygenase pathway. See Ref. 19, pp. 383–90

207. Vick, B. A., Zimmerman, D. C. 1987. In *Biochemistry of Plants*, ed. P. K. Stumpf, Vol. 9, pp. 54–90. New York: Academic

208. Vigh, L., Joo, F., Cseplo, A. 1985. Modulation of membrane fluidity in living protoplasts of *Nicotiana plumbaginifolia* by catalytic hydrogenation. *Eur. J. Biochem.* 146:241–44

209. von Wettstein-Knowles, P. 1979. Genetics and biosynthesis of plant epicuticular waxes. See Ref. 81, pp. 1–26

210. Walker, K. A., Harwood, J. L. 1985. Localisation of chloroplastic fatty acid synthesis *de novo* in the stroma. *Biochem. J.* 226:551–56

211. Walker, K. A., Harwood, J. L. 1986. Evidence for separate elongation enzymes for very long chain fatty acid synthesis in potato (*Solanum tuberosum*). *Biochem. J.* 237:41–46

212. Wang, X., Hildebrand, D. F., Collins, G. B. 1987. Identification of proteins associated with changes in the linolenate content of soybean cotyledons. See Ref. 19, pp. 333–35

213. Weaire, P. J., Kekwick, R. G. O. 1975. The synthesis of fatty acids in avocado mesocarp and cauliflower bud tissue. *Biochem. J.* 146:425–37

214. Wharfe, J., Harwood, J. L. 1978. Fatty acid biosynthesis in the leaves of bar-

ley, wheat and pea. *Biochem. J.* 174: 163–69

215. Willemot, C., Stumpf, P. K. 1967. Development of fatty acid synthesis as a function of protein synthesis in aging potato tuber slices. *Plant Physiol.* 42:391–97

216. Willemot, C., Slack, C. R., Browse, J., Roughan, P. G. 1982. Effect of BASF 13-338, a substituted pyridazinone, on lipid metabolism in leaf tissue of spinach, pea, linseed and wheat. *Plant Physiol.* 70:78–81

217. Williams, J. P., Mitchell, K., Khan, M. 1987. The effect of temperature on desaturation of glycerolipid fatty acids in *Brassica napus*. See Ref. 19, pp. 433–36

218. Wolf, A.-M. A., Perchorowicz, J. T. 1987. The purification of acetyl-CoA:ACP transacylase from *Brassica campestris* leaves. See Ref. 19, pp. 499–504

219. Wood, C., Burgess, N., Thomas, D. R. 1986. The dual location of β-oxidation enzymes in germinating pea cotyldons. *Planta* 167:54–57

220. Yamada, M., Watanabe, S., Takishima, K., Mamiya, G. 1987. Complete amino acid sequence of non-specific lipid transfer protein from castor bean seeds. See Ref. 19, pp. 701–3

Ann. Rev. Plant Physiol. Plant Mol. Biol. 1988. 39:139–55

IMMUNOCYTOCHEMICAL LOCALIZATION OF MACROMOLECULES WITH THE ELECTRON MICROSCOPE

Eliot M. Herman

Plant Molecular Genetics Laboratory, United States Department of Agriculture, Agricultural Research Service, Beltsville, Maryland 20705

CONTENTS

INTRODUCTION 139
GENERAL PRINCIPLES 140
IMMUNOCYTOCHEMICAL METHODS 140
   Fixation and the Retention of Antigenicity 141
   Electron Microscopic Visualization of Bound Antibodies 141
   Pre-embedding Immunocytochemistry 142
   Post-embedding Immunocytochemistry 143
IMMUNOCYTOCHEMICAL OBSERVATIONS, A BRIEF REVIEW 144
   Golgi Apparatus–Mediated Protein Transport 144
   Membranes 147
   Cytoplasmic Proteins 149
   Microbodies 150
   Cell Walls 150
   Heterologous Expressed Proteins 151
CONCLUDING REMARKS 151

"Seeing is believing"

INTRODUCTION

Transmission electron microscopy (EM) resolves intracellular structures of 1–3 nm in thin sections and has thereby revolutionized the understanding of cell architecture. With the development of biological electron microscopy

0066-4294/88/0601-0139$02.00

almost three decades ago, it was recognized that the inherent high resolution is a great asset but that there are also great limitations. For example, conventional EM conveys little compositional information. Early attempts to determine composition involved selective staining with heavy metals or histochemical assays of a few select hydrolases by heavy metal capture of the reaction products. Since the middle 1960s, it has been recognized that immunocytochemical techniques could identify molecules within thin sections. Such procedures would constitute a universal assay technique capable of detecting and localizing macromolecules in any cell type. Recent technical innovations have brought immunocytochemistry to the point of routine laboratory application. Although there have been several recent reviews and monographs on immunocytochemistry (53 and articles within; 64), they have little emphasized applications pertinent to plant research. This review is limited in scope to those subjects that primarily concern plant science investigators. Whenever possible, relevant literature from the plant sciences is cited in preference to papers from other fields.

GENERAL PRINCIPLES

Immunocytochemical localization of an antigen requires several basic steps. The tissue to be assayed must be chemically fixed to immobilize the antigen and preclude its redistribution during subsequent procedures. This fixation of the antigen comes at a price, since many antigenic sites can be destroyed by fixation. As the fixation becomes more stringent, and therefore superior, the loss of antigenicity becomes more severe. The tissue and cells must be made permeable to antibody and electron dense indicator solutions, while at the same time limiting mechanical damage to the cells. After the tissue is prepared for assay it must be incubated with a specific antibody for the antigen of interest. Any contaminating antigens that were present in the immunizations may lead to specific and yet false localizations. The assay must be sensitive enough to mark the antigen of interest without labeling other intracellular sites that do not contain the antigen. The assay must have the resolution necessary to define the location of the antigen. Finally, all of these variables must come together in a single experiment to yield results that have not only scientific merit, but visual impact as well, since microscopic information is presented in a manner that is often as artistic as it is scientific.

IMMUNOCYTOCHEMICAL METHODS

Immunocytochemical experiments may be accomplished with various procedures for pre- or post-embedding labeling. Each method offers distinct advantages and disadvantages with regard to specificity, density of antigen labeling, and structural preservation. These are briefly considered in the following paragraphs and in more detail in cited papers, reviews, and monographs.

Fixation and the Retention of Antigenicity

All immunocytochemical experiments begin with tissue fixation, which serves the dual purpose of preserving the cellular structure and the in vivo distribution of antigens. Nonetheless, immunocytochemical studies depart from all other immunologically based assays in that the antigens are chemically modified by fixation and in many procedures further denatured by dehydration and embedment. Many immunocytochemical projects utilize antisera that are generated against native antigens and that therefore may not recognize fixed antigens. Antigenic sites are often eliminated by stringent fixations (13); however, thorough fixation with extensive crosslinking of cellular constituents is necessary to preserve fine ultrastructural detail. Formaldehyde fixation preserves most antigenic sites (13, 44), but it is reversible (76) and it does not maintain good ultrastructure. Bivalent glutaraldehyde fixation is often used as a compromise between antigenic and structural preservation, maintaining about one half as many antigenic sites as formaldehyde fixation (13, 44). Osmium postfixation is essential to preserve membrane structure and ultrastructural detail; unfortunately, osmium often irreversibly destroys antigenic sites (6).

The inability of the antibodies to recognize the fixed antigen is the most frequent cause of failure to obtain labeling. Labeling may be improved by prefixing antigens prior to immunization by either self-polymerization (39, for example) or by coupling to immunogens (30, for example), thereby selecting antibody populations that will recognize the fixed antigen. The process of eliciting polyclonal and monoclonal antibodies has been extensively reviewed and is not discussed here. However, the importance of establishing the specificity of the antisera cannot be overemphasized. A western blot assay from native- or SDS-PAGE should be the minimum determinant of specificity. One useful test is to fix a replicate blot in the prospective fixative to assay for specificity and for the preservation of antigenicity (40). Nonetheless, immunocytochemical experiments are the only way to ascertain the utility of an antiserum preparation.

Electron Microscopic Visualization of Bound Antibodies

To localize antigens by electron microscopy, it is necessary to impart electron density to the bound antibodies. This is generally accomplished by incubating tissue with an unlabeled primary antibody directed at the antigen of interest and then indirectly localizing the primary antibody with a second label consisting of antibodies or protein A conjugated to an electron-dense material.

Second antibodies coupled to ferritin (67) or peroxidase (72) were formerly the methods of choice (see 53 for reviews). These methods have now been largely displaced by the introduction of colloidal gold labels.

Colloidal gold probes have been extensively adopted for use in post-embedding assays (64 for review) but have also been used in thick- and ultrathin-section pre-embedding assays (24, 25, 46, 56, 57, for examples). Colloidal gold particles of 3–40 nm are formed by the reduction of $HAuCl_4$ (see 53, 64, 69 for reviews; 21, 36, 70, 78 for some preferred methods). Second antibodies (19, 59) or protein A (60, 61) are coupled to colloidal gold particles by electrostatic interaction. Monodisperse and specifically sized colloidal gold particles may be isolated by density gradient centrifugation (68). Double-labeling experiments may be done with two distinctly different sizes of probes (5). Higher labeling densities are usually obtained with smaller gold probes (3–7 nm), since they probably have lower steric hindrance. The advantages of colloidal gold probes are high electron density and visibility with moderate magnification. Unlike ferritin probes (52), the colloidal gold probes have little tendency for nonspecific or pseudospecific binding. The resolution of colloidal gold–based assays is approximately equal to the sum of all the accumulated errors resulting from the physical size of each component of the multiple-step procedure. For example, if a 10-nm gold particle is coupled to a second antibody the total accumulated error is 2×15 nm for the rod-shaped antibodies $+ 10$ nm for the gold particle, giving a maximum error of 40 nm. The significance of the error is dependent on the size of the object and resolution necessary to define the localization.

Pre-embedding Immunocytochemistry

Immunocytochemical localization by thick-section, pre-embedding labeling has been discussed in detail (57). Fixed, cryoprotected tissue is frozen and cryosectioned to a thickness of several micrometers. It is then thawed, and labeled with primary and indirect electron-dense labels which enter the tissue by diffusion. The labeled tissue is then embedded, sectioned, and examined. The primary advantage of this procedure is the excellent antigen retention as the consequence of few prelabeling and processing steps. The disadvantages result from the poor penetration of both primary antibodies and secondary labels into the tissue, limiting the label to a gradient in the superficial few micrometers (see 56, 57 for example). Weakly fixed cellular components may be extracted or redistributed during labeling and intermediate washes. Cellular structures may be damaged by both freezing and thawing (note organelle structure in 46, 56, 57).

A variant of the pre-embedding procedure is to freeze fixed cryoprotected tissues, but then to prepare thin cryosections suitable for electron microscopy (76; see 75 for review). The thawed sections are thin enough that the primary and electron-dense labels penetrate and label the section, which is then stained, dried, and examined. Cryosections offer excellent antigen retention and label density (see 4, for example), yet this technique is not often used because of other disadvantages. Ultrathin cryosectioning can be a difficult art to master and requires costly instrumentation. Plant tissues, especially those rich in reserve oil, can be especially difficult to section (see 26 for comment). The dried sections exhibit very low electron contrast, and subcellular structures may be distorted as a consequence of air drying (see 4, 82, 84 for examples). A procedure has been developed to treat with osmium and embed labeled sections in an ultrathin layer of plastic (42), which corrects the problems of low contrast and structural preservation. Membranes are exceptionally well preserved by this procedure (see 24, 25 for example).

Post-embedding Immunocytochemistry

Advances in post-embedding labeling of plastic sections have made this the technique of choice for most immunocytochemical studies (see 53, 64 for reviews). While other procedures may be used successfully for immunocytochemistry, only post-embedding labeling offers a simple and reproducible routine assay method. In these procedures tissues are fixed, dehydrated, and embedded in plastic using protocols similar to those of conventional EM. Thin plastic sections are labeled by immersion in solutions of primary antibodies followed by electron-dense second label. The main advantages are that the skills and methods are similar to those employed in conventional EM, and no specialized equipment beyond that found in any EM laboratory is required.

The choice of embedding resin determines the processing protocols and the consequent compromises between structural preservation and retention of antigenicity. Embedding resins are generally grouped into two catagories, the hydrophobic epoxy resins known by trivial names such as Epon, Araldite, and Spurrs; and the hydrophilic acrylic resins known as LR White and Lowicryl. The structural preservation of aldehyde- and osmium-fixed tissue embedded in epoxy resins is unsurpassed by other resin choices. Plastic post-embedding label techniques were originally developed with epoxy embedding (61), but these are no longer the first choice for immunocytochemistry owing to the disadvantages of low retention of antigenicity (13), higher background labeling (62), and the emergence of acrylic resins designed for immunocytochemistry.

The introduction of Lowicryl revolutionized post-embedding labeling. Lowicryl is a hydrophilic acrylic resin that tolerates partial dehydration and is processed and photopolymerized at subfreezing temperatures (9, 62). The hydrophilic properties of Lowicryl result in excellent antigen retention and consequently in high label density, specificity, and low background. Plant tissues have been reported to be difficult to embed in Lowicryl (13, 31), possibly as the result of pigment interference with photopolymerization and poor infiltration into cell walls (31) and starch grains (E. M. Herman, unpublished observations). Osmium postfixation is incompatible with Lowicryl embedding, and as a consequence membranes are poorly preserved. The protein body membrane (31, 33), plasmalemma (17, 20, 31, 66, 74), plastid envelope (66, 74), and Golgi apparatus cisternae (8, 27, 31, 33) are in particular poorly preserved in published micrographs.

Many of the problems associated with use of Lowicryl have been corrected with the introduction of LR White. LR White is a commercially available aromatic acrylic resin. Like Lowicryl it tolerates partial dehydration, but unlike Lowicryl it may be used with osmium postfixation and is polymerized by heat cure (48). LR White is reported to be less extractive and easier to infiltrate that Lowicryl, with consequent superior structural preservation (30, for example). However, immunolabeling density of aldehyde-fixed tissues has been reported to be reduced in comparison to parallel Lowicryl-based experiments (30, for example). Osmium-treated plant tissue embedded in LR white has excellent structural preservation (79 for example), but immunolabeling densities are often low (78). Removal of osmium from sections with periodate/HCl may unmask some antigenic sites (14, 16, 78, 79), but labeling densities may be quite low when compared to tissue fixed solely in aldehydes (see 78 for quantitative comparison).

IMMUNOCYTOCHEMICAL OBSERVATIONS, A BRIEF REVIEW

The remainder of this review is devoted to a brief description of some immunocytochemical observations. Much of the emphasis is on the more recent results, with the consequence that some of the pioneering post-embedding localization studies, notably on protein body composition (12, 37, 38, 39, 45, 49, 80), are not considered in detail. The reader is directed to cited reviews and papers for discussions of the rationale for each of the localization studies.

Golgi Apparatus–Mediated Protein Transport

The role of the Golgi apparatus in mediating the formation of intracellular organelles and extracellular secretion has been recently reviewed in this series (28). Immunocytochemical observations are particularly useful in visualizing molecules fixed during transit through the Golgi apparatus or secretion vesicles.

The Golgi apparatus was first proposed to mediate the deposition of legume storage proteins based on interpretation of conventional thin-section micrographs (18). An early attempt to apply electron microscope immunocyto-chemical procedures used immunoferritin-labeled cryosections of developing bean cotyledons to localize phaseolin (4). The endoplasmic reticulum and protein bodies were specifically labeled. However, the Golgi apparatus was not recognized as a labeled structure owing to the technical limitations of low electron contrast and drying artifacts. In subsequent studies, aldehyde-fixed tissue embedded in epoxy and Lowicryl resins was used to localize lectins and storage proteins (31, 32, 50) in developing cotyledons. High density and specific labeling of the Golgi apparatus was observed, but the cellular mem-branes including those of the Golgi apparatus were not well preserved. Further sensitivity of the Lowicryl procedure was demonstrated by localizing a minor protein, α-galactosidase, in the Golgi apparatus of midmaturation soybean cotyledons (33). An example of the labeling of the Golgi apparatus with α-galactosidase antiserum is shown in Figure 1.

Post-embedding immunogold labeling of osmium-treated, epoxy-embed-ded cotyledons was used to localize pea vicilin and legumin in the Golgi apparatus and ER (14). This procedure results in structural preservation comparable to that of conventional electron microscopy. Although antigenic sites were unmasked by removal of the osmium from sections with periodate (6), the density of label on the ER and Golgi apparatus was low, demonstrat-ing the limited sensitivity of this procedure. The removal of the osmium from the sections results in the loss of electron contrast, although intracellular structures are well preserved. A dramatic improvement in the combination of structural preservation and label density was obtained by on-grid osmium treatment and plastic embedding of immunolabeled cryosections (24). Phaseolin and phytohemagglutinin were localized in midmaturation bean cotyledon Golgi apparatus and ER with label density comparable to Lowicryl observations (25). Membranes were well preserved, but the matrix of the protein bodies as well as cytoplasmic and ER bound ribosomes exhibited low contrast.

Monensin inhibits the flow of material through the Golgi apparatus (see 73 for review). Immunogold labeling of osmicated tissue in epoxy resin sections has been used to show that monensin treatment redirects vicilin and legumin (15) and concanavalin A (7) from deposition in the protein bodies to ex-tracellular secretion into the periplasm. The role of the Golgi apparatus in the deposition of legume protein body proteins has been recently reviewed (10, 34).

Monocot seeds contain water-soluble glutelin and alcohol-soluble pro-lamine storage proteins. Observations on osmium-postfixed, epoxy-embedded developing rice embryos assayed after periodate treatment demon-strated that the two major classes of storage proteins are sequestered into distinct classes of protein bodies (43). In a process analogous to legume

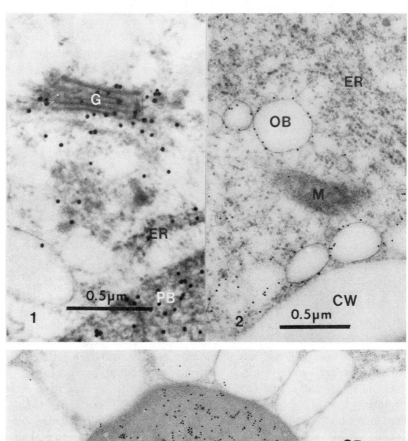

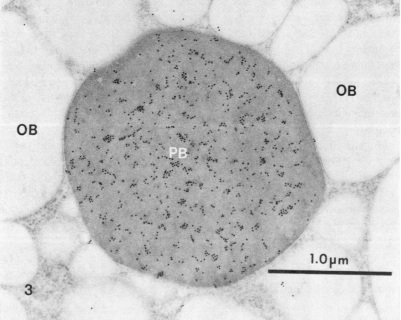

protein body formation, glutelin deposition into vacuolar protein bodies was shown to be mediated by the Golgi apparatus. In contrast, the deposition of prolamine was shown to result by direct fission of protein bodies from the ER (43).

The role of the Golgi apparatus in extracellular protein secretion has been studied by post-embedding immunocytochemistry of α-amylase secreted by barley aleurone layers (27). The Golgi apparatus in Lowicryl sections was observed to be specifically labeled with a low density of gold particles.

Membranes

Immunocytochemistry can provide information about the lateral differentiation of membranes into specialized domains. Preparative techniques usually have adverse consequences on the preservation of membrane structure. The use of detergents and freezing in pre-embedding labeling techniques and the use of organic solvents in post-embedding labeling techniques often result in the loss or degradation of membrane structure.

The thylakoid membranes of chloroplasts consist of stacks of appressed grana thylakoids with interconnecting stroma thylakoids. Biochemical and freeze-fracture studies have suggested that the photosynthetic system proteins are distributed heterogeneously among the grana and stroma thylakoids (see 3 for review).

Lowicryl embedment and immunogold or immunoferritin post-embedding labeling have been used to localize several plastid membrane proteins. Cytochrome b/f and chlorophyll a/b proteins were shown to be homogeneously distributed throughout the grana and stroma thylakoids (2, 17, 66). In contrast, the coupling factor ATPase was shown to be heterogeneously dis-

←―――――――――――――――――――――――――――――――――――――――

Examples of post-embedding immunocytochemical observations are shown in Figures 1–3.

Figure 1:   Shown is the Golgi apparatus, ER, and protein body (PB) of a Lowicryl-embedded maturing soybean cotyledon labeled with anti-α-galactosidase serum and colloidal gold-protein-A. Note although the Golgi apparatus (G) is readily identified, the structural preservation of the cisterna membranes is poor. Observations of this type demonstrate the role of the Golgi apparatus in mediating the transport of specific macromolecules. Reprinted from (33) with permission.

Figure 2:   Shown is the localization of a 24-kDa oil body (OB) membrane protein in a cell of a maturing soybean cotyledon. The indirect protein-A-colloidal gold labeling of the LR White section specifically labeled the oil body membrane while label is absent over the cell wall (CW), ER, and mitochondria (M). Reprinted from (30) with permission.

Figure 3:   Localization of maize zein in the protein body (PB) of a transgenic tobacco embryo. Note that the colloidal gold label is specifically localized on the crystalloid in the interior of the protein body and is absent on the oil bodies (OB) and the matrix of the periphery of the protein body. This observation demonstrates the localization of a heterologously expressed protein. Reprinted from (35) with permission.

tributed along the marginal edges of grana and the entire surface of stroma thylakoids (2, 66). Other Lowicryl studies have shown that NADH-protochlorophyllide oxidoreductase is found in the prolamellar body of etiolated plants but is absent in the thylakoids of mature plastids (17). The distribution of photosystem 1 P700 chlorophyll *a* protein has been probed in epoxy sections by the peroxidase-immunoenzyme technique (81), which labeled the entire grana and stroma thylakoids. The proteins of the photosynthetic oxygen evolving system (23) and cytochrome *f* (22) have been localized in LR White–embedded leaves, but the plastid structural preservation and the label density were poor. Comparison of published micrographs of epoxy (81), LR White (22, 23), and Lowicryl (2, 17, 66) embedment of aldehyde-fixed tissues indicates that Lowicryl is the most effective medium for plastid structural preservation and retention of antigenicity. Lowicryl-embedded plastid thylakoids are easily visible as negatively stained membranes (2, 17, 66, 74), although the plastid fine structure and the envelope are difficult to discern or are lost. Although cytochrome *b/f*, coupling factor, and chlorophyll *a/b* proteins are abundant and highly concentrated in the thylakoid membrane, the density of the observed label was low. Only the chlorophyll *a/b* protein was labeled in every grana stack (17, 66), although even this abundant protein was not labeled in each thylakoid. No more than half of the grana stacks or stroma thylakoids were labeled by antibodies to coupling factor (2, 66) and cytochrome *b/f* (2, 22, 66). These observations illustrate that even very abundant membrane proteins assayed under conditions that compromise structural preservation in favor of retention of antigenicity still may not result in a high density of label.

Lowicryl post-embedding assays have been used to localize membrane proteins of legume root nodule cells (83 for review). The structural preservation of aldehyde-fixed root nodule cells embedded in Lowicryl is poor, although the peribacteroid membrane and *Rhizobium* cells are preserved. In one study the cDNA sequence data of nodulin 24, a major nodule protein, was used to derive a synthetic peptide antigen, which was used to produce specific antisera. The antisera were then used to localize nodulin 24 in the peribacteroid membrane (20). These results demonstrate that sequence data may be used to produce synthetic antigens that can elicit antiserum useful for localization studies. In another study a monoclonal antibody against a membrane protein glycan side chain was used to localize proteins in the peribacteroid membranes, plasma membrane, and Golgi apparatus of infected and uninfected cells (8). These results illustrate how antibodies against a glycan epitope common to several proteins may label several different organelles.

Lowicryl- and LR White–embedded midmaturation soybean cotyledons have been used to localize a 24-kDa membrane protein (mP24) of the reserve oil bodies (30). Direct comparison of the two resins indicated that LR White better preserved fine structure while Lowicryl sections were labeled with a higher density of gold particles. Figure 2 shows the soybean oil body membrane as an example of membrane labeling.

Cytoplasmic Proteins

The localization of cytoplasmic antigens can be a particularly challenging objective. The cytoplasmic protein may be spatially segregated within a cytoplasmic domain such as the "microtrabecular lattice" (see 54 for review). The fixation of cytoplasmic proteins may be hindered by the lack of a boundary membrane, which may permit lightly fixed proteins to redistribute during processing. Many preparative procedures appear to be highly extractive, which may redistribute a more discretely distributed antigen. Even with these cautions, electron microscope immunocytochemistry is the only practical technique to study the intracellular distribution of cytoplasmic proteins.

The red (P_r) to far-red (P_{fr}) conversion of phytochrome is an important developmental regulator in plants (see 55 for review). Post-embedding immunocytochemical observations demonstrated a diffuse distribution of phytochrome (P_r) throughout the cytoplasm. On illumination and conversion to P_{fr} the phytochrome is redistributed and sequestered into cytoplasmic electron-dense aggregations (46, 71). Parallel studies with pre-embedding labeling of thick cryosections (46) and post-embedding labeling of cryofixed, freeze-substituted tissue embedded in LR White (46) confirmed the results of the post-embedding labeling experiments. The observed differential distribution of P_r and P_{fr} demonstrates that fixation followed by either post-embedding or pre-embedding labeling can be used to localize cytoplasmic proteins sequestered into specific domains.

Leghemaglobin of infected legume root nodule cells has been localized in the cytoplasm by post-embedding immunogold labeling of Lowicryl sections (58). Although the tissue preservation exhibited the extracted appearance characteristic of Lowicryl embedding, the leghemaglobin was observed to be diffusely distributed throughout the cytoplasm and to be excluded from other subcellular organelles, bacteroids, and cell wall.

An albumin storage protein was localized by immunogold labeling of LR White–embedded pea seeds (29). Although osmium fixation was not used, preservation of the plasma membrane, mitochondria, and endoplasmic reticulum was enhanced by use of a low temperature chemical catalyst for polymerization of the LR White. The cytoplasm was observed to be densely labeled with gold particles with no apparent label over any organelle or cell wall.

Microbodies

The germination of many oil-storing seeds is accompanied by the synthesis of glyoxysomes utilized in the catabolism of the reserve oil bodies (see 41 for review). Following the completion of oil mobilization the glyoxysomes gradually disappear, to be replaced by peroxisomes. Whether the peroxisomes are a *de novo* synthesized set of organelles or are directly derived from the preexisting glyoxysomes has been debated for many years. Lowicryl embedding and post-embedding immunogold labeling have been used to address directly the question of the glyoxysome-peroxisome transition. Double labeling with the glyoxysomal marker isocitrate lyase and the peroxisomal markers serine:glyoxylate aminotransferase and hydroxypyruvate reductase was used to demonstrate that the glyoxysomal and peroxisomal enzymes are found within the same microbodies (65, 74), substantiating the one-population proposal (77). These studies demonstrate the utility of immunocytochemical techniques to discriminate between competing proposals of organelle ontogeny.

A specific uricase (Uricase II, nodulin 35) is found in nitrogen-fixing legume root nodules. Lowicryl post-embedding immunocytochemistry has been used to localize Uricase II. Although nodule cells are highly extracted by Lowicryl embedding, sufficient structural preservation was obtained to localize Uricase II in the peroxisomes of uninfected cells (51). Subsequently, substantial improvements in structural preservation were obtained by embedding aldehyde- and osmium-fixed nodule tissue in LR White and Spurrs epoxy resin (78, 79). An immunogold labeling assay of uricase with the same antiserum preparation used for the earlier Lowicryl study substantiated the localization in the peroxisomes while better preserving the ultrastructural fine detail of the nodule cells (78, 79).

Cell Walls

The cell wall of plants is a complex layered structure that exhibits a wide variation of cell- and tissue-specific composition (see 1 for review). Immunocytochemistry should be an effective means to probe cell-wall structure as its protein and glycan constituents should be resistant to processing-induced extraction or redistribution. Post-embedding immunogold procedures with Lowicryl-embedded cultured sycamore cell walls have been used to localize cell-wall xyloglucans and rhamnogalacturonans (47). The xyloglucans were observed to be distributed throughout the cell wall while rhamnogalacturonan was shown to be restricted to the middle lamella of the wall. Pollen-tube growth has been studied with a monoclonal antibody against α-L-arabinosyl residues of style glycoproteins. Immunogold post-embedding labeling has demonstrated that the antigen is restricted to the outer layer of the pollen-tube wall (11).

Heterologous Expressed Proteins

The transfer and heterologous expression of plant genes comprise an exciting and active field of biotechnology. Immunocytochemical localization can provide critical information on the site of protein accumulation and can be an important adjunct to heterologous protein expression studies. The difficulty in assaying heterologously expressed proteins results from low levels of expression combined with uncertainly of the site of protein accumulation. Post-embedding immunocytochemistry of osmium-treated (26) and untreated (35) transgenic tobacco embryos embedded in LR White has been used to localize the heterologous expression and accumulation of phaseolin (26) and zein (35) into protein bodies. The localization of zein expressed in transgenic tobacco is shown in Figure 3.

CONCLUDING REMARKS

The currently available immunocytochemical techniques permit the routine and reproducible localization of most moderately abundant antigens. Unfortunately, present immunocytochemical techniques often require compromises between structural and antigenic preservation. The existing stocks of specific antisera in the freezers of plant scientists are an important cytochemical resource. Recent conferences and meetings have included increasing numbers of presentations utilizing electron microscope immunocytochemical procedures. The next few years should see many productive studies on the flow and interaction of membranes, architecture of the cell wall, and targeting of heterologously expressed proteins.

ACKNOWLEDGMENTS

Research in my laboratory is supported by in-house USDA CRIS funds and the USDA Office of Competitive Grants.

Literature Cited

1. Albersheim, P., Darvill, A. G., Davis, K. R., Lau, J. M., McNeil, M., Sharp, J. K., York, W. S. 1984. Why study the structures of biological molecules? In *Structure, Function, and Biosynthesis of Plant Cell Walls*, ed. W. M. Dugger, S. Bartnicki-Garcia, pp. 19–51. Baltimore: Waverley Press. 507 pp.
2. Allred, D. R., Staehelin, L. A. 1985. Lateral distribution of the cytochrome b_6/f and coupling factor ATP synthetase complexes of chloroplast thylakoid membranes. *Plant Physiol.* 88:199–202
3. Anderson, J. M., Andersson, B. 1982. The architecture of photosynthetic membranes: lateral and transverse organization. *Trends Biochem Sci.* 7:288–92
4. Baumgartner, B., Tokuyasu, K. T., Chrispeels, M. J. 1980. Immunocytochemical localization of reserve protein in the endoplasmic reticulum of developing bean *(Phaseolus vulgaris)* cotyledons. *Planta* 150:419–25
5. Bendayan, M. 1982. Double immunocytochemical labeling applying the protein A-gold technique. *J. Histochem. Cytochem.* 30:81–85
6. Bendayan, M., Zollinger, M. 1983. Ultrastructural localization of antigenic sites on osmium fixed tissues applying

152   HERMAN

the protein A-gold technique. *J. Histochem. Cytochem.* 31:101–9

7. Bowles, D. J., Marcus, S. E., Pappin, D. J. C., Findlay, J. B. C., Maycox, P. R., Burgess, J. 1986. Post-translational processing of concanavalin A precursors in jackbean cotyledons. *J. Cell Biol.* 102:1284–97

8. Brewin, N. J., Robertson, J. G., Wood, E. A., Wells, B., Larkins, A. P., Galfre, G., Butcher, G. W. 1985. Monoclonal antibodies to antigens in the peribacteriod membrane from *Rhizobium*-induced root nodules of pea cross-react with plasma membranes and golgi bodies. *EMBO J.* 4:605–11

9. Carlemalm, E., Garavito, R. M., Villiger, W. 1982. Resin development for electron microscopy and an analysis of embedding at low temperature. *J. Microsc.* 126:132–43

10. Chrispeels, M. J. 1985. The role of the golgi apparatus in the transport and post-translational modification of vacuolar (protein body) proteins. *Oxford Surv. Plant Mol. Cell Biol.* 2:43–68

11. Clarke, A. E., Anderson, M. A., Bacic, T., Harris, P. J., Mau, S. L. 1985. Molecular basis of cell recognition during fertilization in higher plants. *J. Cell Sci. Suppl.* 2:261–85

12. Craig, S., Millerd, A. 1981. Pea seed storage proteins-immunocytochemical localization with protein A-gold by electron microscopy. *Protoplasma* 105:333–39

13. Craig, S., Goodchild, D. J. 1982. Post-embedding immunolabeling. Some effects of tissue preparation on the antigenicity of plant proteins. *Eur. J. Cell Biol.* 28:251–56

14. Craig, S., Goodchild, D. J., 1984. Periodic-acid treatment of sections permits on-grid localization of pea seed vicilin in ER and Golgi. *Protoplasma* 122:35–44

15. Craig, S., Goodchild, D. J. 1984. Golgi-mediated vicilin accumulation in pea cotyledon cells is redirected by monensin and nigericin. *Protoplasma* 122:91–97

16. Craig, S., Miller, C. 1984. L R White resin and improved on-grid immunogold detection of vicilin, a pea seed storage protein. *Cell Biol. Int. Rep.* 8:879–86

17. Dehesh, K., van Cleve, B., Ryberg, M., Apel, K. 1986. Light-induced changes in the distribution of the 36000 M_r polypeptide of NADPH-protochlorophyllide oxidoreductase within different cellular compartments of barley *(Hordeum vulgare L.)*. II. Localization by immunogold labeling in ultrathin sections. *Planta* 169:172–83

18. Dieckert, J. W., Dieckert, M. C. 1976. The chemistry and cell biology of the vacuolar proteins of seeds. *J. Food. Sci.* 41:475–82

19. Faulk, W. P., Taylor, G. M. 1971. An immunocolloid method for the electron microscope. *Immunochemistry* 8:1081–83

20. Fortin, M. G., Zelechowska, M., Verma, D. P. S. 1985. Specific targeting of membrane nodulins to bacterioid-enclosing compartment of soybean nodules. *EMBO J.* 4:3041–46

21. Frens, G. 1973. Controlled nucleation for the regulation of particle size in monodisperse gold suspensions. *Nature Phys. Sci.* 241:20–22

22. Goodchild, D. J., Anderson, J. M., Andersson, B. 1985. Immunocytochemical localization of cytochrome b/f complex of chloroplast thylakoid membranes. *Cell Biol. Int. Rep.* 9:715–21

23. Goodchild, D. J., Andersson, B., Anderson, J. M. 1985. Immunocytochemical localization of polypeptides associated with the oxygen evolving system of photosynthesis. *Eur. J. Cell Biol.* 36:294–98

24. Greenwood, J. S., Keller, G. A., Chrispeels, M. J. 1984. Localization of phytohemagglutinin in the embryonic axis of *Phaseolus vulgaris* with ultra-thin cryosections embedded in plastic after indirect immunolabeling. *Planta* 162:548–55

25. Greenwood, J. S., Chrispeels, M. J. 1985. Immunocytochemical localization of phaseolin and phytohemagglutinin in the endoplasmic reticulum and Golgi complex of developing bean cotyledons. *Planta* 164:295–302

26. Greenwood, J. S., Chrispeels, M. J. 1985. Correct targeting of the bean storage protein phaseolin in the seeds of transformed tobacco. *Plant Physiol.* 79:65–71

27. Gubler, F., Jacobson, J. V., Ashford, A. E. 1986. Involvement of the Golgi apparatus in the secretion of α-amylase from gibberellin-treated barley aleurone cells. *Planta* 168:447–52

28. Harris, N. 1986. Organization of the endomembrane system. *Ann. Rev. Plant Physiol.* 37:73–92

29. Harris, N., Croy, R. R. D. 1985. The major albumin protein from pea *(Pisum sativum L.)*. Localisation by immunocytochemistry. *Planta* 165:522–26

30. Herman, E. M. 1987. The immunogold localization and synthesis of an oil body membrane protein in developing soybean seeds. *Planta.* 172:336–45

31. Herman, E. M., Shannon, L. M. 1984.

Immunocytochemical evidence for the involvement of Golgi apparatus in the deposition of seed lectin of *Bauhinia purpurea* (Leguminosae). *Protoplasma* 121:163–70

32. Herman, E. M., Shannon, L. M. 1984. Immunocytochemical localization of concanavalin A in developing jack bean cotyledons. *Planta* 161:97–104

33. Herman, E. M., Shannon, L. M. 1985. Accumulation and subcellular localization of α-galactosidase in developing soybean cotyledons. *Plant Physiol.* 77: 886–90

34. Herman, E. M., Shannon, L. M., Chrispeels, M. J. 1986. The Golgi apparatus mediates the transport and post-translational modification of protein body proteins. In *Molecular Biology of Seed Storage Proteins and Lectins*, ed. L. M. Shannon, M. J. Chrispeels, pp. 163–73. Baltimore: Waverley Press. 229 pp.

35. Hoffman, L. M., Donaldson, D. D., Bookland, R., Rashka, K., Herman, E. M. 1987. Synthesis and protein body deposition of Maize 15 kd zein in transgenic tobacco seeds. *EMBO J.* 6:3213–21

36. Horisberger, M., Rosset, J. 1977. Colloidal gold a useful marker for transmission and scanning electron microscopy. *J. Histochem. Cytochem.* 25: 295–305

37. Horisberger, M., Vonlanthen, M. 1980. Ultrastructural localization of soybean agglutinin on thin sections of *Glycine max* (Soybean) var. Altona by the gold method. *Histochemistry* 65:181–86

38. Horisberger, M., Vonlanthen, M. T. 1983. Ultrastructural localization of Kunitz inhibitor on thin sections of *Glycine max* (Soybean) cv. Maple Arrow by the gold method. *Histochemistry* 77:37–50

39. Horisberger, M., Vonlanthen, M. T. 1983. Ultrastructural localization of Bowman-Birk inhibitor on thin sections of *Glycine max* (Soybean) cv. Maple Arrow by the gold method. *Histochemistry* 77:313–21

40. Hortsch, M., Griffiths, G., Meyer, D. I. 1985. Restriction of docking protein to the rough endoplasmic reticulum: immunocytochemical localization in rat liver. *Eur. J. Cell Biol.* 38:271–79

41. Huang, A. H. C., Trelease, R. N., Moore, T. S. Jr. 1983. *Plant Peroxisomes*. New York: Academic. 252 pp.

42. Keller, G. A., Tokuyasu, K. T., Dutton, A. H., Singer, S. J. 1984. Osmium staining and ultrathin plastic embedding of mounted and immunolabeled frozen sections. *Proc. Natl. Acad. Sci. USA* 81:5744–47

43. Krishnan, H. B., Franceschi, V. R., Okita, T. W. 1986. Immunochemical studies on the role of the Golgi complex in protein-body formation in rice seeds. *Planta* 169:471–80

44. Leenen, P. J. M., Jansen, A. M. A. C., Ewijk, W. van. 1985. Fixation parameters for immunocytochemistry: The effect of glutaraldehyde or paraformaldehyde fixation on the preservation of mononuclear phagocyte differentiation antigens. In *Techniques in Immunocytochemistry*, ed. G. R. Bullock, J. E. Smith, 3:1–24. London: Academic. 241 pp.

45. Manen, J. F., Pusztai, A. 1982. Immunocytochemical localisation of lectins in cells of *Phaseolus vulgaris* L. seeds. *Planta* 155:328–34

46. McCurdy, D. W., Pratt, L. H. 1986. Immunogold electron microscopy of phytochrome in *Avena*: Identification of intracellular sites responsible for phytochrome sequestering and enhanced pelletability. *J. Cell Biol.* 103:2541–50

47. Moore, P. J., Darvill, A. G., Albersheim, P., Staehelin, L. A. 1986. Immunogold localization of xyloglucan and rhamnogalacturonan I in the cell walls of suspension-cultured sycamore cells *Plant Physiol.* 82:787–94

48. Newman, G. R., Jasani, B., Williams, E. D. 1983. A simple post-embedding system for rapid demonstration of tissue antigens under the electron microscope. *Histochem. J.* 15:543–55

49. Nieden, U.-z., Neumann, D., Manteuffel, R., Weber, E. 1982. Electron microscopic immunocytochemical localization of storage proteins in Vicia faba seeds. *Eur. J. Cell Biol.* 26:328–33

50. Nieden, U.-z., Manteuffel, R., Weber, E., Neumann, D. 1984. Dictyosomes participate in the intracellular pathway of storage proteins in developing Vicia faba cotyledons. *Eur. J. Cell Biol.* 34: 9–17

51. Nguyen, T., Zelechowska, M., Foster, V., Bergmann, H., Verma, D. P. S. 1985. Primary structure of the soybean nodulin-35 gene encoding uricase II localized in the peroxisomes of uninfected cell of nodules. *Proc. Natl. Acad. Sci. USA* 82:5040–44

52. Parr, E. L. 1979. Intracellular labelling with ferritin conjugates. A specificity problem due to the affinity of unconjugated ferritin for selected intracellular sites. *J. Histochem. Cytochem.* 27:1095–1102

53. Polak, J. M., Varndell, I. M. 1984. *Immunolabeling for Electron Microscopy,*

ed. J. M. Polak, I. M. Varndell. Amsterdam/New York/Oxford: Elsevier. 370 pp.

54. Porter, K. R. 1984. The cytomatrix: a short history of its study. *J. Cell Biol.* 99:3–12s

55. Pratt, L. H. 1979. Phytochrome: functions and properties. *Photochem. Photobiol. Rev.* 4:59–124

56. Raikhel, N. V., Mishkind, M. L., Palevitz, B. A. 1984. Characterization of wheat germ agglutinin from adult wheat plants. *Planta* 162:55–61

57. Raikhel, N. V., Mishkind, M., Palevitz, B. A. 1984. Immunocytochemistry in plants with collodial gold conjugates. *Protoplasma* 121:25–33

58. Robertson, J. G., Wells, B., Bisseling, T., Farnden, K. J. F., Johnston, A. W. B. 1984. Immunogold localization of leghaemoglobin in nitrogen-fixing root nodules of pea. *Nature* 311:254–56

59. Romano, E. L., Stolinski, C., Hughes-Jones, N. C. 1974. An antoglobulin reagent labelled with colloidal gold for use in electron microscopy. *Immunochemistry* 11:521–22

60. Romano, E. L., Romano, M. 1977. Staphylococcal protein A bound to colloidal gold: A useful reagent to label antigen-antibody sites for electron microscopy. *Immunochemistry* 14:711–15

61. Roth, J., Bendayan, M., Orci, L. 1978. Ultrastructural localization of intracellular antigens by the use of protein A-gold complex. *J. Histochem. Cytochem.* 26:1074–81

62. Roth, J., Bendayan, M., Carlemalm, E., Villiger, W., Garavito, M. 1981. Enhancement of structural preservation and immunocytochemical staining in low temperature embedded pancreatic tissue. *J. Histochem. Cytochem.* 29:663–71

63. Roth, J. 1982. The preparation of 3 nm and 15 nm gold particles and their use in labeling multiple antigens on ultrathin sections. *Histochem. J.* 14:791–801

64. Roth, J. 1986. Post-embedding with gold-labeled reagents: a review. *J. Microsc.* 143:125–37

65. Sautter, C. 1986. Microbody transition in greening watermelon cotyledons. Double immunocytochemical labeling of isocitrate lyase and hydroxypyruvate reductase. *Planta* 167:491–503

66. Shaw, P. J., Henwood, J. A. 1985. Immuno-gold localization of cytochrome f, light-harvesting complex, ATP synthase, and ribulose 1,5-bisphosphate carboxylase/oxygenase. *Planta* 165:333–39

67. Singer, S. J. 1959. Preparation of an electron dense antibody conjugate. *Nature* 183:1523–24

68. Slot, J. W., Gueze, H. J. 1981. Sizing protein A-colloidal gold probes for immunoelectron microscopy. *J. Cell Biol.* 90:533–36

69. Slot, J. W., Gueze, H. J. 1984. Gold markers for single and double immunolabelling of ultrathin cryosections. See Ref. 53, pp. 129–42

70. Slot, J. W., Gueze, H. J. 1985. A new method of preparing gold probes for multiple-labeling cytochemistry. *Eur. J. Cell Biol.* 38:87–93

71. Speth, V., Otto, V., Schafer, E. 1986. Intracellular localisation of phytochrome in oat coleoptiles by electron microscopy. *Planta* 168:299–304

72. Sternberger, L. A., Hardy, P. H. Jr., Cuculis, J. J., Meyer, H. G. 1970. The unlabeled antibody method of immunohistochemistry. *J. Histochem. Cytochem.* 18:315–33

73. Tartakoff, A. M. 1983. Perturbation of vesicular traffic with the carboxylic ionophore monensin. *Cell* 32:1026–28

74. Titus, D. E., Becker, W. M. 1985. Investigation of the glyoxysome-peroxisome transition in germinating cucumber cotyledons using double-label immunoelectron microscopy. *J. Cell Biol.* 101:1288–99

75. Tokuyasu, K. T. 1986. Application of cryoultramicrotomy to immunocytochemistry. *J. Microsc.* 143:139–49

76. Tokuyasu, K. T., Singer, S. J. 1976. Improved procedures for immunoferritin labeling of ultrathin frozen sections. *J. Cell Biol.* 71:894–906

77. Trelease, R. N., Becker, W. M., Gruber, P. J., Newcomb, E. H. 1971. Microbodies (glyoxysomes and peroxisomes) in cucumber cotyledons. Correlative biochemical and ultrastructural study in light- and dark-grown seedlings. *Plant Physiol.* 48:461–75

78. VandenBosch, K. A. 1986. Light and electron microscopic visualization of uricase by immunogold labeling of sections of resin-embedded soybean nodules. *J. Microsc.* 143:187–97

79. VandenBosch, K. A., Newcomb, E. H. 1986. Immunogold localization of nobule-specific uricase in developing soybean root nodules. *Planta* 167:425–36

80. Van Driessche, E., Smets, G., Dejaegere, R., Kanarek, L. 1981. The immunohistochemical localization of lectin in pea seeds (*Pisum sativum* L.). *Planta* 153:287–96

81. Vaughn, K. C., Vierling, E., Duke, S. O., Alberte, R. S. 1983. Immunocytochemical and cytochemical local-

ization of photosystems I and II. *Plant Physiol.* 73:203–7

82. Verbelen, J. P., Pratt, L. H., Butler, W. L., Tokuyasu, K. 1982. Localization of phytochrome in oats by electron microscopy. *Plant. Physiol.* 70:867–71

83. Verma, D. P. S., Fortin, M. G., Stanley, J., Mauro, V. P., Purohit, S., Morrison, N. 1986. Nodulins and nodulin genes of *Glycine max. Plant Mol. Biol.* 7:51–61

84. Vernooy-Gerritsen, M., Leunissen, J. L. M., Veldink, G. A., Vliegenthart, J. F. G. 1984. Intracellular localization of lipoxygenases-1 and -2 in germinating soybean seeds by indirect labeling with protein A-colloidal gold complexes. *Plant Physiol.* 76:1070–79

Ann. Rev. Plant Physiol. Plant Mol. Biol. 1988. 39:157–74

PHOTOCONTROL OF DEVELOPMENT IN ALGAE

M. J. Dring

Department of Biology, Queen's University, Belfast BT7 1NN, Northern Ireland

CONTENTS

INTRODUCTION... 157
PHOTOPERIODIC CONTROL OF ALGAL DEVELOPMENT............................ 158
   Types of Algae Exhibiting Photoperiodic Responses.. 158
   Types of Development Controlled by Photoperiod.. 159
   Physiological Contrasts Between Photoperiodism in Algae and in
      Higher Plants... 160
NONPERIODIC PHOTOCONTROL OF DEVELOPMENT 161
   Types of Algae Whose Development Is Influenced by Light............................... 161
   Types of Development Under Control by Light.. 161
ACTION SPECTRA FOR ALGAL RESPONSES AND POSSIBLE
      PHOTORECEPTORS 166
   Red/Far-Red Reversible Responses—Phytochrome in Algae............................. 166
   Responses Activated by Photoreversible Pigment Systems Other than
      Phytochrome... 167
   Blue Light (Cryptochrome) Responses... 168
   Responses Activated by Both Blue and Red Light.. 169
   Responses Activated by Green Light... 169
SUMMARY 170

INTRODUCTION

Here the term "development" will be treated with what some may regard as excessive freedom in order to include metabolic development (e.g. greening, enzyme synthesis, etc) as well as the more conventional development of vegetative and reproductive structures. This cavalier use of "development" is necessary since many unicellular algae have little morphology to develop, but their responses to light exhibit many similarities to those of macroscopic algae. In effect, this article covers all effects of light on algae that appear to be

157

independent of the photosynthetic apparatus, with the exception of orientation responses [i.e. phototaxis, phototropism, and light-induced chromatophore movement—see reviews by Haupt (45) and Häder (42)] and responses to light that are so rapid that they cannot involve synthesis of proteins or other cell constituents (e.g. changes in membrane potential, activation of existing enzymes, etc). The full breadth of this topic, covering all groups of algae, has not been tackled since Lang presented a "synopsis" of the physiology of growth and development of algae in 1965 (59). However, marine macroalgae have sometimes been singled out for attention (photomorphogenesis—26, 65; blue light effects—22, 23), and photoperiodism in algae was reviewed in 1970 (20) and again in 1984 (21).

The objective of this review is to summarize the state-of-play of such photobiological studies among the algae as a whole, and to derive generalizations that will encourage photobiologists (or photomorphogeneticists) to make wider use of algae as experimental tools. A wide range of responses among algae are controlled by light, and a wide range of different photoreceptors appear to be involved. In the algae, we seem to be looking at the early stages in the evolution of the phytochrome system, which is so well known from (and more widely seen in) flowering plants. In addition, algae exhibit a wider range of cryptochrome responses than is found in vascular plants, and a number of other pigment systems, which are as yet poorly characterized—largely because they are known from only one or a few species. This range of photomorphogenetic pigment systems could be regarded as analagous to the range of photosynthetic pigment systems also found among algae, although there is little indication so far that particular photomorphogenetic pigments are correlated with specific photosynthetic systems and, hence, with phylogenetic groupings of the algae.

PHOTOPERIODIC CONTROL OF ALGAL DEVELOPMENT

This topic was reviewed in detail in 1984 (21) and, although a few interesting new responses have been reported (10, 18, 38), there has been only one significant contribution to the physiology of photoperiodism in algae since then (9). The earlier review requires little revision, therefore, and the present account aims to illustrate the range of response types and algal groups represented, and to summarize the features of greatest physiological interest. The photoreceptors involved in these responses are discussed in the final section of this review.

Types of Algae Exhibiting Photoperiodic Responses

At the latest count, about 55 photoperiodic responses had been reported in macroscopic species from three algal divisions (Chlorophyta, Phaeophyta,

Rhodophyta). The green algae (65, 84) are less well represented than the brown and red algae, but this is almost certainly because of the scarcity of macroscopic green algae in the temperate marine flora of the northern hemisphere, which results in their receiving less attention from experimental phycologists. Freshwater algae (51) are also poorly represented among photoperiodic species, but this probably reflects the numerical distribution of macroscopic algae between marine and freshwater habitats rather than any fundamental insensitivity of freshwater algae to photoperiod.

What may be more fundamental, however, is the complete absence of unicellular algae from the lists of photoperiodic species. This does not mean that a macroscopic morphology is essential for a photoperiodic response, because several of the known responses occur in the microscopic phases of the life histories of macroscopic plants (e.g. 9, 65, 81). The apparent absence of photoperiodic responses among unicellular algae may have more to do with the planktonic habitat that many of them occupy. Factors such as irradiance, temperature, and nutrient supply are often thought to have greater ecological significance than daylength in controlling the seasonal behavior of planktonic species.

Types of Development Controlled by Photoperiod

The majority of the responses reported involve a change of phase in the life history of a species, although this is not always achieved by the formation of reproductive structures in the conventional sense. In both red and brown algae, the formation of erect thalli from a prostrate system (a crust or branching filaments) is commonly controlled by photoperiod (24, 51, 85), but either gametogenesis (10, 41, 105, 107) or sporogenesis (1, 9, 18, 27, 65, 81) may also occur in response to changes in daylength. Most of these responses have been observed in species with a heteromorphic alternation of generations, but a few are found in isomorphic species (e.g. *Dumontia;* 85) and in species with single-phase life histories (e.g. *Ascophyllum;* 107). It is not possible, therefore, to correlate photoperiodic control with any particular type of life history.

Vegetative responses to photoperiod, analogous to the onset or breaking of dormancy in higher plants, have also been observed in a few algae. The short-day (SD) responses of *Laminaria* and *Constantinea* both result in the initiation of a new blade in the autumn or winter (68, 79), while the rate of growth of the young blades of two kelp species (*Pleurophycus* and *Laminaria*) is stimulated by long-day (LD) conditions (38; K. Lüning, unpublished observations). These vegetative responses may be less complex than reproductive responses and are, therefore, potentially easier to study. So far, however, they have been found only in plants whose large size effectively cancels out the experimental advantages of their responses. Nevertheless, two responses offer promising approaches to the biochemistry of photoperiodism. The formation of new blades in *Laminaria* under SD conditions (68) is correlated with changes in the activities of key enzymes (62), while the development of *Derbesia* protoplasts under SD contrasts with that in LD (84).

Physiological Contrasts Between Photoperiodism in Algae and in Higher Plants

Although the first algal responses to be investigated in detail (20, 81) showed a remarkable similarity to photoperiodic responses in vascular plants, it has since become clear that other responses do not share all the classical features of photoperiodic control of flowering. It has, therefore, become necessary to redefine the criteria for accepting a response to daylength as a photoperiodic response in the strict sense of the term—i.e. "the control of some aspect of a life cycle by the timing of light and darkness" (47). The main casualty of this process of redefinition has been the night-break as a diagnostic feature of true photoperidic responses (21).

A few SD responses in algae have been shown to be completely insensitive to night-break treatment (e.g. the red algae *Acrosymphyton* and *Cordylecladia;* 9, 10) even though all other tests (i.e. day extensions, testing different daylengths with the same total daily exposure to light, sharp critical daylengths) show the responses to be controlled by the timing of light and darkness. In another SD red alga *(Rhodochorton),* night-breaks failed to inhibit the response to 8-h days but caused complete inhibition of the response to 10-h days (27). Such insensitivity to night-breaks is unknown among SD flowering plants, although many LD plants show a similar reluctance to respond to night-break treatment (112). Vince-Prue has suggested that such LD plants may be "light-dominant," responding to the length of the day rather than to the shortness of the night, whereas most SD plants are known to respond primarily to the length of the dark period ("long-night"; 114). However, the LD plants that respond to night-breaks are clearly measuring night length, and a few SD plants may conversely be "light-dominant." These two categories of photoperiodic plants can be distinguished by exposing plants to an 8-h day-extension either before or after an 8-h main photoperiod. For long-night responses, both treatments have similar effects but, for light-dominant responses, day-extensions before the main photoperiod inhibit the effects of the SD to a greater extent than day-extensions after the main photoperiod (113).

Recent experiments of this type with *Acrosymphyton* (9) have given precisely the latter result, suggesting that this SD plant is measuring daylength rather than night length. In other respects, however, it does not behave in a way similar to that of Vince-Prue's light-dominant plants. It is far more sensitive to light as a day-extension, and it responds primarily to blue light, with some effect of red but none of far-red (see p. 169; 9). Since phytochrome does not seem to be the photoreceptor, it is hardly surprising that a working hypothesis based on phytochrome action (114) will not fit. We seem to be facing a range of different types of photoperiodic mechanism in algae, and further quantitative data from more species are needed before we can make much sense of the situation.

NONPERIODIC PHOTOCONTROL OF ALGAL DEVELOPMENT

The intention of this admittedly ambiguous heading is to state, in conjunction with the heading of the previous section, that here will be found examples of algal development controlled by light through attributes other than the length of the day. Some aspects of this topic have been reviewed previously (23, 26, 65), but this is probably the first attempt to cover all types of response in all types of algae. A full discussion of the photoreceptors involved will, however, be postponed to the final section of this article so that they can be considered together with the photoreceptors involved in algal photoperiodism.

Types of Algae Whose Development Is Influenced by Light

Light has been shown to affect some aspect of the development of at least one species from every algal division recognized by Bold & Wynne (7; Tables 1–3). In marked contrast to the case with photoperiodic responses, unicellular species are well represented. The familiar laboratory maids-of-all-work from the Chlorophyta (*Chlorella, Chlamydomonas, Scenedesmus)* appear rather frequently, but *Euglena, Cryptomonas,* and the odd dinoflagellate are also present. It is surprising that no response has so far been reported for a diatom, but a recent study of the growth of *Chaetoceros* in blue light found no effects that could not be attributed to differential absorption by the photosynthetic pigments (40). Two divisions of mainly multicellular algae for which no photoperiodic responses have been reported are the Charophyta and the Cyanophyta (or cyanobacteria). The stoneworts (Charophyta) have the most highly differentiated thalli of any algae containing chlorophyll *b,* and it is no surprise, perhaps, to find photoresponses similar to those in mosses and ferns and apparently mediated by phytochrome (Table 2 a,b). The prokaryotic Cyanophyta are, of course, very different both in the types of response they exhibit and the photoreceptors involved (Table 1 b; 2 a,b). That is the main reason for discussing them here, rather than considering them as bacteria and therefore outside the scope of this volume.

Types of Development Under Control by Light

METABOLIC DEVELOPMENT   As is not the case in higher plants, the differentiation of chloroplasts and the synthesis of chlorophyll occur in complete darkness in many algae, but some species or mutants have been found that require light for greening (Table 1 a). *Euglena gracilis* and the mutants of various green algae have received the most attention, and blue light has been shown to precondition the cells so that they respond more rapidly to the wavelengths absorbed by protochlorophyllide (blue and red; 92). The pretreatment with blue light may simply stimulate respiration and thus supply energy or precursors, but a direct influence of blue light on the formation of enzymes that synthesize amino-levulinic acid is often detectable also (95). Recent studies of greening in *Euglena* have suggested that other photoreceptors absorbing orange (55) or green light (30) may be involved in addition to cryptochrome and protochlorophyll. Since all of the detailed work has been done on algae containing chlorophyll *b,* it will be interesting to learn more about the red alga *Delesseria,* which forms phycoerythrin in complete darkness but requires light for chlorophyll formation (67).

The pigment composition of fully "greened" algae is notoriously plastic, and it is often claimed that changes occur in response to either the irradiance

Table 1   Types of metabolic development in algae controlled by light

Type of development	Genus	Algal group[a]	Effective color(s)	Reference
(a) Greening (i.e. chlorophyll synthesis, chloroplast formation, etc)	*Chlorella*	Chl.	blue	95
	Scenedesmus	Chl.	blue	95
	Euglena	Eugl.	various	30, 55, 92
	Delesseria	Rhod.	?	67
(b) Pigment composition	*Chlorella*	Chl.	blue	57
	Scenedesmus	Chl.	blue	49
	Fremyella	Cyan.	green/red	46, 116
	Tolypothrix	Cyan.	green/red	37
	Ochromonas	Chry.	blue-green	53
	Cryptomonas	Cryp	green	53
	Prorocentrum	Pyrr.	green	33
(c) Enzyme synthesis	*Chlorella*	Chl.	blue	88
	Chlorogonium	Chl.	blue	87, 101
	Acetabularia	Chl.	blue	93
	Acrochaetium	Rhod.	blue	110
	Cyanidium	Rhod.	blue	102
(d) Stimulation of cell division	*Chlorella*	Chl.	blue	97
(e) Inhibition of cell division	*Chlamydomonas*	Chl.	blue/yellow	12
	Chlorella	Chl.	blue/yellow	12
	Prototheca	Chl.	blue	32
(f) Cell differentiation	*Volvox*	Chl.	green	56
(g) Osmoregulation	*Chlamydomonas*	Chl.	blue	109

[a] Chl. = Chlorophyta; Chry. = Chrysophyta; Cryp. = Cryptophyta; Cyan. = Cyanophyta; Eugl. = Euglenophyta; Pyrr. = Pyrrophyta; Rhod. = Rhodophyta.

or the quality of the incident light. Regardless of the outcome of the (possibly never-ending) dispute between "intensity adaptation" and "chromatic adaptation," it is clear that pigment synthesis in algae is influenced by light; and it may be more fruitful to enquire about the mechanisms than to continue the argument about the causes. The situation seems to be simplest in those blue-green algae that can vary the ratio between the two phycobilin pigments (Table 1 b). Red light stimulates the formation of phycocyanin and green light the formation of phycoerythrin (37, 46, 116), and the result is chromatic adaptation in the original sense. Outside the Cyanophyta, the phycoerythrin content of *Cryptomonas* also increases in green light (53), as does the peridinin content of *Prorocentrum* (33), but there is no other experimental evidence for adaptive changes in pigments in response to light quality (see 23). All algal groups show a response similar to that of higher plants when grown in low irradiances of white light (i.e. total pigment increases and the ratio of chlorophyll *a* to accessory pigments decreases). In some freshwater green algae, this response appears to be controlled by cryptochrome (whereas in flowering plants this would be a phytochrome response; 49, 57, 96), but the evidence from other algal groups is far less consistent (23) and there is a need for more critical investigation.

There have been numerous reports of blue light affecting (usually stimulating) enzyme activity (see 88, 89), and many of these refer to algal species (Table 1 c). Unicellular green algae are, as usual, favorite subjects, but the giant-celled *Acetabularia* has also been intensively investigated (14, 15, 93). The only non-green algae to appear on these lists so far are the red algae *Acrochaetium* (110) and *Cyanidium* (102). In most of these studies, plants were grown in monochromatic light for an extended period, and enzyme activity was then measured in cell-free extracts under optimal conditions. Higher activity in an extract from a blue-grown plant was usually, therefore, taken to imply that more enzyme had been synthesized in blue light, although direct proof of this interpretation (e.g. through the use of inhibitors) has rarely been obtained. Also, since plants had to be grown for long periods in the different wavelengths, it was not often possible to investigate the action spectrum in detail; and the commonly accepted idea that they are all due to cryptochrome may be questioned.

Almost all of the enzymes investigated (88, 89) control rate-limiting steps in respiratory or photosynthetic pathways, and such photocontrol of enzymes is the probable primary cause of the contrasting patterns of growth and chemical composition of algae cultivated in blue and red light (e.g. 11, 14, 15, 33, 50, 73, 117). The various effects of blue light on cell-cycle events (Table 1 d–f) may also be the indirect result of light affecting enzyme synthesis or activity. Similarly, the swelling and subsequent rupture of a wall-less mutant of *Chlamydomonas* is prevented in blue light, possibly because the cellular components required for osmoregulation are not produced in other wavelengths (109).

DEVELOPMENT OF VEGETATIVE MORPHOLOGY   Studies of the control of seed germination by light provided the basis for much early work on photomorphogenesis in flowering plants, and the development of photomorphogenetic studies on algae may have been delayed by the apparent rarity of similar responses among algal spores. A few exceptions to the rule that algal spores do not require light for germination are now appearing (Table 2 b). These include two charophytes, which have larger and more elaborate spores than many other algae; phytochrome has been implicated in one of these responses (103). There is also some evidence that phytochrome controls akinete germination in *Anabaena fertilissima* (80), although a more detailed action spectrum for the same process in *A. variabilis* (8) indicated phycocyanin as the photoreceptor. Since germination was only slightly reduced by DCMU (dichlorophenyl-dimethyl urea), photosynthesis was not apparently essential for the response; but the opposite result was obtained (and the opposite conclusion reached) in a study of the control of spore germination in *Bangia* by green light (13). The induction of germination in *Scrippsiella* cysts by green light is clearly a photomorphogenetic response, however, since less

Table 2 Types of vegetative development in algae under nonphotoperiodic control by light.

Type of development	Genus	Algal group[a]	Effective color(s)	Reference
(a) Growth responses	*Nostoc*	Cyan.	red/green	60, 86
	Fremyella	Cyan.	red/green	17
	Spirogyra	Chl.	red/far-red	115
	Chara	Char.	red/far-red	82
	Nereocystis	Phae.	red/far-red	28
	Vaucheria	Chry.	blue	54
(b) Spore germination	*Anabaena*	Cyan.	red(/far-red)	8, 80
	Chara	Char.	red/far-red	103
	Nitella	Char.	red	100
	Scrippsiella	Pyrr.	green	3
	Bangia	Rhod.	green	13
(c) Two-dimensional development	*Petalonia*	Phae.	blue	69
	Scytosiphon	Phae.	blue	25
(d) Hair formation	*Acetabularia*	Chl.	blue	94
	Desmotrichum	Phae.	blue	63
	Dictyota	Phae.	blue	73
	Scytosiphon	Phae.	blue	25
(e) Rhizoid formation	*Spirogyra*	Chl.	red/far-red	74

[a] Char. = Charophyta; Phae. = Phaeophyta; other abbreviations as in Table 1.

than 1 s at the standard culturing irradiance was sufficient to induce a 50% response (3). Other dinoflagellates also require light for cyst germination (2), so we may soon be on the trail of other green-light effects in algae.

Up to now, blue-light effects have been most prominent, and a common response to blue light is the formation of hairs (Table 2 d). *Scytosiphon* and *Acetabularia* have been most fully investigated (25, 94), and the responses of both species appear to be typical of cryptochrome, as does the induction of two-dimensional growth in *Scytosiphon* (25). The latter response closely parallels the transition from filamentous to two-dimensional growth in fern gametophytes (48); but there is no obvious parallel to the hair-formation responses among other plants, possibly because hairs of this type only form in aquatic plants. The control of both types of response by blue light may have some ecological significance (see discussion in 23, 26).

The "growth responses" (Table 2 a) are a heterogeneous group. Perhaps their only common feature is that they deserve more detailed investigation. The growth rate of the blue-green alga *Fremyella* (17) and the transition from amorphous to filamentous growth in *Nostoc* appear to be controlled by a red/green photoreversible pigment system (60, 86, 91) similar to that implicated in the chromatic adaptation of pigment composition in other cyanophytes (Table 1 b). Tip growth in the coenocytic chrysophyte *Vaucheria* is stimulated by blue light (54), and this response may be the primary cause of phototropic bending in this species—also a blue-light response. Two very different algae—*Chara* sporelings and the large kelp *Nereocystis*—both show an elongation response to end-of-day treatment with far-red light (28, 82). The response of *Chara* can be reversed by subsequent treatment with red light, but reversal was not convincingly demonstrated for *Nereocystis*. Elongation of cells in *Spirogyra* filaments is also affected by red and far-red light (115), and another apparent phytochrome response is seen in rhizoid formation in this alga (74).

DEVELOPMENT OF REPRODUCTIVE STRUCTURES   The nonphotosynthetic and nonphotoperiodic effects of light on the reproduction of algae may be exerted on the formation of reproductive structures or on the release of propagules from them. In one species of unicellular green algae, zoospore production is inhibited by light, whereas reproduction in another green unicell (*Trebouxia*, the algal symbiont of a lichen species) and in several macroscopic algae is stimulated by light (Table 3 a,b). Release of spores is also more commonly stimulated than inhibited by light (Table 3 c,d). Even in *Laminaria*, which is an apparent exception to this generalization, release is merely delayed for a few hours by blue light and the circadian rhythm of release is inhibited (66). However, keeping ripe receptacles of *Pelvetia* in light prevented release of eggs for up to 450 h, but transfer to darkness for as little as 3

Table 3 Types of reproductive development in algae under nonphotoperiodic control by light.

Type of development	Genus	Algal group[a]	Effective color(s)	Reference
(a) Induction of gamete	*Acetabularia*	Chl.	blue	106
or spore formation	*Dictyota*	Phae.	red	73
	Laminaria	Phae.	blue	64, 70
	Macrocystis	Phae.	blue	71
	Trebouxia	Chl.	red/far-red	39
(b) Inhibition of spore formation	*Protosiphon*	Chl.	blue/yellow	29
(c) Induction of gamete	*Bryopsis*	Chl.	blue	98
or spore release	*Monostroma*	Chl.	blue	99
	Desmotrichum	Phae.	blue	63
	Dictyota	Phae.	blue	58
(d) Inhibition of gamete	*Laminaria*	Phae.	blue	66
or spore release	*Pelvetia*	Phae.	?	52

[a] Abbreviations as in Tables 1 and 2.

min resulted in egg release within 10 min (52). This unusual observation should be followed up.

ACTION SPECTRA FOR ALGAL RESPONSES AND POSSIBLE PHOTORECEPTORS

This section brings together information on the action spectra for all types of nonphotosynthetic effect of light on algae, including where appropriate those nondevelopmental responses (i.e. orientation and rapid effects) excluded from detailed discussion in the rest of this review.

Red/Far-Red Reversible Responses—Phytochrome in Algae

The first indication that phytochrome occurs in algae was obtained when chloroplast movement in the filamentous green alga *Mougeotia* was found to be reversibly influenced by red and far-red light (43). This species has figured prominently in the phytochrome literature ever since. The desmid *Mesotaenium* was soon shown to have a similar response (44), and phytochrome was extracted from this species in 1967 (104). Since the absorption peaks of the two forms of phytochrome from *Mesotaenium* were at slightly shorter wavelengths than those of phytochrome from higher plants, it is possible that green algae possess a different (more primitive?) form of the pigment. There is recent evidence, however, that at least part of the molecule is similar in angiosperms, mosses, and green algae. Three species of green algae gave a positive result when tested with a monoclonal antibody directed to phytochrome from peas (16). The species included *Chlamydomonas* as well as

Mougeotia and *Mesotaenium*, but there is as yet no indication of what phytochrome might do in *Chlamydomonas*.

The evidence for phytochrome in green algae is thus based on action spectra, extraction, and molecular biology. It is surprising, therefore, that only three other species among the Chlorophyta (*Dunaliella*, 61; *Spirogyra*, 74, 115; *Trebouxia*, 39) have been shown to respond to red and far-red light. The stoneworts are included in the Chlorophyta by many authorities, and so the two responses of *Chara* in which red/far-red reversibility has been shown (82, 103) can perhaps be added with reasonable confidence to the list of phytochrome effects. However, the remaining red/far-red effects—all in non-green algae (i.e. from outside the Chlorophyta)—must be treated with more caution.

Such effects have been reported for various types of response in algae from widely separated evolutionary lines—photoperiodic responses (19, 81, 83) and absorbance changes (111) in red algae; growth responses in a brown alga (28), a diatom, a dinoflagellate, and a coccolithophorid (61); a behavioral response in another dinoflagellate (31, 34, 35); and germination in a blue-green alga (80). For many of these responses, red and far-red light were the only wavelengths tested and, in green plants, this is all that would be needed (according to Mohr's operational criteria; 72) to establish phytochrome involvement. As phytochrome has not yet been extracted from a non-green plant, however, better evidence than simple red/far-red reversibility is required, because antagonistic effects of red and far-red light could result from differential absorption by photosystems I and II (5). This possibility can be ruled out, however, if other wavelengths that activate photosystem II have a different morphogenetic effect from red light. Such evidence is available for *Gyrodinium* (31) and for the photoperiodic response of *Porphyra tenera* (19, 81).

Phytochrome, therefore, remains the best candidate for these responses, but its presence and activity outside the Chlorophyta cannot be regarded as proven without more detailed action spectra or successful extraction of the pigment from one of the species that shows a physiological response. There are good reasons why these requirements have not been satisfied already. The photoperiodic responses cannot be conveniently or sensitively quantified (see 21), and it is difficult to grow up sufficient material for extraction. Perhaps we should look to the molecular biologists of the phytochrome world (e.g. 16) to provide us with a probe to identify phytochrome in small quantities and in situ.

Responses Activated by Photoreversible Pigment Systems Other than Phytochrome

In the absence of unequivocal evidence for the occurrence of phytochrome in non-green algae, it is of interest that two distinct photoreversible pigment systems other than phytochrome have been detected in algae. The effects of red light on both the pigment composition (Table 1 b) and the growth responses (Table 2 a) of blue-green algae can be reversed by green light. For a time, there was intense interest in identifying what might prove to be a prokaryotic equivalent of phytochrome (91). One approach has been to fractionate aqueous extracts of the algae and to isolate new pigments (a series of "phycochromes") with photoreversible absorbance properties (4–6). Other workers have attempted to show that the familiar phycobiliproteins (in particular, allophycocyanin) could account for these responses unaided (76, 77). Since none of the absorption spectra exactly matched the action spectra for the physiological responses (6), however, it has not proved possible to settle the claims of the rival pigments, and an uneasy truce exists at present. The structural similarity between phytochrome and allophycocyanin is well known, and it will be valuable to have more information about the role of allophycocyanin or the phycochromes in photomorphogenesis.

A second photoreversible pigment system responding to blue (430 nm) and yellow (580 nm) light has been isolated from three green algae (108), in which it controls zoospore formation (*Protosiphon;* 29) or cell division (*Chlamydomonas, Chlorella;* 12). As is not the case with phytochrome (and, possibly, phycochrome), the photoreversibility of this system appears to result from the interaction of two separable components—a flavoprotein that absorbs blue light, and a plastocyanin that is converted to a form absorbing yellow light after oxidation by the activated flavin (78).

Blue Light (Cryptochrome) Responses

More of the responses in Tables 1–3 are controlled by blue light than by any other waveband. Some of the responses are similar to blue-light effects in higher plants (e.g. enzyme effects, tropic responses, induction of two-dimensional growth), while the induction of reproductive activity (Table 3) is comparable to blue-light-induced sporulation in fungi. Other responses of algae to blue light (e.g. potentiation of greening; pigment changes in low irradiances) are analogous to responses controlled by phytochrome in higher plants (95, 96), but a few responses (e.g. photoperiodic effects, hair formation) appear to be controlled by blue light only in algae.

Perhaps because so many responses have been reported, there are more detailed action spectra available than for other types of response. Those for a wide range of brown algae (24, 25, 58, 66, 70) have many of the characteristic features of cryptochrome spectra from green algae, higher plants, and fungi (see 23 for discussion). It seems reasonable to conclude that cryptochrome—if this really is a single pigment—must be widespread in the Phaeophyta and, possibly, among chromophyte algae in general. However, the situation is far less clear for the Rhodophyta. Although blue light affects enzyme synthesis (110), photoperiodic control (9), and tropic responses (23) in red algae, neither of the two action spectra available (9, 75) is really similar to that of cryptochrome. Large question marks, therefore, hang over both the phytochrome and the cryptochrome of red algae. It is hoped that someone will soon take up the challenge that they pose.

Responses Activated by Both Blue and Red Light

The SD response of *Porphyra tenera* was inhibited only by red light as a night break (19, 81) and that of *Scytosiphon* only by blue light (24), but the SD responses of four other algae (three red algae and one brown) were inhibited to similar extents by blue and red light (27, 79, 85, 107). In two of these species, the effects of red light could not be reversed by far-red (27, 79) so that phytochrome (or its equivalent in red algae) did not appear to be involved; but far-red was not tested on the other species. Blue and red light also seem to be responsible for stimulating dark respiration and starch breakdown in *Dunaliella* (90). The best-known pigment with absorption in blue and red is, of course, chlorophyll, but a simple photosynthetic function of the light can be ruled out for the photoperiodic responses because the same exposure given at a different time would not be inhibitory. The SD response of another red alga, *Acrosymphyton,* could not be inhibited by night-breaks, but day-extensions with low irradiances (0.05 μmol m^{-2} s^{-1}) of white light were completely inhibitory (9). Blue (420 nm) was the most effective wavelength for day-extensions, but 563, 600, and 670 nm were moderately effective. Only far-red (710 and 730 nm) was completely ineffective. These results are also inconsistent with either phytochrome or cryptochrome action, and prompt the suggestion that photoperiodic responses in algae may be controlled by a variety of pigment systems analogous to the variety of pigments involved in algal photosynthesis. There seems, however, to be little correlation between photosynthetic pigments and photoperiodic pigments.

Responses Activated by Green Light

Green light has been reported to affect pigment composition in a cryptomonad (53) and a dinoflagellate (33) as well as chloroplast development in *Euglena* (30), but detailed effectiveness spectra have been determined only for cyst germination in another dinoflagellate (3) and for the differentiation of somatic and reproductive cells in synchronized *Volvox* cultures (56). Both of these spectra show a broad peak around 550 nm. Rhodopsin, which has recently been tentatively identified in *Chlamydomonas* (36), has been proposed as the photoreceptor for the *Volvox* response (56); but the various forms of this pigment all have absorption maxima in the blue-green region (470–510 nm; 36) and so would not fit the effectiveness spectra available so far. However, the appearance of a new candidate for photoreception in plants will certainly revitalize attempts to match action spectra with absorption spectra. This is particularly welcome at a time when the algae are revealing such a variety of new photoresponses.

SUMMARY

Critical points in the life histories of many macroscopic marine algae are controlled by photoperiod, but the physiological mechanisms of their responses are more varied than among flowering plants. At least three pigment systems (responding to red/far-red, blue, or blue + red) appear to be involved, and the variability of other features of the responses (especially their sensitivity to light outside the main photoperiod) may be related to the different photoreceptors. Apart from its photoperiodic effects, light also controls aspects of metabolic, vegetative, and reproductive development in species from every algal division. These include unicellular algae, which appear not to respond to daylength. The commonest photoreceptor found to control algal development is cryptochrome, which is responsible for a wide range of response types. Reversible responses to red and far-red light have been reported for a few species in most algal groups, but phytochrome has been positively identified only in green algae. There is at present no satisfactory alternative explanation for the antagonistic effects of red and far-red light in non-green algae, but the presence of phytochrome in these plants has yet to be confirmed. The action spectra for other responses suggest that at least four other photoreceptors may be found among the algae, including the red/green reversible system of the Cyanophyta. Photobiologists who are saturated by phytochrome and cryptochrome research may, therefore, derive further excitation from the developmental responses of algae.

Literature Cited

1. Abdel-Rahman, M. H. 1982. Photopériodisme chéz *Acrochaetium asparagopsis* (Rhodophycées, Acrochaetiales). Influence de l'interruption de la nyctipériode, par un éclairement blanc ou monochromatique, sur la formation des tetrasporocystes. *C. R. Acad. Sci. Paris, Ser. III* 294:389–92
2. Anderson, D. M., Taylor, C. D., Armbrust, E. V. 1987. The effects of darkness and anaerobiosis on dinoflagellate cyst germination. *Limnol. Oceanogr.* 32:340–51
3. Binder, B. J., Anderson, D. M. 1986. Green light-mediated photomorphogenesis in a dinoflagellate resting cyst. *Nature* 322:659–61
4. Björn, G. S., Björn, L. O. 1976. Photochromic pigments from blue-green algae: phycochromes a, b, and c. *Physiol. Plant.* 36:297–304
5. Björn, L. O. 1979. Photoreversibly photochromic pigments in organisms: properties and role in biological light perception. *Q. Rev. Biophys.* 12:1–23
6. Björn, L. O., Björn, G. S. 1980. Photochromic pigments and photoregulation in blue-green algae. *Photochem. Photobiol.* 32:849–52
7. Bold, H. C., Wynne, M. J. 1985. *Introduction to the Algae*. Englewood Cliffs, NJ: Prentice-Hall. 720 pp. 2nd ed.
8. Braune, W. 1979. C-phycocyanin—the main photoreceptor in the light dependent germination process of *Anabaena* akinetes. *Arch. Mikrobiol.* 122:289–95

9. Breeman, A. M., ten Hoopen, A. 1987. The mechanism of daylength perception in the red alga *Acrosymphyton purpuriferum*. *J. Phycol.* 23:36–42

10. Brodie, J., Guiry, M. D. 1987. Life history and photoperiodic responses in *Cordylecladia erecta* (Rhodophyta). *Br. Phycol. J.* 22:300–1

11. Brown, T. J., Geen, G. H. 1974. The effect of light quality on carbon metabolism and extracellular release of *Chlamydomonas rheinhardtii* Dangeard. *J. Phycol.* 10:213–20

12. Carroll, J. W., Thomas, J., Dunaway, C., O'Kelley, J. C. 1970. Light induced synchronisation of algal species that divide preferentially in darkness. *Photochem. Photobiol.* 12:91–98

13. Charnofsky, K., Towill, L. R., Sommerfeld, M. R. 1982. Light requirements for monospore germination in *Bangia atropurpurea* (Rhodophyta). *J. Phycol.* 18:417–22

14. Clauss, H. 1968. Beeinflussung der Morphogenese, Substanzproduktion und Proteinzunahme von *Acetabularia mediterranea* durch sichtbare Strahlung. *Protoplasma* 65:49–80

15. Clauss, H. 1970. Effect of red and blue light on morphogenesis and metabolism of *Acetabularia mediterranea*. In *Biology of Acetabularia*, ed. J. Brachet, S. Bonotto, pp. 177–91. London: Academic

16. Cordonnier, M.-M., Greppin, H., Pratt, L. H. 1986. Identification of a highly conserved domain on phytochrome from angiosperms to algae. *Plant Physiol.* 80:982–87

17. Diakoff, S., Scheibe, J. 1975. Cultivation in the dark of the blue-green alga *Fremyella diplosiphon*. A photoreversible effect of green and red light on growth rate. *Physiol. Plant.* 34:125–28

18. Dickson, L. G., Waaland, J. R. 1985. *Porphyra nereocystis:* a dual-daylength seaweed. *Planta* 165:548–53

19. Dring, M. J. 1967. Phytochrome in red alga, *Porphyra tenera*. *Nature* 215: 1411–12

20. Dring, M. J. 1970. Photoperiodic effects in microorganisms. In *Photobiology of Microorganisms*, ed. P. Halldal, pp. 345–68. London: Wiley

21. Dring, M. J. 1984. Photoperiodism and phycology. In *Progress in Phycological Research*, ed. F. E. Round, D. J. Chapman, 3:159–92. Bristol: Biopress

22. Dring, M. J. 1984. Blue light effects in marine macroalgae. In *Blue Light Effects in Biological Systems*, ed. H. Senger, pp. 509–16. Berlin: Springer-Verlag

23. Dring, M. J. 1987. Marine plants and blue light. In *Blue Light Responses: Phenomena and Occurrence in Plants and Microorganisms*, ed. H. Senger, 2:121–40. Boca Raton, Fla: CRC Press

24. Dring, M. J., Lüning, K. 1975. A photoperiodic effect mediated by blue light in the brown alga *Scytosiphon lomentaria*. *Planta* 125:25–32

25. Dring, M. J., Lüning, K. 1975. Induction of two-dimensional growth and hair formation by blue light in the brown alga *Scytosiphon lomentaria*. *Z. Pflanzenphysiol.* 75:107–17

26. Dring, M. J., Lüning, K. 1983. Photomorphogenesis of marine macroalgae. In *Encyclopedia of Plant Physiology, Vol. 16B, Photomorphogenesis*, ed. W. Shropshire, H. Mohr, pp. 545–68. Heidelberg: Springer-Verlag

27. Dring, M. J., West, J. A. 1983. Photoperiodic control of tetrasporangium formation in the red alga *Rhodochorton purpureum*. *Planta* 159:143–50

28. Duncan, M. J., Foreman, R. E. 1980. Phytochrome-mediated stipe elongation in the kelp *Nereocystis* (Phaeophyceae). *J. Phycol.* 16:138–42

29. Durant, J. P., Spratling, L., O'Kelley, J. C. 1968. A study of light intensity, periodicity and wavelength on zoospore production by *Protosiphon botryoides* Klebs. *J. Phycol.* 4:356–62

30. Eberly, S. L., Spremulli, G. H., Spremulli, L. L. 1986. Light induction of the *Euglena* chloroplast protein synthesis elongation factors: relative effectiveness of different wavelength ranges. *Arch. Biochem. Biophys.* 245:338–47

31. Ekelund, N. G. A., Björn, L. O. 1987. Photophobic stop-response in a dinoflagellate: modulation by preirradiation. *Physiol. Plant.* 70:394–98

32. Epel, B., Krauss, R. W. 1966. The inhibitory effect of light on growth of *Prototheca zopfii* Kruger. *Biochim. Biophys. Acta* 120:73–83

33. Faust, M. A., Sager, J. C., Meeson, B. W. 1982. Response of *Prorocentrum mariae-lebouriae* (Dinophyceae) to light of different spectral qualities and irradiances: growth and pigmentation. *J. Phycol.* 18:349–56

34. Forward, R. B. 1973. Phototaxis in a dinoflagellate: action spectra as evidence for a two-pigment system. *Planta* 111:167–78

35. Forward, R., Davenport, D. 1968. Red and far-red light effects on a short-term behavioural response of a dinoflagellate. *Science* 161:1028–29

36. Foster, K. W., Saranak, J., Patel, N., Zarilla, G., Okabe, M., et al. 1984. A

rhodopsin is the functional photoreceptor for phototaxis in the unicellular eukaryote *Chlamydomonas. Nature* 311: 756–59

37. Fujita, Y., Hattori, A. 1960. Effect of chromatic lights on phycobilin formation in a blue-green alga, *Tolypothrix tenuis. Plant Cell Physiol.* 1:293–303

38. Germann, I. 1986. Growth phenology of *Pleurophycus gardneri* (Phaeophyceae, Laminariales), a deciduous kelp of the north east Pacific. *Can. J. Bot.* 64: 2538–47

39. Giles, K. L. 1970. The phytochrome system, phenolic compounds, and aplanospore formation in a lichenized strain of *Trebouxia. Can. J. Bot.* 48:1343–46

40. Gostan, J., Lechuga-Deveze, C. 1986. Does blue light affect the growth of *Chaetoceros protuberans* (Bacillariophyceae)? *J. Phycol.* 22:63–71

41. Guiry, M. D., Cunningham, E. M. 1984. Photoperiodic and temperature responses in the reproduction of north eastern Atlantic *Gigartina acicularis* (Rhodophyta: Gigartinales). *Phycologia* 23:357–67

42. Häder, D.-P. 1987. Photomovement. See Ref. 23, 1:101–30

43. Haupt, W. 1959. Die Chloroplastendrehung bei *Mougeotia*. I. Über den quantitativen und qualitativen Lichtbedarf der Schwachlichtbewegung. *Planta* 53:484–501

44. Haupt, W., Thiele, R. 1961. Chloroplastenbewegung bei *Mesotaenium. Planta* 56:388–401

45. Haupt, W. 1982. Light-mediated movement of chloroplasts. *Ann. Rev. Plant Physiol.* 33:205–33

46. Haury, J. F., Bogorad, L. 1977. Action spectra for phycobiliprotein synthesis in a chromatically adapting cyanophyte, *Fremyella diplosiphon. Plant Physiol.* 60:835–39

47. Hillman, W. S. 1979. *Photoperiodism in Plants and Animals* (Carolina Biology Reader 107). Burlington, NC: Carolina Biological Supply Co. 16 pp.

48. Howland, G. P., Edwards, E. E. 1979. Photomorphogenesis of fern gametophytes. In *The Experimental Biology of Ferns*, ed. A. F. Dyer, pp. 393–434. London: Academic

49. Humbeck, K., Schumann, R., Senger, H. 1984. The influence of blue light on the formation of chlorophyll-protein complexes in *Scenedesmus.* See Ref. 22, pp. 359–65

50. Humphrey, G. H. 1983. The effect of the spectral composition of light on the growth, pigments and photosynthetic rate of unicellular marine algae. *J. Exp. Mar. Biol. Ecol.* 66:49–67

51. Huth, K. 1979. Einfluss von Tageslänge und Beleuchtungsstärke auf den Generationswechsel bei *Batrachospermum monoliforme. Ber. Dtsch. Bot Ges.* 92: 467–72

52. Jaffe, L. 1954. Stimulation of the discharge of gametangia from a brown alga by a change from light to darkness. *Nature* 174:743

53. Kamiya, A., Miyachi, S. 1984. Blue-green and green light adaptations on photosynthetic activity in some algae collected from subsurface chlorophyll layer in the western Pacific Ocean. See Ref. 22, pp. 517–28

54. Kataoka, H. 1987. The light-growth response of *Vaucheria.* A *conditio sine qua non* of the phototropic response? *Plant Cell Physiol.* 28:61–71

55. Kaufman, L. S., Lyman, H. 1982. A 600 nm receptor in *Euglena gracilis:* its role in chlorophyll accumulation. *Plant Sci. Lett.* 26:293–99

56. Kirk, M. M., Kirk, D. L. 1985. Translational regulation of protein synthesis, in response to light, at a critical stage of *Volvox* development. *Cell* 41:419–28

57. Kowallik, W., Schürmann, R. 1984. Chlorophyll a/chlorophyll b ratios of *Chlorella vulgaris* in blue or red light. See Ref. 22, pp. 353–58

58. Kumke, J. 1973. Beiträge zur Periodizität der Oogon-Entleerung bei *Dictyota dichotoma* (Phaeophyta). *Z. Pflanzenphysiol.* 70:191–210

59. Lang, A. 1965. Physiology of growth and development in algae. A synopsis. In *Encyclopedia of Plant Physiology*, Vol. 15/1, ed. W. Ruhland, pp. 680–715. Berlin: Springer-Verlag

60. Lazaroff, N., Vishniac, W. 1961. The effect of light on the development cycle of *Nostoc muscorum*, a filamentous blue-green alga. *J. Gen. Microbiol.* 25: 365–74

61. Lipps, M. J. 1973. The determination of the far-red effect in marine phytoplankton. *J. Phycol.* 9:237–42

62. Lobban, C. S., Weidner, M., Lüning, K. 1981. Photoperiod affects enzyme activities in the kelp, *Laminaria hyperborea. Z. Pflanzenphysiol.* 105:81–83

63. Lockhart, J. C. 1982. Influence of light, temperature and nitrogen on morphogenesis of *Desmotrichum undulatum* (J. Agardh) Reinke (Phaeophyta, Punctariaceae). *Phycologia* 21:264–72

64. Lüning, K. 1980. Critical levels of light and temperature regulating the gametogenesis of three *Laminaria* species. *J. Phycol.* 16:1–15

65. Lüning, K. 1981. Photomorphogenesis of reproduction in marine macroalgae. *Ber. Dtsch. Bot. Ges.* 94:401–17

66. Lüning, K. 1981. Egg release in gametophytes of *Laminaria saccharina:* induction by darkness and inhibition by blue light and U.V. *Br. Phycol. J.* 16:379–93

67. Lüning, K. 1984. Growth and lack of chlorophyll *a* in a dark-cultivated *Delesseria sanguinea. Br. Phycol. J.* 19: 196–97

68. Lüning, K. 1986. New frond formation in *Laminaria hyperborea* (Phaeophyta): a photoperiodic response. *Br. Phycol. J.* 21:269–73

69. Lüning, K., Dring, M. J. 1973. The influence of light quality on the development of the brown algae *Petalonia* and *Scytosiphon. Br. Phycol. J.* 8:333–38

70. Lüning, K., Dring, M. J. 1975. Reproduction, growth and photosynthesis of gametophytes of *Laminaria saccharina* grown in blue and red light. *Mar. Biol.* 29:195–200

71. Lüning, K., Neushul, M. 1978. Light and temperature demands for growth and reproduction of laminarian gametophytes in Southern and Central California. *Mar. Biol.* 45:297–309

72. Mohr, H. 1972. *Lectures on Photomorphogenesis.* Heidelberg: Springer-Verlag. 237 pp.

73. Müller, S., Clauss, H. 1976. Aspects of photomorphogenesis in the brown alga *Dictyota dichotoma. Z. Pflanzenphysiol.* 78:461–65

74. Nagata, Y. 1973. Rhizoid differentiation in *Spirogyra.* II. Photoreversibility of rhizoid induction by red and far-red light. *Plant Cell Physiol.* 14:543–54

75. Nultsch, W. 1980. Effects of blue light on movement of microorganisms. In *The Blue Light Syndrome,* ed. H. Senger, pp. 38–49. Berlin: Springer-Verlag

76. Ohad, I., Schneider, H.-J. A. W., Gendel, S., Bogorad, L. 1980. Light-induced changes in allophycocyanin. *Plant Physiol.* 65:6–12

77. Ohki, K., Fujita, Y. 1979. In vitro transformation of phycobiliproteins during photobleaching of *Tolypothrix tenuis* to forms active in photoreversible absorption changes. *Plant Cell Physiol.* 20:1341–47

78. O'Kelley, J. C., Hardman, J. K. 1977. A blue light reaction involving flavin nucleotides and plastocyanin from *Protosiphon botryoides. Photochem. Photobiol.* 25:559–64

79. Powell, J. 1986. A short-day photoperiodic response in *Constantinea subulifera. Am. Zool.* 26:479–87

80. Reddy, P. M., Talpasyi, E. R. S. 1981. Some observations related to red–far red antagonism in germination of spores of the cyanobacterium *Anabaena fertilissima. Biochem. Physiol. Pflanzen* 176: 105–7

81. Rentschler, H.-G. 1967. Photoperiodische Induktion der Monosporenbildung bei *Porphyra tenera* Kjellm. (Rhodophyta–Bangiophyceae). *Planta* 76:65–74

82. Rethy, R. 1968. Red (R), far-red (FR) photoreversible effects on the growth of *Chara* sporelings. *Z. Pflanzenphysiol.* 59:100–2

83. Richardson, N. 1970. Studies on the photobiology of *Bangia fuscopurpurea. J. Phycol.* 6:215–19

84. Rietema, H. 1973. The influence of daylength on the morphology of the *Halicystis parvula* phase of *Derbesia tenuissima* (De Not.) Crn. (Chlorophyceae, Caulerpales). *Phycologia* 12:11–16

85. Rietema, H., Breeman, A. M. 1982. The regulation of the life history of *Dumontia contorta* in comparison to that of several other Dumontiaceae (Rhodophyta). *Bot. Mar.* 25:569–76

86. Robinson, B. L., Miller, J. H. 1970. Photomorphogenesis in the blue-green alga *Nostoc commune* 584. *Physiol. Plant.* 23:461–72

87. Roscher, E., Zetsche, K. 1986. The effects of light quality and intensity on the synthesis of ribulose-1,5-bisphosphate carboxylase and its mRNAs in the green alga *Chlorogonium elongatum. Planta* 167:582–86

88. Ruyters, G. 1984. Effects of blue light on enzymes. See Ref. 22, pp. 283–301

89. Ruyters, G. 1987. Control of enzyme capacity and enzyme activity. See Ref. 23, 2:71–88

90. Ruyters, G., Hirosawa, T., Miyachi, S. 1984. Blue light effects on carbon metabolism in *Dunaliella.* See Ref. 22, pp. 317–22

91. Scheibe, J. 1972. Photoreversible pigment: occurrence in a blue-green alga. *Science* 176:1037–39

92. Schiff, J. A. 1980. Blue light and the photocontrol of chloroplast development in *Euglena.* See Ref. 75, pp. 495–511

93. Schmid, R. 1984. Blue light effects on morphogenesis and metabolism in *Acetabularia.* See Ref. 22, pp. 419–32

94. Schmid, R., Idziak, E.-M., Tünnermann, M. 1987. Action spectrum for the blue-light-dependent morphogenesis of hair whorls in *Acetabularia mediterranea. Planta* 171:96–103

95. Senger, H. 1987. Blue light control of

174 DRING

pigment biosynthesis—chlorophyll biosynthesis. See Ref. 23, 1:75–85
96. Senger, H. 1987. Sun and shade effects of blue light on plants. See Ref. 23, 2:141–49
97. Senger, H., Schoser, G. 1966. Die spektralabhängige Teilungsinduktion in mixotrophen Synchronkulturen von *Chlorella*. *Z. Pflanzenphysiol.* 54:308–20
98. Shevlin, D. E., West, J. A. 1977. Gamete discharge in *Bryopsis hypnoides:* a blue light phenomenon. *J. Phycol.* 13(Suppl.):62
99. Shihara, I. 1958. The effect of light on gamete liberation in *Monostroma*. *Bot. Mag. (Tokyo)* 71:378–85
100. Sokol, R. C., Stross, R. G. 1986. Annual germination window in oospores of *Nitella furcata* (Charophyceae). *J. Phycol.* 22:403–6
101. Stabenau, H. 1972. Aktivitätsänderungen von Enzymen bei *Chlorogonium elongatum* unter dem Einfluss von rotem und blauem Licht. *Z. Pflanzenphysiol.* 67:105–12
102. Steinmüller, K., Zetsche, K. 1984. Photo- and metabolite regulation of the synthesis of ribulose bisphosphate carboxylase/oxygenase and the phycobiliproteins in the alga *Cyanidium caldarium*. *Plant Physiol.* 76:935–39
103. Takatori, S., Imahori, K. 1971. Light reactions in the control of oospore germination of *Chara delicatula*. *Phycologia* 10:221–28
104. Taylor, A. O., Bonner, B. A. 1967. Isolation of phytochrome from the alga *Mesotaenium* and liverwort *Sphaerocarpus*. *Plant Physiol.* 42:762–66
105. ten Hoopen, A., Bos, S., Breeman, A. M. 1983. Photoperiodic response in the formation of gametangia of the long-day plant *Sphacelaria rigidula* (Phaeophyceae). *Mar. Ecol. Prog. Ser.* 13:285–89
106. Terborgh, J. 1965. Effects of red and blue light on the growth and morphogenesis of *Acetabularia crenulata*. *Nature* 207:1360–63

107. Terry, L. A., Moss, B. L. 1980. The effect of photoperiod on receptacle initiation in *Ascophyllum nodosum* (L.) Le Jol. *Br. Phycol. J.* 15:291–301
108. Thomas, J. P., O'Kelley, J. C., Hardman, J. K., Aldridge, E. F. 1975. Flavin as an active component of the photoreversible pigment system of the green alga *Protosiphon botryoides* Klebs. *Photochem. Photobiol.* 22:135–38
109. Thompson, R. J., Davies, J. P., Mosig, G. 1985. "Dark lethality" of certain *Chlamydomonas reinhardtii* strains is prevented by dim blue light. *Plant Physiol.* 79:903–7
110. van der Velde, H. H., Guiking, P., van der Wulp, D. 1975. Glucose-6-phosphate dehydrogenase and 6-phosphogluconate dehydrogenase in *Acrochaetium daviesii* cultured under red, white and blue light. *Z. Pflanzenphysiol.* 76:95–108
111. van der Velde, H. H., Hemrika-Wagner, A. M. 1978. The detection of phytochrome in the red alga *Acrochaetium daviesii*. *Plant Sci. Lett.* 11:145–49
112. Vince-Prue, D. 1975. *Photoperiodism in Plants*. London: McGraw-Hill. 444 pp.
113. Vince-Prue, D. 1976. Phytochrome and photoperiodism. In *Light and Plant Development*, ed. H. Smith, pp. 347–69. London: Butterworths
114. Vince-Prue, D. 1983. Phytochrome and photoperiodic physiology in plants. See Ref. 26, pp. 457–90
115. Virgin, H. I. 1978. Inhibition of etiolation in *Spirogyra* by phytochrome. *Physiol. Plant.* 44:241–45
116. Vogelmann, T. C., Scheibe, J. 1978. Action spectra for chromatic adaptation in the blue-green alga *Fremyella diplosiphon*. *Planta* 143:233–39
117. Wallen, D. G., Green, G. H. 1971. Light quality in relation to growth, photosynthetic rates and carbon metabolism in two species of marine plankton algae. *Mar. Biol.* 10:34–43

Ann. Rev. Plant Physiol. Plant Mol. Biol. 1988. 39:175–219
Copyright © 1988 by Annual Reviews Inc. All rights reserved

THE CONTROL OF FLORAL
EVOCATION AND MORPHOGENESIS

Georges Bernier

Département de Botanique, Université de Liège, Sart Tilman, B-4000 Liège,
Belgium

CONTENTS

INTRODUCTION 175
EXPERIMENTAL SYSTEMS AND MODEL PLANTS 176
ENVIRONMENTAL CONTROL 178
    Genetics of Sensitivity to Environmental Factors...................................... 178
    Control by Day Length.. 178
    Control by Low Temperature.. 180
    Autonomous "Induction" 180
    Comparative Efficiency and Summation of Different Inductive Factors............... 181
    Sites of Perception of Environmental Factors.. 181
CORRELATIVE INFLUENCES 182
FLORAL EVOCATION 184
    Meristem Competence... 184
    Start and End of Evocation... 186
    Component Processes and Nature of Evocation 187
FLORAL MORPHOGENESIS... 192
    Component Processes.. 193
    Genetics... 197
THEORIES OF ENDOGENOUS CONTROL 198
    The Florigen/Antiflorigen Concept 198
    Electrical Signal 201
    The Nutrient Diversion Hypothesis: Control by Assimilates.......................... 201
    The Model of Multifactorial Control: Participation of Plant Growth Regulators.... 203
CONCLUSIONS... 209

In biology we deal with the most complex situations known. . . . Complex situations are
surely going to require complex explanations if the explanation is to be accurate. Or if we
retain simple views should we not honestly admit that they may achieve little when faced
with biological complexity?

A. Trewavas
Aust. J. Plant Physiol. 13:447–57, 1986

175

0066-4294/88/0601-0175$02.00

INTRODUCTION[1]

Research on flowering covers a vast field extending from ecophysiology to biophysics. Despite the fact that flowering is a unitary and integrated process, it is generally divided into the two major phases of flower initiation and development. These phases do not react similarly to environmental and internal variables and thus are not alike. Only the first of these phases is covered here. This field was last reviewed in this series by Zeevaart (258) and Murfet (171). Since then, the proceedings of three conferences on flowering have been published (3, 39, 249), as have a monograph (15, 16, 116) and a handbook in six volumes (88, 89, 90).

This wealth of data has demonstrated the complex nature of flower formation, but the field remains dominated by simplistic and dogmatic ideas. Most exemplary is the idea that the whole process is simply controlled by absence or presence of a single specific hormonal promoter, "florigen." This concept, which celebrated its 50th birthday in 1987, was later complemented with that of a floral inhibitor or "antiflorigen." I aim here to highlight recent developments, to emphasize areas that were sometimes neglected in the past, to argue that the florigen/antiflorigen story is no longer commensurate with the complexity of the phenomenon it was supposed to explain, and to examine alternate hypotheses.

EXPERIMENTAL SYSTEMS AND MODEL PLANTS

The florigen/antiflorigen theory was essentially based on work with particular strains of a very small number of species, the SDP *Xanthium, Perilla, Kalanchoe,* and *Pharbitis;* the LDP *Hyoscyamus, Rudbeckia,* and *Lolium;* the SD and LD tobaccos; and few others (36, 66, 128, 129, 213, 258). These plants were favored because they were considered as absolutely photoperiodic. Work on most other species was often overlooked simply because they were facultatively photoperiodic or DNP. It is clear, however, that absolute and facultative species are basically similar; and most, if not all, species are either absolute or facultative depending on age, history, growing conditions, etc (15, 66, 128).

[1]*Abbreviations used*:

ABA = (±)-abscisic acid; ACC = 1-amino-cyclopropane-1-carboxylic acid; AVG = L-2-amino-4-(2-aminoethoxy)-*trans*-3-butenoic acid; CCC = 2-chloroethyltrimethylammonium chloride; cv = cultivar; 2,4-D = 2,4-dichlorophenoxyacetic acid; DN(P) = day neutral(plant); FR = far-red light; GA = gibberellin; GC–MS = combined gas-liquid chromatography–mass spectrometry; HPLC = high performance liquid chromatography; IAA = indol-3-ylacetic acid; LD(P) = long day (plant); NB = night break; PGR = plant growth regulator; PAR = photosynthetically active radiation; R = red light; SA = salicylic acid; SD(P) = short day(plant); TCL = thin cell layer

A model of multifactorial control of flowering, proposed by Bernier et al (16), based not only on work with the Liège warhorse, *Sinapis,* but also on observations by other workers on a variety of plants, was dismissed because *Sinapis* cannot be regarded as truly photoperiodic (69). Lack of an absolute control hinders, of course, the design of critical experiments; but such a control can often be experimentally devised in nonabsolute species, including DNP and woody perennials (see 152, 205, 221, 227). *Sinapis,* for example, normally a facultative LDP, behaves as an absolute LDP at low irradiance (23). When such a strict control is achieved, these experimental systems then match the best classical model plants. In fact, many observations made on *Sinapis* have been confirmed in several other species, including the "true" photoperiodic plants (see below). I argue that it is necessary to reverse the trend of trying to establish "universal" theories on results obtained with an exclusive small group of "ideal" VIP strains.

Jacobs (99) has emphasized the inadvisability of using plants grown in true LD instead of SD with NB as LD controls. With the SDP *Xanthium,* large differences in leaf $^{14}CO_2$ assimilation, assimilate import, and mitotic index in the apex are indeed found between the two kinds of vegetative controls (102; C. Mirolo, unpublished). In *Chenopodium,* only LD maintain a strict vegetative condition at the apex; NB allows part of the normal changes seen at the floral transition to occur (217). Thus, it is advisable to use both types of LD controls. Another important fact, sometimes overlooked, is that LDP are usually much less sensitive to NB than are SDP. As pointed out by Vince-Prue (246), LDP require NB of one or several hours at higher irradiances. In the LDP *Hyoscyamus,* presence of CO_2 in the air during the NB is required and CO_2 fixation participates in the NB response (254). In other LDP, the energy supplied by the low-intensity photoextension period of LD makes a significant contribution to the CO_2 assimilation (97).

A popular system, well-suited for determining the effects of the chemical environment, is found in the duckweeds (112). These plants present, however, some drawbacks: (*a*) it is unclear whether the mother-frond alone, as against all generations of fronds inside the mother-frond, perceives the photoperiodic signal; (*b*) there is no evidence for transmission of floral promoters or inhibitors from the mother to the younger fronds, which are the only ones capable of flower production; and (*c*) the reacting meristem is extremely small, indistinct, and thus difficult to study.

With few exceptions, the genetics of flowering in most plants is poorly known (171). Clearly, the ideal species is a myth, but *Arabidopsis thaliana* is perhaps the closest to it. It is already well characterized physiologically and genetically (160, 174); it has a number of flowering mutants (159), so an integrated approach is feasible.

ENVIRONMENTAL CONTROL

Genetics of Sensitivity to Environmental Factors

Day-length sensitivity versus insensitivity can be governed by a single gene, as in "mammoth" tobacco and the *ld* mutant of *Arabidopsis* (171, 199), or several genes, as in sorghum and wheat (132, 186). Response to day length is dominant over insensitivity in sorghum, whereas the reverse situation is found in tobacco and wheat. Similarly, a cold requirement is controlled by one gene in *Hyoscyamus, Lunaria,* and Petkus rye (171), but by several in *Arabidopsis* and wheat (132, 174, 175). Cold sensitivity is dominant over insensitivity in *Hyoscyamus* and *Lunaria,* but recessive in rye and wheat. In *Arabidopsis,* vernalization genes can be dominant or recessive. It is commonly thought that dominant alleles cause the formation of a substance that is absent when only the recessive alleles are present (175). If true, the control of flowering appears to involve different substances from plant to plant.

The *Ppd* and *Vrn* genes, controlling respectively day length and cold sensitivity in wheat, proved to be separate and entirely independent genes (132, 175), probably controlling different component processes of flower initiation. An opposite situation was disclosed in pea, where genes *Sn* and *Dne* control two steps in the biosynthetic pathway of a floral inhibitor (172). Joint presence of the dominant alleles at these loci confer a requirement for either LD or vernalization, which here thus both act on the same process.

Interestingly, many of these genes have pleiotropic effects. In wheat *Ppd*$_1$ influences plant height and *Vrn* genes influence tillering (70, 132). In pea, *Sn* and *Dne* affect internode length, leaf morphology, branching, apical senescence, etc (172). These effects on vegetative characters are even clearly apparent in the *veg* mutant, totally unable to initiate flowers (172). Thus, the "floral inhibitor" of pea is not specific for the floral transition. In general, the genes involved in the day-length or cold requirement are not specific. Unfortunately, none of these genes has been cloned and their exact activities remain generally unknown.

Control by Day Length

Inspection of Halevy's *Handbook of Flowering* (88–90) reveals that the physiological evidence is in line with these genetical data. The response to day length has been found to be profoundly altered by several other environmental factors. These multiple interactions vary among different genotypes. In the var. Violet of *Pharbitis,* a model SDP, flower production in LD can be caused by at least six different treatments, including poor nutrition, high irradiance, low temperature, root removal, application of a cytokinin or CCC, etc (179, 232, 247). The dwarf *Pharbitis,* var. Kidachi, reacts like Violet to poor nutrition or low temperature but, unlike Violet, it does not react to high light flux and is inhibited by CCC. On the other hand, culture in small vessels results in flowering in LD in Kidachi, but not in Violet.

Small changes in temperature may suffice to alter the plant response. *Chamelaucium* is an absolute SDP in a 24/16°C day/night regime but is a facultative SDP at 20/10° (221). In sorghum genotypes having the maturity genes Ma_1 and Ma_2, Morgan et al (169) have found that shifting the 30/21°C day/night thermoperiodic regime forward by only 2.5 h relative to the 12/12-h day/night photoperiodic regime causes a considerable acceleration of flower initiation compared to controls with synchronous thermo- and photoperiods.

The situation is not less complex in LDP. *Silene* initiates flowers in SD when subjected to one of the following treatments: low or high temperature, increased CO_2 level in the air, or root removal (256, 260).

LIGHT PERCEPTION AND TIME MEASUREMENT IN PHOTOINDUCTION   In photoinduction, phytochrome is known to interact with a timekeeping mechanism. This topic, outside the scope of this review, was recently discussed (56, 238, 248, 249). The interaction is generally believed to exert its effect by primarily altering membrane properties and/or gene expression. Salicylic acid (SA), a chemical which influences membrane permeability to ions in several animal and plant systems, is a powerful flower-inducing agent in duckweeds (44, 112). As SA is quickly inactivated once it is taken up by the mother-frond and is not translocated to the flowering daughter-fronds, Cleland (44) hypothesized that SA acts by altering membrane permeability in the mother-frond. Curiously, however, the varied flower-inducing activities of a number of SA and benzoic acid analogues are not related to the hydrophobicity of these molecules (255), and SA has no flower-inducing effect in most other plants (16, 259). Other compounds, like lithium, valinomycin, acetylcholine, also affecting membranes, may promote or inhibit duckweed flowering (112), and lithium delays considerably flower initiation in LD-induced spinach (114). In induced leaves of this species, the thickness of plasma membrane increases (5). For Greppin et al (87), these and other results point to membrane modifications as transducers of the photoperiodic signal. Changes in compartmentation of ions, e.g. K^+ and Ca^{2+}, might ensue and these in turn might alter a variety of biochemical and physiological processes (96). However, so many compounds and metabolic systems (including adenine and pyridine nucleotides, pathways of energy metabolism, bioelectrical potential, etc) change early in induced spinach leaves that the disclosed situation defies simple analysis (27, 87, 168). So far, it is impossible to determine which changes are essential and which are simply accompanying and thus incidental.

Gene expression is also altered at induction. In the LDP *Hyoscyamus* and

Sinapis and the SDP *Pharbitis,* analyses of the products of in vitro translated polyA$^+$RNA reveal many early differences between mRNA from induced and noninduced leaves (50, 133, 253). Most differences are quantitative and, except for one increasing mRNA in *Pharbitis,* it is again unclear whether they are related or not to induction. Interestingly, in *Sinapis,* most changes already disappear before the end of the inductive LD.

Control by Low Temperature

The natural low temperature requirement of several winter cereals, perennial grasses, etc can be bypassed by exposure to SD at normal temperature (15, 128). Anaerobiosis and ethylene can also replace the need for vernalization in chicory (107). *Calceolaria* is an absolute LDP at high irradiance, but at low irradiance it flowers only when the LD are preceded by vernalization (205). SD can also substitute here for vernalization and, finally, if the cold treatment is sufficiently long, the requirement for LD is suppressed.

In many cold-requiring species, high temperatures are known as "devernalizing" since they oppose the effect of a chilling treatment (128). To be effective, high temperature must usually be applied immediately after vernalization to suboptimally vernalized plants. However, devernalization is quite effective in supraoptimally vernalized *Cheiranthus cheiri* (63) and easier in *C. allionii* two months after seed vernalization than immediately after it (7). On the other hand, heat devernalization is impossible in several other species (15), whereas in *C. cheiri* it is absolutely required between two periods of flowering at low temperature (63). Unexpectedly, high temperature can substitute for the required cold treatment in *Scrophularia alata* (131) and other species (15).

In several biennials, like beet, *Cheiranthus,* and carrot, the vernalized state disappears more or less rapidly if the plants are grown in SD at moderate temperature after the chilling treatment (7, 15). Opposing effects are common: SD that substitute for vernalization in some species produce a devernalizing effect in others.

Autonomous "Induction"

With time, many plants flower autonomously. All they require is that the environment be conducive to growth. This behavior is often contrasted with that of plants having specific climatic requirements. However, most of the latter species, including *Pharbitis, Kalanchoe, Xanthium, Lolium,* etc, ultimately flower in so-called "non-inductive" conditions (15). On the other hand, flowering in an autonomous DN tobacco could be indefinitely suppressed either by growth in low irradiance and high temperature (16), or by repeated rooting maintaining the distance between the apical meristem and the root system below a minimal number of nodes (152). In pea, the genotype *sn dne* is an early DNP (autonomous) while *Sn Dne* is a late facultative LDP (172). Gene *Hr* magnifies the effect of the *Sn Dne* combination, so that genotype *Sn Dne Hr* is a near-absolute LDP. In old plants of this genotype, a single LD is sufficient to trigger flowering (171). If grown long enough, all pea genotypes will, however, ultimately flower, suggesting that the controlling genes are leaky. This supports the assertion that the difference between autonomous, facultative, and absolute species is only one of degree.

Comparative Efficiency and Summation of Different Inductive Factors

The efficiency of the different factors that cause flower initiation varies enormously. In photoperiodic species, favorable day lengths are in general more efficient than substitute treatments (258). However, vernalized sugar beet and the *ld Arabidopsis* that are LDP flower more rapidly in continuous darkness than in LD (15). An identical flowering response is obtained in *Xanthium* exposed to one long night or to only two LD if part of the light period is at low temperature and if GA_3 is applied (C. Mirolo, unpublished).

In the cold-requiring *Scrophularia,* three weeks at high temperature substitute for seven weeks of chilling (131), while in chicory anaerobiosis or ethylene for half a week has the same effect as eight weeks of cold (107).

In many cases, plants exposed successively to two different inductive factors, each at a subthreshold level, will flower. This indicates summation of whatever changes are caused by each treatment. Curiously, inductive factors of a totally opposite nature may sometimes be summated, as continuous darkness and LD in *Rudbeckia* or low and high temperatures in *Festuca* (see references in 15). In *Silene,* summation occurs irrespective of the kinds of inductive factors combined and of their sequence (256), but in most cases summation is dependent on the sequence of treatments. In Violet *Pharbitis,* for example, subminimal high irradiance and low temperature treatments can only be summated in that sequence (222). Moreover, each of these two treatments, which can substitute for the long night, becomes inhibitory to flowering when combined with a long night (232, 247).

Sites of Perception of Environmental Factors

Knowledge of the site(s) where the environmental factors are perceived is important since when these sites are not shoot apices there is an automatic requirement for long-distance transmission of a signal.

Day length is most effectively perceived in intact plants by young expanded leaves (128, 246). However, various plant fragments, like buds, leaf disks, stem segments and root pieces, grown in vitro were all found to perceive day length (6, 15). Although young leaves and scales transmit part of the incident light (43), photoperiodic treatment of the bud alone in intact plants is usually ineffective. Optical fiber experiments with *Pharbitis,* initiated by Gressel, demonstrate that the apex perceives a R night-break or a FR end-of-day treatment, but much less efficiently than the cotyledons (120, 247). Thus, the role of the apex in day-length perception seems secondary (see also below).

In most plants, vernalization is perceived at the apex (128, 158). Cultured plant fragments are sensitive to cold apparently only when containing dividing cells (15). In chicory, cold is perceived by the root collar whereas anaerobiosis, which can substitute for cold, is perceived by the apex (107). The chilling treatment substituting for the day-length requirement in *Pharbitis, Perilla,* and *Blitum* was reported to be perceived by the leaves, not the apex (see 12), but this awaits confirmation using localized cooling treatments. Vernalization in pea acts in the leaves to repress inhibitor production and in the apex to alter the sensitivity to promoter and inhibitor (172).

In *Brassica pekinensis,* high devernalizing temperature is perceived not by the apex, as expected, but by the roots (195). Similar observations on other cold-requiring plants are badly needed. If confirmed, they will force a dramatic revision of the classical thinking about vernalization/devernalization interactions (128). High temperature, substituting for LD in *Silene,* is also perceived by the roots (256).

Clearly, there are alternate pathways to flowering in all species (genotypes). Environmental factors can interact such that each may change the threshold values for the others. Different factors can be summated, but only following a precise sequence, and they may act at totally different sites within the plant. The resulting matrix of controls forms a stunning array, but such a situation may be essential for the successful reproduction of plants in an often unpredictable environment. Proponents of the florigen/antiflorigen concept suggest that alternate pathways all act through a common mechanism (128, 256, 258, 260). A more realistic hypothesis would imply multiple promotive and inhibitory processes of varied nature occurring concurrently in the various plant parts. These processes are expected to be affected in various ways by different environmental factors and this will explain why the importance of different organs may vary according to the external conditions.

CORRELATIVE INFLUENCES

Two basically different views have been expressed concerning the importance of correlative influences in flower formation. One view, acknowledging their essential role, arose initially from work with perennial plants. Here, internal systems are necessary to counter, at some specific meristems, the effect of environmental factors that favor flowering (202). These systems in several species are seen as a complex network of interactions between different plant parts (31, 124).

The opposite view is essentially based on experiments with red *Perilla* showing that (a) detached leaves can be photoinduced in the absence of buds and roots (but cycling cells are probably present at the petiole base), and (b) these leaves can then be grafted onto derooted receptor shoots where they cause flower formation (262). Thus, the sole organ interaction apparently required in nonperennial photoperiodic plants is the one-way transport of a signal from leaves to apices (128). This latter view is undoubtedly an over-simplification (16). The experiments on red *Perilla* are unique: They are difficult, for technical reasons, to duplicate entirely with other species, even green *Perilla* (262). Further, various organ interactions influencing flower-ing, other than those involved in mediation of the effects of environmental factors, were found in many nonperennial plants. This was reviewed by Miginiac (161, 163) and Krekule (124). Only a few observations are pre-sented here.

In *Xanthium,* the response to photoinduction increases with the number of active buds present (16). This suggests that buds play an essential role in induction. Decapitation inhibits axillary flowering in *Scrophularia* (161). In both species auxin substitutes for the buds. Involvement of expanding leaves is also clearly demonstrated since their removal promotes or hastens flower initiation in various SDP, LDP, and DNP (15).

Excised apices of the SDP *Perilla* and *Chenopodium* and the LDP *Sinapis* and *Scrophularia* initiate flowers in vitro irrespective of the day-length re-gime, indicating that an inhibitory influence is acting in vivo. This influence arises in unfolded leaves in *Perilla* (see 15) and in the roots in *Scrophularia* and *Chenopodium* (161). This root effect can be replaced by a cytokinin treatment. These observations indicate that it is simplistic to disregard the role of roots because some derooted plants form flowers almost normally (260). Plants are well adapted to organ removal. Buds or stems can substitute for roots in the supply of cytokinins and possibly other compounds (33, 41).

A restriction of root growth, obtained by water stress, root pruning, etc, promotes flowering in several plants (163, 190). Culture in small vessels even reduces by 2 h the critical dark period in Violet *Pharbitis* (232). On the other hand, start of root growth correlates with a dramatic drop in the sensitivity of cuttings of *Anagallis* to LD induction (30). Proximity to the root system totally prevents meristems from flowering, not only in annual DN and photoperiodic tobaccos (59, 152), but also in the perennial *Ribes* (214). In this last case, inhibition is attributable to root-generated GAs. The role of roots is not always inhibitory. The rosette LDP *Rudbeckia* is unable to bolt and flower when deprived of roots, even when its leaves are fully photoinduced (121). Since flowering can occur in the absence of bolting, I feel that roots are required here—not for bolting, but for flower initiation (16).

Numerous observations establishing that the root system participates in the determination of inflorescence branching, flower number, sex expression, etc were reviewed by Kinet et al (116).

When plants enter the reproductive phase, other more complex organ interactions develop (116). These may affect flower initiation or early flower morphogenesis. In the woody SDP *Chamelaucium,* for example, flowering is only transient under SD because the plants ultimately return to vegetative growth under the influence of a transmissible inhibitor produced by developing flowers (221). In barley, cessation of spikelet initiation occurs concurrently with stamen initiation in the most advanced spikelet, a situation also explained by the production of an inhibitor (GA?) in stamens (47).

Comparisons of seeded and seedless cvs, as well as other data, show that developing seeds exert a strong inhibitory effect on flower initiation in several fruit trees, such as apple, pear, olive, etc (31, 166, 230). This effect has been implicated in the common phenomenon of alternate bearing of these trees.

It is concluded that the fate of all meristems in perennial and nonperennial plants is under the influence of a network of interorgan long-distance correlations. Although much less complex, the situation in nonperennials is not basically different from that in perennials.

FLORAL EVOCATION

Clonal analysis in corn indicated that about four cells of the meristem at the dry-seed stage are already committed to produce the terminal male inflorescence or tassel (105). No such distinct cell lineages have been found in other plants, such as tobacco, *Petunia,* and sunflower (104), so that except perhaps in corn (*a*) shoot meristems are characterized by the absence of permanent apical initials, and (*b*) the central zone of relatively inactive cells found in many vegetative meristems is not generally a "méristème d'attente" (32) in the sense that there is not a group of cells developmentally restricted to the production of reproductive structures. Nevertheless, the floral transition obligatorily involves the activation of this central zone (see below).

The events occurring in the apex that commit it to flower formation were called "evocation" (66). Not all shoot meristems can react to conditions that otherwise are known to promote flowering. Thus, I consider first the question of meristem competence.

Meristem Competence

In many woody plants, it is usually recognized that juvenile meristems are incompetent—i.e. incapable of responding to floral promoters—because juvenile scions do not flower more rapidly when grafted onto mature trees or vines bearing flowers (251). Wareing (251) emphasized that the changeover to the adult (competent) stage in such plants is thus determined by some mechanism intrinsic to the apex. In many studies, however, it is impossible to know which factor (time, node number, leaf area, distance from roots, or more subtle changes) is critical for the reaching of competence. Moreover, not all cases fit into the above generalization. Even in adult plants, some active meristems are not competent—e.g. axillary meristems of rose under apical dominance (45). Also, juvenile scions of certain plants, fig and pecan for example, can be brought into flowering by grafting onto adult stocks (see 88), indicating that at least in these cases juvenile meristems of woody plants are competent.

In herbaceous plants, it is classically believed that all meristems, young or old, are competent. This idea is based on the observation that juvenile scions of *Bryophyllum,* tobacco, and sweet pea flower very rapidly when grafted onto mature flowering stocks (128, 130, 203). Thus, juvenility in these species seems unrelated to meristem incompetence but is due to physiological limitations in other plant parts, especially the leaves (128). Many experimental data indicate, however, that the situation is not that simple. Differences in photoperiodic sensitivity between different lines of *Xanthium, Pharbitis,* etc can be attributed to differences not only in their leaves but also in their apices (reviewed in 15). In pea, the length of the juvenile phase is controlled by several genes (172). The activities of genes *Sn* and *Dne* (see above), which act in the leaves, normally decrease as the plant ages. Gene *Hr* blocks this temporal decline and thus lengthens the juvenile phase. At the *Lf* locus, several alleles in homozygous condition result in minimum flowering nodes varying from 5 to 15. Grafting studies indicate that the *Lf* alleles act at the apex and influence its competence to respond to the balance between floral promoter(s) and inhibitor(s). The *Veg* gene, when homozygous recessive, totally prevents flower initiation in the most favorable conditions or after grafting to the most promotory donors, but does not hamper normal vegetative growth. A *veg* apex has apparently lost competence definitively, but whether all or part of the evocation process is suppressed is unknown.

As stressed by Chouard (42), competence may be present, in many plants, in young mitotically active, nonzonated meristems and disappear in older meristems. Recent examples that fit this concept are those of the apical meristem in *Bidens* (194) and the axillaries in *Pharbitis* (184). This loss of competence remains unexplained but, as suggested by King & Evans (117), it could be related to the establishment of the phyllotactic pattern (geometry) typical of vegetative morphogenesis.

In *Anagallis,* the meristem is competent to react to the LD stimulus only during a brief period in each plastochron (30). Thus, meristem competence in herbaceous plants is not a fixed character, but we know next to nothing about its molecular and cellular basis. Study of the mode of action of the unique *Veg* gene in pea is perhaps our best hope to gain insight into the molecular mechanisms of competence.

Start and End of Evocation

The meristems of photoperiodic plants, held for protracted periods in unfavorable conditions for flowering, may alter their organization (e.g. zonation is lost) but continue to produce leaves (176). Some of these changes appear similar to, although far less rapid and not as pronounced as, those occurring during the floral transition. Hence, meristems exhibiting these changes were called "intermediate" by Nougarède (176). To me, this situation suggests that evocation has started in noninductive conditions but cannot be completed because only some of the endogenous controlling factors are available. A similar situation may occur in cold-requiring plants (16), so that, depending on the growing conditions, evocation may start at different times in different species.

A way to try to circumvent this problem is to use plants requiring exposure to only one photoinductive cycle and to assume that the time of movement of the leaf-generated floral stimulus (as determined by sequential defoliations) is the start of evocation. But some of these plants—e.g. *Xanthium,* spinach, and *Sinapis*—have an intermediate meristem (10); and in most of them, dramatic changes are observed in the meristem long before the time of stimulus movement (16, 247). Evocation start is thus difficult to determine accurately.

The point at which commitment of the meristem to flower becomes irreversible, i.e. end of evocation or floral determination, is considered to occur in *Sinapis* when inhibitors applied to the apex can no longer prevent flower formation. This point occurs at about the time most histological and morphological changes start and more than a half day before the first signs of flower initiation (12, 16). A reasonable conclusion is therefore that evocation is essentially a molecular and cellular process whose completion triggers the changes at the higher levels of organization.

Floral determination was investigated by McDaniel et al by comparing the developmental fate of buds when grown in situ or in isolation. Determination of the apical bud in both photoperiodic and DN tobaccos occurs, as in *Sinapis,* well before the initiation of flowers (59, 224). In DN tobacco, determination occurs first in the apical bud and only subsequently and independently in the uppermost three axillary buds (224). Tissues from some internodes of flowering plants will form de novo flowers when grown in vitro and are thus also florally determined. Interestingly, the first determined internode tissues appear at about the time of determination of the apical bud but are located some 24 nodes (15cm) below this bud (225). As the plant continues growing, internode tissues progressively higher in the stem become determined. Clearly, determination may occur in cells not organized in meristems and this condition is transmissible through mitotic cycles in tissue culture. There is no evidence, however, that it can be transmitted in grafting experiments (28, 34).

These observations support the conclusion that evocation (determination) is essentially a molecular and cellular process.

In tobacco, cells of determined axillaries and internodes are not clonally derived from determined cells of the apical meristem. Determination is thus believed to result from the action of a transmissible signal on separate competent tissues (224, 225). Not all axillaries and internodes become determined. The basis of this apparent lack of competence is unknown.

Some caution should be exercised here since the timing of irreversible commitment is known to be displaced as a function of the severity of the treatment used for assaying it. Evidence suggests that inhibitor application is a less severe treatment than apex excision and, with cultured apices, the culture technique has much influence on the results (10, 16, 247).

Another complication comes from differences in stability of determination. There is a full range of variation from stability in DN tobaccos to instability in photoperiodic tobaccos (59, 197a). This may explain why the floral condition can be transmitted through several subcultures in DN tobaccos (122) whereas it is weakly or not expressed during the first subculture in most photoperiodic tobaccos (109a, 197a). In several other species, explants from reproductive axes similarly regenerate only vegetative buds or embryos (15). Even in DN tobaccos, determination may be easily lost. The "thin cell layers" (TCL) of Tran Thanh Van (240), which form flowers in culture only when collected from the inflorescence, presumably contain initially determined cells. However, these TCL may be easily forced into other morphogenetic programs (roots, buds, etc) by various manipulations of the culture conditions (239–241).

Component Processes and Nature of Evocation

Two opposite views have been expressed concerning the nature of evocation. The first states that the switch to flowering primarily requires a change in gene expression in meristems, specifically the "turning on" of the genetic program concerned with the control of sexual reproduction (257). This view implies that this primary event sets in motion the complex sequence of all other changes. Another view proposes that evocation is essentially an unspecific activation necessary to eliminate the vegetative pattern of morphogenesis (67). On this clean "slate," the new geometry of reproductive morphogenesis is then drawn, and this will secondarily result in changes in gene expression. The experimental evidence at hand does not allow rejection of either view.

GENE EXPRESSION   Gene expression is, as expected, highly regulated during flower development. In tobacco, for example, almost half of the genes expressed in the anther or ovary are not expressed in the leaf, and vice versa

(109). Each of these floral organs has thousands of unique mRNAs that are undetectable in the other organs, and the anther and ovary each has a distinct leaf mRNA subset. Several specific cDNA clones appear to be expressed in different organs of the tomato flower in a stage- and tissue-specific manner (79). Transcripts of structural genes coding for proteins related to some functions of mature floral organs, such as petal pigmentation, pollen production, or male/female recognition, appear relatively late during flower development (79, 101, 198, 229) and are not detectable in the meristem itself (101). Thus the question is, When and where in the apex are the first changes in gene expression detectable?

Increased synthesis in all RNA fractions is among the earliest events occurring in transitional meristems of single-cycle plants (16, 247, 258); and inhibitors of RNA or protein synthesis reduce the flowering response, especially when applied early after induction (16). So far, however, because of the minute amounts of tissue available, there has been no report of in vitro translation of polyA$^+$RNA from transitional apices. Screening of cDNA libraries has led to the isolation of genomic clones specifically transcribed in flowering tobacco TCL before start of flower initiation (154). One of these clones is expressed later only in subapical tissues of reproductive apices. Further technical refinement is necessary before we shall be able to apply this technology to meristems and identify the mRNAs that may change at evocation.

The published immunological or electrophoretic analyses of the protein complement of transitional apices or meristems are still preliminary (143, 165, 192), but they all tend to show that this complement is altered before flower primordia are initiated. Using 2D-PAGE separation of ^{35}S-labeled proteins, it was found that most of the changes in *Sinapis* meristems are quantitative; but there are a few possible qualitative differences (143; F. Cremer, unpublished). So far, it is unknown whether the changing proteins play an essential role in evocation; but one may speculate that quantitative changes involve enzymes of the transcription/translation machinery, energy metabolism, cell-cycle regulation, etc, while qualitative modifications might be related to the synthesis of some new PGR(s) (GAs?) and/or the changes in sensitivity to these PGRs (see below).

OTHER MOLECULAR CHANGES   In *Sinapis,* rises in sucrose and ATP levels, invertase activity, mitochondrion number, and energy charge are among the earliest recorded events (16, 22, 25); and 2,4-dinitrophenol inhibits flower initiation (23). The increase in sucrose was observed, by Bodson & Outlaw (25), not only after induction by one LD but also after induction by one displaced SD. This change is thus unrelated to photosynthetic input but seems critically involved in evocation.

Similar changes were reported in other species—e.g. early increases in soluble sugar level in several SD- and cold-requiring plants (16, 24, 63) and increases in mitochondrion number in the LDP spinach and the SDP *Xanthium* and tobacco (16, 93, 110). In spinach, Auderset et al (4) found early increases in the activity of enzymes of the pentose phosphate pathway and attached great significance to this activation. However, as no data on other metabolic pathways related to energy metabolism are available, it is unsure that one pathway is preferentially activated. Thus, evocation is characterized in different types of plants by an important stimulation of energy metabolism. However some results, especially in SDP, are at variance. There is an early transient drop in ATP in *Pharbitis* (237) and no early marked changes in soluble sugars, invertase, and fumarase activities in *Xanthium* (M. Bodson & C. Mirolo, unpublished).

Some of the above analyses were not on meristems but on apices bearing leaf primordia and young internodes. In *Pharbitis,* however, it was shown that whole epicotylar buds or smaller apices may yield totally different results in such analyses (1). Thus, in further studies, the meristem and other apical components should be analyzed separately.

Several early ultrastructural changes—for example, the splitting of vacuoles in *Sinapis* and tobacco (16, 110)—suggest that membranes are affected at evocation. Unfortunately, work on membranes is still in its infancy. Cold is known to affect membrane properties profoundly, but in many investigations it is impossible to know whether these effects are essential for thermoinduction of flowering or simply unspecific. In a study comparing a spring and a winter wheat line differing only at the Vrn_1 locus, a higher rate of unsaturated phospholipid synthesis was found during the critical 5–6th weeks of vernalization in the winter genotype (61). There is no difference between the two genotypes during the first five weeks of cold nor during growth at nonvernalizing temperatures. These results suggest that the action of the Vrn_1 gene is related to membrane properties. Whole seedlings were analyzed, and it remains to be shown that this conclusion is valid for the apex.

A claim that stems of flowering tobacco plants contain ten times more DNA than stems of vegetative plants (250) has not been verified (R. B. Goldberg, personal communication).

CELLULAR CHANGES   One of the most conspicuous events of the floral transition is an increase in the rate of cell division (16). This was confirmed in *Silene* (73), sunflower (149), and chrysanthemum (177). In *Sinapis,* Gonthier et al (81) showed that the first change, after exposure to one LD, is a mitotic wave resulting from both a shortening of the G2 phase of cycling cells and a return to cycling of noncycling G2 cells. This is followed by a shortening of G1 and S in cycling cells, resulting in a wave of cells duplicating DNA. A second mitotic wave occurs later and, since the time interval separating the two mitotic waves is identical to the shortened cell cycle, it is clear that cycling cells are synchronized at evocation.

Synchronization was also observed in *Silene,* both in the apical meristem during the floral transition induced by seven LD (73) and in third-order flower meristems of the cyme more than three weeks later (140). In *Lolium,* induced by one LD, synchrony occurs in leaf axils and the meristem summit (i.e. potential spikelet sites) but not in neighboring leaf primordia (180). Remarkably, the floral transition is the only developmental step where a natural synchrony is observed in shoot meristems; and in all cases it is found just before start of flower (spikelet) initiation, suggesting that it is an essential component of the preparation for morphogenesis. In *Sinapis,* indeed, all attempts to dissociate synchrony from flowering have been unsuccessful (16). In *Silene,* however, synchrony at the floral transition can be suppressed without inhibition of flower formation (74); and in *Lolium* caused to flower by an application of 2,2-dimethyl GA_4 in SD, there is so far no evidence for synchronization (J. C. Ormrod and G. Bernier, unpublished). Francis & Lyndon (74) concluded that synchrony is not essential but could be a secondary effect of the sudden arrival of floral promoter(s) at the apex.

The question of what is essential and what is not is difficult to answer (12, 16). In particular cases, it has been possible to achieve evocation in the absence of one or another "universal" evocational event. For example, in *Silene* flower formation may occur in the absence of an increased growth rate of the apex (164). Thus, again, this typical event seems unessential. What may be essential, however, is not so much an absolute increase in growth rate of the apex as an increase relative to the growth rate of other plant parts—e.g. young leaves, axillaries, or roots (10, 137). When one particular event is blocked experimentally, evocation may nevertheless proceed in some cases by a combination of events different from that used in normal conditions. The most general lesson arising from these considerations is that there are probably more ways than one for an apex to achieve the essential changes. The way that is used depends on the environmental and endogenous conditions and in itself is nonessential; but at this stage, the number and nature of the essential events are simply not known.

A special feature in *Silene* is the transient shortening of the cell cycle during the first of the seven inductive LD (72, 74). The reduced cycle is characterized by an accumulation of G2 cells, initially detected 1 h after start of the photoextension of this LD. Work by Ormrod & Francis (74, 181) indicated that these early events could result from the reduced R/FR ratio of the photoextension, since 5 min FR light brought about similar effects. The data seem consistent with a low-irradiance phytochrome response. Interpolation of a dark period of 20 min at the start of photoextension of the first three of seven

LD suppresses both flowering and the characteristic G2 increase but leaves the cycle shortening unaffected (181, 182). Thus, only the G2 increase seems essential for evocation, and it was hypothesized that this event confers competence to the apex to react to the further arrival of floral promoter(s) (72).

A more general interpretation of all cell-cycle data could be that a sufficient proportion of the cell population should be in a critical phase of the cycle for evocation, because only cells in that phase react properly to floral promoters or produce a substance needed for evocation (16, 73).

DIRECT LIGHT EFFECTS   Some changes occur so early in single-cycle plants that the question of direct light/dark effects on the apex can be raised (69). In *Silene,* experiments on cultured apices indicate that the early G2 increase may be due to direct light effects, but this change is also obtained in intact plants with only the leaves exposed to the LD (72). Of the multiple ultrastructural changes seen at evocation in *Sinapis,* only those affecting mitochondria are detected when the apical bud alone is submitted to the LD (A. Havelange and G. Bernier, unpublished). Although both the rate of leaf production and apex enlargement are markedly affected by photoperiod in barley, both are unaffected when LD or SD are given directly to the apex (54). No change in mitotic activity is observed in *Pharbitis* when only the plumule is exposed to the inductive long night (185). Thus, in intact plants, direct light effects seem to play only a secondary role.

RESPONSE OF DIFFERENT APEX COMPONENTS   In general, all apical parts are stimulated, but the degree and timing of activation are not the same in all parts. Activation is generally most marked in the central zone of the meristem, resulting in a definitive or transient loss of the zonate pattern typical of many vegetative meristems (16, 149, 177). This activation is apparently essential for evocation (10, 16, 17, 19, 176). Because sucrose and enzymes of carbohydrate metabolism are uniformly distributed in zonate vegetative meristems (25, 52), the loss of zonation cannot be attributed to changes in carbohydrate level or metabolism.

The most rapid response is recorded in different meristem components in different plants: the peripheral zone in *Sinapis* and *Pharbitis* (16, 247), the central corpus in *Xanthium* and sunflower (102, 148), or the central tunica in tobacco (111). Activation is also seen in leaf primordia: As a rule it occurs as early as in the meristem but is smaller (16, 247).

In particular cases, activation is not general. For example, in *Lolium* caused to flower by a GA application, cell proliferation in leaf primordia is inhibited (J. C. Ormrod and G. Bernier, unpublished). In young seedlings of *Chenopodium,* induced by three long nights, cell activities in all apical parts are first inhibited. Stimulation is only seen after return to continuous light (216, 219). A similar transient inhibition is found in *Sinapis* induced by a displaced SD (16). These general growth inhibitions do not occur in adult *Chenopodium* plants nor in *Sinapis* induced by a LD (16). They can be explained by depletion of reserves during prolonged periods of darkness and are thus incidental to some flowering systems. They are by no means an essential component of evocation.

PARTIAL EVOCATION   Parts of the evocational events in *Sinapis* can be produced by treatments that are unable by themselves to cause flower initiation. Thus, one SD at high irradiance produces the increases in soluble sugar level, invertase activity, and mitochondrion number (94). On the other hand, a single application of a cytokinin in SD causes the splitting of vacuoles, forces the noncycling G2 cells to return to cycling, and shortens S in cycling cells (14, 95; A. Jacqmard and C. Houssa, unpublished). The work of Havelange et al (94, 95) shows that the changes caused by irradiance (presumably via sugar level) are totally different from those caused by cytokinin. Similar observations were made in other plants (11, 145, 164). These cases of "partial" evocation are the basis for the concept of evocation as a process consisting of a limited number of sequences of changes, each sequence being controlled independently of the others (12, 16). Full evocation does not follow automatically the activation of only one of the component sequences but will require activation of all sequences. If true, this implies that there is no single initial evocational event that can set in motion all the subsequent changes, and thus no single controlling agent of evocation.

The different sequences, independent initially, interact at some step of the floral transition, as shown by the fact that the various changes of the cell cycle, observed in LD-induced *Sinapis,* are better mimicked by a combined high-irradiance/cytokinin treatment, than by one of these two treatments alone (A. Jacqmard and C. Houssa, unpublished). Note that this combined treatment does not cause flowering in SD but promotes it in association with a suboptimal LD. I view the interactions between sequences as essential for evocation to proceed and reach completion (12, 16).

FLORAL MORPHOGENESIS

After evocation is completed, the apical growth pattern is profoundly altered. Some changes are common to many, if not all, plants; others are specifically related to the kinds of reproductive structures that will be formed. None of the individual growth changes is in itself flowering. For each species (genotype), flower or inflorescence formation results from a specific spatial and temporal combination of these changes. Research in this area is commonly considered as purely morphological and consequently often ignored by physiologists. This is unfortunate since the information gained may give some insights into the controlling factors of flower initiation. Only the early stages giving rise to flower primordia are considered here. Needless to say, the molecular and cellular study of these stages is still in its infancy. The later processes of growth of floral organs, sex expression, etc were recently reviewed (116).

Component Processes

MERISTEM SHAPE AND SIZE   Doming, the earliest and most common change in meristem shape, is at least partly attributable to the vacuolation and elongation of cells in the pith-rib meristem (16, 176). Later shape and size changes are clearly related to specific features of the potential reproductive structures—e.g. heightening is typical of grasses and cereals (47, 119, 188), and broadening of Compositae (16, 148, 176). This enlargement is essential in *Chrysanthemum,* which will never flower if the meristem does not reach a critical minimal diameter (98). However, the meristem of *Impatiens* keeps the same size (142), and that of *Humulus* becomes smaller at the transition (10, 16). In several barley cvs the transition occurs with widely differing sizes of apical meristems (54). Axillary and nutrient-starved meristems commonly flower at reduced size (10, 142). Thus, except for plants producing a broad receptacle, like *Chrysanthemum,* the absolute size reached by the meristem is not a determinant of flower formation.

A factor involved in the elongation of the grass apex is most probably a GA (47, 188). Applications of GA_3 also increase meristem size in *Xanthium* (145), and there are clearly nutritional factors in the enlargement of many transitional meristems (16, 137).

RATE OF APPENDAGE PRODUCTION   During the floral transition, the rate of appendage initiation generally increases (16, 142). This event is often detectable in advance of reproductive morphogenesis—e.g. several plastochrons before double ridges are seen in wheat (119). It is the only morphological change clearly preceding the completion of evocation in *Sinapis* (12, 13). The most dramatic shortening of plastochron occurs, however, after cessation of leaf initiation, when reproductive structures are formed (16, 142). It is coincident in several species with the reduction in size of nascent primordia of reproductive structures (142).

Plastochron shortening has been obtained by application of various PGRs, such as a GA in *Xanthium, Arabidopsis, Perilla,* and *Rudbeckia* (16, 17, 145); a cytokinin in *Chenopodium, Arabidopsis,* and *Sinapis* (17, 218; G. Bernier, unpublished); or auxin antagonists in *Epilobium* (155).

LEAF GROWTH   The last leaves before the reproductive structures are generally of small size and simple shape. This results from a strong inhibition of primordium growth (16). Again, this change can possibly be traced back to alterations in the balance among several factors that have been found to affect leaf size and/or shape, among which GAs and cytokinins are prominent (65, 91, 145, 187).

PRECOCIOUS INITIATION OF AXILLARY MERISTEMS   This almost universal event of the floral transition is of critical importance. The precocious axillaries grow out, forming either flowers, spikelets, or inflorescence branches. In the flowers themselves, stamens have often been interpreted as homologous to precocious axillaries of sepals or petals (138). Thus, although the growth of leaf (bract) primordia is ultimately much reduced, the growth of axillaries is much promoted and, as emphasized by Seidlová (216, 219), the balance of activities between apical components is critically altered, suggesting profound changes in short-distance interactions within the apex.

This release of axillaries is presumably related to a loss of apical dominance and thus to changes in the factors (mainly auxins, cytokinins, and nutrients) that are known to be normally involved in the correlations between apical and axillary buds. Both the early inhibitory and late promotive effects of exogenous IAA on flowering in *Chenopodium* could be explained on the basis of reinforced apical dominance (216). In grapevine, cytokinins are involved in the control of "anlage" branching, essential for inflorescence formation (170). Other factors are also involved. In grasses, for example, exogenous GA promotes the precocious release of axillary meristems (potential spikelets) and the further development of spikelets (16, 48, 188), and 2,4-D causes ear branching in wheat (220).

INTERNODE GROWTH   Each time a meristem initiates an appendage, it also produces an axis frustum that will give rise to the corresponding node (nonelongating) and internode (elongating) (141). An increased growth of internodes is very common in transitional apices of both caulescent and rosette plants (16). This stimulation is persisting in some species, as *Sinapis*, but only transient in others, as *Xanthium* (16). Genetical and physiological studies have demonstrated that elongation of vegetative internodes and flower initiation are two separate processes in many plants (16, 173, 258). Interestingly, work on a GA-insensitive dwarf of *Silene* has revealed that the regulation of elongation is different for vegetative internodes and flower peduncle (231).

In typical flowers and capitula, there are no internodes at all, and in many species—e.g. tulip, *Silene,* and *Gerbera*—the important elongation of the axis below the flower or inflorescence (scape, peduncle) contrasts with the near cessation of internode growth within these structures themselves. In the sepal frustum in *Silene* all cells are of the internodal type, accounting for the presence of an exceptionally long internode (peduncle) below the flower (141). In the petal-stamen frustum, on the contrary, all cells are of the nodal type, explaining the absence of vertical spacing between the successive whorls of floral organs. At all stages, the flower frusta are smaller at initiation than leaf frusta. This reduction parallels that in the size of floral organs at initiation (see below). A similar situation was found in other species (142). Lyndon (141) hypothesized that young sepals in *Silene,* unlike young leaves, might be a poor source of the auxin needed for elongation of the next internode. It is known indeed that reduced axis (scape) growth, resulting from removal of reproductive structures, is often largely restored to normal by application of IAA and/or GA$_3$ (reviewed in 116).

DECREASE IN PRIMORDIUM SIZE AND PHYLLOTACTIC CHANGES   Generally, there is an increase in the relative size of the meristem to that of its appendages, at their initiation, during the transition to flowering (137, 138, 142). In inflorescences, this increase is often due to an increase of absolute meristem size rather than a reduction in primordium size. When a flower is formed, however, the rule is that the nascent primordia of floral organs are smaller not only relative to the meristem, but also in absolute size. In any event, this decrease in primordium size is a critical change. It permits a closer packing of primordia and is thus basic to the increased complexity in phyllotaxis, which is generally observed during the shift from leaf to inflorescence (flower) formation.

Exogenous substances that can decrease primordium size and alter the phyllotactic arrangement of leaves to a higher-order pattern, in a manner very similar to that appearing during the floral transition, are GA$_3$ in *Xanthium* (145) and auxin antagonists in *Epilobium* (155, 156). A variety of phyllotactic alterations are caused in the wheat ear by 2,4-D (220).

Another perspective on phyllotactic changes is biophysical (84). It is based on observations that the lines of cellulose reinforcement in the surface cells of the meristem run roughly as concentric circles. The organ sites are local irregularities in this pattern. As it bulges, each primordium influences the adjacent meristem surface to align cellulose there so that the reinforcement lines are tangential to the primordium base. Parts of such reinforcement fields combine, where they abut, to generate a new irregularity that produces a subsequent organ. The critical action of a primordium to orient cellulose is thought to be a cytoskeletal response to tangential stress created by rapid growth of the primordium base. Several phyllotactic systems are variations of this biophysical cycle. The variations relate to the differential growth of primordium bases (85).

At the floral transition, due to the decrease in primordium size, this cycle is now restricted to the meristem periphery. The center of reproductive meristems in *Echeveria* and *Catharanthus* exhibits no clear reinforcement pattern (84; P. B. Green, in preparation). Thus, one apparently critical event of the transition is the elimination of the pattern characteristic of the vegetative state. This is essentially achieved by the increase in meristem size, and also perhaps by the elongation of apical internodes, which may contribute to the loss of influence of the last-formed leaves (212).

The successive kinds of floral organs in *Echeveria* are each formed in a specific way (P. B. Green, in preparation). The sepals are unique in that they are produced in the absence of a preexisting reinforcement pattern. Petals and carpels both arise from a band of circumferential reinforcement, in which axiality shifts occur to provide the missing components of hoop reinforcement. Stamens are different since all components of their hoop-reinforced structure are already provided by parts of sepal and petal fields. For Green, these differences determine the subsequent nature of the primordia, and they should precede the expression of organ-specific genes. Thus, he sees sequential mechanical stress pattern as a major controlling factor not only of floral phyllotaxis, but also of the differentiation of floral organs. This view is not without limitations. First, phyllotactic changes, similar to those occurring at flowering, may occur without any shift in the nature of the organs produced— e.g., after application of some PGRs (see above) or in proliferous and virescent "flowers" (2, 8, 116, 139). Also, flowering is not just a change in phyllotaxis but, as indicated previously, results from complex alterations of the growth pattern affecting, in an integrated fashion, all apex parts. Many aspects of this integration are clearly species-specific and thus genetically controlled from the earliest steps of morphogenesis. Finally, specific isoperoxidases appear at the earliest stage of stamen formation in *Mercurialis* (108); and, in *Brassica,* increases in several enzyme activities mark incipient flower primordia (183) (see also speltoid wheats, below). The weight of evidence then suggests that physical parameters interact with gene products to make forms. The way this interaction is achieved is entirely obscure; but, since lateral growth of primordium bases is apparently the decisive factor in phyllotaxis, genes may play a role here in controlling the rate and direction of growth in these localized areas. This is supported by observations made on some morphogenetic mutants (see below). The participation of PGRs (e.g. GA, cytokinins, and ethylene) known to control the orientation of microtubules and cellulose microfibrils in various systems (92) could also be envisaged.

ORGAN NUMBER AND FUSION   The number for each type of floral organ is often considered as a fixed character in the flowers of many species, especially in flowers with a small number of appendages of each sort. Even in these flowers, however, modifications of these numbers can be obtained by a variety of treatments (116)—e.g. after application of cytokinin in tobacco (153) or GA$_3$ in tomato (209).

One of the most fascinating and unique aspects of floral morphogenesis, contrasting it with vegetative morphogenesis, is organ fusion. In *Catharanthus*, fusion of initially free carpel primordia requires not only physical contact but also a diffusible unknown compound (245). 2,4-D has been observed to cause, among other varied abnormalities, fusion of parts in flowers with normally free appendages—e.g. sepals in *Digitalis* or petals and stamens in *Saponaria*—or independent growth of normally concrescent organs—e.g. petals in *Digitalis* (2).

Genetics

Floral morphogenesis is controlled by many genes, and the array of flowering mutants is an invaluable tool, as yet little exploited (151), for unravelling its molecular basis. Modifications in mutants may take a variety of forms that cannot be exhaustively described here. Only a few examples are given.

The *Ppd* and *Vrn* genes in wheat, responsible for day-length and cold sensitivity, respectively, influence various aspects of growth of the reproductive apex—i.e. either the duration or the rate of spikelet initiation (70, 210). In "speltoid" wheat mutants, the basal florets of each spikelet are missing or defective instead of developing until anthesis as is normally the case. A comparison of normal and speltoid wheats showed that the genes controlling floret development exert their effect prior to the initiation of the floret primordium (46).

In hooded barley, a single mutation causes the development on the lemma of two extra florets, the proximal one showing inverted polarization (228). Remarkably, this complex structure appears to result simply from an increase in the rate of cell division accompanied by a loss of the polarity of cell division and elongation in the lemma when it is 300–600 μm long.

The capitulum of *Layia discoidea* is identical to that of its relative *L. glandulosa* from which the ray florets would have been omitted (82). This difference, controlled by two genes, seems also related to well-timed and localized changes in cell proliferation. This interpretation implies that *L. discoidea* has not lost the genetic information to produce ray florets but that this information is not used because conditions are not locally and temporally appropriate (82).

A great number of mutants show the conversion of organs of one type into those of another. These homeotic mutations include, for example, the *ap-2 Arabidopsis* with sepals resembling leaves and petals converted into stamens, and the *pi Arabidopsis* in which petals are transformed into sepals and stamens into carpels (159). In the *blind Petunia*, with petals transformed into stamens, the transformation results again from a very localized change in rate and orientation of cell divisions occurring at a precise stage of petal primordium development (243).

Many other mutants have an increased or decreased number of floral organs of some type. The Do_1 Petunia has extra petals and stamens (62) whereas the ap-1 Arabidopsis has no or rudimentary petals (159). In the pi mutant of tomato with no or rudimentary stamens, the inflorescence continues to grow at its distal end as a typical leafy shoot (200).

Fusion of normally free parts, or independent growth of normally fused parts, may be observed in other mutants. Examples are the ch_3 Petunia with petals not fused (62); or the sf tomato in which sepals, petals, stamens, and carpels all remain free while organs of each floral whorl fuse in the wild type (40). Lack of fusion in the last case seems essentially due to the smaller size of primordia at initiation relative to meristem size. Note that both ch_3 Petunia and sf tomato have leaves of abnormal shape.

The alf Petunia behaves as the wild type until before flowering; then the leaves become smaller, the stem undergoes unlimited branching, and flower structure is very aberrant (80; A. G. M. Gerats et al, unpublished). In old leaves and floral tissues, the level of putrescine and the activity of arginine decarboxylase and its mRNA are 2–4-fold higher in the mutant than in the wild type (80). Isolated tobacco cell lines with altered polyamine synthesis, which could be successfully regenerated into whole plants, also reveal abnormal floral morphologies, including staminoid ovules, stigmoid stamens, etc (146). These observations are substantial in their implication of polyamines in reproductive morphogenesis.

Only in rare occasions do we know where, when, and how the mutant flowering genes act. Moreover, most of them have pleiotropic effects. Their exact modes of action will only be determined by a combination of classical and molecular genetics, physiology, and morphology (196).

THEORIES OF ENDOGENOUS CONTROL

The Florigen/Antiflorigen Concept

SUPPORTING EVIDENCE   The basic evidence for the existence of one or several transmissible leaf-generated promoters and inhibitors, "florigen" and "antiflorigen," comes essentially from grafting experiments. Photoinduced leaves in plants of all photoperiodic response types are supposed to produce florigen. Antiflorigen would be produced in noninduced leaves, at least in some plants. Transmission of the promoter(s), through a graft union, was obtained between different species and different photoperiodic response types in eight plant families, suggesting that florigen is similar, if not identical, in all plants. This evidence was reviewed by Zeevaart (258). Although more recent and limited, similar results concerning the inhibitor(s) were obtained in grafting experiments, leading Lang (129) to suggest that antiflorigen might also be identical in all photoperiodic plant types. Movement of these compounds is generally in the phloem along with assimilates (15, 115, 128, 260), and their action is often believed to be at the meristem [but see Schwabe (213) for evidence on nontransmissible inhibitors acting in leaves]. In plants shown to produce both florigen and antiflorigen, floral evocation would be caused when the balance of these two factors at the meristem is shifted in favor of florigen. Since no known chemical has been found to have a clear promotive or inhibitory effect on flower initiation in all plants, florigen and antiflorigen are usually supposed to be unknown hormones specific for the control of this developmental step.

There are, however, a number of difficulties with this unified theory. First, since grafting experiments can only be made between compatible species, they are totally unable to show, for example, that the promoters of Solanaceae and Crassulaceae are identical, or that the Solanaceae inhibitor is the same as that common to pea and sweet pea (203). Moreover, results of grafting experiments with compatible plants are often negative or anomalous (15, 213, 258), cautioning against hasty generalizations. J. D. Metzger (personal communication) recently found that photoinduced *Sinapis* stocks cause flowering of nonthermoinduced *Thlaspi* scions, whereas thermoinduced *Thlaspi* are unable to cause flowering of nonphotoinduced *Sinapis*. This new case of nonreciprocal transmission suggests that part of the controlling factors in these two Cruciferae is different (15, 16).

Specificity of these factors may also be questioned. The leaves of *Solanum andigenum* produce a transmissible tuber-promoting material when exposed to SD. Chailakhyan et al (38), by grafting florally induced and noninduced scions of either a SD or a LD tobacco onto stocks of this potato, found that the induced scions, but not the noninduced ones, can cause tuber formation in the stocks. This suggests that part at least of the transmissible floral promoters of photoperiodic tobaccos and the tuber promoters of potato are identical and thus unspecific. Similarly, the transmissible floral promoter(s) of cucurbits were shown to be involved in the control of sex expression (233). These results are in line with the observations showing that genes involved in the control of flowering have generally pleiotropic effects.

So-called "translocation" curves for promoter or inhibitor can be obtained, in single-cycle plants, by sequential removals of induced or noninduced leaves, respectively. These curves were also used as supporting evidence for the existence of florigen and antiflorigen (128, 129). However, the significance of these curves can be questioned on several grounds (15, 247). First, if the promoter and/or inhibitor are multifactorial, these curves only tell when their slowest component is exported by the leaves. Indeed, as mentioned above, many dramatic changes are seen in the apex of several plants well before the time of movement out of the leaves of the putative promoter. Second, the immediate destination of exported materials is unknown. The assumption has been that they go directly to the apices, but this is certainly an oversimplification. There is evidence that, in *Xanthium* and *Sinapis,* part at least of the leaf-generated photoperiodic stimulus is transmitted to the roots (134, 252) (see Cytokinins, below).

FLORIGEN AND ANTIFLORIGEN IDENTIFICATION   After decades of extensive work, we still do not have the merest idea of the chemical structure of these hypothetical hormones (15, 213, 258, 260). Progress in this area has been almost nil during the last years. Total ethanolic extracts, prepared in Chailakhyan's lab, from leaves of SD and LD tobaccos, both grown in SD, consistently cause flower formation in seedlings of the SDP *Chenopodium* kept in LD (35, 37). Extracts from leaves of the two types of tobaccos grown in LD elicit flowering in the LDP *Rudbeckia* kept in SD (36). For Chailakhyan (36), these results reinforce his idea that florigen is composed of two complementary materials: an hypothetical "anthesin" produced in SD by both SDP and LDP, and a GA made in LD by both types of plants. Note that so far the active compound(s) or anthesin in the extracts from SD-grown tobaccos have not been identified. Zeevaart (258) reviewed evidence against generalization of this hypothesis. Of particular importance in this rebuttal is the fact that the role of GAs in the floral transition is apparently not the same in all plants (see below).

The exudate (phloem sap) from induced *Perilla* leaves causes no consistent stimulation of flowering of cultured *Perilla* shoot apices (197, 262). Vegetative stem calluses from a DN tobacco—i.e. calluses regenerating only vegetative buds—release inhibitory material(s) whereas reproductive calluses regenerating flowers produce promoter(s) (34). These unknown compounds can influence flower formation in reproductive calluses, but the promoter(s) are unable to cause flowering in vegetative calluses.

Isolation and identification of these factors should be facilitated by analysis of extracts with the powerful techniques of HPLC and GC-MS. So far, however, this approach has not yielded much. *Bis*(2-ethylhexyl)hexane dioate, identified by these techniques in several SDP and claimed to be a floral inhibitor (103), is in fact a contaminant released from plastic materials used for culturing the plants (M. J. Jaffe, personal communication). Analyses of the neutral, acidic, and basic fractions of leaf exudate in *Perilla* failed to reveal significant differences between induced and noninduced plants (262). Active compounds have been identified in SD and LD duckweeds, but they are all present in nearly the same amounts in induced and noninduced plants (235).

The outcome of this work is discouraging and looks like a dead end. The florigen/antiflorigen concept is possibly inadequate for large generalizations. Its major weakness, in my opinion, is that it fails to account for all the complexities of the flowering process. Thus, if we wish to do something more than elaborate on a concept put forward in the 1930s, it is necessary to consider alternate hypotheses.

Electrical Signal

Various stimuli, such as wounding or light, applied to one plant part may exert their effect in other parts so rapidly that transport of a chemical in the symplast or the phloem seems excluded (113). In some such cases, a signal was found to be a fast-propagating wave of electric depolarization (71). Since photoinduction may alter membrane properties and possibly ion fluxes, part at least of the stimulus exported by induced leaves could be a fast-moving electrophysiological signal (114). So far, however, all attempts to detect the transmission of such a signal have failed (167).

The Nutrient Diversion Hypothesis: Control by Assimilates

For many physiologists, assimilates have only a supportive role in flower initiation, simply providing energy for each step of the process (128). A counter "nutrient diversion" hypothesis (208) postulates that induction, whatever the nature of the involved factors, is a means of modifying the source/sink relationships within the plant in such a way that the shoot apex receives a higher concentration of assimilates than under noninductive conditions. Sachs & Hackett (208) raised doubts concerning the possibility of separating the hypothetical florigen and antiflorigen from nonspecific materials, like assimilates, moving along in the phloem sap. They argue that all factors that prevent assimilates from induced leaves from reaching the receptor meristem will appear as floral inhibitors. This is the case, for example, of treatments promoting activity in sinks competing with the meristem for available assimilates, such as repeated rooting which prevents flowering in tobaccos (59, 152). Similarly, the inhibitory effect of noninduced leaves, positioned between induced leaves and the meristem, can be interpreted either by an inhibitor production (213) or on the basis that these leaves provide an alternate source of assimilates (128, 263).

Conversely, agents promoting assimilate supply to the meristem will appear as floral promoters. Indeed, removal or restriction of competing sinks (young leaves, roots) was shown above to promote flower initiation in many plants. Some effects of exogenous PGRs on flowering may also sometimes be explained on the basis that they alter assimilate distribution, as in the case of cytokinin in *Pharbitis* (1, 178, 179).

SUPPORTING EVIDENCE   Many factors controlling flower initiation also influence photosynthesis and/or assimilate availability (16, 206–208). In a variety of plants, the proportion of leaf photosynthates retained as starch is much lower and assimilate export higher under LD than under SD (9, 29). Interestingly, these differences are already detectable 24 h after exposure to a single LD, a single NB, or a single displaced SD in *Digitaria* (29). Increased irradiance or CO_2 level in the air often promotes flowering (16, 75, 207, 208); and we have seen that, in some plants, these treatments may even override the photoperiodic or cold requirement. In the facultative LDP *Brassica campestris,* flower initiation is earlier the higher the photon flux density, and there is a linear relationship between the reciprocal of day length and time to initiation (76). The equal effectiveness of fluorescent and incandescent light given at equal PAR indicates that photosynthesis plays a critical role in this species.

Sucrose or glucose applications promote flowering in several SDP and LDP, sometimes even in noninductive conditions (see 24). In vitro, the optimal sugar concentration for flower production by nonphotosynthetic explants is generally far above that for vegetative bud formation (16). In *Brassica,* again, Friend et al (77) reported that sucrose feeding causes earlier flower initiation in SD and at a lower final leaf number. Thus, the effect is specifically on flowering rather than simply through a general promotion of growth.

ASSIMILATE IMPORT IN THE APEX   Studies on this topic are scanty and inconclusive. In *Pharbitis,* the overall ^{14}C-assimilate distribution between the major plant parts is unchanged after exposure to an inductive long night, but apex import is increased (1). The same treatment in *Xanthium* causes first a decrease and then an increase of apex import (C. Mirolo, unpublished), whereas in *Lolium* and *Sinapis* there is no evidence for an elevated assimilate import into the apex in response to one LD (26, 68). One limitation of these short-term studies is that they are concerned only with recently made assimilates. As the sucrose content of the apex increases early at evocation in *Sinapis* (21, 25), a remobilization of reserve carbohydrates stored within the bud or other plant parts might have occurred (26). Preliminary results on *Pharbitis* indicate that 82% of respiratory CO_2 during the short night in noninduced plants is from CO_2 fixed during the previous light period, whereas only 42% of respiratory CO_2 during a long night is from recently fixed CO_2 (57). In further studies, the extent of reserve remobilization should be explored.

Competition between bud components for assimilates might also occur. In *Pinus,* treatments with $GA_{4/7}$ that promote early and enhanced flowering reduce the flow of assimilates into the terminal bud as a whole, but also cause a significant reallocation of ^{14}C-assimilates within the bud, with more ^{14}C moving to developing seed-cone primordia at the expense of structural tissue (204). Such a situation could be encountered in other plants, so that all bud components should be separately analyzed in further studies (1, 26).

TIMING OF PHOTOSYNTHETIC OR CARBOHYDRATE INPUT   Exposure to high-intensity light during the first half of an inductive LD promotes flowering in *Sinapis* whereas it is inhibitory during the second half (23). Removal of CO_2 from the air reverses both effects, showing that there is an optimal timing for increased photosynthetic input in this species. During the second half of the LD, an intermediate irradiance is best for induction (23). In wheat and sunflower, CO_2 enrichment promotes flowering, but again only when applied during a critical, relatively short, period (147).

There is an optimal timing for sugar application in the SDP *Chenopodium* (53), the LDP *Sinapis* (16, 24), and in cultured tobacco TCL (49); and supraoptimal concentrations inhibit flowering both in *Brassica* (77) and in many cultured explants (16).

An interesting feature of the nutrient diversion hypothesis is that it assigns nonspecific (indirect) roles to many environmental, correlative, and chemical factors that have been shown to control flowering. However, there is probably greater sophistication in control than can be developed from mere assimilate diversion to receptor meristems. Evidence suggests that there is an optimal timing and an optimal input of assimilate supply to promote flower initiation. One may thus wonder whether there is a temporal coordination of supply with some events required for evocation. On the other hand, the sucrose level in the meristem of *Sinapis* increases before most other biochemical and cellular changes (12, 16, 25), ruling out the idea that this initial rise results from a higher demand of the activated meristem. Could then this rise exert a feed-forward control of evocation (207)? Apparently not, because when it is produced by exposing *Sinapis* to one SD at high irradiance only partial evocation ensues (94). Thus, except perhaps in cases like *Brassica*, increased assimilate supply cannot be the sole signal for flower initiation.

The Model of Multifactorial Control: Participation of Plant Growth Regulators

Parts of floral evocation and morphogenesis can be produced by application of various chemicals, including carbohydrates, PGRs, and PGR antagonists. This has led Bernier et al (16) to propose that several factors, promoters and inhibitors, are involved in the control of flower initiation. This developmental step is believed to occur only when all factors are present in the apex at appropriate concentrations and times. While assimilates and PGRs are generally present in most plants, some of these compounds may be absent or present in sub- or supraoptimal amounts. Hence, each factor will not necessarily act in the same direction in all plants. Genetic variation, as well as past and present growing conditions, have resulted in different factors of the complex becoming the critical or "limiting" factor(s) in different species or in a given species in various environments.

Work with exogenous PGRs, if not complemented with other data, should be gauged cautiously (16, 258, 259). No effort is made to cover it comprehensively here. An important component of the response to PGRs, which is given consideration whenever possible, is the sensitivity of the responding tissue (242). For studies on endogenous PGRs, attention is directed mainly to those using present-day analytical techniques and experimental systems in which the temporal and spatial aspects of the flowering process are relatively well known. In duckweeds, only whole-plant analyses have been performed (78), possibly mixing opposite changes in the different plant parts.

CYTOKININS   Exogenous cytokinins cause promotion and inhibition of flower initiation in a variety of species, although promotive effects are much more frequent than inhibitory ones (16). Recent cases of promotion include several SD duckweeds in which flowering can be obtained in the absence of photoinduction by a sequential treatment with benzoic acid and a cytokinin (234). In *Chenopodium* (126), *Anagallis* (20), and *Sinapis* (23), inhibition or promotion has been observed depending on the amount of cytokinin applied and/or timing of treatment. Thus, there is apparently an optimal dose and time of sensitivity. Interestingly, the cytokinin effect is often dependent on presence or absence of other PGRs. In several species—e.g. *Chrysanthemum, Chenopodium,* orchids, etc—a GA reinforces dramatically the promotive effect of cytokinins; and in *Scrophularia,* the concentration of kinetin required to inhibit flowering is far lower in the absence of IAA than in its presence (see references in 16).

In *Arabidopsis,* the stimulation of cell division in the central zone of the meristem and subsequent flowering, by cytokinin (or GA), is possible only in plants having meristems with an "intermediate" configuration—i.e. plants grown in noninductive SD for at least 3 months (17). Thus, sensitivity of the central zone to cytokinin (or GA) increases with age, a feature that may be of critical importance for evocation. In *Pharbitis,* there is evidence that the inductive long night sensitizes the apex to exogenous cytokinin (1).

Cytokinin is also a requirement for flowering in many in vitro systems (16, 223, 240, 244). In tobacco TCL, cytokinin must be present, but if concentration is supraoptimal, vegetative buds are produced instead of flowers (240). Cytokinin is an absolute requirement for flowering of explants from adult *Passiflora* plants, but juvenile explants do not flower even in its presence (215). This suggests that one or more other factors act in conjunction with cytokinin to initiate flowering in this species.

The level and/or metabolism of endogenous cytokinins change markedly, sometimes transiently, in relation to the floral transition—e.g. in *Iris* (86) and chicory root during vernalization (106), as well as in prereproductive buds of *Pinus* (236) and Douglas fir (162). In this last case, where GAs, IAA, and ABA were also examined, the increase in cytokinin is the most obvious difference between vegetative and prereproductive buds. An increase in zeatin riboside occurs in photoinduced leaves of a LD tobacco but not in those of a SD tobacco (136). In *Sinapis*, previous evidence (e.g. from cytokinin application) suggested that the endogenous changes are critical at 16 h after start of the inductive LD (14). Lejeune et al (134, 134a) observe now that, at 16 h, cytokinin levels in roots are lower and levels in leaves are higher in induced than in vegetative plants. At the same time, there is a marked increase of cytokinin activity in both the root and leaf exudates of induced plants.

In *Xanthium*, induced by one long night, there is a marked decline of cytokinin activity in leaves, buds, and root exudate (252). Flowering and decrease in cytokinins are both nullified by a NB, suggesting that they are closely related. Thus, though the trend of changes is opposite in *Xanthium* and *Sinapis*, these changes seem essential in both plants. A similarity between the two plants is the increase in cytokinin content in the phloem sap exported by induced leaves (134a, 191). As postulated by Wareing et al (252), it appears that a signal of unknown nature, produced by induced leaves in *Xanthium* and *Sinapis*, moves rapidly to the roots where it alters the course of cytokinin production and/or release. Bark-ringing experiments suggest that in *Xanthium* this leaf-to-root signal moves in the phloem (252).

GIBBERELLINS   GAs are the most extensively studied class of PGRs because they can elicit a flowering response in many LD- and cold-requiring rosette plants grown in noninductive conditions (16, 128, 259, 261). Work on GAs was reviewed by Pharis & King (189) and is only summarized here.

Although the critical role of GAs in the control of internode elongation is generally recognized (83, 261), their participation in the control of flower initiation is a controversial matter (16, 95a, 189, 261). In some rosette LDP, such as *Samolus, Rudbeckia,* and *Arabidopsis,* growth retardants suppress both bolting and flowering in LD, and GA overcomes these effects, suggesting that in these species GAs are the primary controlling factors of flowering. In other rosette LDP, however—e.g. spinach, *Silene,* and *Agrostemma*— growth retardants suppress bolting, reduce considerably GA levels, but leave flowering unaffected. In addition, not all rosette plants could be induced to flower by GA. Many caulescent species do not flower after GA treatments. In others, however—including the SDP *Pharbitis, Impatiens,* and *Chrysanthemum;* the LSDP *Bryophyllum;* and the perennials grapevine, *Cordyline,* and Conifers—there is evidence that GAs are part of the promoters of flowering. Finally, GAs are clearly inhibitory to flower initiation in some perennial angiosperms, such as the SDP strawberry and *Ribes,* the LDP *Fuchsia,* the SLDP *Poa* and *Bromus,* and the DNP tomato, apple, and *Citrus.*

There are several reasons for this apparent complexity. First, as stressed by Pharis et al (135, 188, 190), the effectiveness of various GAs is different in different species: Pinaceae react to $GA_{4/7}$, not to GA_3; in apple, GA_3 is inhibitory and GA_4 promotive; in *Lolium,* 2,2-dimethyl GA_4, GA_{32}, and GA_5 are more active than GA_3 whereas GA_1 is almost inactive, and so on (16). In many cases, the most effective GAs in eliciting flowering are relatively ineffective for promotion of stem elongation. Thus, different GAs may be involved in the control of these two processes. Second, in some plants, the GA effect is enhanced or even only detectable when GA is associated with an adjunct treatment, such as water stress or root pruning in Pinaceae (190). Third, the timing of GA application is critical. This was best demonstrated in single-cycle plants (16). In dwarf *Pharbitis,* for example, GA_3 is promotive when applied just before the inductive long night and inhibitory just after (118). In grapevine, GA_3 is promotive at the early "anlage" stage of the floral transition but is inhibitory later when cytokinins become the predominant promoter (170). The reverse sequence of GA effects was found in *Fuchsia* and tomato (16, 116). Timing of treatments is also critical in perennial grasses, apple and Pinaceae (95a, 135, 190). Fourth, the GA responses may be much influenced by the growing conditions, the effects being generally much greater in LD than in SD or at moderately low temperatures than at higher temperatures (16, 128, 157). Recall here that low temperature can confer GA sensitivity to insensitive genotypes of wheat aleurone tissue (226). Flowering in response to GA_3 in *Silene* requires the presence of gene F (256) and of an "intermediate" meristem (19). Young F plants behave like f plants that do not flower after GA_3 treatments (18). The F allele seems in fact to accelerate the reaching of the intermediate stage when the central zone of the meristem becomes sensitive to GA_3.

Sorghum genotypes having the Ma_3 or ma_3 allele, treated with GA_3, become very similar to untreated genotypes having the ma^R_3 gene (187). The ma^R_3 allele, which makes plants insensitive to photoperiod, thus seems responsible for increased GA activity.

Several steps of the GA biosynthetic pathways are known to be controlled by day length (55, 83). Low temperature relieves the block(s) somewhere between kaurenoic acid and GA_9 in the non-C-13-hydroxylation pathway in *Thlaspi* (158). Thus, we may expect important changes in the levels of various GAs, together with the formation of new GAs, when plants are subjected to photo- or thermoinduction. There is, indeed, a transient increase in GA level and appearance of new GAs in induced *Hyoscyamus* leaves (127). These changes clearly precede bolting and flower formation. In *Lolium,* polyhydroxylated GAs (GA_{32}?) increase 2–3 fold in the apex on the day following the inductive LD and then subside, whereas the level of a GA_1-like GA does not change (188). In Pinaceae, the cultural treatments that act synergistically with $GA_{4/7}$ applications all tend to increase the level of less polar GA-like substances and decrease GAs of a more polar nature (190).

In pea, allelic differences at the *Sn, Dne, Hr,* and *Lf* loci are clearly expressed in dwarf mutants having almost undetectable levels of GAs (173), suggesting that flowering genes in pea do not act by influencing directly GA metabolism. Note, however, that the dwarfing genes of pea are "leaky" (173) and are not equally operative in all plant parts (193). Genes *na* and *Le,* for example, are operative in vegetative shoots but not in developing seeds. Similarly, GA-insensitive mutations in *Arabidopsis* and *Silene,* expressed during vegetative growth, do not extend to the reproductive phase of the life-cycle (123, 231). Thus, there is the possibility of a rapid alteration in GA metabolism and/or sensitivity at the floral transition. This change might first occur in the evoked apex, as suggested by the observations on *Lolium* (188). Small localized changes in GA levels and/or complement may have escaped detection in studies where whole shoots were analyzed, as was the case in pea (173). Change in apex sensitivity in LD-induced *Lolium* would explain why a GA applied in SD does not mimic the effect of the LD on cell division (see Evocation, above).

In view of all of this, I feel that the role of GAs in flower initiation needs reexamination.

AUXINS   The previous conclusions (16, 259) that (*a*) exogenous auxin can both inhibit and promote flower initiation in SDP and LDP, and (*b*) inhibition is far more widespread than promotion, were verified in recent studies. In *Sinapis,* as in several other plants (16), auxin is promotive at low doses and inhibitory at high doses (23). Inhibition at high doses might simply be related to a general growth inhibition (100) or an auxin-induced ethylene biosynthesis, as in *Chenopodium* (125).

Studies in vitro have confirmed that auxins do not simply oppose flowering; their presence in a certain concentration range is absolutely required if flowers are to be formed in various types of explants (16, 240). For TCL from tobacco flower pedicels, the optimal auxin concentration is ten times higher at some stage of the culture than before and after (244). In some intact photoperiodic plants, auxin may inhibit or promote flowering when applied, at the same concentration, during or after induction, respectively (16, 259). This is perhaps a hint that sensitivity to auxin may change during the process in vitro and in vivo.

Recent work on endogenous IAA confirms the results of earlier studies

(16), showing that levels are generally lower in SDP just before flower initiation—e.g. in *Chenopodium* (125) and in *Sorghum* (64).

These observations tend to implicate auxins in the floral transition. Apparently, high levels inhibit but low levels may be equally limiting. In view of the manifold and dramatic effects that auxins or their antagonists have on different aspects of floral morphogenesis, a reinvestigation of their levels and fluxes in well-controlled systems is long overdue. Not only IAA, but also other auxins, should be considered (100).

ETHYLENE   Exogenous ethylene inhibits flowering in many SDP (16, 259). In *Chenopodium,* induction by long nights results in a decrease of the ACC to ethylene conversion; NB suppress both flowering and this decrease (144). On the contrary, exposure of the LDP spinach to one LD increases markedly this conversion (51). In this case, the plant reacts as if the inductive SD-to-LD shift was a stress.

In *Iris,* ethylene causes flower initiation in bulbs too small to be able to flower (58). After completion of flower formation, however, ethylene induces abortion, indicating a stage-dependent change of sensitivity. In mature *Guzmania,* flowering can be obtained following an ethylene or ACC treatment or by shaking the plants for 15 sec (60). As expected, AVG prevents ethylene release and subsequent flowering in shaken plants, but not in ACC-treated plants. These data are consistent with the view that, in mature *Guzmania* and probably other bromeliads (16, 259), ethylene is the only controlling factor of flowering. Juvenile *Guzmania* plants have a reduced capacity to convert ACC into ethylene (60). Ripeness-to-flower seems related to the development of this capacity at the apex. However, since ethylene applications to juvenile plants do not cause flowering, at least another factor is limiting. Is it absence of sensitivity to ethylene?

Overcrowding stimulates ethylene production by duckweeds, and this in turn might affect their flowering responses (211).

ABSCISIC ACID   Previous observations showed that exogenous ABA is inhibitory in several SDP and LDP grown in inductive conditions, and endogenous ABA levels in several photoperiodic species do not bear consistent relationships with day length (16, 259). Exudates from induced and noninduced *Perilla* leaves contain similar amounts of ABA (197). A late-maturing genotype of *Sorghum* has, however, preflowering levels of ABA at least four-fold greater than an earlier-maturing genotype (64). Thus, ABA does not appear as a major determinant in the floral transition, except perhaps in some species. Another possibility is that ABA would be a general "background" inhibitor, produced more or less constantly irrespective of day length, its effect being overcome in inductive conditions by increased amounts of promoters.

OTHER COMPOUNDS   A range of exogenous unrelated chemicals may affect flowering (16). Prominent among them are phenolics (112, 222), oligosaccharins (241), polyamines, and their conjugates with hydroxycinnamic acids (150, 201, 239). Endogenous levels of polyamines and conjugates change in relation to flowering (16, 80, 146, 150, 239).

CONCLUSIONS

The mechanisms of control of flower initiation defy easy explanations. They appear basically different in different species. They may have evolved independently in a wide range of plant groups (171). As already postulated for woody perennials (202), flowering in all plants seems under the control of several biochemical/physiological systems, all of which must be permissive if reproductive structures are to be formed. The multifactorial model of control (a) best accounts for the complexity that we have encountered at all steps of the process, (b) includes assimilates among the controlling factors and thus incorporates the nutrient-diversion idea, and (c) is not contradictory to the results of grafting experiments showing that transmissible floral promoter(s) and inhibitor(s) are present in plants. It even offers an explanation for the failures and anomalies often seen in these experiments (16).

The evidence concerning participation of known PGRs in the control is still fragmentary, but some recent findings clearly suggest such a participation. The main problem is the lack of comprehensive studies that relate, in well-defined experimental systems, changes in proportions, levels, metabolisms, and fluxes of these substances to the successive steps of the flowering process. Possible changes in sensitivity to PGRs, especially at the apex, have received little attention in the past. This gap should also be filled.

I do not exclude that one or more new regulatory factors remain to be discovered, but I do not expect that these putative new factors will be the exclusive controlling agents of flowering.

Each individual factor of the regulatory complex controls specific events of evocation and/or morphogenesis. The whole process is triggered only when an appropriate balance or sequence of the required factors is achieved. In the harmonious integration of partial processes, giving rise to flower primordia, factors other than regulatory chemicals must also be considered, namely physical forces (85) and, overall, the genetic makeup. The molecular biology and genetics of flowering are still fields of ignorance. A great deal of work in these areas is urgently needed. Flowering and PGR mutants will be the tool of choice in these investigations.

ACKNOWLEDGMENTS

I wish to acknowledge the help and patience of my wife, Martine, during the preparation of this review and the suggestions of Paul B. Green for English improvements. Work by my coworkers and myself was supported by the Belgian Government (Action de Recherche Concertée), F.R.F.C. and I.R.S.I.A.

Literature Cited

1. Abou-Haidar, S. S., Miginiac, E., Sachs, R. M. 1985. [^{14}C]-Assimilate partitioning in photoperiodically induced seedlings of *Pharbitis nil*. The effect of benzyladenine. *Physiol. Plant.* 64:265–70

2. Astié, M. 1962. Tératologie spontanée et expérimentale. Application à l'étude de quelques problèmes de biologie végétale. *Ann. Sci. Nat. Bot.* 12e (sér. 3): 619–844

3. Atherton, J. G. 1987. *Manipulation of Flowering.* London: Butterworths

4. Auderset, G., Gahan, P. B., Dawson, A. L., Greppin, H. 1980. Glucose-6-phosphate dehydrogenase as an early marker of floral induction in shoot apices of *Spinacia oleracea* var. Nobel. *Plant Sci. Lett.* 20:109–13

5. Auderset, G., Sandelius, A. S., Penel, C., Brightman, A., Greppin, H., Morré, J. 1986. Isolation of plasma membrane and tonoplast fractions from spinach leaves by preparative free-flow electrophoresis and effect of photoinduction. *Physiol. Plant.* 68:1–12

6. Badila, P., Lauzac, M., Paulet, P. 1985. The characteristics of light in floral induction in vitro of *Cichorium intybus*. The possible role of phytochrome. *Physiol. Plant.* 65:305–9

7. Barendse, G. W. M. 1964. Vernalization in *Cheiranthus allionii* Hort. *Meded. Landbouwhogesch. Wageningen* 64(14):1–64

8. Battey, N. H., Lyndon, R. F. 1986. Apical growth and modification of the development of primordia during reflowering of reverted plants of *Impatiens balsamina* L. *Ann. Bot.* 58:333–41

9. Baysdorfer, C., Robinson, J. M. 1985. Sucrose and starch synthesis in spinach plants grown under long and short photosynthetic periods. *Plant Physiol.* 79:838–42

10. Bernier, G. 1979. The sequences of floral evocation. See Ref. 39, pp. 129–68

11. Bernier, G. 1984. The factors controlling floral evocation: an overview. See Ref. 249, pp. 277–92

12. Bernier, G. 1986. The flowering process as an example of plastic development. In *Plasticity in Plants*, ed. D. H. Jennings, A. J. Trewavas, pp. 257–86. Cambridge: Company of Biologists

13. Bernier, G. 1986. A quantitative study of morphological changes in the apical bud of *Sinapis alba* in transition to flowering. *Arch. Int. Physiol. Biochim.* 94:PP36

14. Bernier, G., Kinet, J.-M., Jacqmard, A., Havelange, A., Bodson, M. 1977. Cytokinin as a possible component of the floral stimulus in *Sinapis alba*. *Plant Physiol.* 60:282–85

15. Bernier, G., Kinet, J.-M., Sachs, R. M. 1981. *The Physiology of Flowering*, Vol. I. Boca Raton: CRC

16. Bernier, G., Kinet, J.-M., Sachs, R. M. 1981. *The Physiology of Flowering*, Vol. II. Boca Raton: CRC

17. Besnard-Wibaut, C. 1981. Effectiveness of gibberellins and 6-benzyladenine on flowering of *Arabidopsis thaliana*. *Physiol. Plant.* 53:205–12

18. Besnard-Wibaut, C., Cochet, T., Noin, M. 1985. Cell proliferative states and their consequences on apical morphogenesis in *Silene armeria*. *Acta Univ. Agric. Brno.* 33:429–34

19. Besnard-Wibaut, C., Noin, M., Cochet, T. 1985. Apical reactions to gibberellic acid application according to the genotype in *Silene armeria*. *Biol. Plant.* 27: 360–66

20. Bismuth, F., Miginiac, E. 1984. Influence of zeatin on flowering in root forming cuttings of *Anagallis arvensis* L. *Plant Cell Physiol.* 25:1073–76

21. Bodson, M. 1977. Changes in the carbohydrate content of the leaf and the apical bud of *Sinapis* during transition to flowering. *Planta* 135:19–23

22. Bodson, M. 1985. Changes in adenine-nucleotide content in the apical bud of *Sinapis alba* L. during floral transition. *Planta* 163:34–37

23. Bodson, M. 1985. *Sinapis alba*. See Ref. 88, Vol. IV, pp. 336–54

24. Bodson, M., Bernier, G. 1985. Is flowering controlled by the assimilate level? *Physiol. Vég.* 23:491–501

25. Bodson, M., Outlaw, W. H. Jr. 1985. Elevation in the sucrose content of the shoot apical meristem of *Sinapis alba* at floral evocation. *Plant Physiol.* 79:420–24

26. Bodson, M., Remacle, B. 1987. Distribution of assimilates from various source-leaves during the floral transition

of *Sinapis alba* L. See Ref. 3, pp. 341–50

27. Bonzon, M., Degli Agosti, R., Wagner, E., Greppin, H. 1985. Enzyme patterns in energy metabolism during flower induction in spinach leaves. *Plant, Cell Environ.* 8:303–8

28. Bridgen, M. P., Veilleux, R. E. 1985. Studies of de novo flower initiation from thin cell layers of tobacco. *J. Am. Soc. Hort. Sci.* 110:233–36

29. Britz, S. J., Hungerford, W. E., Lee, D. R. 1985. Photoperiodic regulation of photosynthate partitioning in leaves of *Digitaria decumbens* Stent. *Plant Physiol.* 78:710–14

30. Brulfert, J., Fontaine, D., Imhoff, C. 1985. *Anagallis arvensis*. See Ref. 88, Vol. I. pp. 434–49

31. Buban, T., Faust, M. 1982. Flower bud induction in apple trees: internal control and differentiation. *Hort. Rev.* 4:174–203

32. Buvat, R. 1955. Le méristème apical de la tige. *Ann. Biol.* 31:595–656

33. Carmi, A., Van Staden, J. 1983. Role of roots in regulating the growth rate and cytokinin content in leaves. *Plant Physiol.* 73:76–78

34. Chailakhyan, M. Kh., Aksenova, N. P., Konstantinova, T. N. 1982. Bud formation in trapezond tobacco stem explants in culture in vitro under conditions of their contact with explants having a different type of morphogenesis. *Dokl. Akad. Nauk. SSSR* 266:509–12

35. Chailakhyan, M. Kh., Grigor'eva, N. Y., Lozhnikova, V. N. 1977. Effect of leaf extracts from flowering tobacco plants on the flowering of *Chenopodium rubrum* shoots and seedlings. *Dokl. Akad. Nauk. SSSR* 236:773–76

36. Chailakhyan, M. Kh., Lozhnikova, V. N. 1985. The florigen hypothesis and its substantiation by extraction of substances which induce flowering in plants. *Fiziol. Rast.* 32:1172–81

37. Chailakhyan, M. Kh., Lozhnikova, V. N., Grigor'eva, N. Y., Dudko, N. D. 1984. Effect of extracts from leaves of vegetating tobacco plants on flowering of sprouts and seedlings of red goosefoot (*Chenopodium rubrum* L.). *Dokl. Akad. Nauk. SSSR* 279:1276–80

38. Chailakhyan, M. Kh., Yanina, L. I., Devedzhyan, A. G., Lotova, G. N. 1981. Photoperiodism and tuber formation in graftings of tobacco onto potato. *Dokl. Akad. Nauk. SSSR* 257:1276–80

39. Champagnat, P., Jacques, R. 1979. *La Physiologie de la Floraison*. Paris: CNRS

40. Chandra Sekhar, K. N., Sawhney, V.

K. 1987. Ontogenetic study of the fusion of floral organs in the normal and "solanifolia" mutant of tomato (*Lycopersicon esculentum*). *Can. J. Bot.* 65:215–21

41. Chen, C.-M., Petschow, B. 1978. Cytokinin biosynthesis in cultured rootless tobacco plants. *Plant Physiol.* 62:861–65

42. Chouard, P., Tran Thanh Van, M. 1970. L'induction florale et la mise à fleurs en rapport avec la méristématisation et en rapport avec la levée de la dominance apicale. In *Cellular and Molecular Aspects of Floral Induction*, ed. G. Bernier, pp. 449–61. London: Longman

43. Chwirot, W. B., Dygdala, R. S. 1986. Light transmission of scales covering male inflorescences and leaf buds in larch during microsporogenesis. *J. Plant Physiol.* 125:79–86

44. Cleland, C. F. 1984. Biochemistry of induction. The immediate action of light. See Ref. 249, pp. 123–36

45. Cockshull, K. E., Horridge, J. S. 1977. Apical dominance and flower initiation in the rose. *J. Hort. Sci.* 52:421–27

46. Considine, J. A., Knox, R. B., Frankel, O. H. 1982. Stereological analysis of floral development and quantitative histochemistry of nucleic acids in fertile and base-sterile varieties of wheat. *Ann. Bot.* 50:647–63

47. Cottrell, J., Dale, J. E., Jeffcoat, B. 1981. Development of the apical dome of barley in response to treatment with gibberellic acid. *Plant Sci. Lett.* 22:161–68

48. Cottrell, J. E., Dale, J. E., Jeffcoat, B. 1982. The effects of daylength and treatment with gibberellic acid on spikelet initiation and development in Clipper barley. *Ann. Bot.* 50:57–68

49. Cousson, A., Tran Thanh Van, K. 1983. Light- and sugar-mediated control of direct de novo flower differentiation from tobacco thin cell layers. *Plant Physiol.* 72:33–36

50. Cremer, F., Dommes, J., Van de Walle, C., Bernier, G. 1987. Changes in leaf mRNA complement during photoperiodic flower induction in *Sinapis alba*. *14th Int. Bot. Congr.*, West Berlin, p. 129 (Abstr.)

51. Crèvecoeur, M., Penel, C., Greppin, H., Gaspar, Th. 1986. Ethylene production in spinach leaves during floral induction. *J. Exp. Bot.* 37:1218–24

52. Croxdale, J. G. 1983. Quantitative measurements of phosphofructokinase in the shoot apical meristem, leaf primordia, and leaf tissues of *Dianthus chinensis* L. *Plant Physiol.* 73:66–70

53. Cumming, B. G., Seabrook, J. E. A.

1985. *Chenopodium.* See Ref. 88, Vol. II. pp. 196–228

54. Dale, J. E., Wilson, R. G. 1979. The effects of photoperiod and mineral nutrient supply on growth and primordia production at the stem apex of barley seedlings. *Ann. Bot.* 44:537–46

55. Davies, P. J., Birnberg, P. R., Maki, S. L., Brenner, M. L. 1986. Photoperiod modification of [^{14}C]gibberellin A$_{12}$ aldehyde metabolism in shoots of pea, line G$_2$. *Plant Physiol.* 81:991–96

56. Deitzer, G. F. 1987. Photoperiodic processes: induction, translocation and initiation. See Ref. 3, pp. 241–53

57. Deleens, E., Schwebel-Dugue, N., Miginiac, E. 1984. Study of carbon metabolism with stable isotope technique in *Pharbitis nil* induced to flowering by a short photoperiod. *4th Congr. Fed. Eur. Soc. Plant Physiol.,* Strasbourg, pp. 209–10 (Abstr.)

58. De Munk, W. J., Duineveld, Th. L. J. 1986. The role of ethylene in the flowering response of bulbous plants. *Biol. Plant.* 28:85–90

59. Dennin, K. A., McDaniel, C. N. 1985. Floral determination in axillary buds of *Nicotiana silvestris. Dev. Biol.* 112:377–82

60. De Proft, M., Van Dijck, R., Philippe, L., De Greef, J. A. 1985. Hormonal regulation of flowering and apical dominance in bromeliad plants. *12th Int. Conf. Plant Growth Substances,* Heidelberg, p. 93 (Abstr.)

61. De Silva, N. S. 1978. Phospholipid and fatty acid metabolism in relation to hardiness and vernalization in wheat during low temperature adaptation to growth. *Z. Pflanzenphysiol.* 86:313–22

62. De Vlaming, P., Gerats, A. G. M., Wiering, H., Wijsman, H. J. W., Cornu, A., et al. 1984. *Petunia hybrida:* a short description of the action of 91 genes, their origin and their map location. *Plant Mol. Biol. Rep.* 2:21–42

63. Diomaiuto-Bonnand, J., Le Saint, A.-M. 1985. La floraison et les conditions de son renouvellement périodique, chez une espèce vivace polycarpique: la giroflée ravenelle (*Cheiranthus cheiri* L.). II. Analyse des sucres solubles et des acides aminés libres, au niveau des bourgeons. Discussion. *Rev. Cytol. Biol. Vég.-Bot.* 8:63–87

64. Dunlap, J. R., Morgan, P. W. 1981. Preflowering levels of phytohormones in *Sorghum.* II. Quantitation of preflowering internal levels. *Crop Sci.* 21:818–22

65. Engelke, A. L., Hamzi, H. Q., Skoog, F. 1973. Cytokinin-gibberellin regulation of shoot development and leaf form in tobacco plantlets. *Am. J. Bot.* 60:491–95

66. Evans, L. T. 1969. *The Induction of Flowering.* Melbourne: Macmillan

67. Evans, L. T. 1971. Flower induction and the florigen concept. *Ann. Rev. Plant Physiol.* 22:365–94

68. Evans, L. T. 1976. Inhibition of flowering in *Lolium temulentum* by the photosynthetic inhibitor 3(3,4-dichlorophenyl)-1,1-dimethylurea (DCMU) in relation to assimilate supply to the shoot apex. In *Etudes de Biologie Végétale. Hommage au Professeur Pierre Chouard,* ed. R. Jacques, pp. 265–75. Paris

69. Evans, L. T. 1987. Towards a better understanding and use of the physiology of flowering. See Ref. 3, pp. 409–23

70. Flood, R. G., Halloran, G. M. 1986. The influence of genes for vernalization response on development and growth in wheat. *Ann. Bot.* 58:505–13

71. Frachisse, J.-M., Desbiez, M.-O., Champagnat, P., Thellier, M. 1985. Transmission of a traumatic signal via a wave of electric depolarization, and induction of correlations between the cotyledonary buds in *Bidens pilosa. Physiol. Plant.* 64:48–52

72. Francis, D. 1987. Effects of light on cell division in the shoot meristem during floral evocation. See Ref. 3, pp. 289–300

73. Francis, D., Lyndon, R. F. 1979. Synchronization of cell division in the shoot apex of *Silene* in relation to flower initiation. *Planta* 145:151–57

74. Francis, D., Lyndon, R. F. 1985. The control of the cell cycle in relation to floral induction. In *The Cell Division Cycle in Plants,* ed. J. A. Bryant, D. Francis, pp. 199–215. Cambridge: Cambridge Univ. Press

75. Friend, D. J. C. 1984. The interaction of photosynthesis and photoperiodism in induction. See Ref. 249, pp. 257–75

76. Friend, D. J. C. 1985. *Brassica.* See Ref. 88, Vol. II, pp. 48–77

77. Friend, D. J. C., Bodson, M., Bernier, G. 1984. Promotion of flowering in *Brassica campestris* L. cv Ceres by sucrose. *Plant Physiol.* 75:1085–89

78. Fujioka, S., Sakurai, A., Yamaguchi, I., Murofushi, N., Takahashi, N., et al. 1986. Flowering and endogenous levels of plant hormones in *Lemna* species. *Plant Cell Physiol.* 27:1297–1307

79. Gasser, C. S., Smith, A. G., Sachs, K. B., McCormick, S., Hinchee, M. A.

W., et al. 1987. Isolation and characterization of developmentally regulated sequences from tomato flowers. *J. Cell Biochem.* Suppl. 11B, 2

80. Gerats, A. G. M. 1987. Flower developmental mutants in *Petunia hybrida*. *J. Cell Biochem.* Suppl. 11B, 19

81. Gonthier, R., Jacqmard, A., Bernier, G. 1987. Changes in cell-cycle duration and growth fraction in the shoot meristem of *Sinapis* during floral transition. *Planta* 170:55–59

82. Gottlieb, L. D., Ford, V. S. 1987. Genetic and developmental studies of the absence of ray florets in *Layia discoidea*. In *Developmental Mutants in Higher Plants*, ed. H. Thomas, D. Grierson, pp. 1–17. Cambridge: Cambridge Univ. Press

83. Graebe, J. E. 1987. Gibberellin biosynthesis and control. *Ann. Rev. Plant Physiol.* 38:419–65

84. Green, P. B. 1985. Surface of the shoot apex: a reinforcement-field theory for phyllotaxis. *J. Cell Sci.* Suppl. 2:181–201

85. Green, P. B. 1986. Plasticity in shoot development: a biophysical view. In *Plasticity in Plants*, ed. D. H. Jennings, A. J. Trewavas, pp. 211–32. Cambridge: Company of Biologists

86. Gregorini, G. 1983. Cytokinins in *Iris* bulb-scales under flower-inducing conditions. *Sci. Hort.* 21:155–58

87. Greppin, H., Auderset, G., Bonzon, M., Degli Agosti, R., Lenk, R., Penel, C. 1986. Le mécanisme de l'induction florale. *Saussurea* 17:71–84

88. Halevy, A. H. 1985. *Handbook of Flowering*, Vols. I, II, III, IV. Boca Raton: CRC

89. Halevy, A. H. 1986. *Handbook of Flowering*, Vol. V. Boca Raton: CRC

90. Halevey, A. H. 1987. *Handbook of Flowering*, Vol. VI. Boca Raton: CRC. In press

91. Halperin, W. 1978. Organogenesis at the shoot apex. *Ann. Rev. Plant Physiol.* 29:239–62

92. Hardham, A. R. 1982. Regulation of polarity in tissues and organs. In *The Cytoskeleton in Plant Growth and Development*, ed. C. W. Lloyd, pp. 377–403. London: Academic

93. Havelange, A. 1980. The quantitative ultrastructure of the meristematic cells of *Xanthium strumarium* during the transition to flowering. *Am. J. Bot.* 67:1171–78

94. Havelange, A., Bernier, G. 1983. Partial floral evocation by high irradiance in the long-day plant *Sinapis alba*. *Physiol. Plant.* 59:545–50

95. Havelange, A., Bodson, M., Bernier, G. 1986. Partial floral evocation by exogenous cytokinin in the long-day plant *Sinapis alba*. *Physiol. Plant.* 67:695–701

95a. Heide, O. M., Bush, M. G., Evans, L. T. 1987. Inhibitory and promotive effects of gibberellic acid on floral initiation and development in *Poa pratensis* and *Bromus inermis*. *Physiol. Plant.* 69:342–50

96. Hepler, P. K., Wayne, R. O. 1985. Calcium and plant development. *Ann. Rev. Plant Physiol.* 36:397–439

97. Hofstra, G., Ryle, G. J. A., Williams, R. F. 1969. Effects of extending the day length with low-intensity light on the growth of wheat and cocksfoot. *Aust. J. Biol. Sci.* 22:333–41

98. Horridge, J. S., Cockshull, K. E. 1979. Size of the *Chrysanthemum* shoot apex in relation to inflorescence initiation and development. *Ann. Bot.* 44:547–56

99. Jacobs, W. P. 1978. Does the induction of flowering by photoperiod change the polarity or other characteristics of indole-3-acetic acid transport in petioles for the short day plant, *Xanthium*? *Plant Physiol.* 61:307–10

100. Jacobs, W. P. 1985. The role of auxin in inductive phenomena. *Biol. Plant.* 27:303–9

101. Jacqmard, A., Lyndon, R. F., Salmon, J. 1984. Appearance of specific antigenic proteins in the maturing sexual organs of *Sinapis* flowers. *J. Cell Sci.* 68:195–209

102. Jacqmard, A., Raju, M. V. S., Kinet, J.-M., Bernier, G. 1976. The early action of the floral stimulus on mitotic activity and DNA synthesis in the apical meristem of *Xanthium strumarium*. *Am. J. Bot.* 63:166–74

103. Jaffe, M. J., Bridle, K. A., Kopcewicz, J. 1987. A new strategy for the identification of native plant photoperiodically regulated flowering substances. See Ref. 3, pp. 279–87

104. Jegla, D. E., Sussex, I. M. 1987. Clonal analysis of meristem development. See Ref. 3, pp. 101–8

105. Johri, M. M., Coe, E. H. Jr. 1983. Clonal analysis of corn plant development. I. The development of the tassel and the ear shoot. *Dev. Biol.* 97:154–72

106. Joseph, C. 1986. The cytokinins of *Cichorium intybus* L. root: identification and changes during vernalization. *J. Plant Physiol.* 124:235–46

107. Joseph, C., Billot, J., Soudain, P., Côme, D. 1985. The effect of cold, anoxia and ethylene on the flowering abil-

ity of buds of *Cichorium intybus*. *Physiol. Plant.* 65:146–50

108. Kahlem, G. 1976. Isolation and localization by histoimmunology of isoperoxidases specific for male flowers of the dioecious species *Mercurialis annua* L. *Dev. Biol.* 50:58–67

109. Kamalay, J. C., Goldberg, R. B. 1980. Regulation of structural gene expression in tobacco. *Cell* 19:935–46

109a. Kamate, K., Cousson, A., Trinh, T. H., Tran Thanh Van, K. 1981. Influence des facteurs génétique et physiologique chez le *Nicotiana* sur la néoformation in vitro de fleurs à partir d'assises cellulaires épidermiques et sous-épidermiques. *Can. J. Bot.* 59:775–81

110. Kanchanapoom, M. L., Thomas, J. F. 1987. Stereological study of ultrastructural changes in the shoot apical meristem of *Nicotiana tabacum* during floral induction. *Am. J. Bot.* 74:152–63

111. Kanchanapoom, M. L., Thomas, J. F. 1987. Quantitative ultrastructural changes in tunica and corpus cells of the shoot apex of *Nicotiana tabacum* during the transition to flowering. *Am. J. Bot.* 74:241–49

112. Kandeler, R. 1985. Lemnaceae. See Ref. 88, Vol. III, pp. 251–79

113. Karege, F., Penel, C., Greppin, H. 1982. Rapid correlation between the leaves of spinach and the photocontrol of a peroxidase activity. *Plant Physiol.* 69:437–41

114. Karege, F., Penel, C., Greppin, H. 1982. Détection de l'état végétatif et floral de la feuille de l'épinard: emploi d'un indicateur biochimique. *Arch. Sci. Genève* 35:331–40

115. Kavon, D. L., Zeevaart, J. A. D. 1979. Simultaneous inhibition of translocation of photosynthate and of the floral stimulus by localized low-temperature treatment in the short-day plant *Pharbitis nil*. *Planta* 144:201–4

116. Kinet, J.-M., Sachs, R. M., Bernier, G. 1985. *The Physiology of Flowering*, Vol. III. Boca Raton: CRC

117. King, R. W., Evans, L. T. 1969. Timing of evocation and development of flowers in *Pharbitis nil*. *Aust. J. Biol. Sci.* 22:559–72

118. King, R. W., Pharis, R. P., Mander, L. N. 1987. Gibberellins in relation to growth and flowering in *Pharbitis nil* Chois. *Plant Physiol.* 84:1126–31

119. Kirby, E. J. M. 1974. Ear development in spring wheat. *J. Agric. Sci. Camb.* 82:437–47

120. Knapp, P. H., Sawhney, S., Grimmett, M. M., Vince-Prue, D. 1986. Site of perception of the far-red inhibition of flowering in *Pharbitis nil* Choisy. *Plant Cell Physiol.* 27:1147–52

121. Kochankov, V. G., Chailakhyan, M. Kh. 1986. *Rudbeckia*. See Ref. 89, pp. 295–320

122. Konstantinova, T. N., Aksenova, N. P., Bavrina, T. V., Chailakhyan, M. Kh. 1969. On the ability of tobacco stem calluses to form vegetative and generative buds in culture in vitro. *Dokl. Akad. Nauk. SSSR* 187:466–69

123. Koornneef, M., Elgersma, A., Hanhart, C. J., Van Loenen-Martinet, E. P., Van Rijn, L., Zeevaart, J. A. D. 1985. A gibberellin insensitive mutant of *Arabidopsis thaliana*. *Physiol. Plant.* 65:33–39

124. Krekule, J. 1979. Stimulation and inhibition of flowering. See Ref. 39, pp. 19–57

125. Krekule, J., Pavlová, L., Součková, D., Macháčková, I. 1985. Auxin in flowering of short-day and long-day *Chenopodium* species. *Biol. Plant.* 27:310–17

126. Krekule, J., Seidlová, F. 1977. Effects of exogenous cytokinins on flowering of the short-day plant *Chenopodium rubrum* L. *Biol. Plant.* 19:142–49

127. Kriechbaum, D. G., Rau, W. 1985. Changes in gibberellin-content and -composition during flower-induction in *Hyoscyamus niger*. *12th Int. Conf. Plant Growth Substances*, Heidelberg, p. 94 (Abstr.)

128. Lang, A. 1965. Physiology of flower initiation. In *Encyclopedia Plant Physiology*, ed. W. Ruhland, 15 (Pt. 1):1379–1536. Berlin: Springer-Verlag

129. Lang, A. 1980. Inhibition of flowering in long-day plants. In *Plant Growth Substances 1979*, ed. F. Skoog, pp. 310–22. Berlin: Springer-Verlag

130. Lang, A. 1984. Die photoperiodische Regulation von Förderung und Hemmung der Blütenbildung. *Ber. Dtsch. Bot. Ges.* 97:293–314

131. Larrieu, C., Bismuth, F. 1985. *Scrophularia alata* and *S. vernalis*. See Ref. 88, Vol. IV, pp. 283–90

132. Law, C. N. 1987. The genetic control of day-length response in wheat. See Ref. 3, pp. 225–40

133. Lay-Yee, M., Sachs, R. M., Reid, M. S. 1987. Changes in cotyledon mRNA during floral induction in *Pharbitis nil* strain Violet. *Planta* 171:104–9

134. Lejeune, P., Kinet, J.-M., Bernier, G. 1987. Cytokinin level in the LDP *Sinapis alba* during transition to flowering. *14th Int. Bot. Congr.*, West Berlin, p. 105 (Abstr.)

134a. Lejeune, P., Kinet, J.-M., Bernier, G.

1988. Cytokinin fluxes during floral induction in the LDP *Sinapis alba* L. *Plant Physiol.* In press
135. Looney, N. E., Pharis, R. P., Noma, M. 1985. Promotion of flowering in apple trees with gibberellin A₄ and C-3 epigibberellin A₄. *Planta* 165:292–94
136. Lozhnikova, V. N., Krekule, J., Vorob'eva, L. V., Chailakhyan, M. Kh. 1985. Dynamics of activity of endogenous cytokinins in tobacco leaves and roots during photoperiodic induction. *Dokl. Akad. Nauk. SSSR* 282:1021–24
137. Lyndon, R. F. 1977. Interacting processes in vegetative development and in the transition to flowering at the shoot apex. In *Integration of Activity in the Higher Plant*, ed. D. H. Jennings, pp. 221–50. Cambridge: Cambridge Univ. Press
138. Lyndon, R. F. 1978. Phyllotaxis and the initiation of primordia during flower development in *Silene*. *Ann. Bot.* 42:1349–60
139. Lyndon, R. F. 1979. A modification of flowering and phyllotaxis in *Silene*. *Ann. Bot.* 43:553–58
140. Lyndon, R. F. 1987. Synchronization of cell division during flower initiation in third-order buds of *Silene*. *Ann. Bot.* 59:67–72
141. Lyndon, R. F. 1987. Initiation and growth of internodes and stem and flower frusta in *Silene coeli-rosa*. See Ref. 3, pp. 301–14
142. Lyndon, R. F., Battey, N. H. 1985. The growth of the shoot apical meristem during flower initiation. *Biol. Plant.* 27:339–49
143. Lyndon, R. F., Jacqmard, A., Bernier, G. 1983. Changes in protein composition of the shoot meristem during floral evocation in *Sinapis alba*. *Physiol. Plant.* 59:476–80
144. Macháčková, I., Ullmann, J., Krekule, J., Stock, M. 1987. Metabolism of 1-aminocyclopropane-1-carboxylic acid (ACC) in photoperiodic induction of flowering in *Chenopodium rubrum* L. *Proc. Int. Symp. Conjugated Plant Hormones: Structure, Metabolism and Function*, ed. K. Schaeiber, H. A. Schutte, G. Sembdner. East Berlin: Veb. Dtsch. Verlag Wissensch. In press
145. Maksymowych, R., Cordero, R. E., Erickson, R. O. 1976. Long-term developmental changes in *Xanthium* induced by gibberellic acid. *Am. J. Bot.* 63:1047–53
146. Malmberg, R. L., McIndoo, J. 1983. Abnormal floral development of a tobacco mutant with elevated polyamine levels. *Nature* 305:623–25
147. Marc, J., Gifford, R. M. 1984. Floral initiation in wheat, sunflower, and sorghum under carbon dioxide enrichment. *Can. J. Bot.* 62:9–14
148. Marc, J., Palmer, J. H. 1982. Changes in mitotic activity and cell size in the apical meristem of *Helianthus annuus* L. during the transition to flowering. *Am. J. Bot.* 69:768–75
149. Marc, J., Palmer, J. H. 1984. Variation in cell-cycle time and nuclear DNA content in the apical meristem of *Helianthus annuus* L. during the transition to flowering. *Am. J. Bot.* 71:588–95
150. Martin-Tanguy, J. 1985. The occurrence and possible function of hydroxycinnamoyl acid amides in plants. *Plant Growth Regul.* 3:381–99
151. Marx, G. A. 1983. Developmental mutants in some annual seed plants. *Ann. Rev. Plant Physiol.* 34:389–417
152. McDaniel, C. N. 1980. Influence of leaves and roots on meristem development in *Nicotiana tabacum* L. cv. Wisconsin 38. *Planta* 148:462–67
153. McHughen, A. 1982. Inducing organ generation in vitro: sepal-petal structures from tobacco flower buds. *Can. J. Bot.* 60:845–49
154. Meeks-Wagner, D. R., Dennis, E. S., Peacock, W. J. 1987. Molecular genetic analysis of floral differentiation. *J. Cell Biochem.*, Suppl. 11B, 21
155. Meicenheimer, R. D. 1981. Changes in *Epilobium* phyllotaxy induced by N-1-naphthylphthalamic acid and α-4-chloro-phenoxyisobutyric acid. *Am. J. Bot.* 68:1139–54
156. Meicenheimer, R. D. 1982. Change in *Epilobium* phyllotaxy during reproductive transition. *Am. J. Bot.* 69:1108–18
157. Metzger, J. D. 1985. Role of gibberellins in the environmental control of stem growth in *Thlaspi arvense* L. *Plant Physiol.* 78:8–13
158. Metzger, J. D. 1987. *Thlaspi arvense* L. See Ref. 90. In press
159. Meyerowitz, E. M., Pruitt, R. E. 1984. *Genetic Variations of* Arabidopsis *thaliana*. Pasadena: Calif. Inst. Technol.
160. Meyerowitz, E. M., Pruitt, R. E. 1985. *Arabidopsis thaliana* and plant molecular genetics. *Science* 229:1214–18
161. Miginiac, E. 1978. Some aspects of regulation of flowering: role of correlative factors in photoperiodic plants. *Bot. Mag. (Tokyo)*, Spec. Iss. 1:159–73
162. Miginiac, E., Pilate, G., Bonnet-Masimbert, M. 1987. Hormonal regulation of flowering in *Pseudotsuga menziesii* using immunoenzymatic methods. *14th Int. Bot. Congr.*, West Berlin, p. 105 (Abstr.)

163. Miginiac, E., Sotta, B. 1985. Organ correlations affecting flowering in relation to phytohormones. *Biol. Plant.* 27:373–81

164. Miller, M. B., Lyndon, R. F. 1977. Changes in RNA levels in the shoot apex of *Silene* during the transition to flowering. *Planta* 136:167–72

165. Milyaeva, E. L., Kovaleva, L. V., Chailakhyan, M. Kh. 1982. Formation of specific proteins in stem apices of plants in transition from vegetative growth to flowering. *Fiziol. Rast.* 29:253–60

166. Monselise, S. P., Goldschmidt, E. E. 1982. Alternate bearing in fruit trees. *Hort. Rev.* 4:128–73

167. Montavon, M. 1984. *Lumière et biopotentiels chez l'épinard* (Spinacia oleracea *L. cv. Nobel*). PhD thesis. Univ. Genève, Switzerland

168. Montavon, M., Greppin, H. 1985. Potentiel intracellulaire du mésophylle d'épinard (*Spinacia oleracea* L. cv Nobel) en relation avec la lumière et l'induction florale. *J. Plant Physiol.* 118:471–75

169. Morgan, P. W., Guy, L. W., Pao, C.-I. 1987. Genetic regulation of development in *Sorghum bicolor*. III. Asynchrony of thermoperiods with photoperiods promotes floral initiation. *Plant Physiol.* 83:448–52

170. Mullins, M. G. 1980. Regulation of flowering in the grapevine (*Vitis vinifera* L.). In *Plant Growth Substances 1979*, ed. F. Skoog, pp. 323–30. Berlin: Springer-Verlag

171. Murfet, I. C. 1977. Environmental interaction and the genetics of flowering. *Ann. Rev. Plant Physiol.* 28:253–78

172. Murfet, I. C. 1985. *Pisum sativum*. See Ref. 88, Vol. IV, pp. 97–126

173. Murfet, I. C., Reid, J. B. 1987. Flowering in *Pisum*: gibberellins and the flowering genes. *J. Plant Physiol.* 127:23–29

174. Napp-Zinn, K. 1985. *Arabidopsis thaliana*. See Ref. 88, Vol. I, pp. 492–503

175. Napp-Zinn, K. 1987. Vernalization—environmental and genetic regulation. See Ref. 3, pp. 123–32

176. Nougarède, A. 1967. Experimental cytology of the shoot apical cells during vegetative growth and flowering. *Int. Rev. Cytol.* 21:203–351

177. Nougarède, A., Rembur, J., Saint-Côme, R. 1987. Rates of cell division in the young prefloral shoot apex of *Chrysanthemum segetum*. *Protoplasma* 138:156–60

178. Ogawa, Y., King, R. W. 1979. Indirect action of benzyladenine and other chemicals on flowering of *Pharbitis nil* Chois. *Plant Physiol.* 63:643–49

179. Ogawa, Y., King, R. W. 1980. Flowering in seedlings of *Pharbitis nil* induced by benzyladenine applied under a noninductive daylength. *Plant Cell Physiol.* 21:1109–116

180. Ormrod, J. C., Bernier, G. 1987. Cell cycle changes in the shoot apical meristem of *Lolium temulentum* cv. Ceres during the transition to flowering. *Arch. Int. Physiol. Biochim.* 95:PP17

181. Ormrod, J. C., Francis, D. 1986. Cell cycle responses to red or far-red light, or darkness, in the shoot apex of *Silene coeli-raos* L. during floral induction. *Ann. Bot.* 57:91–100

182. Ormrod, J. C., Francis, D. 1987. Effects of interpolated dark periods during the first long day of floral induction on the cell cycle in the shoot apex of *Silene coeli-rosa* L. *Physiol. Plant.* 71:372–78

183. Orr, A. R. 1987. Changes in glyceraldehyde 3-phosphate dehydrogenase activity in shoot apical meristems of *Brassica campestris* during transition to flowering. *Am. J. Bot.* 74:1161–66

184. Owens, V., Paolillo, D. J. Jr. 1986. Effect of aging on flowering of the axillary buds of *Pharbitis nil*. *Am. J. Bot.* 73:882–87

185. Owens, V. A., Paolillo, D. J. 1986. Changes in cell number and mitosis in apices of *Pharbitis nil* subjected to darkness or constant light. *Am. J. Bot.* 73:637–38

186. Pao, C.-I., Morgan, P. W. 1986. Genetic regulation of development in *Sorghum bicolor*. I. Role of the maturity genes. *Plant Physiol.* 82:575–80

187. Pao, C.-I., Morgan, P. W. 1986. Genetic regulation of development in *Sorghum bicolor*. II. Effect of the ma_3^R allele mimicked by GA_3. *Plant Physiol.* 82:581–84

188. Pharis, R. P., Evans, L. T., King, R. W., Mander, L. N. 1987. Gibberellins, endogenous and applied, in relation to flower induction in the long-day plant *Lolium temulentum*. *Plant Physiol.* 84:1132–38

189. Pharis, R. P., King, R. W. 1985. Gibberellins and reproductive development in seed plants. *Ann. Rev. Plant Physiol.* 36:517–68

190. Pharis, R. P., Ross, S. D. 1986. Pinaceae. See Ref. 89, pp. 269–86

191. Phillips, D. A., Cleland, C. F. 1972. Cytokinin activity from the phloem sap of *Xanthium strumarium* L. *Planta* 102:173–78

192. Pierard, D., Jacqmard, A., Bernier, G.,

Salmon, J. 1980. Appearance and disappearance of proteins in the shoot apical meristem of *Sinapis alba* in transition to flowering. *Planta* 150:397–405

193. Potts, W. C., Reid, J. B. 1983. Internode length in *Pisum*. III. The effect and interaction of the *Na/na* and *Le/le* gene differences on endogenous gibberellin-like substances. *Physiol. Plant.* 57:448–54

194. Poulhe, R., Arnaud, Y., Miginiac, E. 1984. Aging and flowering of the apex in young *Bidens radiata*. *Physiol. Plant.* 62:225–30

195. Pressman, E., Negbi, M. 1981. Bolting and flowering of vernalized *Brassica pekinensis* as affected by root temperature. *J. Exp. Bot.* 32:821–25

196. Pruitt, R. E., Chang, C., Pang, P. P.-Y., Meyerowitz, E. M. 1987. Molecular genetics and development of *Arabidopsis*. In *Genetic Regulation of Development*, pp. 327–38. New York: Alan Liss

197. Purse, J. G. 1984. Phloem exudate of *Perilla crispa* and its effects on flowering of *P. crispa* shoot explants. *J. Exp. Bot.* 35:227–38

197a. Rajeevan, M. S., Lang, A. 1987. Comparison of de-novo flower-bud formation in a photoperiodic and a day-neutral tobacco. *Planta* 171:560–64

198. Rall, S., Hemleben, V. 1984. Characterization and expression of chalcone synthase in different genotypes of *Matthiola incana* R. Br. during flower development. *Plant Mol. Biol.* 3:137–45

199. Rédei, G. P. 1962. Supervital mutants of *Arabidopsis*. *Genetics* 47:443–60

200. Rick, C. M., Robinson, J. 1951. Inherited defects of floral structure affecting fruitfulness in *Lycopersicon esculentum*. *Am. J. Bot.* 38:639–52

201. Rohozinski, J., Edwards, G. R., Hoskyns, P. 1986. Effects of brief exposure to nitrogenous compounds on floral initiation in apple trees. *Physiol. Vég.* 24:673–77

202. Romberger, J. A., Gregory, R. A. 1974. Analytical morphogenesis and the physiology of flowering in trees. *Proc. 3rd North Am. Forest Biol. Workshop*, ed. C. P. P. Reid, G. H. Fechner, pp. 132–47. Fort Collins: Colorado State Univ.

203. Ross, J. J., Murfet, I. C. 1985. A comparison of the flowering and branching control systems in *Lathyrus odoratus* L. and *Pisum sativum* L. *Ann. Bot.* 56:847–56

204. Ross, S. D., Bollmann, M. P., Pharis, R. P., Sweet, G. B. 1984. Gibberellin A4/7 and the promotion of flowering in *Pinus radiata*. *Plant Physiol.* 76:326–30

205. Rünger, W. 1975. Flower formation in *Calceolaria* × *herbeohybrida* Voss. *Sci. Hort.* 3:45–64

206. Sachs, R. M. 1979. Metabolism and energetics in flowering. See Ref. 39, pp. 169–208

207. Sachs, R. M. 1987. Roles of photosynthesis and assimilate partitioning in flower initiation. See Ref. 3, pp. 317–40

208. Sachs, R. M., Hackett, W. P. 1983. Source-sink relationships and flowering. In *Strategies of Plant Reproduction*, ed. W. J. Meudt, pp. 263–72. Totowa, NJ: Allanheld, Osmun

209. Sawhney, V. K. 1983. The role of temperature and its relationship with gibberellic acid in the development of floral organs in tomato *(Lycopersicon esculentum)*. *Can. J. Bot.* 61:1258–65

210. Scarth, R., Kirby, E. J. M., Law, C. N. 1985. Effects of the photoperiod genes *Ppd₁* and *Ppd₂* on growth and development of the shoot apex in wheat. *Ann. Bot.* 55:351–59

211. Scharfetter, E., Lesemann, C., Kandeler, R. 1987. Ethylene as a flower-promoting agent in *Lemna*. *Phyton (Austria)* 27:31–37

212. Schwabe, W. W. 1971. Chemical modification of phyllotaxis and its implications. In *Control Mechanisms of Growth and Differentiation. Symp. Soc. Exp. Biol.* 25:301–22. Cambridge: Cambridge Univ. Press

213. Schwabe, W. W. 1984. Photoperiodic induction. Flower inhibiting substances. See Ref. 249, pp. 143–53

214. Schwabe, W. W., Al-Doori, A. H. 1973. Analysis of a juvenile-like condition affecting flowering in the black currant *(Ribes nigrum)*. *J. Exp. Bot.* 24:969–81

215. Scorza, R., Janick, J. 1980. In vitro flowering of *Passiflora suberosa* L. *J. Am. Soc. Hort. Sci.* 105:892–97

216. Seidlová, F. 1980. Sequential steps of transition to flowering in *Chenopodium rubrum* L. *Physiol. Vég.* 18:477–87

217. Seidlová, F., Culafić, L. 1982. The behaviour of the shoot apical meristem in *Chenopodium rubrum* under conditions non-inductive for flowering. *Biol. Plant.* 24:471–73

218. Seidlová, F., Krekule, J. 1977. Effects of kinetin on growth of the apical meristem and floral differentiation in *Chenopodium rubrum* L. *Ann. Bot.* 41:755–63

219. Seidlová, F., Sádlíková, H. 1983. Floral transition as a sequence of growth changes in different components of the shoot apical meristem of *Chenopodium rubrum*. *Biol. Plant.* 25:50–62

220. Sharman, B. C. 1978. Morphogenesis of 2,4-D induced abnormalities of the inflorescence of bread wheat (*Triticum aestivum* L.). *Ann. Bot.* 42:145–53

221. Shillo, R. 1985. *Chamelaucium uncinatum.* See Ref. 88, Vol. II, pp. 185–89

222. Shinozaki, M., Hikichi, M., Yoshida, K., Watanabe, K., Takimoto, A. 1982. Effect of high-intensity light given prior to low-temperature treatment on the long-day flowering of *Pharbitis nil.* *Plant Cell Physiol.* 23:473–77

223. Simmonds, J. 1985. In vitro photoinduction of leaf tissues of *Streptocarpus nobilis.* *Biol. Plant.* 27:318–24

224. Singer, S. R., McDaniel, C. N. 1986. Floral determination in the terminal and axillary buds of *Nicotiana tabacum* L. *Dev. Biol.* 118:587–92

225. Singer, S. R., McDaniel, C. N. 1987. Floral determination in internode tissues of day-neutral tobacco first occurs many nodes below the apex. *Proc. Natl. Acad. Sci. USA.* 84:2790–92

226. Singh, S. P., Paleg, L. G. 1984. Low temperature-induced GA_3 sensitivity of wheat. *Plant Physiol.* 76:139–42

227. Southwick, S. M., Davenport, T. L. 1986. Characterization of water stress and low temperature effects on flower induction in *Citrus.* *Plant Physiol.* 81:26–29

228. Stebbins, G. L., Yagil, E. 1966. The morphogenetic effects of the hooded gene in barley. I. The course of development in hooded and awned genotypes. *Genetics* 54:727–41

229. Stinson, J. R., Eisenberg, A. J., Willing, R. P., Pe, M. E., Hanson, D. D., Mascarenhas, J. P. 1987. Genes expressed in the male gametophyte of flowering plants and their isolation. *Plant Physiol.* 83:442–47

230. Stutte, G. W., Martin, G. C. 1986. Effect of killing the seed on return bloom of olive. *Sci. Hort.* 29:107–13

231. Suttle, J. C., Zeevaart, J. A. D. 1979. Stem growth, flower formation, and endogenous gibberellins in a normal and a dwarf strain of *Silene armeria. Planta* 145:175–80

232. Swe, K. L., Shinozaki, M., Takimoto, A. 1985. Varietal differences in flowering behavior of *Pharbitis nil* Chois. *Mem. Coll. Agric., Kyoto Univ.* 126:1–20

233. Takahashi, H., Saito, T., Suge, H. 1982. Intergeneric translocation of floral stimulus across a graft in monoecious Cucurbitaceae with special reference to the sex expression of flowers. *Plant Cell Physiol.* 23:1–9

234. Takimoto, A., Kaihara, S. 1986. The mode of action of benzoic acid and some related compounds of flowering in *Lemna paucicostata. Plant Cell Physiol.* 27:1309–16

235. Takimoto, A., Kaihara, S., Nishioka, H. 1987. A comparative study of the short-day- and the benzoic acid–induced flowering in *Lemna paucicostata. Plant Cell Physiol.* 28:503–8

236. Taylor, J. S., Koshioka, M., Pharis, R. P., Sweet, G. B. 1984. Changes in cytokinins and gibberellin-like substances in *Pinus radiata* buds during lateral shoot initiation and the characterization of ribosyl zeatin and a novel ribosyl zeatin glycoside. *Plant Physiol.* 74:626–31

237. Thigpen, S. P., Sachs, R. M. 1985. Changes in ATP in relation to floral induction and initiation in *Pharbitis nil. Physiol. Plant.* 65:156–62

238. Thomas, B., Vince-Prue, D. 1987. Photoperiodic control of floral induction in short-day and long-day plants. In *Models in Plant Physiology/Biochemistry/Technology,* ed. D. Newman, K. Wilson. Boca Raton: CRC. In press

239. Tiburcio, A. F., Kaur-Sawhney, R., Galston, A. W. 1987. Flower induction by spermidine in thin-layer tobacco tissue cultures. *14th Int. Bot. Congr.,* West Berlin, p. 130 (Abstr.)

240. Tran Thanh Van, K. 1980. Control of morphogenesis by inherent and exogenously applied factors in thin cell layers. *Int. Rev. Cytol.,* Suppl. 11A: 175–94

241. Tran Thanh Van, K., Toubart, P., Cousson, A., Darvill, A. G., Gollin, D. J., et al. 1985. Manipulation of the morphogenetic pathways of tobacco explants by oligosaccharins. *Nature* 314:615–17

242. Trewavas, A. J. 1982. Growth substance sensitivity: the limiting factor in plant development. *Physiol. Plant.* 55: 60–72

243. Valade, J., Maizonnier, D., Cornu, A. 1987. La morphogenèse florale chez le pétunia. I. Analyse d'un mutant à corolle staminée. *Can. J. Bot.* 65: 761–64

244. Van den Ende, G., Croes, A. F., Kemp, A., Barendse, G. W. M. 1984. Development of flower buds in thin-layer cultures of floral stalk tissue from tobacco: role of hormones in different stages. *Physiol. Plant.* 61:114–18

245. Verbeke, J. A., Walker, D. B. 1986. Morphogenetic factors controlling differentiation and dedifferentiation of epidermal cells in the gynoecium of *Catharanthus roseus.* II. Diffusible morphogens. *Planta* 168:43–49

246. Vince-Prue, D. 1975. *Photoperiodism in Plants*. London: McGraw-Hill
247. Vince-Prue, D., Gressel, J. 1985. *Pharbitis nil*. See Ref. 88, Vol. IV, pp. 47–81
248. Vince-Prue, D., Lumsden, P. J. 1987. Inductive events in the leaves: time measurement and photoperception in the short-day plant, *Pharbitis nil*. See Ref. 3, pp. 255–68
249. Vince-Prue, D., Thomas, B., Cockshull, K. E. 1984. *Light and the Flowering Process*. London: Academic
250. Wardell, W. L. 1977. Floral induction of vegetative plants supplied a purified fraction of deoxyribonucleic acid from stems of flowering plants. *Plant Physiol.* 60:885–91
251. Wareing, P. F. 1987. Juvenility and cell determination. See Ref. 3, pp. 83–92
252. Wareing, P. F., Horgan, R., Henson, I. E., Davis, W. 1977. Cytokinin relations in the whole plant. In *Plant Growth Regulation*, ed. P.-E. Pilet, pp. 147–53. Berlin: Springer-Verlag
253. Warm, E. 1984. Changes in the composition of in vitro translated m-RNA caused by photoperiodic flower induction of *Hyoscyamus niger*. *Physiol. Plant.* 61:344–50
254. Warm, E., Rau, W. 1982. A quantitative and cumulative response to photoperiodic induction of *Hyoscyamus niger*, a qualitative long day plant. *Z. Pflanzenphysiol.* 105:111–18

255. Watanabe, K., Fujita, T., Takimoto, A. 1981. Relationship between structure and flower-inducing activity of benzoic acid derivatives in *Lemna paucicostata* 151. *Plant Cell Physiol.* 22:1469–79
256. Wellensiek, S. J. 1985. *Silene armeria*. See Ref. 88, Vol. IV, pp. 320–30
257. Zeevaart, J. A. D. 1962. Physiology of flowering. *Science* 137:723–31
258. Zeevaart, J. A. D. 1976. Physiology of flower formation. *Ann. Rev. Plant Physiol.* 27:321–48
259. Zeevaart, J. A. D. 1978. Phytohormones and flower formation. In *Phytohormones and Related Compounds: A Comprehensive Treatise*, ed. D. S. Letham, P. B. Goodwin, T. J. V. Higgins, 2:291–327. Amsterdam: Elsevier/North-Holland Biomedical Press
260. Zeevaart, J. A. D. 1979. Perception, nature and complexity of transmitted signals. See Ref. 39, pp. 59–90
261. Zeevaart, J. A. D. 1983. Gibberellins and flowering. In *The Biochemistry and Physiology of Gibberellins*, ed. A. Crozier, 2:333–74. New-York: Praeger
262. Zeevaart, J. A. D., Boyer, G. L. 1987. Photoperiodic induction and the floral stimulus in *Perilla*. See Ref. 3, pp. 269–77
263. Zeevaart, J. A. D., Brede, J. M., Cetas, C. B. 1977. Translocation patterns in *Xanthium* in relation to long day inhibition of flowering. *Plant Physiol.* 60:747–53

Ann. Rev. Plant Physiol. Plant Mol. Biol. 1988. 39:221–44

PHYSIOLOGICAL INTERACTIONS BETWEEN SYMBIONTS IN VESICULAR-ARBUSCULAR MYCORRHIZAL PLANTS

Sally E. Smith

Department of Agricultural Biochemistry, Waite Agricultural Research Institute, University of Adelaide, South Australia, 5001, Australia

Vivienne Gianinazzi-Pearson

Laboratoire de Phytoparasitologie, Station d'Amélioration des Plantes, INRA, BV 1540, 21034 Dijon Cedex, France

CONTENTS

INTRODUCTION 221
THE ENDOPHYTE IN ISOLATION 222
DEVELOPMENT OF INFECTION AND HOST-FUNGUS INTERFACES 224
SYMBIONT PHYSIOLOGY AND THE MYCORRHIZAL CONDITION 227
   Plant Growth and Phosphorus Nutrition 227
   Contribution of the Endophyte 228
   Bidirectional Nutrient Transfer Between Symbionts 230
   Effects on Host-Plant Metabolism 232
   Water Relations 233
   Nonnutritional Modifications of Host Plant Physiology 233
   Carbon Distribution in the Symbiosis 234
PHYSIOLOGICAL COMPATIBILITY AND MYCORRHIZAL EFFICIENCY 235
CONCLUSIONS 236

INTRODUCTION

Research into the physiological bases of symbiont interactions in vesicular-arbuscular (VA) endomycorrhizae has occurred relatively recently. The last review on mycorrhizae in this series (121) was mainly devoted to ectomycor-

221

0066-4294/88/0601-0221$02.00

rhizae, reflecting the lack of knowledge of VA mycorrhizal physiology at that time. Although the fungi had been identified as members of the Endogonaceae and progress had been made in pinpointing the causes of the growth responses of plants to VA mycorrhizal infection, physiological research was only really given impetus at the "Endomycorrhizas" meeting at Leeds in 1974 (147). Many of the hypotheses proposed there have stimulated experimentation and advanced our understanding of the functioning of VA mycorrhizal roots and the physiology of mycorrhizal plants.

It has become clear that mycorrhizae are an integral part of the plant (69) and that in nature most plant species (more than 80%) have a root system that is really a VA mycorrhizal system. Although the overall results of host-fungal interactions in mycorrhizae are usually measured in terms of growth of the host plants, the association contributes to the fitness of both partners. Two features contribute to this mutualism: the persistent biotrophic phase, manifest in structural and physiological compatibility between the symbionts, and the ability of both symbionts to contribute to the nutrition of the association (35, 75, 84).

Our limited knowledge of the physiology of VA mycorrhizal fungi is in striking contrast to the extensive information available on host physiology and biochemistry. Because VA mycorrhizal fungi cannot, as yet, be grown in pure culture for extended periods, information has mainly been obtained from the fungal structures associated with roots, where the fungus is influenced by symbiotic interaction with the higher plant. On the other hand, data on the physiology and biochemistry of hosts largely comes from studies of nonmycorrhizal plants or from plants of unknown mycorrhizal status. Effects of nutrient deficiencies are not well documented, so that interpretation of mycorrhizal effects on growth and metabolism is often difficult.

Here we consider (a) the physiology of VA mycorrhizal fungi in isolation, (b) host-endophyte interactions at the tissue and cellular level, (c) physiology of the symbionts when in association, and (d) mycorrhizal effectiveness in terms of physiological compatibility of the symbionts. We try to analyze data critically and pinpoint areas of research that will be fruitful. References are not exhaustive. Recent papers and reviews have been selected to illustrate specific points and to provide sources of further references.

THE ENDOPHYTE IN ISOLATION

Much of the information on the physiology of VA mycorrhizal fungi in isolation has been gained from attempts to grow them in pure culture rather than in systematic investigations of processes that might be important in a functioning symbiosis.

Spores germinate readily in vitro on water agar, but hyphal growth is restricted to a few centimeters, with a range of organic substrates, vitamins, and sulfur compounds promoting growth to different extents (91). The presence of roots or cell suspensions (45) is also stimulatory, but growth stops if the subtending spore is removed. Exhaustion of spore reserves is probably not responsible for poor or stagnating growth (106).

The recalcitrance of VA fungi in culture has stimulated investigations of the biochemistry of spores following imbibition, to determine whether a vital metabolic pathway is blocked. The operation of glycolytic, TCA, and pentose phosphate pathways has been inferred from enzyme activities determined by cytochemical techniques and electrophoresis (95, 107, 108, 116). Acetate is incorporated into organic and amino acids, confirming that the TCA cycle and pathways of amino acid synthesis operate (21, 22). ATP levels increase slowly up to 45 min after imbibition, but more rapidly between 45 and 60 min, coincident with rapid synthesis of protein, RNA, amino and organic acids, and neutral carbohydrate (23). The dependence of germ-tube growth on RNA and protein synthesis is comparable to that of saprophytic fungi (89, 90, 93). Interestingly, no nuclear DNA synthesis has been detected during germination (23, 37), although EM studies suggest that nuclear division occurs (167).

Lipid reserves in spores are high [about 45% on a dry weight basis (19, 20, 22)] compared to most fungal spores (176). The most abundant neutral lipids are triglycerides; phospholipid content is low (phosphatidyl inositol and glycolipids are absent) and fatty acids are similar to those found in other fungi. Net lipid synthesis occurs during spore germination and early germ-tube growth, followed by net degradation, possibly associated with senescence of the hyphae (19, 22). Spores therefore possess the metabolic machinery and genetic information for initial hyphal growth.

A number of physiological attributes studied in isolated VA fungi may be of significance in the established symbiosis. Nitrate reductase, glutamate oxaloacetate transaminase, and peptidase have been detected in spores (94, 95, 98). Spores and hyphae produce auxin, gibberellin, and cytokinin-like substances endogenously (13, 14). If these are secreted during mycorrhiza formation they might contribute to changes in host physiology (e.g. auxin stimulated activity of membrane-bound ATPase). Germ tubes absorb phosphate (P) by a temperature-sensitive process (36), but mechanisms of uptake and efflux have not been studied, nor has the ability of axenically grown VA mycorrhizal hyphae to translocate nutrients. Rapid bidirectional cytoplasmic streaming, similar to that associated with translocation by extramatrical hyphae, occurs in hyphae arising from spores (124, 134). Streaming is modified by changes in P supply that have no effects on the rates of spore germination or germ-tube growth (134).

The inability to culture VA mycorrhizal fungi saprophytically has focussed research on the physiology of the symbiotic fungus, modified by interactions with the plant. Modification may be fundamental to fungal viability and linked to the supply of a specific nutrient by the host. Alternatively, a signal from the host plant may be required before essential fungal genes can be expressed. The absence in spores of certain isozymes found in internal and external mycelium associated with roots (95) may be an indication of such genetic regulation and merits further investigation. If a system of cell-to-cell signalling operates, the signal molecules must be widespread in the plant kingdom since VA mycorrhizal associations are taxonomically non-specific.

DEVELOPMENT OF INFECTION AND HOST-FUNGUS INTERFACES

Infection is initiated from hyphae growing from soil-borne propagules or from neighboring infected roots. Recognition events must occur between the two organisms, the first visible sign being the formation of appressoria on the root surface (70, 84). Infection units develop within the root cortex, with hyphae growing longitudinally between the cells and intracellular development of arbuscules. Vesicles, containing large amounts of lipid, are formed later in the maturation of an infection unit. At the same time as infection spreads within a root, extramatrical hyphae grow out into the soil.

The spread of infection, measured as the fraction of the root length infected, usually follows a sigmoid curve (171). The length of the lag phase, slope of the phase of rapid spread, and height of the plateau vary depending on host-fungus combination, density of inoculum in soil, and environmental conditions. Models of the spread of infection have been used to analyze the effects of environmental variables on infection processes (38, 146, 157). This approach has emphasized that the progress of infection depends upon growth of the root, as well as upon the rates at which new infection units are formed and grow within the root cortex. It has been postulated that the spread of infection is under host control (40), but the physiological basis for this remains obscure. It is uncertain whether initiation of entry points is restricted to a region immediately behind the growing root apex, to relatively young but differentiated roots, or whether it can occur throughout the length of a root system (40, 92, 157, 160). The question is important, as cell wall synthesis and organization, as well as production of root exudates, vary along a root axis and may interact not only with formation of mycorrhizae, but also with nutrient exchange between the symbionts.

Contact between the fungal endophyte and host cells leads to a sequence of interactions during which compatible cellular relationships are established. The intraradical phase of the association is morphologically complex, and infection patterns can vary depending on the fungus or host species involved (1, 35, 43, 150). Furthermore, the distribution of the hyphal system within

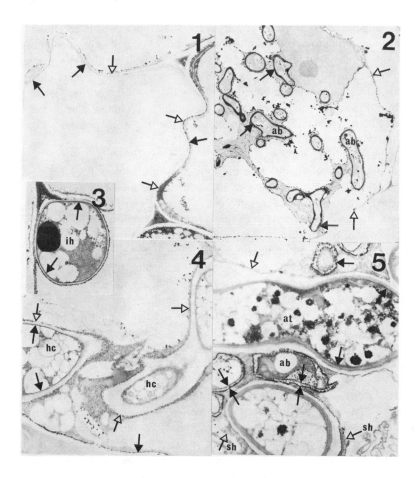

Figures 1–5   Presence (▲) and absence (△) of plasma membrane–bound ATPase activities in fungal and host cells; enzyme localization is indicated by electron-dense precipitates of lead phosphate.

Sporadic ATPase activity occurs along the peripheral plasma membrane of parenchymal cortical cells of a nonmycorrhizal root (1). When an arbuscule develops in these host cells, strong ATPase activity becomes specifically localized along the host plasma membrane around fine arbuscule branches (ab) (2). A weaker activity occurs along the arbuscule trunk hyphae (at) and all activity disappears with hyphal senescence (sh) (5). Specific ATPase localization does not occur around hyphal coils (hc) developing in the outer cell layers (4). Fungal plasma membrane-bound ATPase can be detected in intercellular hyphae (3), hyphal coils (4), the arbuscule trunk and fine branches (2,5).

1: onion (× 3750); 2 and 3: onion with *Glomus* E3 (× 2625 and × 7500); 4 and 5: sycamore with *G. mosseae* (× 4500 and × 7500)

roots and the nature of the host-fungus interfaces formed during cellular interactions are largely determined by the type of host tissue colonized (75).

Fungal penetration of the outer cell layers of the root can be intercellular or intracellular with the formation of a simple unbranched coil in each colonized cell. At this stage host plasma membrane and fungal wall are always separated by host wall material (31, 33, 34, 102, 103); cytoplasmic contents change little and plasma membrane-bound ATPase activities of the host are weak or absent at this interface (Figure 4) (75). In contrast, the fungal plasma membrane of both hyphal coils and intercellular hyphae frequently possess ATPase activity (Figures 3 and 4).

As the infection spreads into cortical parenchyma, intense intracellular fungal development occurs, with the formation of complex much-branched arbuscules. The molecular basis of the induction of arbuscule formation is unknown, but if arbuscules are involved in specific and preferential nutrient transfer, the mechanism must be of fundamental importance for the functional morphology of the symbiosis. Arbuscule formation creates a large surface of contact between the cells of the two organisms, due to proliferation of the host plasma membrane closely around the finely branching hyphae (53, 174).

Fungal wall metabolism is modified as the fungus grows within the root tissues, so that compared to the thickened, fibrillar, chitinous wall of the extraradical hyphae, the fungal wall in the arbuscules is a simplified, thin, amorphous, nonchitinous structure (32). Such alterations probably lead to an increase in plasticity, which could affect water relations of the arbuscule and promote its continued growth. Simultaneously, the amount of wall-like material deposited by the host around the invading hypha diminishes, so that only scattered fibrils are present around the fine arbuscular branches (61, 149). The growing arbuscular branches appear to interfere with the deposition of wall material rather than with synthesis, since host plasma membrane–bound neutral phosphatase activity (considered to be a marker of polysaccharide synthesis) is always present at the interface (99) and wall material accumulates again with arbuscule senescence (61, 77, 149).

Both fungal and host plasma membranes of the arbuscular interface have cytological features suggestive of normal, active membranes (tripartite structure, specific staining properties) and both possess ATPase activities (Figures 2 and 5) (118). High ATPase activity is localized along the host plasma membrane surrounding the arbuscule branches, clearly distinguishing it from the peripheral cell membrane which, as in uninfected parenchyma cells (Figure 1), shows very little ATPase activity (Figure 2). This activity, which is not observed around hyphal coils or degenerated arbuscules (Figures 4 and 5), implies a specialized modification of the host membrane around the arbuscule. Peculiar to VA mycorrhizae is the formation of an intracellular interface in which wall material is reduced to a minimum and membrane-bound enzyme systems exist in both symbionts, capable of generating the necessary energy gradients for active transport. Such extreme specialization is absent in other haustorial host-parasite interactions where nutrient transport is unidirectional towards the parasite (117).

SYMBIONT PHYSIOLOGY AND THE MYCORRHIZAL CONDITION

Plant Growth and Phosphorus Nutrition

The effects of VA mycorrhizal infection on the growth and nutrition of plants have recently been reviewed (2, 48, 73, 84, 87, 154). Briefly, on soils low in available P, mycorrhizal plants have higher rates of growth than nonmycorrhizal plants. Root:shoot ratio is often lower and shoot fresh-weight:dry-weight ratio higher, following infection. The effects are not usually apparent for several weeks after germination. Factors contributing to this delay include a lag phase before the onset of rapid infection of the roots, the masking effect of seed nutrient content, and competition between the symbionts for photosynthate. Reduced or negative growth responses have been observed under a variety of environmental conditions, and the mechanisms by which they occur are receiving attention (48, 84, 105, 154). The rate of plant growth is determined by interactions between mycorrhizal infection and a number of nutritional and nonnutritional aspects of symbiont physiology.

The importance of VA mycorrhizae in the P nutrition of plants has meant that most attention has been given to this aspect of their physiology. Inflow (rate of uptake per unit length of root) of P from soil into the roots of mycorrhizal plants is faster than into nonmycorrhizal plants (81, 145, 148, 155, 158, 168). On soils low in available P this results in increased rates of plant growth and increased concentrations of total P in tissues, at least in the early stages of plant development. The amount of P in the orthophosphate (Pi) fraction in both shoots and roots is increased in mycorrhizal plants (71, 74).

One mechanism underlying the increased rate of uptake of P is the efficiency with which mycorrhizal roots exploit the soil profile, with hyphae extending beyond the depletion zone surrounding the absorbing root and its root hairs (47, 128, 130, 172). This fits with what is known of the factors that determine rates of nutrient supply to and uptake by roots. For nonmobile nutrients such as P (and to a lesser extent K^+ and NH_4^+), root growth, root radius, development of root hairs, and initial concentration in the soil solution are more important determinants of the rate of uptake than are the kinetic properties of the uptake systems of the root (46, 47, 128). Mycorrhizal modification of the nutrient uptake properties of roots depends upon (a) development of extramatrical hyphae in soil, (b) hyphal absorption of phosphate, (c) translocation of P through hyphae over considerable distances, and

(d) transfer of P from the fungus to the root cells. There is clear evidence that all these processes take place. However, the extent of fungal development and the mechanisms and rates of the processes require further investigation before they can be used to compare different fungus-host combinations grown under different environmental conditions. Differences between species and cultivars in root morphology (16, 79, 163), together with differences in growth of hyphae in soil (see below), may all be important in determining the magnitude of the plant growth response.

Contribution of the Endophyte

HYPHAL GROWTH   Hyphae grow extensively in the soil in association with mycorrhizal roots and are an effective means of deploying a limited amount of material in exploitation of the profile. The extent of hyphal production depends upon the species of fungus, the stage of development of the symbiosis, and environmental conditions (3, 5, 26, 27, 80, 148, 173). Table 1 shows data for a number of fungal species, obtained by direct measurement or estimated from dry matter of hyphae associated with roots. There appear to be differences between fungal species, although comparisons are difficult because techniques differed. Whether these variations affect the nutrient acquisition or growth responses of infected plants requires further systematic investigation. The activity, longevity, and turnover rates of external hyphae have received little attention but would be important in determining the biomass of fungus actually involved in nutrient uptake and translocation (107, 108). Fungal development within the root is also important as it determines the surface area across which nutrient transfer occurs (see below) and, together with extramatrical hyphae, provides a measure of the "growth response" of the fungus to symbiotic interaction with a host. Total fungal biomass increases during development of the symbiosis with values ranging from 5 to 20% of root weight. The ratio of internal to external fungus also changes and is influenced by environmental conditions (24, 26, 27, 88, 110).

P UPTAKE AND METABOLISM BY THE FUNGI   Concentrations of Pi in hyphae indicate a gradient of 1000:1 between them and the soil solution (74), so that P must be absorbed actively, against an electrochemical potential gradient. Kinetics of uptake (K_m and V_{max}) have been investigated by comparing mycorrhizal with nonmycorrhizal roots, and conflicting results were obtained (56, 100), possibly owing to differences in the quantity of external hyphae remaining attached to the roots.

The role of mycorrhizal hyphae in tapping sources of soil P chemically unavailable to uninfected roots has not been clearly demonstrated (12, 30, 71, 125, 131, 135). Hyphae may more readily absorb fixed or adsorbed P in high

Table 1 Development of extramatrical hyphae in soil by VA mycorrhizal fungi

Fungus	Host	Hyphal length (m cm^{-1} root)	Reference
On infected root:			
Glomus mosseae	onion	0.79–2.5[c]	145
G. mosseae	onion	0.71[a]	148
G. macrocarpum	onion	0.71[a]	148
G. microcarpum	onion	0.71[a]	148
G. sp (E3)	clover	1.29[b]	173
"	rye grass	1.36[b]	173
G. fasiculatum	clover	2.50[b]	4
G. tenue	clover	14.20[b]	4
Gigaspora calospora	onion	0.71[a]	148
G. calospora	clover	12.30[b]	4
Acaulospora laevis	clover	10.55[b]	4
On whole root system:			
G. fasiculatum	soybean	1.2–2.7[a]	26

[a] Recalculated from hyphal dry weight using a conversion factor of 0.2 m μg^{-1}.
[b] Direct measurement as quoted in the reference.

P-fixing soils (76, 166). However, where increased removal or breakdown of insoluble P reserves in soils by mycorrhizal plants has been observed (76, 179), it has proved impossible to distinguish direct fungal activity either from indirect enhanced root surface or rhizosphere activities, following mycorrhizal infection. In these instances, localized production of siderophores, protons, and enzymes could be important; the appropriate mechanism would clearly depend upon the nature of the P source and the characteristics of the soil.

Following P uptake, polyphosphate (polyP) is accumulated in vacuoles of VA mycorrhizal fungi (53, 54, 115, 150, 165, 177). In *Glomus mosseae,* polyP is probably synthesized via an inducible polyphosphate kinase and has been implicated in both storage (between 16 and 40% of total P) and translocation of P by hyphae (17, 41, 42, 54). Synthesis and breakdown of polyP in vacuoles and transport of Pi between cytoplasm and vacuole is a method of regulating cytoplasmic inorganic Pi concentration (85). Breakdown of polyP could be via polyphosphatases or by reversal of polyphosphate kinase, both of which are present in infected roots (42). The vacuoles of VA mycorrhizal fungi, as well as other fungi that store polyP (119), are also characterized by the presence of an alkaline phosphatase (68). In mycorrhizae the activity of this enzyme is related to the presence of active infection, well-developed arbuscules, and stimulation of plant growth (72, 73); but its role in phosphate metabolism is still a matter for speculation (73).

Cations, in particular calcium and arginine, occur as secondary constituents of polyP (165, 177). Synthesis and breakdown of polyP could therefore have consequences for calcium distribution and nitrogen (N) metabolism in mycor-rhizae. The possibility of a coupled storage and/or translocation system for P and N in hyphae, as in yeast (65), merits further investigation.

TRANSLOCATION TO ROOTS   Mycorrhizal fungi have considerable ability to translocate nutrients, although there is little comparative data for different host-fungus combinations. Elements include P, Zn, S, Ca, and N (11, 50, 51, 133); and the distances over which translocation can take place exceed the radius of any depletion zone likely to develop around an actively absorbing root (139–141).

Rates of translocation of P by *G. mosseae* (determined from tracer fluxes) in association with various host plants are in the range $0.1–2.0 \times 10^{-9}$ mol $cm^{-2} s^{-1}$ (50, 51, 133). The process is temperature sensitive, stopped by cytochalasin B (which inhibits cytoplasmic streaming), and reduced when transpiration of the associated plants is prevented. It probably occurs down a concentration gradient between a P source in the external hyphae and a sink in the roots. Loading and unloading of P from hyphal vacuoles, coupled with cytoplasmic streaming, are presumed to be important in maintaining the high rates of translocation (74, 84, 172).

The contribution of hyphal translocation to P uptake by roots can be estimated from the difference between inflow into mycorrhizal plants and that into nonmycorrhizal plants (81, 145, 148). Values are higher than from in vitro studies based on tracer movement and are likely to be overestimates (84). Even so, they are several orders of magnitude faster than can be accounted for by diffusion of P ions to the root.

Although growth responses to infection frequently do not occur when soil P is high, it is clear that inflow via hyphae can continue and give rise to high P concentrations in shoots as well as roots, the absence of a growth response being due to the operation of a limiting factor other than P availability (S. E. Smith and C. L. Son, unpublished; 155, 156, 158). "Luxury" P accumulation could provide a storage pool, used in later stages of plant growth.

Bidirectional Nutrient Transfer Between Symbionts

The fungus is frequently envisaged as a rapid transit system that delivers P to the root surface, where it can be absorbed by the root cells. This view of the mycorrhizal involvement in P (and other nutrient) acquisition must be an oversimplification, as it ignores transfer between the organisms. The interface is structurally complex, and transfer of nutrients occurs in both directions. This has important implications for the mechanisms that can be envisaged for solutes leaving one cell and being absorbed by the other from the apoplastic space separating them (47, 62, 84, 153, 178).

The simplest concept of bidirectional transport involves opposed (but not necessarily linked) Pi and soluble carbohydrate transport, with the addition of organic N transport from fungus to root. The identity of the molecules transferred is not known, but has important implications for the mechanisms of transport that may operate and the simultaneous movement of other ions (e.g. K^+). The distribution of ATPases on the plasma membranes of both root cells and arbuscules supports the concept of two-way controlled transport in mycorrhizae. It is possible that the ATPases represent proton pumps that would set up proton motive forces driving the uptake of (for example) Pi or hexose contrary to their electrochemical potential gradients, from the interfacial apoplast to root or fungal cell. The uptake of both these molecules could be by proton symport. It is not necessary to postulate special symbiotically modified mechanisms for uptake from the apoplast; what we know of normal uptake mechanisms in fungi and roots is quite adequate for our current knowledge of transport at the interface (153).

Less is known of the mechanisms of loss of solutes from cells, which would maintain the necessary flux into the apoplast. Increased permeability of membranes (e.g. under P deficiency) leading to increased "leakiness" of a range of solutes (84, 137) does not provide a credible mechanism for controlled transport. The mechanism must be specific and must also allow for active and selective uptake into both organisms by the same membranes (153). Cells do not normally have high rates of P loss despite the steep electrochemical gradient in favor of efflux (17, 29). However, fluxes of P across the area of arbuscular interface are of the same order of magnitude [1.3×10^{-14} mol mm^{-2} s^{-1} (53)] as P uptake by fungi and giant algal cells, both much higher than values for P efflux (17, 29). Consequently, modification of fungal P efflux must be envisaged, leading perhaps to rapid reductions in cytoplasmic Pi concentrations and mobilization of polyP reserves in the fine arbuscular branches. The mechanisms that actually trigger or control this process may have most important regulatory roles in the transfer of P between the organisms.

Similar problems apply to carbohydrate transfer from host to fungus. Efflux from leaf mesophyll gives a high concentration of sucrose in the leaf apoplast during phloem loading (78). Similar mechanisms could be important in mycorrhizal roots. Disappearance of starch from infected cells and high activity of invertase in infected roots (60, 102) could be important in maintaining high sugar concentrations within root cells, and could favor high efflux rates. Synthesis of "nonreutilizable" lipid and glycogen by the fungus (see below), together with translocation of carbon compounds away from the site of transfer, would also help to maintain a gradient across the interface (152).

An alternative mechanism facilitating carbohydrate transport in which wall deposition is modified by fungal inactivation of polymerization processes at the host plasma membrane can be envisaged. Root cell wall precursors or components transferred to, but not deposited in, the interface could be utilized by the fungus (63, 99). Furthermore, VA mycorrhizal fungi probably have some limited capacity to degrade cell walls (77, 102; S. Jaquelinet-Jeanmougin, V. Gianinazzi-Pearson, and S. Gianinazzi, unpublished results), but there is no evidence that nutritionally significant amounts of carbohydrate are released. Wall metabolism might provide a fruitful area for future research, especially as small di- and oligosaccharides have been shown to be important intercellular messengers in other symbioses (120).

Effects on Host Plant Metabolism

Once P has reached the plant cells, any process that was previously limited by the availability of P will increase in rate. Unfortunately, there is relatively little information on direct or indirect effects of P deficiency on the activity of specific enzymes or metabolic pathways in plants. Many experiments on mycorrhizal effects on plant growth have shown that the rate of photosynthesis is higher in mycorrhizal than non-mycorrhizal plants (9, 111, 112, 162). The direct involvement of Pi in photosynthesis and in subsequent mobilization or storage of photosynthate has now been clearly demonstrated. Pi availability can limit the rate of photosynthesis in vivo (114, 151). As plant species differ in the sensitivity of their photosynthetic mechanisms to P deficiency (64), this would be a possible basis for differences in response to mycorrhizal infection.

Activities of other enzyme systems can also be directly affected by P supply and are likely to be involved in the response of plants to mycorrhizal infection. Examples include nitrate reductase and glutamine synthetase (44, 129, 159), as well as nitrogen fixation in both legumes and nonlegumes (15). In the case of nitrate reductase, the effects of mycorrhizal infection appear to be entirely due to improved P status of the plants, in line with the requirement of this enzyme for P (18). There is no evidence that fungal nitrate reductase activity (detected in spores) is important in the symbiosis. With glutamine synthetase, direct involvement of the fungus in contributing enzyme activity to roots, as well as a "phosphate" effect on activity in roots and shoots, is indicated. Thus the increased demand for nitrogen following relief of P stress may be partially satisfied by increased activity of enzymes of both ammonium and nitrate assimilation, contributed by both root and fungus. A fungal pathway of uptake, translocation, and transfer of nitrogen to roots may also contribute to higher inflows of N into mycorrhizal roots (10, 11, 158). A recent field study (15a) has confirmed both P-mediated increases in N_2 fixation and enhanced N uptake from soil by mycorrhizal *Hedysarum coronarium*.

Improved P nutrition, as well as direct fungal effects, may be implicated in the enhanced uptake by plants of other macronutrients such as K and S (136, 142), and of micronutrients such as Cu and Zn (50, 101, 170). Nutritional interactions can be complex. For example, K requirement is strongly influenced by the source of N (NO_3^- or NH_4^+) and by the concentration of Na.

These will interact with any effects of mycorrhizal infection and P nutrition on K uptake. The subject has recently been reviewed (48, 84).

The anion:cation balance of wheat and barley is altered on infection, in a way that is apparently not influenced by P nutrition (39). Mycorrhizal plants accumulate lower amounts of organic anions in their vacuoles and may therefore regulate cytoplasmic pH by disposing of a higher proportion of OH^- generated during nitrate reduction to the soil (138). They might therefore be expected to have higher rhizosphere pH than nonmycorrhizal plants assimilating this N source. The reduction in organic anion concentration might also represent a saving in fixed carbon (154). Factors that affect rhizosphere pH will influence nutrient availability and acquisition, fungal and bacterial infection of roots (both pathogens and symbionts), and interactions between microorganisms that operate via pH-sensitive mechanisms.

Water Relations

Alterations in the water relations of plants as a result of mycorrhizal infection have been reviewed recently (48, 67, 84). Interest stems from the possibility that mycorrhizal plants may have improved drought tolerance, but it has proved difficult to distinguish direct mycorrhizal effects from those that could be mediated via improved mineral nutrition. Broadly, mycorrhizal plants have higher hydraulic conductivities, especially when P availability is low. Shoot resistances are frequently reduced, with the main effect being on the stomata, so that both CO_2 and H_2O movement are affected and rates of both transpiration and photosynthesis (on a leaf-area basis) are increased. In many cases these effects can be duplicated by increasing P supply, and must be regarded as indirect (6, 9, 112, 126). Changes in root conductivity can be attributed in part to alterations in root surface area and branching, following relief of P stress (79, 83, 113). A hyphal pathway of water movement from soil to roots may also be involved. Attempts to investigate this directly by removing extramatrical hyphae have not given clear-cut results (82).

In dry soil, P and other nutrients (e.g. NO_3^-) become much less mobile, and a mycorrhizal pathway of nutrient acquisition would become relatively much more important. Relief of nutrient stress might also be followed by increased rates of root growth and more efficient extraction of water from the soil profile (67). Reduction in the severity of symptoms of drought stress such as proline accumulation (48, 112, 127) would follow this improved water availability.

Nonnutritional Modifications of Host Plant Physiology

Hormone accumulation in host tissues is affected by mycorrhizal infection, with changes in the levels of cytokinin, abscisic acid, and gibberellin-like substances (7, 8, 13). It is unclear if this is linked to improved nutrient status of the host, and it seems unlikely that phytohormone synthesis by the fungus could account for the magnitude of the increases. Alterations in biomass partitioning between shoots and roots as well as morphological effects may be hormone mediated (58, 109, 175).

Reduced susceptibility or increased tolerance of roots to certain soil-borne pathogens is frequently associated with an established mycorrhizal infection (59). Increases in synthesis of secondary metabolites like lignin, ethylene, and phenols (59, 60), as well as phytoalexins (122, 123), may all contribute to these "protective" effects.

Carbon Distribution in the Symbiosis

Short-term $^{14}CO_2$ labeling experiments indicate that the carbon requirements of VA mycorrhizal fungi are supplied by photosynthesis in the host plant (28, 48, 55, 97). The fungi convert host metabolites into specific fungal compounds. Lipids are particularly abundant in mature arbuscules, vesicles, and hyphae, but not in young mycelium and arbuscules (31, 49, 55, 102). Glycogen granules are usually associated with these lipid-containing structures and are also found in young vacuolated hyphae and the finest arbuscular branches (31, 77, 102). Polyols and trehalose have been detected in external fungal mycelium (48).

The demand on host photosynthate depends upon the amount of fungus associated with the root and its metabolic rate. Mycorrhizal roots have higher respiratory activity than uninfected roots (84, 161). Part of the increase is likely to be due to fungal respiration itself; and if all the extra were from this source, then fungal metabolic rate would be about 11 mg CO_2 g^{-1} h^{-1} (161). However, the presence of biotrophic pathogens increases the respiratory rate of infected tissue, and we do not know if this occurs with VA infection. As infected root cells have higher cytoplasmic volume, numbers of mitochondria, soluble protein content, and activity of several enzyme systems (52, 60, 129, 159, 174), increased respiratory rates would not be surprising.

Increased metabolic activity in infected roots, together with fungal biomass production and respiration, must be a cost to the mycorrhizal plant. Mycorrhizal plants with low available P export up to 10% more photosynthate to their roots than P-sufficient nonmycorrhizal plants (104, 111, 132, 161). This additional drain could result in reductions in yield, unless compensation mechanisms operate. When P limits the rate of photosynthesis, the investment of carbon in fungal symbiosis leading to increased rates of P acquisition could in turn lead to considerably increased rates of photosynthesis. This would offset the additional carbon utilization in the tissues below ground. Mycorrhizal plants may deploy shoot carbon more effectively than nonmycorrhizal plants, even when all are P sufficient (161). Thus for the same leaf fresh weight, photosynthetic rate (on a leaf-area basis), and relative growth rate (RGR), mycorrhizal plants have a lower percentage dry matter in their shoots (i.e. higher shoot fresh-weight:dry-weight ratios) and higher rates of photosynthesis per unit dry matter (158, 162, 164, 169). Compensation for increased carbon demand via increases in the rate of photosynthesis would not occur under conditions of limiting irradiance or when the fungus spreads rapidly in the roots before the onset of increased P uptake. Under these conditions, significant reductions in mycorrhizal growth response, as well as growth depressions, have often been observed (24, 25, 57, 86, 105, 168).

PHYSIOLOGICAL COMPATIBILITY AND MYCORRHIZAL EFFICIENCY

The relationships between VA mycorrhizal fungi and host plants are nonspecific, and this symbiosis can be found in most terrestrial ecosystems under most soil conditions. The fungus-host interaction is highly compatible at both structural and physiological levels, and at present there is no evidence for genetically controlled resistance mechanisms, such as the gene-for-gene systems that operate in symbioses involving biotrophic shoot pathogens. Mechanisms conferring specific resistance are not to be expected in mutualistic symbioses (84) but it cannot be excluded that some general resistance mechanisms may operate. Despite this lack of absolute specificity considerable variations in symbiotic response occur. These have usually been measured in terms of the growth of the host plant in combination with different fungal strains grown under different environmental conditions. Methods are now available to analyze the responses more precisely in terms of the physiological processes likely to be important determinants of symbiotic efficiency. This analysis is required because the selection of "efficient fungi" for inoculation programs and the prediction of host-fungus combinations likely to show significant responses to inoculation are important goals.

An efficient living organism is one whose physiological and biochemical processes are such that it can successfully cope with limiting environmental constraints. Table 2 lists features of fungi, plants, and symbiotic interactions that may be important in determining symbiotic efficiency. For VA mycorrhizal fungi efficiencies of nutrient uptake and translocation are likely to be important, and these may be determined by efficiency of production of active biomass within and outside the root. In plants, the responses will be affected by nutrient requirements, strategies adopted to cope with nutrient stress, as well as characteristics of the root system (46). In symbiosis, factors that affect the development and function of the interface and rates of transfer of nutrients across it may be important rate-limiting steps.

Mycorrhizal roots are more efficient at absorbing nutrients per unit length than nonmycorrhizal roots, but comparisons of nutrient uptake based on total carbon inputs below ground are yet to be made. Similarly, efficiencies of nutrient use in the whole plant are usually compared on a dry-weight basis, but could also be compared per unit of carbon fixed. In this content, mycor-

Table 2 Possible determinants of symbiotic efficiency

Fungus	Plant	Symbiosis
External hyphae	Roots	Interface
growth rate	thickness	area of contact
nutrient uptake	branching	uptake rates
translocation	root hairs	efflux rates
polyP metabolism	growth rate	wall metabolism
Internal	Nutrient requirements	
infection rate	uptake and deployment	
growth rate	tissue concentrations	
arbuscule production	biochemical sensitivity	
	(e.g. photosynthesis)	
	Stress tolerance	
	RGR	
	root : shoot ratio	

rhizal plants appear less efficient during the early stages of growth and development of infection but more efficient at later stages. Luxury (inefficient) nutrient uptake, root:shoot partitioning, tissue nutrient concentrations on a fresh-weight basis, and compartmentation into metabolic and storage pools need to be considered in comparisons of different host-fungus combinations as well as of mycorrhizal and nonmycorrhizal plants. The most appropriate basis for determination of efficiency will depend upon the context of the research—yield (in a crop), numbers of progeny (in a natural ecosystem), or nutrient transferred per unit carbohydrate utilized (in a physiological comparison)—but must be clearly defined. Modifications of physiological processes during the development of symbiosis must be significant but are difficult to pinpoint in our present state of knowledge.

CONCLUSIONS

Mycorrhizae must be taken into account in whole-plant physiology. The more we learn about interactions between the symbionts the clearer this becomes. Furthermore, understanding how the symbiosis works depends upon knowledge of the details of physiological and biochemical processes occurring in both plants and fungi. More information is required on the fungi, grown both in isolation and symbiotically. Techniques for growing large amounts of external hyphae in association with roots (66) and for determining the processes occurring in fungal structures should be exploited.

Now that we do have some idea of how the symbiosis works, questions about how fungus and plant are modified in the association become relevant.

We need to know much more about the processes taking place during establishment of the mycorrhizal interface and about bidirectional exchange of information and nutrients across it. Another avenue for research [which will require molecular biological approaches (96)] is the possibility that symbiosis may result in specific changes in gene expression and protein synthesis in both organisms, as in other plant:microorganism interactions (143, 144). Such changes could be involved in establishment of the structural and functional dimorphism of the fungus within and outside the root and in promoting bidirectional nutrient exchange between the symbionts. It is at this level that the efficiency of integration of the symbionts, variations in responses of different host-fungus combinations, and effects of different environmental conditions must be analyzed. Exploitation or management of the symbiosis depends on this fundamental information because it is these interactions that determine the final outcome in terms of plant production.

ACKNOWLEDGMENTS

We would particularly like to thank Silvio Gianinazzi and Andrew Smith for many stimulating discussions and for critical appraisal of several drafts of this paper. We are also grateful to Roger Koide for discussions on symbiotic efficiency, Christine Long and Dene Cuthbertson for reading the manuscript, and Gladys Hogg for typing it. Collaboration was funded by the Waite Institute Research Committee and the Institute National de Recherche Agronomique. Research (S.E.S.) is funded by the Australian Research Grants Scheme.

Literature Cited

1. Abbott, L. K. 1982. Comparative anatomy of vesicular-arbuscular mycorrhizas formed on subterranean clover. *Aust. J. Bot.* 30:485–99
2. Abbott, L. K., Robson, A. D. 1984. The effect of mycorrhizas on plant growth. In *VA Mycorrhizae*, ed. C. L. Powell, D. J. Bagyaraj, pp. 113–30. Boca Raton, Fla: CRC Press. 234 pp.
3. Abbott, L. K., Robson, A. D. 1985. The effect of soil pH on the formation of VA mycorrhizas by two species of *Glomus. Aust. J. Soil Res.* 23:253–61
4. Abbott, L. K., Robson, A. D. 1985. Formation of external hyphae in soil by four species of vesicular-arbuscular mycorrhizal fungi. *New Phytol.* 99:245–55
5. Abbott, L. K., Robson, A. D., De Boer, G. 1984. The effect of phosphorus on the formation of hyphae in soil by the vesicular-arbuscular mycorrhizal fungus *Glomus fasiculatum. New Phytol.* 97:437–46
6. Allen, M. F. 1982. Influence of vesicular-arbuscular mycorrhizae on water movement through *Bouteloua gracilis* (H.B.K) Lag ex Steud. *New Phytol.* 91:191–96
7. Allen, M. F., Moore, T. S., Christensen, M. 1980. Phytohormone changes in *Bouteloua gracilis* infected by vesicular-arbuscular mycorrhizae. I. Cytokinin increases in the host plant. *Can. J. Bot.* 58:371–74
8. Allen, M. F., Moore, T. S., Christensen, M. 1982. Phytohormone changes in *Bouteloua gracilis* infected by vesicular-arbuscular mycorrhizae. II. Altered levels of gibberellin-like substances and abscisic acid in the host plant. *Can. J. Bot* 60:468–71

9. Allen, M. F., Smith, W. K., Moore, T. S., Christensen, M. 1981. Comparative water relations and photosynthesis of mycorrhizal and non-mycorrhizal *Bouteloua gracilis* (HBK) Lag ex Steud. *New Phytol.* 88:683–93

10. Ames, R. N., Porter, L., St. John, T. V., Reid, C. P. P. 1984. Nitrogen sources and 'A' values for vesicular-arbuscular and non-mycorrhizal sorghum grown at three rates of ^{15}N ammonium sulphate. *New Phytol.* 97:269–76

11. Ames, R. N., Reid, C. P. P., Porter, L., Cambardella, C. 1983. Hyphal uptake and transport of nitrogen from two ^{15}N-labelled sources by *Glomus mosseae*, a vesicular-arbuscular mycorrhizal fungus. *New Phytol.* 95:381–96

12. Azcon-Aguilar, C., Gianinazzi-Pearson, V., Fardeau, J. C., Gianinazzi, S. 1986. Effect of vesicular-arbuscular mycorrhizal fungi and phosphate solubilising bacteria on growth and nutrition of soybean in a neutral-calcareous soil amended with ^{32}P-^{45}Ca tricalcium phosphate. *Plant Soil* 96:3–15

13. Barea, J. M. 1986. Importance of hormones and root exudates in mycorrhizal phenomena. In *Physiological and Genetical Aspects of Mycorrhizae*, ed V. Gianinazzi-Pearson, S. Gianinazzi, pp. 177–87. Paris: INRA. 832 pp.

14. Barea, J. M., Azcon-Aguilar, C. 1982. Production of plant growth-regulating substances by the vesicular-arbuscular mycorrhizal fungus *Glomus mosseae*. *Appl. Environ. Microbiol.* 43:810–13

15. Barea, J. M., Azcon-Aguilar, C. 1983. Mycorrhizas and their significance in nodulating nitrogen-fixing plants. *Adv. Agron.* 36:1–54

15a. Barea, J. M., Azcon-Aguilar, C., Azcon, R. 1987. Vesicular-arbuscular mycorrhiza improve both symbiotic N_2 fixation and N uptake from soil as assessed with a ^{15}N technique under field conditions. *New Phytol.* 106:717–25

16. Baylis, G. T. S. 1975. The magnolioid mycorrhiza and mycotrophy in root systems derived from it. In *Endomycorrhizas*, ed. F. E. Sanders, B. Mosse, P. B. Tinker, pp. 373–89. London: Academic. 626 pp.

17. Beever, R. E., Burns, D. J. W. 1980. Phosphorus uptake storage and utilisation by fungi. *Adv. Bot. Res.* 8:128–219

18. Beevers, L., Hagerman, R. H. 1980. Nitrate and nitrite reduction. In *The Biochemistry of Plants. A Comprehensive Treatise*, ed. P. K. Stumpf, E. E. Conn, 5:116–68. London: Academic. 670 pp.

19. Beilby, J. P. 1980. Sterol composition of ungerminated and germinated spores of the vesicular-arbuscular mycorrhizal fungus *Glomus caledonium*. *Lipids* 15:375–78

20. Beilby, J. P. 1980. Fatty acid and sterol composition of ungerminated spores of the vesicular-arbuscular mycorrhizal fungus. *Acaulospora laevis*. *Lipids* 15:949–52

21. Beilby, J. P. 1983. Effects of inhibitors on early protein RNA and lipid synthesis in germinating vesicular-arbuscular mycorrhizal fungal spores of *Glomus caledonium*. *Can. J. Microbiol.* 29:596–601

22. Beilby, J. P., Kidby, D. K. 1980. Biochemistry of germinated and ungerminated spores of the vesicular-arbuscular mycorrhizal fungus *Glomus caledonius:* changes in neutral and polar lipids. *J. Lipid Res.* 21:739–50

23. Beilby, J. P., Kidby, D. K. 1982. The early synthesis of RNA protein and some associated metabolic events in germinating vesicular-arbuscular fungal spores of *Glomus caledonius*. *Can. J. Bot.* 28:623–28

24. Bethlenfalvay, G. J., Pacovsky, R. S. 1985. Light effects in mycorrhizal soybeans. *Plant Physiol.* 73:969–72

25. Bethlenfalvay, G. J., Bayne, H. C., Pacovsky, R. S. 1983. Parasitic and mutualistic association between a mycorrhizal fungus and soybean. *Physiologia Plant.* 57:543–49

26. Bethlenfalvay, G. J., Brown, M. S., Pacovsky, R. S. 1982. Relationships between host and endophyte development in mycorrhizal soybeans. *New Phytol.* 90:537–43

27. Bethlenfalvay, G. J., Pacovsky, R. S., Brown, M. S. 1982. Parasitic and mutalistic associations between a mycorrhizal fungus and soybean: development of the endophyte. *Phytopathology* 72:894–97

28. Bevege, D. I., Bowen, G. D., Skinner, M. F. 1975. Comparative carbohydrate physiology of ecto and endomycorrhizas. See Ref. 16, pp. 152–74

29. Bielesky, R. L., Ferguson, I. B. 1983. Physiology and metabolism of phosphate and its compounds. In *Encyclopaedia of Plant Physiology*, Vol. 15a: *Inorganic Plant Nutrition*, ed. A. Lauchli, R. L. Bieleski, pp. 422–49. Berlin: Springer Verlag. 449 pp.

30. Bolan, N. S., Robson, A. D., Barrow, N. J., Aylemore, L. A. G. 1984. Specific activity of phosphorus in mycorrhizal and non-mycorrhizal plants in relation to

the availability of phosphorus to plants. *Soil Biol. Biochem.* 16:299–304

31. Bonfante-Fasolo, P. 1984. Anatomy and morphology of V.A. mycorrhizae. See Ref. 4, pp. 5–32

32. Bonfante-Fasolo, P., Grippiolo, R. 1982. Ultrastructural and cytochemical changes in the wall of a vesicular-arbuscular mycorrhizal fungus during symbiosis. *Can. J. Bot.* 60:2303–12

33. Bonfante-Fasolo, P., Scannerini, S. 1977. A cytological study of the vesicular-arbuscular mycorrhiza in *Ornithogallum umbellatum L. Allionia* 2:5–21

34. Bonfante-Fasolo, P., Dexheimer, J., Gianinazzi, S., Gianinazzi-Pearson, V., Scannerini, S. 1981. Cytochemical modifications in the host-fungus interface during intracellular interactions in vesicular-arbuscular mycorrhizae. *Plant Sci. Lett.* 22:13–21

35. Bonfante-Fasolo, P., Gianinazzi-Pearson, V., Scannerini, S., Gianinazzi, S. 1986. Analyses ultracytologique des interactions entre plante et champignon au niveau cellulaire dans les mycorhizes. *Physiol. Veg.* 24:245–52

36. Bowen, G. D., Bevege, D. I., Mosse, B. 1975. Phosphate physiology of vesicular-arbuscular mycorrhizas. See Ref. 16, pp. 241–60

37. Burggraaf, A. J. P., Beringer, J. E. 1987. Is nuclear division limiting in vitro culture of vesicular-arbuscular mycorrhizal fungi? In *Proceedings of 7th North American Conference on Mycorrhizae*, p. 190

38. Buwalda, J. G., Ross, G. J. S., Stribley, D. P., Tinker, P. B. 1982. The development of endomycorrhizal root systems. III. The mathematical representation of the spread of vesicular-arbuscular mycorrhizal infection in root systems. *New Phytol.* 91:669–82

39. Buwalda, J. G., Stribley, D. P., Tinker, P. B. 1983. Increased uptake of anions by plants with vesicular-arbuscular mycorrhizas. *Plant Soil* 71:463–67

40. Buwalda, J. G., Stribley, D. P., Tinker, P. B. 1984. The development of endomycorrhizal root systems. V. The detailed pattern of infection and the control of development of infection level by host in young leek plants. *New Phytol.* 96:411–27

41. Callow, J. A., Capaccio, L. C. M., Parish, G., Tinker, P. B. 1978. Detection and estimation of polyphosphate in vesicular-arbuscular mycorrhizas. *New Phytol.* 80:125–34

42. Capaccio, L. C. M., Callow, J. A. 1982. The enzymes of polyphosphate metabolism in vesicular-arbuscular mycorrhizas. *New Phytol.* 91:81–91

43. Carling, D. E., Brown, M. F. 1982. Anatomy and physiology of vesicular-arbuscular and non-mycorrhizal roots. *Phytopathology* 72:1108–14

44. Carling, D. R., Riehle, W. G., Brown, M. F., Johnson, D. R. 1978. Effects of a vesicular-arbuscular mycorrhizal fungus on nitrate reductase and nitrogenase activities in nodulating and non-nodulating legumes. *Phytopathology* 68:1590–96

45. Carr, G. R., Hinkley, M. A., Le Tacon, F., Hepper, C. M., Jones, M. G. K., Thomas, E. 1985. Improved hyphal growth of two species of vesicular-arbuscular mycorrhizal fungi in the presence of suspension cultured plant cells. *New Phytol.* 101:417–26

46. Chapin, F. S. 1980. The mineral nutrition of wild plants. *Ann. Rev. Ecol. Syst.* 11:233–60

47. Clarkson, D. T. 1985. Factors affecting mineral nutrient acquisition by plants. *Ann. Rev. Plant Physiol.* 36:77–115

48. Cooper, K. M. 1984. Physiology of VA mycorrhizal associations. See Ref. 2, pp. 155–203

49. Cooper, K. M., Lösel, D. 1978. Lipid physiology of vesicular-arbuscular mycorrhiza. I. Composition of lipids in roots of onion clover and ryegrass infected with *Glomus mosseae. New Phytol.* 80:143–51

50. Cooper, K. M., Tinker, P. B. 1978. Translocation and transfer of nutrients in vesicular-arbuscular mycorrhizas. II. Uptake and translocation of phosphorus, zinc and sulphur. *New Phytol.* 81:43–52

51. Cooper, K. M., Tinker, P. B. 1981. Translocation and transfer of nutrients in vesicular-arbuscular mycorrhizas. IV. Effect of environmental variables on movement of phosphorus. *New Phytol.* 88:327–39

52. Cox, G., Sanders, F. E. 1974. Ultrastructure of the host-fungus interface in a vesicular-arbuscular mycorrhiza. *New Phytol.* 73:901–12

53. Cox, G., Tinker, P. B. 1976. Translocation and transfer of nutrients in vesicular-arbuscular mycorrhizas. I. The arbuscule and phosphorus transfer: a quantitative ultrastructural study. *New Phytol.* 77:371–78

54. Cox, G., Moran, K. J., Sanders, F. E., Nockolds, C., Tinker, P. B. 1980. Translocation and transfer of nutrients in vesicular-arbuscular mycorrhizas. III. Polyphosphate granules and phosphorus translocation. *New Phytol.* 84:649–59

55. Cox, G., Sanders, F. E., Tinker, P. B., Wild, J. A. 1975. Ultrastructural evidence relating to host-endophyte transfer in a vesicular-arbuscular mycorrhiza. See Ref. 16, pp. 297–312

56. Cress, W. A., Throneberry, G. O., Lindsey, D. L. 1979. Kinetics of phosphorus absorption by mycorrhizal and non-mycorrhizal tomato roots. *Plant Physiol.* 64:484–87

57. Daft, M. J., El Giahmi, A. A. 1978. Effects of arbuscular mycorrhiza on plant growth. VIII. Effects of defoliation and light on selected hosts. *New Phytol.* 80:365–72

58. Daft, M. J., Okusanya, B. O. 1973. Effect of *Endogone* mycorrhiza on plant growth. VI. Influence of infection on the anatomy and reproductive development in four hosts. *New Phytol.* 72:1333–39

59. Dehne, H.-W. 1982. Interactions between vesicular-arbuscular mycorrhizal fungi and plant pathogens. *Phytopathology* 72:1115–19

60. Dehne, H.-W. 1986. Influence of V.A. mycorrhizae on host plant physiology. See Ref. 13, pp. 431–35

61. Dexheimer, J., Gianinazzi, S., Gianinazzi-Pearson, V. 1979. Ultrastructural cytochemistry of the host-fungus interface in the endomycorrhizal association *Glomus mosseae/Allium cepa*. *Z. Pflanzenphysiol.* 92:191–206

62. Dexheimer, J., Kreutz-Jeanmaire, C., Gerard, M. J., Gianinazzi-Pearson, V., Gianinazzi, S. 1986. Approche cellulaire du fonctionnement des endomycorhizes à vésicules et arbuscules: les plasmalemmes de l'interface. See Ref. 13, pp. 278–89

63. Dexheimer, J., Marx, C., Gianinazzi-Pearson, V., Gianinazzi, S. 1985. Ultracytological studies on plasmalemma formations produced by host and fungus in vesicular-arbuscular mycorrhizae. *Cytologia* 50:461–71

64. Dietz, K.-J., Foyer, C. 1986. The relationship between phosphate status and photosynthesis in leaves. Reversibility of the effects of phosphate deficiency on photosynthesis. *Planta* 167:376–81

65. Durr, M., Urech, K., Boller, T., Weimken, A., Schwencke, J., Nagy, M. 1979. Sequestration of arginine by polyphosphate in vacuoles of yeast *(Saccharomyces cerevisiae)*. *Arch Microbiol.* 121:161–75

66. Elmes, R. P., Mosse, B. 1984. Vesicular-arbuscular mycorrhizal inoculum production. II. Experiments with maize *(Zea mays)* and other hosts in nutrient flow culture. *Can. J. Bot.* 62:1531–36

67. Fitter, A. H. 1985. Functioning of vesicular-arbuscular mycorrhizas under field conditions. *New Phytol.* 99:257–65

68. Gianinazzi, S., Gianinazzi-Pearson, V., Dexheimer, J. 1979. Enzymatic studies on the metabolism of vesicular-arbuscular mycorrhiza. III. Ultrastructural localisation of acid and alkaline phosphatase in onion roots infected with *Glomus mosseae* (Nicol. & Gerd.) *New Phytol.* 82:127–32

69. Gianinazzi, S., Gianinazzi-Pearson, V., Trouvelot, A., eds. 1982. *Les Mycorhizes Partie Intégrante de la Plante: Biologie et Perspectives d'Utilisation*. (Les Colloques d'INRA13). Paris: INRA. 397 pp.

70. Gianinazzi-Pearson, V. 1984. Host-fungus specificity in mycorrhizae. In *Genes Involved in Plant-Microbe Interactions*, ed. D. P. S. Verma, T. H. Hohn, pp. 225–53. Wien: Springer. 393 pp.

71. Gianinazzi-Pearson, V. 1984. Mycorrhizal effectiveness: How, when and where? In *Proceedings of 6th North American Conference on Mycorrhizae*, ed. R. Molina, pp. 150–54. Oregon State Univ. Corvallis, Ore: For. Res. Lab. 471 pp.

72. Gianinazzi-Pearson, V., Gianinazzi, S. 1978. Enzymatic studies on the metabolism of vesicular-arbuscular mycorrhiza. II. Soluble alkaline phosphatase specific to mycorrhizal infection in onion roots. *Physiol. Plant Pathol.* 12:45–53

73. Gianinazzi-Pearson, V., Gianinazzi, S. 1983. The physiology of vesicular-arbuscular mycorrhizal roots. *Plant Soil* 71:197–209

74. Gianinazzi-Pearson, V., Gianinazzi, S. 1986. The physiology of improved phosphate nutrition in mycorrhizal plants. See Ref. 13, pp. 101–9

75. Gianinazzi-Pearson, V., Gianinazzi, S. 1987. Morphological integration and functional compatibility between symbionts in vesicular-arbuscular endomycorrhizal associations. In *Cell to Cell Signals in Plant, Animal and Microbial Symbiosis*, ed. S. Scannerini, D. C. Smith, P. Bonfante-Fasolo, V. Gianinazzi-Pearson. NATO ASI series. In press

76. Gianinazzi-Pearson, V., Fardeau, J.-C., Asimi, S., Gianinazzi, S. 1981. Source of additional phosphorus absorbed from soil by vesicular-arbuscular mycorrhizal soybeans. *Physiol. Vég.* 19:33–43

77. Gianinazzi-Pearson, V., Morandi, D., Dexheimer, J., Gianinazzi, S. 1981. Ultrastructural and ultracytochemical fea-

tures of a *Glomus tenuis* mycorrhiza. *New Phytol.* 88:633–39
78. Giaquinta, R. T. 1983. Phloem loading of sucrose. *Ann. Rev. Plant Physiol.* 34:347–87
79. Graham, J. H., Syvertsen, J. P. 1984. Influence of vesicular-arbuscular mycorrhiza on the hydraulic conductivity of root of two citrus rootstocks. *New Phytol.* 97:277–84
80. Graham, J. H., Linderman, R. G., Menge, J. A. 1982. Development of external hyphae by different isolates of mycorrhizal *Glomus* spp. in relation to root colonisation and growth of Troyer citrange. *New Phytol.* 91: 183–89
81. Hale, K. A., Sanders, F. E. 1982. Effects of benomyl on vesicular-arbuscular mycorrhizal infection of red clover (*Trifolium pratense* L.) and consequences for phosphorus inflow. *J. Plant Nutr.* 5:1355–67
82. Hardie, K. 1985. The effect of removal of extraradical hyphae on water uptake by vesicular-arbuscular mycorrhizal plants. *New Phytol.* 101:677–84
83. Hardie, K., Leyton, L. 1981. The influence of vesicular-arbuscular mycorrhiza on growth and water relations of red clover. I. In phosphate deficient soil. *New Phytol.* 89:599–608
84. Harley, J. L., Smith, S. E. 1983. *Mycorrhizal Symbiosis.* London/New York: Academic. 483 pp.
85. Harold, F. M. 1966. Inorganic polyphosphates in biology: structure, metabolism and function. *Bacteriol Rev.* 30:772–94
86. Hayman, D. S. 1974. Plant growth responses to vesicular-arbuscular mycorrhiza. VI. The effect of light and temperature. *New Phytol.* 73:71–80
87. Hayman, D. S. 1983. The physiology of vesicular-arbuscular endomycorrhizal symbiosis. *Can. J. Bot.* 61:944–63
88. Hepper, C. M. 1977. A colorimetric method for estimating vesicular-arbuscular mycorrhizal infection in roots. *Soil Biol. Biochem.* 9:15–18
89. Hepper, C. M. 1979. Germination and growth of *Glomus caledonius* spores: the effects of inhibitors and nutrients. *Soil Biol. Biochem.* 11:269–77
90. Hepper, C. M. 1984. Inorganic sulphur nutrition of the vesicular-arbuscular mycorrhizal fungus *Glomus caledonium.* *Soil Biol. Biochem.* 16:669–71
91. Hepper, C. M. 1984. Isolation and culture of VA mycorrhizal (VAM) fungi. See Ref. 2, pp. 95–112
92. Hepper, C. M. 1985. Influence of age of roots on the pattern of vesicular-

arbuscular mycorrhizal infection in leek and clover. *New Phytol.* 101:685–93
93. Hepper, C. M., Jakobsen, I. 1983. Hyphal growth from spores of the mycorrhizal fungus *Glomus caledonius:* effect of amino acids. *Soil Biol. Biochem.* 15:55–58
94. Hepper, C. M., Sen, R. 1986. Biochemical characterisation of vesicular-arbuscular mycorrhizal fungi. See Ref. 13, pp. 611–14
95. Hepper, C. M., Sen. R., Maskell, C. S. 1986. Identification of vesicular-arbuscular mycorrhizal fungi in roots of leek (*Allium porrum* L.) and maize (*Zea mays* L.) on the basis of enzyme mobility during polyacrilimide gel electrophoresis. *New Phytol.* 102:529–39
96. Hirsch, P. R. 1986. Gene cloning and its potential application to mycorrhizal fungi. See Ref. 13, pp. 145–57
97. Ho, I., Trappe, J. M. 1973. Translocation of ^{14}C from *Festuca* plants to their endomycorrhizal fungi. *Nature New Biol.* 244:30–31
98. Ho, I., Trappe, J. M. 1975. Nitrate-reducing capacity of two vesicular-arbuscular mycorrhizal fungi. *Mycologia* 67:886–88
99. Jeanmaire, C., Dexheimer, J., Marx, C., Gianinazzi, S., Gianinazzi-Pearson, V. 1985. Effect of vesicular-arbuscular mycorrhizal infection on the distribution of neutral phosphatase activities in root cortical cells. *J. Plant Physiol.* 119:285–93
100. Karunaratne, R. S., Baker, J. H., Barker, A. V. 1986. Phosphorus uptake by mycorrhizal and non-mycorrhizal roots of soybean. *J. Plant Nutr.* 9:1303–13
101. Killham, K. 1985. Vesicular-arbuscular mycorrhizal mediation of trace and minor element uptake in perennial grasses: relation to livestock herbage. In *Ecological Interactions in Soil,* ed. A. H. Fitter, D. Atkinson, D. J. Read, M. B. Usher, pp. 225–32. Oxford: Blackwell Scientific. 451 pp.
102. Kinden, D. A., Brown, M. R. 1975. Electron microscopy of vesicular-arbuscular mycorrhizae of yellow poplar. II. Intracellular hyphae and vesicles. *Can. J. Microbiol.* 21:1768–80
103. Kinden, D. A., Brown, M. F. 1975. Electron microscopy of vesicular-arbuscular mycorrhizae of yellow poplar. III. Host endophyte interactions during arbuscular development. *Can. J. Microbiol.* 21:1930–39
104. Koch, K. E., Johnson, C. R. 1984. Photosynthate partitioning in split-root citrus seedlings with mycorrhizal and

non-mycorrhizal root systems. *Plant Physiol.* 75:26–30

105. Koide, R. 1985. The nature of growth depressions in sunflower caused by vesicular-arbuscular mycorrhizal infection. *New Phytol.* 99:449–62

106. Koske, R. E. 1981. Multiple germination by spores of *Gigaspora gigantea*. *Trans. Br. Mycol. Soc.* 76:328–30

107. Kough, J., Gianinazzi-Pearson, V. 1986. Physiological aspects of V.A. mycorrhizal hyphae in root tissue and soil. See Ref. 13, pp. 223–26

108. Kough, J. K. L., Gianinazzi-Pearson, V., Ginaninazzi, S. 1987. Depressed metabolic activity of vesicular arbuscular mycorrhizal fungi after fungicide applications. *New Phytol.* 106:707–15

109. Krishna, K. R., Suresh, H. M., Syamsunder, J., Bagyaraj, D. J. 1981. Changes in the leaves of finger millet due to V.A. mycorrhizal infection. *New Phytol.* 87:717–22

110. Kucey, R. M. N., Paul, E. A. 1982. Biomass of mycorrhizal fungi associated with bean roots. *Soil Biol. Biochem.* 14:413–14

111. Kucey, R. M. N., Paul, E. A. 1982. Carbon flow, photosynthesis and N_2 fixation in mycorrhizal and nodulated fababeans (*Vicia faba* L.). *Soil Biol. Biochem.* 14:407–12

112. Levy, Y., Krikun, J. 1980. Effect of vesicular-arbuscular mycorrhiza on *Citrus jambhiri* water relations. *New Phytol.* 85:25–31

113. Levy, Y., Syvertsen, J. P., Nemec, S. 1983. Effect of drought stress and vesicular-arbuscular mycorrhiza on citrus transpiration and hydraulic conductivity of roots. *New Phytol.* 93:61–66

114. Lewis, D. H. 1986. Interrelationships between carbon nutrition and morphogenesis in mycorrhizas. See Ref. 13, pp. 85–100

115. Ling-Lee, M., Chilvers, G. A., Ashford, A. E. 1975. Polyphosphate granules in three different kinds of tree mycorrhiza. *New Phytol.* 75:551–54

116. Macdonald, R. M., Lewis, M. 1978. The occurrence of some acid phosphatases and dehydrogenases in the vesicular-arbuscular mycorrhizal fungus *Glomus mosseae*. *New Phytol.* 80:135–41

117. Manners, J. M., Gay, J. L. 1983. The host-parasite interface and nutrient transfer in biotrophic parasitism. In *Biochemical Plant Pathology*, ed. J. A. Callow, pp. 163–95. Chichester: Wiley. 484 pp.

118. Marx, C., Dexheimer, J., Gianinazzi-Pearson, V., Gianinazzi, S. 1982. Enzymatic studies on the metabolism of vesicular-arbuscular mycorrhiza. IV. Ultracytoenzymological evidence (ATPase) for active transfer processes in the host-arbuscular interface. *New Phytol.* 90:37–43

119. Matile, P., Wiemken, A. 1976. Interactions between cytoplasm and vacuole. In *Encyclopaedia of Plant Physiology, New Series, Vol. III. Transport in Plants III*, ed. C. R. Stocking, U. Heber, pp. 255–87 Berlin: Springer-Verlag

120. McNeil, M., Darville, A. G., Fry, S. C., Albersheim, P. 1984. Structure and function of primary cell walls of plants. *Ann. Rev. Biochem.* 53:625–63

121. Meyer, F. H. 1974. Physiology of mycorrhiza. *Ann. Rev. Plant Physiol.* 25:567–86

122. Morandi, D., Gianinazzi-Pearson, V. 1986. Influence of mycorrhizal infection and phosphate nutrition on secondary metabolite contents of soybean roots. See Ref. 13, pp. 787–91

123. Morandi, D., Bailey, J. A., Gianinazzi-Pearson, V. 1984. Isoflavonoid accumulation in soybean roots infected with vesicular-arbuscular mycorrhizal fungi. *Physiol. Plant Pathol.* 24:357–64

124. Mosse, B. 1959. The regular germination of resting spores and some observations on the growth requirements of an *Endogone* sp. causing vesicular-arbuscular mycorrhiza. *Trans. Br. Mycol. Soc.* 42:273–86

125. Mosse, B., Hayman, D. S., Arnold, D. J. 1973. Plant growth responses to vesicular-arbuscular mycorrhiza. V. Phosphate uptake by three plant species from P-deficient soils labelled with ^{32}P. *New Phytol.* 72:809–15

126. Nelson, C. E., Safir, G. R. 1982. Increased drought tolerance of mycorrhizal onion plants caused by improved phosphorus nutrition. *Planta* 154:407–13

127. Nemec, S., Meredith, F. I. 1981. Amino-acid content of leaves in mycorrhizal and non-mycorrhizal citrus rootstocks. *Ann. Bot.* 47:351–58

128. Nye, P., Tinker, P. B. 1977. *Solute Movement in the Soil-Root System*. Oxford: Blackwell Scientific. 342 pp.

129. Oliver, A. J., Smith, S. E., Nicholas, D. J. D., Wallace, W., Smith, F. A. 1983. Activity of nitrate reductase in *Trifolium subterraneum*: effects of mycorrhizal infection and phosphate nutrition. *New Phytol.* 94:63–79

130. Owusu-Bennoah, E., Wild, A. 1979. Autoradiography of the depletion zone of phosphate around onion roots in the presence of vesicular-arbuscular mycorrhiza. *New Phytol.* 82:133–40

131. Pairunan, A. K., Robson, A. D., Abbott, L. K. 1980. The effectiveness of vesicular-arbuscular mycorrhizas in increasing growth and phosphorus uptake of subterranean clover from phosphorus sources of different solubilities. *New Phytol.* 84:327–38

132. Pang, P. C., Paul, E. A. 1980. Effects of vesicular-arbuscular mycorrhiza on ^{14}C and ^{15}N distribution in nodulated fababeans. *Can. J. Soil Sci.* 60:241–50

133. Pearson, V., Tinker, P. B. 1975. Measurement of phosphorus fluxes in the external hyphae of endomycorrhizas. See Ref. 16, pp. 277–87

134. Pons, F., Gianinazzi-Pearson, V. 1984. Influence du phosphore du potassium de l'azote et du pH sur le comportement de champignons endomyorhizogenes á vésicules et arbuscules. *Cryptogam. Mycol.* 5:87–100

135. Powell, C. L. 1975. Plant growth responses to vesicular-arbuscular mycorrhiza. VIII. Uptake of P by onion and clover infected with different *Endogone* spore types in ^{32}P-labelled soil. *New Phytol.* 75:563–66

136. Powell, C. L. 1975. Potassium uptake by endotrophic mycorrhizas. See Ref. 16, pp. 460–68

137. Ratnayake, R. T., Leonard, R. T., Menge, J. A. 1978. Root exudaton in relation to supply of phosphorus and its possible relevance to mycorrhizal infection. *New Phytol.* 81:543–52

138. Raven, J. A., Smith, F. A. 1976. Nitrogen assimilation and transport in vascular plants in relation to intracellular pH regulation. *New Phytol.* 76:415–31

139. Rhodes, L. H., Gerdemann, J. W. 1975. Phosphate uptake zones of mycorrhizal and non-mycorrhizal onions. *New Phytol.* 75:555–61

140. Rhodes, L. H., Gerdemann, J. W. 1978. Translocation of calcium and phosphate by external hyphae of vesicular-arbuscular mycorrhizae. *Soil Sci.* 126: 125–26

141. Rhodes, L. H., Gerdemann, J. W. 1978. Hyphal translocation and uptake of sulphur by vesicular-arbuscular mycorrhizae of onions. *Soil Biol. Biochem.* 10:355–60

142. Rhodes, L. H., Gerdemann, J. W. 1978. Influence of phosphorus nutrition on sulphur uptake by vesicular-arbuscular mycorrhizae of onions. *Soil Biol. Biochem.* 10:361–64

143. Robertson, J. G., Wells, B., Brewin, N. J., Williams, M. A. 1985. Immunogold localisation of cellular constituents in the Legume-Rhizobium symbiosis. *Oxford Surv. Plant Mol. Cell Biol.* 2:69–89

144. Rossen, L., Johnston, A. W. B., Firmin, J. L., Shearman, C. A., Evans, I. J., Downie, J. A. 1986. Structure function and regulation of nodulation genes of *Rhizobium*. *Oxford Surv. Plant Mol. Cell Biol.* 3:441–47

145. Sanders, F. E. 1975. The effect of foliar-applied phosphate on the mycorrhizal infections of onion roots. See Ref. 16, pp. 261–76

146. Sanders, F. E., Sheikh, N. A. 1983. The development of vesicular-arbuscular mycorrhizal infection in plant root systems. *Plant Soil* 71:223–46

147. Sanders, F. E., Mosse, B., Tinker, P. B., eds. 1975. *Endomycorrhizas.* London: Academic. 626 pp.

148. Sanders, F. E., Tinker, P. B., Black, R. L., Palmerley, S. M. 1977. The development of endomycorrhizal root systems. I. Spread of infection and growth promoting effects with four species of vesicular-arbuscular mycorrhizas. *New Phytol.* 78:257–68

149. Scannerini, S., Bonfante-Fasolo, P. 1979. Ultrastructural cytochemical demonstration of polysaccharides and proteins within the host-arbuscule interfacial matrix in an endomycorrhiza. *New Phytol.* 83:87–94

150. Scannerini, S., Bonfante-Fasolo, P. 1983. Comparative ultrastructural analysis of mycorrhizal associations. *Can. J. Bot.* 61:917–43

151. Sivak, M. N., Walker, D. A. 1986. Photosynthesis *in vivo* can be limited by phosphate supply. *New Phytol.* 102:499–512

152. Smith, D. C., Muscatine, L., Lewis, D. H. 1969. Carbohydrate movement from autotrophs to heterotrophs in parasitic and mutualistic symbioses. *Biol. Rev.* 44:17–90

153. Smith, F. A., Smith, S. E. 1986. Movement across membranes: physiology and biochemistry. See Ref. 13, pp. 75–84

154. Smith, S. E. 1980. Mycorrhizas of autotrophic higher plants. *Biol. Rev.* 55:475–510

155. Smith, S. E. 1982. Inflow of phosphate into mycorrhizal and non-mycorrhizal *Trifolium subterraneum* at different levels of soil phosphate. *New Phytol.* 90:293–303

156. Smith, S. E., Son, C. L. 1987. Inflow of phosphate into onion roots: interactions between photon-irradiance phosphorus supply and mycorrhizal infection. See Ref. 37, p. 265

157. Smith, S. E., Walker, N. A. 1981. A quantitative study of mycorrhizal infection in *Trifolium:* separate determination

of the rates of infection and of mycelial growth. *New Phytol.* 89:225–40

158. Smith, S. E., St. John, B. J., Smith, F. A., Bromley, J.-L. 1986. Effect of mycorrhizal infection on plant growth nitrogen and phosphorus nutrition of glasshouse-grown *Allium cepa* L. *New Phytol.* 103:359–73

159. Smith, S. E., St. John, B. J., Smith, F. A., Nicholas, D. J. D. 1985. Activity of glutamine synthetase and glutamate dehydrogenase in *Trifolium subterraneum* L. and *Allium cepa* L.: effects of mycorrhizal infection and phosphate nutrition. *New Phytol.* 99:211–27

160. Smith, S. E., Tester, M., Walker, N. A. 1986. The development of mycorrhizal root systems in *Trifolium subterraneum* L.: growth of roots and the uniformity of spatial distribution of mycorrhizal infection units in young plants. *New Phytol.* 103:117–31

161. Snellgrove, R. C., Splittstosser, W. E., Stribley, D. P., Tinker, P. B. 1982. The distribution of carbon and the demand of the fungal symbiont in leek plants with vesicular-arbuscular mycorrhizas. *New Phytol.* 92:75–87

162. Snellgrove, R. C., Stribley, D. P., Tinker, P. B., Lawlor, D. W. 1986. The effect of vesicular-arbuscular mycorrhizal infection on photosynthesis and carbon distribution in leek plants. See Ref. 13, pp. 421–24

163. St. John, T. V. 1980. Root size root hairs and mycorrhizal infection: a reexamination of Baylis's hypothesis with tropical trees. *New Phytol.* 84:483–87

164. Stribley, D. P., Snellgrove, R. C. 1985. Physiological changes accompanying mycorrhizal infection in leek. See Ref. 71, p. 395.

165. Strullu, D. G., Gourret, J. P., Garrec, J.-P., Fourcy, A. 1981. Ultrastructure and electron-probe microanalysis of the metachromatic vacuolar granules occurring in *Taxus* mycorrhizas. *New Phytol.* 87:537–45

166. Swaminathan, V. 1979. Nature of the inorganic fraction of soil phosphate fed on by vesicular-arbuscular mycorrhizae of potatoes. *Proc. Indian Acad. Sci.* 88B:423–33

167. Sward, R. J. 1981. The structure of the spores of *Gigaspora margarita*. II. Changes accompanying germination. *New Phytol.* 88:661–66

168. Tester, M., Smith, F. A., Smith, S. E. 1985. Phosphate inflow into *Trifolium subterraneum* L.: effects of photon irradiance and mycorrhizal infection. *Soil Biol. Biochem.* 17:807–10

169. Tester, M., Smith, S. E., Smith, F. A., Walker, N. A. 1986. Effects of photon irradiance on the growth of shoots and roots on the rate of initiation of mycorrhizal infection and on the growth of infection units in *Trifolium subterraneum* L. *New Phytol.* 103:375–90

170. Timmer, L. W., Leyden, R. F. 1980. The relationship of mycorrhizal infection to phosphorus-induced copper deficiency in sour orange seedlings. *New Phytol.* 85:15–23

171. Tinker, P. B. 1975. Effects of vesicular-arbuscular mycorrhizas on higher plants. *Symp. Soc. Exp. Biol.* 29:325–29

172. Tinker, P. B. 1975. Soil chemistry of phosphorus and mycorrhizal effects on plant growth. See Ref. 16, pp. 353–71

173. Tisdall, J. M., Oades, J. M. 1979. Stabilisation of soil aggregates by the root systems of ryegrass. *Aust. J. Soil Res.* 17:429–41

174. Toth, R., Miller, R. M. 1984. Dynamics of arbuscule development and degeneration in a *Zea mays* mycorrhiza. *Am. J. Bot.* 71:449–60

175. Wallace, L. L. 1981. Growth morphology and gas exchange of mycorrhizal and non-mycorrhizal *Panicum coloratum* L., a C4 grass species under different clipping and fertilisation regimes. *Oecologia* 49:272–78

176. Weete, J. D. 1980. *Lipid Biochemistry of Fungi.* New York: Plenum. 388 pp.

177. White, J. A., Brown, M. R. 1979. Ultrastructure and X-ray analysis of phosphorus granules in a vesicular-arbuscular mycorrhizal fungus. *Can. J. Bot.* 57:2812–18

178. Woolhouse, H. W. 1975. Membrane structure and transport problems considered in relation to phosphorus and carbohydrate movements and the regulation of endotrophic mycorrhizal associations. See Ref. 16, pp. 209–39

179. Young, C. C., Juang, T. C., Guo, H. Y. 1986. The effect of inoculation with vesicular-arbuscular mycorrhizal fungi on soybean yield and mineral phosphorus utilisation in subtropical-tropical soils. *Plant Soil.* 95:245–53

Ann. Rev. Plant Physiol. Plant Mol. Biol. 1988. 39:245–65

WATER TRANSPORT IN AND TO ROOTS

J. B. Passioura

CSIRO, Division of Plant Industry, Canberra 2601, Australia

CONTENTS

INTRODUCTION 245
THE UPTAKE OF WATER BY ROOTS FROM SOIL 246
    The Cylindrical Flow Model 246
    Roots in Dry Soil 250
TRANSPORT ACROSS WHOLE ROOT SYSTEMS 252
    Basic Hydraulic and Osmotic Properties 252
    Caveats Based on Possible Experimental Artefacts 256
    Endogenous and Environmental Influences 257
PATHWAYS FOR WATER TRANSPORT 258
    Radial 258
    Axial 260
CONCLUDING REMARKS 261

INTRODUCTION

Plants will not grow unless they are supplied with enough water, in the face of continuing evaporation from their leaves, to maintain those leaves well hydrated. In what follows I review some of the processes that influence the ability of roots to supply leaves with adequate amounts of water, including flow within the roots and flow to them through the soil they occupy. I leave undefined the precise meaning of "well hydrated," for the influence of the water status of its leaves on the performance of a plant is currently controversial (66, 67, 69, 80); but the gross meaning is clear: Wilted plants do not grow.

This review is essentially an expansion and extension of part of Boyer's (4) recent comprehensive coverage of water transport in plants. Other recent

245

reviews (5, 79) and books (37, 75, 81) provide further background. In the world outside the laboratory, the collection of water by roots depends as much on their interactions with the soil—their growth despite mechanical hindrance, their preferential occupation of continuous large pores in the soil, their ability to maintain hydraulic continuity with the soil—as on the processes that influence transport within the root. Accordingly, what follows makes frequent reference to these interactions.

THE UPTAKE OF WATER BY ROOTS FROM SOIL

The Cylindrical Flow Model

Most analyses of the uptake of water by roots from soil are based on the pioneering work of Philip (58) and Gardner (25) and describe the system in terms of cylindrical geometry, with the water flowing radially in response to gradients in pressure, towards a cylindrical, essentially isolated, root. The fundamental equation describing this flow is $F = K \, d\tau/dr$, where F (in $m^3 \, m^{-2} \, s^{-1}$) is the flux density of the water, K (in $m^2 \, s^{-1} \, Pa^{-1}$) is the hydraulic conductivity, τ (in Pa) is the soil water suction, and r (in m) is radial distance from the root. K falls by many orders of magnitude as a soil dries, so that the gradient in τ necessary to induce a given flux correspondingly rises by several orders of magnitude as a soil dries.

Despite the fact that root systems are branched, that the catchments of individual roots overlap in geometrically complicated ways, and that roots growing in real soil are typically not cylindrical because they must weave their ways past obstructions and often have to conform to the shapes of the pores within which they are growing, this model remains useful. It can be extended to complete root systems, despite their complexity, by means of the simple but effective stratagem of imagining that each root has exclusive access to a cylinder of soil whose outer radius, b, is half the average distance between roots (78). Radius b can be calculated as $1/\sqrt{(\pi L)}$, where L is the average rooting density, the length of root in unit volume of soil.

There are two main semiquantitative conclusions that can be drawn from this model. The first is that unless the soil is fairly dry (when its hydraulic conductivity is low), and unless the rooting density, L, is also low, say well below 1 cm cm^{-3}, then the fall in water potential across the roots, from the epidermis to the xylem, will be much greater than $\Delta\tau$, the difference between the average value of τ in the soil and its value at the surface of the root (44, 51, 61). The second is that once the soil is so dry that the flow of water through it is largely limiting the rate of uptake, then the time constant, or approximate half time, $t_{1/2}$ (in days), for the withdrawal of water from the soil, is approximately $1/\pi L$, or b^2, if b is measured in centimeters (53).

There have been many attempts to test this model, both in detail, for single roots (28), and more generally, for root systems (17, 18, 30, 51, 87). The latter have been bedeviled by ignorance about what proportion of the total root length is actually involved in taking up water. Some of these attempts have clearly supported the model, at least for first-order effects (28, 51). Others have conspicuously failed to support it (17, 18, 30, 31, 87), including a recent comparative study of sunflower and sorghum in the field, which showed that sunflower was much more effective at extracting water from the subsoil than was sorghum despite their root systems' having apparently similar distributions (6). There are several reasons why the model, at least in its simplest form, may fail to describe accurately the uptake of water by roots. These are discussed below.

THE EFFECT OF CLUMPING OF ROOTS   The model assumes that the roots are evenly distributed through the soil. This assumption is a good one in disturbed soil, but in undisturbed soil it is common for roots to grow in preexisting continuous large pores, so that they have a clumped distribution (16, 73, 74, 85). The effect of this clumping on the rate of uptake of water is likely to be large. If the roots are clumped into large pores, say wormholes, then it is no longer appropriate to use L as the average length of root in unit volume of soil. The appropriate variable is now L', the length of occupied large pores in unit volume of soil. That is, we assume that the clumped roots in a large pore act as though they were effectively a single root having access to a catchment whose outer boundary b' is given by $1/\sqrt{(\pi L')}$. Thus, by analogy with the single-root model, the half time for the removal of water by the roots from the soil is b'^2 days, and since b' may be many centimeters (73), $t_{1/2}$ may be of the order of weeks rather than of days as it would be were the roots evenly distributed (53).

INTERFACIAL RESISTANCE BETWEEN ROOT AND SOIL   In addition to the clumping effect described above, roots growing in large pores may suffer another impediment to taking up water. The intimate contact between root and soil that results when a root creates its own pore is missing when the root is growing in a large preexisting pore. Channels for the flow of water as liquid are sparse, and much of the flow may occur as vapor. The flux of water as vapor in response to a given gradient in water potential is many times less than the corresponding flux of liquid water in soil that is wet enough to support the growth of plants. Thus if there is poor hydraulic continuity between root and soil there will be a large fall in water potential at the interface between the two whenever there is a substantial rate of uptake of water by the root (78).

There has been some speculation (58), and some evidence (19, 32), that roots may shrink during times of low water potential in the plant, so that even with roots that have created their own pores in the soil, a large interfacial resistance may develop. The clearest evidence comes from observations by Huck et al (32) of cotton roots, which they found would shrink diurnally by up to 40% in diameter, in phase with diurnal changes in plant water potential. These observations are convincing for roots growing in large pores, which are the only ones visible using the glass-walled observation chamber of Huck et al (32); but it is not at all clear that roots that are in intimate contact with soil will also shrink. Indeed, evidence from neutron radiography suggests that they do not (76). The likelihood that they will shrink depends on where in the root the main resistance to the radial flow of water is. If, as has long been popularly assumed, the main resistance is at the endodermis, it follows that the cortex of the root, which is the tissue most likely to shrink, will have a water potential much closer to that of the soil than to that of the rest of the plant; and hence it will be buffered against the diurnal changes experienced by the rest of the plant. In a root growing in a large pore, however, the main resistance to flow will be at the interface between root and soil, and hence the cortex will not be buffered in the same way.

Several papers that have explored the relation between the flow of water through the roots and the fall in water potential across plant and soil have concluded that there is a large resistance to flow that does not appear to be in either the plant or the soil, and by implication is at the interface between the two (7, 18, 19, 31). In contrast, other apparently similar experiments have shown no such large unexplained resistance (2, 51). Neither does it seem necessary to invoke such a resistance in field experiments (61), unless they be on very sandy soils (87). These contradictory results may have resulted from differing rates of change of conditions in the soil among the different experiments. It is common in laboratory experiments to have large plants in small pots. Once water is withheld from such pots the rate of drying of the soil is fast—typically much faster than occurs in the field. There is little time for the roots to adjust, either physiologically or morphologically, to the changing conditions—for example, by increasing the osmotic pressure of their sap, so that their cells remain turgid as the soil water potential falls (26, 68), or by molding and gluing themselves to adjacent soil particles as they continue to grow into the drying, and usually hardening, soil (51).

Thus although it is conceivable that roots do shrink, and thereby induce an interfacial resistance even when growing in pores of their own making, such behavior is likely to be rare in the field, where the rate of change of soil water potential is much smaller than in the typical laboratory experiment, except perhaps in very sandy soils, as discussed in the next section.

However, the sharply increasing resistance that sometimes appears as the soil dries may originate within the plant. Blizzard & Boyer (2) found such a resistance: The difference in water potential between root and shoot in their transpiring plants remained remarkably constant as the soil dried, despite the transpiration rate's falling to only about 20% of its original value. Bristow et al (7) found similar behavior; but because they measured only the difference in water potential between shoot and soil, they attributed the increase in the apparent hydraulic resistance as the soil dried to the development of an interfacial resistance between root and soil, implicitly assuming a constant resistance in the plant. Had they measured the water potential of the roots they might have reached the same conclusion as did Blizzard & Boyer (2).

The special case of sand   Sands and very sandy soils are worth discussing specifically, in relation to the flow of water to roots. They are a popular medium for root growth in laboratory experiments, yet they have properties that distinguish their behavior so markedly from the general run of soils in the field, that extrapolation from laboratory experiments to the field can easily be misleading on this account alone. Field soils are typically composed of aggregates of primary particles of sand, silt, and clay, and roots usually grow between these aggregates. Water can flow through the aggregates as well as around them. But water cannot flow through sand particles, which are the effective units between which roots weave in a very sandy soil. As roots remove water from a wet sand, the water-filled pores steadily drain until they fairly suddenly become discontinuous. From then on flow is strongly hindered, an increasing proportion of it being as vapor. By contrast, in aggregated soils at the same stage, water-filled pores between the aggregates remain hydraulically continuous via the small pores within the aggregates.

   Thus the rapid drop in hydraulic conductivity in a sand that coincides with the water-filled pores' becoming largely discontinuous may well be responsible for the interfacial resistances that many have observed (19, 31, 87).

MODIFICATION BY ROOTS OF THE RELATION BETWEEN SUCTION AND WATER CONTENT OF THE SOIL   This is a distinct yet so far unexplored possibility. The theoretical basis for relating the suction to the water content of a soil is epitomized by the Laplace-Young equation (14), which describes the pressure drop, ΔP, across a spherical air-water meniscus as $2\gamma \cos\theta/a$, where γ (N m^{-1}) is the surface tension, θ is the angle of contact between liquid and solid, and a (m) is the radius of curvature of the meniscus. It is common to assume that, in soil, θ is zero (that is, the soil is strongly hydrophilic), and that the surface tension is that of pure water, namely, 70 mN m^{-1}. These assumptions lead to the conclusion that the suction of the water in a soil at equilibrium is inversely related to the size of the largest water-filled pores in the soil, and that if such a pore were cylindrical and of diameter 3 μm, then the suction would be 100 kPa (or 1 MPa if the diameter were 0.3 μm, and so on).

But there is recent evidence that soil contains surface-active materials, presumably of biological origin, that change the surface tension (47). Common surfactants such as soaps and detergents typically halve the surface tension of an air-water meniscus to about 30 mN m^{-1} (14). If roots exuded similar surfactants (or stimulated the production of them by microorganisms) they would halve the suction of the water in the soil at a given water content. At low suctions such an effect would be negligible in influencing the ability of the roots to take up water, but once the suction has risen to a few hundred kilopascal, the effect could be substantial.

Roots in Dry Soil

When a soil becomes so dry that its water potential falls below that of the roots, flow into the roots necessarily stops and may even reverse (1, 62). If the soil is dry for a long time the roots suberize and may develop new structures (83). It is of interest to know how fast roots can respond in their ability to take up water once the soil is rewet, and also whether the putative reverse flow is substantial.

RECOVERY OF UPTAKE BY ROOTS AFTER WATERING DRY SOIL   Surprisingly little is known about this despite its evident agricultural and ecological importance. A light to medium fall of rain, say less than 10 mm, will penetrate no more than about 50 mm into dry soil (unless the soil is very sandy), and is therefore little protected against rapid loss by direct evaporation from the soil surface. Can roots absorb much of this water before it evaporates, which could be within two days at a moderate potential evaporation rate of say 5 mm per day?

Nobel & Sanderson (48) explored the behavior of the roots of desert succulents and found that the permeability of the roots fell drastically when they were exposed for some days to a dry environment, and that, although there was some recovery in permeability within a few hours after rewatering, substantial recovery, which was augmented by the growth of new roots, took at least two days. Dirksen & Raats (15) claimed that alfalfa roots recovered their permeability immediately on rewetting, but their published data refer only to uptake between two and five days after watering. Thus the available evidence suggests that it is unlikely that roots can recover fast enough, if they have been exposed to very dry soil, to capture much of the water entering the soil after a fall of rain of less than about two days of potential evaporation from the soil surface. Sala & Lauenroth (63) have shown transient rises in leaf water potential and conductance in droughted *Bouteloua gracilis,* lasting about two days, in response to a simulated light fall of rain. They did not measure transpiration; but from the response in leaf conductance it seems likely that little of the rain was transpired, and that the bulk of it was lost by evaporation from the soil.

REVERSE FLOW: FROM ROOTS TO DRY SOIL   Interest in reverse flow has waxed and waned over many years. The evidence is variable. Of four recent papers, two find no reverse flow, in alfalfa (15) and in desert succulents (48), while the other two find much, in couch (1), and in *Artemisia tridentata* (62), with the roots transferring enough water from wet to dry soil during the night to account for much of the following day's transpiration.

There may be various reasons for the various results. There is the obvious one that different species may behave differently. Also, reverse flow is much less likely into very dry soil, in which the permeability of the roots may have fallen drastically as was the case with the desert succulents (48), than it is into moderately dry soil from which the roots are still managing to extract substantial amounts of water, as was the case with the couch grass (1) and the *Artemisia* (62).

If roots can act as rectifiers, what could the mechanism be? Nobel & Sanderson (48) conclude that it is simply that the permeability of the roots becomes very low when they are in dry soil, presumably because of suberization. Dirksen & Raats (15) speculate that a large interfacial resistance develops between root and soil (but see the discussion of this above). An interesting physiological explanation suggested by C. B. Tanner (personal communication) is that water can flow out of the roots, but in so doing solutes in the xylem sap are filtered out, thereby increasing in concentration. These solutes cannot escape back up the xylem—the distance is much too great for diffusion to disperse them—so eventually the osmotic pressure of the xylem sap rises until the flow stops. One can imagine many complications of this hypothesis—for example, that the filtered solutes might be removed in the phloem or sequestered in vacuoles, but in principle it could explain rectification where this occurs.

Would there be any advantage to a plant whose roots did allow reverse flow? Richards & Caldwell (62) argue that where a plant has but a few roots tapping a deep but moist part of the soil profile, these roots may be unable to support the maximum daytime transpiration rate of the plant, but their contribution to the daily transpiration is substantially increased if they continue to take up water during the night, transferring it to shallower, drier parts of the soil, where it is available for uptake during the day. This argument makes a lot of sense where the receiving soil is deep enough to be protected against loss of water by direct evaporation from the soil surface, as was the case with their *Artemisia*. Where the receiving soil is close to the surface, this process may be wasteful if it results in much of the transferred water's being lost by evaporation from the surface of the soil. If the process does occur there, its benefit to the plant may be through allowing the roots that are growing in this dry but probably nutritionally rich soil to take up some nutrients that would otherwise be inaccessible (10, 59).

TRANSPORT ACROSS WHOLE ROOT SYSTEMS

Basic Hydraulic and Osmotic Properties

Experiments for exploring the hydraulic properties of whole root systems typically involve placing a detached root system inside a pressure chamber, with the basal end of the main root protruding through a gland in the top of the chamber. Applying a pressure in the chamber induces a flow of solution through the roots which can be measured by collecting exudate from the cut stump of the main root. The relation between the applied pressure, ΔP (Pa), and the exudation rate, Q (m³ s⁻¹), is typified by Line A in Figure 1, the main features of which are that the relation is linear at high flow rates but is markedly curved at low flow rates, with a substantial intercept on the flow axis even when no pressure is applied.

TWO-COMPARTMENT MODEL   The shape of Line A is explicable, at least qualitatively, in terms of a two-compartment model developed independently

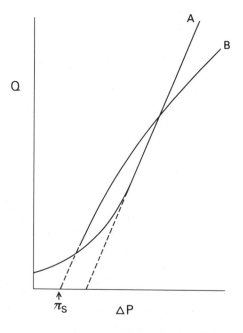

Figure 1 Flow rate of water through a root *(Q)* as a function of a difference in hydrostatic pressure (ΔP) applied across the root between the external medium, of osmotic pressure π_s, and the xylem. Line A represents typical experimental data. Note that the extrapolation of the straight portion of this line intercepts the ΔP axis at a value greater than π_s. Line B is the expected shape of $Q(\Delta P)$ if there is a buildup of solutes at the membrane in the two-compartment model. Adapted with permission from Passioura (52).

some years ago by Dalton et al (13) and Fiscus (20). This model is based on an equation for describing the flow of water across an individual semipermeable membrane, which, in a form suitable for describing the flow across a system of membranes as complex as those in a root system, is:

$$Q = k(\Delta P - \sigma \Delta \pi)$$

1.

where k ($m^3 \, s^{-1} \, Pa^{-1}$) is the hydraulic conductance, ΔP (Pa) is the difference in hydrostatic pressure, σ is an effective reflection coefficient for weighting any osmotically induced flow, and $\Delta \pi$ is the difference in osmotic pressure across the roots (52).

According to the model, the curvature of Line A at low flow rates arises from the active transport of solutes into the xylem where they manifest themselves osmotically and generate a flow of water through the root even when ΔP is zero. When ΔP is large, however, the resulting high flow rates so dilute these solutes that their effect on Q becomes negligible, and a linear relation between Q and ΔP results whose slope is k, the hydraulic conductance of the root system, and whose intercept on the pressure axis would equal π_s, the osmotic pressure of the external solution, were $\sigma = 1$ and the osmotic pressure in the xylem sap zero.

This model is attractive and deserves the wide acceptance it has by those working in this area. But there is a common feature of the experimental results that it fails to explain. This difficulty, first pointed out by Newman (45) and elaborated on by several others (38, 40, 52, 54, 55), is that the intercept of Line A on the pressure axis often exceeds the external osmotic pressure. This can only be explained by the model in its simplest form either by having σ exceed 1, which is impossible, or by assuming that k increases with Q in such a strangely contrived way that it allows Q versus ΔP to be linear while having a large intercept on the pressure axis; k in such circumstances would be given, for any particular point on the linear part of Line A, by the slope of the line joining that point to the pressure axis at π_s.

Fiscus (21) has pointed out that at high flow rates the osmotic pressure of the xylem sap is usually less than that of the external solution, so that solutes must be filtered from the flowing solution at some semipermeable membrane, and that therefore there may be a buildup of solutes at that membrane that could induce the strange offset in Figure 1—that is, that the external concentration experienced by the roots would be higher than that in the external solution. This is not a convincing explanation for both theoretical and empirical reasons.

Theoretically, one would expect any buildup of solutes to give not a constant offset, but one which increased with increasing flow rate. The buildup would be the result of a balance between the convection of solutes in the flow of solution towards the membrane, and the diffusion of solutes in the reverse direction. In the extreme in which all of the solutes are filtered out by the membrane, there will develop a steady state in which the diffusive and convective flows are equal, giving: $D \, dC/dx = VC$, where D (m^2 s^{-1}) is a diffusion coefficient, dC/dx is the gradient of concentration C (mol m^{-3}) with respect to distance x (m) from the membrane, and V is the velocity of the solution. The solution to this equation shows that C at $x = 0$ (i.e. at the filtering membrane) increases exponentially with increasing V, by a factor of exp (Va/D), where a (m) is the thickness of the layer in which the buildup occurs, and D (m^2 s^{-1}) is the diffusion coefficient for the solutes (11). Thus if there were a substantial buildup, $Q(\Delta P)$ would curve away from the Q axis as illustrated by Line B in Figure 1 (52). That $Q(\Delta P)$ is not curved at high flow rate is good evidence against there being a substantial buildup.

This conclusion is reinforced by some observations on barley plants whose roots had been exposed to various concentrations of NaCl up to 200 mol m^{-3} (43). The roots filtered out virtually all of the NaCl at all concentrations, and one would therefore expect that if there were a substantial buildup of solutes at the filtering membrane, then $Q(\Delta P)$ would become more markedly curved the higher was the external NaCl concentration. In fact $Q(\Delta P)$ was almost perfectly linear at all concentrations, which is strong evidence against any significant buildup.

THREE-COMPARTMENT MODEL   Several have pointed out that water must pass at least two membranes in flowing across the root, once to get into the symplast, and once to get out again into the xylem (12, 40, 45). It therefore makes sense to use a three-compartment model for exploring the flow of water across roots, rather than the two-compartment one discussed above. Miller (40) has also pointed out that the concentration of solutes in the symplast is generally much greater than that outside the plasma membranes, so that a convectional shift of solutes within the symplast, from the downstream side of the first membrane to the upstream side of the second, is likely to generate a substantial difference in osmotic pressure opposing the flow of water; this, he argues, could give rise to the discrepancy shown in Figure 1 between π_s and the intercept of Line A. But as with the two-compartment model it is hard to see how such a difference could be utterly independent of flow rate. Furthermore, the value of the average diffusion coefficient of the solutes that is needed to skew their distribution significantly within the cells is about 100 times smaller than typical values in water, as the following argument shows: Consider a typical cortical cell of width 20 μm. A high value of the flux density of water entering a root is 10^{-7} m s^{-1} (46). If this water is flowing largely within the cells, then we may assume that V within the cell is 10^{-7} m s^{-1}. The effective diffusion coefficient for the solutes within the cell is presumably similar to that of common solutes in water, namely 10^{-9} m^2 s^{-1}. Thus if the ratio of concentrations in the cell at its upstream and downstream wall is of the order of exp (Va/D), as discussed above, this ratio is very close to one, for Va/D is only 0.002 if we take a to be the width of the cell, namely 20 μm. It is hard to see why D within a cell should not be close to that in water, in which case any buildup would be trivial.

CONCLUSION   The models described above have about them such an appealing simplicity, and describe the shape of $Q(\Delta P)$ so well, that it seems sensible to accept them despite the unexplained offset on the pressure axis. In interpreting this offset, the best working hypothesis is that it arises from an unexplained jump in P or π somewhere along the flow path that is independent of flow rate, rather than from a variable hydraulic conductance. The linearity of $Q(\Delta P)$ is so strong except at low Q that it is hard to believe that the conductance varies with Q. If we accept this hypothesis then it is appropriate to modify Equation 1 to:

$$Q = k(\Delta P - P_0 - \sigma \Delta \pi) \qquad\qquad 2.$$

where P_0 is the offset on the pressure axis.

What in principle could induce the offset? I have already discussed Miller's (40) suggestion of a jump in osmotic pressure based on the three-compartment model. In a simple hydraulic system it is easy to imagine a structure that could induce an offset. A gravity-operated clack valve, for example, will produce an offset whose value is given by the weight of the valve divided by its area. There is a minimum pressure required to lift it, and the drop in pressure across it remains equal to its weight divided by its area irrespective of the flow rate; as the flow rate increases the valve simply opens more widely. It is faintly conceivable that the endoplasmic reticulum, where it passes through a plasmodesma, could operate as a valve in this way, with the tension in the membrane inducing a constant force analogous to that produced by gravity on the clack valve (54). Both of these suggestions are unlikely, and there seems no alternative for the moment but to accept the offset without explanation. Certainly Equation 2 is the most economical way of summarizing the experimental data.

Note that it is common to define the "hydraulic resistance," R, of a root system as $\Delta \psi / Q$, where $\Delta \psi$ is the difference in water potential across the root system. R equals the reciprocal of k in Equation 2 only if $\sigma = 1$ and $P_0 = 0$, (whence $\Delta \psi = \Delta P - \Delta \pi$). If these conditions are not true, and if we take the true hydraulic resistance to be $1/k$, then R is only an apparent hydraulic resistance; and variation in it, through time or with Q, need not imply that k varies (24, 52, 54).

Finally, it is worthwhile reiterating Dainty's (12) point that analyses based on these models, particularly experiments with maize, often conclude that σ is much less than one (39, 40, 72), thus implying that the roots are so leaky that one wonders how they can function properly. These low estimates of σ could imply that the roots are damaged, or are conducting solution artefactually in the intercellular spaces of the cortex, as discussed in the next section; but even when explicit care is taken to avoid damage, as in Miller's (39, 40) experiments, σ is low. However these results need not imply that the membranes of individual root cells are leaky; rather they could simply imply that there are some breaks in the continuity of the endodermis (72), such as those that occur where lateral roots emerge, and which have been shown to be permeable to apoplastic dyes (57). There are evident differences among species in the leakiness of the roots. Experiments with apoplastic tracers in mangrove (42) and red pine (29) showed only minute amounts of tracer getting into the xylem; similarly, experiments with barley roots exposed to large concentrations of NaCl showed that little NaCl got into the xylem (43), and that σ was very close to one.

Caveats Based on Possible Experimental Artefacts

When roots are decapitated, placed in nutrient solution, and then pressurized, there is a danger that experimental artefacts may lead to results that are irrelevant to the behavior of intact plants growing at normal pressures.

The most obvious problem is that the intercellular spaces in the cortex of the roots will become infiltrated thereby allowing them to become conduits for the longitudinal transport of water that normally occurs only in the xylem (36, 64). Although such infiltration must always occur when roots immersed in nutrient solution are pressurized, the extreme form of artefactual longitudinal transport observed by Salim & Pitman (64), in which external solution gains access to the intercellular spaces and moves through them, does not seem to be common. Perhaps this extreme form is associated with breakages in the cortex that arise when unsupported roots are transferred from one container to another or are jostled by too vigorous aeration; both Miller (39) and Moon et al (42) have clear evidence of this as a likely problem. Artefactual longitudinal transport is not a problem in roots growing in soil with a large air-filled porosity, for then most of the gas-filled intercellular spaces survive the application of pressure (54).

Another possible problem associated with decapitated roots is that they are starved of carbohydrate. The effect on potassium transport is rapid and profound (3). That on water transport is unknown, but could be troublesome. One of the attractions of using detached root systems is that both the pressure drop across them and the flow rate through them can be determined with great precision. However, one can achieve comparable precision with intact plants

(54), and given that these are much less likely to be beset by artefacts it is preferable to use them.

Finally, there is uncertainty about what a large hydrostatic or pneumatic pressure might do to roots in those experiments in which the roots are enclosed within a pressure chamber. The possibility is remote that hydrostatic pressure in itself is a problem; the pressures applied, generally <2 MPa, are small compared with those known to affect cell physiology in plants (84). Of more concern is that high partial pressures of the pressurizing gases, typically a mixture of oxygen and nitrogen, will damage the cells in some way. Certainly, a high partial pressure of oxygen is damaging (77), and it is important to control this at about 20 kPa, its value in normal air. A high partial pressure of nitrogen could also be damaging, by inducing the high concentration of nitrogen in cell membranes that is thought to cause narcosis in deep-sea divers (41). However such effects appear to be troublesome only in nerve cells. The fact that $Q(\Delta P)$ is so strongly linear over a wide range of pressures, at least up to 2 MPa, suggests that the hydraulic properties of the roots are independent of the pressure.

Endogenous and Environmental Influences

The hydraulic and osmotic properties of roots respond, in ways that are generally poorly understood, to various environmental attributes, including circadian changes that can persist as endogenous rhythms.

CIRCADIAN RHYTHMS   These have been known for some time, both as variation in the rate of free exudation by decapitated roots (82) and as variation in apparent hydraulic resistance, R (i.e. $\Delta\psi/Q$) (49). But although R may vary, it does not follow that k in Equation 2 varies. Passioura & Munns (54) found a marked daily variation in R in intact barley plants that was entirely attributable to changes in P_0, with k remaining quite constant; P_0 was close to zero in the morning, rose to about 100 kPa in the afternoon, and fell back to zero during the night. They found a more complex pattern in lupin, in which P_0 varied from about 100 kPa in the morning to 250 kPa in the afternoon, while k simultaneously fell by about 30%. Fiscus (23) has proposed a complex model for describing such changes. It is not at all clear whether this circadian behavior has any functional significance.

ENVIRONMENTAL INFLUENCES   Several conditions in the soil strongly affect the permeability of roots to water. Low temperatures (34, 37), anoxia (16a, 37), nutrient deficiency (60), and dehydration as discussed above, all markedly decrease the permeability. The mechanisms involved are largely unknown, although with dehydration the laying down of suberin in the cell walls presumably affects the permeability.

Of the nutrients, phosphorus especially has a large effect (55, 60). Indeed the effect is so large that it has led Radin & Eidenbock (60) to propose that one of the reasons for the poor growth of phosphorus-deficient plants is that the low permeability of their roots induces such low water potentials in the shoot during transpiration that growth is inhibited. This is an interesting suggestion that awaits critical testing. There has been some discussion of the effects of mycorrhizae on the uptake of water by roots (35a), but their possible direct effect on the uptake of water is confounded by their great influence on the phosphorus status of the plant.

PATHWAYS FOR WATER TRANSPORT

Radial

The peculiar anatomy of the endodermis of roots, with its Casparian strip signalling that no water can pass through its cell walls en route to the stele, persuaded several generations of plant physiologists that water flowed predominantly in the cell walls of the root except where it was forced to flow through the interior of the endodermal cells. Logically, this conclusion was too strong, for the impermeability of the endodermal cell walls signals only that water must pass a membrane (in fact two membranes) before entering the xylem.

Newman (46) has been an influential critic of this popular view, arguing that the permeability of cell walls was likely to be rather lower than that of the symplast, and that therefore water was likely to enter the cells near the epidermis and travel within the cells towards the xylem, possibly emerging briefly to traverse tangential walls. Two recent lines of evidence appear to support this view. One is anatomical, and shows the blockage of apoplastic tracers at the exodermis of the root, a layer of cortical cells just inside the epidermis that has a suberized band in its cell walls similar to the Casparian strip (42, 56). The other is physiological, and concerns direct measurements of the hydraulic conductivites of the plasma membranes of the root cells (33, 71) made using a pressure-probe. These measurements showed that the impediment to the putative passage of water into and out of each cell during radial flow across the root was approximately equal to that across the root as a whole. This agreement could be fortuitous, for the cellular measurements were made only on the cortical cells; a large resistance to flow through the endodermis, coupled with a much smaller resistance in the cell walls (i.e. the traditional view) could produce the same overall effect. Nevertheless the agreement is intriguing.

This longstanding debate about the relative importance of transport within the cells, and along their walls, will not be resolved until adequate measurements have been made of the permeability of the walls to water, not only for flow across the walls, but also and perhaps more importantly for flow along them, for the walls could well be anisotropic. There are copious intercellular spaces within the cortex, and if films of water occupied these their permeability could be very high. Admittedly almost all of these spaces are oriented longitudinally, and also, the surfaces of the walls bordering these spaces are somewhat hydrophobic, having contact angles with water of about 100° (my own unpublished observations). This reluctance to wet is to be expected, for simple capillary theory (see 86 and the comments above in the section on modification by roots of the relation between suction and water content of the soil) predicts that gas in a small pore would be under considerable pressure were the walls of that pore hydrophilic (having a contact angle with an air-water meniscus of close to zero), and the gas would soon dissolve. For example, a cylindrical pore of diameter 3 μm, which is a common approximate diameter of the intercellular spaces, would submit any bubble of gas within it to a pressure of about 100 kPa. That the gas in the intercellular spaces persists is, therefore, strong evidence that the walls are not hydro–philic. Nevertheless we cannot discount the possibility that there are radially oriented water-filled pores within the cell walls of the cortex that are substantially larger than those within the normal cell-wall matrix and that could transmit water readily. The puzzling spatial variation in water content in the cortex of *Pelargonium* roots observed by Brown et al (9) could signify preferential paths on the scale of a few cell diameters through the cortex, which might be associated with different properties of the cell walls.

An issue related to that of what is the pathway across the root is where along that pathway does the main resistance to flow lie? My own view is that there is unlikely to be a uniformly distributed hydraulic resistance across the whole of the cortex, as implied in the work of Steudle & Jeschke (71) and Jones et al (33). I believe, in accordance with the traditional view, that there is likely to be a major resistance near or in the endodermis, for the reasons discussed above in the section on the interfacial resistance to flow between root and soil. On functional grounds one would expect the cortex to be buffered against the daily swings in water potential experienced by the xylem; otherwise, substantial shrinkage of it might occur with detrimental effects on the root's ability to extract water from the soil. As noted earlier, there is little evidence that roots shrink diurnally when growing in nonsandy soils through pores of their own making. If Peterson and colleagues (42, 56) are right that water must pass through the exodermal cells, this need not imply that the endodermis is not the site of the main resistance; the measurements with the pressure probe discussed above attribute only a small resistance to any given layer of the cortical cells (33, 71).

UPTAKE OF WATER BY DIFFERENT REGIONS OF ROOTS   Brouwer (8) showed, many years ago, that there was substantial variation in the uptake of water along roots of *Vicia faba,* and that the pattern of uptake varied with transpiration rate. Fiscus (22) provided a plausible explanation of this behavior in terms of the two-compartment model discussed above. It is generally found that roots take up water most rapidly in the region from about 10–100 mm behind their tips (4, 37); but, in barley at least, the uptake remains substantial even in old regions of the root, much further than 100 mm from the tips, where the tertiary, heavily suberized endodermis has formed (65).

These measurements were made on roots in nutrient solution; the behavior of roots growing in soil may be quite different, but is difficult to measure. When exploring the uptake of water by roots growing in soil, as discussed earlier, it is important to know what proportion of the whole root system is effectively involved in uptake. Little is known about this, but for wheat in a drying soil there is circumstantial evidence that about one third of the root system is involved in uptake (51).

Axial

Axial flow of water in roots occurs almost entirely in the xylem, and is described fairly well by Poiseuille's equation (27). In dicots, with their facility for secondary growth, there is abundant xylem, so one would not expect large axial gradients in water potential to occur, unless the vessels were diseased, or unless cavitation emptied most of the vessels. The latter seems unlikely except in very water-deficient plants, although Blizzard & Boyer (2) proposed this as an explanation for the sharply increased hydraulic resistance they found in plants in a drying soil (discussed above in the section on interfacial resistance).

In monocots, however, secondary growth is rare. Extra axial carrying capacity is provided only by new adventitious roots growing from the crown of the plant. Normally there seem to be enough roots for axial gradients in water potential to be small; but in the grasses especially, a dry topsoil after establishment can result in plants having no or few adventitious roots so that they are relying entirely on their seminal roots—i.e. those arising from the seed—to extract water from a moist subsoil. When that happens, flow through the xylem may be seriously hindered (27, 35, 50). A wheat plant, for example, typically has 3–5 seminal axes each containing only one large metaxylem vessel of about 65 μm diameter (50). A plant with no nodal roots might therefore be relying on as few as three xylem vessels to transport its entire water supply from a moist subsoil through a dry topsoil to the transpiring leaves. Such a plant would seem to be particularly at risk from cavitation, but evidently cavitation in these xylem vessels is rare. Plants that have been allowed to grow with only one seminal root and no nodal roots, and hence have essentially only one xylem vessel to carry water from the roots into the crown, survive to maturity even when their growth is strongly restricted by water supply (50).

It is interesting to note that grasses relying solely on their seminal roots may have enormous mean speeds of water in the xylem of these roots—far higher than has been reported for any other system. For example, a large wheat plant can grow well with a root system comprising only one seminal axis plus associated laterals, and the mean speed of water in the xylem of that axis can reach the astonishing value of 0.8 m s^{-1} (50).

It is common to assume that the late metaxylem elements so obvious to the eye in transections of roots are mature—that is, that they have lost their endwalls and contain no cytoplasm. But this may not be generally true. St. Aubin et al (70), working with field-grown maize roots, have shown that immature elements persist as far as 300 mm from the tip. They review earlier but largely forgotten work, not only in maize but also in other species, pointing to the same conclusion. They calculate, using Poiseuille's Law, that the carrying capacity of the xylem increases by a factor of about 100 when the late metaxylem matures, and they suggest that roots with immature xylem may have little part to play in taking up water. It is hard to reconcile this suggestion with the common observation, described in the previous section, that roots generally take up water most rapidly near their tips. The discrepancy is perplexing. Perhaps the carrying capacity of the xylem remains large despite the immaturity of the late metaxylem.

CONCLUDING REMARKS

The processes I have been discussing largely concern the provision of water by roots to shoots. They would therefore seem to be of critical importance to the growth and survival of plants, which will die if their shoots dehydrate. Yet apart from the obvious requirements that the roots colonize moist soil and establish a conducting system for transferring water from soil to shoot, it is difficult to gauge the importance of the various phenomena. Does the resistance to radial or axial flow in roots, or cavitation in the xylem, or shrinkage of the roots ever influence the growth of a plant? Possibly they do, but there is little evidence of this, especially in field-grown plants. More generally, when a soil dries so far that the growth of a plant is affected, is the effect mediated by a lowered water potential in the shoot, so that if the resistance to flow through the roots and adjacent soil were less, the plant would grow faster? A few years ago it would have been taken for granted that the answer was yes, but evidence is accumulating that roots often sense conditions in the soil and send information to the shoot that influences its behavior so much that effects of water status may be overridden (79). It will be interesting to see how our views of the relative importance of these two effects develop over the next few years.

ACKNOWLEDGMENT

I thank Ernst Steudle for helpful comments on the manuscript.

Literature Cited

1. Baker, J. M., van Bavel, C.H.M. 1986. Resistance of plant roots to water loss. *Agron J.* 78:641–44
2. Blizzard, W. E., Boyer, J. S. 1980. Comparative resistance of the soil and the plant to water transport. *Plant Physiol.* 66:809–14
3. Bowling, D.J.F., Watson, B. T., Ehwald, R. 1985. The effect of ringing on root growth and potassium uptake by *Helianthus annuus*. *J. Exp. Bot.* 36: 290–97
4. Boyer, J. S. 1985. Water transport. *Ann. Rev. Plant Physiol.* 36:473–516
5. Bradford, K. J., Hsiao, T. C. 1982. Physiological responses to moderate water stress. In *Encyclopedia of Plant Physiology* (NS), Vol. 12B: *Physiological Plant Ecology II,* ed. O. L. Lange, P. S. Nobel, C. B. Osmond, H. Ziegler, pp. 264–324. Berlin/Heidelberg/New York: Springer-Verlag
6. Bremner, P. M., Preston, G. K., Fazekas de St. Groth, C. 1986. A field comparison of sunflower *(Helianthus annuus)* and sorghum *(Sorghum bicolor)* in a long drying cycle. I. Water extraction. *Aust. J. Agric. Res.* 37:483–93
7. Bristow, K. L., Campbell, G. S., Calissendorff, C. 1984. The effects of texture on the resistance to water movement within the rhizosphere. *Soil Sci. Soc. Am. J.* 48:266–70
8. Brouwer, R. 1954. The regulating influence of transpiration and suction tension on the water and salt uptake by the roots of intact *Vicia faba* plants. *Acta Bot. Neerl.* 3:264–312
9. Brown, J. M., Johnson, G. A., Kramer, P. J. 1986. In vivo magnetic resonance microscopy of changing water content in *Pelargonium hortorum* roots. *Plant Physiol.* 82:1158–60
10. Cornish, P. S., So, H. B., McWilliam, J. R. 1984. Effects of soil bulk density and water regimen on root growth and uptake of phosphorus by ryegrass. *Aust. J. Agric. Res.* 35:631–44
11. Dainty, J. 1963. Water relations of plant cells. *Adv. Bot. Res.* 1:279–326
12. Dainty, J. 1985. Water transport through the root. *Acta Hortic.* 171:21–31
13. Dalton, F. N., Raats, P.A.C., Gardner, W. R. 1975. Simultaneous uptake of water and solutes by plant roots. *Agron. J.* 67:334–39
14. Davies, J. T., Rideal, E. K. 1963. *Interfacial Phenomena.* New York: Academic
15. Dirksen, C., Raats, P.A.C. 1985. Water uptake and release by alfalfa roots. *Agron. J.* 77:621–26
16. Ehlers, W., Köpke, U., Hesse, F., Böhm, W. 1983. Penetration resistance and root growth of oats in tilled and untilled loess soil. *Soil Till. Res.* 3: 261–75
16a. Everard, J. D., Drew, M. C. 1987. Mechanisms of inhibition of water movement in anaerobically treated roots of *Zea mays* L. *J. Exp. Bot.* 38:1154–65
17. Faiz, S.M.A., Weatherley, P. E. 1977. The location of the resistance to water movement in the soil supplying the roots of transpiring plants. *New Phytol.* 78: 337–47
18. Faiz, S.M.A., Weatherley, P. E. 1978. Further investigations into the location and magnitude of the hydraulic resistances in the soil:plant system. *New Phytol.* 81:19–28
19. Faiz, S.M.A., Weatherley, P. E. 1982. Root contraction in transpiring plants. *New Phytol.* 92:333–43
20. Fiscus, E. L. 1975. The interaction between osmotic- and pressure-induced water flow in plant roots. *Plant Physiol.* 55:917–22
21. Fiscus, E. L. 1977. Determination of hydraulic and osmotic properties of soybean root systems. *Plant Physiol.* 59:1013–20
22. Fiscus, E. L. 1977. Effects of coupled solute and water flow in plant roots with special reference to Brouwer's experiments. *J. Exp. Bot.* 28:71–77
23. Fiscus, E. L. 1986. Diurnal changes in volume and solute transport coefficients of *Phaseolus* roots. *Plant Physiol.* 80:752–59

24. Fiscus, E. L., Klute, A., Kaufmann, M. R. 1983. An interpretation of some whole plant water transport phenomena. *Plant Physiol.* 71:810–17
25. Gardner, W. R. 1960. Dynamic aspects of water availability to plants. *Soil Sci.* 89:63–73
26. Greacen, E. L., Oh, J. S. 1972. Physics of root growth. *Nature New Biol.* 235:24–25
27. Greacen, E. L., Ponsana, P., Barley, K. P. 1976. Resistance to water flow in the roots of cereals. In *Water and Plant Life,* ed. O. L. Lange, L. Kappen, E.-D. Schulze, pp. 86–100. Berlin:Springer-Verlag
28. Hainsworth, J. M., Aylmore, L.A.G. 1986. Water extraction by single plant roots. *Soil Sci. Soc. Am. J.* 50:841–48
29. Hanson, P. J., Sucoff, E. I., Markhart, A. H. 1985. Quantifying apoplastic flux through red pine root systems using trisodium, 3-hydroxy-5, 8, 10-pyrenetrisulfonate. *Plant Physiol.* 77:21–24
30. Herkelrath, W. N., Miller, E. E., Gardner, W. R. 1977. Water uptake by plants. I. Divided root experiments. *Soil Sci. Soc. Am. J.* 41:1033–38
31. Herkelrath, W. N., Miller, E. E., Gardner, W. R. 1977. Water uptake by plants. II. The root contact model. *Soil Sci. Soc. Am. J.* 41:1039–43
32. Huck, M. G., Klepper, B., Taylor, H. M. 1970. Diurnal variations in root diameter. *Plant Physiol.* 45:529–30
33. Jones, H., Tomos, A. D., Leigh, R. A., Wyn Jones, R. G. 1984. Water-relation parameters of epidermal and cortical cells in the primary root of *Triticum aestivum* L. *Planta* 158:230–36
34. Kaufmann, M. R. 1975. Leaf water stress in Engelmann spruce: influence of the root and shoot environments. *Plant Physiol.* 56:841–44
35. Klepper, B. 1983. Managing root systems for efficient water use: axial resistances to flow in root systems—anatomical considerations. See Ref. 75, pp. 115–25
35a. Koide, R. 1985. The effect of VA mycorrhizal infection and phosphorus status on sunflower hydraulic and stomatal properties. *J. Exp. Bot.* 36:1087–98
36. Kozinka, V., Luxova, M. 1971. Specific conductivity of conducting and nonconducting Zea mays root. *Biol. Plant.* 13:257–66
37. Kramer, P. J. 1983. *Water Relations of Plants.* Orlando: Academic. 489 pp.
38. Michel, B. E. 1977. A model relating root permeability to flux and potentials. *Plant Physiol.* 60:259–64
39. Miller, D. M. 1985. Studies of root function in *Zea mays*. III. Xylem sap composition at maximium root pressure provides evidence of active transport into the xylem and a measurement of the reflection coeffient of the root. *Plant Physiol.* 77:162–67
40. Miller, D. M. 1985. Studies of root function in *Zea mays*. IV. Effects of applied pressure on the hydraulic conductivity and volume flow through the excised root. *Plant Physiol.* 77:168–74
41. Miller, K. W. 1972. Inert gas narcosis and animals under high pressure. In *The Effects of Pressure on Organisms,* pp. 363–78. Cambridge: Cambridge Univ. Press
42. Moon, G. J., Clough, B. F., Peterson, C. A., Allaway, W. G. 1986. Apoplastic and symplastic pathways in *Avicennia marina* (Forsk.) Vierh. roots revealed by fluorescent tracer dyes. *Aust. J. Plant Physiol.* 13:637–48
43. Munns, R., Passioura, J. B. 1984. Hydraulic resistance of plants. III. Effects of NaCl in barley and lupin. *Aust. J. Plant Physiol.* 11:351–59
44. Newman, E. I. 1969. Resistance to water flow in soil and plant. II. A review of experimental evidence on the rhizosphere resistance. *J. Appl. Ecol.* 6:261–72
45. Newman, E. I. 1976. Interaction between osmotic- and pressure-induced water flow in plant roots. *Plant Physiol.* 57:738–39
46. Newman, E. I. 1976. Water movement through root systems. *Philos. Trans. R. Soc. London Ser. B.* 273:463–78
47. Nimmo, J. R., Miller, E. E. 1986. The temperature dependence of isothermal moisture vs. potential characteristics of soils. *Soil Sci. Soc. Am. J.* 50:1105–13
48. Nobel, P. S., Sanderson, J. 1984. Rectifier-like activities of roots of two desert succulents. *J. Exp. Bot.* 35:727–37
49. Parsons, L. R., Kramer, P. J. 1974. Diurnal cycling in root resistance to water movement. *Physiol. Plant.* 30:19–23
50. Passioura, J. B. 1972. Effect of root geometry on the yield of wheat growing on stored water. *Aust. J. Agric. Res.* 23:745–52
51. Passioura, J. B. 1980. The transport of water from soil to shoot in wheat seedlings. *J. Exp. Bot.* 31:333–45
52. Passioura, J. B. 1984. Hydraulic resistance of plants. I. Constant or variable? *Aust. J. Plant Physiol.* 11:333–39

53. Passioura, J. B. 1985. Roots and water economy of wheat. In *Wheat Growth and Modelling*, ed. W. Day, R. K. Atkin, pp. 185–98. New York: Plenum

54. Passioura, J. B., Munns, R. 1984. Hydraulic resistance of plants. II. Effects of rooting medium, and time of day, in barley and lupin. *Aust. J. Plant Physiol.* 11:341–50

55. Passioura, J. B., Tanner, C. B. 1985. Oscillations in apparent hydraulic conductance of cotton plants. *Aust. J. Plant Physiol.* 12:455–61

56. Perumalla, C. J., Peterson, C. A. 1986. Deposition of Casparian bands and suberin lamellae in the exodermis and endodermis of young corn and onion roots. *Can. J. Bot.* 64:1873–78

57. Peterson, C. A., Emanuel, G. B., Humphreys, G. B. 1981. Pathway of movement of apoplastic fluorescent dye tracers through the endodermis at the site of secondary root formation in corn *(Zea mays)* and broad bean *(Vicia faba)*. *Can. J. Bot.* 59:618–25

58. Philip, J. R. 1957. The physical principles of soil water movement during the irrigation cycle. *Proc. Int. Congr. Irrig. Drain.* 8:125–54

59. Pinkerton, A., Simpson, J. R. 1986. Interactions of surface drying and subsurface nutrients affecting plant growth on acidic soil profiles from an old pasture. *Aust. J. Exp. Agric.* 26:681–89

60. Radin, J. W., Eidenbock, M. P. 1984. Hydraulic conductivity as a factor limiting leaf expansion of phosphorus deficient cotton plants. *Plant Physiol.* 75:372–77

61. Reicosky, D. C., Ritchie, J. T. 1976. Relative importance of soil resistance and plant resistance in root water absorption. *Soil Sci. Soc. Am. Proc.* 40:293–97

62. Richards, J. H., Caldwell, M. M. 1987. Hydraulic lift: substantial water transport between soil layers by *Artemisia tridentata* roots. *Oecologia.* In press

63. Sala, O. E., Lauenroth, W. K. 1982. Small rainfall events: an ecological role in semiarid regions. *Oecologia* 53:301–4

64. Salim, M., Pitman, M. G. 1984. Pressure-induced water and solute flow through plant roots. *J. Exp. Bot.* 35:869–81

65. Sanderson, J. 1983. Water uptake by different regions of the barley root. Pathways of radial flow in relation to development of the endodermis. *J. Exp. Bot.* 34:240–53

66. Schulze, E.-D. 1986. Whole-plant responses to drought. *Aust. J. Plant Physiol.* 13:127–41

67. Shackel, K. A., Matthews, M. A., Morrison, J. C. 1987. Dynamic relation between expansion and cellular turgor in growing grape *(Vitis vinifera* L.) leaves. *Plant Physiol.* 84:1166–71

68. Sharp, R. E., Davies, W. J. 1979. Solute regulation and growth by roots and shoots of water-stressed maize plants. *Planta* 147:43–49

69. Sinclair, T. R., Ludlow, M. M. 1985. Who taught plants thermodynamics? The unfulfilled potential of plant water potential. *Aust. J. Plant Physiol.* 12: 213–17

70. St. Aubin, G., Canny, M. J., McCully, M. E. 1986. Living vessel elements in the late metaxylem of sheathed maize roots. *Ann. Bot.* 58:577–88

71. Steudle, E., Jeschke, W. D. 1983. Measurements of root pressure and hydraulic conductivity of root cells. *Planta* 158:237–48

72. Steudle, E., Oren, R., Schulze, E.-D. 1987. Water transport in maize roots. *Plant Physiol.* 84:1220–32

73. Tardieu, F., Manichon, H. 1986. Caractérisation en tant que capteur d'éau de l'enracinement du maïs en parcelle cultivée. II. Une méthode d'étude de la répartition verticale et horizontale des racines. *Agronomie* 6:415–25

74. Taylor, H. M. 1983. Managing root systems for efficient water use: an overview. See Ref. 75, pp. 87–113

75. Taylor, H. M., Jordan, W. R., Sinclair, T. R., eds. 1983. *Limitations to Efficient Water Use in Crop Production.* Madison: Am. Soc. Agronomy

76. Taylor, H. M., Willatt, S. T. 1983. Shrinkage of soybean roots. *Agron. J.* 75:818–20

77. Termaat, A., Passioura, J. B., Munns, R. 1985. Shoot turgor does not limit shoot growth of NaCl-affected wheat and barley. *Plant Physiol.* 77:869–72

78. Tinker, P. B. 1976. Transport of water to plant roots in soil. *Philos. Trans. R. Soc. London Ser. B.* 273:445–61

79. Turner, N. C. 1986. Adaptation to water deficits: a changing perspective. *Aust. J. Plant Physiol.* 13:175–90

80. Turner, N. C. 1987. Crop water deficits: a decade of progress. *Adv. Agron.* 39:1–51

81. Turner, N. C., Passioura, J. B., eds. 1986. *Plant Growth, Drought and Salinity.* Melbourne: CSIRO. 201 pp.

82. Vaadia, Y. 1960. Autonomic diurnal fluctuations in rate of exudation and root pressure of decapitated sunflower plants. *Physiol. Plant.* 13:701–17

83. Vartanian, N. 1981. Some aspects of

structural and functional modifications induced by drought in root systems. *Plant Soil* 63:83–92

84. Vidaver, W. 1972. Effects of pressure on the metabolic processes of plants. In *The Effects of Pressure on Organisms,* pp. 159–74. Cambridge: Cambridge Univ. Press

85. Wang, J., Hesketh, J. D., Woolley, J. T. 1986. Preexisting channels and soy-

bean rooting patterns. *Soil Sci.* 141:432–37

86. Woolley, J. T. 1983. Maintenance of air in intercellular spaces of plants. *Plant Physiol.* 72:989–91

87. Zur, B., Jones, J. W., Boote, K. J., Hammond, L. C. 1982. Total resistance to water flow in field soybeans. II. Limiting soil moisture. *Agron. J.* 74:99–105

Ann. Rev. Plant Physiol. Plant Mol. Biol. 1988. 39:267–95

THE CONTROL OF LEAF EXPANSION

J. E. Dale

Department of Botany, University of Edinburgh, King's Buildings, Edinburgh, EH9 3JH, Scotland

CONTENTS

INTRODUCTION 267
CELL DIVISION AND CELL ENLARGEMENT 269
    Cell Division 269
    Cell Expansion 272
    The Grass Leaf 273
THE MECHANICS OF ENLARGEMENT 273
    The Role of the Epidermis 273
    Intercellular Spaces and Their Significance 275
THE MECHANISM OF LEAF CELL GROWTH 276
    Biophysical Considerations 276
    The Phaseolus Primary Leaf System 278
REGULATION BY LIGHT 279
    The Light Environment 279
    Effects of Light Quantity 280
    Effects of Light Quality 282
    The Role of Phytochrome 283
WATER AND LEAF EXPANSION 284
    Methodological Difficulties 284
    Effects on Cell Division 285
    Cell Enlargement and Turgor Control 285
    Abscisic Acid and Leaf Growth 288
CONCLUSIONS 289

INTRODUCTION

The importance of leaves to the productivity and stability of most terrestrial ecosystems needs little emphasis, and yet substantial questions about the control and regulation of leaf growth remain unanswered. The topic was

267

0066-4294/88/0601-0267$02.00

reviewed in the early days of this series (72), and a number of recent reviews have relevance for leaf growth studies (22, 61, 67). The book by Maksymowych (103) summarizes much of his work on leaf growth in *Xanthium,* and two recent multiauthor works (7, 36) cover a variety of topics relevant to the more limited themes of this article.

This review is selective. Canopy expansion is covered only in passing, and I concentrate on the control of expansion of angiosperm leaves that are dorsiventrally flattened. The initiation of foliar primordia is covered elsewhere (61, 67, 98), and the formidably difficult topic of the control of leaf form and shape are considered only in passing.

The leaf is an organ of limited growth and limited life span. The ontogeny of individual leaves is often abbreviated; and different parts of the same expanding leaf may be at different developmental stages, with significant consequences for analysis and sampling. Variation is also important when successive leaves on a stem are studied since ontogenetic drifts in leaf development are frequently found, most obviously in heteroblastic development but also in features such as size and anatomy (108). There are also smaller, more subtle differences that cannot be neglected, since the development of successive leaves does not necessarily follow an identical course, qualitatively or quantitatively. If leaves are not the same within a plant it is axiomatic that they will differ even more between species. Obvious morphological differences may well go hand in hand with differences in regulatory mechanisms.

Defining growth as the irreversible accretion of dry mass means that a leaf "grows" almost to the onset of senescence, since dry mass can increase long after area expansion has ceased (37). Increase in dry mass is the result of a highly complex series of biochemical events, and numerous attempts have been made to model leaf growth (15, 16, 89, 164). Inevitably these rely heavily on the biochemical factors involved, particularly on those aspects of nitrogen and carbon metabolism easily identified with cell growth. Although these models offer no major new insight into the mechanisms of leaf expansion, they highlight the enormous complexity that has to be incorporated into any but the most trivial mathematical simulation. Since the review of Humphries & Wheeler (72) there has been a welcome and significant reassessment of the importance of biophysical parameters in growing systems. The interaction between biophysical and biochemical events and the significance of these is a running theme here.

Because the plant acquires carbon and loses water mainly through its leaves, it is not surprising that light and the availability of water are major determinants of leaf growth. They are considered in some detail. To do this requires an understanding of the cellular basis of leaf growth. Consideration of recent work on cell division and expansion in developing leaves is a natural starting point.

CELL DIVISION AND CELL ENLARGEMENT

Growth and morphogenesis of the primordium and the unfolding lamina are intimately associated with both cell enlargement and cell division; therefore following changes in cell size or cell number alone cannot adequately describe leaf expansion. Examining cell numbers and volumes of successive primordia of *Lupinus albus,* Sunderland & Brown (156) showed that while primordium volume increased by about 50-fold, mean cell volume increased by only 3–4-fold over seven plastochrons, the difference being accounted for by increase in cell number. Superficially, this argues for the importance of cell multiplication as the major determinant of primordial enlargement. There is also a massive increase in cell number in the unfolding lamina. However, as pointed out by Haber & Foard (64), "Simple geometrical considerations alone demand that expansion of an organ be the summation of the expansions in all the individual cells that comprise it [taking account also of intercellular spaces—see the section below on intercellular spaces], irrespective of the presence or extent of cell divisions." It is often but not always the case that large leaves have large numbers of cells while cell size can vary widely depending upon ontogenetic factors and the environment. It is therefore sterile to argue for one or the other being of key importance in the determination of leaf size; growth cannot occur without cell expansion. Nevertheless, it is convenient, if arbitrary, to separate the processes of division and cell expansion here.

Cell Division

Recent work, comprehensively and critically reviewed by Cusset (26), has modified our views of the contribution of cell division to leaf morphogenesis. The widely accepted interpretation of Avery (6) from studies on the tobacco leaf, that "the addition of cells at its apex [i.e. that of the young primordium] may be traced to the activity of a single sub-epidermal cell," is no longer tenable following elegant work by Dulieu and his collaborators (48–51) and by Poethig & Sussex (132, 133). The Dijon workers treated tobacco seed with ethyl methanesulfonate and induced sectorial chimeras of albino genotype. Careful analysis of these showed that several layers of cells contributed to the growing primordium and that from the earliest stages of primordial growth there was generalized division along its length, with the distal region of the primordium gradually differentiating into the nonmeristematic condition. The method of clonal analysis of cell lineages, widely used in study of morphogenesis in animal systems (see 131), was employed to follow cell fates in developing leaves of Xanthi Nc tobacco (132) and supports this interpretation. It also led to the conclusion that at least 18 cells in the epidermis of the apical meristem contribute to leaf development, probably with a similar number in each of the two or three subepidermal layers, giving a total of between 50 and 100 initial cells.

It is not known what determines this number, or whether more cells contribute to primordia formed as the apical meristem enlarges, although since such primordia often appear to be larger at initiation, this seems likely. Of more importance is whether final cell number in the lamina is influenced by the number in the primordium. Studies on *Cucumis sativus* suggest that neither final leaf size nor cell number correlates with the rate of cell division in the primordium or with the number of cells in the primordium at the onset of unfolding (114). Environmental factors, in this case light intensity, appear to override any affects attributable to primordium size.

The work on tobacco also showed that files of cells tend to run longitudinally from base to margin in the young leaf rather than originating from the apex as would be expected on Avery's interpretation of an apical initial. The early differentiation of epidermal hairs seen in SEM of the Xanthi Nc material (132) confirms Dulieu's work showing that cells in the tip of the extending primordium differentiate and stop dividing at an early stage and that gradually intercalary growth comes to predominate. The cause of this early maturation of the primordial tip is not known. Growth in volume and dry mass is exponential over this period (183) and only begins to decline much later; thus effects at the tip do not reflect any general trend in the growing primordium.

Analysis of cell lineage patterns has also contributed to a reassessment of the origin of the lamina. Avery's view (6) was that a row of subepidermal cells, appearing as a single cell in transverse section, formed a marginal meristem from which the tissues of the lamina were derived by a pattern of regular divisions cutting off alternately palisade and spongy mesophyll mother cells. Poethig & Sussex (133) concluded that the lamina is derived from a minimum of six cells in the transverse section of the Xanthi Nc leaf, with no evidence for specialized initial cells in the marginal position.

This is not to say that *initiation* of the lamina does not involve localized growth responses, perhaps with locally enhanced meristematic activity, although evidence for this is not strong. The frequency of mitotic figures in sectioned or cleared leaf material is often used as evidence for local effects, but mitotic indexes do not necessarily indicate the rate of cell division unless it can be demonstrated that the duration of mitosis is the same in material being compared (123). There is little information on temporal and spatial variations in cell cycle length and the duration of stages of the cycle for leaf tissue (34, 92). The acquisition of such data will be necessary if the advantages of the clonal analysis technique (131) are to be fully realized.

When expansion begins, the leaf shows a number of cell layers in the horizontal plane, usually 6 or 7 but occasionally up to 10. Within these layers the plane of division is predominantly but not exclusively anticlinal; their continuity thus tends to be maintained, notably in the epidermis, where intrusion of underlying tissues does not normally occur. However, this trend is far from universal. In periclinal chimeras, the contribution of genetically different underlying layers of the leaf frequently varies from point to point along and across the leaf, to be seen as apparently random variegation where one or more of the layers is genetically free of chlorophyll (34, 48, 131). An important implication of this is that since leaf form in variegated plants does not usually differ significantly from that of the nonvariegated type, cell division has only a limited role in determining leaf form in the later stages of expansion; wide variation in the detailed pattern of division in different parts of the leaf has little effect on shape. As in many animal systems, a cell's position appears to determine its fate. The constant allometric relationship between leaf width and length shown by Haber & Foard (64) for expanding leaves of tobacco, even when divisions had stopped, is further support for this view.

In the *early* stages of lamina formation, differences in frequency of mitosis and in the planes of division and direction of cell expansion can be important in the generation of leaf form in a number of species with lobed or palmate leaves (59, 60, 76, 163), although what determines these differences remains unknown. In the *Knotted* mutant of maize additional divisions in the epidermal layers appear to be induced by the passage of an unknown substance from the underlying mesophyll cells (66). The role of cell-to-cell communication in leaves, particularly as it may affect the control of shape, is largely unknown and could prove a fruitful research area.

The main features of leaf shape and structure are already established by the time that unfolding and rapid expansion of the lamina begin. Up to 99% of the cells in a leaf arise from divisions that occur after this time (see 34). Although cell division can continue until the leaf is up to 95% of its final size, earlier cessation is not uncommon (31, 34, 114, 155).

In general, cell divisions cease first in the distal portions of the lamina, and continue longest at the leaf base (6, 102). Superimposed on this basipetal gradient is a difference in the behavior of the various tissues of the developing leaf since cells of the epidermis, with the exception of stomatal guard cell initials (45, 46, 145), cease dividing before those of the mesophyll, with the palisade cells the last to stop dividing (6, 31). This raises the interesting possibility that, depending on conditions during expansion, the ratio of epidermal to mesophyll cells may vary, with consequences for photosynthetic potential. There are few data on this point. Dengler (43) found leaves of sunflower to be more than twice as large and to contain 2.3 times as many cells when grown in full daylight compared with 25% daylight. The two epidermises accounted for 38% of all the cells in the larger leaves but 41% in the smaller, the difference being at the expense of the spongy tissue. The smaller proportion of mesophyll cells formed under shade conditions may reflect a general effect that persists over the period during which divisions occur rather than to specific effects on the mesophyll itself.

The very rapid rates of cell division in the growing leaf mean a large demand on the associated biochemical machinery. The mechanisms that direct metabolite and precursor flow into cell division rather than enlargement are not understood, and consequently these complex spatial and temporal trends in development across and along the leaf remain unexplained. Nor is it known what leads to the cessation of cell division in this complicated pattern. There is limited evidence, unfortunately based only on total leaf cell number, that the number of cells continuing into successive divisions follows a negative logistic relationship with time (20, 33), with declining numbers of cells going into successive cycles of division. In at least one species, *Cucumis,* high photon fluence rate leads to faster and more prolonged cell division, and involvement of photosynthetic products in this has been suggested (34).

Cell Expansion

The basipetal trends and tissue differences seen for cell division are also seen for cell enlargement and physiological maturation. As the lamina unfolds, mean cell volume continues to increase so that the final value when expansion is completed is, typically, around 20–40 times that at the start of unfolding (37). Invariably there is a final period, sometimes short, when lamina expansion is accounted for entirely by cell enlargement. Cells complete expansion and differentiation in basipetal sequence from the tip (6, 102); the leaf base matures last. This pattern of basipetal development masks major differences between tissues. Detailed work on expansion of the epidermal and palisade layers of leaf 7 of *Xanthium strumarium* by Maksymowych (102) has shown that cells in the epidermis begin to expand in the plane of the lamina at least 1 LPI unit (53, 91) earlier than the palisade cells, which in contrast extend vertically faster and earlier (by about 2 LPI units) than the epidermal cells. The absolute growth rate of area expansion for the palisade cells is negative between LPI 0 and 1.5, meaning that the cells are getting smaller in this plane; this is caused by the rapid rate of cell division's outstripping the rate of cell enlargement.

Differences in rates of cell division and expansion are reflected in differences in expansion rate across the lamina. Avery's data (6, 142), obtained by marking the leaf into a grid of 5 mm squares and following growth of these, show that relative growth rates tend to be highest in the center of the leaf, closest to the main veins, the local source of water and metabolites, and lowest at the leaf perimeter and tip. This has been confirmed by studies on *Xanthium* (52, 101), *Cucumis,* and *Vitis* (151), where strain-rate analysis has been used (152).

The Grass Leaf

In the dicot leaf, meristematic activity and differentiation cease last at the base. In some species, meristematic activity may continue at the base of the lamina for several days, as in spinach (146), or for many months, as in the unifoliate *Streptocarpus wendlandii* (78), where the leaf-like phyllomorph continues to grow as a result of activity at the basal meristem.

The grass leaf also extends as the result of activity at the lamina base (81, 150). At an early stage in primordial development an intercalary meristem is established that subsequently separates into two, the distal giving rise to the lamina, the proximal to the sheath. All cells in the leaf originate from these meristems, and there is no generalized division over the expanding leaf surface as occurs in the dicotyledon. Cells expand rapidly on leaving the meristem; thus the extension zone which includes the meristems is short (9, 41, 83, 106, 171) and totally enclosed in the subtending sheaths of older leaves. It follows that the cells in the emerged blade are completely expanded.

This has a number of important consequences. First, the extension zone is insulated from any direct effect of light, except insofar as a small amount, tending to be rich in far-red because of absorption of red, will pass through the surrounding sheath tissue. Second, the extending cells will grow in an atmosphere with a higher water vapor pressure than that surrounding the emerged lamina; hence evaporation from the expanding cells is likely to be low. Third, the pathway for exported assimilate from the more mature distal part of the blade must inevitably pass through the extending zone; in principle, assimilate may be subject to loss to it. Conversely, cells of the extending zone must have first call on organic materials, minerals, and especially water entering the leaf and destined for the distal portion; this may represent an important difference from leaves of dicotyledons, where the transpiring and expanding regions are not separated to the same degree (136).

THE MECHANICS OF ENLARGEMENT

The Role of the Epidermis

Because the epidermal cells represent an interface with the aerial environment and commence expansion so early, it can be asked whether growth of these cells governs lamina expansion initially, or throughout the whole of its period of growth. Over a century ago, Sachs (144) showed that for a number of organs the epidermis contained the internal pressures generated by the tissues within, since when the epidermal layers, probably including also one or two outer cortical layers (111, 129), were removed, a substantial increase in length of the remaining inner tissues resulted. Such effects can be easily demonstrated using, for example, pieces of rhubarb petiole. This suggests that the epidermis restrains tissue enlargement and certainly does not drive it. However, recent work, mainly using stem pieces (12, 40, 56, 158), has emphasized the importance of the epidermis especially in relation to responses to auxin, to which inner tissues appear to be largely insensitive. In other words, here growth centers on the epidermis. The conclusion that, "although the inner tissue apparently plays a role, the epidermis primarily contributes to auxin-induced elongation of stem tissue" (158) has been widely accepted without the implications being always fully appreciated (61, 88).

A current view (61) is that yielding properties of the outer epidermal walls are of critical importance in morphogenesis since it is these walls that contain the stresses generated by the inner cells and the turgor of the epidermal cells themselves. Little is known about the chemistry and rheological properties of the outer wall. In segments of maize coleoptiles treated with IAA, osmiophil-ic granules (0.3 μm diameter) accumulated specifically at the interface be-tween the plasma membrane and the outer epidermal wall (88). The nature of these particles remains to be established; but if, as has been suggested, they contain protein or phenolic compounds, they could be involved in modifying wall properties by changing the extent and perhaps the pattern of phenolic crosslinks (58, 185). The view that "steady growth involves the coordinated action of wall loosening in the epidermis and regeneration of tissue tensions by the inner tissues" (88) is attractive.

Does this interpretation hold also for the dorsiventrally flattened di-cotyledonous leaf? Clearly the situation is less straightforward than for a centric structure such as a pea hypocotyl or a maize coleoptile, where the tissue tensions are directed towards a single outer cell layer. In the leaf there are two epidermises, of different character and abutting on different classes of underlying tissues; and the outer walls of both would have to be load-bearing and able to loosen in concert, and to similar extents, if lamina expansion is to proceed smoothly and without distortion and wrinkling. There are no quan-titative data to show how general and close the contacts are between the two epidermises and the underlying palisade and spongy mesophyll, especially in the later stages of lamina expansion, although published electron micrographs (44, 74) suggest that areas of contact may be relatively small. Internally generated pressures may therefore be confined to discrete points and not distributed universally and equally.

The convoluted jigsaw puzzle appearance of some epidermal cells has been associated with localized stretching due to the underlying tissues (6). An alternative suggestion (42) is that the waviness represents an "expansion reserve" of wall substance, since in leaves of *Avicennia* the convolutions found early in development disappear as the leaf grows. More data on the connections between epidermises and adjacent tissues and on the properties of the inner and outer epidermal cell walls are urgently needed. Species with readily strippable epidermises could be useful here.

Intercellular Spaces and Their Significance

Expansion of the epidermal layers occurs concurrently with the formation of intercellular spaces, and the two events have been linked (6). In the mature, fully expanded leaf, intercellular spaces constitute from under 10% to more than 40% of the total volume. As a percentage of the mesophyll volume the figure is clearly much larger, and in the palisade of leaves of *Helianthus* spaces account for 42% of the volume, with a figure of 58% for the mesophyll (43). As a result of space development, the internal surface area of a leaf can be from about 7–32 times its external area, depending on species and on ecological factors such as light intensity or quality (97, 126, 168).

Except in a few cases (42, 179), spaces are schizogenous in origin and result from the partial separation of cells following wall breakdown rather than from the lysis of cells of the mesophyll. Published sections of leaves indicate rapid development of spaces as the lamina unfolds (31, 45, 74, 104). The detailed mechanism of space formation and wall breakdown varies between species and tissues (74, 86, 143), but in primary leaves of *Phaseolus vulgaris* (74), as in various other tissues (105), spaces begin to form in predictable positions at points where three or more mesophyll cells abut. These points correspond with sites where the cell plate and daughter-cell wall meet the wall of the mother cell. It is possible that the mechanism of separation begins to operate as the new wall is being formed. Cell separation begins early in lamina expansion, when cell division is frequent, and is associated with breakdown of part of the primary wall (74, 105), almost certainly including both cellulose and pectin. The likely involvement of one or more hydrolytic enzymes in the initiation of separation (82) offers a possible point of regulation through control of their synthesis, amount, or activity. Immunocytochemistry offers an approach that could yield important data here.

The initiation, at least, of the separation process is not the result of stretching forces generated by the expanding epidermis, as suggested by Avery (6), although the expansion of the epidermis may help to continue the process of separation once wall dissolution has begun. On the other hand it is probably incorrect to consider the continued formation as a purely passive process. Scanning electron microscopy (74, 75) shows the spongy mesophyll cells to be stellate and to form a reticulate meshwork in the plane of the leaf, with arms connecting to cells above; there is no sign of arms being ruptured, nor does their appearance, which is plump and rounded, suggest that they are being subjected to passive pulling (75). On the contrary, the arms give the appearance of being sites of extension growth. They may extend as small rams, helping to drive expansion and generate the inner tensions found in many expanding tissues (45). It is also hard to reconcile with passive pulling the formation of large substomatal cavities, which develop concurrently with differentiation of guard cells.

Although in the fully expanded leaf most of the surface of mesophyll cells borders intercellular space, nevertheless some areas of wall abut on adjacent cells where separation has not occurred. What governs the siting of these regions is unknown, and it is not clear whether the walls are resistant to hydrolysis because of differences in composition, or whether hydrolysis is checked some other way, perhaps by progressive inactivation of the enzymes responsible. Thus expansion of the cells of the spongy tissues is intimately linked to space formation and involves extension in certain directions governed by local wall characteristics.

THE MECHANISMS OF LEAF CELL GROWTH

Biophysical Considerations

The expansion of any cell depends upon the expansion of the cell wall, which has been characterized as a form of biochemically controlled creep (148). Although not perfectly elastic, the cell wall can be stretched elastically, and in the growing cell irreversibly, in response to turgor. Cell volume, V, and turgor, P, are related through the bulk modulus of elasticity, ϵ, thus (169):

$$\epsilon = (dP/dV) \cdot V \qquad\qquad 1.$$

It follows that ϵ is not constant but varies such that as turgor rises, resistance to elastic stretching of the cell wall becomes progressively greater (28, 85, 188). Bulk values of ϵ appear to increase (i.e walls become less elastic) as leaves age (8), consistent with the idea that rheological properties of the wall change as leaf cells expand—perhaps as a result of increased wall thickening with concomitant increase in the number of wall bonds that will restrict elastic stretching and the capacity to deform irreversibly.

Accepting that turgor is the driving force for cell expansion (141) the relative rate of volume increase is given empirically as the product of the effective turgor—i.e. turgor less a yield threshold term, Y, and a wall yielding coefficient, ϕ, (22), often referred to as wall or plastic extensibility:

$$1/V \cdot dV/dt = \phi(P - Y) \qquad\qquad 2.$$

This form of one of Lockhart's (96) equations is often combined with a second equation that relates volume increase to the product of the gradient of water potential between the cell and its surroundings, $\Delta\psi$, and the hydraulic conductivity of the cell wall and membranes, L_p, to give:

$$1/V \cdot dV/dt = (L_p \cdot \phi/L_p + \phi) \cdot [\Delta\psi + (P - Y)] \qquad 3.$$

Since $\Delta\psi = \pi_e - \pi - P$, an alternative form is

$$1/V \cdot dV/dt = (L_p \cdot \phi/L_p + \phi) \cdot [\sigma\Delta\pi - Y], \qquad 4.$$

which takes into account the fact that cell membranes may not be truly semipermeable by including the reflection coefficient, σ. Y, π, ϕ, and L_p have all been implicated in the control of leaf cell expansion.

Strictly speaking, Lockhart's treatment applies only to single cells. The problems of applying the equations to multicellular tissues and organs such as leaves have been assessed and discussed by several workers (22, 71, 166). One of the difficulties concerns the estimation of L_p since in the intact plant the water pathway in multicellular tissues, and hence its characteristics, is often not known with any certainty. For example, in phosphorus-deficient cotton plants, reduction in leaf growth appears to be due to a decrease in hydraulic conductance of the roots (137). Indirect effects on leaf growth through changes in hydraulic conductivity outside the leaves occur in sunflower plants acclimated to low water potentials (107). In neither of these cases was there any evidence of a direct effect of treatment on leaf hydraulic conductivity—i.e on the L_p term in Equation 4.

Another difficulty in applying the Lockhart approach to growing multicellular systems concerns the wall-related parameters, ϕ and Y, which have been widely implicated in the control of leaf cell expansion. Wall extension (as opposed to elastic stretching) must involve the breaking of bonds within the wall and the reestablishment and reformation of at least some of these in different locations and conformations. Cleland (18, 19) has argued that this is unlikely to be an instantaneous process and that the three methods that have been used to measure wall mechanical properties—i.e. analysis of creep, stress relaxation, and the Instron technique, all of which use isolated cell walls—probably measure previously induced capacity for wall extension and not the future capacity. Obviously care needs to be taken in the interpretation of data using these approaches, the more so because the forces used are not the same as those generated through turgor, where expansion is multiaxial, even where it appears to be directional as in the cells of the growing leaf.

Examples are known where cell extension occurs without concomitant synthesis of new wall material so that in consequence the wall becomes thinner (140). This is not normal, though; and in sustained growth of the cell, stretching of previously formed wall and synthesis of new wall proceed side by side. That is to say, parallel with the biochemical events associated with wall loosening are those leading to formation of new wall. So far, we are unable to disentangle the two. In vivo, all these processes will be reflected in ϕ and Y, which themselves are not independent since changes in the wall that affect the one may also affect the other (71).

The Phaseolus Primary Leaf System

Recent important work on the control of expansion of primary leaves of *P. vulgaris* (172–175) is now considered in detail. Plants were grown in low-intensity red light (3–4 μmol m^{-2} s^{-1}) for 10 days until cell division in the leaves had ceased (172). Such plants show greatly elongated stems; but the primary leaves, though green, remain unexpanded and grow only very slowly. Bringing leaves, or leaf pieces cultured on 10 mM sucrose and 10 mM KCl, into white light (250–400 μmol m^{-2} s^{-1}) led to a rapid increase in the rate of leaf growth after 10–20 min (173). The pH of solutions added to the light abraded surface of the leaves fell in illuminated leaves and proton secretion into the solution began some 10 minutes before the growth response, continuing for some hours for tissue exposed to white light but quickly ceasing for tissue returned to low-intensity red light. The secretion of protons was inhibited, after a short lag, by cycloheximide and immediately by the ATPase inhibitor, dicyclohexylcarbodiimide. It was concluded that in white light on ATP-mediated proton pump was activated and that proton secretion was intimately associated with the growth response.

Support for this hypothesis came from the finding that fusicoccin promoted proton secretion and growth in leaf pieces, while infiltration with neutral buffer prevented growth in white light. Wall extensibility, measured by the Instron technique, significantly increased in the white-light treatment, while frozen-thawed material subjected to a constant load, extended significantly more when bathed with a solution at pH 4 than at pH 6. It was a short step to argue that the acid-growth theory could be applied to the bean leaf system, with white light substituting for auxin as the promotor of acid-induced wall loosening.

Indirect indications that changes in wall extensibility are major determinants of growth were also obtained by examining other parameters of the Lockhart equations (174). In pieces of leaf tissue treated with white light, water potential, osmotic pressure, and turgor all fell, although the differences from material in red light were not statistically significant. These values were obtained by isopiestic psychrometry following prolonged incubation in the dark at 25°C so that stress relaxation (21, 24) would have occurred. Estimates of wall yield threshold, Y, following incubation of leaf pieces on solutions of polyethylene glycol, ignoring possible effects of wall solutes (23), gave values for the effective turgor for growth, P_e (i.e $P - Y$), of about 0.05 MPa in the white-light material compared with nearly 0.4 MPa in the red light material, making it very unlikely that an increase in P_e could explain the enhanced growth. Later work (175) using the method of stress relaxation (24) showed that white-light treatment caused an immediate fall in P, Y, and P_e, while as growth decreased Y also continued to decrease although P_e rose. These results support the idea that the light-stimulated growth is controlled primarily by changes in wall extensibility.

The increase in ϵ as tissues age, (see the section on biophysical considerations, above), could arise from an increase in wall thickness, perhaps with parallel changes in wall composition. These changes might also be expected to affect ϕ. Work with the *Phaseolus* leaf system has confirmed that values of wall plastic extensibility fall as the rate of leaf area expansion goes down (177). The decline in leaf expansion coincides with the period when maximum rates of photosynthesis are achieved; if high concentrations of locally produced assimilate enhance the rate of wall synthesis so that this exceeds the rate at which wall loosening can occur, reduced extensibility could result.

Although wall extensibility appears to be a major factor governing leaf expansion in *Phaseolus,* evidence from other species (e.g. 160) does not always support this view, as will be seen below.

REGULATION BY LIGHT

Since leaves are primarily photosynthetic organs the plant's developmental strategy might be expected to optimize this function, while at the same time maintaining and optimizing water availability (25, 55). The twin requirements impose conflicting restrictions on leaf development as conditions of high light, which favor photosynthesis, must also impose a high evaporational load on the leaf.

The Light Environment

In full sun, the photon fluence rate (400–800 nm) may reach a maximum of about 1900 μmol m^{-2} s^{-1}, while in deep shade within a forest or crop canopy it can fall to values 1/50 of this or less, close to or even below the light compensation point for photosynthesis (115, 153). The photomorphogenetically important red:far-red ratio (R:FR) can vary from about 1.15 in full sun to as low as 0.12 in canopies of mixed deciduous woodland (154). Although for most of the day the R:FR ratio is constant, at dawn and dusk, with low solar angles and a longer atmospheric pathlength, there is a fall in the ratio of 20–30% and a substantial increase in the blue component of the spectrum.

For a leaf, the light environment can vary spatially, according to aspect and the position of other plants or other leaves on the same plant: it can also vary with time—systematically, owing to the position of the sun, or randomly, because of variable cloud cover. Even within a plant, differences in leaf development can be ascribed to the light environment, as for example in sun and shade leaves, which can vary substantially in size, morphology, and physiology (10, 93, 130).

Distinguishing between effects of light quantity and quality on leaf growth is not straightforward. Natural shade conditions inevitably involve changes in both fluence rate and R : FR ratio; the response of plants to *natural* conditions of shade will reflect this and involve complex interactions between photomorphogenic and photosynthetic effects that are not easily, or perhaps realistically, disentangled. Furthermore, as pointed out by Smith (153), with present technology it is difficult to vary R : FR ratio experimentally at photon fluence rates that do not limit photosynthesis, although it is possible to change the ratio by adding FR to light of relatively low fluence rates.

Effects of Light Quantity

Where light quality has been kept constant, leaf number and final size of individual leaves are often positively correlated with photon fluence rate (32, 43, 95, 124, 184). There are also differences in how different species distribute dry mass in response to shade treatments. Shading the tolerant *Impatiens parviflora* causes leaf dry mass to rise slightly as a proportion of total plant mass, at the expense of roots; however, specific leaf area—i.e. leaf area per unit leaf weight—rises markedly under shade, resulting in larger but thinner leaves (54). In the shade-intolerant *Helianthus annuus*, leaf weight ratio falls slightly in shade with a much smaller rise in specific leaf area (68); in consequence leaf area ratio changes little, although shade reduces the area of individual leaves (43). *P. vulgaris* shows intermediate responses, shading having little effect on leaf weight ratio but causing an increase in specific leaf area and leaf area ratio (32).

Photoperiod has only small effects on leaf growth. Evidence suggests that increased photoperiod reduces specific leaf area slightly while leaf weight ratio is unaffected (32, 95). Where sugar beet plants were subjected to day-length extension by 4 h of low-intensity incandescent light (113), specific leaf area was increased by about 20% coincident with a massive increase in leaf area per plant of 47%; these are probably end-of-day responses, ascribable to light quality (79) rather than photoperiod.

Leaf weight ratio and specific leaf area are whole-plant based and represent average values integrated for different numbers of leaves and different plant dry weights. The response of individual leaves to photon fluence rate is not always the same, even on the same plant. In *Phaseolus vulgaris*, the area of primary leaves and trifoliates showed a parabolic relationship with increasing total daily fluence (32). Cell number per leaf did not vary for primary leaves despite considerable differences in area, implying substantial effects of light on cell size of the epidermal cells at least; but a doubling of trifoliate leaf area in high photon fluence rates was associated with an increase in cell number of about 75% and greater development of the palisade mesophyll (32). Leaf size and cell number were also greater for *Cucumis sativus* grown under high photon fluence rate (114).

The work with *Cucumis* also showed that the total amount of radiation per day had much greater effects on leaf expansion than photon fluence rate or day length; this has also been shown for leaves of *Fragaria virginiana* (14) and *Hyptis emoryi* (125). These results suggest that the production of photoassimilate could be a major factor governing leaf size. However, some at least of these studies used photon fluence rates that at their highest were less than one third, and at their lowest less than one thirtieth of those encountered in full sunlight. Since the species used can be classified as shade avoiding, the results must be interpreted with some caution. Under low photon fluence rates the availability of metabolites deriving directly or indirectly from photosynthetic activity may be greatly restricted, with effects on all facets of growth.

Increases in thickness are often found for leaves grown under high total fluence (14, 125). These are due to a great increase in dimensions of the mesophyll and may involve differentiation of additional palisade from cells that would form spongy tissue in lower light intensities. This finding can be interpreted as follows. Under conditions of high total fluence, accumulation of assimilate leads to a fall in π in the mesophyll cells, with a corresponding rise in P and maintenance of the potential for growth, which is also linked to light-induced wall loosening (see the section on the *Phaseolus* primary leaf system, above); the greater concentration of soluble carbohydrate in the mesophyll could also favor the formation of new walls there. For the epidermal cells, on the other hand, the greater availability of potential substrate for wall building may not be matched by the capacity for wall extension, especially in the outer walls, so that wall synthesis occurs without concomitant cell expansion.

Young expanding primary leaves of glasshouse-grown plants of *Phaseolus vulgaris* contain high concentrations of hexose sugars which fall as the leaves begin to act as sources of assimilate (121). These leaves also show a high specific activity of soluble acid invertase which correlates with rates of leaf and epidermal cell enlargement, as found for other species (62, 134).

Using field-grown crops of wheat, Kemp (83, 84) found that the concentration of carbohydrate (hexoses, sucrose, and fructans) in the extension zone of leaves was correlated with leaf extension only under conditions of extreme shade. He concluded that under normal field conditions carbohydrate supply is unlikely to limit growth; that does not mean that rate of utilization may not be limiting under low irradiances. Leaves on shaded plants extended more rapidly than those in full daylight, and the extension zone was longer (84) in keeping with the well-known finding that parallel-veined monocotyledon leaves tend to grow longer under low-light conditions (99). Since the extension zone in wheat is not exposed to direct light the effect of shading on its size must be indirect. It is not known whether this effect is attributable to more cells being meristematic or in the expanding condition, and consequently the effects of shade on rates of division and extension cannot be distinguished.

With the exception of the work on *Phaseolus* (see the section on the *Phaseolus* primary leaf system, above) comparatively few studies have analyzed the biophysics of leaf growth in response to light, and fewer still have used trees. Recently, Taylor & Davies (159–161) made detailed comparisons of the effect of light on leaf growth in *Betula pendula* and *Acer pseudoplatanus*. They found that on illumination with white light, wall extensibility was increased for leaves of the shade-avoiding *Betula,* and the surface pH of abraded leaves fell; this species showed a higher leaf growth rate in light than in darkness. In contrast, light had no effect on wall extensibility of the shade-tolerant sycamore, where leaf growth was greater in darkness than in light. Measurements of Y, for a single size range of growing leaves only, showed this to be unaffected by illumination but to be much greater in sycamore (0.25 MPa compared with 0.07 MPa for birch). It was concluded that while wall extensibility limited growth in *Betula,* the higher value of Y made turgor a limitation of expansion of *Acer* leaves. It seems that differences in wall metabolism are involved in both species modifying the related parameters ϕ and Y but in different ways. The involvement of low-energy responses, especially in the case of *Acer,* remains to be explored.

Effects of Light Quality

At the ecological level, plants have long been arbitrarily classified into shade-avoiding and shade-tolerant species (63). The important role of light quality in avoidance and tolerance reactions is now better understood. Broadly speaking, for shade-avoiding herbaceous plants, petiole and stem growth are greatly and rapidly enhanced by shade, which at the same time leads to greatly reduced leaf expansion (118, 120). This strategy offers the maximum opportunity for growth into more favorable light environments where leaf expansion can proceed normally. In shade-tolerant species, growth in shade is often very slow and leaf expansion is favored, with concomitant adaptation at the biochemical level to optimize photosynthetic efficiency under unfavorable light conditions (10, 153). There are no clear taxonomic differences between the two types of plant (119), and species within a genus may vary significantly in their response to light and tolerance to shade (57).

Evidence from a number of herbaceous species grown under a range of experimental conditions shows that varying light quality while maintaining photon fluence rate constant can have major effects on leaf morphogenesis. Morgan & Smith (119) found the ratio of leaf to stem dry weight to rise markedly in shade avoiding species as the R:FR ratio fell, whereas for woodland species the dry weight ratio was unaffected. Changing the R:FR ratio was also associated with an increase in total chlorophyll concentration for all species examined except *Mercurialis perennis,* a plant of deep shade. Experiments with *Rumex obtusifolius* showed leaf area per plant to increase with photon fluence rate but to decrease as the R:FR ratio was lowered; leaf thickness decreased with low photon fluence rate and also with low values of R:FR ratio (110). For *Impatiens,* too, both photon fluence rate and R:FR ratio are important in determining the leaf response (182).

The Role of Phytochrome

There is little doubt that the phytochrome system is involved in the rapid shade-avoiding growth responses of stems exposed to enhanced levels of FR light as well as in at least some of the slower responses shown by leaves (117).

Phytochrome is also implicated in the control of leaf growth in etiolated leaves. Most work here has used leaves or disks of dark-grown plants of *Phaseolus vulgaris* (38, 39, 47, 94, 139, 178). The enhanced leaf expansion that follows a short exposure to red light, and which is far-red reversible, is due to an increase in cell number with little effect on cell size (38). The sites of perception of the red-light stimulus are in the primary leaves and the cotyledons, and the breakdown of cotyledonary reserves and the utilization of these in the leaf appear to be involved in the response (39). Although a single treatment with red light is insufficient to increase cell number to that found in plants grown in high-fluence white light, growing plants continuously in red light for 10 days gives leaves whose cell number does not increase when exposed to white light (39).

Whether phytochrome is involved in the control of cell division in plants in the field is unknown. Within many buds the light environment will have a low R:FR ratio due to attenuation of red light by surrounding leaves and bud scales. Prolonged exposure to these conditions could have morphogenic effects comparable to those found in bean, and may affect the initiation and development of the photosynthetic apparatus.

Phytochrome is known to be involved in the control of a large number of enzymes at transcription and by posttranslational mechanisms (87, 90, 147, 169). Among the genes whose transcription is controlled by phytochrome are those coding for the synthesis of the chlorophyll *a/b* binding protein and for the small subunit of RuBISCO, as well as others also concerned with the development of the photosynthetic machinery. However, it is unclear which enzymes are involved in the control of leaf growth by phytochrome. Of interest is that phytochrome can control at least two enzymes known to be involved in cell wall metabolism—peroxidase (128, 149) and UDP-galactosyl transferase (170)—as well as a number involved in the synthesis of lignin (see 65).

There remains a paradox. For shade-avoiding species there is an increase in individual leaf area as light increases, whereas for shade-tolerant species leaf areas tend to be enhanced at low photon fluence rates. One interpretation is that in the shade-avoiders, leaf expansion is closely linked to photosynthesis and the availability of carbon skeletons either from older leaves or from concurrent photosynthesis; the response is substrate-determined. In contrast, in shade-tolerant species, light-quality responses appear to be the main determinants of leaf development, with regulation occurring through the pattern of substrate utilization. With such a variety of responses, mediated through phytochrome and other photoreceptors, including chlorophyll, it seems that subtly different controls exist between species. Thus only broad-brush generalizations to explain these can be made. Analysis of the metabolic and biophysical basis of these differences remains a major task.

WATER AND LEAF EXPANSION

Methodological Difficulties

Two difficulties hinder any but the most superficial study of the effects of water stress on leaf growth. The first concerns the application of stress. This is often done by allowing plants to dry out and suffer progressive stress. Under field conditions, the rate of drying out is usually slow and the irradiance high; but where controlled environment chambers have been used with pot-grown plants, not only is the irradiance much less but the rate of drying out of pot and plant is much faster. This is also the case where plants are subjected to sharp and rapid stress by changing the osmotic pressure of the rooting medium by adding polyethylene glycols (PEG) or similar osmotica. For operational reasons the experimenter usually welcomes rapid responses in his material, but in water-stress studies these have to be interpreted with more than usual caution.

The second difficulty is equally important. Because the pattern of growth in leaves of the linear-leaved monocotyledons differs from that of the broad-leaved dicots spatially and temporally (see the section on the grass leaf, above), analysis of cell growth and of the responses to water stress will differ between the two groups (see 71). There are several reports that water potential in the meristematic and expanding zones of grass leaves differs from that of the emerged, fully expanded blade, although turgor is often comparable (28, 29, 71, 106, 109, 112), implying lower values of π in the growing zone. In a grass leaf, the expanding region is enclosed by the sheaths of older leaves which shield it from light and maintain an environment with low evaporational demand (see the section on the grass leaf, above). This is not the case in the expanding broadleaf where, although in the bud the primordium is protected, once unfolding begins the new leaf is directly exposed to light and to loss of water from epidermal cells that are initially poorly cuticularized. Even under favorable conditions for growth, these outermost cells will be vulnerable to water stress. Most workers using in vitro psychrometry to measure water potentials of intact growing leaves have ignored spatial differences (11, 159) and differences between cell types, although these are known to exist (73). With the wider availability of the micropressure probe more data on differences in ψ and P across and within the leaf will become available.

Effects on Cell Division

While major water stress will inevitably cause extensive metabolic disturbance with attendant disruption of mitosis, milder stress is generally believed to affect cell expansion, with much smaller effects on division. Clough & Milthorpe (20) found that witholding water for up to 12 days from young tobacco plants led to a gradual fall in leaf water potential, ψ_{leaf}, to about -1.1 MPa. Even at that value cell division continued in young leaves, although at a reduced rate compared with watered controls. In contrast, leaf expansion and area increase of palisade cells in the paradermal plane were reduced significantly at values of ψ_{leaf} of about -0.4 MPa and ceased entirely when ψ_{leaf} fell to -0.75 MPa, the continuing cell division leading to a reduction in cell area.

However, it is possible that not all species respond similarly. Reductions in soil water potential as small as -0.0026 MPa have been reported to affect cell division but not expansion in leaves of sugar beet (162). In these experiments, plants were stressed either using PEG6000 that had been dialyzed to remove toxic impurities, or using a tension plate. The apparent sensitivity of cell division is hard to explain. Unfortunately, no data on ψ_{leaf} are given so that it is impossible to assess the degree of stress at the leaves themselves. Further work on this species is desirable, particularly in view of recent work on it (180) suggesting that leaf extension is extremely sensitive to small and rapid changes in atmospheric humidity, thereby implying effects on cell expansion rather than division.

Cell Enlargement and Turgor Control

Expanding leaves are extremely sensitive to water stress. Even small falls in water potential and turgor may be sufficient to cause expansion to cease or to be greatly reduced (69, 70). Acevedo et al (1), using pot-grown maize plants, showed that reductions in soil water potential, ψ_{soil}, of 0.01–0.02 MPa, led to measurable reductions in leaf extension rate. Extension rate also fell away linearly as ψ_{leaf}, measured in the expanded part of the blade only, dropped from -0.2 to -0.7 MPa. Even more remarkable was the finding that removal of stress at the roots was followed by recovery in leaf elongation rate within a few seconds. This is compatible with the concept that the hydraulic pathway from root to expansion zone is continuous (138). Furthermore, the rate after the relief of stress was often greater than that of the unstressed control for a short while so that, with periods of stress of a day or two, a measure of compensation was achieved and stressed leaves approached the controls in length.

Small reductions in relative water content (RWC) can lead to significant falls, at least in physiological terms, in the bulk leaf water potential. The magnitude of the fall depends on the value of ϵ, since $dV/d\psi = V/(\epsilon + \pi)$ (30), V being approximated by RWC. As an example, for values of ϵ and π of 4 and 1 MPa, respectively, a fall of 5% in relative water content will cause a drop in ψ of 0.25 MPa. In the absence of any osmotic adjustment, this change will lead to a fall in turgor, bringing P closer to Y (Equation 2). For many plants zero turgor is reached when RWC falls to around 90%, corresponding to a fall in ψ of 0.5 MPa on the values quoted above.

There are three main ways in which turgor-driven growth can be maintained in a leaf suffering water stress: by lowering the osmotic potential such that turgor remains high and above the yield threshold (see eqn 2), by alteration of ϵ such that turgor is maintained, or by adapting to lower turgors by lowering the yield threshold and maintaining the effective turgor (i.e $P - Y$).

There is substantial and growing evidence that osmotic adjustment (see 116, 167) is a major regulatory mechanism for leaf extension (2, 29, 106, 112, 122). In experiments using field-grown maize stressed by witholding water over several days, Acevedo et al (2) found that although evaporative demand for water was greatest during the day when ψ_{leaf} tended to be lowest, osmotic potential fell significantly such that bulk leaf turgor was maintained or even rose. This osmotic adjustment was interpreted as a major factor in the maintenance of high extension rates in otherwise unfavorable conditions. Other experiments (71) showed that slow drying enabled leaf growth to continue for longer and with a more substantial fall in π than where drying was rapid and the fall in π smaller. Greater maintenance of turgor following slow as opposed to fast drying has been found for several other species (167).

In the studies on maize, the fall in π was largely accounted for by the production of carbohydrates; pronounced diurnal trends were found in π, presumably as photosynthetically produced solutes were either exported or else used locally in growth overnight. Other studies have given evidence for osmotic adjustment and turgor maintenance in the expanding zone of grass leaves and not only in the mature portion of the lamina (106, 109, 112, 122).

Interestingly, in stressed wheat, the solutes accumulated in the blade are mainly accounted for by sucrose, while in the expanding region glucose accumulates (122). This suggests that apart from transport costs for accumulating solutes in epidermal cells, there may be no major energy cost involved in osmotic adjustment of this kind since both sucrose and glucose would normally be found in those tissues and there is no evidence for enhanced synthesis of them; however, there may be an opportunity cost since compounds required for osmotic adjustment cannot be utilized in other ways. The frequent, although far from exclusive and universal, involvement of photosynthesis in the osmotic-adjustment mechanism means that results from water-stress experiments conducted under low-light conditions must be assessed with caution.

Although osmotic adjustment is known to occur in mature leaves of a large number of dicots, there have been few studies to quantify effects in growing leaves. By inference, expanding leaves might be expected to show osmotic adjustment (27), and this has been shown for slowly stressed plants of sunflower (77). If local photosynthesis is involved in the production of the necessary solutes then behavior of the distal tissues, which mature first, might differ from that of the basal regions; but this has not been examined.

The possibility that water stress in growing leaves can be accommodated through changes in ϵ has been examined in a few cases. The work on sunflower gave no evidence for changes in ϵ during stress (77), nor was this parameter found to vary in studies on rice (28). On the other hand, values of ϵ fell when field-grown plants of *Vicia faba* were stressed by natural drought conditions (80); pronounced increases in ϵ over the season are known (169), and the changes found for *V. faba,* although in the right direction to maintain turgor, cannot be specifically ascribed to growing as opposed to mature leaves (85). In the absence of convincing evidence to the contrary it seems that large falls in ϵ do not occur. Indeed, bearing in mind that the half time for rehydration is dependent upon ϵ (22), too large a fall, with concomitant fall in P, could be a disadvantage since it would delay recovery from water stress.

In theory, effects of water stress and turgor reduction could also be alleviated by walls becoming more extensible (a rise in ϕ) or by a fall in the yield threshold to maintain $P - Y$. This does not appear to happen except perhaps transitorily over short periods (71). Instead the opposite seems to be true, for in a number of cases (29, 107, 112, 171) the maintenance of turgor by osmotic adjustment, although allowing some growth under stress conditions, is not adequate to sustain growth rates at control levels when stress is relieved by re-watering. For sunflower, prolonged water stress caused a fall in ϕ to about 60% of the value in controls and also a small fall (0.05 MPa) in the value of Y (107). The responses may be rapid. Indirect evidence from maize (71) suggests that changes in wall characteristics can occur with stress periods as short as 2 h. Using up to 5 days of stress, Van Volkenburgh & Boyer (171) found a progressive fall in wall extensibility that correlated with a decline in leaf extension rate; the ability of leaf tissue to acidify the apoplast by proton extrusion was also reduced in stressed leaves, suggesting a basis for the reduction in wall loosening.

There are conflicting reports concerning diurnal variation in leaf growth rates (13, 17, 35, 127, 136, 180, 181). It is important to note that rates are seldom constant throughout either day or night. During the day some water stress may occur, perhaps only in the epidermal cells (100), such that turgor may fall close to P_e, effectively reducing or even stopping growth; at night, rise in turgor will lead to recovery in growth rate, provided that wall extension is possible. The rapid recovery in leaf extension rate when stressed maize plants were rewatered (see the section on cell enlargement and turgor control, above) has led to the suggestion that the capacity of leaf tissue to grow can be stored (1) to a limited extent. This implies that a fall in turgor may not immediately curtail wall loosening, which can be used to allow limited later growth. Whether this 'stored' growth is of general importance is not clear, but it could provide a mechanism for maintaining expansion following short-term, ephemeral stress.

Regulation of leaf growth under conditions of water stress is thus extremely complex. The low values of ϵ for growing cells mean that small short-term fluctuations in water potential, measured over minutes, can be accommodated without major falls in turgor. Osmotic adjustment to maintain turgor represents a second line of defense that will enable growth to continue, although maintenance of turgor is no guarantee that *maximum* rates will occur since alteration in wall properties inimical to extension will eventually follow.

Abscisic Acid and Leaf Growth

The accumulation of ABA in water-stressed leaves, including those still expanding, (3, 139) is well-known. Is ABA implicated in causing the reduced growth of leaves suffering water stress, either directly or through stomatal closure? The evidence remains equivocal. The *flacca* mutant of tomato, which does not produce ABA (157), develops normal leaves, suggesting that this growth regulator may not be involved here. In contrast, "normal" expanding unstressed leaves contain higher concentrations of ABA than mature leaves (187), suggesting that it is not inhibitory to growth in these circumstances. On the other hand, ABA can affect leaf form in aquatic plants (4, 5, 186), and in the *Phaseolus* system leaf growth was reduced, although only when ABA was applied at high concentration (176); here, although incubation with ABA did not affect proton extrusion and the acidification of the incubating medium, the capacity for acid-induced wall loosening appears to be reduced. Thus a possible mechanism exists whereby water stress could inhibit extension growth by enhancing ABA concentration. More data are needed about the effects of ABA on wall biochemistry.

CONCLUSIONS

Over the past 25 years substantial progress has been made in understanding the factors that affect leaf growth, but large gaps remain in understanding the complex and multifaceted processes of regulation. Because the leaf is a highly plastic structure varying in its development between and within species (35), simple models to describe leaf growth are unlikely to provide any profound general insight into regulation, which must operate through a graded series of controls.

At the coarsest level, since growth in the post-seedling stage depends upon the input of new dry matter, photosynthesis and nitrogen metabolism and the partition of products between root and shoots must limit leaf initiation and development. Within the leaf, regulation of vascular development will determine the local availability of water, minerals, and the organic metabolites necessary for growth, while local photosynthesis, itself dependent on stomatal initiation, mesophyll differentiation, and intercellular space formation, may provide materials to affect and perhaps control the later stages of expansion.

Finer controls determine the extent and duration of cell division in the different tissues of the leaf; how these interact with those regulating the direction, rate, and duration of cell expansion to determine form and size remains largely unknown. But the ultimate level of regulation lies in the expanding cell itself. A theme of this article is that the biochemistry and biophysics of the growing leaf cell are inseparable. The twin needs of maintaining turgor and allowing the wall to loosen and extend irreversibly in preferred or predetermined directions are paramount requirements for the growth of any cell. While we are beginning to understand mechanisms of turgor maintenance in the growing leaf, those that regulate the properties and growth of the wall remain less well understood. Unravelling them, in the context of the complexities of the expanding leaf, is a challenging and exciting task requiring imagination and ingenuity in the use of the powerful techniques now available from cell and molecular biology.

Literature Cited

1. Acevedo, E., Hsiao, T. C., Henderson, D. W. 1971. Immediate and subsequent growth responses of maize leaves to changes in water status. *Plant Physiol.* 48:631–36
2. Acevedo, E., Fereres, E., Hsiao, T. C., Henderson, D. W. 1979. Diurnal growth trends, water potential and osmotic adjustment of maize and sorghum leaves in the field. *Plant Physiol.* 64:476–80
3. Ackerson, R. C. 1982. Synthesis and movement of abscisic acid in water-stressed cotton leaves. *Plant Physiol.* 69:609–13
4. Anderson, L. W. J. 1978. Abscisic acid induces formation of floating leaves in the heterophyllous aquatic angiosperm *Potamogeton nodosus. Science* 201:1135–38
5. Anderson, L. W. J. 1982. The effects of abscisic acid on growth and development in the American pondweed (*Potamogeton nodosus* Poir.). *Aquat. Bot.* 13:29–44

6. Avery, J. S. 1933. Structure and development of the tobacco leaf. *Am. J. Bot.* 20:565–92

7. Baker, N. R., Davies, W. J., Ong, C. K. 1985. *Control of Leaf Growth.* Soc. Exp. Biol. Seminar Ser. 27. Cambridge: Cambridge Univ. Press

8. Barlow, E. W. R. 1983. Water relations of the mature leaf. See Ref. 36, pp. 315–45

9. Begg, J. E., Wright, M. J. 1962. Growth and development of leaves from intercalary meristems of *Phalaris arundinacea. Nature* 194:1097–98

10. Boardman, N. K. 1977. Comparative photosynthesis of sun and shade plants. *Ann. Rev. Plant Physiol.* 28:355–77

11. Boyer, J. S. 1968. Relationship of water potential to growth of leaves. *Plant Physiol.* 43:1056–62

12. Brumell, D. A., Hall, J. L. 1980. The role of the epidermis in auxin-induced and fusicoccin-induced elongation of *Pisum sativum* stem segments. *Planta* 150:371–79

13. Bunce, J.A. 1977. Leaf elongation in relation to leaf water potential in soybean. *J. Exp. Bot.* 28:156–61

14. Chabot, B. F., Jurik, T. W., Chabot, J. R. 1979. Influence of instantaneous and integrated light flux density on leaf anatomy and photosynthesis. *Am. J. Bot.* 66:940–45

15. Charles-Edwards, D. A. 1979. A model for leaf growth. *Ann. Bot.* 44:523–35

16. Charles-Edwards, D. A. 1983. Modelling leaf growth and function. See Ref. 36, pp. 489–97

17. Christ, R. A. 1978. The elongation rate of wheat leaves. I. Elongation rates during day and night. *Exp. Bot.* 28:156–61

18. Cleland, R. E. 1981. Wall extensibility; hormones and wall extension. In *Encyclopedia of Plant Physiology (NS), Plant Carbohydrates.* II. *Extracellular Carbohydrates,* ed. W. Tanner, F. A. Loewus, 13B:255–73. Berlin: Springer-Verlag

19. Cleland, R. E. 1984. The Instron technique as a measure of immediate-past wall extensibility. *Planta* 160:514–20

20. Clough, B. F., Milthorpe, F. L. 1975. Effect of water deficit on leaf development in tobacco. *Aust. J. Plant Physiol.* 2:291–300

21. Cosgrove, D. J. 1985. Cell wall yield properties of growing tissue. Evaluation by *in vitro* stress relaxation. *Plant Physiol.* 78:347–56

22. Cosgrove, D. J. 1986. Biophysical control of plant cell growth. *Ann. Rev. Plant Physiol.* 37:377–405

23. Cosgrove, D. J., Cleland, R. E. 1983. Solutes in the free space of growing stem tissues. *Plant Physiol.* 72:326–31

24. Cosgrove, D. J., Van Volkenburgh, E., Cleland, R. E. 1984. Stress relaxation of cell walls and the yield threshold for growth. Demonstration and measurement by micro-pressure probe and psychrometer techniques. *Planta* 162:46–54

25. Cowan, I. R., Farquhar, G. D. 1977. Stomatal function in relation to leaf metabolism and environment. In *Integration of Activity in the Higher Plant,* ed. D. H. Jennings. Soc. Exp. Biol. Symp. 31:471–505. Cambridge: Cambridge Univ. Press

26. Cusset, G. 1986. La morphogenese du limbe des Dicotyledones. *Can. J. Bot.* 64:2807–39

27. Cutler, J. M., Rains, D. W. 1977. Effects of irrigation history on responses of cotton to subsequent water stress. *Crop. Sci.* 17:329–35

28. Cutler, J. M., Shahan, K. W., Steponkus, P. L. 1980. Alteration of the internal water relations of rice in response to drought hardening. *Crop Sci.* 20:307–10

29. Cutler, J. M., Shahan, K. W., Steponkus, P. L. 1980. The influence of water deficits and osmotic adjustment on leaf elongation in rice. *Crop Sci.* 20:314–18

30. Dainty, J. 1976. Plant-cell water relations. In *Encyclopedia of Plant Physiology,* ed. U. Lüttge, M. G. Pitman, 2A:12–35. Berlin: Springer-Verlag

31. Dale, J. E. 1964. Leaf growth in *Phaseolus vulgaris* L. I. Growth of the first pair of leaves under controlled conditions. *Ann. Bot.* 28:579–89

32. Dale, J. E. 1965. Leaf growth in *Phaseolus vulgaris.* II. Temperature effects and the light factor. *Ann. Bot.* 29:293–308

33. Dale, J. E. 1970. Models of cell number increase in developing leaves. *Ann. Bot.* 34:267–73

34. Dale, J. E. 1976. Cell division in leaves. In *Cell Division in Higher Plants,* ed. M. M. Yeoman, pp. 315–45. London: Academic

35. Dale, J. E. 1986. Plastic responses of leaves. In *Plasticity in Plants,* ed. D. H. Jennings, A. J. Trewavas, Soc. Exp. Biol. Symp. 40:287–305. Cambridge: Company of Biologists

36. Dale, J. E., Milthorpe, F. L. 1983. *The Growth and Functioning of Leaves.* Cambridge: Cambridge Univ. Press

37. Dale, J. E., Milthorpe, F. L. 1983. General features of the production and growth of leaves. See Ref. 36, pp. 151–78

38. Dale, J. E., Murray, D. 1968. Photomorphogenesis, photosynthesis and ear-

ly growth of primary leaves of *Phaseolus vulgaris*. *Ann. Bot.* 32:767–80

39. Dale, J. E., Murray, D. 1969. Light and cell division in primary leaves of *Phaseolus Proc. R. Soc. London Ser. B* 173:541–55

40. Darvill, A. G., Smith, C. J., Hall, M. A. 1978. Cell wall structure and elongation growth in *Zea mays* coleoptile tissue. *New Phytol.* 80:503–16

41. Davidson, J. L., Milthorpe, F. L. 1966. Leaf growth in *Dactylis glomerata* following defoliation. *Ann. Bot.* 30: 173–84

42. De Chalain, T. M. B., Berjak, P. 1979. Cell death as a functional event in the development of the leaf intercellular spaces in *Avicennia marina* (Forskal) Vierh. *New Phytol.* 83:147–55

43. Dengler, N. G. 1980. Comparative histological basis of sun and shade leaf dimorphism in *Helianthus annuus*. *Can. J. Bot.* 58:717–30

44. Dengler, N. G., Mackay, L. B. 1975. The leaf anatomy of beech, *Fagus grandifolia*. *Can. J. Bot.* 53:2202–11

45. Dengler, N. G., Mackay, L. B., Gregory, L. M. 1975. Cell enlargement during leaf expansion in beech, *Fagus grandifolia*. *Can. J. Bot.* 53:2846–65

46. Denne, M. P. 1966. Leaf development in *Trifolium repens*. *Bot. Gaz.* 127:202–10

47. Downs, R. J. 1955. Photoreversibility of leaf and hypocotyl elongation of dark-grown red kidney bean seedlings. *Plant Physiol.* 30:468–73

48. Dulieu, H. 1968. Emploi des chimeres chlorophylliennes pour l'etude de l'ontogenie foliare. *Bull. Sci. Bourgogne* 25:1–60

49. Dulieu, H. 1974. Somatic variations on a yellow mutant in *Nicotiana tabacum* (al+/al, al2+/al2). 1. Non-reciprocal events occurring in leaf cells. *Mutation Res.* 25:283–304

50. Dulieu, H. 1975. Somatic variations on a yellow mutant in *Nicotiana tabacum* (al+/al, al2+/al2). 2. Reciprocal events occurring in leaf cells. *Mutation Res.* 28:69–77

51. Dulieu, H., Bugnon, F. 1966. Chimeres chlorophylliennes et ontogenie foliare chez le tabac (*Nicotiana tabacum* L.) *C. R. Acad. Sci. Ser. D* 263:1714–17

52. Erickson, R. O. 1966. Relative elemental rates and anisotropy of growth in area: a computer programme. *J. Exp. Bot.* 17:390–403

53. Erickson, R. O., Michelini, F. J. 1957. The plastochron index. *Am. J. Bot.* 44:297–305

54. Evans, G. C., Hughes, A. P. 1961. Plant growth and the aerial environment. I. Effect of artificial shading on *Impatiens parviflora* DC. *New Phytol.* 60:150–86

55. Farquhar, G.D., Sharkey, T. D. 1982. Stomatal conductance and photosynthesis. *Ann. Rev. Plant Physiol.* 33:317–45

56. Firn, R. D., Digby, J. 1977. The role of the peripheral cell layers in the geotropic curvature of sunflower hypocotyls; a new model of shoot geotropism. *Aust. J. Plant Physiol.* 4:337–47

57. Fitter, A. H., Ashmore, C. J. 1974. Responses of two *Veronica* species to a simulated woodland light climate. *New Phytol.* 73:997–1001

58. Fry, S. C. 1983. Oxidative phenolic coupling reactions cross-link hydroxyproline-rich glycoprotein molecules in plant cell walls. *Curr. Top. Plant Biochem. Physiol.* 2:59–72

59. Fuchs, C. 1972. Croissance de la feuille et acquisition de la forme chez le *Tropaeolum peregrinum* L. L'activite mitotique. *C. R. Acad. Sci. Ser. D* 274:3206–9

60. Fuchs, C., Broniatowski, M., Delarue, Y. 1985. Etude de la distribution spatiale des mitoses dans la feuille du *Tropaeolum peregrinum* L. *Rev. Cytol. Biol. Veg. Bot.* 8:255–64

61. Green, P. B. 1980. Organogenesis—a biophysical viewpoint. *Ann. Rev. Plant Physiol.* 31:51–82

62. Greenland, A. J., Lewis, D. H. 1981. The acid invertases of the developing third leaf of oat. I. Changes in activity of invertase and concentrations of ethanol-insoluble carbohydrates. *New Phytol.* 88:265–77

63. Grime, J. P. 1965. Shade tolerance in flowering plants. *Nature* 208:161–63

64. Haber, A. H., Foard, D. E. 1963. Nonessentially of concurrent cell divisions for degree of polarisation of leaf growth. II. Evidence from untreated plants and from chemically induced changes of the degree of polarisation. *Am. J. Bot.* 50:937–44

65. Hahlbrock, K., Grisebach, H. 1979. Enzymic controls in the biosynthesis of lignin and flavonoids. *Ann. Rev. Plant Physiol.* 30:105–30

66. Hake, S., Freeling, M. 1986. Analysis of genetic mosaics shows that the extra epidermal cell divisions in *knotted* mutant maize plants are induced by adjacent mesophyll cells. *Nature* 320:621–23

67. Halperin, W. 1978. Organogenesis at the shoot apex. *Ann. Rev. Plant Physiol.* 29:239–62

292 DALE

68. Hiroi, T., Monsi, M. 1963. Physiological and ecological analyses of shade tolerance of plants. 3. Effect of shading on growth attributes of *Helianthus annuus. Bot. Mag.* 76:121–29
69. Hsiao, T. C. 1973. Plant response to water stress. *Ann. Rev. Plant Physiol.* 24:519–70
70. Hsiao, T. C., Acevedo, E. 1974. Plant responses to water stress, water use efficiency and drought resistance. *Agric. Meteorol.* 14:59–84
71. Hsiao, T. C., Silk, W. K., Jing, J. 1985. Leaf growth and water deficits; biophysical effects. See Ref. 7, pp. 239–66
72. Humphries, E. C., Wheeler, A. W. 1963. The physiology of leaf growth. *Ann. Rev. Plant Physiol.* 14:385–410
73. Husken, D., Steudle, E., Zimmermann, U. 1978. Pressure probe technique for measuring water relations of cells in higher plants. *Plant Physiol.* 61:158–63
74. Jeffree, C. E., Dale, J. E., Fry, S. C. 1986. The genesis of intercellular spaces in developing leaves of *Phaseolus vulgaris* L. *Protoplasma* 132:90–98
75. Jeffree, C. E., Read, N. D., Smith, J. A. C., Dale, J. E. 1987. Water droplets and ice deposits in leaf intercellular spaces: redistribution of water during cryofixation for scanning electron microscopy. *Planta* 172:20–37
76. Jeune, B. 1984. Position et orientation des mitoses dans la zone organogene de jeunes feuilles de *Fraxinus excelsior, Glechoma hederacea* et *Lycopus europaeus. Can. J. Bot.* 62:2861–64
77. Jones, M. M., Turner, N. C. 1980. Osmotic adjustment in expanding and fully expanded leaves of sunflower in response to water deficits. *Aust. J. Plant Physiol.* 7:181–92
78. Jong, K., Burtt, B. L. 1975. The evolution of morphological novelty exemplified in the growth patterns of some Gesneriaceae. *New Phytol.* 75:297–311
79. Kasperbauer, M. J. 1971. Spectral distribution of light in a tobacco canopy and effects of end of day light quality on growth and development. *Plant Physiol.* 47:775–78
80. Kassam, A. H., Elston, J. F. 1974. Seasonal changes in the status of water and tissue characteristics of leaves of *Vicia faba* L. *Ann. Bot.* 38:419–29
81. Kaufman, P. B. 1959. Development of the shoot of *Oryza sativa* L. II. Leaf histogenesis. *Phytomorphology* 9:277–311
82. Kawase, M. 1979. Role of cellulase in aerenchyma development in sunflower. *Am. J. Bot.* 66:183–90
83. Kemp, D. R. 1980. The location and size of the extension zone of emerging wheat leaves. *New Phytol.* 84:729–37
84. Kemp, D. R. 1981. The growth rate of wheat leaves in relation to the extension zone sugar concentration manipulated by shading. *J. Exp. Bot.* 32:141–50
85. Kim, J. H., Lee-Stadelman, O. Y. 1984. Water relations and cell wall elastic quantities in *Phaseolus vulgaris* leaves. *J. Exp. Bot.* 35:841–58
86. Kolloffel, C., Linssen, P. W. T. 1984. The formation of intercellular spaces in the cotyledons of developing and germinating pea seeds. *Protoplasma* 120:12–19
87. Kuhlemeier, C., Green, P. J., Chua, N.-H. 1987. Regulation of gene expression in higher plants. *Ann. Rev. Plant Physiol.* 38:221–57
88. Kutschera, U., Bergfeld, R., Schopfer, P. 1987. Cooperation of epidermis and inner tissues in auxin-mediated growth of maize coleoptiles. *Planta* 170:168–80
89. Lainson, R. A., Thornley, J. H. M. 1982. A model for leaf expansion in cucumber. *Ann. Bot.* 50:407–25
90. Lamb, C. J., Lawton, M. A. 1983. Photocontrol of gene expression. In *Encyclopedia of Plant Physiology,* ed. W. Shropshire, H. Mohr, 16A:213–57. Berlin: Springer-Verlag
91. Lamoreaux, R. J., Chaney, W. R., Brown, K. M. 1978. The plastochron index; a review after two decades of use. *Am. J. Bot.* 65:586–93
92. Landre, P. 1976. Teneurs en DNA nucleaire de quelque types cellulaires de l'epiderme de la morelle noire (*Solanum nigrum* L.) au cours du developpement de la feuilles. Etude histologique et cytophotometrique. *Ann. Sci. Nat. Bot.* (Paris) 17:5–104
93. Lichtenthaler, H. K. 1985. Differences in morphology and chemical composition of leaves grown at different light intensities and qualities. See Ref. 7, pp. 201–21
94. Liverman, J. L., Johnson, M. P., Starr, L. 1955. Reversible photoreaction controlling expansion of etiolated bean-leaf disks. *Science* 121:440–41
95. Loach, K. 1970. Shade tolerance in tree seedlings. II. Growth analysis of plants raised under artificial shade. *New Phytol.* 69:273–86
96. Lockhart, J. A. 1985. An analysis of irreversible plant cell elongation. *J. Theor. Biol.* 8:264–75
97. Longstreth, D. J., Bolanos, J. A., Goddard, R. D. 1985. Photosynthetic rate and mesophyll surface area in expanding leaves of *Alternanthera philoxeroides* grown at two light levels. *Am. J. Bot.* 72:14–19

98. Lyndon, R. F. 1983. The mechanism of leaf initiation. See Ref. 36, pp. 3–24

99. MacDougal, D. T. 1903. The influence of light and darkness upon growth and development. *Mem. NY Bot. Gard.* 2:1–319

100. Maier-Maercker, U. 1979. "Peristomatal transpiration" and stomatal movement: a controversial view. III. Visible effects of peristomatal transpiration on the epidermis. *Z. Pflanzenphysiol.* 91:225–38

101. Maksymowych, R. 1962. An analysis of leaf elongation in *Xanthium pennsylvanicum* presented as relative elemental rates. *Am. J. Bot.* 49:7–13

102. Maksymowych, R. 1963. Cell division and cell elongation in leaf development of *Xanthium pennsylvanicum. Am. J. Bot.* 50:891–901

103. Maksymowych, R. 1973. *Analysis of Leaf Development.* Cambridge: Cambridge Univ. Press

104. Maksymowych, R., Erickson, R. O. 1960. Development of the lamina of *Xanthium italicum* represented by the plastochron index. *Am. J. Bot.* 47:451–59

105. Martens, P. 1937. L'origine des espaces intercellulaires. *La Cellule* 46:355–88

106. Matsuda, K., Riazi, A. 1981. Stress-induced osmotic adjustment in growing regions of barley leaves. *Plant Physiol.* 68:571–76

107. Matthews, M. A., Van Volkenburgh, E., Boyer, J. S. 1984. Acclimation of leaf growth to low water potentials in sunflower. *Plant Cell Environ.* 7:199–206

108. Maximov, N. A. 1929. *The Plant in Relation to Water.* London: Unwin Bros. Ltd.

109. Maxwell, J. O., Redman, R. E. 1978. Water potential and component potentials in expanded and unexpanded leaves of two xeric grasses. *Physiol. Plant.* 44:383–87

110. McLaren, J. S., Smith, H. 1978. Phytochrome control of the growth and development of *Rumex obtusifolius* under simulated canopy light environments. *Plant Cell Environ.* 1:61–67

111. Mentze, J., Raymond, B., Cohen, J. D., Rayle, D. L. 1977. Auxin-induced H$^+$-secretion in *Helianthus* and its implications. *Plant Physiol.* 60:509–12

112. Michelena, V. A., Boyer, J. S. 1982. Complete turgor maintenance at low water potentials in the elongation region of maize leaves. *Plant Physiol.* 69:1145–49

113. Milford, G. F. J., Lenton, J. R. 1976. Effect of photoperiod on growth of sugarbeet. *Ann. Bot.* 40:1309–15

114. Milthorpe, F. L., Newton, P. 1963. Studies on the expansion of the leaf surface. III. The influence of radiation on cell division and leaf expansion. *J. Exp. Bot.* 14:483–95

115. Morgan, D. C. 1981. Shadelight quality effects on plants growth. In *Plants and the Daylight Spectrum,* ed. H. Smith, pp. 205–21. London: Academic

116. Morgan, J. M. 1984. Osmoregulation and water stress in higher plants. *Ann. Rev. Plant Physiol.* 35:299–319

117. Morgan, D. C., O'Brien, T., Smith, H. 1980. Rapid photomodulation of stem extension in *Sinapis alba* L. Studies on kinetics, site of perception and photoreceptor. *Planta* 150:95–101

118. Morgan, D. C., Smith, H. 1978. The relationship between phytochrome photoequilibrium and development in light grown *Chenopodium album. Planta* 142:187–93

119. Morgan, D. C., Smith, H. 1979. A systematic relationship between phytochrome-controlled development and species habitat for plants grown in simulated natural radiation. *Planta* 145:253–58

120. Morgan, D. C., Smith, H. 1981. Control of development in *Chenopodium album* L. by shadelight: the effect of light quantity (total fluence rate) and light quality (red:far-red ratio). *New Phytol.* 88:239–48

121. Morris, D. A., Arthur, E. D. 1984. An association between acid invertase activity and cell growth during leaf expansion. *J. Exp. Bot.* 35:1369–79

122. Munns, R., Weir, R. 1981. Contribution of sugars to osmotic adjustment in elongating and expanded zones of wheat leaves during moderate water deficits at two light levels. *Aust. J. Plant Physiol.* 8:93–105

123. Nachtwey, D. S., Cameron, I. L. 1968. Cell cycle analysis. In *Methods in Cell Physiology,* ed. D. M. Prescott, 3:213–59. New York:Academic

124. Newton, P. 1963. Studies on the expansion of the leaf surface. II. The influence of light intensity and photoperiod. *J. Exp. Bot.* 13:458–82

125. Nobel, P. S. 1976. Photosynthesis of sun versus shade leaves of *Hyptis emoryi* Torr. *Plant Physiol.* 58:218–33

126. Nobel, P. S., Zaragosa, L. J., Smith, W. K. 1975. Relation between mesophyll surface area, photosynthetic rate, and illumination level during development for leaves of *Plectranthus parviflorus. Plant Physiol.* 55:1067–70

127. Parrish, D. J., Wolf, D. D. 1983. Kinetics of tall fescue leaf elongation; re-

sponse to changes in illumination and vapour pressure. *Crop Sci.* 23:659–63

128. Penel, C., Greppin, H., Boisard, J. 1976. In vitro photomodulation of a peroxidase activity through membrane-bound phytochrome. *Plant Sci.* 6:117–21

129. Penny, D., Miller, K. F., Penny, P. 1972. Studies on the mechanism of cell elongation of lupin hypocotyl segments. *NZ J. Bot.* 10:97–111

130. Pieters, G. A. 1974. The growth of sun and shade leaves of *Populus euramericana* 'Robusta' in relation to age, light intensity and temperature. *Meded. Landbouwhogesch. Wageningen* 74:1–106

131. Poethig, R. S. 1987. Clonal analysis of cell lineage patterns in plant development. *Am. J. Bot.* 74:581–94

132. Poethig, R. S., Sussex, I. M. 1985. The developmental morphology and growth dynamics of the tobacco leaf. *Planta* 165:158–69

133. Poethig, R. S., Sussex, I. M. 1985. The cellular parameters of leaf development in tobacco; a clonal analysis. *Planta* 165:170–84

134. Pollock, C. J., Lloyd, E. J. 1977. The distribution of acid invertase in developing leaves of *Lolium temulentum*. *Planta* 113:197–200

135. Powell, R. D., Griffiths, M. M. 1960. Some anatomical effects of red light and kinetin on disks of bean leaves. *Plant Physiol.* 35:273–75

136. Radin, J. W. 1983. Control of plant growth by nitrogen; differences between cereals and broadleaved species. *Plant Cell Environ.* 6:65–68

137. Radin, J. W., Eidenbock, M. P. 1984. Hydraulic conductance as a factor limiting leaf expansion of phosphorus deficient cotton plants. *Plant Physiol.* 75:372–77

138. Raschke, K. 1970. Leaf hydraulic system; rapid epidermal and stomatal responses to changes in water supply. *Science* 167:189–91

139. Raschke, K., Zeevart, J. A. D. 1976. Abscisic acid content, transpiration and stomatal conductance as related to leaf age in plants of *Xanthium strumarium*. *Plant Physiol.* 58:169–74

140. Ray, P. M. 1962. Cell wall synthesis and cell elongation in oat coleoptile tissues. *Am. J. Bot.* 49:928–39

141. Ray, P. M., Green, P. B., Cleland, R. E. 1972. Role of turgor in plant cell growth. *Nature* 239:163–64

142. Richards, O. W., Kavanagh, A. J. 1943. The analysis of the relative growth gradients and changing form of growing

organisms; illustrated by the tobacco leaf. *Am. Nat.* 77:385–99

143. Roland, J. C. 1978. Cell wall differentiation and stages involved with intercellular gas space opening. *J. Cell Sci.* 32:325–36

144. Sachs, J. 1887. *Lectures on the Physiology of Plants.* Transl. M. Ward. Oxford: Clarendon

145. Sachs, T. 1978. The development of spacing patterns in leaf epidermis. In *The Clonal Basis of Development*, ed. S. S. Subtelny, I. M. Sussex, pp. 161–83. New York: Academic

146. Saurer, W., Possingham, J. V. 1970. Studies on the growth of spinach leaves (*Spinacia oleracea*). *J. Exp. Bot.* 21:151–58

147. Schopfer, P. 1977. Phytochrome control of enzymes. *Ann. Rev. Plant Physiol.* 28:223–52

148. Sellen, D. C. 1980. The mechanical properties of plant cell walls. In *The Mechanical Properties of Biological Materials*, ed. J. F. Vincent, J. D. Currey. Soc. Exp. Biol. Symp., 345: 315–29. Cambridge: Cambridge Univ. Press

149. Sharma, R., Sopory, S. K., Guha-Mukherjee, S. 1976. Phytochrome regulation of peroxidase activity in maize. *Plant Sci. Lett.* 6:69–75

150. Sharman, B. C. 1945. Leaf and bud initiation in the Gramineae. *Bot. Gaz.* 106:269–89

151. Silk, W. K. 1983. Kinematic analysis of leaf expansion. See Ref. 36, pp. 89–108

152. Silk, W. K., Erickson, R. O. 1979. Kinematics of plant growth. *J. Theor. Biol.* 76:481–501

153. Smith, H. 1982. Light quality, photoreception and plant strategy. *Ann. Rev. Plant Physiol.* 33:481–518

154. Stoutjesdijk, P. 1972. Spectral transmission curves of some types of leaf canopy with a note on seed germination. *Acta Bot. Neerl.* 21:185–91

155. Sunderland, N. 1960. Cell division and expansion in the growth of the leaf. *J. Exp. Bot.* 7:126–45

156. Sunderland, N., Brown, R. 1956. Distribution of growth in the apical region of the shoots of *Lupinus albus*. *J. Exp. Bot.* 7:126–45

157. Tal, M., Imber, D., Erez, A., Epstein, E. 1979. Abnormal stomatal behavior and hormonal imbalance in *flacca*, a wilty mutant of tomato. *Plant Physiol.* 63:1044–48

158. Tanimoto, E., Masuda, Y. 1971. Role of the epidermis in auxin-induced elongation of light-grown pea stem segments. *Plant Cell Physiol.* 12:663–73

159. Taylor, G., Davies, W. J. 1985. The control of leaf growth of *Betula* and *Acer* by photoenvironment. *New Phytol.* 101:259–68

160. Taylor, G., Davies, W. J. 1986. Yield turgor of growing leaves of *Betula* and *Acer*. *New Phytol.* 104:347–53

161. Taylor, G., Davies, W. J. 1986. Leaf growth of *Betula* and *Acer* in simulated shadelight. *Oecologia* 69:589–93

162. Terry, N., Waldron, L. J., Ulrich, A. 1971. Effects of moisture stress on the multiplication and expansion of cells in leaves of sugar beet. *Planta* 97:281–89

163. Thomasson, M. 1970. Quelques observations sur la repetition des zones de croissance de le feuille du *Jasminum nudiflorum* Lindley. *Candollea* 25:297–340

164. Thornley, J. H. M., Hurd, R. G., Pooley, A. 1981. A model of growth of the fifth leaf of tomato. *Ann. Bot.* 48:327–40

165. Tobin, E. M., Silverthorne, J. 1985. Light regulation of gene expression in higher plants. *Ann. Rev. Plant Physiol.* 36:569–93

166. Tomos, A. D. 1985. The physical limitations of leaf cell expansion. See Ref. 7, pp. 1–33

167. Turner, N. C., Jones, M. M. 1980. Turgor maintenance by osmotic adjustment; a review and evaluation. In *Adaptation of Plants to Water and High Temperature Stress,* ed. N. C. Turner, P. J. Kramer, pp. 87–104. New York: Wiley

168. Turrill, F. M. 1936. The area of the internal exposed surface of dicotyledon leaves. *Am. J. Bot.* 23:255–64

169. Tyree, M. T., Jarvis, P. G. 1982. Water in tissues and cells. In *Encyclopedia of Plant Physiology (NS), Physiological Plant Ecology II,* ed. O. L. Lang, P. S. Nobel, C. B. Osmond, H. Ziegler, 12B:36–77. Berlin: Springer-Verlag

170. Unser, G., Masoner, M. 1972. Kinetics of monogalactosyl-transferase in mustard seedlings. *Naturwissenschaften* 59:39

171. Van Volkenburgh, E., Boyer, J. S. 1985. Inhibitory effects of water deficit on maize leaf elongation. *Plant Physiol.* 77:190–94

172. Van Volkenburgh, E., Cleland, R. E. 1979. Separation of cell enlargement and division in bean leaves. *Planta* 146:245–47

173. Van Volkenburgh, E., Cleland, R. E. 1980. Proton excretion and cell expansion in bean leaves. *Planta* 148:273–78

174. Van Volkenburgh, E., Cleland, R. E. 1981. Control of light induced bean leaf expansion; role of osmotic potential, wall yield stress and hydraulic conductivity. *Planta* 153:572–77

175. Van Volkenburgh, E., Cleland, R. E. 1986. Wall yield threshold and effective turgor in growing bean leaves. *Planta* 167:37–43

176. Van Volkenburgh, E., Davies, W. J. 1983. Inhibition of light-stimulated leaf expansion by abscisic acid. *J. Exp. Bot.* 34:835–45

177. Van Volkenburgh, E., Schmidt, M. G., Cleland, R. E. 1985. Loss of capacity for acid-induced wall-loosening as the principle cause of the cessation of cell enlargement in light grown bean leaves. *Planta* 163:500–5

178. Verbelen, J.-P., De Greef, J. A. 1979. Leaf development of *Phaseolus vulgaris* L. in light and darkness. *Am. J.Bot.* 66:970–76

179. Veres, J. S., Williams, G. J. III. 1985. Leaf cavity size differentiation and water relations in *Carex eleocharis. Am. J. Bot.* 72:1074–77

180. Waldron, L. J., Terry, N. 1987. The influence of atmospheric humidity on leaf expansion in *Beta vulgaris* L. *Planta* 170:336–42

181. Waldron, L. J., Terry, N., Nemson, J. A. 1985. Diurnal cycles of leaf expansion in unsalinised and salinised *Beta vulgaris. Plant Cell Environ.* 8:207–11

182. Whitelam, G. C., Johnson, C. B. 1982. Photomorphogenesis in *Impatiens parviflora* and other plant species under simulated natural canopy radiations. *New Phytol.* 90:611–18

183. Williams, R. F. 1975. *The Shoot Apex and Leaf Growth.* Cambridge; Cambridge Univ. Press

184. Wilson, G. L. 1966. Studies on the expansion of the leaf surface. V. Cell division and expansion in a developing leaf as influenced by light and upper leaves. *J. Exp. Bot.* 17:440–51

185. Wilson, L. G., Fry, J. C. 1986. Extensin—a major glycoprotein. *Plant Cell Environ.* 9:239–60

186. Young, J. P., Horton, R. F. 1985. Heterophylly in *Ranunculus flabellaris* Raf.; the effect of abscisic acid. *Ann. Bot.* 55:899–902

187. Zeevaart, J. A. D. 1977. Sites of ABA synthesis and metabolism in *Ricinus communis* L. *Plant Physiol.* 59:788–91

188. Zimmermann, U., Steudle, E. 1975. The hydraulic conductivity and volumetric elastic modulus of cells and isolated cell walls of *Nitella* and *Chara* spp.; pressure and volume effects. *Aust. J. Plant Physiol.* 2:1–13

Ann. Rev. Plant Physiol. Plant Mol. Biol. 1988. 39:297–319

GENETIC ANALYSIS OF LEGUME NODULE INITIATION

Barry G. Rolfe

Plant Molecular Biology Group, Research School of Biological Sciences, Australian National University, Canberra, A.C.T. 2600 Australia

Peter M. Gresshoff*

Botany Department, Faculty of Science, Australian National University, Canberra, A.C.T. 2600 Australia

CONTENTS

INTRODUCTION 297
RHIZOBIUM INFECTION AND LEGUME NODULE ONTOGENY 298
REGULATION OF THE *RHIZOBIUM* NODULATION GENES 301
GENETIC ORGANIZATION OF *RHIZOBIUM* GENES AFFECTING
    NODULATION 304
COMPATIBILITY OF INFECTING RHIZOBIA 306
HOST FACTORS CONTROLLING NODULATION 307
ISOLATION OF PLANT NODULATION MUTANTS 310
INDUCIBLE PLANT DEFENSE SYSTEMS 311
EARLY NODULIN GENE EXPRESSION 313
CONCLUSIONS 313

INTRODUCTION

Through a symbiotic association with a soil bacterium [*Rhizobium* or *Bradyrhizobium* (78)] leguminous plants have the ability to utilize (fix) atmo-

*New address: Plant Molecular Genetics (OHLD), College of Agriculture, University of Tennessee, Knoxville, TN 37901

0066-4294/88/0601-0297$02.00

spheric dinitrogen gas. Rhizobia are able to induce the formation of morphologically defined structures called nodules on legume roots. The rhizobia
within these nodules reduce atmospheric nitrogen to ammonia.

This review examines the bacterial and plant roles in the initiation of
nodulation. We aim to provide not an extensive literature review but an
assessment of some of the biological aspects of the nodulation process and
possible future developments. There are many recent reviews and articles on
different aspects of *Rhizobium* infection of legumes. These cover the infection
process (4, 5, 7, 128), the signals involved (41, 56, 74, 91, 98, 103), the
process as a "controlled disease" (43, 120, 129, 130), the genetics of *Rhizo-*
bium species (33, 40, 42, 47, 52, 57, 71, 82, 107, 113, 115), and plant host
genetics (64, 66). The last review on the infection process by rhizobia in the
Annual Review of Plant Physiology appeared in 1981 (4).

RHIZOBIUM INFECTION AND LEGUME NODULE ONTOGENY

Rhizobia occupy the whole legume root surface and can attach to epidermal
cells, particularly the root hairs. Most of the bacteria that do so do not initiate
infections of the plant cells. Some rhizobia specifically interact with newly
emerging root hairs, however, and initiate a pronounced curling of these
growing hair cells. To initiate this interaction, rhizobia have evolved to use
flavonoid compounds. These compounds are released by the plant into the
root rhizosphere and act as positive regulation of the infection (nodulation,
nod) genes. Following the initiation of an infection process, rhizobia entrap-
ped within curled root-hair cells begin the invasion of these plant cells.
Invasion occurs via the induction of an infection thread that penetrates into the
plant tissue and continues to grow and ramify in the root cortex (106).

The infection thread eventually invades a focus of dividing plant cells and
rhizobia are released into these cells following "packaging" within a plant
membrane (106). The bacteria continue to grow and ultimately differentiate
into bacteroids capable of fixing nitrogen (107). A specific signal from the
plant coupled with the appropriate physiological environment is thought to
stimulate the derepression of the nitrogenase and other bacterial genes involved in nitrogen fixation. In addition, the plant is able to regulate further the
growth of the microsymbiont by affecting oxygen and nutrient availability
(66). Thus the plant is actively involved in restricting and preventing a
parasitic invasion by *Rhizobium*.

Legumes form a variety of nodule types. Figure 1 summarizes the variation
in infection and nodulation strategies that have been found (6, 13, 20, 21, 48,
62, 92, 96, 101, 126, 127, 128). Entry of the bacterium occurs via a variety of
routes (through root hairs, into cracks or middle lamellae) and relies in all

cases on plant cell divisions. These are either induced by invading bacteria, as seen in soybean or clovers, or are preexisting cells that have not undergone the complete differentiation to a highly vacuolated cortical cell. It is possible that the bacterium interferes with an existing positional gradient controlling the cell cycle, so that cell division is induced in otherwise latent tissue regions (for literature on positional information in plants, 132). Collins (24, 63) showed that there exist in uninoculated white clover plants preinfection foci near the xylem poles within the root cortex. Further, these sites are apparently the sites of cellular division after bacterial infections. In soybean, cortical cell division is initiated by a diffusible factor and does not require contact between plant and bacterium (5). Thus pseudoinfections are induced and, as demonstrated through the use of plant mutants (see below), these cell foci allow further infection to occur on the developing root hairs just above them.

A large degree of variation in the final nodule structure occurs with different host species. Some nodules maintain a root meristem (i.e. they are indeterminate), others are determinate, or have a central vascular system, reminiscent of the lateral root structure. Bacterial release from the infection thread is not a prerequisite for the development of a nitrogen-fixing bacteriod. In the *Parasponia* symbiosis the microsymbiont is retained in modified infection threads, called fixation threads (101). Recently, similar structures were observed in *Andira* nodules induced by *Rhizobium* (34, 35).

Temperate legumes (such as *Pisum sativum, Medicago sativa,* and *Trifolium* species) develop indeterminate nodules characterized by (*a*) a defined meristem during nodule growth, (*b*) an open vascular system connecting the root's vascular system with the nodule meristem, (*c*) vacuolated infected cells (92), (*d*) asparagine/glutamine as the major translocation products of nitrogen fixation (62, 66, 107), and (*e*) enlarged, single bacteroids. The nodules are cylindrical in shape and tend to be infected with rhizobia belonging to the *Rhizobium* genus rather than the slow-growing *Bradyrhizobium* group (78). These nodules arise from cortical tissue close to the endodermis just near xylem poles in the root (99) as well as concomitant pericycle divisions (24).

Tropical legumes like soybean, peanut, *Vigna* species, and lupins form spherical, determinate nodules, which arise from the outer cortical cell regions just below the epidermis (128). Meristematic activity is not persistent and is limited to an early stage of nodule growth, to be followed by a period of extension growth. Other legumes can form more primitive nodules similar to those seen on *Parasponia* and *Frankia,* which are nodulated nonlegumes (Figure 1). Nodules on *Andira* species (legumes) (34, 35) resemble those of *Parasponia* (a nonlegume), being coralloid, branched, and meristematic with a central vascular cylinder.

Some invasions utilize the emergence of a lateral root, such as the stem nodulation of *Sesbania* (48) and the nodulation of peanut, *Parasponia,* and

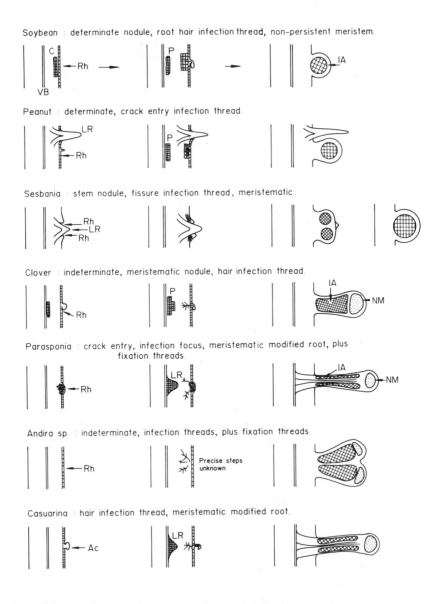

Figure 1   Schematic representations of different nodulation ontogenies: VB, vascular bundle; Rh, *Rhizobium* or *Bradyrhizobium* inoculant; Ac, actinomycete (i.e. *Frankia*) inoculant; C, cortex; P, pericycle-associated cell divisions; IA, infected area and region of nitrogen fixation; LR, lateral root (emerging or dormant) with associated root meristem; NM, nodule meristem.

Casuarina (126). The continued meristematic activity always involves a secondary effect on the pericycle tissue proximal to the initial cell-division center. This is the tissue source *(Anlage)* for lateral root formation, and primitive nodules possess this lateral root ontogeny, while more advanced nodule types rely on de novo induced cell-division centers in the cortex. A more detailed and perhaps speculative description of the nodulation ontogeny as we now believe it to occur in *Glycine max* is shown in Figure 2 (based in part on 13, 87).

REGULATION OF THE *RHIZOBIUM* NODULATION GENES

The model in Figure 3 summarizes the various interactions and steps during the infection of legume root cells by *Rhizobium*. Variòus plant-derived compounds have been identified in root exudates that can either activate or antagonize *nod* gene expression (41, 56, 74, 91, 98, 103). Stimulatory compounds now have been isolated from clovers, 7,4'-dihydroxyflavone (DHF) (103), and from alfalfa, luteolin (98). For peas, apigenin-7-O-glucoside and eriodictyl have been suggested as the chief stimulatory compounds (56, 124, 133). These stimulatory compounds are active at low concentrations and are derived from the phenylpropanoid biosynthetic pathway (Figure 4) (46, 50, 90). For the tropical legume soybean, the isoflavones (daidzein, genistein, and coumestrol) have been isolated and shown to stimulate the expression of *Rhizobium nod : lacZ* gene fusions (R. Kosslak, E. Appelbaum; G. Stacey; H. Hennecke; B. Bassam, personal communications). These stimulatory hydroxyflavones and isoflavones, which are released predominantly from the zone of emerging root hairs, have been shown to induce bacterial *nod* genes rapidly (41, 56, 98, 103). Recent studies also have identified compounds (coumarins and isoflavones) secreted by legumes that antagonize the induction of *nod* genes by stimulatory compounds (41). While the stimulator is chiefly released behind the growing root tip in the zone of emerging root hairs, the release of the inhibitor occurs in the region between this zone and the root tip and elsewhere on the root (41). The target for the action of the antagonizing and stimulatory compounds is the nodD product. In vitro experiments have shown that the activation of *nod* genes is determined by the ratio of stimulator (S) to inhibitor (i.e. antagonist) (I) (41). As this S : I ratio varies over the developing root, it may be the chief determinant of the sites of nodule initiation. Fluctuations in the concentrations of these compounds in the plant could affect *nod* gene expression in the infection thread. An excess of antagonizing compounds may cause the infection thread to cease growing because the bacteria are not able to stimulate the plant to continue either the synthesis of the infection-thread cell wall or cortical cell division.

SOYBEAN NODULE INITIATION

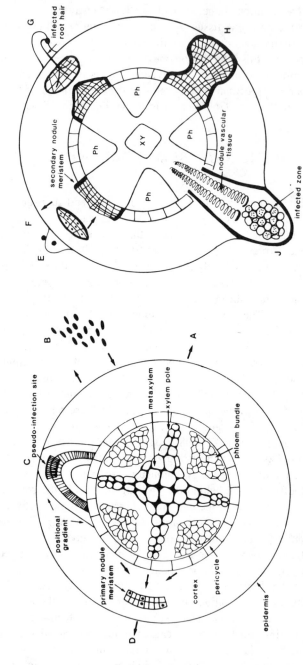

Figure 2 Ontogeny of nodule initiation in soybean: sequence of events is indicated by letters A to J. Plant roots excrete substances (stage A), which interact with bacteria (stage B), which produce subepidermal cell division stimulating factors. These interact with specific hypodermal cells near the xylem poles, suggesting a possible positional gradient emanating from the xylem and perhaps the phloem (stage C). The hypodermal division focus forms the primary nodule meristem, which potentiates the developing root hair just above it to become a target site for bacterial infection (stage D). Bacteria attach and invade the root hair (stage E) while the primary meristem induces pericycle divisions, again near the xylem pole (stage F). Infection threads are clearly visible as the root hair grows. The two meristematic foci grow together (stage G), giving rise to a fused cluster of dividing and invaded cell types (stage H). The infection thread ramifies, and bacteria increase rapidly in number within the cortex. Subsequent differentiation of the nodule yields vascular connections and the variety of cell types needed for nodule function (stage J).

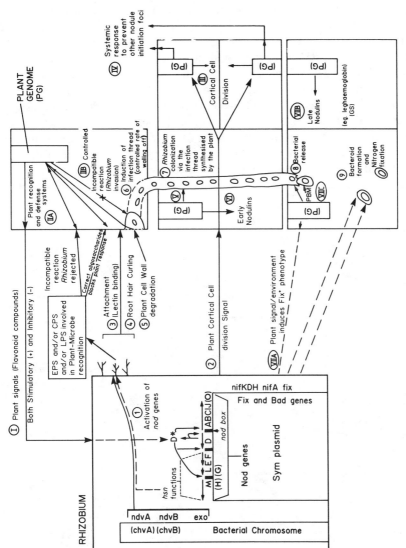

Figure 3 Diagrammatic representation of the interaction steps during the formation of a nitrogen-fixing nodule. Numbered circles indicate bacterial gene reactions, and circled roman numerals show plant responses. *hsn* genes *nodFELM* for *R. trifolii* and *nodFEGH* for *R. meliloti*. Reproduced with permission from the *Annual Review of Phytopathology* 1987, 25:145–68.

GENETIC ORGANIZATION OF *RHIZOBIUM* GENES AFFECTING NODULATION

The genetic determinants for *Rhizobium* invasion of a plant host are located on the bacterial chromosome and indigenous plasmids, particularly the symbiotic (Sym) plasmid (Figure 3). A set of four contiguous and highly conserved genes designated *nodDABC* have been identified (52, 113, 115). Generally, mutations in these genes resulted in strains that were unable to curl root hairs (Hac⁻ phenotype) of their respective hosts (42, 52, 82, 115). Furthermore, interspecies complementation studies have shown that the *nodDABC* genes are functionally interchangeable between most species (42, 82).

The *nodA* gene of *R. meliloti* encodes a 21.8-kDa protein and the *nodC* gene a protein of 44 kDa (77, 113). The nodA protein is located in the cytosol, and the *nodC* protein is an integral membrane protein in *R. meliloti*

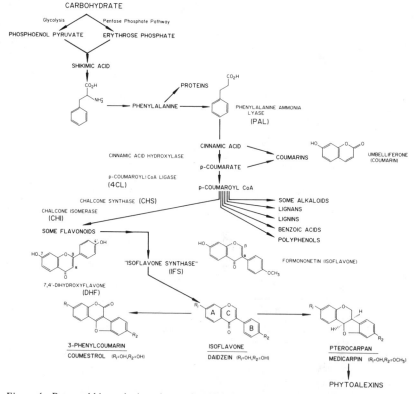

Figure 4   Proposed biosynthetic pathways for shikimic acid, phenolic compounds, flavonoids, and isoflavonoid-derived phytoalexins. Modified after Dickinson & Lucas (39).

cells. In addition, nodA and nodC proteins are detected in mature bacteroids (77).

The main regulatory gene, the *nodD* gene, in many rhizobia is linked to the *nodABC* genes but is transcribed divergently (51). The *nodD* gene product is probably a type of "environmental sieve," sensing the concentrations of stimulatory and antagonizing compounds released by a particular legume (41, 70, 108, 124). Some species of *Rhizobium* contain multiple copies of the *nodD* gene. This suggests that specific nodD product-flavone interactions may occur (41, 43, 70, 74, 108, 124). The expression of the *R. trifolii nodD* gene has been shown to be constitutive (41). The current working hypothesis (Figure 3) is that the nodD product is regulatory and requires the presence of the plant stimulatory compound to convert to an active form that promotes the expression of the *nodABC* and *nodFE* operons and probably the other inducible *nod* genes (*nodH* or *nodM*).

Two genes, *nodI* and *nodJ,* have been identified downstream of the *nodABC* genes and are probably part of the *nodABC* operon, which is regulated through the *nodA* promoter (53, 125, 133). These two genes are involved but are not—at least, in some rhizobia—obligatory for nodulation (42, 53). Mutations in the *nodIJ* genes of *R. trifolii* result in defective infection thread formation in clover plants (72).

A third group of legume nodulation genes in *Rhizobium* species are the host–specific nodulation *(hsn)* genes. In *R. meliloti* these are represented by the *nodFEGH* genes (71), in *R. trifolii* and *R. leguminosarum* by the *nod-FELM* cluster of genes (40, 47, 114). The *nodF* may be an acyl-carrier protein involved in fatty acid biosynthesis (121). An investigation of *Tn5* induced mutants and various subclones of these *hsn* genes in *R. trifolii* has shown that (*a*) the *nodFE* genes are intimately involved in conferring host range, (*b*) white clover plants can prevent the *R. trifolii nodFE* mutants from nodulating, (*c*) pea plants can prevent the wild-type strain (but not the *nodFE* mutants) from nodulating, (*d*) the *nodL* gene is involved in root-hair curling, and (*e*) *nodO* (J. M. Watson, personal communication) is involved in host-range nodulation of Afghanistan peas *(P. sativum)* (C. A. Wijffelman, unpublished).

Finally, a reiterated DNA sequence of about 26–28 bases called the "*nod*-box" sequences form at least part of the promoters of several flavonoid-induced *nod* gene operons (*nodABC, nodFE, nodH,* and *nodM*) (109, 115, 125, 133). The nodD product is thought to interact with the *nod*-box sequences to initiate *nod* gene transcription (109, 115).

At present, with the exception of the *nodD* gene there are no known functions for the different *nod* genes of *Rhizobium* strains. Perhaps they represent a hierarchy of regulatory genes starting with the *nodD* gene, which is constitutively expressed (41, 57, 74, 91, 108). Once flavonoid plant signals have bound to the nodD product, other *nod* genes are induced and they then regulate other non-Sym plasmid located genes. Such functions that may be induced are the pectinases and lytic enzymes required for degrading plant cell walls (12, 105, 128); the possible control of enzymes that "decorate" the *Rhizobium* surface polysaccharides with noncarbohydrate substitutions (31, 32, 38, 67); the synthesis of bacterial signals causing cortical cell division in the legume roots (5); the production of factors that cause changes in plant root growth [thick and short roots (Tsr)] phenotype (11, 15); and the possible control of ethylene synthesis. Ethylene production may be induced during the degradation of root-hair cell walls (plant wound response), and may cause the activation of the phenylpropanoid pathway (Figure 4) (10, 112).

COMPATIBILITY OF INFECTING RHIZOBIA

Recent studies with *Rhizobium* mutants defective in polysaccharide synthesis have generally supported the idea that these polymers play an important role in legume root infection (8, 22, 54, 86). A number of genes that affect polysaccharide synthesis have been located in either of several replicons (Figure 3) (8, 9, 22, 49, 54, 55, 73, 102). Many genes are involved in the synthesis of these complex polysaccharide polymers, and mutants defective in their formation are often characterized by poor infectivity and nodule formation (22, 54, 73, 86). The plant may respond unfavorably to these mutants by developing a nodule-like structure but with little evidence of root-hair curling and infection-thread formation having occurred (54, 86). Microscopic analysis of these aberrant nodules often demonstrates the presence of osmiophilic inclusions in the nodule tissue (22). These inclusions could indicate plant rejection of invading microbes (equivalent to an induced-incompatible reaction). They also suggest that many of the Exo⁻ mutants (defective in acidic exopolysaccharide synthesis) are recognized early and blocked during infection (22). In a number of bacterial plant pathogens the exopolysaccharides (EPSs) are critical factors for compatibility and are thought to function by masking the initial recognition of these bacteria from agglutinating plant lectins (3, 117, 118). The precise role of the various surface polysaccharides [such as capsular polysaccharides (CPS), EPS, lipopolysaccharides (LPS), and cyclic β-1,2-glucans] produced by *Rhizobium* species is still unknown (14, 31, 44, 68, 86, 102).

Recently it has been shown that the exogenous addition of the EPS or specific oligosaccharide repeat-unit molecules isolated from the parent strain to plants together with some Exo⁻ mutants could result in the restoration of the ability to induce nitrogen-fixing nodules (45). This finding provides evidence for a direct role of EPS in the nodulation process but still requires confirmation in other systems before a more specific functional role in infection may be proposed for EPS polymers. However, the addition of cyclic β-1,2, glucans, LPS, or CPS has increased infection-thread formation and invasion in clover root hairs (1, 2, 68). The plant response of infection-thread formation can be likened to a "controlled-incompatible" host-pathogen reaction, with the plant cell being stimulated to "wall off" the invading microbe (Figure 3). This implies that some of the *Rhizobium* polysaccharides may act as elicitors of plant function.

HOST FACTORS CONTROLLING NODULATION

The number of nodules forming on a root after *Rhizobium* infection is influenced by a plant process called autoregulation (94, 100). This phenomenon has been investigated in clovers (*Trifolium* species), soybeans *(Glycine max)*, and siratro *(Macroptilium atropurpureum)* (93, 94, 95, 100, 105, 111). Infections are predominantly possible in these legumes in a distinct region of the root just behind the root tip. This region is characterized by the presence of newly emerging root hairs, which are susceptible to curling and invasion via an infection thread. If two genetically marked strains were inoculated 15–30 hr apart, and if the position of the root tip at each inoculation point was noted, the first inoculated strain prevented the nodulation of the second strain in the region of the root tip at time point 2 (97, 100, 111).

These findings generate two concepts: first, of a moving "window of infectivity;" and second, restriction of the successful transition of infections into nodules. Infection on lateral roots was similar to the tap root (5, 7, 100). Microscopic studies by Calvert et al (13) and Mathews (87) showed that "pseudoinfections" (i.e. cortical cell divisions just below the epidermis) were more numerous than actual infections. The induction did not require direct contact, as experimental separation of plant and bacterium by a millipore membrane resulted in the same phenomenon (5). Similarly, pseudoinfections could be induced by an exogenous application of cytokinins such as benzyladenine (5). Mathews (87) found that 80 mm long sections of the tap root of soybean cultivar Bragg developed approximately 200 pseudoinfections and about 75 actual infections—i.e. cortical cell division centers associated with a bacterial infection. Yet only six nodules developed in this region of the root. The ontogenetic process of both types of division centers was stalled predominantly at the transition point from stage 3 to 4, being cell clusters of about 20 to 30 cells (13).

Thus the plant controls the extent of cell division in the root cortex and its associated pericycle (and thus nodule development) by (*a*) limiting infectivity to certain regions of the root, (*b*) restricting the infectability of root hairs within these regions (i.e. only those developing off the xylem poles are primary infection targets) (99), and finally (*c*) regulating the degree of general cortical cell division (87, 100). Autoregulation of nodulation means that developing infection centers, prior to the onset of nitrogen fixation, trigger a general plant response that inhibits further cell division activity in newly developing root tissue. Regulation of flavone, coumarin, and isoflavone synthesis (S:I ratios) in the vicinity of developing infection threads or by translocation from other plant parts (Figure 4) could be a mechanism by which legumes prevent overnodulation. Several other physiological and environmental factors also affect nodulation, such as nitrate levels, acidity, and several micronutrients (16, 69, 116, 131). The inhibitory activity of these factors may be modulated through effects on these fundamental autoregulatory systems.

The phenomenon of autoregulation in soybeans also has been examined using a split-root system (69, 83, 97) where the inoculation of the first side with a *Bradyrhizobium* suppressed the nodulation on the second side. This process gave maximum suppression if the two inoculation events were separated by 4–7 days. Observations in the laboratory of Calvert and Bauer, and in ours in Canberra, confirm that autoregulation seems to work via the arrest of developing nodule foci, rather than by a prevention of further infection and nodule initiation.

Nitrate inhibition of nodulation and nitrogen fixation has been investigated using split-root systems in both soybeans (69) and clovers (16). Nitrate inhibited the formation of nodules only on the side of exposure, while its effect of lowering the specific nitrogenase activity was expressed on both sides of the split-root system. Singleton & van Kessel (123) used split-root systems of soybean to study the allocation of photosynthetic resources to differentially treated root systems. Shoots were exposed to $^{14}CO_2$ to monitor current photosynthate. Using uninoculated root systems first, they found the root that received nitrate received most of the labeled carbon. In teleological terms this makes sense: The plant recognizes which root side is most important for its fitness and supports the growth of this side through the shoot. These studies were extended by nodulating both sides of the root system equally but then exposing one side to a constant stream of air and the other side to argon/oxygen for 11 days. This resulted in nitrogen fixation in one root portion but not the other. The half receiving air and thus being able to fix nitrogen was allocated most of the available photosynthate. Singleton & van Kessel did not compare the allocation of carbon between a nodulated root portion and a portion that receives nitrate alone.

Olsson et al (97) investigated autoregulation responses in split-root systems of the soybean cultivar Bragg and its supernodulating mutant nts382 (18). Potassium nitrate (0.5 mM), supplied to both root systems, has no inhibitory effect on nodulation or nitrogen fixation, and hence was used to allow both root portions to remain functional. Concurrently inoculated roots nodulated equally well; root growth was nearly equivalent. If, however, roots were inoculated one week apart, then nodulation on the side inoculated first totally suppressed nodulation on the second side in the case of wild-type plants but not the mutant nts382. Suppression was detectable already with a 24-hr inoculation delay. Root weights were nearly identical, suggesting that under the experimental conditions of Olsson et al, both root systems, whether nodulated or not, received similar amounts of photosynthate. This was verified by translocation studies using ^{14}C-sucrose (97). This was probably the result of the inclusion of the low-level nitrate in the nutrient medium.

Not all soybean cultivars possess the same ability to autoregulate. Olsson et al (97) illustrated that cultivar Bragg has a greater suppression ability than cultivar Clark or Williams. This may represent an adaptation to the local environment and environmental factors, such as the intensity and amount of light (83). This again stresses that, even within a species with a small genetic base such as soybean, one finds diversity in the ability to regulate the nodulation response.

The involvement of nitrogen fixation in autoregulation is still not entirely resolved [see the conclusions of Singleton & van Kessel (123)]. Although Kosslak & Bohlool (83) and Olsson et al (97) find that autoregulation suppression occurs before the onset of nitrogen fixation (about day 13 to 14 after germination and inoculation for soybean grown under the described conditions), the experimental design did not rule out an alternative explanation. A 7-day headstart may allow the symbiosis to advance to the nitrogen fixation stage, at which time preferential allocation of carbon may occur, so that the lagging nodule development on the delayed-inoculation side is arrested in its symbiotic development prior to the emergence of visible nodules. In contrast, preliminary data from soybean split roots infected with the *Bradyrhizobium sm5* mutant (Fix⁻) still showed (although less efficiently than wild type) the ability to suppress nodulation on the delayed-inoculation root side (7a).

A *nifA* mutant of strain USDA110 gave increased nodulation and distributed nodule pattern on soybean (104). Nodule weight, however, was drastically reduced. This type of observation is common, i.e. nonfixing nodules do not always suppress further nodulation. The plant may recognize that a particular nodule is "nonfunctional" and hence may curtail translocation to it, leading to a senescence of the nodule meristem and the release from the symbiotic inhibition loop. It also demonstrates that there is a distant separation between nodule initiation and nodule growth. Often the latter is monitored as an indication of nodule initiation.

Split-root assays also have been used to examine the initial infection events on subterranean clovers (111). Specific mutants derived from *R. trifolii* strain ANU843 were used to investigate the systemic plant response induced by infective *Rhizobium* strains. The ability to invade root hairs, not just curl them, was required to initiate the plant inhibitory response. Moreover, plants could discriminate between infections initiated by the parent strain or by mutants subtly impaired in their ability to nodulate. In addition, the split-root assay was successfully used as a simple system for ranking the rapid infectivity of various rhizobia. Preexposure of some poorly competitive strains to the flavone plant signal for 4 hr enables them to outcompete a normally more successful strain (L. Sargent, unpublished results).

ISOLATION OF PLANT NODULATION MUTANTS

Over the last few years several laboratories have reported the isolation of symbiotically altered plant mutants. The most studied species are those in soybean *(Glycine max)* and pea *(Pisum sativum)* (64, 66). Nonnodulation and supernodulation mutants have been isolated in pea (75, 76, 80, 81), soybeans (17, 18, 19), and chickpeas (28). The soybean mutants were induced with ethyl methyl sulfonate (EMS).

Carroll et al (17, 18) described the isolation of 12 nitrate-tolerant symbiotic (nts) mutants that supernodulated—i.e. their nodulation regulation mechanisms allow excessive nodulation along the entire root system. The nts mutants have been described in detail (29, 30, 36, 37, 65, 116). The essential features of this material are: (*a*) they have a normal nitrate and possibly altered ureide metabolism; (*b*) they are capable of near normal growth on nitrate in the absence of inoculum; (*c*) they have elevated acetylene reduction activity on 5 mM nitrate compared to Bragg controls, but their specific nitrogenase activity on 0 mM nitrate (i.e. fully symbiotic plants) was about 25–35% that of wild-type nodules; (*d*) they show a high level of nodulation (up to 4000 nodules per plant was recorded) both in the presence and absence of nitrate; (*e*) they were symbiotically sensitive to extremely high levels of nitrate (above 10 mM); (*f*) the phenotype was caused by a single recessive Mendelian inherited allele, and (*g*) the phenotype could be suppressed by the injection into the hypocotyl of plant extracts from inoculated wild-type plants.

Mathews (87) demonstrated that the total number of pseudo- and actual infections in the supernodulation mutant nts382 was within the range found for Bragg, yet the number of nodules was increased by a factor of 10 (data from 80-mm sections from plastic growth pouch grown plants). Symbiotic arrest after stage 3 did not occur, and more effective transition to larger nodule centers was pronounced. Grafting studies (36, 37), split-root experiments (97), and refeeding data (65) led to the postulate that the supernodulation mutants lack a shoot-derived inhibitory substance, which targets cortical cell divisions irrespective of whether they are pseudoinfections or actual infections. Furthermore, the level of this substance would increase in

Bragg plants following inoculation by a wild-type *Bradyrhizobium* (A. Krotzky, unpublished data).

Induced mutagenesis in soybean also was used to isolate nonnodulating mutants (nod49, nod139, and nod772), which were unable to curl root hairs (87, 88, 89). Mathews found nod49 and nod772 to be allelic with the naturally occurring nonnodulation allele *rj₁*, while *nod139* represents a different gene locus (87). All nonnodulation mutants have the ability to show normal nodule development (but not numbers), if inoculated with the appropriate strain at high ($>10^8$ cells plant^{-1}) bacterial titers. Such occasional nodulation follows a normal ontogeny leading to a nitrogen-fixing nodule. Nodule numbers per plant, however, are extremely low, being in the range of 1–5 nodules. Under field conditions this occasional nodulation was never observed. All three induced nodulation mutants are controlled by single Mendelian recessive alleles. Linkage analysis showed that the *nod* alleles are unlinked to *nts382* and *purple flower*.

Serial sections of inoculated taproots of the mutants showed that the nodulation phenotype is controlled not only by the plant genotype and the bacterial inoculant titer, but also by the bacterial strain. Nod49 was able to develop pseudoinfections and some actual infections, if inoculated with *Bradyrhizobium* strain CB 1795 at high (10^9 cells plant^{-1}) cell titers, but not with strain USDA110 at similar titers or strain 1795 at low (10^5 cell plant^{-1}) titers (87). The same phenotype was found for nod772 and *rj₁* Lee (84, 87); nod139, in contrast, showed no pseudoinfections with either strain at either inoculant level. The observation that pseudoinfections are possible but root-hair curling and infection threads are is absent means that the cortical cell division stimulus may precede the actual infection mechanism. The dividing cortical cells may potentiate the developing root hair to become an infection target. For uninfected plants, the analysis of root exudates and extracts of supernodulation, nonnodulation, and wild-type lines of soybean for their interactive and/or crossfeeding ability showed that the lectin, extracellular protein, and *nod* gene activation were similar in all cases (88).

It is thus possible to use a genetic approach in the analysis of the host contribution to the nodule symbiosis. Clearly, multiple genes exist within the legume that are exclusively responsible for a symbiotic function; this mirrors the situation in the bacterium, where likewise nodulation-specific genes are found.

INDUCIBLE PLANT DEFENSE SYSTEMS

The development of disease resistance in plants involves inducible defense mechanisms produced in response to pathogen attack (79, 119). A number of enzyme systems accumulate, such as those involved in the synthesis of phytoalexin antibiotics and those leading to deposition of lignin-like material, accumulation of hydroxy-proline-rich glycoproteins (HRGP) and proteinase inhibitors, and increased activity of certain hydrolytic enzymes (26, 27, 50, 122). Readers are referred to a number of reviews on various aspects of these responses (39, 46, 50, 68, 90, 110, 119).

Generally, nodulation does not cause cell death or induction of the plant hypersensitive reaction. This type of plant reaction has been studied intensely because it is a clear example of the dynamic role of the infected host in the early stages of a pathogen attack; this reaction confers a high degrees of resistance on the plant host. The hypersensitive reaction that is widespread among plants occurs during incompatible combinations of host plant and a variety of pathogens (39). A number of physiological changes occur in the infected cells when the hypersensitive reaction is induced, such as loss of permeability of cell membranes, increased respiration, accumulation and oxidation of phenolic compounds, and production of phytoalexins (39) The end result is the death and collapse of the infected and perhaps a few surrounding cells and the containment of the pathogen. For example, aviru-lent mutants of the bacterial pathogen *Pseudomonas solanacearum* cause rapid localized cell death and fail to multiply when infiltrated into tobacco leaves. In contrast, virulent strains cause no pronounced initial reaction and multiply in the intercellular spaces. The initial host response to the invading avirulent strains is the hypersensitive reaction. This reaction was not triggered by the virulent strain, and its induction suggests that the plant has "recog-nized" the presence of the avirulent pathogen (5, 6).

The interesting question is why rhizobia are so different from many other soil bacteria in being able to initiate successfully such a complex set of interactions with a plant without eliciting a major host response. Presumably, it is because the bacteria have acquired genes for regulating the recognition and defense responses of a host plant. *Rhizobium* workers are interested in plant defense systems and the synthesis of certain phenylpropanoid com-pounds because (*a*) flavonoid compounds (flavones, isoflavones) and coumar-ins (Figure 4) affect the expression of the *Rhizobium* nodulation genes (41, 56, 98, 103), (*b*) preincubation of rhizobia with flavones increases the infectivity of such strains (unpublished results), and (*c*) the levels of com-pounds detected from the phenylpropanoid pathway after *Rhizobium* infec-tion are influenced by the host-range genes of rhizobia (B. G. Rolfe, unpub-lished results). Investigators also realize that the *Rhizobium* bacterium behaves like a parasite in the early stages in the infection process (43, 107, 120, 129).

Characterization of the host-specific nodulation (*hsn*) genes of R. *trifolii* suggested that these genes might be involved in determining avirulence of *Rhizobium* strains (40, 42). In addition, when white clovers are infected with

R. leguminosarum strain 300 (a pea nodulating strain), a rapid induction of the synthesis of flavones occurs (an incompatible response). *R. trifolii* strains cause a less pronounced induction, which decreases with time (a compatible response). The nodF and nodE mutants of *R. trifolii* on white clovers cause a rapid and prolonged induction of flavonoid compounds like strain 300. These mutants behave as avirulent strains on white clovers but as virulent strains on peas (42). Thus *Rhizobium hsn* genes appear to be involved in the induction of the plant phenylpropanoid biosynthetic pathway (B. G. Rolfe, unpublished results).

EARLY NODULIN GENE EXPRESSION

The presence of nodule-specific plant gene products (termed nodulins) has been demonstrated in several *Rhizobium*-legume symbioses (59, 60, 61, 85). During nodule development at least 20 nodule-specific (or nodule-amplified) genes are expressed. Most of these genes (including the leghemoglobin genes) are expressed after the initiation of the visible nodule (61). Hybridization studies, however, have also shown the presence of an early nodulin (ENOD2) in both soybeans and peas (58, 61). DNA sequence analysis of the ENOD2 cloned gene indicated that it is a hydroxy-proline-rich glycoprotein (HRGP) (58). The synthesis of HRGPs is developmentally regulated in plants and is associated with cell walls. HRGPs are thought to be involved in host defense responses, accumulating upon wounding and pathogen attack (23, 25, 58, 122). Plant molecular analyses of early nodulation steps would be aided by a coupling of the genetic variability (i.e. plant and bacterial mutants) with existing knowledge of the biological chemistry and molecular genetics of root infection.

CONCLUSIONS

A wealth of physiological, biochemical, and molecular biological data is now available about *Rhizobium* and legumes. Available are particular mutants and specific strain constructions that can be used to investigate the resistance mechanism of infected plants. The use of reporter gene systems, such as the *E. coli lacZ* gene fusions, have helped to identify and characterize plant substances used in the molecular communication between *Rhizobium* and its host legume. A simple rapid-staining technique for detecting cell division foci in the roots is needed to identify the bacterial signals and their corresponding genes that stimulate cortical cell division. New plant gene probes for different key steps of the phenylpropanoid pathway are available for a coordinated investigation of how the *Rhizobium hsn* genes may be involved in the regulation of this plant defense system. Although our knowledge is still incomplete, many significant observations about the bacterial and plant contributions to the initial steps of nodulation have been made since the publication of Bauer's review in 1981 (4).

ACKNOWLEDGMENTS

We thank our colleagues, research staff, and students for their contribution to the analysis of the bacterial/plant interaction, Val Rawlings for typing of the manuscript, Tessa Reath for Figure 2.

Literature Cited

1. Abe, M., Amemura, A., Higashi, S. 1982. Studies on cyclic β-1,2-glucan obtained from periplasmic space of *Rhizobium trifolii* cells. *Plant Soil.* 64:315–24

2. Abe, M., Sherwood, J. E., Hollingsworth, R. I., Dazzo, F. B. 1984. Stimulation of clover root hair infection by lectin-binding oligosaccharides from the capsular and extracellular polysaccharides of *Rhizobium trifolii. J. Bacteriol.* 160:517–20

3. Ayers, A. R., Ayers, S. B., Goodman, R. N. 1979. Extracellular polysaccharide of *Erwinia amylovora:* a correlation with virulence. *Appl. Environ. Microbiol.* 38:659–66

4. Bauer, W. D. 1981. The infection of legumes by rhizobia. *Ann. Rev. Plant. Physiol.* 32:407–49

5. Bauer, W. D., Bhuvaneswari, T. V., Calvert, H. E., Law, I. J., Malik, N. A. S., Vesper, S. J. 1985. Recognition and infection by slow-growing rhizobia. In *Nitrogen Fixation Research Progress,* ed. H. J. Evans, P. J. Bottomley, W. E. Newton, pp. 247–53. Amsterdam: Martinus Nijhoff

6. Bender, G. L., Nayudu, M., Goydych, W., Rolfe, B. G. 1987. Early infection events in the nodulation of the non-legume *Parasponia andersonii* by *Bradyrhizobium. Plant Sci.* 51:285–93

7. Bhuvaneswari, T.V., Turgeon, B. G., Bauer, W. D. 1980. Early events in the infection of soybean (*Glycine max* L. Merr.). *Plant Physiol.* 66:1027–31

7a. Bohlool, B. B., Nakao, P., Singleton, P. W. 1986. Ecological determinants of interstrain competition in *Rhizobium/* legume symbiosis. *Aust. Inst. Agri. Sci. (Occas. Publ.)* 25:145–48

8. Borthakur, D., Barber, C. E., Lamb, J. W., Daniels, M. J., Downie, J. A., Johnston, A. W. B. 1986. A mutation that blocks exopolysaccharide synthesis prevents nodulation of peas by *Rhizobium leguminosarum* but not of beans by *R. phaseoli* and is corrected by cloned DNA from *Rhizobium* or the phytopathogen *Xanthomonas. Mol. Gen. Genet.* 203:320–23

9. Borthakur, D., Johnston, A. W. B. 1987. Sequence of *Psi,* a gene on the symbiotic plasmid of *Rhizobium phaseoli* which inhibits exopolysaccharide synthesis and nodulation and demonstration that its transcription is inhibited by *psr,* another gene on the symbiotic plasmid. *Mol. Gen. Genet.* 207:149–54

10. Broglie, K. E., Gaynor, J. J., Broglie, R. M. 1986. Ethylene-regulated gene expression: molecular cloning of the genes encoding an endochitinase from *Phaseolus vulgaris. Proc. Natl. Acad. Sci. USA* 83:6820–24

11. van Brussel, A. A. N., Zaat, S. A. J., Canter Cremers, H. C. J., Wijffelman, C. A., Pees, E., et al. 1985. Role of plant root exudate and Sym plasmid-localized nodulation genes in the synthesis by *Rhizobium leguminosarum of Tsr* factor, which causes thick and short roots on common vetch. *J. Bacteriol.* 165:517–22

12. Callaham, D. A., Torrey, J. G. 1981. The structural basis for infection of root hairs of *Trifolium repens* by *Rhizobium. Can. J. Bot.* 59:1647–64

13. Calvert, H. E., Pence, M. K., Pierce, M., Malik, N. S. A., Bauer, W. D. 1984. Anatomical analysis of the development and distribution of *Rhizobium* infections in soybean roots. *Can. J. Bot.* 62:2375–84

14. Cangelosi, G. A., Hung, L., Puvanesarajah, V., Stacey, G., Ozga, S. D., et al. 1987. Common loci for *Agrobacterium tumefaciens* and *Rhizobium meliloti* exopolysaccharide synthesis and their roles in plant interactions. *J. Bacteriol.* 169: 2086–91

15. Canter Cremers, H. C. J., van Brussel, A. A. N., Rolfe, B. G. 1985. Sym plas-

mid and chromosomal gene products of *Rhizobium trifolii* elicit developmental responses on various legume roots. *J. Plant Physiol.* 122:25–40

16. Carroll, B. J., Gresshoff, P. M. 1983. Nitrate inhibition of nodulation and nitrogen fixation in white clover. *Z. Pflanzenphysiol.* 110:77–88

17. Carroll, B. J., McNeil, D. L., Gresshoff, P. M. 1985. Isolation and properties of soybean *(Glycine max)* mutants that nodulate in the presence of high nitrate concentrations. *Proc. Natl. Acad. Sci. USA* 82:4162–66

18. Carroll, B. J., McNeil, D. L., Gresshoff, P. M. 1985. A supernodulation and nitrate tolerant symbiotic *(nts)* soybean mutant. *Plant Physiol.* 78:34–40

19. Carroll, B. J., McNeil, D. L., Gresshoff, P. M. 1986. Mutagenesis of soybean *(Glycine max* (L.) Merr.) and the isolation of non-nodulating mutants. *Plant Sci.* 47:109–14

20. Chandler, M. R. 1978. Some observations on infection of *Arachis hypogaea.* L. by *Rhizobium. J. Exp. Bot.* 29:749–55

21. Chandler, M. R., Date, R. A., Roughley, R. J. 1982. Infection and root nodule development in *Stylosanthes* sp. by *Rhizobium. J. Exp. Bot.* 33:45–57

22. Chen, H., Batley, M., Redmond, J. W., Rolfe, B. G. 1985. Alteration of the effective nodulation properties of a fast-growing broad host range *Rhizobium* due to changes in exopolysaccharide synthesis. *J. Plant Physiol.* 120:331–49

23. Chen, J. A., Varner, J. E. 1985. Isolation and characterization of cDNA clones for carrot extension and a proline-rich 33-kDa protein. *Proc. Natl. Acad. Sci. USA* 82:4399–403

24. Collins, J. 1983. *Anatomical investigations of nodule initiation in white clover.* Honours Degree Dissertation. Australian Natl. Univ., Canberra, Australia

25. Cooper, J. B., Chen, J. A., van Holst, G.-J., Varner, J. E. 1987. Hydroxyproline-rich glycoproteins of plant cell walls. *Trends Biochem. Sci.* 12:24–29

26. Cramer, C. L., Bell, J. N., Ryder, T. B., Bailey, J. A., Schuch, W., et al. 1985. Coordinated synthesis of phytoalexin biosynthetic enzymes in biologically-stressed cells of bean *(Phaseolus vulgaris* L.). *EMBO J.* 4:285–89

27. Darvill, A. G., Albersheim, P. 1984. Phytoalexins and their elicitors—a defense against microbial infection in plants. *Ann. Rev. Plant Physiol.* 35:243–75

28. Davies, T. M., Foster, K. W., Phillips, D. A. 1985. Non-nodulation mutants of chickpea. *Crop Sci.* 25:345–48

29. Day, D. A., Lambers, H., Bateman, J., Carroll, B. J., Gresshoff, P. M. 1986. Growth comparisons of a supernodulating soybean mutant and its wildtype parent. *Physiol. Plant.* 68:375–82

30. Day, D. A., Price, G. D, Schuller, K. A., Gresshoff, P. M. 1987. Nodule physiology of a supernodulating soybean *(Glycine max)* mutant. *Aust. J. Plant Physiol.* 14: In press

31. Dazzo, F. B., Hollingsworth, R. I., Sherwood, J. E., Abe, M., Hrabak, E. M. et al. 1985. Recognition and infection of clover root hairs by *Rhizobium trifolli.* In *Nitrogen Fixation Research Progress,* ed. H. J. Evans, P. J. Bottomley, W. E. Newton, pp. 239–45. Amsterdam: Martinus Nijhoff

32. Dazzo, F. B., Hollingsworth, R. I., Philip, S., Smith, K. B., Welsch, M. A., et al. 1987. Involvement of *pSym* nodulation genes in production of surface and extracellular components of *Rhizobium trifolii* which interact with white clover root hairs. In *Molecular Genetics of Plant-Microbe Interactions,* ed. D. P. S. Verma, N. Brisson, pp. 171–72. Dordrecht: Martinus Nijhoff

33. Debelle, F., Sharma, S. R. 1986. Nucleotide sequence of *Rhizobium meliloti* RCR 2011 genes involved in host specificity of nodulation. *Nucl. Acids Res.* 14:7453–72

34. de Faria, S. M., McInroy, S. G., Sprent, J. I. 1987. The occurrence of infected cells, with persistent infection threads, in legume root nodules. *Can. J. Bot.* 65:553–58

35. de Faria, S. M., Sutherland, J. M., Sprent, J. I. 1986. A new type of infected cell in root nodules of *Andira* Spp. (Leguminosae). *Plant Sci.* 45:143–47

36. Delves, A. C., Mathews, A., Day, D. A., Carter, A. S., Carroll, B. J., Gresshoff, P. M. 1986. Regulation of the soybean-*Rhizobium* nodule symbiosis by shoot and root factors. *Plant Physiol.* 82:588–90

37. Delves, A. C., Higgins, A., Gresshoff, P. M. 1987. Supernodulation in interspecific grafts between *Glycine max (soybean)* and *Glycine soja. J. Plant Physiol.* 128:473–78

38. Diaz, C. L., Van Spronsen, P. C., Bakhuizen, R., Logman, G. J. J., Lugtenberg, E. J. J., Kijne, J. W. 1986. Correlation between infection by *Rhizobium leguminosarum* and lectin on the

surface of *Pisum sativum L. roots. Planta* 168:350–59

39. Dickinson, C. H., Lucas, J. A. 1982. In *Plant Pathology and Plant Pathogens.* Oxford: Blackwell Scientific

40. Djordjevic, M. A., Innes, R. W., Wijffelman, C. A., Schofield, P. R., Rolfe, B. G. 1986. Nodulation of specific legumes is controlled by several distinct loci in *Rhizobium trifolii. Plant Mol. Biol.* 6:389–403

41. Djordjevic, M. A., Redmond, J. W., Batley, M., Rolfe, B. G. 1987. Clovers secrete specific phenolic compounds which stimulate or repress *nod* gene expression in *Rhizobium trifolii. EMBO J.* 6:1173–79

42. Djordjevic, M. A., Schofield, P. R., Rolfe, B. G. 1985. Tn5 mutagenesis of *Rhizobium trifolii* host-specific nodulation genes result in mutants with altered host range ability. *Mol. Gen. Genet.* 200:463–71

43. Djordjevic, M. A., Gabriel, D.W., Rolfe, B. G. 1987. *Rhizobium*—the refined parasite of legumes. *Ann. Rev. Phytopathol.* 25:145–68

44. Djordjevic, S. P., Batley, M., Redmond, J. W., Rolfe, B. G. 1986. The structure of the exopolysaccharide from *Rhizobium* sp. strain ANU280 (NGR 234). *Carbohydr. Res.* 148:87–99

45. Djordjevic, S. P., Chen, H., Batley, M., Redmond, J. W., Rolfe, B. G. 1987. Nitrogen fixation ability of exopolysaccharide synthesis mutants of *Rhizobium* sp. strain NGR234 and *Rhizobium trifolii* is restored by the addition of homologous exopolysaccharides. *J. Bacteriol.* 169:53–60

46. Douglas, C., Hoffman, H., Schulz, W., Hahlbrock, K. 1987. Structure and elicitor or U.V.-light-stimulated expression of two 4-coumarate: CoA ligase genes in parsley. *EMBO J.* 6:1189–95

47. Downie, J. A., Hombrecher, G., Ma, Q. S., Knight, C. D., Wells, B., Johnston, A. W. B. 1983. Cloned nodulation genes of *Rhizobium leguminosarum* determine host-range specificity *Mol. Gen. Genet.* 190:359–56

48. Dreyfus, B., Dommergues, Y. R. 1980. Nitrogen fixing nodules induced by *Rhizobium* on the stems of the tropical legume *S. rostrata. FEMS Microbiol. Lett.* 10:313–17

49. Dyan, T., Ielpi, L., Stanfield, S., Kashyap, L., Douglas, C., et al. 1986. *Rhizobium meliloti* genes required for nodule development are related to chromosomal virulence genes in *Agrobacterium tumefaciens. Proc. Natl. Acad. Sci. USA* 83:4403–7

50. Edwards, K., Cramer, C. L., Bolwell, G. P., Dixon, R. A., Schuch, W., Lamb, C. J. 1985. Rapid transient induction of phenylalanine ammonia-lyase mRNA in elicitor-treated bean cells. *Proc. Natl. Acad. Sci. USA* 82:6731–35

51. Egelhoff, T. T., Fisher, R. F., Jacobs, T. W., Mulligan, J. T., Long, S. R. 1985. Nucleotide sequence of *Rhizobium meliloti* 1021 nodulation genes: *nodD* is read divergently from *nodABC. DNA* 4:241–48

52. Egelhoff, T. T., Long, S. R. 1985. *Rhizobium meliloti* nodulation genes: identification of *nodDABC* gene products, purification of *nodA* protein, and expression of *nodA* in *Rhizobium meliloti. J. Bacteriol.* 164:591–99

53. Evans, I. J., Downie, J. A. 1986. The *nodI* gene product of *Rhizobium leguminosarum* is closely related to ATP-binding bacterial transport proteins; nucleotide sequence analysis of the *nodI* and *nodJ* genes. *Gene* 35‹01

54. Finan, T. M., Hirsch, A. M., Leigh, J. A., Johansen, E., Kuldau, G. A., et al. 1985. Symbiotic mutants of *Rhizobium meliloti* that uncouple plant from bacterial differentiation. *Cell* 40:869–87

55. Finan, T. M., Kunkel, B., de Vos, G. F., Signer, E. R. 1986. Second symbiotic megaplasmid in *Rhizobium meliloti* carrying exopolysaccharide and thiamine synthesis genes. *J. Bacteriol.* 167:66–72

56. Firmin, J. L., Wilson, K. E., Rossen, L., Johnston, A. W. B. 1986. Flavonoid activation of nodulation genes in *Rhizobium* reversed by other compounds present in plants. *Nature* 324:90–92

57. Fisher, R. F., Brierley, H. L., Mulligan, J. T., Long, S. R. 1987. Transcription of *Rhizobium meliloti* nodulation genes. Identification of a *nodD* transcription initiation site in vitro and in vivo. *J. Biol. Chem.* 262:6849–55

58. Franssen, H. J., Nap, J.-P., Gloudemans, T., Stiekema, W., van Dam, H., et al. 1987. Characterization of cDNA for nodulin-75 of soybean: a gene product involved in early stages of root nodule development. *Proc. Natl. Acad. Sci. USA* 84:4495–99

59. Fuller, F. F., Verma, D. P. S. 1984. Appearance and accumulation of nodulin mRNAs and their relation to the effectiveness of root nodules. *Plant Mol. Biol.* 3:21–28

60. Govers, F., Gloudemans, T., Moerman, M., van Kammen, A., Bisseling, T. 1985. Expression of plant genes during the development of pea root nodules. *EMBO J.* 4:861–67

61. Govers, F., Moerman, M., Downie, J. A., Hooykaas, P., Franssen, H. J., Louwerse, J., van Kammen, A., Bisseling, T. 1986. *Rhizobium nod* genes are involved in inducing an early nodulin gene. *Nature* 323:564–66

62. Gresshoff, P. M. 1981. Luftstickstoff als Pflanzendünger. *Umschau Wissensch. Techn.* 15:465–67

63. Gresshoff, P. M., Mohapatra, S. S. 1981. Legume cell and tissue culture. In *Tissue Culture of Economically Important Crop Plants,* ed. A. N. Rao, pp. 11–24. Singapore: Univ. Singapore Press

64. Gresshoff, P. M., Day, D. A., Delves, A. C., Mathews, A., Olsson, J. A., et al. 1986. Plant host genetics of nodulation and symbiotic nitrogen fixation in pea and soybean. In *Nitrogen Fixation Research Progress,* ed. H. J. Evans, P. J. Bottomley, W. E. Newton, pp. 19–25. Amsterdam: Nijhoff

65. Gresshoff, P. M., Krotzky, A., Mathews, A., Day, D. A., Schuller, K. A., et al. 1987. Suppression of the symbiotic supernodulation symptoms of soybean. *J. Plant Physiol.* In press

66. Gresshoff, P. M., Delves, A. C. 1986. Plant genetic approaches to symbiotic nodulation and nitrogen fixation in legumes. In *Plant Gene Research 3,* ed. A. D. Blonstein, P. J. King, pp. 159–206. Vienna: Springer-Verlag

67. Halverson, J. L., Stacey G. 1985. Host recognition in the *Rhizobium*-soybean symbiosis. Evidence for the involvement of lectin in nodulation. *Plant Physiol.* 77:621–25

68. Halverson, J. L., Stacey, G. 1986. Signal exchange in plant-microbe interactions. *Microbiol. Rev.* 50:193–225

69. Hinson, K. 1975. Nodulation response from nitrate applied to soybean half-root systems. *Agron. J.* 67:799–804

70. Homna, M. A., Ausubel, F. M. 1986. Multiple *nodD* genes in *Rhizobium meliloti.* In *Third Int. Symp. on the Mol. Genet. Plant-Microbe Interactions,* Montreal, Canada. Abstr., p. 112

71. Horvath, B., Kondorosi, E., John, M., Schmidt, J., Torok, I., et al. 1986. Organisation, structure and symbiotic function of *Rhizobium meliloti* nodulation genes determining host specificity for alfalfa. *Cell* 46:335–43

72. Huang, S. Z., Rolfe, B. G., Djordjevic, M. A. 1987. Mutation of *Rhizobium trifolii nodIJ* genes affects infection thread initiation and synthesis in *Trifolium subterraneum. J. Plant Physiol.* In press

73. Hynes, M. F., Simon, R., Muller, P., Niehaus, K., Labes, M., Pühler, A.

1986. The two megaplasmids of *Rhizobium meliloti* are involved in the effective nodulation of alfalfa. *Mol. Gen. Genet.* 202:356–62

74. Innes, R. W., Kuempel, P. L., Plazinski, J., Canter-Cremers, H., Rolfe, B. G., Djordjevic, M. A. 1985. Plant factors induce expression of nodulation and host-range genes in *Rhizobium trifolii. Mol. Gen. Genet.* 201:426–32

75. Jacobsen, E., Feenstra, W. J. 1984. A new pea mutant with efficient nodulation in the presence of nitrate. *Plant Sci. Lett.* 33:337–44

76. Jacobsen, E. 1984. Modification of symbiotic interaction of pea *(Pisum sativum)* and *Rhizobium leguminosarum* by induced mutation. *Plant Soil* 82:427–38

77. John, M., Schmidt, J., Weineke, U., Kondorosi, E., Kondorosi, A., Schell, J. 1985. Expression of nodulation gene *nodC* of *Rhizobium meliloti* in *Escherichia coli:* role of the nodC product in nodulation. *EMBO J.* 4:2425–30

78. Jordan, D. C. 1982. Transfer of *Rhizobium japonicum* Buchanan 1980 to *Bradyrhizobium japonicum* gen. nov., a genus of slow-growing, root nodule producing bacteria from leguminous plants. *Int. J. Syst. Bacteriol.* 32:136–39

79. Klement, Z., Goodman, R. N. 1967. The hypersensitive response by bacterial pathogens. *Ann. Rev. Phytopathol.* 5:17–44

80. Kneen, B. E., LaRue, T. A. 1984. Nodulation resistant mutant of *Pisum sativum* (L.). *J. Hered.* 75:238–40

81. Kneen, B. E., LaRue, T. A. 1984. Peas *(Pisum sativum* L.) with strain specificity for *Rhizobium leguminosarum. Heredity* 52:383–89

82. Kondorosi, E., Banfalvi, Z., Kondorosi, A. 1984. Physical and genetic analysis of a symbiotic region of *Rhizobium meliloti:* identification of nodulation genes. *Mol. Gen. Genet.* 193:445–52

83. Kosslak, R. M., Bohlool, B. B. 1984. Suppression of nodule development of one side of a split root system of soybeans caused by prior inoculation of the other side. *Plant Physiol.* 75:125–30

84. LaFavre, J. S., Eaglesham, A. R. J. 1984. Increased nodulation of "nonnodulating" (rj, rj,) soybeans by high dose inoculation. *Plant Soil* 80:297–300

85. Lang-Unasch, N., Ausubel, F. M. 1985. Nodule-specific polypeptides from effective alfalfa root nodules and from ineffective nodules lacking nitrogenase. *Plant Physiol.* 77:833–38

86. Leigh, J. A., Signer, E. R., Walker, G. C. 1985. Exopolysaccharide-deficient mutants of *Rhizobium meliloti* that form

ineffective nodules. *Proc. Natl. Acad. Sci. USA* 82:6231–35

87. Mathews, A. 1987. *Host involvement in nodule initiation in the soybean-*Brady-rhizobium *symbiosis*. PhD dissertation. Australian Natl. Univ., Canberra, Australia

88. Mathews, A., Kosslak, R., Sengupta-Gopalan, C., Appelbaum, E., Carroll, B. J., Gresshoff, P. M. 1987. Soybean *(Glycine max)* non-nodulation and supernodulation mutants have wildtype root exudates and extracts. *Mol. Gen. Genet.* (submitted)

89. Mathews, A., Carroll, B. J., Gresshoff, P. M. 1987. Characterization of non-nodulation mutants of soybean (*Glycine max* (L) Merr): *Bradyrhizobium* effects and absence of root hair curling. *J. Plant Physiol.* 131:349–61

90. Mehdy, M. C., Lamb, C. J. 1987. Chalcone isomerase cDNA cloning and mRNA induction by fungal elicitor, wounding and infection. *EMBO J.* 6:1527–33

91. Mulligan, J. T., Long, S. R. 1985. Induction of *Rhizobium meliloti nodC* expression by plant exudate requires *nodD. Proc. Natl. Acad. Sci. USA* 82:6609–13

92. Newcomb, W. 1976. A correlated light and electron microscopic study of symbiotic growth and differentiation in *Pisum sativum* root nodules. *Can. J. Bot.* 54:2163–86

93. Nutman, P. S. 1949. Physiological studies on nodule function. II. The influence of delayed nodulation on the rate of nodulation in red clover. *Ann. Bot.* 13:261–83

94. Nutman, P. S. 1952. Studies on the physiology of nodule formation. III. Experiments on the excision of root-tips and nodules. *Ann. Bot.* 16:81–102

95. Nutman, P. S. 1953. Symbiotic effectiveness in nodulated red clover. I. Variation in host and bacteria. *Heredity* 8:35–46

96. Olsson, J. E., Rolfe, B. G. 1985. Stem and root nodulation of the tropical legume *Sesbania rostrata* by *Rhizobium* strains ORS-571 and WE7. *J. Plant Physiol.* 121:199–210

97. Olsson, J. E., Nakao, P., Bohlool, B. B., Gresshoff, P. M. 1988. Host genetic control of soybean nodulation in split root systems. *Plant Physiol.* (Submitted)

98. Peters, K. N., Frost, J. W., Long, S. R. 1986. A plant flavone, luteolin, induces expression of *Rhizobium meliloti* nodulation genes. *Science* 223:977–79

99. Phillips, D. A. 1971. A cotyledonary inhibitor of root nodulation in *Pisum sativum. Physiol. Plant.* 25:482–87

100. Pierce, M., Bauer, W. D. 1983. A rapid regulatory response governing nodulation in soybean. *Plant Physiol.* 73:286–90

101. Price, G. D., Mohapatra, S. S., Gresshoff, P. M. 1985. Structure of nodules formed by *Rhizobium* strain ANU289 in the non-legume *Parasponia* and the legume siratro *(Macroptilium atropurpureum). Bot. Gaz.* 145:444–51

102. Puvanesarajah, V., Schell, J. M., Stacey, G., Douglas, C. J., Nester, E. W. 1985. Role for 2-linked-β-D-glucan in the virulence of *Agrobacterium tumefaciens. J. Bacteriol.* 164:102–6

103. Redmond, J. W., Batley, M., Djordjevic, M. A., Innes, R. W., Kuempel, P. L., Rolfe, B. G. 1986. Flavones induce the expression of the nodulation genes in *Rhizobium. Nature* 323:632–35

104. Regensburger, B., Meyer, L., Filser, M., Weber, J., Studer, D., Lamb, J. W., et al. 1986. *Bradyrhizobium japonicum* mutants defective in root-nodule bacteroid development and nitrogen fixation. *Arch. Microbiol.* 144:355–66

105. Ridge, R. W., Rolfe, B. G. 1986. Sequence of events during the infection of the tropical legume *Macroptilium atropurpureum* Urb. by the broad-host-range, fast-growing *Rhizobium* ANU 240. *J. Plant Physiol.* 122:121–37

106. Robertson, J. G., Lyttleton, P. 1982. Coated and smooth vesicles in the biogenesis of cell walls, plasma membranes, infection threads and peribacteroid membranes in root hairs and nodules of white clovers. *J. Cell Sci.* 58:63–78

107. Rolfe, B. G., Shine, J. 1984. *Rhizobium-Leguminosae* symbiosis: the bacterial point of view. In *Genes Involved in Microbe-Plant Interaction, Plant Gene Research*, ed. D. P. S. Verma, Th. Hohn, 1:95–128. Wien/New York: Springer

108. Rossen, L., Shearman, C. A., Johnston, A. W. B., Downie, J. A. 1985. The *nodD* gene of *Rhizobium leguminosarum* is autoregulatory and in the presence of plant exudate induces the expression of *nodABC* genes. *EMBO J.* 4:3369–73

109. Rostas, K., Kondorosi, E., Horvath, B., Simoncsits, A., Kondorosi, A. 1986. Conservation of extended promoter regions of nodulation genes in *Rhizobium. Proc. Natl. Acad. Sci. USA* 83:757–61

110. Ryan, C. A., Bishop, P. D., Walker-Simmons, M., Brown, W. E. Graham, J. S. 1985. Pectic fragments regulate the expression of proteinase inhibitor genes in plants. In *Cellular and Molecular Biology of Plant Stress*, pp. 319–34.

UCLA Symp. Mol. Biol., New Ser. 22. New York: Alan R. Liss

111. Sargent, L., Huang, S. Z., Rolfe, B. G., Djordjevic, M. A. 1987. Split-root assays using *Trifolium subterraneum* show that *Rhizobium* infection induces a systemic response that can inhibit nodulation of another invasive *Rhizobium* strain. *Appl. Environ. Microbiol.* 53:1611–19

112. Sarkar, S. K., Phan, C. T. 1974. Effect of ethylene on the phenylalanine ammonia-lyase activity of carrot tissues. *Physiol. Plant.* 32:318–21

113. Schmidt, J., John, M., Kondorosi, E., Kondorosi, A., Weineke, U., et al. 1984. Mapping of the protein coding regions of *Rhizobium meliloti* common nodulation genes. *EMBO J.* 3:1705–11

114. Schofield, P. R., Ridge, R. W., Rolfe, B. G., Shine, J., Watson, J. M. 1984. Host-specific nodulation is encoded on a 14kb DNA fragment in *Rhizobium trifolii*. *Plant Mol. Biol.* 3:3–11

115. Schofield, P. R., Watson, J. M. 1986. DNA sequence of the *Rhizobium trifolii* nodulation genes reveals a reiterated and potentially regulatory sequence preceding the *nodABC* and *nodFE* genes. *Nucl. Acids Res.* 14:2891–905

116. Schuller, K. A., Day, D. A., Gibson, A. H., Gresshoff, P. M. 1986. Enzymes of ammonia assimilation and ureide biosynthesis in soybean nodules: effect of nitrate. *Plant Physiol.* 80:646–50

117. Sequeira, L. 1984. Recognition systems in plant-pathogen interactions. *Biol. Cell* 51:281–86

118. Sequeira, L., Graham, T. L. 1977. Agglutination of avirulent strains of *Pseudomonas solanacearum* by potato lectin. *Physiol. Plant Pathol.* 10:43–54

119. Sequeira, L. 1983. Mechanisms of induced resistance in plants. *Ann. Rev. Microbiol.* 37:51–79

120. Sharifi, E. 1984. Parasitic origins of nitrogen-fixing *Rhizobium*-legume symbioses. A review of the evidence. *BioSystems* 16:269–89

121. Shearman, C. A., Rossen, L., Johnston, A. W. B., Downie, J. A. 1986. The *Rhizobium leguminosarum* nodulation gene *nodF* encodes a polypeptide similar to acyl-carrier protein and is regulated by *nodD* plus a factor in pea root exudate. *EMBO J.* 5:647–52

122. Showalter, A. M., Bell, J. N., Cramer, C. L., Bailey, J. A., Varner, J. E., Lamb, C. J. 1985. Accumulation of hydroxyproline-rich glycoprotein

mRNAs in response to fungal elicitor and infection. *Proc. Natl. Acad. Sci. USA* 82:6551–55

123. Singleton, P. W., van Kessel, C. 1987. Effect of localized nitrogen availability to soybean half-root systems on photosynthate partitioning to roots and nodules. *Plant Physiol.* 83:552–56

124. Spaink, H. P., Wijffelman, C. A., Pees, E., Okker, R. J. H., Lugtenberg, B. J. J. 1987. *Rhizobium* nodulation gene *nodD* as a determinant of host specificity. *Nature* 328:337–40

125. Spaink, H. P., Okker, R. J. H., Wijffelman, C. A., Pees, E., Lugtenberg, B. J. J. 1987. Promoters in the nodulation region of the *Rhizobium leguminosarum* Sym plasmid *pRL1J1*. *Plant Mol. Biol.* 9:27–39

126. Torrey, J. G. 1976. Initiation and development of root nodules of *Casuarina* (Casuarinacae) *Am. J. Bot.* 63:335–44

127. Trinick, M. J. 1979. Structure of nitrogen-fixing nodules by *Rhizobium* on roots of *Parasponia andersonii* Planch. *Can. J. Microbiol.* 25:565–78

128. Turgeon, B. G., Bauer, W. D. 1982. Early events in the infection of soybean by *Rhizobium japonicum*. Time course and cytology of the early infection process. *Can. J. Bot.* 60:152–61

129. Vance, C. P. 1983. *Rhizobium* infection and nodulation: a beneficial plant disease? *Ann. Rev. Microbiol.* 37:399–424

130. Vance, C. P., Boylan, K. L. M., Stade, S. 1987. Host plant determinants of legume nodule function: similarities to plant disease situations. In *Molecular Determinants of Plant Diseases*, ed. S. Nishimura, pp. 271–87. Tokyo: Jpn. Sci. Press; Berlin: Springer-Verlag

131. Vessey, J. K., Layzell, D. B. 1987. Regulation of assimilate partitioning in soybean: initial effects following change in nitrate supply. *Plant Physiol.* 83:341–48

132. Warren Wilson, J., Warren Wilson, P. M. 1984. Control of tissue differentiation in normal development and in regeneration. In *Positional Control of Plant Development*, ed. P. W. Barlow, D. J. Carr, pp. 225–80. Cambridge: Cambridge Univ. Press

133. Zaat, S. A. J., Wijffelman, C. A., Spaink, H. P., van Brussel, A. A. N., Okker, R. J. H., Lugtenberg, B. J. J. 1987. Induction of the *nodA* promoter of *Rhizobium leguminosarum* Sym plasmid pRL1J1 by plant flavanones and flavones. *J. Bacteriol.* 169:198–204

Ann. Rev. Plant Physiol. Plant Mol. Biol. 1988. 39:321–53

CELL WALL PROTEINS

Gladys I. Cassab and Joseph E. Varner

Plant Biology Program, Department of Biology, Washington University, St. Louis Missouri 63130

CONTENTS

INTRODUCTION 321
STRUCTURAL PROTEINS 322
   History 322
   Biochemical Characterization 326
   Molecular Biology Studies 328
   Cellular Localization and Possible Function in the Cell Wall 331
   Interaction with Other Cell Wall Components 337
CELL WALL ENZYMES 340
CONCLUDING REMARKS 346

INTRODUCTION

The presence of extraprotoplasmic walls is one of the outstanding characteristics distinguishing the cells of plants from those of animals. Few plant cells lack walls, and few animal cells (these belong to the lower organisms) have extraprotoplasmic envelopes even roughly comparable to the walls of plant cells. The cell wall is a complex entity with unique characteristics related to the developmental stage of a given plant cell type. Most commonly, each cell within a tissue has its own wall. Moreover, the function of each cell is largely determined by its wall. A plant cannot long exist unless its body is firmly knit together and its organs possess the collective mechanical strength of the cell walls. The many variations observed in plant cells reflect quantitative and qualitative changes in their walls. Cell walls differ in thickness and constituents. Wall models have been proposed from studies in plant tissue cultures (37). However, because the cell wall is not, like the mitochondrion,

321

an organelle with relatively constant duties, but rather a cell component subject to the continuous developmental processes that govern cell size, shape, and function, these models may lack the detail necessary for an understanding of how the wall affects cell form and function.

In general, the cell walls are comprised of cellulose, hemicelluloses, pectic compounds, lignin, suberin, proteins, and water. Water is one of the most important and most variable components of the various walls. The polysaccharide components of the wall have been reviewed extensively (37, 108, 115, 121). Here, we emphasize only the protein component of the wall. Cell walls may contain structural proteins as well as enzymes. So far, the best-characterized structural protein is extensin (32, 158, 169). Extensin is one member of a class of hydroxyproline-rich glycoproteins present in a wide variety of plants and algae (18, 84). It has been proposed that extensins are the major protein component of the primary wall and that they play a role in cell wall architecture (89). However, we do not know that extensin is present in all plant cell types, nor is there yet an elucidation of its role in the architecture of any cell. Recently, Cassab & Varner (17) demonstrated that extensin is most abundantly localized in cells that belong to the sclerenchyma tissue. The sclerenchyma cells act as the skeletal elements of the plant body. These cells enable the plant body to withstand various strains, such as stretching, bending, compression, and tension (46, 65). Thus, the presence of extensin in the sclerenchyma cell walls together with other wall components, may determine the unique characteristics of these cells. Extensin is probably not the only structural protein of the cell wall. Furthermore, structural proteins may vary from plant to plant, and from one cell type to another.

Several enzymes appear to be associated with cell walls. However, we cannot yet distinguish enzymes in the wall from those in the wall space. Cell wall enzymes—e.g. peroxidases and glycosidases—presumably play a role in the modification of macromolecules in muro.

STRUCTURAL PROTEINS

History

Tupper-Carey & Prestly (1924) were apparently the first to report the presence of protein in the cell wall (the primary wall of meristematic cells) (120). Tripp et al (159) found in primary walls of cotton seed hairs the amino acids serine, glycine, and aspartic acid. Lamport & Northcote (94) and Dougall & Shimbayashi (40) discovered hydroxyproline (Hyp) as a major amino acid constituent of hydrolysates of cell walls from tissue cultures. Since that time many other workers, using a wide variety of plants (77, 117), algae (122, 156), and fungi (6, 36, 116), have reported Hyp in cell wall hydrolysates. The evidence that these amino acids were in peptide linkage has been summarized by Lamport (84, 86), who, in view of the presumed function of the hydroxyproline-containing protein in cell wall extensibility, named it extensin. In some algal walls the protein contains hydroxylysine (155), another amino acid, which like Hyp is typical of collagen, the major component of animal extracellular matrixes.

For several years, what was known about extensin was extrapolated from the characterization of a series of hydroxyproline-rich glycopeptides and peptides obtained by enzymatic digestion and partial hydrolysis of tomato cell walls (85, 86). These all contain arabinose, galactose, and Hyp with a number of other amino acids (valine, serine, threonine, lysine, and tyrosine). By alkaline hydrolysis of these glycopeptides and whole walls, Lamport (86) was able to recover hyp-O-arabinosides, confirming a glycosidic link with the 4-OH of the amino acid. The serine is also O-substituted with galactose (92). The first amino acid sequences characteristic of extensin were the sequences of three peptides resulting from a tryptic digestion of cell walls: Ser-Hyp-Hyp-Hyp-Hyp-Thr-Hyp-Val-Tyr-Lys, Ser-Hyp-Hyp-Hyp-Hyp-Lys, and Ser-Hyp-Hyp-Hyp-Hyp-Val("Tyr"-Lys-Lys) (87). The solubilization of extensin from the wall was unsuccessful, so it was concluded that extensin was characteristically insoluble in the wall. The presence of wall-bound protein was generally accepted except by Steward et al. Steward's laboratory was the first to find Hyp in plants (147), and they concluded that the Hyp-rich material was present only in the cytoplasm (119). Autoradiographic studies, with ^{14}C proline and ^{3}H proline, showed that the accumulation of the labeled Hyp was in the cytoplasmic protein fraction and not in the wall of carrot cells in tissue culture (73, 145). Further autoradiographic studies made in the giant algae *Valonia ventricosa* (146) revealed that no radioactivity could be detected in the wall that develops during culture. Taken together, these results appear to speak against the presence of Hyp-protein in the wall. The results may be explained, however, by the fact that Steward et al did not attempt to show chemically the presence or absence of protein in the walls they examined. Sadava & Chrispeels (131) did experiments similar to those of Israel et al (73) but with carrot explants. The autoradiography of phloem-parenchyma tissue from carrot roots labeled with ^{3}H proline and then plasmolyzed indicated that much of the incorporated proline was associated with the cell wall.

Chrispeels (23) was the first to identify a salt-extractable Hyp-containing protein in carrot roots, and suggested that it might be the precursor to the covalently bound cell wall extensin (12, 24). The synthesis and secretion of extensin in carrot roots were shown to be enhanced by slicing and aeration of the tissue (25). This process may be involved in structural reformation of the wall, or perhaps in disease resistance following wounding (25, 149). The first reported purification of extensin from carrot root was by Stuart & Varner (149). Since then, several laboratories have purified extensin from different plants and tissues, such as potato tuber (96), tobacco callus (109), tomato cell suspension cultures (138), and soybean seed coats (16).

The only physiological role proposed for extensin is in the protection of the plant from pathogen attack and desiccation at the wounded surface (32). Extensin may have a structural role in the development and maturation of sclereids in soybean seed coats (17), cells that are involved in the protective function of the seed coat. Moreover, an extremely hydroxyproline-rich sulfated glycoprotein is expressed under strict developmental control in inverting *Volvox* colonies, which supports the idea of a functional role of this glycoprotein in inversion (133). This sulfated glycoprotein, I-SG, whose main sugar constituents are arabinose and galactose, contains the highest amount of Hyp (62%) known so far. It is synthesized for less than 10 min within the 48-hr life cycle of *Volvox*. Peptide sequences from this I-SG glycoprotein revealed repeating clusters of Hyp residues such as Leu-Arg-Hyp-Hyp-Hyp-Hyp-Hyp-Hyp-Hyp-Hyp-Arg-Hyp-Hyp-Hyp-Hyp-Leu-Leu-Hyp-Gln-Ala-Hyp-Phe and Val-Ala-Ser-Asn-Hyp-Ser-Hyp-Hyp-(Hyp)$_{>6}$- (133).

Perhaps the best example of a structural role for extensin in the wall comes from studies of *Chlamydomonas* cell walls in which several hydroxyproline-rich glycoproteins constitute the major structural components (61, 125). The structure and assembly of cell walls in *Chlamydomonas reinhardtii* have been analyzed in vitro (61, 125). Each wall consist of chaotrope-insoluble (W1, W2) and chaotrope-soluble (W4, W6) layers. The W6 is of particular interest in that it displays a crystalline lattice (61, 127). Moreover, dialysis of chaotrope-soluble W6 components leads to their in vitro reassembly into native crystalline arrays (70). The four major glycoproteins of the complex GP1, GP1.5, GP2, and GP3, have been purified and characterized morphologically by transmission electron microscopy and by amino acid composition. Three of the four are hydroxyproline-rich glycoproteins that co-polymerize to form the W6 layer. The fourth (GP1.5) is a glycine-rich glycoprotein apparently found within the W4 layer (61).

A system to study W6 assembly in a quantitative fashion has been developed (1) that employs perchlorate-extracted *Chlamydomonas* cells as nucleating agents. Wall reconstitution by biotinylated W6 monomers was monitored by FITC (fluorescein isothiocyanate)-streptavidin fluorescence and quick-freeze/deep-etch electron microscopy. Assembly occurred from multiple nucleation sites, and reflected the structure of the intact W6 layer. The W6 components cannot assemble onto two wall-less mutants (*cw-2* and *cw-15*); nonetheless, W6 glycoproteins secreted into the medium by *cw-2* are capable of assembly onto wild-type cells (1, 70). W6 sublayers can be assembled from purified components: GP2 and GP3 coassembled to form the inner (W6A) sublayer; this sublayer then served as a substrate for self-assembly of GP1 into the outer (W6B) sublayer (1). Interestingly, evolutionary relationships between *C. eugametos* and *Volvox carteri* have been studied by performing interspecific reconstitutions. Hybrid walls can be obtained between *C. reinhardtii* and *Volvox* but not with *C. eugametos,* which confirms taxonomic assignments based on structural criteria (1).

Primary cell walls, the first to be formed by the cell, from dicotyledoneous plants contain 5–10% protein and 2% hyp (87, 151). The content of protein in wood is relatively low (84), but it is approximately 20% in sclereids, cells with thick secondary walls of developing soybean seed coats (G. I. Cassab, J. E. Varner, unpublished results). In these seed coat sclereids, extensin represents approximately 7% of the dry weight of the cells (17). An earlier report (128), suggested the presence of a proline-rich substance in wound vessel members. Wound vessel members are xylem cells (with secondary walls) differentiated from the parenchymatous cells of the pith, formed when incisions are made in a stem. It was proposed that this proline-rich substance may be the "xylogenic factor" in initiating wound vessel member differentiation. However, no further studies have sought to detect Hyp-containing extensin in wound vessel members, studies that might reveal a role of extensin in wounding and cell differentiation.

Extensin may not be the only structural protein in the cell wall. With the use of molecular biology techniques, some investigators have reported other structural cell wall proteins, like the glycine-rich protein in petunia (30), the proline-rich protein in carrot (21, 157), and the proline-rich protein in soybean (4, 71). On the other hand, a 28-kD glycoprotein that is accumulated at low water potentials has been reported in the cell wall of the growing stems of soybean (10, 11). This 28-kd glycoprotein does not contain Hyp and is normally associated with early wall growth, in either a catalytic or structural role. Moreover, a 70-kD protein, mainly extractable from mature cell walls of soybean stems, appeared to decrease at low water potentials (10, 11). Further, an arabinogalactan fraction from the extracellular polysaccharides of suspension cultured sycamore cells contains approximately 3.7% protein (g protein/100 g dry weight) (144). This protein is characterized by large amounts of histidinyl residues (25 mole%), tryptophan (17 mole%), and no detectable residues of Hyp, an amino acid usually present in arabinogalactan proteins (50). Recently, a hydroxyproline-rich glycoprotein from the cell walls of maize cell suspension cultures was purified and characterized. This glycoprotein is unusually rich in threonine (25 mole%), proline (15 mole%), and hydroxyproline (25 mole%) (76). In general, monocotyledon cell walls contain low levels of Hyp, with the exception of seed coats and pericarps (161). A glycoprotein rich in Hyp and threonine was also isolated, but from maize pericarp (9).

Our knowledge of structural cell wall proteins has only begun to develop. Future studies on the characterization, localization, assembly, and interaction with other cell.and/or cell wall components should provide surprising insights into how cell walls affect the growth, development, and function of plant cells.

Biochemical Characterization

So far, the best-characterized salt-extractable extensin is the one present in carrot root cell walls. The phloem-parenchyma tissue from carrot root synthesizes large amounts of peptidyl-Hyp following wounding and aeration (23, 149). This Hyp accumulates in a salt-extractable glycoprotein in the cell wall (149, 162). Another hydroxyproline-rich glycoprotein, called extensin-2, has been reported in aerated carrot roots (141). The extensin reported by Van Holst & Varner (162) is not a special wound protein, but a protein normally present in the root that accumulates in wounded tissue. The amino acid composition of the soluble carrot extensin is mostly Hyp, Ser, His, Tyr, Lys, and Val; and many amino acids are not present at all. The abundance of Lys and low content of Asp and Glu contribute to the high isoelectric point observed in this molecule. This glycoprotein consists of 35% protein and 65% carbohydrate. Arabinose represents 97% of the sugar present, and galactose only 3%. The arabinose is attached to Hyp in short side chains of mainly four and three residues, and galactose is presumably linked to serine (92).

The secondary structure of extensin from carrot root has been studied by circular dichrometry (162). The spectra show that extensin is completely in the polyproline II conformation (an extended left-handed helix). If extensin is deglycosylated, much of the polyproline II conformation of the peptide backbone is lost, which indicates that the glycosylated portion reinforces this conformation. The secondary structure of extensin is consistent with the information obtained from electron micrographs of the glycoprotein. A protein in a polyproline II conformation with a molecular mass for the polypeptide portion of approximately 30kD will be a narrow rod-like molecule 80 nm long. Extensin proves to be too narrow to be detectable by negative staining with uranyl acetate, but it can be detected by rotary shadowing with platinum/carbon (69, 160). The micrographs of pure extensin showed thin rod-like molecules with an average length of 80–84 nm (140, 162).

Extensin has also been isolated from different plants and tissues, such as potato tuber (96), tobacco callus (109), tomato cell suspension cultures (138), and soybean seed coats (16). In all these glycoproteins, Hyp is the major amino acid representing 33–42 mole% of the total amino acids. Other abundant amino acids are Ser, His, Lys, Tyr, Val, and Pro. Arabinose and galactose are the only carbohydrates present in the protein (16, 96, 109). The relative amounts of oligoarabinose substituents of the Hyp seem to be constant for a given species (93). In tomato cell suspension cultures, Hyp-tetra-arabinosides and triarabinosides predominate in both extensins isolated

(138). In soybean seed coat extensin, arabinose is mainly bound to Hyp in side chains of three arabinosyl residues (16).

It has been proposed that extensin is slowly insolubilized in the cell wall by a covalent link (33, 138). One proposed covalent link is the isodityrosine formed between two tyrosine residues from different extensin molecules (55, 57). There is evidence that isodityrosine is present in tomato extensin from cell suspension cultures (45) and in cell wall hydrolyzates from callus of several plants (55). Soluble dimers of extensin have been isolated from carrot root cell walls, suggesting intermolecular linking, but it was not determined whether the link was isodityrosine (140). Carrot root cell walls can insolubilize soluble monomeric extensin in vitro in parallel with the formation of isodityrosine (34), and wall-bound peroxidase has been implicated in this process (34, 57). However, isodityrosine is absent in *Chlamydomonas* cell walls (126) and in soybean seed coat cell walls (G. I. Cassab, J. E. Varner, unpublished results), where hydroxyproline-rich glycoproteins and extensin are abundant. Moreover, in a tomato tryptic peptide from partially hydrolyzed cell walls, isodityrosine is formed between two tyrosine residues of the same polypeptide and not between tyrosines from different polypeptides (45). Thus to date no intermolecular crosslink has been characterized. During seed development, extensin is highly accumulated in the cell walls of soybean seed coats. The most (about 30% of the total cell wall protein) salt-extractable extensin is seen in seeds at 21 days after anthesis (17). However, once the seed is desiccated (mature dry stage), no extensin can be extracted (G. I. Cassab, J. E. Varner, unpublished results). The insolubilization of extensin may not be the consequence of a covalent link but of an irreversible change in the conformation of the wall during desiccation.

Amino acid sequences of two different extensin monomers from tomato cell suspension cultures, P1 and P2, have been reported (139). The tryptic peptide maps showed that both extensin precursors are highly periodic structures. P1 contains primarily two different peptide blocks: Ser-Hyp-Hyp-Hyp-Hyp-Thr-Hyp-Val-Tyr-Lys and Ser-Hyp-Hyp-Hyp-Hyp-Val-Lys-Pro-Tyr-His-Pro-Thr-Hyp-Val-Tyr-Lys; and P2 may consist entirely of a single repeating decapeptide, Ser-Hyp-Hyp-Hyp-Hyp-Val-Tyr-Lys-Tyr-Lys. These sequences of P1 and P2 show two different repeated domains, one of glycosylated Ser-$(Hyp)_4$ sequences and the other a nonglycosylated and repeated domain. The significance of two different domains in the extensin molecule has been discussed (139). Lamport proposed that the glycosylated domain is relatively rigid and the nonglycosylated one flexible, a structure that might allow the weaving of the cellulose microfibrils of the primary wall with an extensin network of defined porosity, the so-called "warp-weft" model (warp, cellulose; weft, extensin).

In summary, all the extensins that have been characterized are highly basic molecules with much lysine, little aspartate and glutamate, and much Hyp and arabinose. Most of the Hyp is found in Ser-$(Hyp)_4$ peptide sequences. The Hyp-arabinosylation pattern varies from species to species, as does the content of non-Hyp amino acids. The variability observed in the different extensins may be plant specific and tissue specific. In some cases, there are two different extensins in the same tissue or cell type. This may tell us that extensin can have different functions according to the cell type, and even in the same cell type or tissue.

Molecular Biology Studies

The existence of multiple forms of extensin raises questions concerning the number of different genes and mRNAs. Answers to these questions are most readily obtained by recombinant DNA methods. Such methods also offer the simplest means of determining the primary sequences of extensins. Although the repetitive segments of the primary sequences of two different extensins from tomato cell suspension cultures (139) have been determined, completion of the sequence by protein chemical methods may prove difficult because of the presence of many imino acid residues and of many posttranslational modifications. Chen & Varner isolated and sequenced a partial cDNA clone for carrot root extensin, which provided the sequence of the carboxyl-terminal (21). This partial cDNA clone isolated from a cDNA library from wounded carrot root mRNA encodes a peptide containing Ser-$(Pro)_4$ repeats and Tyr-Lys-Tyr-Lys (21) sequences also found in the tomato extensin (45, 139). Using extensin cDNA clones as probes, six different clones from carrot genomic libraries were isolated (22). One of the genomic clones was characterized and found to contain an open reading frame possibly encoding extensin, and a single intron in the 3'-noncoding region. The derived amino acid sequence contained a putative signal peptide, and 25 Ser-$(Pro)_4$ sequences. Two different extensin RNA transcripts were found corresponding to the genomic clone with different 5' start sites. Both transcripts increase markedly upon wounding, which correlates with the extensin accumulation seen in the cell wall after wounding in carrot roots (22, 25, 149). However, immunoprecipitation with specific antibodies against extensin of the hybrid-release RNA translation products are necessary to confirm that this particular gene encodes the extensin isolated from carrot root.

The effect of ethylene on the expression of the gene(s) encoding extensin has been analyzed (42). The accumulation of mRNA for extensin following ethylene was different from that of wounding. Ethylene induces two extensin mRNAs (1.8 and 4.0 kb), whereas wounding of carrot root produces the accumulation of an additional extensin mRNA (1.5 kb). These results suggest that the two signals, ethylene and wounding, are distinct (42).

A tomato extensin genomic clone has been isolated with the genomic clone for carrot extensin as a probe (136). The sequence of this clone encodes a polypeptide with numerous Ser-(Pro)$_4$ repeats, which are usually followed by Val-His or Val-Ala. It has been reported that there is an increase in possible extensin mRNAs in elicitor-treated bean cells in suspension culture and infected bean hypocotyls during race : cultivar-specific interactions with *Colletotrichum lindemunthianum,* causal agent of anthracnose (136). On the other hand, this clone has been used to study probable extensin mRNA levels in unwounded and wounded tomato stems and leaf tissue (137). When tomato stems are wounded, the expression of a new and larger extensin transcript (1.7 kb) occurred, together with the disappearance of two extensin transcripts (both about 1.2 kb) that were present in unwounded stems. The accumulation of a larger transcript (4.9 kb), presumably analogous to the one present in carrot root Northern blots, was also detected. Leaf tissue has very low levels of extensin mRNA, even after wounding, which is consistent with the trace amount of Hyp (as determined by chemical analysis) of leaves (16). The fact that there are multiple RNAs present in the same plant may reflect the presence of a multigene family; on the other hand, these multiple mRNAs may all be encoded by a single gene and arise by alternative splicing, as in the case of fibronectin (134).

Nucleic acid studies have indicated the possible presence of other classes of structural proteins in the cell wall: the proline-rich and the glycine-rich proteins. A cDNA from carrot root has been isolated (21) that encodes a proline-rich 33kD protein, whose RNA is also markedly accumulated upon wounding. The predicted peptide sequence of the cDNA that encodes for the proline-rich protein contains 24 repeat units of Pro-Pro-Val-Xa-Xaa. Subsequent studies have shown that this proline-rich protein is present in the cell wall of carrot root (157). An expressed gene from petunia encoding a glycine-rich protein has been characterized (30). The predicted amino acid sequence of the glycine-rich gene contains 67% glycine residues. The entire amino acid sequence of this clone can be represented as (Gly-X)$_n$, where X is frequently glycine. The product of this gene is presumably secreted because it begins with a transit peptide sequence. It has been proposed that the protein encoded by this gene is likely to function as a cell wall structural protein (30). Recently, it has been shown that there is differential expression of this gene according to the developmental stage of leaves and stems, young tissues having higher mRNA levels (31). Moreover, it was found that this mRNA increases within 5 min after wounding. The increased accumulation of this transcript coding for a glycine-rich structural protein by wounding appears to be one of the earliest events of the plant wound response.

There is preliminary evidence that glycine-rich proteins are present in the cell walls of pumpkin seed coats (163). In some plant tissues, where the Hyp content in the cell wall is low, glycine is a major fraction of the total protein nitrogen. These include the gourd *(Cucurbita ficifolia)* seed coat (21% glycine); the pumpkin *(Cucurbita pepo)* seed coat, where the major protein of the cell walls contains more than 47% glycine; milkweed *(Periploca graeca)* stem (cell walls, 31% glycine); and oat *(Avena sativa)* coleoptiles (epidermal cell walls, 27% glycine; D. Pope, personal communication). On the other hand, a glycine-rich glycoprotein has been characterized in the cell wall of *Chlamydomonas reinhardtii* (61). The discovery and characterization of a gene encoding a protein rich in glycine and most probably secreted, and the characterization of a cDNA for a proline-rich protein also probably localized in the wall may mean that the cell wall contains other kinds of structural proteins in addition to or instead of the extensins. These different proteins would reflect and determine the diverse functions of plant cell walls.

Recently Franssen et al (51) reported the characterization of a cDNA for nodulin-75 of soybean, a gene product involved in early stages of root development. The cDNA clone hybridizes to nodule-specific RNA. The RNA that was hybrid selected with this clone produced two nodulins when translated in vitro. These two nodulins with an apparent molecular mass of 75 kD differ slightly in charge, and one contains no methionine. The amino acid composition derived from the DNA sequence shows that proline accounts for 45% of the residues, and the sequence of one of the nodulins contains at least 20 repeating heptapeptide units, Pro-Pro-His-Glu-Lys-Pro-Pro. Whether any of these prolines is hydroxylated is not known. The amino acid composition of the nodulins-75, especially with respect to the high glutamic acid and the low serine content, does not resemble any of the hydroxyproline-rich glycoproteins already reported. This suggests that the nodulins belong to an as yet unidentified class of presumably structural proteins.

In addition, it has been reported that in soybean root nodules, the level of hydroxyproline-containing molecules is developmentally regulated (15). Hydroxyproline accumulates in early nodulation, and it is found later in development in large amounts in the cortex and in the medulla. In the cortex, the Hyp is mainly localized in the wall, presumably as extensin; but in the medulla it is mainly in the soluble fraction as an arabinogalactan protein, a class of hydroxyproline-rich glycoproteins (15). These results suggest that different (hydroxy)proline-rich proteins present in nodules may play a role in nodule formation and/or in the maintenance of the plant-bacteria interaction during nodule development.

Another gene that encodes a proline-(Hyp)-rich protein has been isolated from soybean (71). The cDNA and its corresponding genomic clone have been characterized. The nucleotide sequence of the gene corresponds to a proline-rich highly basic protein, designated SbPRP1, composed primarily of 43 repeat units of Pro-Pro-Val-Tyr-Lys and a putative signal sequence of 26 amino acids. The hybrid-select translation analyses produced a basic protein of 29 kD on an acid-urea/SDS-urea two-dimensional gel system. The presence of signal sequence on this SbPRP protein suggests that it may be a secreted extracellular cell wall protein. A comparison of the hydropathy profiles of the SbPRP1 gene with carrot 33-kD proline-rich protein derived from partial cDNA sequence analysis (21), and carrot extensin gene pDC5A1 (22) showed that each protein is highly hydrophilic, has a hydrophobic NH$_2$-terminal region, and has a repetitive structure (71).

Although there is no evidence that the encoded proline-rich protein is localized in the wall, it has some properties similar to those of other isolated wall proteins, i.e. it is highly basic, with high levels of Pro, Lys, and Tyr, and its repeating unit shares homology with that of extensins. On the other hand, the message for this SbPRP1 gene (1220 nucleotides) is observed in the mature section of the soybean hypocotyl tissue. In addition, treatment of the mature region with auxin leads to switching in size of the hybridizing RNA from the larger 1220-nucleotide form to a smaller 1050-nucleotide message. It has been proposed (71) that the SbPRP1 protein may play a role in cell differentiation and maturation, and that a change in cell wall protein gene expression in response to auxin could be related in these developmental transitions. The fact that some growth regulators—e.g. auxin and ethylene—affect the expression of cell wall proteins (42, 71) suggests that further studies on the effect of hormones on cell wall proteins are needed in order to understand the role of wall protein in cell growth.

Finally, another cDNA that codes for a proline-rich protein has been found (4). This clone hybridizes to an mRNA that accumulates in the axis of 5-day soybean seedlings. The derived amino acid sequence of this clone indicates a pentameric repeat of Pro-Pro-Val-Tyr-Lys, as reported by Hong et al (71), that occurs 24 times and lacks Ser and His. A 33-kD protein was isolated from cell walls of soybean cells in suspension culture that has an amino acid composition similar to that coded for by the cDNA clone (4). The protein contains 20% Hyp, 22% Pro, 17% Val, 16% Lys, 13% Tyr, and lacks Ser and His. This is clearly a novel hydroxyproline-rich protein. Although it is a basic protein, only half of the prolines are hydroxylated, and it has no Ser and His.

Cellular Localization and Possible Function in the Cell Wall

For several years it was thought that extensin was the major protein component of the primary wall of plant cells. Nonetheless, when the Hyp content was determined in several organs and tissues of the soybean plant, it was found that seed coats and root nodules contain the highest ratio of Hyp to dry weight compared to roots, leaves, stems, and flowers (15, 16). Furthermore, extensin was primarily localized in two of the external layers of the seed coat, palisade epidermal and hourglass cells, and in the cortex of soybean root nodules. These results allow the suggestion that the distribution of extensin in plant cells is not general but rather tissue specific. In order to assign a possible function to a protein, it is necessary to know in what type of cell it is present, and its cellular location. Thus, localization of extensin in different plant tissues is needed to answer this question. Specific antibodies to soybean seed coat extensin were used for immunocytolocalization, since preliminary efforts to separate the palisade layer from the hourglass cells by mechanical means and to discern where extensin is localized were unsuccessful (17). Immunogold-silver localization of extensin in the seed coat revealed marked deposition of the glycoprotein in the walls of both palisade and hourglass cells. The distribution of extensin among the different cells of the seed coat changes during seed development. Extensin is not detected at early developmental stages (1–6 days after anthesis). At this seed stage, the palisade cell layer has ceased division and begun its maturation process, and precursors of the hourglass cells can be seen. Seeds at 19 days after anthesis start accumulating extensin, primarily in the walls of the palisade cells, but also in the cytoplasm and walls of the hourglass cells, as well as in the parenchyma. At this developmental stage, immature hourglass cells begin a marked differentiation process, where the walls in the region of the cell equator become heavily thickened, preventing further expansion, while the cell ends retain their thin walls and continue to expand (67). Extensin is even more concentrated in the walls of palisade cells in seeds at 21 days after anthesis. More label is also detected in the walls of the hourglass cells, which at this stage are fully differentiated. In mature green seeds, extensin is heavily concentrated in the palisade cell walls close to the cuticle following the secondary wall deposit, and in the upper part of the wall of hourglass cells. In the hilum region, extensin is also heavily deposited, particularly in the counter-palisade and palisade cell walls.

A new simple technique, called "tissue-printing" on nitrocellulose paper, was developed to immunolocalize extensin in different plant tissues as well as plant species (17). Tissue-printing of developing soybean seeds shows that extensin is mainly localized in the seed coat, hilum, and vascular regions of the seed. These tissue-prints differ from the control prints stained with india ink, where the seed coat and cotyledons show uniform protein stain and the vascular region of the seed cannot be distinguished. The fact that extensin is localized in the sclerenchyma cells of the vascular elements of the seed as well as in the two types of sclereids present in the leguminous seed coat (macrosclereids and osteosclereids) suggests that extensin is a marker for the sclerenchyma tissue of the plant. The presence of extensin in the sclerenchyma could be related to the mechanical function of this tissue in the plant. As Haberlandt pointed out (65), there is harmony between the histological structure of mechanical cells and their functions. He showed at some length that the main physical properties of the cell walls of mechanical elements fully qualify them to act as the skeletal elements of the plant body. Schwender (1884, see 65), was the first to undertake an investigation of the elasticity and tensile strength of sclerenchyma cells, particularly fibers (Table 1).

The tensile strength of fibers below the limit of elasticity is remarkable. Their tensile strength lies usually between 15 and 20 kg/mm^2 and is thus equal to that of wrought iron; the fibers of *Nolina recurvata* match steel. Nonetheless, fibers differ from metals in two important characteristics: they are more extensible than any metal, and the extension at the limit of elasticity varies from 1–1.5% for fibers, while for metals it is less than 1%. In addition, only a very slight increase of tension over the limit of elasticity will cause fibers to break. In metals the difference is much larger; the breaking strength of wrought iron is almost three times as great as its elastic limit. Schwender remarks that "Nature has evidently concentrated her attention upon the tensile strength of mechanical cells, since great breaking strength would be of no value in the case of structures which cannot be stretched beyond the elastic limit without suffering injury".

There appears to be no correlation between the strength of fibers and the degree of lignification of their walls (Sonntag 1894, see 65). This may mean that components other than lignin contribute to the tensile strength of

Table 1 Comparison of tensile strength, breaking strength, and elongation at the elastic limit between plant fibers and metals[a]

Plant or Metal	Tensile strength (kg/mm^2)	Breaking strength (kg/mm^2)	Elongation at the elastic limit (per 1000 linear units)
Dianthus capitatus	14.3		7.5
Dasylirion longifolium	17.8	21.6	13.3
Dracaena indivisa	17	21.8	17
Phormium tenax	20	25	13
Hyacinthus orientalis	12.3	16.3	50
Allium porrum	14.7	17.6	38
Lillium auratum	19		7.6
Nolina recurvata	25		14.5
Papyrus antiquorum	20		15.2
Molinia coerulea	22		11
Secale cereale	15–20		4.4
Cibotium schiedei	18–20		10
Silver	11	29	
Copper (wire)	12.1		1
Brass (wire)	13.3		1.4
Wrought iron	13.1	41	0.7
Steel (German)	24.6	82	1.2

[a] From Haberlandt (65).

schlerenchyma cells. Extensin is a major component of sclerenchyma cell walls. It could be that extensin contributes, with other wall components, to the tensile strength of mechanical cells. Sclereids from soybean root nodules are also immunolabeled intensely with extensin antiserum (G. I. Cassab, J. E. Varner, unpublished results). The sclereids are localized in the cortex of nodules. The cortex of soybean root nodules contains large amounts of Hyp (15), which is present as extensin (G. I. Cassab, J. E. Varner, unpublished results). Hydroxyproline can be detected in very young nodules (one week after inoculation); thus extensin may be a marker of early nodulation. It is interesting that the cDNA for nodulin-75 of soybean, which encodes for two (hydroxy)proline-rich proteins, is expressed in nodule-like structures devoid of intracellular bacteria and infection threads. This indicates that these nodulins do not function in the infection process but more likely in nodule morphogenesis (51). Light microscopic studies on nodule initiation have shown (13) that at the stage in which the meristems emerge through the epidermis, nodule development can stop. Nonetheless, when a meristem has reached the emergence stage it will continue to develop into a mature nodule. The expression of the nodulin-75 genes coincides with the moment the soybean nodule meristems have reached a presumably critical developmental stage. This expression may reflect the commitment of meristems to develop into a nodule. The nature of the involvement of the (hydroxy)proline-rich nodulins-75 in this commitment remains to be established, as well as their precise cellular location in the nodule. This function resembles the function of extensin found in palisade and hourglass cells, where it plays a role in cell maturation and differentiation.

The collenchyma cells, which are also part of the mechanical system of the plant, have an absolute strength lower than that of fiber strands. On an average, collenchymatous walls break under a strain of not less than 10–12 kg/mm^2. The main difference between collenchyma and sclerenchyma is the fact that the limit of elasticity is much lower in the former tissue, where a load of 1.5–2 kg/mm^2 is sufficient to produce a permanent elongation (3). In other words, fibers are elastic and collenchyma is plastic. This physical peculiarity of collenchymatous walls is related to the special function of this tissue—that is, to give mechanical support during intercalary growth, without preventing longitudinal extension. If fibers were to differentiate in growing organs, they would hinder tissue elongation because of their tendency to regain their original length when stretched. Collenchyma is the first supporting tissue that occurs in peripheral position in stems, leaves, and floral parts (46). The structure of the wall is the most distinctive feature of collenchyma cells. The wall thickenings are depostied unevenly, particularly in the corners where several cells are joined together. This peculiarity is closely connected with the fact that collenchyma serves as the mechanical tissue of growing organs. The wall thickness in collenchyma is increased if during development the plant is exposed to motion by wind (166). Loss of wall material in collenchyma was also induced experimentally by etiolation (166).

The walls of collenchyma consist mainly of cellulose and pectic compounds and contain much water—over 60% water, based on fresh weight. It is not known what kind of structural protein(s), if any, is present in the walls of collenchyma cells. Collenchyma may be a suitable system for studying what type of structural proteins are present in the wall, whether they are localized in the wall thickenings, and whether they contribute to the mechanical function of these cells. On the other hand, it will be interesting to know whether structural proteins accumulate when plants are exposed to motion by wind, or decrease by etiolation. Such studies, together with what it is known about sclerenchyma cell walls, will help to reveal the role of wall proteins in the architecture of the plant body.

To appreciate the materials comprising plant skeletal systems, compare their mechanical constants with those of ordinary cellulose walls, such as parenchyma or cells in culture. According to Schewender, the tensile strength of the thin walls of parenchymatous and cortical cells of young stems is 1 kg/mm^2. There is no doubt that these cells are not specialized for mechanical purposes, since in the walls of sclereids, the breaking strain is 10–25 times as great. The tensile strength of parenchyma cells may be related to the very low level of extensin found in their walls. In the parenchymatous cell walls of soybean seed coats, extensin is almost absent (16, 17). Further studies may enable us to correlate the presence of a structural protein with a particular function of a plant cell.

Extensin has also been immunolocalized in cell walls of carrot roots (142). In carrot phloem parenchyma walls, extensin-1 is distributed uniformly across the wall but is absent from the expanded middle lamella at the intersection of three or more cells, and is reduced in the narrow middle lamella between two cells. This distribution is the same as that of cellulose. Results from immunolocalization studies in carrot root suggest that: (a) newly synthesized extensin-1 is added to the wall by intussusception—that is, by intercalation of new particles among those existing in the wall; (b) extensin cannot cross the middle lamella; and (c) incorporation of extensin is a late event in the development of carrot phloem parenchyma walls, since it is absent in walls of young root tissue. Extensin immunolocalization in carrot root was done in unwounded tissue. Wounding enhances accumulation of extensin (25, 149), but the variability of immunolabeling seen from one root to the next made it difficult to determine directly where in the wall newly synthesized extensin becomes incorporated (142).

The role of extensin in plant growth is unknown. Extensin accumulation in the wall of growing pea epicotyls was coincident with the cessation of cell elongation (132). On the other hand, it has been reported that in elongating tissue of pea stems the protein content of the wall increases as well as the content of Hyp and Hyp-arabinosides; these changed only slightly once elongation was complete (78). There was an inverse relation between the rate of elongation and the concentration of the hydroxyproline-rich wall proteins, which suggested that extensin stiffens the wall during growth, thus reducing the rate of elongation (78).

Other immunocytochemical studies of extensin have been reported (107). Extensins accumulated in the walls of living, uninfected cells close to sites where fungal and bacterial growth is restricted by plants, an observation supporting their proposed role in disease resistance. Extensins also accumulated in plant papillae, which may present a physical barrier to penetration by fungi (107).

Several studies suggest a role for extensins in plant defense. Extensins accumulate in melon plants infected with the fungus *Colletotrichum lagenarium* (106). Artificial enhancement or suppression of extensin levels in melon plants results in increased or decreased resistance to *C. lagenarium* (48). Cell wall Hyp levels increase more rapidly in resistant than in susceptible cultivars of cucumber infected with the fungus *Cladosporium cucumerinum* (66). Elicitor treatment of melon and soybean hypocotyls results in the stimulation of extensin synthesis (129). There is rapid accumulation of extensin mRNAs in resonse to fungal infection and elicitor treatment (136). The exact role of extensins in the defense response is not clear, but they may act as structural barriers, provide matrixes for the deposition of lignin, and/or act as nonspecific agglutinins of microbial pathogens. Furthermore, extensin mRNAs accumulated rapidly in response to ethylene. The earliest detectable event during plant-pathogen interaction is a rapid increase in ethylene biosynthesis. It has been proposed that ethylene produced in response to biological stress is a signal for plants to activate defense mechanisms against invading pathogens (42).

Cell wall proteins unrelated to extensin have been shown to accumulate in response to a variety of developmental and stress signals (10, 11, 157). The 33-kD cell wall protein was first identified in carrot root (21, 157). The mRNA encoding the proline-rich protein accumulates in carrot root within 1 hr after wounding, and the level of this protein increases in the wall within 24 hr after wounding, as detected by Western blot analysis using an antibody against a synthetic peptide (157). The sequence of the synthetic peptide is present in the partial cDNA clone that codes for the 33-kD protein (21). Two other cell wall proteins from soybean seedlings have been reported recently (10, 11) that are differentially accumulated in young and mature tissue, and also respond to low water potentials. Dark-grown soybean seedlings showed decreased stem growth when the roots were exposed to low water potentials

(19). After a time, growth resumed but at a reduced rate relative to the controls. A 28-kD protein is accumulated at lower water potentials, particularly in the dividing and elongating region of the seedling. The amount of the 28-kD protein increased in the wall of the dividing region after the initial growth inhibition, and it appeared in the elongation region when growth recovered (10, 11). On the other hand, a 70-kD protein extractable from mature cell walls seemed to decrease slightly at low water potentials. Because this 70-kD protein is associated with mature tissue, it may play a role in the maturation of walls. These results indicate that biochemical changes occur in the cell walls after periods of low water potentials (10, 11, 150).

The expression of the glycine-rich protein, presumably a structural protein of the wall, in petunia has been investigated (31). This gene is expressed on a single RNA species of 1600 bases in leaves, stems, and flowers but not in roots. The steady-state mRNA levels were highest in stems and leaves and lowest in flowers. Furthermore, the steady-state level of the glycine-rich protein message is greater in young than in old tissue. This pattern differs from that of the tomato extensin gene *tom5* (136), which is expressed in higher levels in roots and stems than in leaves and is higher in old than in young tissue (137). These results suggest that the glycine-rich protein is involved in early development of primary wall and presumably is not associated with wall strength (31). The glycine-rich mRNA also responds to wounding. The mRNA accumulates within 5 min after wounding. The enhancement of this message by wounding seems to be one of the earliest events of the wound response in petunia (31).

As more genes that encode cell wall proteins are isolated and characterized, it will be possible to discover their function by using genetic engineering techniques, particularly by altering their expression either by manipulating regulatory sequences or by inactivation of transcripts with antisense RNA (41).

Interaction with Other Cell Wall Components

When we consider that different cell types in the plant have different cell wall compositions and that this composition is determined by developmental stage and by external signals such as light, flexion, rubbing, wounding, infection, etc, it seems self-evident that the wall components interact with each other differently in different cells. The challenge is to find out exactly what these interactions are in muro.

It is generally agreed that much of the xyloglucan of the dicot wall is strongly associated by noncovalent bonds to cellulose microfibrils (68). Xyloglucans are therefore logical candidates for maintaining the integrity of the wall framework. Other polysaccharides not so firmly bound to cellulose might interact with the xyloglucans or might simply form a gel of sufficient strength to prevent protoplasmic blow-out between the cellulose microfibrils held in place by xyloglucans. The gel properties of wall polysaccharides in vitro have been reviewed recently (5, 49). The gel properties of the polyanionic polysaccharides are dependent upon degree and pattern of esterification, pH, and Ca^{2+} (or Mg^{2+}) concentration.

Extensins are basic proteins because of their high lysine content, and the lysine residues are regularly spaced along the extended peptide backbone. Therefore an interaction in the wall between extensin and the block polyanion regions of pectin seems inevitable. However, antibodies against rhamnogalacturonan 1, the major pectic polysaccharide of sycamore suspension cell walls, find it only in the middle lamella (110). In contrast, antibodies against xyloglucan label the sycamore primary cell wall, including the middle lamella, uniformly. Antibodies against carrot root extensin do not label the middle lamella of carrot root cells but do stain all other regions of the primary wall (142). As discussed above, extensin is not uniformly distributed across the walls of the soybean seed coat macrosclereid and osteosclereid cells. These results taken together raise the possibility that layering of various components occurs.

If extensin and acidic blocks of pectin come together an electrovalent zippering up of opposite charges might well occur. Because extensin is 80 nm long with about 30 lysine residues, a single extensin molecule might fasten three or four pectin molecules together. Lowering the pH or increasing [Ca^2] would unzipper the complex. Such a reversible assembly of wall macromolecules could be useful in regulating wall yield strength, in regulating wall pore size to allow insertion of new wall material or the secretion of macromolecules through the wall, and in positioning wall components for irreversible crosslinking to stop cell expansion.

If the distribution of the polycationic wall proteins and the polyanionic pectins is such that they cannot interact, reversible covalent binding of protein to polysaccharide might nonetheless occur through reaction of the ϵ-amino groups of the lysines with the reducing ends of the polysaccharides (Schiff's base). Thus one protein molecule might orient several neutral polysaccharide molecules without irreversibly crosslinking them. And the report of an oxohexuronic acid in the walls of several plant species (118) allows the possibility for extensin to crosslink polysaccharides by Schiff's base linkages.

Some of the extensins are histidine rich; some are histidine poor. The histidine imidazole nitrogen has a pK of about 6. Therefore the charge on the nitrogen would vary during physiological changes in the wall pH, as would the opportunities for interaction with wall polyanions. It seems within the realm of possibilities that the histidine residues are attacked by oxygen free radicals to open the imidazole ring and generate an aldehyde (124). This would allow for a reversible polymerization, through Schiff's base linkages, of extensin molecules.

Complete elucidation of the distribution of the various components within the walls of each cell type will require the use of probes for specific components. These probes will include component specific antibodies, labeled enzymes that bind specifically to a wall component, carbohydrate hybridization probes (Figure 1), probes for specific amino acid residues (e.g. isodityrosine), probes for free aldehyde groups, etc.

Localization of each of the wall components needs to be followed by information about possible changes in conformation of these components under transient changes in ionic strength, pH, $[Ca^{2+}]$, water potential, degree of methylation, and electropotential across the wall. The kinetics of enzyme activity have been used to follow changes in the environment of wall acid phosphatase (35). At low ionic strength, wall-bound acid phosphatase shows negative cooperativity while in free solution it shows Michaelis-Menten kinetics. At high ionic strength the bound enzyme shows the same kinetics as the free enzyme at low ionic strength. Removal of Ca^{2+} from the wall inhibits the wall-bound enzyme; readdition of the Ca^{2+} restores the original activity. Phosphatase in free solution is unaffected by changes in $[Ca^{2+}]$ or changes in ionic strength. It is concluded that in the wall compartments, loading and unloading Ca^{2+} during cell division and cell elongation may regulate the activity of cell wall enzymes such as phosphatase (35) and pectin methylesterase (111, 113) in a way that regulates and coordinates wall extension. These observations and conclusions have been elaborated into an intriguing model for an ionic control of cell wall expansion (123). Even if the model is incorrect it is clear that acid phosphatase and probably other wall enzymes can be used as "reporters" about conditions within the wall. Ongoing characterization of several enzymes extracted from purified walls (112) offers additional opportunities to use the properties of wall-bound enzymes as reporters of their environment.

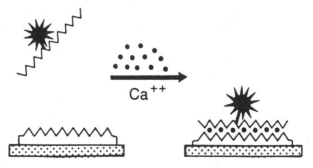

Figure 1   Alginate G blocks self-associate in the presence of calcium or certain other divalent cations to form dimers and aggregates. When a purified G block is labeled in vitro with a fluorochrome or other marker and added back to cells, it can identify another G block in bound alginate. From Vreeland et al (165).

There are many elegant studies of the in vitro physical changes in the gel-forming components of cell walls. For example, the chain association of pectic molecules with different levels and patterns of esterification during calcium-induced gelation has been studied by light scattering, circular dichrometry, viscometry, and determination of Ca^{2+} activity and Ca^{2+} transport parameters (20, 153, 154). These experiments tell us what is possible without telling us exactly what goes on in any given wall at any given stage of development. We now need new methods of examining in muro, in situ, and in vivo the gel properties of specific regions of specific cell walls.

CELL WALL ENZYMES

Several enzymes have been detected in the cell walls of higher plants, and the possibility has been considered that they may play a part in cell wall metabolism. These enzymes, usually glycoproteins, can be extracted from the wall to varying degrees. Of all glycoenzymes, peroxidase is the most studied, though it is misleading to term such diverse peroxidatic activities a single enzyme when they comprise a family of several. The oligosaccharide substituent of horseradish peroxidase isozyme C contains the ubiquitous branched core mannan, attached via a chitobiosyl unit to aspargine (29) as found in extracellular enzymes, which may be related to increased solubility, thermal stability, and resistance to proteases (89). Peroxidases can catalyze phenolic crosslinks between macromolecules such as lignin (63), protein (81), hemicellulose (167), and ferulic acid (56). Peroxidase may also crosslink proteins by oxidative deamination of lysine (143).

The polymers of growing plant cell walls contain phenolic side chains, such as the tyrosine residues of extensin (55), and ferulic acid and p-coumaric acid residues attached to polysaccharides, pectic arabinogalactan in dicotyledons (56), and hemicellulosic arabinoxylans in monocotyledons (114). These phenolic side chains are quantitatively minor components of the wall, but they might nonetheless play a role in wall architecture by participating in the formation of inter-polymer crosslinks (57). The formation of this type of crosslink has been demonstrated in vitro. Horseradish peroxidase plus H_2O_2 at low pH will dimerize tyrosine residues of ionically bound extensin present in isolated cell walls to isodityrosine (34, 57). Similarly, a gel of feruloyl-polysaccharides in the presence of peroxidase plus H_2O_2 can be oxidized from feruloyl side chains to diferuloyl crosslinks (60). In addition, diferulate and isodityrosine have been found in vivo in some plant tissues (55, 57), which suggests that the oxidative coupling reaction may occur in muro. These results taken together suggest that wall-bound peroxidase can crosslink extensin via isodityrosine, or pectic fragments by diferuloyl bridges. This cannot be a firm conclusion because we lack proof that isodityrosine or diferuloyl bridges are subtended by either tyrosine residues in different extensin molecules, or by a pair of different pectic fragments. It has been recently reported (75) that a salt-elutable enzyme from cell walls of tomato cell suspension cultures has the ability to crosslink extensin monomers. The crude activity eluted from intact 7–8-day-old cultures was heat labile, H_2O_2 dependent, and had a pH optimum between pH 6 and 6.5. The crude activity did not crosslink bovine serum albumin, and horesradish peroxidase did not crosslink extensin monomers.

Because phenolic crosslinks between matrix polymers would probably reduce the extensibility of the growing cell wall, their appearance may be related to the control of cell growth rate. There is a negative correlation between wall peroxidase activity and growth rate. For instance, gibberellic acid promoted the expansion of spinach cells in suspension culture, and simultaneously suppressed peroxidase secretion, reduced the activity of phenylalanine ammonia-lyase, and favored the accumulation of wall-esterified ferulate (53, 54). Peroxidase may restrict growth by rigidifying the cell wall by catalyzing the covalent conversion of feruloyl side chains into diferuloyl crosslinks, and by the noncovalent conversion of soluble phenolics into hydrophobic quinones. Feruloyl side groups bestow certain soluble poly-saccharides with the property of producing gels upon oxidation with H_2O_2 plus peroxidase. It has been speculated that this "oxidative gelation" of the matrix polysaccharides in the cell wall may affect its extensibility and thus its growth rate (88, 102). Wall rigidification may be prevented in the presence of gibberellic acid by inhibiting peroxidase secretion (53). Peroxidase is an abundant enzyme in plants, and it can be argued that its activity in the wall is unlikely to be a growth rate–limiting factor.

It has been proposed that isoperoxidase may be an early marker of the differentiation in vitro of *Zinnia* mesophyll cells into tracheary elements (27). Isoperoxidase is induced in the cell wall and soluble fraction 36 hr after induction of mesophyll cells to tracheary elements. Inhibitors that block or delay tracheary differentiation reduce the expression of this isozyme, which suggests that modulation of the level of the differentiation-specific isoperox-idase may be related to the formation of tracheary elements. How this isoperoxidase is involved in the differentiation of mesophyl cells into trache-ary elements is not known, but it means that the cell wall machinery can regulate cell differentiation.

Stylar peroxidase activity has been involved in incompatibility reactions in *Petunia* (14). Wall peroxidases are absent in cross-pollinated styles, but they are seen on the cell walls of the outer portion of the transmitting tissue in self- and nonpollinated styles. In accordance with these observations it has been suggested that wall peroxidases present in the cells of the outer portion of the transmitting tissue are involved in the gametophytic self-incompatibility of *Petunia*, and that nonpollinated styles, which are characterized by the presence of wall peroxidases in the outer portion of the transmitting tissue, can reject incompatible pollen tubes. The disappearance of peroxidase activity seems to be an important step in the compatibility pollination process (14). On the other hand peroxidase has also been proposed to be an indicator of differential response of corn isolines to heat stress (2). Peroxidase leakage from leaves was observed to be greater in heat stress–sensitive than tolerant cultivars of corn. Thus, a change in cell wall enzymes upon environmental stimuli is observed.

Peroxidases are also involved in the induction of new cell wall biosynthesis, which may be an important defensive response to pathogen attack (58, 135). They are induced by wounding and are presumably implicated in the repair of damaged walls (7) in which they in part form a watertight barrier over the wound by the deposition of polymeric aliphatic and aromatic compounds (47).

Plants contain several peroxidase isozymes whose pattern of expression is tissue specific, developmentally regulated, and controlled by environmental stimuli (83). Tissue samples from leaf, root, pith, and callus from tobacco showed a different isozyme fingerprint. In tobacco there are 12 isozymes classified in three subgroups; the anionic, the moderately anionic, and the cationic. Each subgroup seems to have a different function in the cells. The cationic peroxidase isozymes (pI 8.1–11) have been localized recently in the central vacuole (98) and catalyze the synthesis of H_2O_2 from NADH and H_2O (100). These isozymes have also an indoleacetic acid-oxidase activity in the absence of H_2O_2 (62), and may provide H_2O_2 to other peroxidase isozymes (100). Their actual function in the plant is unclear.

The moderately anionic peroxidase isozymes (pI 4.5–6.5) have been localized in the cell walls, and they posses a moderate activity toward lignin precursors. They are highly accumulated upon wounding in tobacco stems (83). However, the function of the anionic isoperoxidases (pI 3.4–4.0) is better understood. These isozymes are also associated with the cell wall and have a high activity in the polymerization of cinnamyl alcohols in vitro (99). They accumulate in wound-healing potato tuber (47). Immunocytochemical studies on wound-healing potato tuber with an antibody specific to an anionic peroxidase associated with suberization showed that only the inner side of cell walls of the periderm were labeled in the tuber tissue (47). The formation of suberized periderm seems to be the usual response to wounding in any plant organ (80). The aromatic domain may be similar to that of lignin, and the deposition of the aromatic polymeric domain of suberin involved an anionic-isoperoxidase. Suberin forms a barrier and consequently seals off the wound from water loss.

Thus peroxidases may be an important wall component involved in the defense mechanisms of plants. They are also involved in lignification and the crosslinking of extensin monomers and feruloylated polysaccharides (90). Recently the first cDNA encoding a plant peroxidase has been cloned (82). This may help in the elucidation of the physiological role that each of the isozymes plays in plant development. The cloned peroxidase from tobacco was an anionic isozyme. The predicted amino acid sequence showed a putative signal sequence, and the coding sequence when compared to that of the cationic peroxidases from horseradish and turnip was found to be 52% and 46% homologous, respectively. The message for the anionic isozyme was highly expressed in stems, but it was expressed at very low levels in leaves and roots (82).

Peroxidases are not the only cell wall enzymes that change their expression upon different environmental stimuli. Hydrolases are another class of plant proteins with a potential role in defense (8). Plant hydrolases are usually localized in the "lytic compartment" of cells, which includes the vacuoles and the cell wall space (104), although they are frequently found in the cytoplasmic fraction (39). A cell wall location seems to be appropriate for a function in defense, since pathogens necessarily attack cells from the outside and often have to penetrate the host wall in order to start the infection. Some hydrolases are secreted in the apoplastic space and can be readily extracted from the cell walls (β-glucosidase) with high salt treatment, while in other cases their activity (α-mannosidase) cannot be removed even with sequential salt treatment (112). The hydrolases that have been found in the wall compartment are: β-1,3-glucanase, cellulase, arabinosidases, β-fructofuranosidases, α-galactosidases, β-galactosidases, α-mannosidases, β-mannosidases, trehalases, β-glucosidase, β-glucuronidase, β-xylosidases, and acid phosphatases. However, many hydrolases are found almost exclusively in the central vacuole—i.e. chitinase (8), N-acetylglucosaminidase, and proteinase inhibitors. Thus, the vacuole may be considered as a depository for a defense arsenal. For instance, in the so-called hypersensitive reaction one or few cells close to the invading pathogen collapse and then set free their vacuolar contents, thus overwhelming the pathogen.

β-1,3-Glucanase is an abundant enzyme in higher plants (28). Callose (β-1,3-glucan) is a substrate of this enzyme that is present in sieve tubes, in cell appositions formed in response to wounding, and in primary cell walls. β-1,3-Glucanases have been studied primarily with regard to their function in the turnover and degradation of these structures (28). Nonetheless, β-1,3-glucanase is induced in response to infection (105), at the same time as chitinase. Purified β-1,3-glucanase from tomato has been shown to digest partially the cell walls of pathogenic fungus (171). β-1,3-Glucanase purified from soybean can release fragments from *Phytophthora* cell walls, which can activate the defense reactions in plants (74).

Cellulase plays a major role in cell wall degradation in plants, and has been much studied with regard to abscission. Two kinds of cellulase have been found, one that is abscission specific, and the other that is constitutively present in plant tissues and is of unknown function. However, some fungi contain cellulose rather than chitin in their walls, and it may be that the walls of such fungi can be attacked by this cellulase (8). Cellulase has also been implicated in the ripening of fleshy fruits. In avocados, the mRNA coding for cellulase is barely detectable in preclimacteric fruit but it is abundant in climacteric fruit (26).

α-Mannosidase shows high activity in most plants. It probably functions in the turnover of glycoproteins and in the processing of oligosaccharide derivatives that are implicated in the synthesis of many diverse glycoproteins. However, it may also attack specific wall structures of plant pathogens (8). Little is known about β-mannosidases from plants (39). The enzyme has been shown to be cell-wall bound in *Convolvulus arvensis* callus (79). Numerous β-glucosidases exist in the plant kingdom. Nonetheless, the precise role of β-glucosidases in plants has not been well explored (39). β-Glucosidase accumulation can be increased to 80 times the control level in tobacco that has been infected with the fungus *Peronospora tabacina* (43). The activity did not increase in response to abiotic stress or during senescence. The increased β-glucosidase activity was found histochemically in the mesophyll cells adjacent to the invading mycelium. It has been suggested that the increase in glucosidase activity is part of the defense mechanism of tobacco against this fungus (43). The ability of some β-glucosidase to hydrolyze cinnamyl alcohol glucosides is probably related to the biosynthesis of lignin. Several of these glucosides and enzymes have been detected in lignifying tissues of gymnosperms (52). The enzyme may also contribute to lignification of angiosperms (72) and may play a role in the phenylpropanoid metabolism.

Arabinosidases, namely arabinofuranosidases, are more commonly known. They have been detected in lupin seeds, where a fraction of the enzyme was cell wall bound (103). In lupin seeds, germination brings about modification of the chemical structure of the polysaccharides of the primary cell wall. This also occurs in fruit softening, where pectin-degrading enzymes have not been detected (39).

β-Fructofuranosidase (invertase) is one of the oldest known enzymes, isolated first from yeast by Berthelot in 1860. Invertase has an important function in storage organs, where it breaks down sucrose. Sucrose is the most abundant transportable free carbohydrate in plants. Invertases play an important role in the regulation of sucrose metabolism in plants. A high level of invertase activity is located in tissues with rapid growth (59). Multiple forms of invertase have been found in several plants (39). The presence of soluble and cell wall–bound forms of the enzyme have been reported (95, 97). β-Fructofuranosidase is a glycoprotein with high mannose and complex fucosylated glycans (95).

α-Galactosidase is widely distributed in plants, and its presence can be predicted in tissues that contain α-D-galactosyl-containing oligo- or polysaccharides, particularly in storage organs (39). Another possible function may be to protect plants from α-galactosidic phytotoxic substances (148). A relationship has been shown between α-galactosidase production and virulence of some plant pathogens (44). The physiological role of β-galactosidases is unclear. They are involved in the degradation of galactolipids, and thus they may change membrane characteristics. On the other hand, the fact that they are wall bound and present in root tips of several plants may have physiological significance (152).

Trehalase has been found in several plant species. However, its substrate, trehalose, has never been conclusively shown in higher plants, but it is common in bacteria and fungi (164). Trehalose can be toxic for plants with low trehalase activity. It has been suggested that trehalase activity may protect plants from the potentially harmful effect of microbial trehalose (164). The pollen of various plants contains trehalase, and such pollen can even germinate in media containing trehalose as a sole carbon source. However, the function of trehalase in pollen is unknown (64).

β-Glucuronidases have not been much studied, but they have been reported in the pollen walls of *Portulaca grandiflora,* where their level increases upon pollen germination. The role of this enzyme, as in the case of other glycosidases, is unknown (39). β-Xylosidases, also named hemicellulases, may play a role in the degradation of xyloglucans, which as already noted are important constituents of cell walls (37). β-Xylosidases were found to be involved in the mobilization of cell wall components of the wilting flowers of *Ipomea tricolor* (168), and they have also been implicated in host-pathogen interactions.

Finally, acid phosphatase is the most abundant wall-bound hydrolase in some tissues (91). The significance of the presence of acid phosphatase in the cell wall is not known, since cellular phosphate esters usually do not leave the cells. The possible role of acid phosphatase in plant-microbe interactions has been discussed (8).

Almost all the hydrolases discussed above have endogenous substrates as well as the potential exogenous substrates present in cell walls or products of pathogens. Most plant hydrolases are present in multiple forms. One set of isozymes may be involved in primary metabolism while the others may have evolved new properties useful in defense against pathogens. However, for some of the hydrolases (e.g. trehalase), no endogenous substrate is known. The possibility that this enzyme contributes to defense against pathogens cannot be excluded.

Pectinesterases have been found in all species of higher plants tested and are also produced by a number of plant pathogenic fungi and bacteria. Plant pectinesterases participate in the conversion of protopectin to soluble pectin and pectate, and are involved in plant maturation processes as well as in mechanisms that protect plants from infection. The enzyme catalyzes the deesterification of pectin and is highly specific for the D-galacturonan structure of the substrate. It has been suggested that by producing changes in gelling properties the gradual deesterification of pectin by pectinesterase may affect interactions between cell wall polymers associated with plant growth (170). Much attention has been devoted to tomato pectinesterase because it is so plentiful in the ripening fruit. The primary structure of the tomato enzyme has been determined (101).

Polygalacturonase is also widely distributed in plants. The richest plant source of polygalacturonase, as in the case of pectinesterase, is ripening tomato fruit. Both enzymes may be widely distributed since pectin occurs in most plant materials but is particularly abundant in young tissues and fruits. The role of this enzyme in cell wall degradation has been firmly established (130). The levels of polygalacturonase are developmentally regulated during tomato fruit ripening (38). Polygalacturonase activity and immunologically detectable protein are entirely absent from mature green fruits, and increase dramatically as ripening proceeds. Polygalacturonidase cDNA clones have been isolated from ripe tomato fruit to examine the accumulation of the enzyme mRNA during fruit ripening. Red-ripe tomato fruit contain at least 2000 times the polygalacturonase mRNA of immature-green fruit (38).

Most cell wall enzymes may be involved in protecting the plant from wounding or pathogen attack. However, it is difficult to evaluate the extent to which these enzymes contribute to plant resistance. Nonetheless, cell wall enzymes may also be important components of the wall that help determine the wall's structure throughout plant development.

CONCLUDING REMARKS

Wall components are more complex than anticipated. Part of the observed complexity derives from our habit of lumping together many different cell types before we do our biochemistry. Much of the complexity, however, is real. Because the processes of evolution tend to pare away unused form and discard unneeded function, most of the real complexity is likely necessary. If this is true our task is straightforward. We must identify all of the wall components of a given cell and assemble them (or let them self-assemble) into a wall that is functional for that particular cell.

ACKNOWLEDGMENTS

This work was supported by a grant from the US Department of Energy (DE-FG02-84ER13255) to J. E. Varner. G. I. Cassab is supported by a predoctoral fellowship from the Division of Biology and Biomedical Sciences at Washington University and a program training grant from the Monsanto Company.

Literature Cited

1. Adair, W. S., Steinmetz, S. A., Mattson, D. M., Goodenough, U. W., Heuser, J. E. 1987. Nucleated assembly of *Chlamydomonas* and *Volvox* cell walls. *J. Cell Biol.* In press

2. Akhtar, M., Garraway, M. O. 1987. Peroxidase: indicator of differential response of corn isolines to heat stress. *Plant Physiol. Suppl.* 83:422 (Abstr.)

3. Ambronn, H. 1881. Über die Entwicklungsgeschichte und die mechanischen Eigenschaften des Collenchyms. Ein Beitrag zur Kenntnis des mechanischen Gewebesystems. *Jahrb. Wiss. Bot.* 12: 473–541

4. Averyhart-Fullard, V., Datta, K., Marcus, A. 1987. A new hydroxyproline-rich protein in the soybean cell wall. *Proc. Natl. Acad. Sci. USA.* In press

5. Bacic, A., Harris, P. J., Stone, B. A. 1987. Structure and function of plant cell walls. In *Carbohydrate Structure and Function,* Vol. 12, ed. J. Preiss. New York: Academic

6. Bartnicki-Garcia, S. 1966. Chemistry of hyphal walls of Phytophtora. *J. Gen. Microbiol.* 42:57–69

7. Birecka, H., Miller, A. 1974. Cell wall and protoplast isoperoxidases in relation to injury, indoleacetic acid and ethylene effects. *Plant Physiol.* 53:569–74

8. Boller, T. 1987. Hydrolytic enzymes in plant disease resistance. In *Plant-Microbe Interactions,* ed. T. Kosuge, E. W. Nester, pp. 385–414. New York: Macmillan

9. Boundy, J. A., Wall, J. S., Turner, J. E., Woychik, J. H., Dimler, R. J. 1967. A mucopolysaccharide containing hydroxyproline from corn pericarp. *J. Biol. Chem.* 242:2410–15

10. Bozarth, C. S., Boyer, J. S. 1987. Characterization of a cell wall protein which is altered at low water potentials. *Suppl. Plant Physiol.* 83:282 (Abstr.)

11. Bozarth, C. S., Boyer, J. S. 1987. Cell wall proteins at low water potentials. *Plant Physiol.* 85:261–67

12. Brysk, M. M., Chrispeels, M. J. 1972. Isolation and partial characterization of a hydroxyproline-rich cell wall glycoprotein and its cytoplasmic precursor. *Biochem. Biophys. Acta* 251:421–32

13. Calvert, H. E., Pence, M., Pierce, M., Malik, N. S. A., Bauer, W. D. 1984. Anatomical analysis of the development and distribution of *Rhizobium* infections in soybean roots. *Can. J. Bot.* 62:2375–84

14. Carraro, L., Lombardo, G., Gerola, F. M. 1986. Stylar peroxidase and incompatibility reactions in *Petunia hybrida. J. Cell Sci.* 82:1–10

15. Cassab, G. I. 1986. Arabinogalactan proteins during the development of soybean root nodules. *Planta* 168:441–46

16. Cassab, G. I., Nieto-Sotelo, J., Cooper, J. B., Van Holst, G. J., Varner, J. E. 1985. A developmentally regulated hydroxyproline-rich glycoprotein from the cell walls of soybean seed coats. *Plant Physiol.* 77:532–35

17. Cassab, G. I., Varner, J. E. 1987. Immunocytolocalization of extensin in developing soybean seed coats by immunogold-silver staining and by tissue printing on nitrocellulose paper. *J. Cell Biol.* 105:2581–88

18. Catt, J. W., Hills, G. J., Roberts, K. 1976. A structural glycoprotein, containing hydroxyproline, isolated from the cell wall of *Chlamydomonas reinhardtii. Planta* 131:165–71

19. Cavalieri, A. J., Boyer, J. S. 1982. Water potentials induced by growth in soybean hypocotyls. *Plant Physiol.* 69: 492–96

20. Cesaro, A., Ciana, A., Delben, F., Manzine, G., Paoletti, S. 1982. Physiochemical properties of pectic acid. I. Thermodynamic evidence of a pH-induced conformational transition in aqueous solution. *Biopolymers* 21:431–49

21. Chen, J. A., Varner, J. E. 1985. Isolation and characterization of cDNA clones for carrot extensin and a proline-rich 33-kDa protein. *Proc. Natl. Acad. Sci. USA* 82:4399–4403

22. Chen, J. A., Varner, J. E. 1985. An extracellular matrix protein in plants: characterization of a genomic clone for carrot extensin. *EMBO J.* 4:2145–51

23. Chrispeels, M. J. 1969. Synthesis and secretion of hydroxyproline containing macromolecules in carrots. I. Kinetic analysis. *Plant Physiol.* 44:1187–93

24. Chrispeels, M. J. 1970. Synthesis and secretion of hydroxyproline-containing macromolecules in carrots. *Plant Physiol.* 45:223–27

25. Chrispeels, M. J., Sadava, D., Cho, Y. P. 1974. Enhancement of extensin biosynthesis in aging discs of carrot storage tissue. *J. Exp. Bot.* 25:1157–66

26. Christoffersen, R. E., Tucker, M. L., Laites, G. G. 1984. Gene expression during fruit ripening in avocado fruit: the accumulation of cellulase in RNA and protein as demonstrated by cDNA hybridization and immunodetection. *Plant Mol. Biol.* 3:385–91

27. Church, D. L., Galston, A. W. 1987.

Expression of isoperoxidases in *Zinnia* mesophyl cells differentiating to tracheary elements *in vitro*. *Plant Physiol. Suppl.* 83:446 (Abstr.)

28. Clarke, A. E., Stone, B. A. 1962. β-1,3-Glucan hydrolases from the grape vine *(Vitis vinifera)* and other plants. *Phytochemistry* 1:175–88

29. Clarke, J. A., Shannon, L. M. 1976. The isolation and characterization of the glycopeptides from horseradish peroxidase isoenzyme C. *Biochim. Biophys. Acta* 427:428–42

30. Condit, C. M., Meagher, R. B. 1986. A gene encoding a novel glycine-rich structural protein of petunia. *Nature* 323:178–81

31. Condit, C. M., Meagher, R. B. 1987. Expression of a gene encoding a glycine-rich protein in petunia. *Mol. Cell. Biol.* 7. In press

32. Cooper, J. B., Chen, J. A., Van Holst, G. J., Varner, J. E. 1987. Hydroxyproline-rich glycoproteins of plant cell walls. *Trends Biochem. Sci.* 12:24–27

33. Cooper, J. B., Varner, J. E. 1983. Insolubilization of hydroxyproline-rich glycoprotein in aerated carrot root slices. *Biochem. Biophys. Res. Commun.* 112:161–67

34. Cooper, J. B., Varner, J. E. 1984. Cross-linking of soluble extensin in isolated cell walls. *Plant Physiol.* 76:414–17

35. Crasnier, M., Moustacas, A. M., Ricard, J. 1985. Electrostatic effects and calcium ion concentration as modulators of acid phosphatase bound to plant cell walls. *Eur. J. Biochem.* 151:187–90

36. Crook, E. M., Johnston, I. R. 1962. The qualitative analysis of the cell walls of selected species of fungi. *Biochem. J.* 83:325–31

37. Darvill, A., McNeil, M., Albersheim, P., Delmer, D. P. 1980. The primary cell walls of flowering plants. In *The Biochemistry of Plants*, ed. N. E. Tolbert, 1:91–162. New York: Academic. 705 pp.

38. Della Penna, D., Alexander, D. C., Bennett, A. B. 1986. Molecular cloning of tomato fruit polygalacturonase: analysis of polygalacturonase mRNA levels during ripening. *Proc. Natl. Acad. Sci. USA* 83:6420–24

39. Dey, P. M., Campillo, E. D. 1984. Biochemistry of the multiple forms of glycosidases in plants. *Adv. Enzymol.* 56:141–249

40. Dougall, D. K., Shimbayashi, K. 1960. Factors affecting growth of tobacco callus tissue and its incorporation of tyrosine. *Plant Physiol.* 35:396–404

41. Ecker, J. R., Davis, R. W. 1986. Inhibition of gene expression in plant cells by expression of antisense RNA. *Proc. Natl. Acad. Sci. USA* 83:5372–76

42. Ecker, J. R., Davis, R. W. 1987. Plant defense genes are regulated by ethylene. *Proc. Natl. Acad. Sci. USA* 84:5202–6

43. Edreva, A. M., Georgieva, I. D. 1980. Biochemical and histochemical investigations of α- and β-glucosidase activity in an infectious disease, a physiological disorder and in senescence of tobacco leaves. *Physiol. Plant Pathol.* 17:237–43

44. English, P. D., Albersheim, P. 1969. Host pathogen interactions: I. A correlation between α-galactosidase production and virulence. *Plant Physiol.* 44:217–24

45. Epstein, L., Lamport, D. T. A. 1984. An intramolecular linkage involving isodityrosine in extensin. *Phytochemistry* 23:1241–46

46. Esau, K. 1965. *Plant Anatomy.* New York: Wiley. 767 pp.

47. Espelie, K. E., Franceschi, V. R., Kolattukudy, P. 1986. Immunocytochemistry localization and time course of appearance of an anionic peroxidase associated with suberization in wound-healing potato tuber tissue. *Plant Physiol.* 81:487–92

48. Esquerré-Tugayé, M. T., Lafitte, C., Mazau, D., Toppan, A., Touzé, A. 1979. Cell surfaces in plant-microorganism interactions. II. Evidence for the accumulation of hydroxyproline-rich glycoproteins in the cell wall of diseased plants as a defense mechanism. *Plant Physiol.* 64:320–26

49. Fincher, G. B., Stone, B. A. 1987. Cell walls and their components in grain technology. *Adv. Cereal Sci. Technol.* 8:207–95

50. Fincher, G. B., Stone, B. A., Clarke, A. E. 1983. Arabinogalactan-proteins: structure, biosynthesis, and function. *Ann. Rev. Plant Physiol.* 34:47–70

51. Franssen, H. J., Nap, J. P., Gloudemans, T., Stiekema, W., Van Dam, H., et al. 1987. Characterization of cDNA for nodulin-75 of soybean: a gene product involved in early stages of root nodule development. *Proc. Natl. Acad. Sci. USA* 84:4495–99

52. Freudenberg, K., Harkin, J. M. 1963. The glucosides of cambial sap of spruce. *Phytochemistry* 2:189–93

53. Fry, S. C., 1979. Phenolic components of the primary cell wall and their possible role in the hormonal regulation of growth. *Planta* 146:343–51

54. Fry, S. C. 1980. Gibberellin-controlled

pectinic acid and protein secretion in growing cells. *Phytochemistry* 19:735–40

55. Fry, S. C. 1982. Isodityrosine, a new crosslinking amino acid from plant cell-wall glycoprotein. *Biochem. J.* 204:449–55

56. Fry, S. C. 1983. Feruloylated pectins from the primary cell wall: their structure and possible functions. *Planta* 157:111–23

57. Fry, S. C. 1986. Cross-linking of matrix polymers in the growing cell walls of Angiosperms. *Ann. Rev. Plant Physiol.* 37:165–86

58. Gaspar, T., Penel, C., Thorpe, T., Greppin, H. 1982. *Peroxidases. A Survey of Their Biochemical and Physiological Roles in Higher Plants.* Geneva: Univ. Geneva Press

59. Gayler, K. R., Glasziou, K. T. 1972. Physiological functions of acid and neutral invertases in growth and sugar storage in sugar cane. *Physiol. Plant.* 27:25–31

60. Geissmann, T., Neukom, H. 1973. On the composition of water-soluble wheat flour pentosans and their oxidative gelation. *Lebensm. Wiss. Technol.* 6:59–62

61. Goodenough, U. W., Gebhart, B., Mecham, R. P., Heuser, J. E. 1986. Crystals of the *Chlamydomonas reinhardtii* cell wall: polymerization, depolymerization, and purification of glycoprotein monomers. *J. Cell Biol.* 103:405–17

62. Grambow, H. J., Langenbeck-Schwich, B. 1983. The relationship between oxidase activity, peroxidase activity, hydrogen peroxide, and phenolic compounds in the degradation of indole-3-acetic acid *in vitro*. *Planta* 157:131–37

63. Gross, G. G. 1977. Biosynthesis of lignin and related monomers. *Recent Adv. Phytochem.* 11:141–84

64. Gussin, A. E. S., McCormack, J. H., Waung, L. Y., Gluckin, D. S. 1969. Trehalase: a new pollen enzyme. *Plant Physiol.* 44:1163–68

65. Haberlandt, G. 1914. *Physiological Plant Anatomy.* New Delhi: Today & Tomorrow's Book Agency. 755 pp. Reprinted

66. Hammerschmid, R., Lamport, D. T. A., Muldoon, E. P. 1984. Cell wall hydroxyproline enhancement and lignin deposition as an early event in the resistance of cucumber to *Cladosporium cucumeriunum*. *Physiol. Plant Pathol.* 24:43–47

67. Harris, W. M. 1984. On the development of osteosclereids in seed coats of *Pisum sativum* L. *New Phytol.* 98:135–41

68. Hayashi, T., MacLachlan, G. 1984. Pea xyloglucan and cellulose. III. Metabolism during lateral expansion of pea epicotyl cells. *Plant Physiol.* 76:739–42

69. Heuser, J. E. 1983. Procedure for freeze drying molecules absorbed to mica flakes. *J. Mol. Biol.* 169:155–95

70. Hills, G. J., Phillips, J. M., Gay, M. R., Roberts, K. 1975. Self-assembly of a plant cell wall *in vitro*. *J. Mol. Biol.* 96:431–41

71. Hong, J. C., Nagao, R. T., Key, J. L. 1987. Characterization and sequence analysis of a developmentally regulated putative cell wall protein gene isolated from soybean. *J. Biol. Chem.* 262:8367–76

72. Hösel, W., Surholt, E., Borgman, E. 1978. Characterization of β-glucosidase isoenzymes possibly involved in lignification from chick pea (*Cicer arietinum* L.) cell suspension cultures. *Eur. J. Biochem.* 84:487–92

73. Israel, H. W., Salpeter, M. M., Steward, F. C. 1968. The incorporation of radioactive proline into cultured cells. *J. Cell Biol.* 39:698–715

74. Keen, N. T., Yoshikawa, M. 1983. β-1,3-Endoglucanase from soybean releases elicitor-active carbohydrates from fungus cell walls. *Plant Physiol.* 71:460–65

75. Kiefer, S., Everdeen, D., Lamport, D. T. A. 1987. In vitro crosslinking of extensin precursors. *Plant Physiol.* 83:654 (Abstr.)

76. Kieliszewski, M., Lamport, D. T. A. 1987. Purification and partial characterization of an HRGP in a Graminaceous monocot, *Zea mays*. *Plant Physiol.* 85:823–27

77. King, N. J., Bayley, S. T. 1965. A preliminary analysis of the proteins of the primary walls of some plant cells. *J. Exp. Bot.* 16:294–303

78. Klis, F. M. 1976. Glycosylated seryl residues in wall protein of elongating pea stems. *Plant Physiol.* 57:224–26

79. Klis, F. M., Dalhuizen, R., Sol, K. 1974. Wall-bound enzymes in callus of *Convolvulus arvensis*. *Phytochemistry* 13:55–57

80. Kolattukudy, P. E. 1981. Structure, biosynthesis, and biodegradation of cutin and suberin. *Ann. Rev. Plant Physiol.* 32:539–67

81. Labella, F., Waykole, P., Queen, G. 1968. Formation of insoluble gels and dityrosine by the action of peroxidase on soluble collagens. *Biochem. Biophys. Res. Commun.* 30:333–38

82. Lagrimini, L. M., Burkhart, W., Moyer, M., Rothstein, S. 1987. Molecular cloning of complementary DNA encoding the lignin-forming peroxidase from tobacco: molecular analysis and tissue specific expression. *Proc. Natl. Acad. Sci. USA* 84:7542–46

83. Lagrimini, L. M., Rothstein, S. 1987. Tissue specificity of tobacco peroxidase isozymes and their induction by wounding and tobacco mosaic virus infection. *Plant Physiol.* 84:438–42

84. Lamport, D. T. A. 1965. The protein component of primary cell walls. *Adv. Bot. Res.* 2:151–218

85. Lamport, D. T. A. 1967. Hydroxyproline-O-glycosidic linkage of the plant cell wall glycoprotein extensin. *Nature* 216:1322–24

86. Lamport, D. T. A. 1969. The isolation and partial characterization of hydroxyproline-rich glycopeptides obtained by enzymatic degradation of primary cell walls. *Biochemistry* 8:1155–63

87. Lamport, D. T. A. 1974. *The role of hydroxyproline-rich proteins in the extracellular matrix of plants.* Presented at 30th Symp. Soc. Dev. Biol., pp. 113–30. New York: Academic

88. Lamport, D. T. A. 1978. Cell wall carbohydrates in relation to structure and function. In *Frontiers of Plant Tissue Culture,* ed. T. A. Thorpe, pp. 235–44. Calgary: Int. Assoc. Plant Tissue Culture

89. Lamport, D. T. A. 1980. Structure and function of plant glycoproteins. In *The Biochemistry of Plants,* ed. P. K. Stumpf, E. E. Conn, 3:501–36. New York: Academic. 639 pp.

90. Lamport, D. T. A. 1986. Roles for peroxidases in cell wall genesis. In *Molecular and Physiological Aspects of Plant Peroxidases,* ed. H. Greppin, C. Penel, T. Gaspar, pp. 199–208. Geneva: Univ. Geneva Press

91. Lamport, D. T. A., Catt, J. W. 1981. Glycoproteins and enzymes of the cell wall. In *Encyclopedia of Plant Physiology* (NS), *Plant Carbohydrates II,* ed. W. Tanner, F. A. Holmes, 13B:133–65. New York: Springer-Verlag

92. Lamport, D. T. A., Katona, L., Roering, S. 1973. Galactosylserine in extensin. *Biochem. J.* 133:125–31

93. Lamport, D. T. A., Miller, D. H. 1971. Hydroxyproline arabinosides in the plant kingdom. *Plant Physiol.* 48:454–56

94. Lamport, D. T. A., Northcote, D. H. 1960. Hydroxyproline in primary cell walls of higher plants. *Nature* 118:665–66

95. Laurière, C., Sturm, A., Chrispeels, M.

J. 1987. Characteristics of the soluble and cell wall-associated β-fructosidase from suspension-cultured carrot cells. *Plant Physiol. Suppl.* 83:545 (Abstr.)

96. Leach, J. E., Cantrell, M. A., Sequeira, L. 1982. Hydroxyproline-rich bacterial agglutinin from potato. *Plant Physiol.* 70:1353–58

97. Little, G., Edelman, J. 1973. Solubility of plant invertases. *Phytochemistry* 12: 67–71

98. Mäder, M. 1986. Cell compartmentation and specific roles of isoenzymes. See Ref. 90, pp. 247–60

99. Mäder, M., Nessel, A., Bopp, M. 1977. On the physiological significance of the isoenzyme groups of peroxidase from tobacco demonstrated by biochemical properties. *Z. Pflanzenphysiol.* 82:247–60

100. Mäder, M., Ungemach, J., Schloss, P. 1980. The role of peroxidase isoenzyme groups of *Nicotiana tabacum* in hydrogen peroxide formation. *Planta* 147: 467–70

101. Markovic, O., Jörnvall, H. 1986. Pectinesterase. The primary structure of the tomato enzyme. *Eur. J. Biochem.* 158: 455–62

102. Markwalder, H. U., Neukom, H. 1976. Diferulic acid as a possible crosslink in hemicellulases from wheat germ. *Phytochemistry* 15:836–37

103. Matheson, N. K., Saimi, H. S. 1977. Polysaccharide and oligosaccharide changes in germinating lupin cotyledons. *Phytochemistry* 16:59–66

104. Matile, P. 1975. *The Lytic Compartment of Plant Cells.* Vienna: Springer-Verlag

105. Mauch, F., Hadwiger, L. A., Boller, T. 1984. Ethylene: symptom, not signal for the induction of chitinase and β-1, 3-glucanase in pea pods by pathogens and elicitors. *Plant Physiol.* 76:607–11

106. Mazau, D., Esquerré-Tugayé, M. T. 1986. Hydroxyproline-rich glycoprotein accumulation in the cell walls of plants infected by various pathogens. *Physiol. Mol. Plant Pathol.* 29:147–57

107. Mazau, D., Rumeau, D., Esquerré-Tugayé, M. T. 1987. Biochemical characterization and immunocytochemical localization of hydroxyproline-rich glycoproteins in infected plants. *Plant Physiol. Suppl.* 83:658 (Abstr.)

108. McNeil, M., Darvill, A. G., Fry, S. C., Albersheim, P. 1984. Structure and function of the primary cell walls of plants. *Ann. Rev. Biochem.* 53:625–63

109. Mellon, J. E., Helgeson, J. P. 1982. Interaction of a hydroxyproline-rich glycoprotein from tobacco callus with

potential pathogens. *Plant Physiol.* 70: 401–5

110. Moore, P., Darvill, A. G., Albersheim, P., Staehelin, L. A. 1986. Immunogold localization of xyloglucan rhamnogalacturonan I in the cell walls of suspension-cultured sycamore cells. *Plant Physiol.* 82:787–94

111. Moustacas, A. M., Nari, J., Diamantidis, G., Noat, G., Crasnier, M., et al. 1986. Electrostatic effects and the dynamics of enzyme reactions at the surface of plant cells. 2. The role of pectin methylesterase in the modulation of electrostatic effects in soybean cell walls. *Eur. J. Biochem.* 155:191–97

112. Nagahashi, G., Seibles, T. 1986. Purification of plant cell walls: isoelectric focusing of $CaCl_2$ extracted enzymes. *Protoplasma* 134:102–10

113. Nari, J., Noat, G., Diamantidis, G., Wouastra, M., Ricard, J. 1986. Electrostatic effects and the dynamics of enzyme reactions at the surface of plant cells. 3. Interplay between cell-wall autolysis, pectin methyl esterase activity and electrostatic effects in soybean cell walls. *Eur. J. Biochem.* 155:199–202

114. Neukom, H. 1976. Chemistry and properties of non-starchy polysaccharides (NSP) of wheat flour. *Lebensm. Wiss. Technol.* 6:143–48

115. Northcote, D. H. 1972. Chemistry of the plant cell wall. *Ann. Rev. Plant Physiol.* 23:113–32

116. Novaes-Ledieu, M., Jiménez-Martinez, A., Villanueva, J. R. 1966. Chemical composition of hyphae wall of Phycomycetes. *J. Gen. Microbiol.* 47:237–45

117. Olson, A. C. 1964. Proteins and plant cell walls. Proline to hydroxyproline in tobacco suspension cultures. *Plant Physiol.* 39:543–50

118. Painter, T. J. 1983. Residues of D-lyxo-5-hexosulopyranuronic acid in *Sphagnum* holocellulose, and their role in cross-linking. *Carbohydr. Res.* 124:18–21

119. Pollard, J. K., Steward, F. C. 1959. The use of ^{14}C proline by growing cells: its conversion to protein and to hydroxyproline. *J. Exp. Bot.* 10:17–32

120. Preston, R. D. 1974. *The Physical Biology of Plant Cell Walls.* London: Chapman & Hall. 491 pp.

121. Preston, R. D. 1979. Polysaccharide conformation and cell wall function. *Ann. Rev. Plant Physiol.* 30:55–78

122. Punnett, T., Derrenbacker, E. C. 1965. The amino acid composition of algal cell walls. *J. Gen. Microbiol.* 44:105–14

123. Ricard, J., Noat, G. 1986. Electrostatic effects and the dynamics of enzyme reactions at the surface of plant cells. 1. A theory of the ionic control of a complex multi-enzyme system. *Eur. J. Biochem.* 155:183–90

124. Rivett, A. J. 1986. Regulation of intracellular protein turnover: covalent modification as a mechanism of marking proteins for degradation. *Curr. Top. Cell. Regul.* 28:291–337

125. Roberts, K. 1974. Crystalline glycoprotein cell-walls of algae: their structure, composition and assembly. *Philos. Trans. R. Soc. London Ser. B* 268:129–46

126. Roberts, K., Grief, C., Hills, G. J., Shaw, P. J. 1985. Cell wall glycoproteins: structure and function. *J. Cell Sci. Suppl.* 2:105–27

127. Roberts, K., Gurney-Smith, M., Hills, G. J. 1972. Structure, composition, and morphogenesis of the cell wall of *Chlamydomonas reinhardtii.* I. Ultrastructure and preliminary analysis. *J. Ultrastruct. Res.* 40:599–613

128. Roberts, L. W., Baba, S. 1968. Effect of proline on wound vessel member formation. *Plant Cell Physiol.* 9:353–60

129. Roby, D., Toppan, A., Esquerré-Tugayé, M. T. 1985. Cell surface in plant-microorganisms interactions. V. Elicitors of fungal and of plant origin trigger the synthesis of ethylene and of cell wall hydroxyproline-rich glycoprotein in plants. *Plant Physiol.* 77:700–4

130. Rushing, J. W., Huber, D. J. 1984. *In vitro* characterization of tomato fruit ripening. *Plant Physiol.* 75:891–94

131. Sadava, D., Chrispeels, M. J. 1969. Cell wall protein in plants: autoradiographic evidence. *Science* 165:299–300

132. Sadava, D., Walker, F., Chrispeels, M. J. 1973. Hydroxyproline-rich cell wall protein (extensin): biosynthesis and accumulation in growing pea epicotyls. *Dev. Biol.* 30:42–48

133. Schlipfenbacher, R., Wenzl, S., Lottspeich, F., Sumper, M. 1986. An extremely hydroxyproline-rich glycoprotein is expressed in inverting *Volvox* embryos. *FEBS Lett.* 209:57–62

134. Schwarzbauer, J. E., Tamkun, J. W., Lemischka, I. R., Hynes, R. O. 1983. Three different fibronectin mRNAs arise by alternative splicing within the coding region. *Cell* 35:421–31

135. Sequeira, L. 1983. Mechanisms of induced resistance in plants. *Ann. Rev. Microbiol.* 37:51–79

136. Showalter, A. M., Bell, J. N., Cramer, C. L., Bailey, J. A., Varner, J. E., et al. 1985. Accumulation of hydroxyproline-

rich glycoprotein mRNAs in response to fungal elicitor and infection. *Proc. Natl. Acad. Sci. USA* 82:6551–55

137. Showalter, A. M., Varner, J. E. 1987. Molecular details of plant cell wall hydroxyproline-rich glycoprotein expression during wounding and infection. *UCLA Symp. Mol. Cell. Biol.* NS 48:375–92

138. Smith, J. J., Muldoon, E. P., Lamport, D. T. A. 1984. Isolation of extensin precursors by direct elution of intact tomato cell suspension cultures. *Phytochemistry* 23:1233–39

139. Smith, J. J., Muldoon, E. P., Willard, J. J., Lamport, D. T. A. 1986. Tomato extensin precursors P1 and P2 are highly periodic structures. *Phytochemistry* 25:1021–30

140. Stafstrom, J. P., Staehelin, L. A. 1986. Cross-linking patterns in salt-extractable extensin from carrot cell walls. *Plant Physiol.* 81:234–41

141. Stafstrom, J. P., Staehelin, L. A. 1987. A second extensin-like hydroxyproline-rich glycoprotein from carrot cell walls. *Plant Physiol.* 84:820–25

142. Stafstrom, J. P., Staehelin, L. A. 1987. Antibody localization of extensin in carrot cell walls. *Planta*. In press

143. Stahmann, M. A., Spencer, A. K. 1977. Deamination of protein lysyl. ε-amino groups by peroxidase *in vitro*. *Biopolymers* 16:1299–1306

144. Stevenson, T. T., McNeil, M., Darvill, A. G., Albersheim, P. 1986. Structure of plant cell walls. *Plant Physiol.* 80:1012–19

145. Steward, F. C., Chang, L. O. 1963. The incorporation of ^{14}C proline into the proteins of growing cells. II. A note on evidence from cultured carrot explants by acrylamide gel electrophoresis. *J. Exp. Bot.* 14:379–86

146. Steward, F. C., Mott, R. L., Israel, H. W., Ludford, P. M. 1970. Proline in the vesicles and sporelings of *Valonia ventricosa* and the concept of cell wall protein. *Nature* 225:760–63

147. Steward, F. C., Thompson, J. F., Millar, F. K., Thomas, M. D., Hendricks, R. H. 1951. The amino acids of alfalfa as revealed by paper chromatography with special reference to compounds labelled with ^{35}S. *Plant Physiol.* 26:123–35

148. Strobel, G. A. 1974. Phytotoxins produced by plant parasites. *Ann. Rev. Plant Physiol.* 25:541–66

149. Stuart, D. A., Varner, J. E. 1980. Purification and characterization of a salt-extractable hydroxyproline-rich glycoprotein from aerated carrot discs. *Plant Physiol.* 66:787–92

150. Taiz, L. 1984. Plant cell expansion: regulation of cell wall mechanical properties. *Ann. Rev. Plant Physiol.* 35:585–657

151. Talmadge, K. W., Keegstra, K., Bauer, W. D., Albersheim, P. 1973. The structure of plant cell walls. *Plant Physiol.* 51:158–73

152. Tanimoto, E., Pilet, P. E. 1978. α- and β-glycosidases in maize roots. *Planta* 138:119–22

153. Thibault, J. F., Rinaudo, M. 1985. Interactions of macro- and divalent counterions with alkali- and enzyme-deesterified pectins in salt-free solutions. *Biopolymers* 24:2131–43

154. Thibault, J. F., Rinaudo, M. 1986. Chain association of pectic molecules during calcium-induced gelation. *Biopolymers* 25:455–68

155. Thompson, E. W., Preston, R. D. 1967. Proteins in the cell walls of some green algae. *Nature* 213:684–85

156. Thompson, E. W., Preston, R. D. 1968. Evidence for a structural role of protein in algal cell walls. *J. Exp. Bot.* 19:690–97

157. Tierney, M. L. 1986. Accumulation of a new proline-rich protein in carrot root cell walls after wounding. *Plant Physiol. Suppl.* 80:365 (Abstr.)

158. Tierney, M. L., Varner, J. E. 1987. The extensins. *Plant Physiol.* 84:1–2

159. Tripp, V. W., Moore, A. T., Rollins, M. L. 1951. Some observations on the constitution of the primary wall of the cotton fiber. *Text. Res. J.* 21:886–94

160. Tyler, J. M., Branton, D. 1980. Rotary shadowing of extended molecules dried from glycerol. *J. Ultrastruct. Res.* 71:95–102

161. Van Etten, C. H., Miller, R. W., Earle, F. R., Wolff, I. A., Jones, Q. 1961. Hydroxyproline content of seed meals and distribution of the amino acid in the kernel, seed coat, and pericarp. *J. Agric. Food Chem.* 9:433–35

162. Van Holst, G. J., Varner, J. E. 1984. Reinforced polyproline II conformation in a hydroxyproline-rich cell wall glycoprotein from carrot root. *Plant Physiol.* 74:247–51

163. Varner, J. E., Cassab, G. I. 1986. A new protein in petunia. *Nature* 323:110

164. Veluthambi, K., Mahadevan, S., Maheshwari, R. 1981. Trehalose toxicity in *Cuscuta reflexa*. *Plant Physiol.* 68:1369–74

165. Vreeland, V., Zablackis, E., Baboszewski, B., Laetsch, W. M. 1987. Molecular markers for marine algal

polysaccharides. *Proc. 12th Int. Seaweed Symp. Hydrobiologia.* 7421 São Paulo, Brazil

166. Walker, W. S. 1960. The effect of mechanical stimulation and etiolation on the collenchyma of *Datura stramonium. Am. J. Bot.* 47:717–24

167. Whitmore, F. W. 1976. Binding of ferulic acid to cell walls by peroxidases of *Pinus elliottii. Phytochemistry* 15: 375–78

168. Wiemken-Gehrig, V., Wiemken, A., Matile, P. 1974. Cell wall breakdown in wilting flowers of *Ipomea tricolor* Cav. *Planta* 115:297–307

169. Wilson, L. G., Fry, J. C. 1986. Extensin—a major cell wall glycoprotein. *Plant Cell Environ.* 9:239–60

170. Yamaoka, T., Chiba, N. 1983. Changes in the coagulation ability of pectin during growth of soybean hypocotyls. *Plant Cell Physiol.* 24:1281–90

171. Young, D. H., Pegg, G. F. 1982. The action of tomato and *Verticillium alboatrum* glycosidases on the hyphal wall of *V. albo-atrum. Physiol. Plant Pathol.* 21:411–23

Ann. Rev. Plant Physiol. Plant Mol. Biol. 1988. 39:355–78
Copyright © 1988 by Annual Reviews Inc. All rights reserved

METABOLISM AND COMPARTMENTATION OF IMPORTED SUGARS IN SINK ORGANS IN RELATION TO SINK STRENGTH

Lim C. Ho

Department of Physiology, Division of Plant Science, AFRC Institute of Horticultural Research, Littlehampton, West Sussex BN17 6LP, United Kingdom

CONTENTS

INTRODUCTION 356
SINK STRENGTH AND IMPORT OF ASSIMILATE 357
    Definitions of Sink Strength 357
    Sink Strength and Sink Competition 359
CHARACTERISTICS OF SINK ORGANS 360
    Utilization and Storage Sink Organs 360
    Reversible and Irreversible Sink Organs 361
REGULATION OF IMPORT BY SINK ORGANS 362
    Unloading of Sugars from Phloem Conducting Tissues 362
    Uptake of Sugars by Sink Cells 365
    Chemical Conversion of Sugars Inside Sink Cells 367
    Physical Compartmentation of Sugars Inside Sink Cells 368
DETERMINATION OF SINK STRENGTH 370
    Genetic Determination 370
    Genetic Expression 371
CONCLUDING REMARKS 372

355

0066-4294/88/0601-0355$02.00

INTRODUCTION

In the last two centuries, the yield of economically important crops has been substantially increased through plant breeding and optimization of growing conditions. The improvement of yield was made possible because both dry-matter production in the leaves and accumulation of dry matter by harvestable organs have been improved. For instance in potato, the modern cultivar *(Solanum tuberosum)* has a plant dry weight 10 times higher than the wild species *(S. demissum);* tuber dry weight, as a proportion of plant weight (i.e. harvest index) has increased from 7% to 81% (65). Through plant breeding, the genetic yield potential of wheat, soya bean, corn, and peanuts has been improved by 40–100% within this century (46). The increased grain yield of winter wheat and spring barley in England can be entirely accounted for by the increase of harvest indexes, as the biomass yield of the modern varieties is the same as that of ones bred before modern breeding was practiced (1).

In general, the higher photosynthetic capacity of the modern cultivars of economic crops has been achieved mainly by increasing the light-intercepting area of leaves; this has been achieved by either increasing the number of leaves, as in corn (108); producing a more erect leaf posture, as in rice (109); or attaining a larger individual leaf area, as in wheat (26). The higher dry-matter accumulation capacity of the harvestable organs has been accomplished mainly by increasing either the number of grains, as in rice (109), or the size of individual grains, as in wheat (26). The capacity of dry-matter production in leaves may either be higher or lower than the capacity of dry-matter accumulation in other parts of the plant. Therefore, at different times, either source- or sink-limiting situations may exist in crop production. In essence, better yield is achieved by successful regulation of source-sink relationships in the production and utilization of photoassimilate within a plant. Neither source nor sink manipulation alone can improve crop yield indefinitely. In this review, I deal only with the regulation of dry-matter accumulation by sink organs, as a means for crop yield improvement, under sink-limiting conditions.

While the production and the utilization of dry matter within a plant depend on each other, the regulation of the partitioning of dry matter into different organs is independent from the production of assimilate. The competition among sink organs for assimilate is determined partly by the intrinsic ability of the sink organs to receive assimilate, relative to each other. Therefore, studies on the determination of potential sink strength may provide better strategies for improving crop productivity.

Recent progress of phloem transport studies indicates that the loading of sucrose into sieve tubes in leaves of a number species may be controlled by an energy-requiring proton/sucrose co-transport process at the plasmalemma of the sieve tube/companion cell (se/cc) complex (44). In contrast, the unloading of sucrose from the sieve tube in the sink organs has been demonstrated to be either an active energy-requiring process at the plasmalemma or a leakage sensitive to turgor pressure in the apoplast (119). However, there is no evidence that the rate of import is mainly limited by the rate of unloading. If unloading is not the controlling step in the import of assimilate, then possible roles of processes inside sink cells related to the utilization or storage of imported sugars must be examined.

In this review, I limit discussion of the metabolism and compartmentation of imported sugars with respect to the regulation of assimilate import, despite the fact that amino acids and organic acids are also imported by sink organs; proteins and lipids, as well as carbohydrates, are accumulated in sink organs. The relatively well-documented studies on the transport and accumulation of sucrose in sink organs should give a fair assessment of the regulation of photoassimilate import. I hope this review provides the essential links between crop productivity and the biochemical and biophysical processes at the cellular level within sink organs. This should enable agronomists and membrane physiologists alike to appreciate both the complexity and the similarity of their mutual quest for the controlling mechanism of assimilate partitioning.

SINK STRENGTH AND IMPORT OF ASSIMILATE

Definitions of Sink Strength

Sink organs for assimilate are net importers of assimilate. Essentially, all plant organs at some stages of plant development would act as *sinks,* receivers of assimilate. In terms of assimilate transport, the ability of a sink organ to import assimilate is the *sink strength.* The proportion of imported assimilate used for respiration by sink organs can be substantial (28). Thus, sink strength of a sink organ, measured as an absolute growth rate or net accumulation rate of dry matter, fails to assess the true ability of a sink organ to receive assimilate and is a measure of *apparent sink strength.* The import rate of assimilate, measured as the sum of the net carbon gain and respiratory carbon loss by a sink organ (128), should give a more appropriate estimate of the *actual sink strength.*

Although actual sink strength would be affected by the availability of assimilate supply and the proximity of the sink to the source (i.e. transport conductivity), the most critical determinant is the intrinsic ability of the sink to receive or attract assimilate (16). This intrinsic ability of a sink is the *potential sink strength.* It is genetically determined and can be fully expressed when the supply of assimilate is sufficient to meet the demand and the environmental conditions for the metabolic activity of the sink organ are optimal. Therefore, potential sink strength is the maximum attainable ability of a sink organ to obtain assimilate. When more than one sink exists in the same transport system and the supply of assimilate is less than the potential demand, the import rate is an estimate of the actual sink strength.

Sink strength has been considered as a product of sink size and sink activity (134). However, the early attempts to quantify sink strength by defining organ weights as sink size and relative growth rates as sink activity did not improve our understanding of the mechanism regulating sink strength (133). Instead, sink size should be defined as the *physical constraint* and sink activity as the *physiological constraint* upon a sink organ's assimilate import.

The number of cells in the sink organ may be a suitable measure of sink size. For instance, among varieties of wheat, the final grain dry weight is potentially determined by the number of cells in the endosperm and the number of plastids in each endosperm cell (48). Apart from genetic determination, cell number can also be regulated by plant growth regulators during cell division of the sink organs (13). Indeed, cell division activity is crucial in attracting assimilate to sink organs in the early stages of development (73, 81). Therefore, by measuring the number of storage cells or the storage organelles within, sink size can be quantified.

At present, factors determining cell number in sink organs have not been studied thoroughly. In tomato pericarp, the cell number appears to be associated with cambial activity of the vascular bundles and thus with the number of vascular bundles (13). Although most cells of a tomato fruit are derived during cell division following pollination, the difference in cell number of a fruit at various positions of the same truss already exists in the pre-anthesis ovary (8). Similarly, the higher cell number in mutant fruit was mainly due to a higher cell number in the ovary, as the cell division activity after pollination was the same in both the mutant and its parent, *Lycopersicon pimpinelliforium* (8). How the cell number of an ovary is determined is not well understood. However, larger tomato fruit can be induced by applying gibberellins (GA) to the apex just before floral initiation to increase the number of locules (103). This observation suggests that the cell number of the carpels may be determined at the initiation of the floral primordium.

However, import rate is not entirely determined by potential sink size but is regulated by the metabolic activity of the sink organ during development. It has long been recognized that the import rate of a sink can be reduced by local application of low temperature, anoxia, or metabolic inhibitors, resulting in reduced growth (132). The transport processes within a sink, such as the unloading of sugars from the phloem tissue, the retrieval of sugars by sink cells from the cell wall, and utilization or storage of sugars within the cell, may be energy-dependent processes. Import of assimilate may be controlled by these metabolic activities (37). It has been suggested that, when any of these processes is affected, the sucrose concentration in the free space at the point of unloading will be raised. As a result, the sucrose concentration gradient between the source and the sink required to sustain the rate of import would decrease. As has been demonstrated in tomato, the import rate of a fruit can be reduced by lowering fruit temperature to decrease both respiration and sucrose hydrolysis (130). If the sucrose concentration at the point of unloading is raised, then reduced import rate into the cooled fruit may indeed be due to a reduced sucrose concentration gradient between the leaves and the cooled fruit (129, 131). In wheat, the gradient of solute concentration between the sieve tube sap at the crease and the endosperm cavity sap was substantial (33). Thus the sucrose gradient at the point of sucrose unloading may regulate the rate of import.

Sink Strength and Sink Competition

The availability of mobile assimilate to a particular sink may be determined by the competition between sink organs. When dry-matter production in flowering tomato plants was reduced, the limited amount of mobile assimilate was mainly imported by the developing young leaves to sustain growth, and the initiating inflorescence was aborted (71). The higher priority of importing assimilate by developing leaves over inflorescences remained, even when extra assimilate was made available. Import to the inflorescence was only improved after the demand by the developing leaves had been met (72).

Although sink competition may be amplified when assimilate supply is limited, the priority of partitioning is consistent at each stage of plant development. In tomato, a developing inflorescence is a weaker sink for assimilate than the expanding leaves, but a truss with growing fruit is a stronger sink than young leaves and roots (64). The potential sink strength of the inflorescence increased from flowering to fruiting stage, and the priority between sinks for assimilate in the order of roots > young leaves > inflorescence in a flowering plant changed to the order of fruit > young leaves > flowers > roots in a fruiting tomato plant (58).

By altering the potential sink strength of an individual sink, the priority of assimilate partitioning can be changed. In normal truss development of tomato, the import rate of the early-set fruit in the proximal position is greater than that of later-set fruit in the distal position (62). By altering the fruit-set sequence, early-set fruit in a distal position had a higher assimilate import rate than later-set fruit in a proximal position (2). The cell number of the pre-anthesis ovary at the proximal position is higher than that at the distal position (8), but the early multiplication and enlargement of cells in the distal ovary after induction may exert a greater demand for assimilate than the later-induced ovaries at the proximal position.

The import rate for a specific sink organ can be altered either by changing its sink strength or by changing the strength of its competing sink. This kind of manipulation can be achieved without altering the production of dry matter in the plant. For seed crops, increasing sink number rather than sink size is an effective way to improve seed yield (46). Winter wheat grain yield was successfully improved by introducing the Norin 10 dwarfing gene *Gai/Rht*, which shortened the stem. The higher yield of the semi-dwarf winter wheat was due to a greater number of grains, which may in turn be due to reduced maximal stem growth when 90% of the ear dry weight was normally accumulated (11). It appears that the reduced sink strength of the stem with respect to the ears may be due to reduced cell division activity in the stem (92).

CHARACTERISTICS OF SINK ORGANS

Utilization and Storage Sink Organs

All sink organs use imported assimilates for growth, maintenance, and storage. However, the allocation of imported assimilates is substantially different from one sink organ to another. Thus the rate of import by a sink would be mainly affected by the predominant allocation process, which might be different from that in other sinks. Based on the fate of the imported assimilates, there are in general two kinds of sinks—utilization and storage sinks (59, 60).

In meristem tissue, most of the imported assimilate would be used for growth, and only a small amount would be stored temporarily. Such sink organs can be classified as *utilization sinks*. For instance, the fibrous roots of barley use 40% and 55% of imported sucrose for respiration and structural growth, respectively (28), and stored sugars account for only 1% of the root weight (27). The storage of sugars is unlikely to be the controlling step for utilization sinks. In most of the utilization sinks investigated, such as developing leaves or root tips (18, 45, 104, 125), the unloading of assimilate from the phloem to the surrounding tissue occurs via intercellular cytoplasmic connections. There are sufficient plasmodesmatal connections between the conducting tissue, and no hydrolysis of imported sucrose is found in the cell wall. Thus, the use of imported sucrose for respiration and for cellular-structure biosynthesis will sustain a steep sucrose concentration gradient between the sink cell and the phloem tissue and will facilitate further transport (36).

The main metabolic activity in utilization sinks is respiration. Thus, it is expected that the rate of import in these sinks would follow the rate of respiration. Although the relative growth rate is linearly related to the respiration rate of the roots of chickpea (70), there is no direct proof that import rate is regulated by the rate of respiration. However, there is indirect evidence to support such a notion. For instance, the import of assimilate into the nodules of soya bean was retarded by low oxygen, which also inhibited the respiration of the nodules (49). Furthermore, the resumption of cell division and the increase of assimilate import coincide in an aborting inflorescence of tomato after growth regulator treatment (72, 81). This also suggests that the import of assimilate is regulated by metabolic activities associated with cell division and/or cell enlargement within the sink.

In storage organs such as fruit, stem, tuber, and root, substantial amounts of imported assimilate are stored in different forms. For instance, more than 70% of the dry matter accumulated by sugar beet tap roots is sucrose (88), while more than 80% of the dry matter accumulated by the wheat grain endosperm is starch (69). These organs are classified as *storage sinks* because the storage process may be the controlling step for assimilate import.

Among the storage sinks that have been well investigated, the principal carbohydrate reserves are sugars (i.e. sucrose, glucose, or fructose) and starch or fructosans. For convenience, these are termed *sugar-* or *starch-storage sinks,* respectively. Although most storage sinks would store principally one kind of reserve, some storage sinks, such as tomato fruit, store equal amounts of hexoses and starch in the early stage of fruit development. Later, starch is depleted and only hexoses are stored (62). If the form of reserve stored in the sink organs is significant in controlling the import rate, then sinks with a transition of storage substances may be termed *starch/sugar-storage sinks.*

Detailed studies of the anatomical structure of the vascular tissue and surrounding tissue in a number of storage sinks demonstrates that there are sufficient cytoplasmic connections for the observed rate of sucrose transfer from the phloem to the adjacent cells (15, 93). However, the subsequent transport of sugars frequently involves an apoplastic route between maternal and embryonic tissue of the reproductive sinks, such as cereal grains or legume seeds, or in vegetative sinks such as sugarcane stalk. Therefore, storage sinks share two features: storage of a substantial amount of imported sugars as reserves, and the involvement of membrane transport of sugars by sink cells.

In starch-storage sinks, such as wheat grains, the rate of starch accumulation in the endosperm is normally limited not by the supply of sucrose (68) but by the activity of the starch-synthesizing enzymes. Therefore, the amount of starch accumulation in wheat grains is determined by both the rate of starch synthesis and the duration of accumulation in the endosperm (6, 35). Apart from the enzymatic regulation of starch synthesis, the amount of starch accumulation in a wheat grain may also be determined by the number of plastids in the endosperm (48).

Reversible and Irreversible Sink Organs

Some storage sinks, such as fruit, store reserves continuously and permanently during the life cycle of the plant while others, such as unmodified stem, have their reserves remobilized at times. The former has an irreversible import of assimilate and the unloading is symplastic (93), while the latter has a reversible import of assimilate and its unloading can be switched from symplastic to apoplastic (53). They are termed *irreversible* and *reversible sinks*, respectively (94).

This kind of classification of sink organs is useful, as there are substantial functional differences between permanent and temporary storage sinks. There is a need to understand how the continuous import to the irreversible sinks and the reversible transport in the reversible sinks is made possible. The reversibility of the import is assumed to be controlled by the mode of the unloading processes; entirely symplastic unloading is essential for irreversible sinks, and the ability to switch from import to export by minimizing the symplastic unloading is essential for reversible sinks (94). All these assumptions have to be tested to justify this interesting hypothesis.

REGULATION OF IMPORT BY SINK ORGANS

Unloading of Sugars from Phloem Conducting Tissues

The transfer of sugars from the se/cc complex to the adjacent cells can occur via cytoplasmic connections, the plasmodesmata, or transmembrane transport at the plasmalemma in both the se/cc complex and the sink cells, with an apoplastic route in between the membranes. The possibility that one of these two modes is operating has been frequently tested by examining whether the surface area of the plasmalemma (for apoplastic unloading) or the cross-sectional area of the plasmodesmata (for symplastic unloading) is consistent with the rate of assimilate import (93). This possibility was further tested by comparing the rates of unloading with and without the interference of metabolic inhibitors of membrane transport (93), by applying plasmolyzing agents to restrict the cytoplasmic connections (94), or by tracing the route of the symplastic tracer fluorescein from the conducting tissue to sink cells (15). Symplastic unloading has been identified in roots, where a diffusional barrier, the endodermis, exists (18), while apoplastic unloading has been identified in seeds, where a symplastic discontinuity exists between the seed coat and the embryonic cotyledon (117).

The symplastic route of unloading has been commonly and convincingly proven for utilization sinks. In corn seedlings, despite the presence of a cell wall–bound invertase in their roots, imported sucrose is not hydrolyzed prior to storage in the vacuole. As there are sufficient plasmodesmata between phloem conducting cells and adjacent cortical cells, sucrose is likely to be unloaded symplastically in the roots (45). Similarly, the import of assimilate into a developing sugar beet leaf is not affected by exogenously applied PCMBS (p-chloromercuribenzene sulfonic acid), a nonpermeable sulfhydryl-group modifier, in the free space. Thus, apoplastic unloading, which is normally inhibited by PCMBS, is unlikely to be the principal route for sucrose unloading from phloem to mesophyll in the sink leaf of sugar beet (104). Apparently, unloading of assimilate in developing tobacco leaves continued for a short time, when the import of assimilate was inhibited (124). These observations indicate that unloading in a sink leaf may be a passive symplastic movement from the phloem conducting tissue to adjacent cells.

Because of the inaccessibility of the site of unloading, the control of symplastic unloading in utilization sinks is still poorly understood. However, the cessation of unloading appears to be controlled by the maturation of the minor veins in the sink leaves (105, 123). In contrast, by changing the source-sink relationships in the plant, mature leaves can import assimilate and metabolize imported sugars in cells adjacent to the conducting tissue (34). Most likely, import in utilization sinks is controlled by the metabolic activity of the sink cells, rather than the transfer processes between conducting cells and sink cells.

The control of apoplastic unloading has been studied mainly in French bean stem, sugar beet storage roots, sugarcane stalk, and legume seed coats. In the bean stem, the transfer of assimilate from the conducting tissue to the adjacent storage cells occurs via the apoplast (96). However, this apoplastic unloading of assimilate may not occur entirely by means of simple diffusion (90), but also by facilitated movement, for it is affected by plant growth regulators (54). Nevertheless, the plasmodesmatal frequency between conducting tissue and the adjacent tissue is sufficient for symplastic unloading in the bean stem (53). It has been suggested that symplastic movement may be controlled by pressure-sensitive "valves" in the plasmodesmatal connections (3). Once membrane transport is saturated, as a result of high sugar concentration in the free space, symplastic unloading would begin and become the principal route of unloading (55). These observations imply that assimilate may be unloaded by either apoplastic or symplastic routes, depending on the conditions.

It has been assumed that unloading in sugar beet storage roots is apoplastic because (a) sucrose can be taken up by the sink cells, (b) both the symplastic connections between conducting tissue and sink cells are few (142), and (c) the invertase activity in the storage roots is low (43). However, anatomical studies of sugar beet storage root demonstrated that there may be sufficient plasmodesmatal connections between the conducting tissue and the adjacent cells (87), particularly between the storage cells away from the conducting tissue, to transport sugars throughout the root (101).

Recently, it has been reported that glucose rather than sucrose was pre-ferentially taken up by protoplasts isolated from the bundle region of red beets (39). Sucrose may be unloaded apoplastically from the conducting tissue and hydrolyzed in the cell wall prior to the uptake by sink cells. Although randomization of asymmetrically labelled [^{14}C-fructosyl]-sucrose taken up by the sugar beet root discs was very low (41), the hydrolysis of sucrose was estimated to be 30% (141). Although acid invertase activity was reduced substantially when sucrose was stored in sugar beet tap roots, the activity of sucrose synthetase increased markedly (43). As the activity of sucrose synthetase is evenly distributed in both the core and peripheral tissues of the beet and is high enough to hydrolyze the imported sucrose (32), sucrose synthetase may be responsible for sucrose hydrolysis and thus partly determines the sink strength in sugar beet (78).

In the last few years, the characteristics of apoplastic unloading have been investigated by using attached empty seed coats of soya bean (120), garden pea (140), broad bean (139), cowpea (97), and maize (99). The efflux of assimilate into the seed coat, after seed removal, is comparable to that occurring during normal growth of these seeds. The regulation of the efflux revealed by such studies has been regarded as the regulation of apoplastic unloading in these organs. In soya bean and garden pea, the efflux appears to be an energy-dependent, perhaps carrier-mediated, transport process (118, 138). On the other hand, efflux in the maize pedicel was not affected by metabolic inhibitors and may be a passive diffusional process (99). As the unloading of assimilate is reduced by high concentrations of slowly (rather than rapidly) penetrating solutes in the pedicel cavity, a reduction of turgor of the unloading cells in the pedicel has been identified as the cause of reduced unloading (100). Among the cations, potassium profoundly stimulated the unloading of assimilate from the seed coats, owing either to a direct activation of a putative sucrose/proton symport from the unloading cells (127), or to an inhibition of reabsorption of assimilate by the cells (140). Either could be possible, because the movement of K^+ into the unloading cells will depolarize the membrane potential (107). It has been proposed that the variation of K^+ concentration in the free space of the seed coat may serve as a signal to integrate sucrose uptake by the embryo with sucrose unloading from the seed coat (95).

However, it is obvious from the anatomical study of the seed coat of *Phaseolus vulgaris* that assimilate is likely to be transferred from the sieve element to adjacent vascular parenchyma symplastically and subsequently unloaded apoplastically to the cell wall, prior to efflux from the seed coat (93). The intracellular conversion of imported glutamine to glutamate in the seed coat of soya bean also suggests that assimilate is not apoplastically unloaded from the phloem tissue (63). Therefore, the kinetics of the efflux in the seed coat are most likely a combination of the symplastic unloading from the sieve element, the apoplastic unloading from the branch parenchyma, and the leakage from the seed coat. Thus, the energy-dependent characteristics of the efflux may be mainly due to the symplastic transfer and the metabolic activity of the seed coat, and the diffusional characteristics may be due to the apoplastic leakage from the branch parenchyma or from the seed coat.

Sieve elements, regardless of their position in the plant, should possess an energy-dependent active loading facility to absorb sucrose or other assimilates from the adjacent free space while releasing some of the sucrose back to the apoplast. This apoplastic unloading would occur only in the sink tissue if the reloading process is retarded, the unloaded sucrose is hydrolyzed at the point of unloading, or the leakage is enhanced (25).

Uptake of Sugars by Sink Cells

Whether apoplastic unloading from the sieve elements occurs or not, some sink cells do retrieve imported sucrose or hexoses from the adjacent free space. The biphasic uptake of amino acids and sugars by developing soya bean embryos suggests that there are saturating and nonsaturating uptake processes at low and high substrate concentrations, respectively. The main component of the nonsaturating uptake may be a passive diffusion process (5). In sugarcane stalk, the uptake of sugar by storage cells is a selective process; only glucose is taken up by the protoplast (75). The uptake of sucrose by cell suspensions confirm that hydrolysis of sucrose by cell wall–bound acid invertase is the essential step for sucrose storage in sugarcane stalk (9, 47, 50). Furthermore, the uptake of glucose appears to be a proton symport with stoichiometries of 1 H^+ and 1 K^+ per sugar for charge compensation. It has been suggested that the establishment of the membrane potential of the plasmalemma may result from a series of redox reactions in sugarcane (113). However, further examination failed to demonstrate that the observed redox reactions are mechanistically and stoichiometrically related to form a plasmalemma-bound electron-transport system for membrane energization (77).

On the other hand, discs of sugar beet root take up sucrose, glucose, and fructose, although the uptake of sucrose is nonsaturating with increasing substrate concentration and is greater than that of glucose or fructose. The uptake of the latter saturates with increasing substrate concentration (141). However, the rate of sugar uptake by isolated protoplasts from red beet roots favors the hexoses (39). While the uptake of sucrose by sugar beet root discs is not affected by the pH between 4 and 8, the optimal pH for glucose and fructose uptake is 5 (141). The uptake of sucrose or glucose by both the discs and protoplasts of red beet root is optimal at pH 5 (39). Whether these contradicting observations are due to varietal or tissue differences has not been clarified. Nevertheless, the uptake of sugars by both sugar beet and red beet root tissue, or isolated protoplasts, is inhibited by a number of metabolic inhibitors, uncouplers, and ionophores (39, 40, 141). Recent studies of sucrose uptake by sugar beet root discs demonstrated that the cell turgor may regulate the uptake process (143). When the cell turgor was increased by low osmotic concentrations in the bathing solution, the uptake of sucrose was reduced. As the acidification of the medium is greatly reduced at high cell turgor, the plasma-membrane ATPase activity was presumably inhibited. On the other hand, the membrane electrical potential difference, *Em,* of the cells with lower turgor was higher than those with higher turgor and the proton extrusion was more vigorous (74). Therefore, the regulation of cell turgor of the sugar-storage tissue may provide a means to regulate the uptake of sugars by the sink cells.

However, it is not always true that uptake of sugars by the sink cells is a selective process. Although sucrose is hydrolyzed in the pedicel cavity of the maize kernel (106), the uptake of sugars, such as sucrose and glucose, is not selective (31, 116). The hydrolysis of sucrose in the pedicel cavity may increase the total solute concentration gradient and facilitate the diffusion of sugars along the cell-wall path for subsequent uptake by the endosperm cells. However, in the other starch-storage sinks, such as wheat (67) and barley (30), imported sucrose was not hydrolyzed prior to the uptake by sink cells.

Studies of the regulation of sugar uptake by sink cells have provided a deeper insight when comparisons were made between the uptake kinetics of tissue discs, free cell suspensions, and isolated protoplasts (20, 38, 75, 76, 141). By deduction, the possible role of processes associated with cell wall or intercellular free space can be identified. However, there are indications that the uptake of sugars by isolated protoplasts may be affected by the changes in membrane properties or by disruption of the existing plasmodesmatal connections during protoplast preparation. If the low sucrose concentration of the isolated protoplast of red beet root (137) is due to leakage of sucrose during preparation, the subsequent sugar uptake may not be the same as it is in situ.

The discussion above deals mainly with the cellular uptake of sugars from the adjacent apoplast. Even in sink organs where sugars are unloaded apoplastically, the further transport of sugars to sink cells away from the conducting tissue may occur by means of an intercellular movement, as suggested in sugar beet roots (38). The regulation of this symplastic transport may be crucial for overall assimilate accumulation in the sink organs.

It is appropriate to draw attention to the phenomenon of the intercellular mass extrusion of disassembled protoplasm in both rapidly growing and wilting tissue of garlic scales (83). Apparently, this kind of intercellular movement is particularly important in rapidly growing tissue as the new cells receive not just the normal mobile assimilate, such as sugars, but also structural and other materials of high molecular weight. The intercellular transport of this ready-made protoplasm may be beneficial for new cell development. Detailed studies of this transport within developing wheat ovules revealed that a high frequency of mass intercellular transport occurs first between disintegrating nucellar tissue and the embryo sac and coincides with the proliferation of antipodal cells (144), and then in the degenerating antipodal cells during the establishment of endosperm tissue (145). Such intercellular transport takes place via enlarged modified plasmodesmata by contraction and expansion of the disassembled protoplasm (146). As this mass intercellular movement would supplement the normal assimilate transport, the regulation of this kind of transport may be crucial for embryo or meristem development. Undoubtedly, both phloem transport and intercellular transport are part of the integrated transport system for assimilate. Mass intercellular transport may occur widely in the region of sink organs where degenerating and developing tissues are adjacent. Further investigation of the significance and the regulation of such transport is needed.

Chemical Conversion of Sugars Inside Sink Cells

Even when sucrose is the principal reserve stored in sink cells, there may still be a number of chemical conversions occurring en route from the sieve elements to the vacuole. If sucrose is hydrolyzed outside storage cells, as in the sugarcane stalk, hexoses in the cytoplasm will then be transported by UDP-glucose-dependent, tonoplast-bound group translocators; and sucrose will subsequently be made inside the vacuole (112). In a study using isolated tonoplast vesicles of sugarcane stem storage cells, UDP-glucose was found to be converted to sucrose and sucrose phosphate by a nucleotide pyrophosphorylase (114).

For sugar beet storage roots, there may be resynthesis of sucrose in the cytoplasm, if indeed sucrose is hydrolyzed in the apoplast and hexoses are taken up by sink cells adjacent to the conducting tissue (39). Apparently, there is sucrose phosphate synthetase (SPS) in sugar beet storage roots, and the activity of this enzyme is higher in the core tissue (where sucrose content is higher) than in the peripheral tissue (32). However, SPS in beet root tissue may not be entirely in the cytoplasm for sucrose resynthesis, because this enzyme has also been identified as one of the enzymes in the group translocator at the tonoplast of sugar cane stem (112). Neither the extent nor the location of sucrose resynthesis in the beet root tissue is certain.

In grape berry, hexoses are the principal reserves in the vacuoles, and the chemical conversion of imported sucrose inside the cells can be complex. It has been suggested that hexoses are phosporylated for sucrose phosphate synthesis, after the hydrolysis of imported sucrose. Subsequently, sucrose is resynthesized prior to hydrolysis in the cytosol (51). It has recently been suggested that tonoplast-bound group translocators may synthesize sucrose phosphate, which is then used for hexose synthesis inside the vacuole (12).

For some storage organs, such as the leaf bases of onion (17) and the tubers of Jerusalem artichoke (22), the principal reserves are fructosans. The synthesis and remobilization of fructosans are closely linked with the metabolism of sucrose and starch (98), but the possible role of fructosans in the regulation of import rate has not yet been studied.

For many legume seeds, cereal kernels, and stem- or root-tubers, the imported sucrose is stored mainly as starch. The rate of starch accumulation undoubtedly regulates the rate of import. The import rate of assimilate was reduced when potato tubers were treated with GA, while the conversion of sugar to starch was reduced (85). There are strong correlations between the rate of growth and the activity of ADPG-pyrophosphorylase, on the one hand, and the ratio of ADPG-pyrophosphorylase to starch phosphorylase in the tubers on the other (52). In wheat grains, the enzyme activity of starch synthesis appears to be ontogenetically controlled (126) and the synthesis of starch is not entirely regulated by the supply of sucrose (4, 66).

The compartmentation of imported sucrose for sugar or starch accumulation in sink organs is genetically determined, but it can be osmotically regulated. In the early stages of tomato fruit development, starch can be as much as 20% of the fruit dry matter (57, 62). However, when water import by the fruit was restricted by high salinity in the root environment, the accumulation of starch increased and accounted for 40% of the dry matter (23). In these fruit, the proportion of the imported assimilate accumulated as hexoses was decreased, and the hexose concentration in the fruit juice was similar to that from plants grown at low salinity. The accumulation of sugars in sink cells may be osmotically regulated, and an enhanced accumulation of starch may keep the osmotic pressure of the cell constant (24). In this case, the enhanced accumulation of starch did not increase the dry-matter accumulation of the fruit.

However, there are relationships between starch accumulation in green fruit and total soluble solids content (TSS) of ripe fruit of different tomato cultivars (19). Indeed, fruit with high TSS did have a higher conversion rate of imported sugars to starch, while the import rate of assimilate is higher than those with low TSS. If the enhanced starch synthesis reduces the sucrose concentration in the fruit, a higher import rate would be expected, owing to a steeper sucrose concentration gradient between the leaves and the fruit (56). Whether the import rate of assimilate can be increased by raising the rate of starch accumulation in tomato fruit is still not certain. However, it is interesting that the decline of starch accumulation in tomato fruit late in fruit development is related not to an increased activity of starch-hydrolyzing enzymes but to a decreased activity of starch-synthesizing enzymes (102).

Physical Compartmentation of Sugars Inside Sink Cells

For sugar-storage sinks, such as beet roots, most of the sugars are stored in the vacuole of the storage sink cells (79). Because the sugar concentration inside the vacuole is much greater than that in the cytosol, the transport of sugars across the tonoplast must be an energy-dependent process. In red beet roots, transport of sucrose across the tonoplast of mechanically isolated vacuoles is sucrose specific and against a sucrose concentration gradient. It is stimulated by ATP and the promotors of ATPase, such as K^+ and Mg^{2+} (20). The optimal sucrose uptake by these vacuoles occurs at pH 5.6, and there is a correlation between sucrose uptake and proton concentration in the uptake medium. Thus, transport of sucrose across the tonoplast appears to be related to proton extrusion from the vacuole. In this system, the rate of sucrose uptake was temperature dependent and the substrate concentration for half maximum velocity, K_m, was reduced from 32 mM to 16 mM as the temperature increased from 10°C to 25°C. Furthermore, the uptake rate of sucrose was substrate dependent, and the uptake process consists of both saturation (active uptake at low concentration) and nonsaturation (passive diffusion at high concentration) components (136, 137). The mechanism of sucrose uptake by vacuoles had also been studied by tonoplast vesicles from sugar beet tap roots. As the uptake of sucrose is stimulated by the addition of ATP, through the establishment of a pH gradient, the transport of sucrose into the vacuole may be driven by the protonmotive force through the action of a sucrose/proton antiport carrier (10).

Recent sugar uptake studies, using vacuoles derived from protoplasts rather than vacuoles isolated by mechanical means, demonstrated that both sucrose and UDP-glucose were taken up by vacuoles (38). In the presence of Mg-ATP, the uptake of sucrose by the vacuoles was greater than the uptake of UDP-glucose. However, the rate of sucrose uptake was similar throughout the beet root tissue, while the rate of UDP-glucose uptake by vacuoles from the vascular bundle regions was greater than from those isolated from storage tissue away from conducting tissue. These observations suggest that UDP-glucose may be the main sugar taken up by vacuoles in cells adjacent to vascular tissue. As these cells may use ATP for glucose uptake from the apoplast, glucose may then be taken up by vacuoles by UDP-glucose-dependent group translocators at the tonoplast. On the other hand, if storage cells away from the conducting tissue mainly import sucrose via the symplastic route, the imported sucrose will be taken up by vacuoles by means of sucrose/proton antiports, as more ATP is available in the cytoplasm (135). Therefore, the principal mechanism for sugar uptake by vacuoles may vary depending on tissue type. This investigation also demonstrated that only vacuoles derived from protoplasts, not those mechanically isolated from tissue, take up UDP-glucose and have an enhanced uptake of sucrose in the presence of Mg-ATP (38). These observations imply that some of the membrane properties of the tonoplast related to membrane transport can be seriously damaged during improper vacuole preparations.

Detailed studies comparing the uptake of sugars by isolated vacuoles from sugarcane stem with that measured in situ or in cell suspension cultures have also demonstrated the complexity of sugar accumulation in vacuoles. During cell growth, the sugar concentration in the vacuole increased and accounted for all sugars in the cell (115). Both glucose and sucrose are taken up by isolated vacuoles. While the uptake rate of hexoses by isolated vacuoles is comparable to the transfer rate of hexoses from the cytosol to the vacuole in situ, the uptake rate of sucrose by isolated vacuoles is below the in situ rate. Although sucrose is synthesized in the cytosol from the imported glucose, the uptake rate of hexoses by the isolated vacuole is much greater than that of sucrose. It appears that the sucrose transport system in the tonoplast of the isolated vacuoles may have been damaged during preparation (111). Furthermore, the protonmotive potential differences across the tonoplast of isolated vacuoles were qualitatively different from those from isolated cells (76). It appears that the uptake of sucrose by the vacuoles of sugarcane stem storage cells is by a sugar/proton antiport process (110), while the uptake of hexoses is facilitated by tonoplast-bound UDP-glucose-dependent group translocators (112). It is quite likely that the accumulation of sucrose in vacuoles of sugarcane is operated mainly by the group-translocator mechanism (86).

The accumulation of sugars in the vacuole is also osmotically controlled. In sugar beet tap roots, the sucrose concentration is inversely affected by potassium salts and nitrogen compounds, which may account for 30% of the osmotic potential of the root sap (T. O. Pocock, personal communication). It has been reported that large cells of high-yield cultivars (21) or large cells at the central zone between cambial rings (89) have a lower sucrose concentration than small cells of low-yield cultivars or those cells adjacent to the conducting tissue. Furthermore, tomato fruit grown in high-nutrient solution accumulated less sugar (23). The high solute potential of these fruit is mainly due to high levels of potassium accumulated in a small amount of vacuole sap (61). This observation suggests that sugar concentration inside the vacuoles may not be determined by the supply of sugars outside the cells, but may be osmotically regulated by the osmotic potential of the sap within the vacuoles.

DETERMINATION OF SINK STRENGTH

Genetic Determination

Both the potential sink size and the potential sink activity are genetically determined. During the evolution of wheat, grain weights have increased with increasing ploidy, from diploid to hexaploid (26). The increase of grain weight is mainly due to increase in cell number of the endosperm and in plastid number per cell, resulting in a greater capacity to accumulate starch (48). However, it is not clear how the endosperm cell number or plastid number is genetically determined. In tomato fruit, the low cell number of the ovary of a parthenocarpic isogenic line in comparison to the seeded ovary is clearly related to low endogenous cytokinin levels and high GA levels at the onset of ovary development (84). Together with the effect of different growth regulators on the fruit size of the induced parthenocarpic fruit (13), cell numbers in the tomato fruit may be regulated by the interactions of endogenous growth regulators, the levels of which are genetically determined.

Apart from the physical constraint, metabolic activity is also predetermined genetically. For instance, the low level of starch in the shrunken mutants of corn is due to mutations in either the *sh* or the *sh2* locus, causing a severe reduction in starch-synthesis-related enzymes such as sucrose synthetase (14) or ADPG-pyrophosphorylase, respectively (122). Whether the change of storage from glucose to sucrose during the growth of melon (82) and the compartmentation of imported sucrose into starch and hexoses in young tomato fruit (62) are genetically determined is worthy of further investigation.

In cereals, a low grain weight is not necessarily due to a low capacity of the endosperm to synthesize starch, but may result from a restriction of assimilate supply. For instance, the small grains of the shrunken endosperm mutant of barley *segl* is due to a premature cessation of grain-filling caused by genetic defects in the maternal spike or grain tissue (30), rather than a low capacity of the endosperm to synthesize starch (29).

The examples described above indicate that it is feasible to improve grain yield by genetically improving the capacity of starch accumulation in the endosperm cells. On the other hand, grain yield can also be improved by altering the priority in assimilate partitioning as in semi-dwarf wheats (11). As both sink strength and sink competition for assimilate are genetically determined, further improvement in crop yield can be achieved by genetic engineering if the genetic determinants are well defined.

Genetic Expression

Although the potential sink strength is genetically determined, the actual sink strength is determined by factors affecting rate-limiting processes within the sinks during development. The low yield of wheat can be caused by drought and high temperature during the grain-filling period. Both the maximum cell and starch granule number were reduced by these growing conditions (91). Starch synthesis in wheat grains may be reduced by temperatures higher than 30°C (7). Therefore, optimization of the rate-limiting processes by manipulating growing conditions is an effective way to increase crop yield.

There is no simple relationship between endogenous plant growth regulators and the growth of a sink organ. For instance, if abscisic acid (ABA) is applied to barley grains when the endogenous ABA level is low at the early stage of grain development, the grain weight can be increased (121). It appears that when the endogenous ABA level becomes limiting, application of ABA will increase the rate of critical processes. In tomato, the endogenous cytokinin level of the developing inflorescence is also related to the rate of assimilate transport (80). Therefore, it is just as important to identify the critical period as it is to determine the critical level of the growth regulator to control the metabolic activities in sink organs.

CONCLUDING REMARKS

The precise roles of chemical conversion and physical compartmentation of imported sugars and the rate-limiting steps in regulating the import of assimilate into sink organs are still uncertain. Considering the complexity and diversity among the very few and mainly sucrose-importing sink organs discussed in this review, a great variety of controlling mechanisms in sink organs, with different mobile assimilate and reserves, is expected. It would thus be futile to apply present knowledge, obtained from a few case studies, to explain the regulation of assimilate import in other sink organs without a proper individual investigation. Lessons learned from these case studies should, however, help future investigations on sink organs other than those described here. Much information on the control of import has been obtained from studies using isolated organs, tissues, cells, protoplasts, vacuoles, and membrane vesicles. Apart from artifacts caused by damage during preparation, the value of such studies should also be assessed by comparing information obtained from different levels of plant organization. Short-distance transport studies, based on short-term radioactive tracer kinetics, should also be assessed with long-term mass transport studies.

Membrane transport processes and the hormonal and enzymatic regulation of rate-limiting processes within sink organs are the most important future research areas. Such research will be more fruitful if it investigates sink strength. The understanding of the mechanism of sink competition can provide a means to increase the harvest index as well as to change yield components. The quality of products can also be improved by manipulating compartmentation and metabolism of imported sugars in sinks. As partitioning of assimilate is genetically determined, a close collaboration between plant physiologists and plant geneticists in this common quest is most desirable.

ACKNOWLEDGMENTS

I thank Prof. J. Warren Wilson, Dr. R. Grange, and Dr. D. Hendrix for their valuable comments on the manuscript.

Literature Cited

1. Austin, R. B., Ford, M. A., Morgan, C. L. 1987. Physiological changes associated with genetic improvement in English cereal yields. *OECD Workshop: Genet. Physiol. Photosynth. Crop Yield.* Abstr. 5

2. Bangerth, F., Ho, L. C. 1984. Fruit position and fruit set sequence in a truss as factors determining final size of tomato fruits. *Ann. Bot.* 53:315–19

3. Barclay, G. F., Fensom, D. S. 1984. Physiological evidence for the existence of pressure-sensitive valves in plasmodesmata between internodes of *Nitella*. In *Membrane Transport in Plants,* ed. W. J. Cram, K. Janacek, R. Rybova, K. Sigler, pp. 316–17. New York: Wiley

4. Barlow, E. W. R., Donovan, G. R., Lee, J. W. 1983. Water relations and composition of wheat ears grown in liquid culture: effect of carbon and nitrogen. *Aust. J. Plant Physiol.* 10:99–108

5. Bennett, A. B., Damon, S., Osteryoung, K., Hewett, J. 1986. Mechanisms of retrieval and metabolism following phloem unloading. In *Phloem Transport,* ed. J. Cronshaw, W. J. Lucas, R. T. Giaquinta, pp. 307–16. New York: Alan R. Liss

6. Bhullar, S. S., Jenner, C. F. 1983. Responses to brief period of elevated temperature in ear and grains of wheat. *Aust. J. Plant Physiol.* 10:549–60

7. Bhullar, S. S., Jenner, C. F. 1985. Differential responses to high temperatures of starch and nitrogen accumulation in the grain of four cultivars of wheat. *Aust. J. Plant Physiol.* 12:263–75

8. Bohner, J. 1986. *Fruchtgrösse und Konkurrenzverhalten bei Tomatenfrüchten Beziehungen zwischen Zellzahl, Zellgrosse und Phytohormonen.* Dokt. dissertation, Universität Hohenheim

9. Bowen, J. C., Hunter, J. E. 1972. Sugar transport in immature internodal tissue of sugarcane. II. Mechanism of sucrose transport. *Plant Physiol.* 49:789–93

10. Briskin, D. P., Thornley, W. R., Wyse, R. E. 1985. Membrane transport in isolated vesicles from sugar beet taproot. II. Evidence for a sucrose/H antiport. *Plant Physiol.* 78:871–75

11. Brooking, I. R., Kirby, E. J. M. 1981. Interrelationships between stem and ear development in winter wheat: the effects of a Norin 10 dwarfing gene, *Gai/Rht2*. *J. Agric. Sci. Cambridge* 97:373–81

12. Brown, S. C., Coombe, B. G. 1982. Sugar transport by an enzyme complex at the tonoplast of grape pericarp cells. *Naturwissenschaften* 69:43–45

13. Bunger-Kibler, S., Bangerth, F. 1983. Relationship between cell number, cell size and fruit size of seeded fruits of tomato (*Lycopersicon esculentum* Mill.) and those induced parthenocarpically by the application of plant growth regulators. *Plant Growth Regul.* 1:143–54

14. Chourey, P. S., Nelson, O. E. 1976. The enzymatic deficiency conditioned by shrunken-1 mutation in maize. *Biochem. Genet.* 14:1041–55

15. Cook, H., Oparka, K. J. 1983. Movement of fluorescein into isolated caryopses of wheat and barley. *Plant Cell Environ.* 6:239–42

16. Cook, M. G., Evans, L. T. 1983. The roles of sink size and location in the partitioning of assimilates in wheat ears. *Aust. J. Plant Physiol.* 10:313–27

17. Darbyshire, B., Henry, R. J. 1978. The distribution of fructan in onions. *New Phytol.* 81:29–34

18. Dick, P. S., ap Rees, T. 1975. The pathway of sugar transport in roots of *Pisum sativum*. *J. Exp. Bot.* 26:305–14

19. Dinar, M., Stevens, M. A. 1981. The relationship between starch accumulation and soluble solids content of tomato fruits. *J. Am. Soc. Hortic. Sci.* 106:415–18

20. Doll, S., Rodier, F., Willenbrink, J. 1979. Accumulation of sucrose in vacuoles isolated from red beet tissue. *Planta* 144:407–11

21. Doney, D. L., Wyse, R. E., Theurer, J. C. 1981. The relationship between cell size, yield and sucrose concentration of the sugar beet root. *Can. J. Plant Sci.* 61:447–53

22. Edelman, J., Jefford, T. G. 1968. The mechanism of fructosan metabolism in higher plants as exemplified in *Heliathus tuberosum*. *New Phytol.* 67:517–31

23. Ehret, D. L., Ho, L. C. 1986. Effects of salinity on dry matter partitioning and fruit growth in tomatoes grown in nutrient film culture. *J. Hortic. Sci.* 61:361–67

24. Ehret, D. L., Ho, L. C. 1986. Effects of osmotic potential in nutrient solution on diurnal growth of tomato fruit. *J. Exp. Bot.* 37:1294–302

25. Eschrich, W. 1986. Mechanisms of phloem unloading. See Ref. 5, pp. 225–30

26. Evans, L. T., Dunstone, R. L. 1970.

Some physiological aspects of evolution in wheat. *Aust. J. Biol. Sci.* 23:725–41

27. Farrar, J. F. 1981. Respiration rate of barley roots: its relation to growth, substrate supply and the illumination of the shoot. *Ann. Bot.* 48:53–63

28. Farrar, J. F. 1985. Fluxes of carbon in roots of barley plants. *New Phytol.* 99:57–69

29. Felker, F. C., Peterson, D. M., Nelson, O. E. 1983. Growth characteristics, grain filling, and assimilate transport in a shrunken endosperm mutant of barley. *Plant Physiol.* 72:679–84

30. Felker, F. C., Peterson, D. M., Nelson, O. E. 1984. [^{14}C]sucrose uptake and labelling of starch in developing grains of normal and *seg1* barley. *Plant Physiol.* 74:43–46

31. Felker, F. C., Shannon, J. C. 1980. Movement of ^{14}C-labelled assimilates into kernels of *Zea mays* L. III. An anatomical examination and microautoradiographic study of assimilate transfer. *Plant Physiol.* 65:864–70

32. Fieuw, S., Willenbrink, J. 1987. Sucrose synthase and sucrose phosphate synthase in sugar beet plants (*Beta vulgaris* L. spp. altissima). *J. Plant Physiol.* 131:153–61

33. Fisher, D. B., Gifford, R. M. 1986. Accumulation and conversion of sugars by developing wheat grains. VI. Gradients along the transport from the peduncle to the indosperm cavity during grain filling. *Plant Physiol.* 82:1024–30

34. Fisher, D. G., Eschrich, W. 1985. Import and unloading of ^{14}C assimilate into mature leaves of *Coleus blumei*. *Can. J. Bot.* 63:1700–7

35. Ford, M. A., Pearman, I., Thorne, G. N. 1976. Effects of variation in ear temperature on growth and yield of spring wheat. *Ann. Appl. Biol.* 82:317–33

36. Geiger, D. R., Fondy, B. R. 1980. Phloem loading and unloading: pathways and mechanisms. *What's New in Plant Physiology*, ed. G. J. Fritz, 11:25–28. Gainesville: Univ. Florida

37. Geiger, D. R., Sovonick, S. A. 1975. Effects of temperature, anoxia and other metabolic inhibitors on translocation. In *Encyclopaedia of Plant Physiology. Transport in Plants. I. Phloem Transport.* ed. M. H. Zimmermann, J. A. Milburn, 1:256–86. Berlin: Springer

38. Getz, H. P. 1987. Accumulation of sucrose in vacuoles released from isolated beetroot protoplasts by both direct sucrose uptake and UDP-glucose-dependent translocation. *Plant Physiol. Biochem.* 25:573–79

39. Getz, H. P., Knauer, D., Willenbrink, J. 1987. Transport of sugars across the plasma membrane of beetroot protoplasts. *Planta* 171:185–96

40. Getz, H. P., Schulte-Altedorneburg, M., Willenbrink, J. 1987. Effects of fusicoccin and abscisic acid on glucose uptake into isolated beetroot protoplasts. *Planta* 171:235–40

41. Giaquinta, R. T. 1977. Sucrose hydrolysis in relation to phloem translocation in *Beta vulgaris*. *Plant Physiol.* 60:339–43

42. Giaquinta, R. T. 1978. Source and sink leaf metabolism in relation to phloem translocation—carbon partitioning and enzymology. *Plant Physiol.* 61:380–85

43. Giaquinta, R. T. 1979. Sucrose translocation and storage in the sugar beet. *Plant Physiol.* 63:828–32

44. Giaquinta, R. T. 1983. Phloem loading of sucrose. *Ann. Rev. Plant Physiol.* 34:347–87

45. Giaquinta, R. T., Lin, W., Sadler, N. L., Franceschi, V. R. 1983. Pathway of phloem unloading of sucrose in corn roots. *Plant Physiol.* 72:362–67

46. Gifford, R. M., Thorne, J. H., Hitz, W. D., Giaquinta, R. T. 1984. Crop productivity and photoassimilate partitioning. *Science* 225:801–8

47. Glasziou, K. T., Gayler, K. R. 1972. Sugar accumulation in sugar cane. Role of cell walls in sucrose transport. *Plant Physiol.* 49:912–13

48. Gleadow, R. M., Dalling, M. J., Halloran, G. M. 1982. Variation in endosperm characteristics and nitrogen content in six wheat lines. *Aust. J. Plant Physiol.* 9:539–51

49. Gordon, A. J., Ryle, G. J. A., Mitchell, D. F., Powell, C. E. 1985. The flux of ^{14}C-labelled photosynthate through soybean root nodules during N_2 fixation. *J. Exp. Bot.* 46:756–59

50. Hatch, M. D., Sacher, J. A., Glasziou, K. T. 1963. Sugar accumulation cycles in sugarcane. I. Studies on enzymes of the cycle. *Plant Physiol.* 38:338–43

51. Hawker, J. S. 1969. Changes in the activities of enzymes concerning with sugar metabolism during the development of grape berries. *Phytochemistry* 8:9–17

52. Hawker, J. F., Marschner, H., Krauss, A. 1979. Starch synthesis in developing potato tubers. *Physiol. Plant.* 46:25–30

53. Hayes, P. M., Offler, C. E., Patrick, J. W. 1985. Cellular structures, plasma membrane surface areas and plasmodesmatal frequencies of the stem of *Phaseolus vulgaris* L. in relation to radial photosynthate transfer. *Ann. Bot.* 56:125–38

54. Hayes, P. M., Patrick, J. W. 1985.

Photosynthate transport in stems of *Phaseolus vulgaris* L. treated with gibberellic acid indole-3-acetic acid or kinetin. *Planta* 166:371–79

55. Hayes, P. M., Patrick, J. W., Offler, C. E. 1987. The cellular pathway of radical transfer of photosynthates in stems of *Phaseolus vulgaris* L. Effects of cellular plasmolysis and p-chloromercuribenzene sulphonic acids. *Ann. Bot.* 59:635–42

56. Ho, L. C. 1979. Regulation of assimilate translocation between leaves and fruits in the tomato. *Ann. Bot.* 43:437–48

57. Ho, L. C. 1980. Control of import into tomato fruits. *Ber. Dtsch. Bot. Ges.* 93:315–25

58. Ho, L. C. 1984. Partitioning of assimilates in fruiting tomato plants. *Plant Growth Regul.* 2:277–85

59. Ho, L. C. 1986. Metabolism and compartmentation of translocates in sink organs. See Ref. 5, pp. 317–24

60. Ho, L. C., Baker, D. A. 1982. Regulation of loading and unloading in long distance transport systems. *Physiol. Plant.* 56:225–30

61. Ho, L. C., Grange, R. I., Picken, A. J. 1987. An analysis of the accumulation of water and dry matter in tomato fruit. *Plant Cell Environ.* 10:157–62

62. Ho, L. C., Sjut, V., Hoad, G. V. 1983. The effect of assimilate supply on fruit growth and hormone levels in tomato plants. *Plant Growth Regul.* 1:155–71

63. Hsu, F. C., Bennett, A. B., Spanswick, R. M. 1984. Concentrations of sucrose and nitrogenous compounds in the apoplast of developing soybean seed coats and embryos. *Plant Physiol.* 75:181–86

64. Hurd, R. G., Gay, A. P., Mountifield, A. C. 1979. The effect of partial flower removal on the relation between root, shoot and fruit growth in the indeterminate tomato. *Ann. Appl. Biol.* 93:77–89

65. Inoue, H., Tanaka, A. 1978. Comparison of source and sink potentials between wild and cultivated potatoes. *J. Sci. Soil. Man. Jpn.* 49:321–24

66. Jenner, C. F. 1970. Relationship between levels of soluble carbohydrate and starch synthesis in detached ears of wheat. *Aust. J. Biol. Sci.* 23:991–1003

67. Jenner, C. F. 1974. An investigation of the association between the hydrolysis of sucrose and its absorption by grains of wheat. *Aust. J. Plant Physiol.* 1:319–29

68. Jenner, C. F. 1986. End product storage in cereals. See Ref. 5, pp. 561–72

69. Jennings, A. C., Morton, R. K. 1963. Changes in carbohydrate, protein and non-protein nitrogenous compounds of developing wheat grains. *Aust. J. Biol. Sci.* 16:318–31

70. Kallarackal, J., Milburn, J. A. 1985. Respiration and phloem translocation in the roots of chickpea *(Cicer arietinum)*. *Ann. Bot.* 56:211–18

71. Kinet, J. M. 1977. Effect of light condition on the development of the inflorescence in tomato. *Sci. Hortic.* 6:15–26

72. Kinet, J. M. 1977. Effect of defoliation and growth substances on the development of the inflorescence in tomato. *Sci. Hortic.* 6:27–35

73. Kinet, J. M., Zime, V., Linotte, A., Jacqmard, A., Bernier, G. 1986. Resumption of cellular activity induced by cytokinin and gibberellin treatments in tomato flowers targeted for abortion in unfavourable light conditions. *Physiol. Plant.* 64:67–73

74. Kinraide, T. B., Wyse, R. E. 1986. Electrical evidence for turgor inhibition of proton extrusion in sugar beet taproot. *Plant Physiol.* 82:1148–50

75. Komor, E., Thom, M., Maretzki, A. 1981. The mechanism of sugar uptake by sugarcane suspension cells. *Planta* 153:181–92

76. Komor, E., Thom, M., Maretzki, A. 1982. Vacuoles from sugarcane suspension cultures. III. Protonmotive potential difference. *Plant Physiol.* 69:1326–30

77. Komor, E., Thom, M., Maretzki, A. 1987. The oxidation of extracellular NADH by sugarcane cells: coupling to ferricyanine reduction, oxygen uptake and pH change. *Planta* 170:34–43

78. Kursanov, A. L. 1974. Assimilattransport und Zuckerspeicherung in der Zuckerrube. *Z. Zuckerindust.* 24:478–87

79. Leigh, R. A., ap Rees, T., Fuller, W. A., Banfield, J. 1979. The location of acid invertase activity and sucrose in vacuoles isolated from storage roots of red beet *(Beta vulgaris* L.). *Biochem. J.* 178:539–47

80. Leonard, M., Kinet, J. M. 1982. Endogenous cytokinin and gibberellin levels in relation to inflorescence development in tomato. *Ann. Bot.* 50:127–30

81. Leonard, M., Kinet, J. M., Bodson, M., Bernier, G. 1983. Enhanced inflorescence development in tomato by growth substance treatments in relation to ^{14}C-assimilate distribution. *Physiol. Plant.* 57:85–89

82. Lester, G. E., Dunlap, J. R. 1985. Physiological changes during development and ripening of 'Perlita' musk melon fruits. *Sci. Hortic.* 26:323–31

83. Lou, C. H., Wu, S. H., Chang, W. C., Shao, L. M. 1956. Intercellular movement of protoplasm as a means of translocation of organic material in garlic. *Acta Bot. Sinica* 5:345–62

84. Mapelli, S., Frova, C., Torti, G., Soressi, G. P. 1978. Relationship between set, development and activities of growth regulators in tomato berries. *Plant Cell Physiol.* 19:1281–88

85. Mares, D. J., Marschner, H. 1980. Assimilate conversion in potato tubers in relation to starch deposition and cell growth. *Ber. Dtsch. Bot. Ges.* 93:299–314

86. Maretzki, A., Thom, M. 1987. UDP-glucose-dependent sucrose translocation in tonoplast vesicles from stalk tissue of sugarcane. *Plant Physiol.* 83:235–37

87. Mierzwa, R. J., Evert, F. 1984. Plasmodesmatal frequency in the root of sugar beet. *Am. J. Bot.* 71(S):39

88. Milford, G. F. J. 1973. The growth and development of the storage root of sugar-beet. *Ann. Appl. Biol.* 75:427–38

89. Milford, G. F. J., Thorne, G. N. 1973. The effect of light and temperature late in the season on the growth of sugarbeet. *Ann. Appl. Biol.* 75:419–25

90. Minchin, P. E. H., Ryan, K. G., Thorpe, M. R. 1984. Further evidence of apoplastic unloading into the stem of bean: identification of the phloem buffering pool. *J. Exp. Bot.* 35:1744–53

91. Nicolas, M. E., Gleadow, R. M., Dalling, M. J. 1984. Effects of drought and high temperature on grain growth in wheat. *Aust. J. Plant Physiol.* 11:553–66

92. Nilson, E. B., Johnson, V. A., Gardner, C. O. 1957. Parenchyma and epidermal cell length in relation to plant height and culm internode length in winter wheat. *Bot. Gaz.* 119:38

93. Offler, C. E., Patrick, J. W. 1984. Cellular structures, plasma membrane surface areas and plasmodesmatal frequencies of seed coats of *Phaseolus vulgaris* L. in relation to photosynthate transfer. *Aust. J. Plant Physiol.* 11:79–99

94. Offler, C. E., Patrick, J. W. 1986. Cellular pathway and hormonal control of short-distance transfer in sink regions. See Ref. 5, pp. 295–306

95. Patrick, J. W. 1987. Effects of potassium on photosynthate unloading from seed coats of *Phaseolus vulgaris* L. Specificity and membrane location. *Ann. Bot.* 59:181–90

96. Patrick, J. W., Turvey, P. M. 1981. The pathway of radical transfer of photosynthate in decapitated stem of *Phaseolus vulgaris* L. *Ann. Bot.* 47:611–21

97. Peoples, M. B., Atkins, C. A., Pate, J. S., Murray, D. R. 1985. Nitrogen nutrition and metabolic interconversions of nitrogenous solutes in developing cowpea fruits. *Plant Physiol.* 77:382–88

98. Pollock, C. J. 1986. Fructans and the metabolism of sucrose in vascular plants. *New Phytol.* 104:1–24

99. Porter, G. A., Knievel, D. P., Shannon, J. C. 1985. Sugar efflux from maize (*Zea mays* L.) pedicel tissue. *Plant Physiol.* 77:524–31

100. Porter, G. A., Knievel, D. P., Shannon, J. C. 1987. Assimilate unloading from maize (*Zea mays* L.) pedicel tissue. 1. Evidence for regulation of unloading by cell turgor. *Plant Physiol.* 83:131–36

101. Richter, E., Ehwald, R. 1983. Apoplastic mobility of sucrose in storage parenchyma of sugar beet. *Physiol. Plant.* 58:263–68

102. Robinson, N. L., Hewitt, J., Bennett, A. B. 1987. Carbohydrate metabolism during development in tomato fruit. *Plant Physiol.* 83(S):3

103. Sawhney, V. K. 1984. Gibberellins and fruit formation in tomato: a review. *Sci. Hortic.* 22:1–8

104. Schmalstig, J., Geiger, D. R. 1985. Phloem unloading in developing leaves of sugar beet. I. Evidence for pathway through the symplast. *Plant Physiol.* 79:237–41

105. Schmalstig, J. G., Geiger, D. R. 1987. Phloem unloading in developing leaves of sugar beet. II. Termination of phloem unloading. *Plant Physiol.* 83:49–52

106. Shannon, J. C. 1972. Movement of ^{14}C-labelled assimilates into kernels of *Zea mays* L. I. Pattern and rate of sugar movement. *Plant Physiol.* 49:198–202

107. Spanswick, R. M. 1981. Electrogenic ion pumps. *Ann. Rev. Plant Physiol.* 32:267–89

108. Tanaka, A., Yamaguchi, J. 1972. Dry matter production, yield components and grain yield of the maize plant. *J. Fac. Agric. Hokkaido Univ.* 57:71–132

109. Tanaka, A., Kawano, K., Yamaguchi, J. 1966. Phytosynthesis, respiration and plant type of the tropical rice plant. *IRRI Tech. Bull.* No. 7

110. Thom, M., Komor, E. 1984. H-sugar antiport as the mechanism of sugar uptake by sugarcane vacuoles. *FEBS Lett.* 173:1–4

111. Thom, M., Komor, E., Maretzki, A. 1982. Vacuoles from sugarcane suspension cultures. II. Characterization of sugar uptake. *Plant Physiol.* 69:1320–25

112. Thom, M., Maretzki, A. 1985. Group translocation as a mechanism for sucrose transfer into vacuoles from sugarcane cells. *Proc. Natl. Acad. Sci. USA* 82:4697–701
113. Thom, M., Maretzki, A. 1985. Evidence for a plasmalemma redox system in sugarcane. *Plant Physiol.* 77:871–76
114. Thom, M., Maretzki, A. 1987. UDPglc breakdown by the group translocator for sucrose. *Plant Physiol.* 83(S):3
115. Thom, M., Maretzki, A., Komor, E. 1982. Vacuoles from sugarcane suspension cultures. I. Isolation and partial characterization. *Plant Physiol.* 69:1315–19
116. Thomas, P. A., Knievel, D. P., Shannon, J. 1987. Characterisation of sugar movement in kernels of maize. *Plant Physiol.* 83(S):6
117. Thorne, J. H. 1981. Morphology and ultrastructure of maternal seed tissues of soybean in relation to the import of photosynthate. *Plant Physiol.* 67:1016–25
118. Thorne, J. H. 1985. Phloem unloading of C and N assimilates in developing seeds. *Ann. Rev. Plant Physiol.* 36:317–43
119. Thorne, J. H. 1986. Sieve tube unloading. See Ref. 5, pp. 211–24
120. Thorne, J. H., Rainbird, R. M. 1983. An in vivo technique for the study of phloem unloading in seed coats of developing soybean seeds. *Plant Physiol.* 72:268–71
121. Tietz, A., Dingkuhn, M. 1981. Regulation of assimilate transport in barley (*Hordeum vulgare* cv. Sorte Claudia) by abscisic acid content of young cryopses. *Z. Pflanzenphysiol.* 104:475–79
122. Tsai, C. Y., Nelson, O. E. 1966. Starch deficient maize mutants lacking adenosine diphosphate glucose pyrophosphorylase activity. *Science* 151:341–43
123. Turgeon, R. 1984. Termination of nutrient import and development of vein loading capacity in albino tobacco leaves. *Plant Physiol.* 76:45–48
124. Turgeon, R. 1987. Phloem unloading in tobacco sink leaves: insensitivity to anoxia indicates a symplastic pathway. *Planta* 171:73–81
125. Turgeon, R., Webb, J. A. 1975. Leaf development and phloem transport in *Cucurbita pepo*: maturation of minor veins. *Planta* 121:265–69
126. Turner, J. F. 1969. Starch synthesis and changes in uridine diphosphate glucose pyrophosphorylase and adenosine diphosphate glucose pyrophosphorylase in developing wheat grain. *Aust. J. Biol. Sci.* 22:1321–27
127. van Bel, A. J. E., Patrick, J. W. 1984. No direct linkage between proton pumping and photosynthate unloading from seed coats of *Phaseolus vulgaris* L. *Plant Growth Regul.* 2:319–26
128. Walker, A. J., Ho, L. C. 1977. Carbon translocation in the tomato: carbon import and fruit growth. *Ann. Bot.* 41:813–23
129. Walker, A. J., Ho, L. C. 1977. Carbon translocation in the tomato: effect of fruit temperature on carbon metabolism and the rate of translocation. *Ann. Bot.* 41:825–32
130. Walker, A. J., Ho, L. C., Baker, D. A. 1978. Carbon translocation in the tomato: Pathways of carbon metabolism in the fruit. *Ann. Bot.* 43:437–48
131. Walker, A. J., Thornley, J. H. M. 1977. The tomato fruit: import, growth, respiration and carbon metabolism at different fruit sizes and temperature. *Ann. Bot.* 41:977–85
132. Wardlaw, I. F. 1968. The control and pattern of movement of carbohydrates in plants. *Bot. Rev.* 34:79–105
133. Wareing, P. F., Patrick, J. 1975. Source-sink relations and the partition of assimilates in the plant. In *Photosynthesis and Productivity in Different Environments,* ed. J. P. Cooper, pp. 481–99. Cambridge: Cambridge Univ. Press
134. Warren Wilson, J. 1972. Control of crop processes. In *Crop Processes in Controlled Environments,* ed. A. R. Rees, K. E. Cockshull, D. W. Hand, R. G. Hurd, pp. 7–30. London/New York: Academic
135. Willenbrink, J. 1987. Die pflanzliche Vakuole als Speicher. *Naturwissenschaften* 74:22–29
136. Willenbrink, J., Doll, S. 1979. Characteristics of sucrose uptake system of vacuoles isolated from red beet tissue: kinetics and specificity of the sucrose uptake system. *Planta* 147:159–62
137. Willenbrink, J., Doll, S., Getz, H. P., Meyer, S. 1984. Uptake of sugars into vacuoles and protoplasts isolated from storage tissue of *Beta vulgaris*. *Ber. Dtsch. Bot. Ges.* 97:27–39
138. Wolswinkel, P. 1985. Phloem unloading and turgor-sensitive transport: factors involved in sink control of assimilate partitioning. *Physiol. Plant.* 65:331–39
139. Wolswinkel, P., Ammerlaan, A. 1983. Phloem unloading in developing seeds of *Vicia faba* L. The effect of several inhibitors on the release of sucrose and amino acids by the seed coat. *Planta* 158:205–15
140. Wolswinkel, P., Ammerlaan, A. 1985.

Effect of potassium on sucrose and amino acid release from the seed coat of developing seeds of *Pisum sativum*. *Ann. Bot.* 56:35–43

141. Wyse, R. E. 1979. Sucrose uptake by sugar beet taproot tissue. *Plant Physiol.* 64:837–41

142. Wyse, R. E. 1985. Membrane transport of sucrose: possible control site for assimilate allocation. In *Frontiers of Membrane Research in Agriculture*, ed. J. B. St. John, E. Berlin, P. B. Jackson, pp. 257–71. Totowa: Rowman & Allanheld

143. Wyse, R. E., Zamski, E., Tomos, A. D. 1986. Turgor regulation of sucrose

transport in sugar beet taproot tissue. *Plant Physiol.* 81:478–81

144. Zhang, W. C., Yan, W. M., Wu, S. H. 1980. Intercellular migration of protoplasm in its relation to the development of embryo sac in nucellus of wheat. *Acta Bot. Sinica* 22:32–36

145. Zhang, W. C., Yan, W. M., Lou, C. H. 1984. Transport of disassembled protoplasm from degenerated nucellus into embryo sac and its role in feeding the proliferating antipodals in wheat. *Acta Bot. Sinica* 28:11–18

146. Zhang, W. C., Yan, W. M., Lou, C. H. 1985. Mechanism of intercellular movement of protoplasms in wheat nucellus. *Sci. Sinica (Ser. B)* 28:1175–88

Ann. Rev. Plant Physiol. Plant Mol. Biol. 1988. 39:379–411

ELECTRON TRANSPORT IN PHOTOSYSTEMS I AND II

L.-E. Andréasson and T. Vänngård

Department of Biochemistry and Biophysics, University of Göteborg and Chalmers University of Technology, S-412 96 Göteborg, Sweden

CONTENTS

INTRODUCTION 379
PHOTOSYSTEM I 380
    Polypeptide Composition 381
    P700 382
    The Acceptors A_0 and A_1 383
    The Iron-Sulfur Centers F_X, F_A, and F_B 384
    Path and Rates of Electron Transfer 385
PHOTOSYSTEM II 387
    The Acceptor Side 387
    The Donor Side 390
    Oxygen Evolution 391
CONCLUDING REMARKS 396

INTRODUCTION

Photosynthetic electron transport in plants, algae, and cyanobacteria converts light energy to useful chemical energy. Light is absorbed in the photosynthetic membranes by the chlorophyll antennas of the two photosystems, I and II (PSI and PSII), which cooperate in the transfer of electrons from water to NADP and of protons across the photosynthetic membrane. The two photosystems function in series to permit the electrons to flow against the thermodynamic barrier associated with these reactions. The electron transport components in oxygenic photosynthesis are organized in membrane-bound supramolecular complexes, the water-splitting PSII complex and the NADP-

379

0066-4294/88/0601-0379$02.00

reducing PSI complex with the interconnecting b_6/f complex, which is responsible for the vectorial transfer of hydrogen ions. Small electron carriers connect these complexes; plastoquinone carries electrons from PSII to the b_6/f complex, which in turn is coupled to PSI by the copper-containing blue protein plastocyanin. PSI does not reduce NADP directly but via a ferredoxin-dependent dehydrogenase. Ferredoxin may also participate in cyclic electron transport by routing electrons back to the b_6/f complex.

Much interest has lately been focused on the analogies between the structure and function of the complexes mentioned above and those in photosynthetic bacteria. The reaction center in PSII is strikingly similar to that in Rhodopseudomonads both in composition and mechanism of reaction of the components. The similarities with bacteria also include PSI, which shows parallels with the reaction center in green bacteria, represented by *Chlorobium,* and the b_6/f complex, which is almost identical to the bacterial b/c complex.

The reaction centers in PSI and PSII have been treated in a recent volume in this series (110). The reader is also referred to recent books (3, 16, 251) and articles (164, 281), which cover most of this field, and to the proceedings of the 6th (256) and 7th (28) International Congresses in Photosynthesis.

Neither the b_6/f complex nor the control of photosynthetic electron transport in relation to the lateral distribution of the components is discussed here. Excellent reviews provide more information on these subjects (17, 57, 110, 215).

PHOTOSYSTEM I

During recent years, more interest has been devoted to PSII and oxygen evolution than to PSI. Progress has been made concerning the identity of the electron transfer components of PSI; but important questions, such as the sequence of components in the chain, are still under debate. Figure 1 summarizes some of the present ideas about PSI [for further details, see the recent reviews (160, 165, 219, 241, 299)]. Some selected room-temperature halftimes for forward and recombination electron transfer are given in the Figure.

The whole chloroplast genome from two organisms has been sequenced (192, 244), which already has helped in defining cofactor binding. Two groups have reported crystallization of cyanobacterial reaction centers (91, 298), giving great promise for a better understanding of PSI in the future.

Since the acceptor side of PSI contains many components (at least five), one approach has been to reduce these successively with chemical reductants in order to study their spectroscopic properties. Also, one then follows the flash-induced kinetics of the unreduced acceptors, which is made easier by the slowness of the recombination reaction compared to the forward reactions.

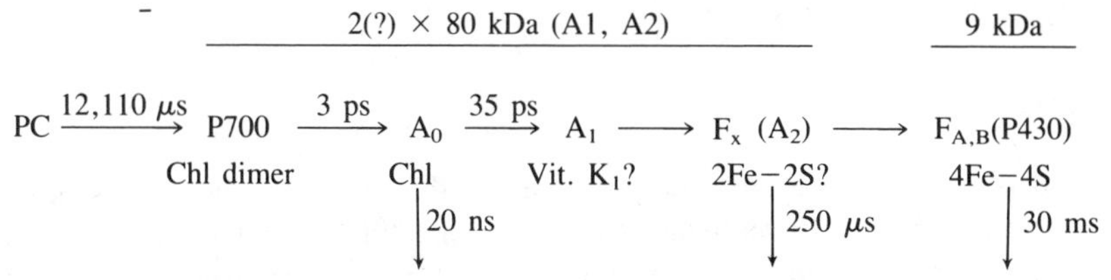

$$2(?) \times 80 \text{ kDa (A1, A2)} \qquad\qquad 9 \text{ kDa}$$

$$PC \xrightarrow{12,110\ \mu s} P700 \xrightarrow{3\ ps} A_0 \xrightarrow{35\ ps} A_1 \longrightarrow F_x (A_2) \longrightarrow F_{A,B}(P430)$$

Chl dimer   Chl   Vit. K_1?   2Fe−2S?   4Fe−4S

$$\Big\downarrow 20 \text{ ns} \qquad\qquad \Big\downarrow 250\ \mu s \qquad\qquad \Big\downarrow 30 \text{ ms}$$

Figure 1   Electron transfer reactions in PSI and some room-temperature half-times.

However, the very low reduction potentials of some acceptors make this approach difficult (just as the high potential at the donor side of PSII has essentially prohibited the use of chemical oxidants for this system) and often requires alkaline conditions. Additional reduction of PSI acceptors can be achieved by illumination during freezing, followed by studies of the frozen samples. Such reduction, however, may have limited relevance for the situation in vivo.

Polypeptide Composition

All photochemically active preparations of PSI contain the polypeptides PSI-A1 and PSI-A2 (or subunits A1 and A2, not to be confused with the acceptors A_1 and A_2—in Nelson's terminology, subunits I_a and I_b; 183) coded by the chloroplast genes *pslAl* and *pslA2*, respectively (90, 147, 157, 192, 244). Their nucleotide sequences show great homology and correspond to molecular masses of about 83 and 82 kDa. Using electrophoresis many groups found a molecular mass of around 65 kDa, but Fish & Bogorad showed (89) that the gene products are not extensively processed. Thus these proteins show an abnormal migration, although more recent analyses yield the proper molecular mass (131). Molecular mass determinations, often forming the basis for subunit stoichiometry of more intact preparations, must therefore be interpreted with great caution. This difficulty might be one reason why the smallest photochemically active preparation, CP-1 [also named P700-Chl *a*-protein, or CCI (core complex I); 270], containing only the subunits A1 and/or A2, has been reported to contain one to four polypeptides per P700 (see discussion in 160, 183, 241, 270, 299, and the section on the iron-sulfur centers, below). The number of chlorophylls in such preparations also varies from 10 to 100 per P700, in part reflecting whether the goal was to obtain a preparation as "intact" or as small as possible.

A full complement of electron acceptors requires additional smaller polypeptides in the range 8–25 kDa (131, 155) (Nelson's subunits II–VII; 183), most of these without known function. In the complete PSI there are additional antenna chlorophyll proteins (molecular masses 20–30 kDa), raising the Chl/P700 ratio to about 200. Many of these smaller nuclear-coded polypeptides have now been cloned (273).

P700

P700 is bound to the large polypeptides of the reaction center. It is often depicted as bound to both the A1 and A2 peptides, but there is no hard evidence that this is the case. The nature of P700 is under debate with respect both to its chemical identity and to whether it is a monomeric or dimeric molecule. It was for many years thought to consist of Chl a (for a review of the earlier literature, see 158), but Hiyama et al (129) suggested that it consists of a C10 epimer of Chl a (Chl a'), which in reconstitution experiments can restore some of the optical (but not the redox) properties of P700. Dörnemann & Senger (82) detected a chlorinated chlorophyll, 13-hydroxy-20-chloro-Chl a (chlorophyll RC I), in a constant ratio to P700 in many diverse organisms; but detailed characterization of this chlorophyll (87) gave no additional support to its assignment to P700. In addition, the early suggestion (295) that P700 contains Chl a enolized in ring V (which decreases the reduction potential from about 900 to 500 mV, more similar to that of P700/P700$^+$) has not been refuted, but this species has not been found in chemical analyses.

The question whether P700 is a chlorphyll monomer or dimer has been attacked by spectroscopic techniques, although the observed amounts of chlorophyll RC I and Chl a', one per P700 (82, 129), lead to obvious suggestions. The dimer hypothesis rested primarily on interpretations of the hyperfine couplings between the unpaired electron of oxidized P700 (P700$^+$) and Chl protons observed using electron paramagnetic resonance (EPR). More refined studies have shown that these data can no longer be considered as strong evidence for a dimer structure (187). Similarly, the parameters characterizing the EPR spectrum of ^{3}P700, the excited triplet state of P700, which is formed in a back reaction after a photoinduced charge separation, are more consistent with a monomeric than a dimeric structure (220). On the other hand, the absorption maximum at 700 nm is at longer wavelength than expected for a monomer, indicating that the neutral ground state of P700 has the properties of a rather strongly coupled dimer (30). In ^{3}P700 and P700$^+$ the chlorophyll molecules could be less well coupled, thus giving essentially monomeric-type properties. The chlorophyll molecules in a dimer need not be physically separated to show monomeric behavior, but IR spectra do indicate some changes in the pigment environment on charge separation (267). Recent so-called hole-burning experiments (109) have also indicated the presence of a dimer with a charge-transfer excited state having a large electric dipole moment. Summing up, the dimer hypothesis seems again most plausible.

The Acceptors A_0 and A_1

The first two known acceptors are named A_0 and A_1. The earlier literature (before about 1983) is somewhat confusing since only one acceptor was recognized, termed A_1, whereas in the actual experiment both A_0 and A_1 might have been observed. A_0 and probably also A_1 (219, 237) are associated with the two larger polypeptides in PSI.

In photoaccumulation experiments under continuous illumination of frozen samples in the presence of dithionite (which reduces the iron-sulfur centers and eventually also reduced $P700^+$, thereby preventing recombination), A_0^- and A_1^- had EPR g values typical of a chlorophyll anion and a semiquinone, respectively (see 161, 271). The same conclusion was reached in room-temperature flash experiments, particularly those performed at a higher EPR microwave frequency (99, 272). The detailed interpretation of these experiments has met with some difficulty (45, 271), but the results are similar to those from the well-defined quinone-containing *Rhodopseudomonas sphaeroides* reaction center (272). Chlorophyll and quinone in a hydrophobic environment can probably also have the very low reduction potentials of A_0 and A_1—below -800 mV.

The chemical nature of A_0 is still not established with certainty, and it cannot be excluded that A_0 consists of the previously mentioned variations of chlorophyll, RC I or Chl a'. Photoaccumulation experiments show that it has a strong absorption at 670 nm, indicating a chlorophyll monomer (161). However, an intermediate observed in room-temperature picosecond (ps)-flash kinetic measurements, having a recombination half-life of 20–30 nanoseconds (ns) when forward electron transfer was eliminated (245) and absorbing at 693 nm with only a shoulder at 675 nm, has been associated with a chlorophyll dimer (185). In both cases the iron-sulfur centers, and presumably also A_1, were reduced, the only difference being the oxidation state of P700 and the temperature of observation. The conflicting results could be caused by different conformations of the acceptor or by the existence of two separate chlorophyll acceptors, the EPR species possibly being on a sidetrack. In favor of the dimer proposition, Ikegami & Itoh reported the occurrence of 3 dimeric and 2 monomeric chlorophylls in a low-chlorophyll preparation (136).

Likewise, the discussion on the chemical nature of A_1 is intense. Its reduced-minus-oxidized optical absorption spectrum has been observed both in photoaccumulation (162) and low-temperature (10 K) transient experiments (43), supporting its quinone nature. Its spectrum is in fact similar to that of vitamin K_1 (phylloquinone, 2-methyl-3-phytyl-1,4-naphthoquinone) (43), which has been detected by many groups in PSI in the ratio of two vitamin K_1 per P700 (159, 237). The molecules are apparently different, however, one being easier to extract with organic solvents (159) and to destroy with UV light (307). (The latter difference disappears in a more purified reaction-center preparation.) Also, at most one A_1^- per P700 has been observed by EPR spectroscopy (31).

Further evidence that A_1 is vitamin K_1 has been obtained by Itoh et al (141), who investigated the effects of extraction of vitamin K_1. This led to a loss of the photoaccumulated A_1 EPR signal, which, under conditions where previously only A_1 was reduced, was replaced by the A_0^- signal. Also, removal of vitamin K_1 decreased the lifetime of flash-induced $P700^+$, suggesting that the fast recombination from A_0^- dominated. Addition of vitamin K_1 again increased the lifetime to that corresponding to the slow recombination from iron-sulfur centers.

Some workers find that iron-sulfur centers can be reduced in low-temperature photoaccumulation experiments even when vitamin K_1 has been extracted (141, 159; but see 163). This is not a strong argument against the idea that A_1 is vitamin K_1, since electron transfer could take place over the longer distance albeit at a slower rate. However, in direct contrast to the data reported above, when vitamin K_1 was destroyed by UV light, the lifetime of $P700^+$ was not shortened (202), and the electron transfer through A_1 appeared to occur in the normal 100-μs range. Both experimental approaches have shortcomings; in extraction experiments one might cause other modifications in addition to the removal of vitamin K_1, while photodestruction with UV light could give a product, of as yet unknown nature, that replaces vitamin K_1 in electron transfer.

The Iron-Sulfur Centers F_X, F_A, and F_B

There are three membrane-bound iron-sulfur centers in PSI—F_X, F_A, and F_B (also labelled Centers X, A, and B). Consistent with this notion, quantitative analyses yield 10–14 Fe per P700 and about the same amount of acid-labile sulfur. The EPR properties and reduction potentials probably exclude the presence of 3Fe-3S centers; thus the only possibilities are 2Fe-2S or 4Fe-4S centers. As with other iron-sulfur clusters, the photosynthetic clusters all seem to absorb at around 430 nm, but relatively weakly compared to chlorophyll. The term P430 is usually reserved for F_A and F_B (Hiyama identifies P430 with F_X; 128)

F_A and F_B, with reduction potentials -500 to -600 mV, have long been considered to be 4Fe-4S centers from their EPR and Mössbauer spectra (see 219 for a discussion of earlier results). Studies by X-ray absorption of the intact system (168) and by ^{19}F NMR of the extruded centers (112) confirm that most of the iron resides in clusters of the 4Fe-4S type. Whereas earlier studies assigned the F_A and F_B centers to a polypeptide with molecular mass around 15 kDa, only a 9-kDa polypeptide has enough cysteines to accommodate these centers (155). From sequence information (131, 189, 192, 244) this polypeptide is a typical 2(4Fe-4S) protein, certainly consistent with the earlier observed magnetic interactions seen in EPR (see 219). The closeness of the two clusters also explains why it is difficult to modify one center selectively. The more easily altered F_B center may reside in a more hydrophilic part of the protein (131).

The F_X center is more difficult to define. Its EPR properties are unusual for an iron-sulfur cluster but might be explained in terms of a unique geometry with small interaction between the iron atoms. The low reduction potential, -700 mV, makes it difficult to reduce, and the Mössbauer study that suggested F_X to be 4Fe-4S (85) suffered from incomplete reduction of the sample. Extended X-ray absorption fine-structure (EXAFS) studies of the oxidized centers revealed the presence of some 2Fe-2S (168; K. Sauer, personal communication), as did the ^{19}F NMR experiments (112); but none gave an accurate quantitative result.

Most likely, F_X is localized on the large A1 and A2 polypeptides since these contain the required cysteines (130, 155). Further evidence is provided by analyses for ^{59}Fe labels and so-called zero-valence sulfur (130, 224), a signature of SDS-treated iron-sulfur centers. From both optical and EPR kinetic experiments on samples treated with LDS, which disconnects the smaller polypeptides from the larger ones, Golbeck and coworkers (111, 293), too, could assign the F_X cluster to the larger polypeptides.

For each iron-sulfur center four cysteines are required; the exceptional coordination to nitrogen (268) seems connected with centers with high reduction potential and is not supported by the EXAFS data (K. Sauer, personal communication). The A2 subunit contains only two cysteines, and the A1 unit from some organisms has three cysteines, two in positions homologous to those in A2. Thus, an A1-A2 pair can bind at most one normal iron-sulfur center. EPR studies indicate the presence of two F_X per P700 (32), but according to polypeptide determinations (183; B. Lindberg Møller, personal communication) there are only two large and one 9-kDa polypeptides per P700. Thus, at present it seems plausible that each reaction center contains one 2Fe-2S bridged by one each of subunit A1 and A2; however, future work on the number and nature of F_X certainly requires an open-minded attitude.

Path and Rates of Electron Transfer

The construction of an electron transfer chain and measurements of transfer rates are met with many difficulties. Rates depend critically on distances, and rather small modifications of the reaction center from one preparation to another may greatly affect rates. A given step often also depends strongly on the acceptors not directly involved in the transfer (whether they are oxidized, reduced, destroyed, or removed). In addition, most rates show a strong temperature dependence with different activation energies, so that an electron transfer path might be different at room and cryogenic temperatures.

Figure 1 above gives some selected values for rates obtained. The excited state ^{1}P700* is reached within 1.5 ps after a flash (294). At room temperature the first charge separation to the state P700$^+$-A$_0^-$ takes place in 14 ps based on absorption measurements (Govindjee and coworkers; 294) and in 3 ps based on fluorescence-decay studies (201). The latter value is more similar to the one found for the bacterial reaction center. The half-life at 1.6 K has been estimated to be 90 ps (109). Shuvalov et al find that the next step, transfer to A$_1$, has a half-life of 35 ps at room temperature (245), whereas Govindjee's data (294) indicate a much longer time, more like the 200 ps Shuvalov found earlier for a damaged preparation.

The further transfer to the iron-sulfur centers is still not well defined (see discussion in 219, 241). F$_A$ and F$_B$ are situated close to each other on the same polypeptide, and indeed it has been difficult to determine if they act in parallel or in series. With both A$_1$ and F$_X$ on the same polypeptides, it seems reasonable to think of an easy transfer from A$_1$ to F$_X$ and further to the other iron-sulfur centers. Room-temperature optical absorption data are scarce for this part of the chain (see 219). The low-temperature EPR determinations of the forward rates have been used against the linear scheme above, but the arguments are beset with difficulties.

Rates for recombination between reduced acceptors and P700$^+$ have been studied in detail, in part because they are slower and can be followed via the strong absorptions of P700. The electron leaves A$_0^-$ in about 20 ns at room temperature and in 100–200 ns at cryogenic temperatures (42a, 239). Some centers return to the ground state of P700 via an excited ^{3}P700 state, but the fraction doing so is still somewhat uncertain (42a) and is predicted to vary also with the flash frequency (114). Recombination from A$_1$ at low temperatures occurs in 120 μs with only F$_A$ and F$_B$ prereduced (242) and in 20 μs if F$_X$ is reduced also (240), an acceleration presumably due to the repulsive effect of the extra charge. Electrons return from F$_X$(A$_2$) in 0.25 ms at room temperature (F$_A$ and F$_B$ reduced), but this time increases to 1.2 ms if F$_{A,B}$ are removed (111), again suggesting an electrostatic effect.

Finally, the reduction of P700$^+$ by plastocyanin should be considered (for a review, see 121). This water-soluble protein is better characterized than any other in plant photosynthesis. It shows interesting behavior at low pH, where it is redox incompetent in isolation and yet functions well in transferring electrons to P700 (see 120). There is a binding site for plastocyanin on PSI, from which electrons can be transferred to P700$^+$ in about 10–20 μs (36). The binding is sufficiently strong that the site is also partly occupied in chloroplasts. However, this high rate is never observed for all reaction centers—in chloroplasts possibly because the binding is incomplete because of the low internal concentration of plastocyanin, and in PSI particles because some centers may have been modified in the preparation. In a second phase, $P700^+$ is reduced at a rate that increases with plastocyanin concentration but reaches a limiting value around 100 μs; this was interpreted as binding to a second, more "distant" site, from which the electron was transferred to $P700^+$ in 100 μs (36). The situation is more complicated than that, however. Bottin & Mathis (37) showed that under conditions where the slow phase after a single flash has a half-time of 200 μs, the transfer after the second of two closely spaced flashes is faster—100 μs. This suggests the presence of a second *occupied* binding site for plastocyanin. The half-life of this phase is slowed down severalfold by the addition of glycerol, which is evidence against a direct intramolecular transfer from the second site. In the light of these findings, many of the earlier interpretations of the effect of pH and chemical modifications of plastocyanin (see 120) should be reconsidered. Similarly, the suggested involvement of subunit III in the electron transfer to $P700^+$ is questioned, since this subunit is missing in maize (182).

PHOTOSYSTEM II

To a considerable degree, PSII research during recent years has progressed closely in step with the development of methods for the separation of the PSII complex from other components in the photosynthetic membrane and for its resolution into subcomponents. A score of highly purified PSII preparations have been isolated from plants (97, 104, 116, 138, 180, 181, 304, 305) and cyanobacteria (12, 154, 233).

The elucidation of the complete 3-D structure of the reaction center from *Rps. viridis* (64) has given PSII research new impetus and allowed results from fields as diverse as molecular biology and picosecond spectroscopy to be condensed into a coherent picture of the PSII reaction center. Similarities with bacterial systems have been emphasized, and old results have been put into a new perspective.

Another area of PSII research dealt with here is the oxygen-evolving reaction. A flood of articles dealing with this process has been published during the last few years, as well as several reviews of general character (13, 67, 77, 117, 211, 228, 280, 286, 309) or specializing in specific areas such as the role of manganese (46, 79, 80) or protein and cofactors (7, 107, 132). The energetics of water-splitting has also received attention (152, 248, 291).

The Acceptor Side

ELECTRON TRANSFER REACTIONS   On excitation of P680, the primary donor in PSII, charge separation takes place with the formation of a radical pair comprised of the oxidized primary donor chlorophyll and the reduced pheophytin intermediate acceptor. (The electron transfer reactions in PSII are summarized in Figure 2.) The electron may then be passed on to the first stable acceptor plastoquinone, Q_A, in 250–300 ps, or recombine with the hole on the primary donor in 2–30 ns, depending on whether Q_A is reduced or absent (62, 186, 259), to form the ground or triplet states of P680. The orientation dependence of the triplet EPR signal implies that the ring plane of the chlorophyll is parallel to the membrane and not perpendicular as in photosynthetic bacteria (217).

The flash-induced reduction of the pheophytin in PSII particles is not electrogenic, which indicates that pheophytin is located on the same side of the photosynthetic membrane as P680 (277). Two pheophytin molecules have been detected in plant and cyanobacterial reaction centers by chemical methods (178). No evidence has been found for the participation of accessory reaction center chlorophylls.

The magnetic interaction of Q_A^- with Fe^{2+} in QFe results in an EPR signal at $g = 1.82$ or $g = 1.90$ depending on the pH; this indicates the protonation of a group close to the quinone site (223, 287), as has been found in photosynthetic bacteria (218). A similar EPR signal may have been observed from the interaction with the next quinone, Q_B (218). The reduced Q_A and Q_B both exhibit typical semiquinone optical spectra dominated by absorption in the UV (232). Although kinetic methods show only one stable quinone acceptor between pheophytin and Q_B (42), the presence of an additional Fe-quinone (86) has been proposed (for further details about additional acceptors, see 77, 165).

Electron transfer to Q_B takes place in the ms regime. The doubly reduced Q_BH_2 may then transfer its electrons to the next component in the chain. The reactions involving Q_B are regulated by HCO_3^-, which facilitates electron transfer through the two-electron gate (reviewed in 83). The charge on Q_B can

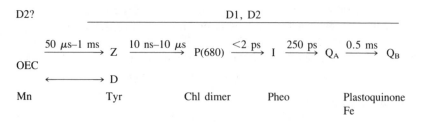

Figure 2   Summary of electron transfer reactions in PSII with some half-times at room temperature indicated.

also recombine with the positive hole on the donor side (127, 221). The recombination reaction is influenced by the conditions on the acceptor and donor sides and has been studied by thermoluminescence (70, 148, 151; also reviewed in 227), which gives information about the lifetime of the S states of the oxygen-evolving system (118) and the temperature dependence of the transitions between these states (150).

Q_{400} The electron acceptor Q_{400} has been identified as the iron of FeQ by Mössbauer, EPR, and redox studies (205). Not only ferricyanide but also various quinones commonly used as acceptors oxidize the iron via the Q_B site (77a, 310). The replacement of the Q_B plastoquinone with other quinones (300) or herbicides (77a, 142) distorts the Q_B site. With the iron oxidized, two electrons may be accommodated on the acceptor side by $Q_A Fe^{3+}$ (42), with Q_A reduced first in light (205a), although only Q_A appears to be active in electron transport (71).

THE ROLE OF THE D1 AND D2 PROTEINS   The 32-kDa D1 protein, which binds the Q_B quinone, is coded by the *psbA* gene; this gene has been sequenced (312). The redox state of the bound quinone is determined by its pH-dependent midpoint potential (221, 283). The D1 protein binds herbicides in competition with quinones and is covalently modified by herbicide analogs (206). Herbicide resistance, which frequently affects the electron transfer properties on the acceptor side of PSII (11, 144), can be traced to alterations in the amino acids of the D1 protein (11, 84, 146, 276). Modification of the D1 protein by mutation (171) or iodination (see below), as well as the binding of DCMU and several herbicide-like compounds (113, 134, 292) to the protein, also affects the donor side of PSII.

The 34-kDa protein, also known as the D2 protein, is a hydrophobic subunit coded by the *psbD* gene (100, 184), which also has been sequenced (216). This protein may function on the donor side, possibly in Mn binding (172).

Extensive sequence homologies have been noted between the D1 and D2 proteins (216) but also between these and the L and M subunits in bacterial reaction centers (see, for example, 306), while no sequence homologies were found between the latter and the 47-kDa protein (275) that was earlier believed to harbor the reaction center (reviewed in 77, 110, 228). Hydropathy plots of the D1 and D2 proteins indicate the presence of five membrane-spanning helixes (231, 276) as in the corresponding bacterial subunits. Since the latter harbor the reaction center in *Rps. viridis,* the D1 and D2 proteins may carry the reaction center in PSII. The presence of conserved amino acids in D1 and D2 relative to the bacterial L and M proteins points at possible binding sites for the electron-transferring components Fe, Q_A, Q_B, primary donor chlorophyll, and pheophytin (276). These predictions concerning the roles of the D1 and D2 proteins have recently been verified with the observation of light-induced charge separation and the spin-polarized triplet state of P680 (62, 193, 259) in a PSII reaction-center preparation containing only these subunits and cytochrome b_{559} (181). Thus, there is now definite proof that the reaction center of photosystem II is located in the D1 and D2 polypeptides in plants and that it is very similar to that in photosynthetic purple bacteria with regard to its structure and function. This similarity includes the acceptor side also (218).

CYTOCHROME b_{559} Cytochrome b_{559} has been shown to consist of two protein subunits of 9 and 4 kDa (58). The amino acid sequences of both subunits are known (126). The heme appears to be crosslinked between histidines in each subunit of the heterodimer. The cytochrome is always copurified with the reaction center polypeptides as a result of an association with the D1 and D2 subunits (18).

The function of cytochrome b_{559} in PSII is still clouded by uncertainty. It may be involved in the electron flow around P680 by accepting electrons from the Q_B plastoquinone (88, 279).

Removal of the extrinsic 23- and 16-kDa polypeptides (see below), which leads to the loss of oxygen-evolving activity, also results in the conversion to the low-potential form (44, 108). However, any connection between the high-potential form of the cytochrome and oxygen-evolving activity seems excluded as the activity can be almost completely revived by $CaCl_2$ without change in the midpoint potential (27, 44, 108).

PSII HETEROGENEITY   The two populations of PSII centers known as α and β centers (reviewed in 29, 77, 165, 280) differ with respect to the size of the chlorophyll antenna (282). The smaller antenna size in β units [which limits excitation energy transfer between PSII units (212)], together with the absence of a functional plastoquinone pool, indicates that these entities are precursors that, when complemented with the missing functions, are incorporated into the grana region as α units (170). Different populations of PSII centers also appear to be responsible for the recent observation of nonelectrogenic charge recombination (169).

The Donor Side

P680 REDUCTION KINETICS   The reduction of P680$^+$ by the secondary donor, Z, can be followed at 680 or 820 nm. Under repetitive flash excitation the reduction is multiphasic in the ns range, which reflects the Coulomb interaction with the charge on the different states of the oxygen-evolving complex (41, 234). Also in the microsecond range, where the transfer from the oxygen-evolving complex is rate-limiting, multiphasic behavior is observed (235). The reduction of P680$^+$ is pH dependent and influenced by several groups with pK-values in the 4.5–8 pH range (56, 236).

The donor Z is associated with the EPR signal II_{Vf} in intact PSII units (34) and with signal II_f when the oxygen-evolving site is inactivated (35). In contrast, the component D, associated with signal II_{slow}, may react with the oxygen-evolving complex by reducing the higher S states and oxidizing the less stable manganese in the S_0 state (222). This function may be important during photoactivation when the newly incorporated manganese needs to be stabilized (255).

THE NATURE OF Z AND D   The optical properties of the oxidized secondary donor, Z$^+$, resemble those of plastosemiquinone (68, 78, 296), possibly influenced by an additional, unidentified carrier on the donor side. Although the identification of Z and D with plastoquinone (188) is supported to some extent by the high midpoint potentials of the Z/Z$^+$ (about 1 V) and D/D$^+$ (500–760 mV) (38, 278) couples, the low amount of plastoquinone in some highly purified PSII preparations (191, 257, 260) poses a problem. Sequence homologies with the L and M subunits in bacterial reaction centers instead suggest that Z and D may be tyrosines in the D1 and D2 proteins (222). This view has received support from experiments in which the use of I$^-$ as an artificial electron donor to Z and D resulted in the iodination of the D1 and D2 proteins, most likely of tyrosines (137, 261, 262). These experiments have located Z and D on the D1 and D2 subunits, respectively. The simultaneous modification of these proteins and the donors to P680 in photoinhibited material (47) has led to similar conclusions.

The strongest arguments for a role of tyrosine on the donor side come from experiments in which cyanobacteria are grown in the presence of deuterated tyrosine which is incorporated into newly synthesized proteins (20). This leads to a structureless EPR signal II_{slow}, which clearly shows that the original hyperfine structure of the signal arises from proton couplings and positively identifies D and most likely also Z as tyrosines. Incorporation of deuterium in the methyl group of plastoquinone via deuterated methionine has no effect.

Oxygen Evolution

THE BINDING OF MANGANESE   Studies using highly purified, well resolved PSII preparations have confirmed that a minimum of four manganese ions are required for full water-splitting activity. Light-dependent restoration of oxygen evolution with added manganese after removal of the metal is Ca^{2+}-dependent (149, 264). Radiation-inactivation experiments indicate that binding sites are inequivalent (258). The released 33-kDa extrinsic polypeptide has been reported to retain metal (2), but this appears to be accidental as the protein can also be removed with the manganese still retained by PSII (95, 174). There are, however, some indications that manganese binding is stabilized by the 33-kDa protein (135, 174), and a 15-kDa active fragment has been identified (289). Other observations link manganese with the D2 protein (172). Both proteins may contribute to the stabilization of the manganese during photoactivation (48).

SPECTROSCOPY OF MANGANESE   Spectroscopic methods have, during the last few years, provided rich information about the involvement of manganese in oxygen evolution. Particularly important was the observation of a light-induced multiline EPR signal from the S_2 state of the oxygen-evolving system, which was attributed to a mixed-valence cluster of manganese ions (see 79).

The S_2 state is associated with an additional EPR signal centered around g = 4.1, originally thought to arise from an intermediate in the oxygen-evolving system (51, 308; but see 290). The S_2 species associated with this signal may be catalytically inactive (23). However, the signal shows the same type of oscillating behavior as a response to light flashes as the multiline signal (311), which is evidence of cyclic interconversion of the S states. Charge recombination between Q^-_A and the g = 4.1 species may induce the Z_V thermoluminescence band (311, but see 198).

The temperature dependence and saturation properties of the multiline EPR signal suggested that it originated from a antiferromagnetically coupled pair of manganese ions (123). The essentially isotropic nature of the EPR spectrum has later been confirmed by spectra recorded at other microwave frequencies (122), although attempts have been made to simulate the spectrum assuming various degrees of anisotropy (72, 73). Owing to the unusual temperature dependence of the g = 4.1 and the multiline EPR signals (74, 80), these have also been proposed to arise from the ground and the thermally populated excited states, respectively, of a tetrameric manganese cluster where the two signals are associated with different conformation of the tetramer (72, 311).

The temperature-independent ratio of the two S_2 signals (122) and difficulties in reproducing the peculiar temperature dependence of the multiline EPR signal (1) have lent some support to an alternative model involving a dimeric or possibly trimeric mixed-valence cluster of manganese ions responsible for the multiline signal in redox equilibrium with an isolated tetravalent manganese ion (122).

X-ray absorption edge spectroscopy (XAES) shows that the manganese in the S_1 state is mainly Mn (III) and that the transition to the S_2 state leads to an oxidation to Mn (IV) (115). On further oxidation, the oxidizing equivalent is not stored on the manganese but on another donor or possibly water (119,

301). The reaction of the S_2 state with hydroxylamine leads to a reduction of manganese (301). Other observations firmly associate the $g = 4.1$ signal with manganese (54).

EXAFS studies of spinach PSII place manganese atoms at a distance of 2.7 Å from each other with indications of another Mn at 3.3 Å (301) and O or N ligands at 1.75 and 2.0 Å from the manganese, some of these possibly as bridging ligands (301, 302). The EXAFS data do not support a regular structure of cubane type but rather two manganese-containing centers at close distance or a highly distorted tetranuclear center. The structure appears largely unaffected by the oxidation of the S_1 state to S_2 and by the removal of the extrinsic 24- and 16-kDa polypeptides (301).

Proton relaxation measurements also support an oxidation of manganese on the S_1–S_2 transition (249) with no further oxidation on the transition to the S_3 state. Also this method indicates that the S_0 state is more reduced than the S_1 state (250).

The relaxation properties of Signal II$_{slow}$, which are influenced by the manganese of the oxygen-evolving system, also indicate changes in the oxidation state of manganese; but different results have been presented by various groups (63, 222).

Further evidence for oxidation-state changes of the manganese has been found in the optical absorption changes—in the UV (68, 69) or in the near IR (81) regions—that accompany the transitions between the S states. The kinetics of the optical changes (66) show excellent agreement with the kinetics of the oxidation of the oxygen-evolving system measured by EPR spectroscopy as reduction of Z^+ (14). There are differences in opinion about the spectral changes associated with the various S-state transitions (156, 213, 214) and about the S_0–S_1 transition in particular. The optical changes on this transition have been reported (156a, 230a) to deviate from those on the S_1–S_2 and S_2–S_3 transitions; others found no such difference (207).

MODELS OF THE MANGANESE CENTER   Binuclear, mixed-valence complexes with di-μ-oxo (252) or μ-oxo bridging in combination with one or more carboxylate bridges (243) show multiline EPR spectra similar to that of the S_2 state of the oxygen-evolving system, some with unusual temperature dependence. The metal-metal and metal-ligand distances in di-μ-oxo complexes (252) closely match those found from EXAFS spectroscopy (302), but the exchange coupling between the manganese ions in mixed-valence complexes of this type is much stronger (243, 252) than that of the photosynthetic multiline species (123). In μ-oxo-carboxylate-coupled dimers the distance between the metal ions is significantly larger (243) than that found in the oxygen-evolving system, with extremely weak magnetic interaction between the metal centers in the Mn(III)$_2$ state. Trinuclear manganese complexes have also been considered as models for the photosynthetic oxygen-evolving system (204).

REACTIONS OF THE OXYGEN-EVOLVING SYSTEM   The inactivation of oxygen evolution at high concentrations of amines is closely correlated with the solubilization of manganese and extrinsic polypeptides (263) and with the loss of the S_2 EPR signals (9). At lower concentrations, hydroxylamine and hydrazine are oxidized by the oxygen-evolving system causing a shift in the flash-induced oxygen and proton release patterns (92) and a delay in the formation of the multiline EPR signal (24). The binding to the S_1 state is cooperative and involves more than three molecules of hydroxylamine (92) and at least two molecules of hydrazine (94, 125). Hydroxylamine also reacts rapidly with the S_2 and S_3 states (125), evidently by direct reduction of the manganese (9).

The change in the multiline EPR signal upon binding of ammonia to the S_2 state suggests a direct interaction with the manganese (25). Of several amines, only ammonia modifies the multiline EPR signal, possibly for steric reasons (21, 22). Binding of ammonia to a second site in the S_1 state, favors the form with the $g = 4.1$ EPR signal in the S_2 state (21) and alters the shape and position of the signal (9). Other inhibitory amines and chloride, which interfere with ammonia binding (225, 226), may also bind at this site (23), as well as anions, such as F^- and NO^-_3, which also enhance the formation of the $g = 4.1$ EPR signal (198). A direct binding of anions to manganese is contradicted by the absence of anion effects on the shape of the multiline EPR signal (61, 303) and by EXAFS, which excludes chloride from the first coordination sphere of manganese in the S_1 and S_2 states (301, 302).

Studies of the binding of water to the oxygen-evolving system by mass spectrometry of ^{18}O indicate that no nonexchangeable water is bound to the S_2 or S_3 states (210). The broadening of the S_2 multiline EPR signal in a medium containing ^{17}O-labeled water demonstrates that water or oxygen atoms derived from water are directly ligated to the manganese in the native S_2 state (124) also when ammonia is bound (9).

On oxidation of water to molecular oxygen by the photosynthetic oxygen-evolving system, protons are released in a $1:0:1:2$ pattern at the successive oxidation steps starting with the S_0 state (76, 93, 211, 230) in accordance with the pH-independent formation of the multiline EPR signal on the S_1–S_2 transition (60). The kinetics of proton release show reasonable agreement with the rates expected from the formation of the different S states but may be influenced by protonation/deprotonation reactions of the secondary donor, Z (93).

With hydroxylamine, which sets back the oxygen release pattern by two steps, two protons are released at the formation of S_0 on the first flash, originating from the binding of two OH^- to the oxygen-evolving system when the S_0 state is generated (94, 230a).

THE ROLE OF PROTEIN AND COFACTORS   Photosystem II contains several extrinsic water-soluble proteins that regulate the function of the oxygen-evolving system. The 23- and 16-kDa proteins are present in plants and green algae (10, 19) but appear to be absent in cyanobacteria (253). Amino acid compositions and partial amino acid sequences of the proteins have been determined (153, 167, 179) and, based on the sequence, an analysis of the secondary structure of the 24-kDa protein has been made (285). The 33-kDa protein is present in all oxygen-evolving organisms. It is coded in the nucleus and expression of the gene is required for functional oxygen evolution (166). The complete amino-acid sequence has recently been published (190).

Salt-washing exposes the manganese site to reductants (106), mainly owing to the loss of the 23-kDa polypeptide, and allows H_2O_2 to act as an electron donor (26, 98, 238). The removal of the polypeptides also eliminates the ns P680 reduction kinetics (5) and stimulates charge recombination (6).

Although depletion of the 23- and 16-kDa proteins may inhibit oxygen evolution, the S_1–S_2 transition is still allowed, as evidenced by the reduction of added acceptor (297), suppressed oxidation of external donor (265), and rapid re-reduction of the secondary donor, Z^+ (96, 145). The latter observation has also been interpreted as indicating the presence of an intermediate donor between the oxygen-evolving complex and Z (33, 53). Thermoluminescence studies, however, support S-state advancement in polypeptide-depleted PSII membranes (227). There is some evidence that depletion of the 16-, 24-, and 33-kDa proteins specifically inhibits (196, 284) or at least retards (177) the S_3–S_4/S_0 transition—i.e. the oxygen release step. Readdition of the 33-kDa protein alone can reactivate oxygen evolution in undamaged PSII units (139, 195, 266, 274).

To a large extent the effects, induced by the removal of the extrinsic proteins, can be accounted for by the role of these in the regulation of the affinity for Cl^- and Ca^{2+}, which are essential for the oxygen-evolving activity (59, 102). The 24-kDa protein, for example, lowers the requirement for Cl^- of oxygen evolution severalfold (8, 175). The role of the 16-kDa polypeptide in Cl^--binding is unclear at present (4, 8, 175).

The depletion of the 16- and 23-kDa proteins also leads to a decreased affinity for Ca^{2+}: Only very high concentrations of Ca^{2+} can restore oxygen evolution (39, 173), the rapid decay of Z^+ (65), and the ns kinetics in P680 reduction (288) and make the multiline EPR signal reappear (75), provided Cl^- is present (49, 102, 208, 254). Other divalent cations with the exception of Sr^{2+} are ineffective (101, 288). The loss of Ca^{2+} is accelerated in light (69, 176). Lanthanum replaces Ca^{2+} and induces a release of extrinsic polypeptides in intact PSII membranes (103). The 16- and 23-kDa polypeptides may provide a suitable environment for tight binding of Ca^{2+} (and Cl^-) (105), possibly by acting on a Ca^{2+}-binding protein (208). The presence of such a protein is indicated by the sensitivity of PSII to calmodulin-type inhibitors (50). In fact, a protein with increased affinity for Ca^{2+} has been isolated from PSII membranes (247).

It has also been proposed that the 33-kDa protein provides binding sites for Cl^- and Ca^{2+} (55). The lower stability of the manganese (174), the inhibition of oxygen evolution (197), and the loss of the multiline EPR signal (254) after removal of this protein all can be counteracted by high (200 mM) concentrations of chloride (but see 135). However, the Cl^--reactivated PSII units may be structurally modified as evidenced by abnormalities in the S_3-S_0 transition and in the stability of the $S_2Q_A^-$ charge pair (197).

In Ca^{2+} (40) or Cl^--depleted PSII membranes partial S-state advancement to the S_2 state is still possible (143, 194, 246, 269), although the S_2 state may show abnormal properties such as high stability (200), loss of multiline EPR signal (but not of the $g = 4.1$ signal; 199), and different redox properties (133). Several other monovalent anions can replace Cl^- and at least partially reactivate the oxygen-evolving complex (61), but OH^- (52) or acetate (229) is inhibitory and blocks electron transfer from the oxygen-evolving complex.

The exchange of bound Cl^- with the medium is rapid (140) and allows the use of $^{35}Cl^-$ NMR to probe the binding of Cl^- to the photosynthetic membrane. Some early experiments suggested that binding has already occurred in the dark (15), while others found that binding takes place only in the S2 and S3 states (209). A recent detailed NMR study shows a complex mode of Cl^- binding involving several distinct sites (55).

CONCLUDING REMARKS

During the last few years we have witnessed a rapid evolution in photosynthesis research, highlighted by the unveiling of the molecular details of a bacterial reaction center. As with the bacterial systems, progress in plant research has depended on the development of preparative methods for the photosystem complexes. Studies of simple PSII complexes have stressed their similarities to bacterial systems, at which molecular genetic investigations had already hinted. The new preparative methods and progress in crystallization techniques, exemplified by the recent successful crystallization of a PSI reaction center, should soon resolve longstanding questions about the chemical identity and localization of the components of the two photosystems and reveal whether components are duplicated as in photosynthetic bacteria. Detailed structural information is also lacking about the water-splitting system. A successful marriage of molecular genetic and X-ray crystallographic techniques should allow a deeper understanding of photosynthetic electron transfer and the role of protein dynamics in these processes.

ACKNOWLEDGMENTS

The authors express their gratitude to those who have communicated their manuscripts prior to publication. We are also indebted to Drs. T. Wydrzynski and S. Styring for their critical reading of the manuscript. This work was supported by a grant from the Swedish Natural Science Research Council.

Literature Cited

1. Aasa, R., Andréasson, L.-E., Lagenfelt, G., Vänngård, T. 1987. A comparison between the multiline EPR signals of spinach and *Anacystis nidulans* and their temperature dependence. *FEBS Lett.* 221:245–48

2. Abramowicz, D. A., Dismukes, G. C. 1984. Manganese proteins isolated from spinach thylakoid membranes and their role in O_2 evolution. II. A binuclear manganese-containing 34 kilodalton protein, a probable component of the water dehydrogenase. *Biochim. Biophys. Acta* 765:318–28

3. Amesz, J., ed. 1987. *New Comprehensive Biochemistry*, Vol. 15, *Photosynthesis*. Amsterdam: Elsevier

4. Akabori, K., Imaoka, A., Toyoshima, Y. 1984. The role of lipids and 17-kDa protein in enhancing the recovery of O_2 evolution in cholate-treated thylakoid membranes. *FEBS Lett.* 173: 36–40

5. Åkerlund, H.-E., Brettel, K., Witt, H. T. 1984. Reversible alteration of nanosecond reduction of chlorophyll a_{II}^+ in inside-out thylakoids correlated to inhibition and reconstitution of oxygen-evolving activity. *Biochim. Biophys. Acta* 765:7–11

6. Åkerlund, H.-E., Renger, G., Weiss, W., Hagemann, R. 1984. Effect of partial removal and readdition of a 23 kilodalton protein on oxygen yield and flash-induced absorbance changes at 320 nm of inside-out thylakoids. *Biochim. Biophys. Acta* 765:1–6

7. Andersson, B., Åkerlund, H.-E., 1987. Proteins of the oxygen-evolving complex. See Ref. 16, pp. 379–420

8. Andersson, B., Critchley, C., Ryrie, I. J., Jansson, C., Larsson, C. et al. 1984. Modification of the chloride requirement for photosynthetic O_2 evolution. The role of the 23 kDa polypeptide. *FEBS Lett.* 168:113–17

9. Andréasson, L.-E., Hansson, Ö. 1987.

EPR studies of the oxygen-evolving system. The interaction with amines. See Ref. 28, 1:503–10

10. Aoki, K., Ideguchi, T., Kakuno, T., Yamashita, J., Horio, T. 1986. Effects of NaCl and glycerol on photosynthetic oxygen-evolving activity with thylakoid membranes from halophilic green alga *Dunaliella tertiolecta. J. Biochem. (Tokyo)* 100:1223–30

11. Astier, C., Meyer, I., Vernotte, C., Etienne, A. L. 1986. Photosystem II electron transfer in highly herbicide resistant mutants of *Synechocystis* 6714. *FEBS Lett.* 207:234–38

12. Astier, C., Styring, S., Maison-Peteri, B., Etienne, A.-L. 1986. Preparation and characterization of thylakoid membranes and photosystem II particles from the facultative phototrophic cyanobacterium *Synechocystis* 6714. *Photobiochem. Photobiophys.* 11:37–47

13. Babcock, G. T. 1987. The photosynthetic oxygen-evolving process. See Ref. 3, pp. 125–58

14. Babcock, G. T., Blankenship, R. E., Sauer, K. 1976. Reaction kinetics for positive charge accumulation on the water side of chloroplast photosystem II. *FEBS Lett.* 61:286–89

15. Baianu, I. C., Critchley, C., Govindjee, Gutowsky, H. S. 1984. NMR study of chloride ion interactions with thylakoid membranes. *Proc. Natl. Acad. Sci. USA* 81:3713–17

16. Barber, J., ed. 1987. *The Light Reactions (Topics in Photosynthesis,* Vol. 8). Amsterdam: Elsevier

17. Barber, J. 1987. Composition, organisation and dynamics of the thylakoid membrane in relation to its function. In *The Biochemistry of Plants*, Vol. 14, ed. M. D. Hatch, N. K. Boardman. New York: Academic. In press

18. Barber, J., Gounaris, K., Chapman, D. J. 1987. Isolation of the photosystem two reaction centre and the location and

function of cytochrome b-559. See Ref. 203

19. Barber, J., Pick, U., Gounaris, K. 1987. Variable and conserved characteristics of photosystem 2 of spinach and of the halotolerant green alga *Dunalliela salina*. See Ref. 28, 2:97–100

20. Barry, B. A., Babcock, G. T. 1987. Characterization of the tyrosine radical involved in photosynthetic oxygen evolution. *Proc. Natl. Acad. Sci. USA*. In press

21. Beck, W. F., Brudvig, G. W. 1986. Binding of amines to the O_2-evolving center of photosystem II. *Biochemistry* 25:6479–86

22. Beck, W. F., Brudvig, G. W. 1987. Coordination of ammonia, but not larger amines, to the manganese site of the O_2-evolving center in the S2 state. See Ref. 28, 1:499–502

23. Beck, W. F., Brudvig, G. W. 1987. Ligand-substitution reactions of the O_2-evolving center of photosystem II. *Chem. Scr.* In press

24. Beck, W. F., Brudvig, G. W. 1987. Reactions of hydroxylamine with the electron-donor side of photosystem II. *Biochemistry*. In press

25. Beck, W. F., de Paula, J. C., Brudvig, G. W. 1986. Ammonia binds to the manganese site of the O_2-evolving complex of photosystem II in the S_2 state. *J. Am. Chem. Soc.* 108:4018–22

26. Berg, S. P., Seibert, M. 1987. Evidence for the role of functional manganese in hydrogen-peroxide-stimulated oxygen production on the first flash in $CaCl_2$-washed photosystem II membranes. See Ref. 28, 1:589–92

27. Bergström, J., Franzén, L.-G. 1987. Restoration of high-potential cytochrome b-559 in salt-washed photosystem II-enriched membranes as revealed by EPR. *Acta Chem. Scand. B* 41:126–28

28. Biggins, J., ed. 1987. *Progress in Photosynthesis Research* (Proc. 7th Int. Congr. Photosynth.). Dordrecht: Nijhoff

29. Black, M. T., Brearley, T. H., Horton, P. 1986. Heterogeneity in chloroplast photosystem II. *Photosynth. Res.* 8: 193–207

30. Blanken, H. J. den, Hoff, A. J. 1983. High-resolution absorbance-difference spectra of the triplet state of the primary donor P-700 in photosystem I subchloroplast particles measured with absorbance-detected magnetic resonance at 1.2 K. Evidence that P-700 is a dimeric chlorophyll complex. *Biochim. Biophys. Acta* 724:52–61

31. Bonnerjea, J., Evans, M. C. W. 1982. Identification of multiple components in the intermediary electron carrier complex of photosystem I. *FEBS Lett.* 148:313–16

32. Bonnerjea, J. R., Evans, M. C. W. 1984. Evidence that the low-potential (-700 mV) electron acceptor (X) in photosystem I has two iron-sulphur centres. *Biochim. Biophys. Acta* 767:153–59

33. Boska, M., Blough, N. V., Sauer, K. 1985. The effect of mono- and divalent salts on the rise and decay kinetics of EPR signal II in photosystem II preparations from spinach. *Biochim. Biophys. Acta* 808:132–39

34. Boska, M., Sauer, K. 1984. Kinetics of EPR signal II_{vf} in chloroplast photosystem II. *Biochim. Biophys. Acta* 765:84–87

35. Boska, M., Sauer, K., Buttner, W., Babcock, G. T. 1983. Similarity of EPR signal II_f rise and P-680$^+$ decay kinetics in Tris-washed chloroplast photosystem II preparations as a function of pH. *Biochim. Biophys. Acta* 722:327–30

36. Bottin, H., Mathis, P. 1985. Interaction of plastocyanin with the photosystem I reaction center: a kinetic study by flash spectroscopy. *Biochemistry* 24:6453–60

37. Bottin, H., Mathis, P. 1987. Turn-over of electron donors in photosystem I: double-flash experiments with pea chloroplasts and photosystem I particles. *Biochim. Biophys. Acta* 892:91–98

38. Boussac, A., Etienne, A. L. 1984. Midpoint potential of signal II (slow) in Tris-washed photosystem-II particles. *Biochim. Biophys. Acta* 766:576–81

39. Boussac, A., Maison-Peteri, B., Etienne, A.-L., Vernotte, C. 1985. Reactivation of oxygen evolution of NaCl-washed photosystem-II particles by Ca^{2+} and/or the 24 kDa protein. *Biochim. Biophys. Acta* 808:231–34

40. Boussac, A., Rutherford, A. W. 1987. S-state formation after Ca^{2+} depletion in the photosystem-II oxygen evolving complex. *Chem. Scr.* In press

41. Brettel, K., Schlodder, E., Witt, H. T. 1984. Nanosecond reduction kinetics of photooxidized chlorophyll-a_{II} (P-680) in single flashes as a probe for the electron pathway, H^+-release and charge accumulation in the O_2-evolving complex. *Biochim. Biophys. Acta* 766:403–15

42. Brettel, K., Schlodder, E., Witt, H. T. 1985. Evidence for only one stable electron acceptor in the reaction center of photosystem II in spinach chloroplasts. *Photobiochem. Photobiophys.* 9:205–13

42a. Brettel, K., Sétif, P. 1987. Magnetic-field effects on primary reactions in photosystem I. *Biochim. Biophys. Acta* 893:109–12

43. Brettel, K., Sétif, P., Mathis, P. 1986. Flash-induced absorption changes in photosystem I at low temperature: evidence that the electron acceptor A_1 is vitamin K_1. *FEBS Lett.* 203:220–24

44. Briantais, J.-M., Vernotte, C., Miyao, M., Murata, N., Picaud, M. 1985. Relationship between O_2 evolution capacity and cytochrome b-559 high-potential form in photosystem II particles. *Biochim. Biophys. Acta* 808:348–51

45. Broadhurst, R. W., Hoff, A. J., Hore, P. J. 1986. Interpretation of the polarized electron paramagnetic resonance signal of plant photosystem I. *Biochim. Biophys. Acta* 852:106–11

46. Brudvig, G. W. 1987. The tetranuclear manganese complex of photosystem II. *J. Bioenerg. Biomembr.* 19:91–103

47. Callahan, F. E., Becker, D. W., Cheniae, G. M. 1986. Studies on the photoactivation of the water-oxidizing enzyme. II. Characterization of weak light photoinhibition of PSII and its light-induced recovery. *Plant Physiol.* 82:261–69

48. Callahan, F. E., Cheniae, G. M. 1985. Studies on the photoactivation of the water-oxidizing enzyme. I. Processes limiting photoactivation in hydroxylamine-extracted leaf segments. *Plant Physiol.* 79:777–86

49. Cammarata, K., Cheniae, G. 1987. PS II Ca abundance and interaction of the 17,24 kD proteins with the Cl^-/Ca^{2+} essential for oxygen evolution. See Ref. 28, 1:617–20

50. Carpentier, R., Nakatani, H. Y. 1985. Inhibitors affecting the oxidizing side of photosystem II at the Ca^{2+}- and Cl^--sensitive sites. *Biochim. Biophys. Acta* 808:288–92

51. Casey, J. L., Sauer, K. 1984. EPR detection of a cryogenically photogenerated intermediate in photosynthetic oxygen evolution. *Biochim. Biophys. Acta* 767:21–28

52. Cole, J., Boska, M., Blough, N. V., Sauer, K. 1986. Reversible and irreversible effects of alkaline pH on photosystem II electron-transfer reactions. *Biochim. Biophys. Acta* 848:41–47

53. Cole, J., Sauer, K. 1987. The flash number dependence of EPR signal II decay as a probe for charge accumulation in photosystem II. *Biochim. Biophys. Acta* 891:40–48

54. Cole, J., Yachandra, V. K., Guiles, R.

D., McDermott, A. E., Britt, R. D. et al. 1987. Assignment of the $g = 4.1$ EPR signal to manganese in the S2 state of the photosynthetic oxygen-evolving complex: an X-ray absorption edge spectroscopy study. *Biochim. Biophys. Acta* 890:395–98

55. Coleman, W. J., Govindjee, Gutowsky, H. S. 1987. Involvement of Ca^{2+} in Cl^- binding to the oxygen evolving complex of photosystem II. See Ref. 28, 1:629–32

56. Conjeaud, H., Mathis, P. 1986. Electron transfer in the photosynthetic membrane. Influence of pH and surface potential on the P-680 reduction kinetics. *Biophys. J.* 49:1215–21

57. Cramer, W. A., Widger, W. R., Black, M. T., Girvin, M. E. 1987. Structure and function of the photosynthetic cytochrome b-c_1 and b_6-f complexes. See Ref. 16, pp. 447–94

58. Cramer, W. A., Theg, S. M., Widger, W. R. 1986. On the structure and function of cytochrome b-559. *Photosynth. Res.* 10:393–403

59. Critchley, C. 1985. The role of chloride in photosystem II. *Biochim. Biophys. Acta* 811:33–46

60. Damoder, R., Dismukes, G. C. 1984. pH dependence of the multiline, manganese EPR signal for the 'S2' state in PS II particles. Absence of proton release during the $S_1 \rightarrow S_2$ electron transfer step of the oxygen evolving system. *FEBS Lett.* 174:157–61

61. Damoder, R., Klimov, V. V., Dismukes, G. C. 1986. The effect of Cl^- depletion and X^- reconstitution on the oxygen-evolution rate, the yield of the multiline manganese EPR signal and EPR Signal II in the isolated photosystem-II complex. *Biochim. Biophys. Acta* 848:378–91

62. Danielius, R. V., Satoh, K., van Kan, P. J. M., Plijter, J. J., Nuijs, A. M., et al. 1987. The primary reaction of photosystem II in the D1-D2-cytochrome b_{559} complex. *FEBS Lett.* 213:241–44

63. de Groot, A., Plijter, J. J., Evelo, R., Babcock, G. T., Hoff, A. J. 1986. The influence of the oxidation state of the oxygen-evolving complex of photosystem II on the spin-lattice relaxation time of signal II as determined by electron spin-echo spectroscopy. *Biochim. Biophys. Acta* 848:8–15

64. Deisenhofer, J., Epp, O., Miki, K., Huber, R., Michel, H. 1984. X-ray structure analysis of a membrane protein complex: electron density map at 3 Å resolution and a model of the chromophores of the photosynthetic reaction

center from *Rhodopseudomonas viridis*. *J. Mol. Biol.* 180:385–98

65. Dekker, J. P., Ghanotakis, D. F., Plijter, J. J., van Gorkom, H. J., Babcock, G. T. 1984. Kinetics of the oxygen-evolving complex in salt-washed photosystem II preparations. *Biochim. Biophys. Acta* 767:515–23

66. Dekker, J. P., Plijter, J. J., Ouwehand, L., van Gorkom, H. J. 1984. Kinetics of manganese redox transitions in the oxygen-evolving apparatus of photosynthesis. *Biochim. Biophys. Acta* 767: 176–79

67. Dekker, J. P., van Gorkom, H. J. 1987. Electron transfer in the water-oxidizing complex of photosystem II. *J. Bioenerg. Biomembr.* 19:125–42

68. Dekker, J. P., van Gorkom, H. J., Brok, M., Ouwehand, L. 1984. Optical characterization of photosystem II electron donors. *Biochim. Biophys. Acta* 764: 301–9

69. Dekker, J. P., van Gorkom, H. J., Wensink, J., Ouwehand, L. 1984. Absorbance difference spectra of the successive redox states of the oxygen-evolving apparatus of photosynthesis. *Biochim. Biophys. Acta* 767:1–9

70. Demeter, S., Rózsa, Zs., Vass, I., Sallai, A. 1985. Thermoluminescence study of charge recombination in photosystem II at low temperatures. I. Characterization of the Z_V and A thermoluminescence bands. *Biochim. Biophys. Acta* 809:369–78

71. Dennenberg, R. J., Jursinic, P. A. 1985. A comparison of the absorption near 325 nm and chlorophyll *a* fluorescence characteristics of the photosystem II acceptors Q_a and Q_{400}. *Biochim. Biophys. Acta* 808:192–200

72. de Paula, J. C., Beck, W. F., Brudvig, G. W. 1986. Magnetic properties of manganese in the photosynthetic O_2-evolving complex. 2. Evidence for a manganese tetramer. *J. Am. Chem. Soc.* 108:4002–9

73. de Paula, J. C., Beck, W. F., Miller, A.-F., Wilson, R. B., Brudvig, G. W. 1987. Studies of the manganese site of photosystem II by electron paramagnetic resonance spectroscopy. *J. Chem. Soc. Faraday.* In press

74. de Paula, J. C., Brudvig, G. W. 1985. Magnetic properties of manganese in the photosynthetic O_2-evolving complex. *J. Am. Chem. Soc.* 107:2643–48

75. de Paula, J. C., Li, P. M., Miller, A.-F., Wu, B. W., Brudvig, G. W. 1986. Effect of the 17- and 23-kilodalton polypeptides, calcium, and chloride on

electron transfer in photosystem II. *Biochemistry* 25:6487–94

76. Dietrich-Glaubitz, R., Völker, M., Renger, G., Gräber, P. 1987. Proton release by photosynthetic water oxidation. See Ref. 28, 1:519–22

77. Diner, B. A. 1987. The reaction center of photosystem II. See Ref. 251, pp. 422–36

77a. Diner, B. A., Petrouleas, V. 1987. Light-induced oxidation of the receptor-side Fe(II) of Photosystem II by exogenous quinones acting through the Q_B binding site. II. Blockage by inhibitors and their effects on the Fe(III) EPR spectra. *Biochem. Biophys. Acta* 893:138–48

78. Diner, B. A., de Vitry, C. 1984. Optical spectrum and kinetics of the secondary donor, Z, of photosystem II. See Ref. 256, I:407–11

79. Dismukes, G. C. 1986. The metal centers of the photosynthetic oxygen-evolving complex. *Photochem. Photobiol.* 43:99–115

80. Dismukes, G. C. 1986. The organization and function of manganese in the water-oxidizing complex of photosynthesis. In *Manganese in Metabolism and Enzyme Function*, ed. F. C. Wedler, V. L. Schram, pp. 275–309. New York: Academic

81. Dismukes, G. C., Mathis, P. 1984. A near infrared electronic transition associated with conversion between S-states of the photosynthetic O_2-evolving complex. *FEBS Lett.* 178:51–54

82. Dörnemann, D., Senger, H. 1986. The structure of chlorophyll RC I, a chromophore of the reaction center of photosystem I. *Photochem. Photobiol.* 43: 573–81

83. Eaton-Rye, J. J., Blubaugh, D. J., Govindjee. 1986. Action of bicarbonate on photosynthetic electron transport in the presence or absence of inhibitory anions. In *Ion Interactions in Energy Transport Systems*, ed. J. Barber, S. Papa, G. Papageorgiou, pp. 263–78. New York: Plenum

84. Erickson, J. M., Rahire, M., Rochaix, J.-D., Mets, L. 1985. Herbicide resistance and cross-resistance: changes at three distinct sites in the herbicide-binding protein. *Science* 228:204–7

85. Evans, E. H., Dickson, D. P. E., Johnson, C. E., Rush, J. D., Evans, M. C. W. 1981. Mössbauer spectroscopic studies of the nature of centre X of photosystem I reaction centres from the cyanobacterium *Chlorogloea fritschii*. *Eur. J. Biochem.* 118:81–84

86. Evans, M. C. W., Ford, R. C. 1986.

Evidence for two tightly bound iron-quinones in the electron acceptor complex of photosystem II. *FEBS Lett.* 195:290–94

87. Fajer, J., Fujita, E., Frank, H. A., Chadwick, B., Simpson, D. et al. 1987. Are chlorinated chlorophylls components of photosystem I reaction centers? See Ref. 28, 1:307–10

88. Falkowski, P. G., Fujita, Y., Ley, A., Mauzerall, D. 1986. Evidence for cyclic electron flow around photosystem II in *Chlorella pyrenoidosa. Plant Physiol.* 81:310–12

89. Fish, L. E., Bogorad, L. 1986. Identification and analysis of the maize P700 chlorophyll *a* apoproteins PSI-A1 and PSI-A2 by high pressure liquid chromatography analysis and partial sequence determination. *J. Biol. Chem.* 261:8134–39

90. Fish, L. E., Kück, U., Bogorad, L. 1985. Two partially homologous adjacent light-inducible maize chloroplast genes encoding polypeptides of the P700 chlorophyll *a*-protein complex of photosystem I. *J. Biol. Chem.* 260:1413–21

91. Ford, R. C., Picot, D., Garavito, R. M. 1987. Crystallization of the photosystem I reaction centre. *EMBO J.* 6:1581–86

92. Förster, V., Junge, W. 1985. Cooperative and reversible action of three or four hydroxylamine molecules on the water-oxidizing complex. *FEBS Lett.* 186:153–57

93. Förster, V., Junge, W. 1985. Stoichiometry and kinetics of proton release upon photosynthetic water oxidation. *Photochem. Photobiol.* 41:183–90

94. Förster, V., Junge, W. 1986. On the action of hydroxylamine, hydrazine and their derivatives on the water-oxidizing complex. *Photosynth. Res.* 9:197–210

95. Franzén, L.-G., Andréasson, L.-E. 1984. Studies on manganese binding by selective solubilization of photosystem-II polypeptides. *Biochim. Biophys. Acta* 765:166–70

96. Franzén, L.-G., Hansson, Ö., Andréasson, L.-E. 1985. The roles of the extrinsic subunits in photosystem II as revealed by EPR. *Biochim. Biophys. Acta* 808:171–79

97. Franzén, L.-G., Styring, S., Etienne, A.-L., Hansson, Ö., Vernotte, C. 1986. Spectroscopic and functional characterization of a highly oxygen evolving photosystem II reaction center complex from spinach. *Photobiochem. Photobiophys.* 13:15–28

98. Frasch, W. D., Mei, R. 1987. Hydrogen peroxide as an alternate substrate for the oxygen-evolving complex. *Biochim. Biophys. Acta* 891:8–14

99. Furrer, R., Thurnauer, M. C. 1983. Resolution of signals attributed to photosystem I primary reactants by time-resolved EPR at K band. *FEBS Lett.* 153:399–403

100. Geiger, R., Berzborn, R. J., Depka, B., Oettmeier, W., Trebst, A. 1987. Site directed antisera to the D-2 polypeptide subunit of photosystem II. *Z. Naturforsch. Teil C* 42:117–24

101. Ghanotakis, D. F., Babcock, G. T., Yocum, C. F. 1984. Calcium reconstitutes high rates of oxygen evolution in polypeptide depleted photosystem II preparations. *FEBS Lett.* 167:127–30

102. Ghanotakis, D. F., Babcock, G. T., Yocum, C. F. 1985. On the role of water-soluble polypeptides (17, 23 kDa), calcium and chloride in photosynthetic oxygen evolution. *FEBS Lett.* 192:1–3

103. Ghanotakis, D. F., Babcock, G. T., Yocum, C. F. 1985. Structure of the oxygen-evolving complex of photosystem II: calcium and lanthanum compete for sites on the oxidizing side of photosystem II which control the binding of water-soluble polypeptides and regulate the activity of the manganese complex. *Biochim. Biophys. Acta* 809:173–80

104. Ghanotakis, D. F., Demetriou, D. M., Yocum, C. F. 1987. Isolation and characterization of an oxygen-evolving photosystem II reaction center core preparation and a 28 kDa Chl-*a*-binding protein. *Biochim. Biophys. Acta* 891:15–21

105. Ghanotakis, D. F., Topper, J. N., Babcock, G. T., Yocum, C. F. 1984. Water-soluble 17 and 23 kDa polypeptides restore oxygen evolution activity by creating a high-affinity binding site for Ca^{2+} on the oxidizing side of photosystem II. *FEBS Lett.* 170:169–73

106. Ghanotakis, D. F., Topper, J. N., Yocum, C. F. 1984. Structural organization of the oxidizing side of photosystem II. Exogenous reductants reduce and destroy the Mn-complex in photosystem II membranes depleted of the 17 and 23 kDa polypeptides. *Biochim. Biophys. Acta* 767:524–31

107. Ghanotakis, D. F., Yocum, C. F. 1985. Polypeptides of photosystem II and their role in oxygen evolution. *Photosynth. Res.* 7:97–114

108. Ghanotakis, D. F., Yocum, C. F., Babcock, G. T. 1986. ESR spectroscopy demonstrates that cytochrome b_{559} remains low potential in Ca^{2+}-reactivated, salt-washed PSII particles. *Photosynth. Res.* 9:125–34

109. Gillie, J. K., Fearey, B. L., Hayes, J. M., Small, G. J. 1987. Persistent hole burning of the primary donor state of photosystem I: strong linear electron-phonon coupling. *Chem. Phys. Lett.* 134:316–22

110. Glazer, A. N., Melis, A. 1987. Photochemical reaction centers: structure, organization, and function. *Ann. Rev. Plant Physiol.* 38:11–45

111. Golbeck, J. H., Cornelius, J. M. 1986. Photosystem I charge separation in the absence of centers A and B. I. Optical characterization of center 'A$_2$' and evidence for its association with a 64-kDa peptide. *Biochim. Biophys. Acta* 849:16–24

112. Golbeck, J. H., McDermott, A. E., Jones, W. K., Kurtz, D. M. 1987. Evidence for the existence of [2Fe-2S] as well as [4Fe-4S] clusters among F_A, F_B and F_X. Implications for the structure of the photosystem I reaction center. *Biochim. Biophys. Acta* 891:94–98

113. Golbeck, J. H., Warden, J. T. 1985. Site of salicylaldoxime interaction with photosystem II. *Photosynth. Res.* 6:371–80

114. Goldstein, R. A., Boxer, S. G. 1987. Effects of nuclear spin polarization on reaction dynamics in photosynthetic bacterial reaction centers. *Biophys. J.* 51:937–46

115. Goodin, D. B., Yachandra, V. K., Britt, R. D., Sauer, K., Klein, M. P. 1984. The state of manganese in the photosynthetic apparatus. 3. Light-induced changes in X-ray absorption (K-edge) energies of manganese in photosynthetic membranes. *Biochim. Biophys. Acta* 767:209–16

116. Gounaris, K., Barber, J. 1985. Isolation and characterisation of a photosystem II reaction centre lipoprotein complex. *FEBS Lett.* 188:68–73

117. Govindjee, Kambara, T., Coleman, W. 1985. The electron donor side of photosystem II: the oxygen evolving complex. *Photochem. Photobiol.* 42:187–210

118. Govindjee, Koike, H., Inoue, Y. 1985. Thermoluminescence and oxygen evolution from a thermophilic blue-green alga obtained after single-turnover light flashes. *Photochem. Photobiol.* 42:579–85

119. Guiles, R. D., Yachandra, V. K., McDermott, A., Britt, R. D., Dexheimer, S. L., et al. 1987. Structural features of the manganese cluster in different states of the oxygen evolving complex of photosystem II: an X-ray absorption spectroscopy study. See Ref. 28, 1:561–64

120. Guss, J. M., Harrowell, P. R., Murata, M., Norris, V. A., Freeman, H. C. 1986. Crystal structure analyses of reduced (Cu^I) poplar plastocyanin at six pH values. *J. Mol. Biol.* 192:361–87

121. Haehnel, W. 1986. Plastocyanin. See Ref. 251, pp. 547–59

122. Hansson, Ö., Aasa, R., Vänngård, T. 1987. The origin of the multiline and g = 4.1 electron paramagnetic resonance signals from the oxygen-evolving system of photosystem II. *Biophys. J.* 51:825–32

123. Hansson, Ö., Andréasson, L.-E., Vänngård, T. 1984. Studies on the multiline EPR signal associated with state S2 of the oxygen-evolving system. See Ref. 256, I:307–10

124. Hansson, Ö., Andréasson, L.-E., Vänngård, T. 1986. Oxygen from water is coordinated to manganese in the S$_2$ state of photosystem II. *FEBS Lett.* 195:151–54

125. Hanssum, B., Renger, G. 1985. Studies on the interaction between hydroxylamine and hydrazine as substrate analogues and the water-oxidizing enzyme system in isolated spinach chloroplasts. *Biochim. Biophys. Acta* 810:225–34

126. Herrmann, R. G., Alt, J., Schiller, B., Widger, W. R., Cramer, W. A. 1984. Nucleotide sequence of the gene for apocytochrome *b*-559 on the spinach plastid chromosome: implications for the structure of the membrane protein. *FEBS Lett.* 176:239–44

127. Hideg, É., Demeter, S. 1985. Binary oscillation of delayed luminescence: evidence of the participation of Q^-_B in the charge recombination. *Z. Naturforsch. Teil C* 40:827–31

128. Hiyama, T. 1986. Iron-sulfur clusters bound to photosynthetic membranes. In *Iron-Sulfur Protein Research*, ed. H. Matsubara, et al, pp. 254–59. Berlin: Springer-Verlag

129. Hiyama, T., Watanabe, T., Kobayashi, M., Nakazato, M. 1987. Interaction of chlorophyll *a*′ with the 65 kDa subunit protein of photosystem I reaction center. *FEBS Lett.* 214:97–100

130. Høj, P. B., Møller, B. L. 1986. The 110-kDa reaction center protein of photosystem I, P700-chlorophyll *a*-protein 1, is an iron-sulfur protein. *J. Biol. Chem.* 261:14292–300

131. Høj, P. B., Svendsen, I., Scheller, H. V., Møller, B. L. 1987. Identification of a chloroplast encoded 9 kDa polypeptide as a 2[4Fe-4S] protein carrying centers A and B of photosystem I. *J. Biol. Chem.* 262:12676–84

132. Homann, P. H. 1987. The relations between the chloride, calcium and

polypeptide requirements of photosynthetic water oxidation. *J. Bioenerg. Biomembr.* 19:105–23

133. Homann, P. H., Gleiter, H., Ono, T., Inoue, Y. 1986. Storage of abnormal oxidants 'Σ_1', 'Σ_2', and 'Σ_3' in photosynthetic water oxidases inhibited by Cl⁻-removal. *Biochim. Biophys. Acta* 850:10–20

134. Hsu, B.-D., Lee, J.-Y., Pan, R.-L. 1986. The two binding sites for DCMU in photosystem II. *Biochem. Biophys. Res. Commun.* 141:682–88

135. Hunziker, D., Abramowicz, D. A., Damoder, R., Dismukes, G. C. 1987. Evidence for an association between a 33 kDa extrinsic protein, manganese and photosynthetic oxygen evolution. I. Correlation with the S_2 multiline EPR signal. *Biochim. Biophys. Acta* 890:6–14

136. Ikegami, I., Itoh, S. 1986. Chlorophyll organization in P-700-enriched particles isolated from spinach chloroplasts. CD and absorption spectroscopy. *Biochim. Biophys. Acta* 851:75–85

137. Ikeuchi, M., Inoue, Y. 1987. Specific ¹²⁵I labeling of D1 (herbicide-binding protein). An indication that D1 functions on both the donor and acceptor sides of photosystem II. *FEBS Lett.* 210:71–76

138. Ikeuchi, M., Yuasa, M., Inoue, Y. 1985. Simple and discrete isolation of an O_2-evolving PS II reaction center complex retaining Mn and the extrinsic 33 kDa protein. *FEBS Lett.* 185:316–22

139. Imaoka, A., Yanagi, M., Akabori, K., Toyoshima, Y. 1984. Reconstitution of photosynthetic charge accumulation and oxygen evolution in CaCl₂-treated PSII particles. I: Establishment of a high recovery of O_2 evolution and examination of the 17-, 23- and 34-kDa proteins, focusing on the effect of Cl⁻ on O_2 evolution. *FEBS Lett.* 176:341–45

140. Itoh, S., Iwaki, M. 1986. Rate of exchange of Cl⁻ between the aqueous phase and its action site in the O_2 evolving reaction of photosystem II studied by rapid, ionic-jump-induced Cl⁻ depletion. *FEBS Lett.* 195:140–44

141. Itoh, S., Iwaki, M., Ikegami, I. 1987. Extraction of vitamin K-1 from photosystem I particles by treatment with diethyl ether and its effects on the A_1^- EPR signal and system I photochemistry. *Biochim. Biophys. Acta.* 893:508–16

142. Itoh, S., Tang, X.-S., Satoh, K. 1986. Interaction of the high-spin Fe atom in the photosystem II reaction center with the quinones Q_A and Q_B in purified oxygen-evolving PS II reaction center

complex and in PS II particles. *FEBS Lett.* 205:275–81

143. Itoh, S., Yerkes, C. T., Koike, H., Robinson, H. H., Crofts, A. R. 1984. Effects of chloride depletion on electron donation from the water-oxidizing complex to the photosystem II reaction center as measured by the microsecond rise of chlorophyll fluorescence in isolated pea chloroplasts. *Biochim. Biophys. Acta* 766:612–22

144. Jansen, M. A. K., Hobé, J. H., Wesselius, J. C., van Rensen, J. J. S. 1986. Comparison of photosynthetic activity and growth performance in triazine-resistant and susceptible biotypes of *Chenopodium album. Physiol. Vég.* 24:475–84

145. Jansson, C., Hansson, Ö., Åkerlund, H.-E., Andréasson, L.-E. 1984. EPR studies on the photosystem II donor side in salt-washed and reconstituted inside-out thylakoids. *Biochem. Biophys. Res. Commun.* 124:269–76

146. Johanningmeier, U., Bodner, U., Wildner, G. F. 1987. A new mutation in the gene coding for the herbicide binding protein in *Chlamydomonas. FEBS Lett.* 211:221–24

147. Kirsch, W., Seyer, P., Herrmann, R. G. 1986. Nucleotide sequence of the clustered genes for two P700 chlorophyll *a* apoproteins of the photosystem I reaction center and the ribosomal protein S14 of the spinach plastid chromosome. *Curr. Genet.* 10:843–55

148. Klimov, V. V., Allakhverdiev, S. I., Shafiev, M. A., Demeter, S. 1985. Effect of complete extraction and readdition of manganese on thermoluminescence of pea photosystem II preparations. *Biochim. Biophys. Acta* 809:414–20

149. Klimov, V. V., Ganago, I. B., Allakhverdiev, S. I., Shafiev, M. A., Ananyev, G. M. 1987. Structural and functional aspects of electron transfer in photosystem 2 of oxygen-evolving organisms. See Ref. 28, 1:581–84

150. Koike, H., Inoue, Y. 1987. Temperature dependence of the S-state transition in a thermophilic cyanobacterium measured by thermoluminescence. See Ref. 28, 1:645–48

151. Koike, H., Siderer, Y., Ono, T., Inoue, Y. 1986. Assignment of thermoluminescence A band to $S_3 Q_A^-$ charge recombination: sequential stabilization of S_3 and Q_A^- by a two-step illumination at different temperatures. *Biochim. Biophys. Acta* 850:80–89

152. Krishtalik, L. I. 1986. Energetics of multielectron reactions. Photosynthetic

404 ANDRÉASSON & VÄNNGÅRD

oxygen evolution. *Biochim. Biophys. Acta* 849:162–71
153. Kuwabara, T., Murata, T., Miyao, M., Murata, N. 1986. Partial degradation of the 18-kDa protein of the photosynthetic oxygen-evolving complex: a study of a binding site. *Biochim. Biophys. Acta* 850:146–55
154. Lagenfelt, G., Hansson, Ö., Andréasson, L.-E. 1987. Spectroscopic characterization of a photosystem II preparation from the blue-green alga (cyanobacterium) *Anacystis nidulans*. *Acta Chem. Scand. B* 41:123–25
155. Lagoutte, B., Sétif, P., Duranton, J. 1984. Tentative identification of the apoproteins of iron-sulfur centers of photosystem I. *FEBS Lett.* 174:24–29
156. Lavergne, J. 1986. Stoichiometry of the redox changes of manganese during the photosynthetic water oxidation cycle. *Photochem. Photobiol.* 43:311–17
156a. Lavergne, J. 1987. Optical-difference spectra of the S-state transitions in the photosynthetic oxygen-evolving complex. *Biochim. Biophys. Acta* 894:91–107
157. Lehmbeck, J., Rasmussen, O. F., Bookjans, G. B., Jepsen, B. R., Stummann, B. M., et al. 1986. Sequence of two genes in pea chloroplast DNA coding for 84 and 82 kDa polypeptides of the photosystem I complex. *Plant Mol. Biol.* 7:3–10
158. Malkin, R. 1982. Photosystem I. *Ann. Rev. Plant Physiol.* 33:455–79
159. Malkin, R. 1986. On the function of two vitamin K₁ molecules in the PS I electron acceptor complex. *FEBS Lett.* 208:343–46
160. Malkin, R. 1987. Photosystem I. See Ref. 16, pp. 495–520
161. Mansfield, R. W., Evans, M. C. W. 1985. Optical difference spectrum of the electron acceptor A₀ in photosystem I. *FEBS Lett.* 190:237–41
162. Mansfield, R. W., Evans, M. C. W. 1986. UV optical difference spectrum associated with the reduction of electron acceptor A₁ in photosystem I of higher plants. *FEBS Lett.* 203:225–29
163. Mansfield, R. W., Hubbard, J. A. M., Nugent, J. H. A., Evans, M. C. W. 1987. Extraction of electron acceptor A₁ from pea photosystem I. *FEBS Lett.* 220:74–78
164. Mathis, P. 1986. Structural aspects of vectorial electron transfer in photosynthetic reaction centers. *Photosynth. Res.* 8:97–111
165. Mathis, P., Rutherford, A. W. 1987. The primary reactions of photosystems I

and II of algae and higher plants. See Ref. 3, pp. 63–96
166. Mayfield, S. P., Bennoun, P., Rochaix, J.-D. 1987. Expression of the nuclear encoded OEE1 protein is required for oxygen evolution and stability of photosystem II particles in *Chlamydomonas reinhardtii*. *EMBO J.* 6:313–18
167. Mayfield, S. P., Rahire, M., Frank, G., Zuber, H., Rochaix, J.-D. 1987. Expression of the nuclear gene encoding oxygen-evolving enhancer protein 2 is required for high levels of photosynthetic oxygen evolution in *Chlamydomonas reinhardtii*. *Proc. Natl. Acad. Sci. USA* 84:749–53
168. McDermott, A. E., Yachandra, V. K., Guiles, R. D., Britt, R. D., Dexheimer, S. L., et al. 1987. Iron X-ray absorption spectra of acceptors in PS I. See Ref. 28, 1:249–52
169. Meiburg, R. F., van Gorkom, H. J., van Dorssen, R. J. 1984. Non-electrogenic charge recombination in photosystem II as a source of sub-millisecond luminescence. *Biochim. Biophys. Acta* 765:295–300
170. Melis, A. 1985. Functional properties of photosystem IIβ in spinach chloroplasts. *Biochim. Biophys. Acta* 808:334–42
171. Metz, J. G., Pakrasi, H. B., Seibert, M., Arntzen, C. J. 1986. Evidence for a dual function of the herbicide-binding D1 protein in photosystem II. *FEBS Lett.* 205:269–74
172. Metz, J. G., Seibert, M. 1984. Presence in photosystem II core complexes of a 34-kilodalton polypeptide required for water photolysis. *Plant Physiol.* 76:829–32
173. Miyao, M., Murata, N. 1984. Calcium ions can be substituted for the 24-kDa polypeptide in photosynthetic oxygen evolution. *FEBS Lett.* 168:118–20
174. Miyao, M., Murata, N. 1984. Role of the 33-kDa polypeptide in preserving Mn in the photosynthetic oxygen-evolution system and its replacement by chloride ions. *FEBS Lett.* 170:350–54
175. Miyao, M., Murata, N. 1985. The Cl⁻ effect on photosynthetic oxygen evolution: interaction of Cl⁻ with 18-kDa, 24-kDa and 33-kDa proteins. *FEBS Lett.* 180:303–8
176. Miyao, M., Murata, N. 1986. Light-dependent inactivation of photosynthetic oxygen evolution during NaCl treatment of photosystem II particles: the role of the 24-kDa protein. *Photosynth. Res.* 10:489–96
177. Miyao, M., Murata, N., Lavorel, J., Maison-Peteri, B., Boussac, A., et al.

1987. Effect of the 33-kDa protein on the S-state transitions in photosynthetic oxygen evolution. *Biochim. Biophys. Acta* 890:151–59

178. Murata, N., Arakai, S., Fujita, Y., Suzuki, K., Kuwabara, T. et al. 1986. Stoichiometric determination of pheophytin in photosystem II of oxygenic photosynthesis. *Photosynth. Res.* 9:63–70

179. Murata, N., Kajiura, H., Fujimura, Y., Miyao, M., Murata, T., et al. 1987. Partial amino acid sequences of the proteins of pea and spinach photosystem II complex. See Ref. 28, 1:70–74

180. Murata, N., Miyao, M., Omata, T., Matsunami, H., Kuwabara, T. 1984. Stoichiometry of components in the photosynthetic oxygen evolution system of photosystem II particles prepared with Triton X-100 from spinach chloroplasts. *Biochim. Biophys. Acta* 765:363–69

181. Nanba, O., Satoh, K. 1987. Isolation of a photosystem II reaction center consisting of D-1 and D-2 polypeptides and cytochrome b-559. *Proc. Natl. Acad. Sci. USA* 84:109–12

182. Nechushtai, R., Schuster, G., Nelson, N., Ohad, I. 1986. Photosystem I reaction centers from maize bundle-sheath and mesophyll chloroplasts lack subunit III. *Eur. J. Biochem.* 159:157–61

183. Nelson, N. 1987. Structure and function of protein complexes in the photosynthetic membrane. See Ref. 3, pp. 213–31

184. Nixon, P. J., Dyer, T. A., Barber, J., Hunter, C. N. 1986. Immunological evidence for the presence of the D1 and D2 proteins in PS II cores of higher plants. *FEBS Lett.* 209:83–86

185. Nuijs, A. M., Shuvalov, V. A., van Gorkom, H. J., Plijter, J. J., Duysens, L. N. M. 1986. Picosecond absorbance difference spectroscopy on the primary reactions and the antenna-excited states in photosystem I particles. *Biochim. Biophys. Acta* 850:310–18

186. Nuijs, A. M., van Gorkom, H. J., Plijter, J. J., Duysens, L. N. M. 1986. Primary-charge separation and excitation of chlorophyll a in photosystem II particles from spinach as studied by picosecond absorbance-difference spectroscopy. *Biochim. Biophys. Acta* 848:167–75

187. O'Malley, P. J., Babcock, G. T. 1984. Electron nuclear double resonance evidence supporting a monomeric nature for P700$^+$ in spinach chloroplasts. *Proc. Natl. Acad. Sci. USA* 81:1098–101

188. O'Malley, P. J., Babcock, G. T. 1984.

EPR properties of immobilized quinone cation radicals and the molecular origin of signal II in spinach chloroplasts. *Biochim. Biophys. Acta* 765:370–79

189. Oh-oka, H., Takahashi, Y., Wada, K., Matsubara, H., Ohyama, K., et al. 1987. The 8 kDa polypeptide in photosystem I is a probable candidate of an iron-sulfur center protein coded by the chloroplast gene *frxA*. *FEBS Lett.* 218:52–54

190. Oh-oka, H., Tanaka, S., Wada, K., Kuwabara, T., Murata, N. 1986. Complete amino acid sequence of 33 kDa protein isolated from spinach photosystem II particles. *FEBS Lett.* 197:63–66

191. Ohno, T., Satoh, K., Katoh, S. 1986. Chemical composition of purified oxygen-evolving complexes from the thermophilic cyanobacterium *Synechococcus* sp. *Biochim. Biophys. Acta* 852:1–8

192. Ohyama, K., Fukuzawa, H., Kohchi, T., Shirai, H., Sano, T., et al. 1986. Chloroplast gene organization deduced from complete sequence of liverwort *Marchantia polymorpha* chloroplast DNA. *Nature* 322:572–74

193. Okamura, M. Y., Satoh, K., Isaacson, R. A., Feher, G. 1987. Evidence of the primary charge separation in the D_1D_2 complex of photosystem II from spinach: EPR of the triplet state. See Ref. 28, 1:379–81

194. Ono, T., Conjeaud, H., Gleiter, H., Inoue, Y., Mathis, P. 1986. Effect of preillumination on the P-680$^+$ reduction kinetics in chloride-free photosystem II membranes. *FEBS Lett.* 203:215–19

195. Ono, T., Inoue, Y. 1984. Reconstitution of photosynthetic oxygen evolving activity by rebinding of 33 kDa protein to $CaCl_2$-extracted PS II particles. *FEBS Lett.* 166:381–84

196. Ono, T., Inoue, Y. 1985. S-state turnover in the O_2-evolving system of $CaCl_2$-washed photosystem II particles depleted of three peripheral proteins as measured by thermoluminescence. Removal of 33 kDa protein inhibits S_3 to S_4 transition. *Biochim. Biophys. Acta* 806:331–40

197. Ono, T., Inoue, Y. 1986. Effects of removal and reconstitution of the extrinsic 33, 24, and 16 kDa proteins on flash oxygen yield in photosystem II particles. *Biochim. Biophys. Acta* 850:380–89

198. Ono, T., Nakayama, H., Gleiter, H., Inoue, Y., Kawamori, A. 1987. Modification of the properties of S_2 state in photosynthetic O_2-evolving center by replacement of chloride with other an-

ions. *Arch. Biochem. Biophys.* 256:618–24

199. Ono, T., Zimmermann, J. L., Inoue, Y., Rutherford, A. W. 1986. EPR evidence for a modified S-state transition in chloride-depleted photosystem II. *Biochim. Biophys. Acta* 851:193–201

200. Ono, T., Zimmermann, J. L., Inoue, Y., Rutherford, A. W. 1987. Abnormal S_2 state formed in chloride depleted photosystem II as revealed by manganese EPR multiline signal. See Ref. 28, 1:653–56

201. Owens, T. G., Webb, S. P., Mets, L., Alberte, R. S., Fleming, G. R. 1987. Antenna size dependence of fluorescence decay in the core antenna of photosystem I: estimates of charge separation and energy transfer rates. *Proc. Natl. Acad. Sci. USA* 84:1532–36

202. Palace, G. P., Franke, J. E., Warden, J. T. 1987. Is phylloquinone an obligate electron carrier in photosystem I? *FEBS Lett.* 215:58–62

203. Papa, S., ed. 1987. *Cytochrome Systems. Molecular Biology and Bioenergetics.* New York: Plenum. In press

204. Pecoraro, V. L., Kessissoglou, D. P., Li, X., Butler, W. M. 1987. Models for manganese centers in metalloenzymes. See Ref. 28, 1:725–28

205. Petrouleas, V., Diner, B. A. 1986. Identification of Q_{400}, a high-potential electron acceptor of photosystem II, with the iron of the quinone-iron acceptor complex. *Biochim. Biophys. Acta* 849:264–75

205a. Petrouleas, V., Diner, B. A. 1987. Light-induced oxidation of the acceptor-side Fe(II) of photosystem II by quinones acting through the Q_B binding site. I. Quinones, kinetics and pH-dependence. *Biochim. Biophys. Acta* 893:126–37

206. Pfister, K., Steinback, K. E., Gardner, G., Arntzen, C. J. 1981. Photoaffinity labeling of an herbicide receptor protein in chloroplast membranes. *Proc. Natl. Acad. Sci. USA* 78:981–85

207. Plijter, J. J., de Groot, A., van Dijk, M. A., van Gorkom, H. J. 1986. Destabilization by high pH of the S_1-state of the oxygen-evolving complex in photosystem II particles. *FEBS Lett.* 195:313–18

208. Preston, C., Critchley, C. 1985. Ca^{2+} requirement for photosynthetic oxygen evolution of spinach and mangrove photosystem II membrane preparations. *FEBS Lett.* 184:318–22

209. Preston, C., Pace, R. J. 1985. The S-state dependence of Cl^- binding to plant

photosystem II. *Biochim. Biophys. Acta* 810:388–91

210. Radmer, R., Ollinger, O. 1986. Do the higher oxidation states of the photosynthetic O_2-evolving system contain bound H_2O?. *FEBS Lett.* 195:285–89

211. Renger, G. 1987. Mechanistic aspects of photosynthetic water cleavage. *Photosynthetica* 21:203–24

212. Renger, G., Schulze, A. 1985. Quantitative analysis of fluorescence induction curves in isolated spinach chloroplasts. *Photobiochem. Photobiophys.* 9:79–87

213. Renger, G., Weiss, W. 1986. Functional and structural aspects of photosynthetic water oxidation. *Biochem. Soc. Trans.* 14:17–20

214. Renger, G., Weiss, W. 1986. Studies on the nature of the water oxidizing enzyme system. III. Spectral characterization of the intermediary redox states in the water oxidizing enzyme system Y. *Biochim. Biophys. Acta* 850:184–96

215. Rich, P. R. 1985. Mechanisms of quinol oxidation in photosynthesis. *Photosynth. Res.* 6:335–48

216. Rochaix, J.-D., Dron, M., Rahire, M., Malnoe, P. 1984. Sequence homology between the 32K dalton and the D2 chloroplast membrane polypeptides of *Chlamydomonas reinhardii. Plant Mol. Biol.* 3:363–70

217. Rutherford, A. W. 1986. How close is the analogy between the reaction centre of photosystem II and that of bacteria? *Biochem. Soc. Trans.* 14:15–17

218. Rutherford, A. W. 1987. How close is the analogy between the reaction centre of PSII and that of purple bacteria? 2. The electron acceptor side. See Ref. 28, 1:277–83

219. Rutherford, A. W., Heathcote, P. 1985. Primary photochemistry in photosystem-I. *Photosynth. Res.* 6:295–316

220. Rutherford, A. W., Mullet, J. E. 1981. Reaction center triplet states in photosystem I and photosystem II. *Biochim. Biophys. Acta* 635:225–35

221. Rutherford, A. W., Renger, G., Koike, H., Inoue, Y. 1984. Thermoluminescence as a probe of photosystem II. The redox and protonation states of the secondary acceptor quinone and the O_2-evolving enzyme. *Biochim. Biophys. Acta* 767:548–56

222. Rutherford, A. W., Styring, S. 1987. EPR signal II in photosystem II: redox and paramagnetic interactions with the O_2 evolving enzyme. See Ref. 203

223. Rutherford, A. W., Zimmerman, J. L.

1984. A new EPR signal attributed to the primary plastosemiquinone acceptor in photosystem II. *Biochim. Biophys. Acta* 767:168–75

224. Sakurai, H., San Pietro, A. 1985. Association of Fe-S center(s) with the large subunit(s) of photosystem I particles. *J. Biochem. (Tokyo)* 98:69–76

225. Sandusky, P. O., Yocum, C. F. 1984. The chloride requirement for photosynthetic oxygen evolution. Analysis of the effects of chloride and other anions on the amine inhibition of the oxygen-evolving complex. *Biochim. Biophys. Acta* 766:603–11

226. Sandusky, P. O., Yocum, C. F. 1986. The chloride requirement for photosynthetic oxygen evolution: factors affecting nucleophilic displacement of chloride from the oxygen-evolving complex. *Biochim. Biophys. Acta* 849:85–93

227. Sane, P. V., Rutherford, A. W. 1986. Thermoluminescence from photosynthetic membranes. In *Light Emission by Plants and Bacteria,* ed. Govindjee, J. Amesz, D. C. Fork, pp. 329–60. New York: Academic

228. Satoh, K. 1985. Protein-pigments and photosystem II reaction center. *Photochem. Photobiol.* 42:845–53

229. Saygin, Ö., Gerken, S., Meyer, B., Witt, H. T. 1986. Total recovery of O_2 evolution and nanosecond reduction kinetics of chlorophyll-a^+_{II} (P-680$^+$) after inhibition of water cleavage with acetate. *Photosynth. Res.* 9:71–78

230. Saygin, Ö., Witt, H. T. 1985. Evidence for the electrochromic identification of the change of charges in the four oxidation steps of the photoinduced water cleavage in photosynthesis. *FEBS Lett.* 187:224–26

230a. Saygin, Ö., Witt, H. T. 1987. Optical characterization of intermediates in the water-splitting enzyme system of photosynthesis—possible states and configurations of manganese and water. *Biochem. Biophys. Acta* 893:452–69

231. Sayre, R. T., Andersson, B., Bogorad, L. 1986. The topology of a membrane protein: the orientation of the 32 kd Qb-binding chloroplast thylakoid membrane protein. *Cell* 47:601–8

232. Schatz, G. H., van Gorkom, H. J. 1985. Absorbance difference spectra upon charge transfer to secondary donors and acceptors in photosystem II. *Biochim. Biophys. Acta* 810:283–94

233. Schatz, G. H., Witt, H. T. 1984. Extraction and characterization of oxygen-evolving photosystem II complexes from a thermophilic cyanobacterium *Synechococcus* spec. *Photobiochem. Photobiophys.* 7:1–14

234. Schlodder, E., Brettel, K., Schatz, G. H., Witt, H. T. 1984. Analysis of the Ch1-a^+_{II} kinetics with nanosecond time resolution in oxygen-evolving photosystem II particles from *Synechococcus* at 680 and 824 nm. *Biochim. Biophys. Acta* 765:178–85

235. Schlodder, E., Brettel, K., Witt, H. T. 1985. Relation between microsecond reduction kinetics of photooxidized chlorophyll a_{II} (P-680) and photosynthetic water oxidation. *Biochim. Biophys. Acta* 808:123–31

236. Schlodder, E., Meyer, B. 1987. pH dependence of oxygen evolution and reduction kinetics of photooxidized chlorophyll a_{II} (P-680) in photosystem II particles from *Synechococcus* sp. *Biochim. Biophys. Acta* 890:23–31

237. Schoeder, H.-U., Lockau, W. 1986. Phylloquinone copurifies with the large subunit of photosystem I. *FEBS Lett.* 199:23–27

238. Schröder, W. P., Åkerlund, H.-E. 1986. H_2O_2 accessibility to the photosystem II donor side in protein-depleted inside-out thylakoids measured as flash-induced oxygen production. *Biochim. Biophys. Acta* 848:359–63

239. Sétif, P., Bottin, H., Mathis, P. 1985. Absorption studies of primary reactions in photosystem I. Yield and rate of formation of the P-700 triplet state. *Biochim. Biophys. Acta* 808:112–22

240. Sétif, P., Mathis, P. 1986. Photosystem I photochemistry: a new kinetic phase at low temperature. *Photosynth. Res.* 9:47–54

241. Sétif, P., Mathis, P. 1986. Photosystem I reaction center and its primary electron transfer reactions. See Ref. 251, pp. 476–86

242. Sétif, P., Mathis, P., Vänngård, T. 1984. Photosystem I photochemistry at low temperature. Heterogeneity in pathways for electron transfer to the secondary acceptors and for recombination processes. *Biochim. Biophys. Acta* 767:404–14

243. Sheats, J. E., Czernuszewicz, R. S., Dismukes, G. C., Reingold, A. L., Petrouleas, V., et al. 1987. Binuclear manganese(III) complexes of potential biological significance. *J. Am. Chem. Soc.* 109:1435–44

244. Shinozaki, K., Ohme, M., Tanaka, M., Wakasugi, T., Hayashida, N., et al. 1986. The complete nucleotide sequence of the tobacco chloroplast genome: its

gene organization and expression. *EMBO J.* 5:2043–49

245. Shuvalov, V. A., Nuijs, A. M., van Gorkom, H. J., Smit, H. W. J., Duysens, L. N. M. 1986. Picosecond absorbance changes upon selective excitation of the primary electron donor P-700 in photosystem I. *Biochim. Biophys. Acta* 850:319–23

246. Sinclair, J. 1984. The influence of anions on oxygen evolution by isolated spinach chloroplasts. *Biochim. Biophys. Acta* 764:247–52

247. Sparrow, R. W., England, R. R. 1984. Isolation of a calcium-binding protein from an oxygen-evolving photosystem II preparation. *FEBS Lett.* 177:95–98

248. Spencer, L., Sawyer, D. T., Webber, A. N., Heath, R. L. 1987. Thermodynamic constraints to photosynthetic water oxidation. See Ref. 28, 1:717–20

249. Srinivasan, A. N., Sharp, R. R. 1986. Flash-induced enhancements in the proton NMR relaxation rate of photosystem II particles. *Biochim. Biophys. Acta.* 850:211–17

250. Srinivasan, A. N., Sharp, R. R. 1986. Flash-induced enhancements in the proton NMR relaxation rate of photosystem II particles: response to flash trains of 1–5 flashes. *Biochim. Biophys. Acta.* 851:369–76

251. Staehelin, L. A., Arntzen, C. J., eds. 1986. *Encyclopedia of Plant Physiology; Photosynthesis III* (NS), Vol. 19. Berlin: Springer-Verlag. 802 pp.

252. Stebler, M., Ludi, A., Bürgi, H.-B. 1986. $[(phen)_2Mn^{IV}(\mu\text{-}O)_2Mn^{III}(phen)_2]$ $(PF_6)_3 \cdot CH_3CN$ and $[(phen)_2Mn^{IV}(\mu\text{-}O)_2Mn^{IV}(phen)_2](ClO_4)_4 \cdot CH_3CN$ (phen = 1, 10-phenanthroline): Crystal structure analyses at 100 K, interpretation of disorder, and optical, magnetic and electrochemical results. *Inorg. Chem.* 25:4743–50

253. Stewart, A. C., Ljungberg, U., Åkerlund, H.-E., Andersson, B. 1985. Studies on the polypeptide composition of the cyanobacterial oxygen-evolving complex. *Biochim. Biophys. Acta* 808:353–62

254. Styring, S., Miyao, M., Rutherford, A. W. 1987. Formation and flash-dependent oscillation of the S_2-state multiline EPR-signal in an oxygen evolving photosystem-II preparation lacking the three extrinsic proteins in the oxygen-evolving system. *Biochim. Biophys. Acta* 890:32–38

255. Styring, S., Rutherford, A. W. 1987. In the oxygen-evolving complex of photo-system II the S_0-state is oxidized to the S_1-state by D^+ (signal II_{slow}). *Biochemistry* 26:2401–5

256. Sybesma, C., ed. 1984. *Advances in Photosynthesis Research* (Proc. 6th Int. Congr. Photosynth.). The Hague: Martinus Nijhoff/Dr. W. Junk

257. Tabata, K., Itoh, S., Yamamoto, Y., Okayama, S., Nishimura, M. 1985. Two plastoquinone A molecules are required for photosystem II activity: analysis in hexane-extracted photosystem II particles. *Plant Cell Physiol.* 26:855–63

258. Takahashi, M., Asada, K. 1986. Sizes of Mn-binding sites in spinach thylakoids. *J. Biol. Chem.* 261:16923–26

259. Takahashi, Y., Hansson, Ö., Mathis, P., Satoh, K. 1987. Primary radical pair in the photosystem II reaction center. *Biochim. Biophys. Acta* 893:49–59

260. Takahashi, Y., Katoh, S. 1986. Numbers and functions of plastoquinone molecules associated with photosystem II preparations from *Synechococcus* sp. *Biochim. Biophys. Acta* 848:183–92

261. Takahashi, Y., Styring, S. 1987. A comparative study of the reduction of EPR signal II_{slow} by iodide and the iodolabeling of the D2-protein in photosystem II. *FEBS Lett.* 223:371–75

262. Takahashi, Y., Takahashi, M., Satoh, K. 1986. Identification of the site of iodide photooxidation in the photosystem II reaction center complex. *FEBS Lett.* 208:347–51

263. Tamura, N., Cheniae, G. 1985. Effects of photosystem II extrinsic proteins on microstructure of the oxygen-evolving complex and its reactivity to water analogs. *Biochim. Biophys. Acta* 809:245–59

264. Tamura, N., Cheniae, G. 1987. Photoactivation of the water-oxidizing complex in photosystem II membranes depleted of Mn and extrinsic proteins. I. Biochemical and kinetic characterization. *Biochim. Biophys. Acta* 890:179–94

265. Tamura, N., Radmer, R., Lantz, S., Cammarata, K., Cheniae, G. 1986. Depletion of photosystem II–extrinsic proteins. II. Analysis of the PS II/water-oxidizing complex by measurement of N,N,N',N'-tetramethyl-*p*-phenylenediamine oxidation following an actinic flash. *Biochim. Biophys. Acta* 850:369–79

266. Tang, X.-S., Satoh, K. 1986. Reconstitution of photosynthetic water-splitting activity by the addition of 33 kDa polypeptide to urea-treated PS II re-

action center complex. *FEBS Lett.* 201:221–24

267. Tavitian, B. A., Nabedryk, E., Mäntele, W., Breton, J. 1986. Light-induced Fourier transform infrared (FTIR) spectroscopic investigations of primary reactions in photosystem I and photosystem II. *FEBS Lett.* 201:151–57

268. Telser, J., Hoffman, B. M., LoBrutto, R., Ohnishi, T., Tsai, A.-L., et al. 1987. Evidence for N coordination to Fe in the [2Fe-2S] center in yeast mitochondrial complex III. Comparison with similar findings for analogous bacterial [2Fe-2S] proteins. *FEBS Lett.* 214:117–21

269. Theg, S. M., Jursinic, P. A., Homann, P. H. 1984. Studies on the mechanism of chloride action on photosynthetic water oxidation. *Biochim. Biophys. Acta* 766:636–46

270. Thornber, J. P. 1986. Biochemical characterization and structure of pigment-proteins of photosynthetic organism. See Ref. 251, pp. 98–142

271. Thurnauer, M. C., Gast, P. 1985. Q-band (35 GHz) EPR results on the nature of A_1 and the electron spin polarization in photosystem I particles. *Photobiochem. Photobiophys.* 9:29–38

272. Thurnauer, M. C., Gast, P., Petersen, J., Stehlik, D. 1987. EPR evidence that the photosystem I acceptor A_1 is a quinone molecule. See Ref. 28, 1:237–40

273. Tittgen, J., Hermans, J., Steppuhn, J., Jansen, T., Jansson, C., et al. 1986. Isolation of cDNA clones for fourteen nuclear-encoded thylakoid membrane proteins. *Mol. Gen. Genet.* 204:258–65

274. Toyoshima, Y., Akabori, K., Imaoka, A., Nakayama, H., Ohkoushi, N., et al. 1984. Reconstitution of photosynthetic charge accumulation and oxygen evolution in $CaCl_2$-treated PSII particles. II: EPR evidence for reactivation of the $S_1 \rightarrow S_2$ transition in $CaCl_2$-treated PSII particles with the 17-, 23-, and 34-kDa proteins. *FEBS Lett.* 176:346–50

275. Trebst, A., Depka, B. 1985. The architecture of photosystem II in plant photosynthesis. Which peptide subunits carry the reaction center of PSII? In *Antennas and Reaction Centers of Photosynthetic Bacteria—Structure, Interactions and Dynamics* (Springer Series in Chemical Physics, Vol. 42), ed. M. E. Michel-Beyerle, pp. 216–24. Berlin: Springer-Verlag

276. Trebst, A., Draber, W. 1986. Inhibitors of photosystem II and the topology of the herbicide and Q_B binding polypeptide in the thylakoid membrane. *Photosynth. Res.* 10:381–92

277. Trissl, H.-W., Kunze, U. 1985. II. Primary electrogenic reactions in chloroplasts probed by picosecond flash-induced dielectric polarization. *Biochim. Biophys. Acta* 806:136–44

278. Tso, J., Hunziker, D., Dismukes, G. C. 1987. The effects of chemical oxidants on the electron transport components of photosystem II and the water-oxidizing complex. See Ref. 28, 1:487–90

279. Tsujimoto, H. Y., Arnon, D. I. 1985. Differential inhibition by plastoquinone analogues of photoreduction of cytochrome b-559 in chloroplasts. *FEBS Lett.* 179:51–54

280. van Gorkom, H. J. 1985. Electron transfer in photosystem II. *Photosynth. Res.* 6:97–112

281. van Gorkom, H. J., Nuijs, A. M. 1987. The primary photochemical reactions in systems I and II of photosynthesis. In *Primary Processes in Photobiology, Proc. 12th Int. Symp. Biophys.*, Physics Ser., ed. T. Kobayashi, pp. 61–69. Berlin: Springer-Verlag

282. van Rensen, J. J. S., Spätjens,, E. E. M. 1987. Photosystem II heterogeneity in triazine-resistant and susceptible biotypes of *Chenopodium album. Z. Naturforsch. Teil C.* 42:794–97

283. Vass, I., Inoue, Y. 1986. pH dependent stabilization of $S_2Q_A^-$ and $S_2Q_B^-$ charge pairs studied by thermoluminescence. *Photosynth. Res.* 10:431–36

284. Vass, I., Koike, H., Inoue, Y. 1985. High pH effect on S-state turnover in chloroplasts studied by thermoluminescence. Short-time alkaline incubation reversibly inhibits S_3-to-S_4 transition. *Biochim. Biophys. Acta* 810:302–9

285. Vater, J., Salnikow, J., Jansson, C. 1986. N-terminal sequence determination and secondary structure analysis of extrinsic membrane proteins in the water-splitting complex of spinach. *FEBS Lett.* 203:230–34

286. Velthuys, B. R. 1987. The photosystem II reaction center. See Ref. 16, pp. 341–78

287. Vermaas, W. F. J., Rutherford, A. W. 1984. EPR measurements on the effect of bicarbonate and triazine resistance on the acceptor side of photosystem II. *FEBS Lett.* 175:243–48

288. Völker, M., Eckert, H.-J., Renger, G. 1987. Effects of trypsin and bivalent cations on P-680$^+$-reduction, fluorescence induction and oxygen evolution in photosystem II membrane fragments

from spinach. *Biochim. Biophys. Acta* 890:66–76

289. Völker, M., Ono, T., Inoue, Y., Renger, G. 1985. Effect of trypsin on PS-II particles. Correlation between Hill-activity, Mn-abundance and peptide pattern. *Biochim. Biophys. Acta* 806:25–34

290. Völker, M., Renger, G., Rutherford, A. W. 1986. Effects of trypsin upon EPR signals arising from components of the donor side of photosystem II. *Biochim. Biophys. Acta* 851:424–30

291. Volkov, A. G. 1986. The molecular mechanism of functioning of photosystem II in higher plants: a hypothesis. *Photobiochem. Photobiophys.* 11:1–7

292. Warden, J. T., Csatorday, K. 1987. On the mechanism of linolenic acid inhibition in photosystem II. *Biochim. Biophys. Acta* 890:215–23

293. Warden, J. T., Golbeck, J. H. 1986. Photosystem I charge separation in the absence of centers A and B. II. ESR spectral characterization of center 'X' and correlation with optical signal 'A$_2$'. *Biochim. Biophys. Acta* 849:25–31

294. Wasielewski, M. R., Fenton, J. M., Govindjee. 1987. The rate of formation of P700$^+$-A$_0^-$ in photosystem I particles from spinach as measured by picosecond transient absorption spectroscopy. *Photosynth. Res.* 12:181–90

295. Wasielewski, M. R., Norris, J. R., Shipman, L. L., Lin, C.-P., Svec, W. A. 1981. Monomeric chlorophyll *a enol:* evidence for its possible role as the primary electron donor in photosystem I of plant photosynthesis. *Proc. Natl. Acad. Sci. USA* 78:2957–61

296. Weiss, W., Renger, G. 1986. Studies on the nature of the water-oxidizing enzyme. II. On the functional connection between the reaction-center complex and the water-oxidizing enzyme system Y in photosystem II. *Biochim. Biophys. Acta* 850:173–83

297. Wensink, J., Dekker, J. P., van Gorkom, H. J. 1984. Reconstitution of photosynthetic water splitting after salt-washing of oxygen-evolving photosystem-II particles. *Biochim. Biophys. Acta* 765:147–55

298. Witt, I., Witt, H. T., Gerken, S., Saenger, W., Dekker, J. P., et al. 1987. Crystallization of reaction center I of photosynthesis. *FEBS Lett.* 221:260–64

299. Wollman, F.-A. 1986. Photosystem I proteins. See Ref. 251, pp. 487–95

300. Wydrzynski, T., Inoue, Y. 1987. Modified photosystem II acceptor side properties upon replacement of the quinone at the Q$_B$ site with 2,5-dimethyl-*p*-

benzoquinone and phenyl-*p*-benzoquinone. *Biochim. Biophys. Acta* 893:33–42

301. Yachandra, V. K., Guiles, R. D., McDermott, A., Britt, R. D., Cole, J., et al. 1987. The state of manganese in the photosynthetic apparatus determined by X-ray absorption spectroscopy. *J. Phys.* C8:1121–28

302. Yachandra, V. K., Guiles, R. D., McDermott, A., Britt, R. D., Dexheimer, S. L., et al. 1986. The state of manganese in the photosynthetic apparatus. 4. Structure of the manganese complex of photosystem II studied using EXAFS spectroscopy. The S$_1$ state of the O$_2$-evolving photosystem II complex from spinach. *Biochim. Biophys. Acta* 850:324–32

303. Yachandra, V. K., Guiles, R. D., Sauer, K., Klein, M. P. 1986. The state of manganese in the photosynthetic apparatus. 5. The chloride effect in photosynthetic oxygen evolution. Is halide coordinated to the EPR-active manganese in the O$_2$-evolving complex? Studies of the substructure of the low-temperature multiline EPR signal. *Biochim. Biophys. Acta* 850:333–42

304. Yamada, Y., Tang, X.-S., Itoh, S., Satoh, K. 1987. Purification and properties of an oxygen-evolving photosystem II reaction-center complex from spinach. *Biochim. Biophys. Acta* 891:129–37

305. Yamamoto, Y., Tabata, K., Isogai, Y., Nishimura, M., Okayama, S., et al. 1984. Quantitative analysis of membrane components in a highly active O$_2$-evolving photosystem II preparation from spinach chloroplasts. *Biochim. Biophys. Acta* 767:493–500

306. Youvan, D. C., Bylina, E. J., Alberti, M., Begusch, H., Hearst, J. E. 1984. Nucleotide and deduced polypeptide sequences of the photosynthetic reaction center, B870 antenna, and flanking polypeptides from *R. capsulata. Cell* 37:949–57

307. Ziegler, K., Lockau, W., Nitschke, W. 1987. Bound electron acceptors of photosystem I. Evidence against the identity of redox center A$_1$ with phylloquinone. *FEBS Lett.* 217:16–20

308. Zimmermann, J. L., Rutherford, A. W. 1984. EPR studies of the oxygen-evolving enzyme of photosystem II. *Biochim. Biophys. Acta* 767:160–67

309. Zimmermann, J.-L., Rutherford, A. W. 1985. The O$_2$-evolving system of photosystem II. Recent advances. *Physiol. Vég.* 23:425–34

310. Zimmermann, J.-L., Rutherford, A. W.

1986. Photoreductant-induced oxidation of Fe^{2+} in the electron-acceptor complex of photosystem II. *Biochim. Biophys. Acta* 851:416–23

311. Zimmermann, J.-L., Rutherford, A. W. 1986. Electron paramagnetic resonance properties of the S_2 state of the oxygen-evolving complex of photosystem II. *Biochemistry* 25:4609–15

312. Zurawski, G., Bohnert, H. J., Whitfeld, P. R., Bottomley, W. 1982. Nucleotide sequence of the gene for the M_r 32,000 thylakoid membrane protein from *Spinacia oleracea* and *Nicotiana debneyi* predicts a totally conserved primary translation produce of M_r 28,950. *Proc. Natl. Acad. Sci. USA* 79:7699–703

Ann. Rev. Plant Physiol. Plant Mol. Biol. 1988. 39:413–37

THE CHROMOSOMAL BASIS OF SOMACLONAL VARIATION

Michael Lee

Department of Agronomy, Iowa State University, Ames, Iowa 50011

Ronald L. Phillips

Department of Agronomy and Plant Genetics, University of Minnesota, St. Paul, Minnesota 55108

CONTENTS

INTRODUCTION 414
LESSONS FROM PAST REVIEWS 415
OBSERVING CHROMOSOME ABERRATIONS 416
CHROMOSOME ABERRATIONS IN CULTURED CELLS
    AND REGENERATED PLANTS 417
POSSIBLE ORIGINS OF CHROMOSOME REARRANGEMENTS
    IN TISSUE CULTURE 420
    Late-Replicating Heterochromatin 420
    Nucleotide Pool Imbalance 422
    Mitotic Recombination 424
ACTIVATION OF CRYPTIC TRANSPOSABLE ELEMENTS 425
    Chromosome Breakage 425
    Transposable-Element Activity in Tissue Culture 425
    Origin of Qualitative Variants 426
CHROMOSOME REARRANGEMENTS IN RELATION TO
    OTHER FORMS OF SOMACLONAL VARIATION 427
    Morphology of Regenerated Plants 427
    Position Effects 428
    Qualitative Variation and Chromosome Rearrangements 428
    Changes in Copy Number 429
    Gene Amplification 429
CONCLUDING REMARKS 430

INTRODUCTION

In 1939, three reports were published describing the first successful methods for in vitro growth of carrot (42, 93) and tobacco (140) tissues. Tissue-culture techniques have subsequently been developed and refined for hundreds of plant species. Scientists have incorporated these techniques into the mainstream of many fundamental and applied aspects of plant biology. As new in vitro systems are designed, refined, and applied, investigators should be cognizant of two facts established during preliminary studies of cultured cells and regenerated plants: 1. genetic and cytogenetic variation commonly increases during exposure of cells to in vitro growth, and 2. while much phenomenology has been recorded, the underlying mechanisms generating the variation remain to be described.

Variation among plants regenerated from tissue culture has been termed "somaclonal variation" (65). The nature, extent, and possible origin of somaclonal variation have been reviewed on numerous occasions from such various viewpoints as chromosomal variation in cultured cells (6, 131) and regenerated plants (24), cytoplasmic variation (102), the biological basis and possible role in plant improvement programs (36, 56, 64, 95, 96), and the epigenetic basis (84). Variation has been recorded using virtually every tool available to the geneticist, from the light microscope to the nucleic acid sequencer. Somaclonal variation has been recorded for most species capable of plant regeneration from tissue culture, and variation has been detected in all components of the plant genome. In this review, we consider somaclonal variation from the standpoint of numerical and structural chromosomal variation in cell culture and regenerated plants, its relationship to other forms of variation, and its possible role in generating other forms of somaclonal variation.

Investigations of cytological variation among cultured cells and regenerated plants have been limited for a number of reasons. Earlier studies were confined to species poorly suited to precise cytological and genetic analysis. Indeed, many of the data were initially collected from carrot and tobacco culture systems, both severely limited for cytological studies by small, numerous chromosomes (131). As tissue-culture technology advanced, other species such as maize and the small grains (44), with large and distinctive chromosomes and a rich tradition of cytogenetic investigations, became available for study. Finally, scientists have generally been unaware, until recently, of the breadth and scope of somaclonal variation. Therefore, experiments have not been planned accordingly. At this point, we have accumulated enough information to ask the right questions and to suggest which experiments should be performed. The results should be rewarding.

Certain aspects of somaclonal variation may have important ramifications for basic and applied branches of science. To basic plant scientists, somaclonal variation may be the avenue through which somatic genetic changes may best be studied. This will depend on the mechanisms and frequencies of the events. Certainly the frequency of changes appears to be high, perhaps much higher than normally occurs in vivo. Other basic scientists, especially those interested in in vitro gene transfer, may view somaclonal variation with some disdain because it represents biological complexity. Alternatively, some scientists interested in in vitro selection methods for generating variation may desire more frequent or more extensive variation. For those interested in somaclonal variation for crop improvement, it will be important to know whether the variation is truly novel or is obtainable through other sources. For these reasons, it is important to study somaclonal variation and attempt to identify the causal mechanisms.

LESSONS FROM PAST REVIEWS

Because considerable data have been accumulated on cytological variation among cultured cells and regenerated plants, it has become possible to formulate some general "truths" about in vitro chromosomal variation and factors affecting its occurrence:

1. Some cytological variation may preexist in the explant, especially for polysomatic[1] species (23, 24). In these species, and perhaps nonpolysomatic species, choice of explant may be an important determinant of in vitro cytological variation.
2. Within a species, explant genotype may be an important determinant of chromosomal variation (83).
3. Culture regimes (subculture interval, physical condition of media) (5, 124) and media components, especially hormones and growth regulators, have been shown to influence the cytological status of cultured cells (135).
4. Chromosomal variation tends to increase with the duration of in vitro growth.
5. A positive correlation exists between chromosomal stability and regenerative capacity of the cultured cells (73, 90, 143).
6. Chromosome instability is more frequently associated with disorganized callus growth (6, 25) as opposed to the relative stability of organized cultures derived from meristems (23, 24).

Most of these generalities are based upon numerical chromosome data obtained from the classical carrot and tobacco tissue-culture systems. As

---

[1]Polysomatic = tissues or individuals containing diploid and polyploid cells side by side as shown by the occurrence of both diploid and endopolyploid mitotic nuclei (108).

Sunderland (131) points out, the limitations of these systems were plainly evident; in search of better species several groups switched to species containing fewer and more distinct chromosomes [*Haplopappus gracilis,* 2n = 2x = 4 (85, 86) and *Crepis capillaris,* 2n = 2x = 6 (114, 115)]. Sunderland also suggested this pattern should logically continue and encouraged "diminishing use of the more traditional species in favor of low-chromosome monocots." In this view, such a switch would reduce the "preoccupation with chromosome numbers (which) has tended to divert attention from the far more important source of variation in tissue and cell cultures, namely that of structural change and loss of genetic material." Bayliss (6) reemphasized the desirability of conducting studies with "plant species with easily checked karyotypes" and encouraged coupling detailed cytological studies with genetic analysis of regenerated plants. By 1978, tissue-culture systems including plant regeneration were developed for most of the major cereals including maize (44a). Once those systems were available, a full battery of elegant genetic techniques could be used to evaluate somaclonal variation, its chromosomal basis, and mechanisms that may generate genetic variation in cultured somatic cells.

OBSERVING CHROMOSOME ABERRATIONS

A variety of techniques may be employed to study chromosomal variation in cultured cells and regenerated plants. Thus far, the most penetrating analyses have been accomplished with light microscopy. In most cases, information has been collected from stained squash preparations of mitotic cells. Using simple staining techniques, this approach permits observations of chromosome number and other cytological landmarks such as centromere position, arm length ratio, chromosome length, and location of nucleolus organizer regions (NOR). Banding techniques help to elucidate the linear differentiation of chromosomes and therefore permit more sophisticated evaluations, such as the distribution of heterochromatic regions.

Analysis of meiotic cells has also made an important contribution to our understanding of the cytological variation associated with tissue culture. Meiotic analysis offers the additional advantage of evaluating pairing relationships among homologous chromosomes and chromosome behavior through several stages of meiosis. This approach often permits detection of subtle variation that would be more difficult to observe through analysis of mitotic cells.

The optimal cytological strategy is determined by the species under investigation. For example, Giemsa C-banding has been particularly informative with *Vicia faba* (53, 97) and certain cereal species (3), but the technique has not been advantageous for studying *Zea mays* L. In contrast, meiotic analysis has been extremely useful in many grass species (especially maize) because of their large and distinct meiotic chromosomes. Until a complete battery of cytological techniques becomes available for plant species with advanced tissue-culture systems, we must rely on data collected from several culture systems using a variety of cytological tools.

CHROMOSOME ABERRATIONS IN CULTURED CELLS AND REGENERATED PLANTS

Thorough characterization and classification of tissue culture–induced chromosome aberrations have led to a more complete understanding of somaclonal variation. Such analyses have provided specific information on the types and frequencies of aberrations and the specific chromosomes and chromosome regions affected by the rearrangements. Collectively, these studies have been useful in identifying patterns of variation and, subsequently, a possible underlying mechanism generating some of the structural changes.

Variation in chromosome number and structure has been observed among cultured cells and regenerated plants. The most thorough studies have indicated that structural chromosome changes most accurately reflect the frequency and extent of karyotypic changes. Karyotypic analysis of 40 protoplasts from a celery cell suspension culture showed that 68% of the protoplasts contained an abnormal chromosome number, and greater than 95% of the cells exhibited structural changes relative to the standard karyotype of root-tip cells (87). The most striking and consistently observed alteration was the appearance of chromosomes exhibiting two or more centromere-like constrictions, presumably the result of chromosome fusion events. Extensive rearrangements were also detected through Giemsa C-banding and karyotyping of cells from a *Crepis capillaris* callus culture (4). The karyotype was characterized by the presence of seven chromosomes, one above the normal complement. Chromosome length measurement and C-banding pattern showed all chromosomes had undergone several rearrangements involving chromatin loss, gain, and exchange among nonhomologous chromosomes.

Table 1 contains summaries of cytological data collected from regenerated plants of several species. As with cultured cells, the predominant type of aberration was the result of changes in chromosome structure. Therefore, events leading to chromosome breakage, and in some instances subsequent exchange or reunion of fragments, appear to be of fundamental importance. Most events result in chromatin loss. The extent of the loss has ranged from nearly complete elimination of chromosome arms (yielding telocentric or acrocentric chromosomes) (3, 83) to deletions of 10–20% of chromosome arm length (68). Another significant fraction of the events results in exchange of large chromosome segments among nonhomologous chromosomes to produce reciprocal translocations. Observations of chromatin gain, either through increases in chromosome number or duplication of chromosome segments, have been less frequent. Perhaps events leading to chromatin duplication occur less frequently. Alternatively, products of such events may not readily proliferate in culture (5) and may not participate in plant regeneration (94) owing to a functional advantage of other karyotypes.

There are probably many ways to generate such variation, but most explanations may be placed in two categories: 1. events leading to direct exchange of chromosome or chromatid fragments (such as sister-chromatid exchange or forms of somatic crossing-over); and 2. events generating acentric chromosome or chromatid fragments accompanied by subsequent reunion, exchange, or loss of the fragments. Presently, the balance of the evidence supports the latter types of events.

Specific chromosome rearrangements are known to be biologically significant in certain instances, i.e. some human cancers (111) and phase variation in *Salmonella* (126). Their role in the response of the plant genome to tissue-culture conditions, however, has been difficult to define. Analysis of chromosome breakpoint location and distribution suggests rearrangement events in cultured plant cells may involve specific chromosomes and chromosome regions, particularly regions containing heterochromatin. For example, a high incidence of structural alterations involving the chromosome containing the nucleolus organizer region (NOR) has been reported on two occasions. In *Crepis capillaris* (114) 82% of the rearrangements, and in *Zea mays* L. (68) 51% of the translocations involved the NOR chromosome. Another striking example of nonrandom chromosome rearrangement, also from maize, was a report of three independent recoveries of reciprocal translocations involving chromosomes 7 and 8 (7). Nonrandom chromosome rearrangement has also been reported for triticale (12) and celery (87).

Surveys of chromosome breakpoint location for spontaneous and induced breakage unrelated to tissue culture have also indicated a nonrandom distribution. In tomato *(Lycopersicon esculentum)*, breakpoints for reciprocal translocations (43) and deletions (60) preferentially occur within heterochromatin. In both maize and *Drosophila melanogaster* (52) breakpoints of reciprocal translocations and inversions were preferentially found in regions near the centromere, and within the centromere for maize. Also, Longley (71) concludes, from a survey of breakpoints in maize chromosomes, that heterochromatic areas have been involved in translocations more frequently than euchromatic regions. Heterochromatic regions have also been sites of a high incidence of chromatid breakage following mutagenesis in *Vicia faba* (109).

Breakpoint locations in regenerated maize plants also tend to be placed on chromosome arms containing large blocks of heterochromatin (7, 68, 101), such as knobs and the NOR. Generally, the breakpoints occur between the

Table 1 Summary of chromosome aberrations observed in regenerated plants

Reference	Species	Number of regenerated plants	Analysis	Aberration[a]					
				Polyploidy	Aneuploidy	Translocation	Deficiency	Duplication	Other[b]
83	Avena sativa L.	799	meiotic	0	43	48	180	0	2
7	Zea mays L.	370	meiotic	6	0	23	13	0	1
68	Zea mays L.	267	meiotic	2	1	45	59	0	1
107	Zea mays L.	257	meiotic	37	14	32	36	0	0
—[c]	Zea mays L.	142	meiotic	1	1	7	0	0	0
33	Zea mays L.	110	mitotic	1	1	0	0	0	0
3	Triticosecale	51	mitotic	0	12	3	29	0	0
63	Wheat-rye hybrid	10	mitotic	0	0	4	10	6	0

[a] Aberration = number of plants containing that aberration.
[b] Other = includes inversions, centric fusions, and more complex, unclassified rearrangements.
[c] — = C. L. Armstrong and R. L. Phillips, manuscript in review.

heterochromatin and the centromere. Heterochromatic regions also seem to be sites of a high frequency of chromosome breakage and rearrangement in triticale (3, 12) and wheat-rye hybrid (63) callus cultures and regenerated plants. Lapitan et al. (63) reported 12 of 13 translocation or deficiency breakpoints occurred within heterochromatin. Deletion or amplification of heterochromatic bands, primarily telomeric, were also observed in all regenerated plants; changes in euchromatic regions were not generally observed. Variation in heterochromatic blocks was also the predominant chromosome aberration reported by Brettell et al. (12) in regenerated triticale plants with the majority of the rearrangements being reductions of heterochromatic bands on chromosome 7RL, the same location of several rearrangements observed by Lapitan et al.

Breakage events within or near heterochromatin have also been implicated as a primary source of chromosome rearrangements in tissue cultures and regenerated plants of oats (83) and celery (88). In regenerated oat plants, telocentric or near-telocentric chromosomes occur at high frequency, presumably owing to breakage near the centromere, a region characterized by large blocks of pericentromeric heterochromatin (54, 55). In celery, chromosome fusion events leading to formation of chromosomes containing Robertsonian translocations and multiple constrictions have been observed as a primary type of chromosome rearrangement in tissue cultures and regenerated plants (13, 88). The origin of such events could be attributed to breakage in or near telomeric heterochromatin found on all celery chromosomes (89).

The structural and functional features of heterochromatin and specific chromosome regions responsible for their conspicuous involvement in gross chromosome rearrangements have not been identified. One established characteristic of heterochromatin, late replication during the mitotic cell cycle (70), may be a source of many of the observed changes in chromosome structure during in vitro culture.

POSSIBLE ORIGINS OF CHROMOSOME REARRANGEMENT IN TISSUE CULTURE

Gross chromosome rearrangement during cell culture has been well documented in plants and animals (48, 50). The next step, identifying and describing mechanisms that trigger the rearrangements, has been a more elusive task. In this review, we present two possible mechanisms: the first, based on late replication of heterochromatin; and the second, an outcome of nucleotide pool imbalance.

Late-Replicating Heterochromatin

The mitotic cell cycle of higher organisms consists of four phases, G1 (gap), S (synthesis of DNA), G2 (gap), and M (mitosis, consisting of prophase, metaphase, anaphase, and telophase), each with a species-specific and cell-type specific duration (15, 142). Any perturbation affecting the synchrony between chromosome replication during S phase and cell division would likely result in chromosome aberrations. Because heterochromatic regions replicate later than euchromatic segments (70), their integrity may be particularly vulnerable to fluctuations in the cycle.

The potential of late replication of DNA as a mechanism for tissue culture–induced chromosome rearrangement was first recognized by Sacristan (114). In her study of *Crepis capillaris* callus cultures, the SAT-chromosome (SAT=satellite; 108) was involved in 82% of the rearrangements, with the breakpoint(s) corresponding to a region of late DNA synthesis. In 1982, Mc-Coy et al (83) hypothesized that the high frequency of telocentric chromosomes in regenerated oat plants could result from late or delayed replication of pericentromeric heterochromatin. According to their hypothesis, "if late-replicating, pericentromeric heterochromatic regions exist in oat chromosomes, they occasionally may replicate even later in cultured cells leading to a bridge and breakage at anaphase." The presence of pericentromeric heterochromatic regions and evidence for their late replication in root tip cells have subsequently been established for oats (55). Important aspects of the hypothesis, most notably a greater delay in replication of heterochromatin during tissue culture and a relationship between sites of breakage and late replication, have not been fully resolved. However, the association between sites of late-replicating heterochromatin and in vivo chromosome breakage has been documented in maize (103) and *Vicia faba* (109).

In maize, Rhoades & Dempsey proposed a model of late-replicating heterochromatin as the basis for chromosome breakage in A chromosomes in the presence of B chromosomes (104, 105). According to their model, replication of knob heterochromatin in A chromosomes is delayed, thus preventing separation of chromatids and causing a bridge to form at anaphase. Cytological and genetic observations have shown that breakage occurs between the knob and the centromere (105). Labeling studies have subsequently demonstrated that knob heterochromatin was the last to replicate in root-tip cells containing B chromosomes (103). In *Vicia faba*, late-replicating, heterochromatic chromosome regions were also observed to undergo rearrangement at a high frequency following various mutagenic treatments (109).

The Rhoades & Dempsey model, although developed for haploid cells, could be used to account for many of the observations of chromosome rearrangement in tissue culture (Figure 1). In cultures of somatic cells, simultaneous breakage of homologous chromosomes (presumably with similar replication patterns) could lead to duplications and deficiencies whereas simultaneous breakage in nonhomologous chromosomes could lead to reciprocal interchanges (68), although these may not be exact (118). The basic model is easily extended to account for the changes previously discussed. It is obvious that the types and frequencies of rearrangements will depend on the species-specific distribution of heterochromatic blocks. For example, oat chromosomes have large blocks of pericentromeric heterochromatin, so one may anticipate a higher frequency of breaks in those regions. In contrast, maize chromosomes generally lack extensive regions of pericentromeric heterochromatin; instead, large heterochromatic blocks typically occur at knob sites, telomeric regions, and the nucleolus organizer region (80). Differences in heterochromatin distribution may, therefore, explain the predominance of telocentric chromosomes in regenerated oat plants and the general absence of such aberrations in regenerated maize plants. Chromosome breakage in regions of late DNA synthesis (likely heterochromatic) seems consistent with observations of tissue culture–induced chromosome rearrangement in many cases (3, 4, 7, 12, 53, 54, 63, 68, 83, 88, 97, 114, 125).

Late-replicating heterochromatin has also been implicated as a mechanism generating chromosome aberrations in humans (127). In several other systems, including *Drosophila* (47, 117), Chinese hampster ovary cells (17, 28), *Mus, Bos* (cattle), and *Myopus* (wood lemming) cell cultures (137) a high incidence of chromosome aberrations has been located at heterochromatic regions. Therefore, it seems plausible that a model invoking late-replicating DNA may have general applicability in generating genetic variation in somatic cell cultures.

Nucleotide Pool Imbalance

A growing body of evidence suggests that intracellular deoxyribonucleotide (dNTP) pools have an important influence on the fidelity of several components of prokaryotic and eukaryotic DNA metabolism, including precursor biosynthesis, replication, repair, recombination, and possibly degradation (49). Imbalances in the dNTP pools may have serious genetic consequences; these include nuclear chloroplast and mitochondrial mutation, mitotic recombination, chromosome (structural) aberrations, aneuploidy, sister-chromatid exchange, increased sensitivity to mutagens, and oncogenic transformation (61). The effects of nucleotide pool imbalances have been documented for a wide variety of species, including many prokaryotes (61), yeast (62), and mammals (139). It is tempting to speculate that such imbalances may occur in cells of higher plants with comparable genetic consequences.

Plant cells in tissue culture may be especially susceptible to dNTP pool imbalances because they can be serially transferred from depleted to fresh media almost indefinitely. Do media components reach critically low levels toward the end of subculture intervals? Do key metabolic processes fluctuate with subculture intervals? These areas have not been evaluated critically for

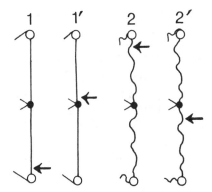

Figure 1 Chromosome bridges owing to late replication of knob heterochromatin (solid circles) in a diploid cell. Open circles represent centromeres and arrows indicate breakpoints. If breaks occur in homologous chromosomes 1 and 1' on opposite sides of equatorial plate, a duplicate or deficient homologue and a normal homologue could result in each daughter cell by exchange of segments distal to the breakpoints. Interchanges could result if simultaneous breaks occurred in chromosomes 1' and 2 on the same side of the plate and in 1 and 2' on the other side. (See references 55 and 68 for further discussion and illustration of possible outcomes.)

plant tissue culture, but an example from human cell culture suggests the answer may be affirmative for at least the first question.

Human cytogeneticists have used lymphocyte culture media deficient in thymidine and folic acid to enhance expression of fragile sites on chromosomes. Fragile sites are heritable chromosome regions characterized as having nonstaining gaps at metaphase and a high susceptibility to breakage. Chromosomes containing fragile sites, specifically the X chromosome, have been associated with a common form of X-linked mental retardation (130). Jackey et al (51) have reported an increased incidence of chromosome breakage in lymphocyte cultures from normal individuals on media deficient in thymidine and folic acid. They hypothesize that the expression of fragile sites in lymphocytes from carriers, the afflicted, and normal individuals depends on a reduction in thymidine monophosphate during the late stages of DNA synthesis. Furthermore, they propose that a general class of chromosome breakage exists that results from deficiencies in DNA precursors. Deficiencies in thymidine or folate have also been associated with fragile-site expression in the Indian mole rat (*Nesokia indica;* 133). In the case of the Indian mole rat, the folate-sensitive fragile sites corresponded to sites of deletions and other structural polymorphisms involving constitutive heterochromatin.

Unlike most animal cells, plant cell cultures do not generally require exogenous thymidine and folic acid for growth. Perhaps some unique, yet undefined, features of nucleotide biosynthesis and DNA metabolism in plants may be sensitive to fluctuations or stresses in the tissue-culture environment.

Mitotic Recombination

Various forms of mitotic recombination, including somatic crossing-over and sister-chromatid exchange could produce several types of chromosome rearrangements observed in tissue culture, especially if the exchanges were asymmetric or between nonhomologous chromosomes (65). Evidence supporting the importance of mitotic recombination in generating somaclonal variation has generally been lacking, with a few noteworthy exceptions.

Mitotic recombination in tobacco *(Nicotiana tabacum)* has been estimated by the frequency of dark green spots or dark green/yellow twin spots in *Su/su* heterozygotes at the *aurea* locus (18, 34). Lörz & Scowcroft (72) report that the frequency of dark green and twin spots on leaves of regenerated plants heterozygous for *Su/su* increased after culture regimes including both several subcultures and mutagenesis. They speculate that genetic events occurring in cell culture increased mitotic recombination during leaf development. The cytogenetic consequences of enhanced mitotic recombination were not evaluated in that study. However, the authors noted that 37% of all regenerated plants, and a disproportionate fraction (63%) of regenerated plants with abnormal morphology, had progenies segregating in ratios that deviated from Mendelian expectations. Also, variation in protoplast-derived tobacco callus colonies and regenerated plants, presumably owing to somatic recombination involving the *al* and *yg* loci, has been reported by Barbier and Dulieu (4a, 32a).

The recovery of homozygous sectors from heterozygous progenitor tissue is one possible outcome of mitotic recombination. Such events, although rare, have occurred in regenerated plants of maize (81) and wheat (66). McCoy observed a self-pollinated, fully fertile regenerated maize plant that contained all sugary kernels even though the callus tissue was heterozygous for the recessive sugary allele. He speculated this could have resulted from a point mutation or somatic recombination. Regenerated wheat plants homozygous for nonparental genotypes at loci affecting grain (testa) color and awn development have been observed by Larkin et al. They suggest several possible causes including cycles of point mutation followed by monosomy/disomy or trisomy/disomy, nonreciprocal transfer of genetic information between repeated DNA sequences, and gene conversion. Tests for somatic crossing-over, including the use of flanking markers, are needed to evaluate critically the importance of mitotic recombination in generating chromosome rearrangement in tissue culture.

Data obtained from light-microscope studies of cultured cells and regenerated plants lead to the conclusion that gross chromosome rearrangement represents a common form of somaclonal variation. We believe this perception to be conservative, because even the best cytological technique detects only a fraction of the rearrangements. Because chromosome rearrangement, through either breakage and exchange or some form of mitotic recombination, can lead to so many changes in the structure and possibly function of the genome, it may occupy a central role in generating several forms of somaclonal variation.

ACTIVATION OF CRYPTIC TRANSPOSABLE ELEMENTS

In their landmark review of somaclonal variation, Larkin & Scowcroft (65) list possible causes of variation, including the involvement of transposable elements. Subsequent reports in maize (37, 38, 98) and alfalfa (45, 46) lend support to their hypothesis. In these instances, presumed activation of previously silent transposable elements was detected among regenerated plants and their progeny. In the case of maize, activation may have originated in chromosome breakage and rearrangement in tissue culture.

Chromosome Breakage

Chromosome breakage has been a means for initiating activity of maize transposable elements (40). This has been initiated in several ways: chromosome stocks with special arrangement (31, 75), exposure to radiation (92), and the presence of B chromosomes (106). The steps leading to activation through chromosome breakage have not been described, but suggested mechanisms include DNA rearrangements within element-related sequences (40, 99); change in location of intact, cryptic elements residing within inactive chromatin regions (i.e. heterochromatin; 77); and change in DNA methylation patterns (19a) perhaps resulting from DNA damage and repair (43a). Once activated, the elements may generate a wide array of genomic changes with many manifestations (39, 76, 99).

Transposable-Element Activity in Tissue Culture

Activation of maize transposable elements following tissue culture has been documented for the *Ac* (38, 98) and *Spm (En)* (37) systems. Peschke et al (98) detected active *Ac* elements in tissue culture–derived plants by crossing these materials as males to tester plants containing the receptor, *Ds,* and appropriate endosperm marker alleles. Plants with an active *Ac* traced back to three of 94 independent, embryo-derived callus cultures. In a similar study, Evola et al (37) observed reactivation of an *Spm (En)* element among half of the plants regenerated; activation of *Ac* was also noted in a later report (38). To the best of our knowledge, other maize transposable-element systems (at least 14 known; 99) have not been tested for activation following tissue culture. In alfalfa, phenomena typical of transposable-element activity—frequent reversions of unstable alleles, and recovery of new alleles at a locus—have been recorded for an anthocyanin pigmentation locus in regenerated plants (45, 46).

Origin of Qualitative Variants

The discovery of activation of maize transposable elements in tissue culture suggested a possible relationship between somaclonal variation and mobile elements. A common manifestation of somaclonal variation is the occurrence of single-gene variants segregating among progeny of regenerated plants. Observations in maize (7, 33, 69, 82), tomato (35), and rice (41) indicate that the majority of the variants exhibit classic recessive Mendelian inheritance patterns. What portion, if any, of these variants result from insertion or excision of transposable elements? Answers to this question may be obtained by two approaches: monitoring the phenotypic expression of such variants, or molecular analysis of variant alleles.

The occurrence of unstable alleles has been one of the hallmarks associated with transposable-element activity. Our experience with tissue cultured–induced, single-gene variants has been that most variants exhibit stable expression over several generations. However, two noteworthy examples of instability have been reported. In maize, an unstable phenotype for cob color has been observed among progeny of regenerated plants (141). Also, variation in alfalfa flower anthocyanin levels represents a dramatic example of unstable gene expression (45).

Molecular analysis of single-gene Mendelian variants has been achieved for alleles at the *Adh1* (11) and *Sh* (37) loci in maize. An electrophoretic variant allele of *Adh1* was recovered from a population of 645 regenerated plants. Sequencing the genomic DNA demonstrated that a single base substitution has occurred in exon 6. Restriction mapping of six tissue culture–induced mutants at the *Sh* locus did not detect any differences between progenitor and variant alleles. Evola et al (37) concluded the mutants were not attributable to insertions or deletions of greater than 100 base pairs but more subtle variation could not be ruled out.

Insertion of plant transposable elements result in perfect host sequence duplication. Excision can restore the original allele, leave the host sequence duplication intact, or result in mutations (i.e. base substitutions, complementary transversions) of the duplicated target region (116, 121). Sequence analyses of progenitor alleles, insertional mutants, and revertant (or null) alleles have demonstrated that insertion site mutations also include small deletions of host DNA adjacent to the insertion site. The deletion may (97a) or may not (132a) accompany excision of the element. Until more sequencing data have been collected from insertion and excision sites of different transposable-element systems our current expectations may be a bit speculative. Transposable-element "footprints" do come in various sizes and levels of complexity. Also, the elements may not behave as expected in tissue culture. For example, mobile elements in *Drosophila* cell cultures show an increased rate of transposition over in vivo rates (112).

CHROMOSOME REARRANGEMENTS IN RELATION TO OTHER FORMS OF SOMACLONAL VARIATION

As we propose, one way chromosome breakage and rearrangement function in the generation of somaclonal variation occurs in the following scenario: late replication of heterochromatin → bridge formation and breakage → rearrangement → activation of transposable elements → insertion and exci-sion → mutation. Of course, there are other avenues of chromosome rearrangement and many forms of somaclonal variation. For a moment, let us disregard transposable elements and consider the possible relationships be-tween chromosome rearrangements per se and some examples of somaclonal variation.

Morphology of Regenerated Plants

Departures from normal chromosome number and structure have been known to be associated with morphological, physiological, and developmental abnormalities in plants (16, 59) and animals (134). Regenerated plants often have an abnormal appearance, which may or may not reflect karyotypic changes.

Plant morphology and chromosome number of regenerated plants appear to be loosely correlated, depending on the species; whereas morphology and chromosome structural change seem unrelated. In alfalfa, regenerated plants with a doubled chromosome number ($2n = 4x = 32$) have been easy to identify owing to their larger leaves and flowers (10). In one study, five variant protoclones (plants from protoplasts) with elevated levels of resistance to verticillium wilt contained doubled or partly doubled chromosome numbers (67). Regenerated plants have also been observed with variation not attribut-able to doubled chromosome numbers. Aneuploidy and structural chromo-some changes among regenerated alfalfa plants have been observed, but their association with other forms of somaclonal variation has not been strong (9). In contrast, morphology of regenerated potato plants ($2n = 4x = 48$) has been reported to be a reliable indicator of chromosome number (57), but the reliability depends on the degree of aneuploidy. For example, all regenerated plants with 49–96 chromosomes were grossly aberrant (20). Associations between abnormal morphology and structural chromosome changes could not be assessed owing to the small and numerous chromosomes of potato. In *Crepis capillaris,* a change in chromosome structure was closely associated with abnormal morphology of regenerated plants (115); perhaps this reflects the sensitivity of a diploid to fluctuations in the genome. As predicted by the model (Figure 1), chromosome breakage and reunion could produce duplications, deletions, and both reciprocal and nonreciprocal interchanges. Therefore, portions of the *Crepis* genome may have been represented in an abnormal manner ranging from single copies (deletions) to several copies (duplications and interchanges).

Position Effects

Changes in the phenotypic effect of one or more genes as a result of a change in their position with respect to other genes (i.e. position effects; 108) appears to be a logical phenomenon resulting from chromosome rearrangements in tissue culture. Position effects have been documented in *Drosophila* (129) and the evening primrose, *Oenothera* (19). Further evaluation of position effects in plants has been limited; such effects have not yet been reported in plant tissue culture.

Qualitative Variation and Chromosome Rearrangements

Two hallmarks of somaclonal variation have been karyotypic changes in cultured cells and regenerated plants and the occurrence of qualitative variants segregating in progeny of regenerated plants. Simultaneous study of these two aspects of somaclonal variation has been most thoroughly accomplished in wheat (27, 66, 113) and maize (7, 33, 68, 69, 82).

In the initial report of heritable somaclonal variation in wheat, Larkin et al (66) observed variation for several morphological and developmental characters. After routine mitotic analysis of root tips of regenerated plants, they concluded the variation was not attributable to aneuploidy. A wide range in fertility of regenerated plants suggests more subtle changes in chromosome structure may have been carried by some plants. Davies et al (27) employed more rigorous cytological techniques of N-banding of root-tip chromosomes and meiotic examination in the analysis of *Adh1* variants. Among 551 regenerated plants, 17 had progeny with variant zymograms; 13 of these were aneuploid, while 3 were directly associated with changes in chromosome structure. Examination of β-amylase zymograms (113) of progeny of 149 plants identified one plant with progeny characterized by new bands. Mitotic and meiotic analyses of variant progeny suggested the new bands were not associated with an obvious karyotypic change.

Cytogenetic analysis of regenerated maize plants and progeny evaluation have been coupled on several occasions. In each study, cytogenetically normal regenerated plants have produced progeny that subsequently segregated for qualitative variants. While cytogenetic and qualitative variation increased with culture age in some cases, (7, 68; C. L. Armstrong and R. L. Phillips, manuscript in review), thus far they appear to be indirectly related phenomena.

Changes in Sequence Copy Number

Many characteristics of the plant genome appear to be capable of saltatory change (138). One aspect of the genome prone to such change has been the copy number of DNA sequences. For example, inbred lines of maize have characteristic copy numbers of rDNA cistrons (100) and various other repeated sequences (110). The environmentally induced genomic changes in flax dramatically illustrated the capacity of a plant genome to respond to its surroundings. Most of the known repetitive portion of the flax genome may change in a single generation when plants are grown under stressful conditions (21).

Similar variation has also been observed in tissue culture. Again, in flax, most of the reiterated sequences exhibited changes in copy number during tissue culture (22); the only exception was a satellite which was also invariant in previous studies of flax genotrophs. The cytogenetic basis of these changes has not been established. In triticale, Brettell et al (12) have reported an 80% reduction of 1R rDNA spacer sequences and other significant but unquantified reductions in terminal heterochromatin. These regions have been shown to consist of well-characterized repeated sequences (2). The observations from triticale tissue culture could be the result of unequal crossing-over, as proposed for *Drosophila* (132), or a mechanism similar to that in Figure 1 involving late replication of terminal heterochromatin and subsequent breakage within that region. Changes in the proportion of heterochromatin (30) and total DNA content (29) were also observed among anther-derived tobacco plants. The elevation in DNA content may trace to amplification events occurring in the vegetative nucleus of the microspore prior to embryonic development in tissue culture (26). The cytogenetic basis of sequence amplification in plants has not been described (91), but several mechanisms have been proposed for gene amplification in mammalian cell culture (119).

Gene Amplification

Gene amplification in response to herbicide selection schemes has been reported in alfalfa (32) and petunia (123) cells in culture. Alfalfa suspension cells selected for resistance to i-phosphinothricin, an inhibitor of glutamine synthase (GS), were characterized by a 4–11-fold amplification of a GS gene and an 8-fold elevation in mRNA levels. Selection of petunia cells for tolerance to N-(phosphonomethyl) glycine (glyphosate), an inhibitor of 5-enolpyruvylshikimate-3-phosphate (EPSP) synthase resulted in a 20-fold amplification of the EPSP gene and enzyme levels (128).

Gene amplification in response to chemical agents, usually a drug that binds to a "housekeeping" enzyme, has been demonstrated in mammalian cell culture (122). The laboratory of R. T. Schimke has proposed several mechanisms for amplification of dihydrofolate reductase (DHFR) in methotrexate-resistant cell cultures (1, 120, 120a): 1. uptake, extrachromosomal replication, and site-specific integration of DNA from killed cells; 2. unequal sister-chromatid exchange; and 3. disproportionate replication resulting in free DNA strands, tandem arrays of the DHFR gene, or double-minute chromosomes containing this gene. Cytological manifestations of gene amplification include the appearance of an expanded chromosome (homogeneously staining regions; 8), possible translocation of amplified genes to different chromosomes, and double-minutes in unstable cell lines (58).

Methotrexate-resistant cell lines of maize have also been selected (136). Total DHFR activity and sensitivity of DHFR to methotrexate were not altered, but five of the six selected lines had increased ploidy levels. Cellular selection has been remarkably successful in plants (74), but in general the response to selection has not been characterized by obvious karyotypic changes.

CONCLUDING REMARKS

As predicted by Sunderland, chromosome rearrangement has emerged as a prevalent form of variation in plant tissue culture. The rearrangements observed among cells and plants of diverse species may originate through a common mechanism: late or delayed replication of heterochromatin. The mechanism rests on a series of events, starting with a deviation in the timing of the mitotic cell cycle, late-replication of heterochromatin leading to bridge formation and breakage, and perhaps, continuing through many cell and sexual cycles via activation of transposable elements. The proposed activation of transposable elements as early as the first division following chromosome breakage may represent a response by the plant genome to imposed "trauma" or "stress," as suggested by McClintock (78, 79). This aspect of the model fits the tissue-culture data nicely because the induction of qualitative variants (events we propose result primarily from insertion and excision by transposons) in tissue culture appears to occur very early during in vitro growth (4a, 7, 72).

Of course, our model contains several components in need of critical evaluation. These include: the occurrence of late-replicating heterochromatin, a need for further documentation of cytological events, activation of cryptic transposable elements, and the molecular basis of single-gene, qualitative variants. Thus, while confirmation or modification of the model awaits results of current and future investigations, the premise (late or delayed replication of heterochromatin leading to chromosome breakage and activation of transposable elements), established in a classical series of experiments in maize genetics, rests on a foundation of biological precedent and scattered data collected from cell cultures, regenerated plants, and their progeny of various species.

The direct involvement of gross chromosome rearrangement with other manifestations of somaclonal variation remains a vague issue. In a few cases (i.e. wheat *Adh1* variants; 27) the relationship was strong, while in many other instances (i.e. maize qualitative variants) the direct association was apparently nonexistent. It seems the direct relationship will have to be evaluated on an individual-case basis.

Further molecular investigations of somaclonal variations will likely identify additional cases. For example, Brown & Lörz (14) have reported rapid and dramatic changes in methylation patterns among regenerated plants of the same inbred line of maize. Future studies may succeed in defining the consequences of the novel patterns.

As more data become available, our perception of somaclonal variation could require revision. Much of our current understanding of somaclonal variation has been derived from differentiated cell cultures. As protoplast technology becomes developed and refined for more species, the nature and extent of the variation may change. Some forms of somaclonal variation may result from mechanisms that also occur in somatic tissue in vivo; cultured cells may show an accelerated and perhaps controllable rate of variation. Therefore, tissue-culture cells may be useful in analyzing somatic genetic change. Identification of the mechanisms, especially if some prove to be unique to tissue culture, should have a profound impact on our understanding of somaclonal variation and the fluidity of the plant genome.

ACKNOWLEDGMENTS

The authors express their appreciation to Mary Lents for her careful typing and to Debra Metzger Lee, Elizabeth Lee, Sharon Lockhart, Virginia M. Peschke and Wendy Woodman for their editorial contributions. The drawing was prepared by Kris Kohn.

Literature Cited

1. Alt, F. W., Kellems, R. E., Bertino, J. R., Schimke, R. T. 1978. Selective multiplication of dihydrofolate reductase genes in methotrexate-resistant variants of cultured murine cells. *J. Biol. Chem.* 253:1357–70
2. Appels, R., Moran, L. B. 1984. Molecular analysis of alien chromatin introduced into wheat. In *Gene Manipulation in Plant Improvement,* ed. J. P. Gustafson, pp. 529–58. New York: Plenum. 668 pp.
3. Armstrong, K. C., Nakamura, C., Keller, W. A. 1983. Karyotype instability in tissue culture regenerants of Triticale (x *Triticosecale* Wittmack) cv. 'Welsh'

from 6-month-old callus cultures. *Z. Pflanzenzuecht.* 91:233–45
4. Ashmore, S. E., Gould, A. R. 1981. Karyotype evolution in a tumour derived plant tissue culture analysed by Giemsa C-banding. *Protoplasma* 106:297–308
4a. Barbier, M., Dulieu, H. 1983. Early occurrence of genetic variants in protoplast cultures. *Plant Sci. Lett.* 29:201–6
5. Bayliss, M. W. 1977. Factors affecting the frequency of tetraploid cells in a predominantly diploid suspension culture of *Daucus carota. Protoplasma* 92:109–15
6. Bayliss, M. W. 1980. Chromosomal

variation in plant tissue culture. *Int. Rev. Cytol.* Suppl. 11A:113–44

7. Benzion, G., Phillips, R. L., Rines, H. W. 1986. Case histories of genetic variability *in vitro:* oats and maize. In *Cell Culture and Somatic Cell Genetics of Plants,* ed. I. K. Vasil, 3:435–48. New York: Academic

8. Biedler, J. L. 1982. Evidence for transient or prolonged extrachromosomal existence of amplified DNA sequences in antifolate-resistant, vincristine-resistant, and human neuroblastoma cells. In *Gene Amplification,* ed. R. T. Schimke, pp. 39–45. New York: Cold Spring Harbor Laboratory

9. Bingham, E. T., McCoy, T. J. 1986. Somaclonal variation in alfalfa. *Plant Breed. Rev.* 4:123–52

10. Bingham, E. T., Saunders, J. W. 1974. Chromosome manipulation in alfalfa: scaling the cultivated tetraploid to seven ploidy levels. *Crop Sci.* 14:474–77

11. Brettell, R. I. S., Dennis, E. S., Scowcroft, W. R., Peacock, W. J. 1986. Molecular analysis of a somaclonal mutant of maize alcohol dehydrogenese. *Mol. Gen. Genet.* 202:235–39

12. Brettell, R. I. S., Pallotta, M. A., Gustafson, J. R., Appels, R. 1986. Variation at the *Nor* loci in triticale derived from tissue culture. *Theor. Appl. Genet.* 71:637–43

13. Browers, M. A., Orton, T. J. 1982. Transmission of gross chromosomal variability from suspension cultures into regenerated celery plants. *J. Hered.* 73: 159–62

14. Brown, P. T. H., Lorz, H. 1986. Molecular changes and possible origins of somaclonal variation. In *Somaclonal Variations and Crop Improvement,* ed. J. Semal, pp. 148–59. Boston: Martinus Nijhoff

15. Bryant, J. A. 1976. The cell cycle. In *Molecular Aspects of Gene Expression in Plants,* ed. J. A. Bryant, pp. 117–216. New York: Academic

16. Burnham, C. R. 1962. *Discussions in Cytogenetics.* Minneapolis: Burgess Publishing Co. 393 pp.

17. Campbell, C. E., Worton, R. G. 1977. Chromosome replication patterns in an established cell line (CHO). *Cytogenet. Cell. Genet.* 19:303–19

18. Carlson, P. S. 1974. Mitotic crossing-over in a higher plant. *Genet. Res.* 24:109–12

19. Catcheside, D. G. 1939. A position effect in *Oenothera. J. Genet.* 38:345–52

19a. Chandler, V. L., Walbot, V. 1986. DNA modification of a maize transpos-

able element correlates with loss of activity. *Proc. Natl. Acad. Sci. USA* 83:1767–71

20. Creissen, C. P., Karp, A. 1985. Karyotypic changes in potato plants regenerated from protoplasts. *Plant Cell Tiss. Organ Cult.* 4:171–82

21. Cullis, C. A. 1983. Environmentally induced DNA changes in plants. *CRC Crit. Rev. Plant Sci.* 1:117–31

22. Cullis, C. A., Cleary, W. 1985. Fluidity of the flax genome. In *Plant Genetics,* ed. M. Freeling, pp. 303–10. New York: Alan R. Liss. 861 pp.

23. D'Amato, F. 1975. The problem of genetic stability in plant tissue and cell cultures. In *Crop Genetic Resources for Today and Tomorrow,* ed. O. H. Frankel, J. G. Hawkes, pp. 333–48. New York: Cambridge Univ. Press

24. D'Amato, F. 1977. Cytogenetics of differentiation in tissue and cell culture. In *Applied and Fundamental Aspects of Plant Cell, Tissue and Organ Culture,* ed. J. Reinert, V. P. S. Bajaj, pp. 343–464. New York: Springer/Verlag

25. D'Amato, F. 1978. Genetic stability and cell preservation. In *Frontiers of Plant Tissue Culture; Proc. 4th Int. Congr. Plant Tiss. and Cell Cult.,* ed. T. A. Thorpe, pp. 287–95. Calgary: Univ. Calgary

26. D'Amato, F., Devreux, M., Scarascia, G. T. 1965. The DNA content of the nuclei of the pollen grains in tobacco and barley. *Caryologia* 18:377–82

27. Davies, P. A., Pallotta, M. A., Ryan, S. A., Scowcroft, W. R., Larkin, P. J. 1986. Somaclonal variation in wheat: genetic and cytogenetic characterization of alcohol dehydrogenase 1 mutants. *Theor. Appl. Genet.* 72:644–53

28. Deaven, L. L., Peterson, D. F. 1973. The chromosomes of CHO an aneuploid chinese hampster cell line: G-band, C-band, and autoradiographic analyses. *Chromosoma* 41:129–44

29. DePaepe, R., Prat, D., Huguet, T. 1982. Heritable nuclear DNA changes in doubled haploid plants obtained by pollen culture of *Nicotiana sylvestris. Plant Sci. Lett.* 28:11–28

30. Dhillon, S. S., Wernsman, E. A., Miksche, J. P. 1982. Evaluation of nuclear DNA content and heterochromatin changes in anther-derived dihaploids of tobacco *(Nicotiana tabacum)* cv. Coker 139. *Can. J. Genet. Cytol.* 25:169–73

31. Doerschug, E. B. 1973. Studies of *Dotted,* a regulatory element in maize. *Theor. Appl. Genet.* 43:182–89

32. Donn, G., Tischer, E., Smith, J. A., Goodman, H. M. 1984. Herbicide-

resistant alfalfa cells: an example of gene amplification in plants. *J. Mol. Appl. Genet.* 2:621–35

32a. Dulieu, H., Barbier, M. 1982. High frequency of genetic variant plants regenerated from cotyledons of tobacco. In *Variability in Plants Regenerated from Tissue Culture*, ed. E. D. Earle, Y. Demarly, pp. 211–27. New York: Praeger. 392 pp.

33. Edallo, S., Zucchinali, C., Perenzin, M., Salamini, F. 1981. Chromosomal variation and frequency of spontaneous mutation associated with *in vitro* culture and plant regeneration in maize. *Maydica* 26:39–56

34. Evans, D. A., Paddock, E. F. 1976. Comparison of somatic crossing over frequency in *Nicotiana tabacum* and three other crop species. *Can. J. Genet. Cytol.* 18:57–65

35. Evans, D. A., Sharp, W. R. 1983. Single gene mutations in tomato plants regenerated from tissue culture. *Science* 221:949–51

36. Evans, D. A., Sharp, W. R., Medina-Filho, H. P. 1984. Somaclonal and gametoclonal variation. *Am. J. Bot.* 71(6):759–74

37. Evola, S. V., Burr, F. A., Burr, B. 1984. *11th Ann. Aharon Katzir-Katchalsky Conf.*, Jerusalem (Abstr.)

38. Evola, S. V., Tuttle, A., Burr, F., Burr, B. 1985. *First Int. Congr. Plant Mol. Biol.*, Savannah (Abstr.)

39. Fedoroff, N. 1983. Controlling elements in maize. In *Mobile Genetic Elements*, ed. J. Shapiro, pp. 1–63. New York: Academic. 688 pp.

40. Freeling, M. 1984. Plant transposable elements and insertion sequences. *Ann. Rev. Plant Physiol.* 35:277–98

41. Fukui, K. 1983. Sequential occurrence of mutations in a growing rice callus. *Theor. Appl. Genet.* 65:225–30

42. Gautheret, R. J. 1939. Sur la possibilité de réaliser la culture indefinie des tissus de tubercules de carotte. *C. R. Acad. Sci.* 208:218–20

43. Gill, B. S., Burnham, C. R., Stringam, G. R., Stout, J. T., Weinheimer, W. H. 1980. Cytogenetic analysis of chromosomal translocations in the tomato: preferential breakage in heterochromatin. *Can. J. Genet. Cytol.* 22:333–41

43a. Grafstrom, R. H., Hamilton, D. L., Yuan, R. 1984. DNA methylation: DNA replication and repair. In *DNA Methylation: Biochemistry and Biological Significance*, ed. A. Razin, H. Cedar, A. D. Riggs, pp. 111–26. New York: Springer-Verlag. 392 pp.

44. Green, C. E. 1978. *In vitro* plant regeneration in cereals and grasses. In *Frontiers of Plant Tissue Culture; Proc. 4th Int. Congr. Plant Tiss. Cell Cult.*, ed. T. A. Thorpe, pp. 411–18. Calgary: Univ. Calgary

44a. Green, C. E., Phillips, R. L. 1975. Plant regeneration from tissue cultures of maize. *Crop. Sci.* 15:417–21

45. Groose, R. W., Bingham, E. T. 1986. An unstable anthocyanin mutation recovered from tissue culture of alfalfa *(Medicago sativa)*. 1. High frequency of reversion upon reculture. *Plant Cell Rep.* 5:104–7

46. Groose, R. W., Bingham, E. T. 1986. An unstable anthocyanin mutation recovered from tissue culture of alfalfa *(Medicago sativa)*. 2. Stable nonrevertants derived from reculture. *Plant Cell Rep.* 5:108–10

47. Halfer, C., Privitera, E., Barigozzi, C. 1980. A study of spontaneous chromosome variations in seven cell lines derived from *Drosophila melanogaster* stocks marked by translocations. *Chromosoma* 76:201–18

48. Harris, M. 1964. *Cell Culture and Somatic Variation*. London/New York: Holt, Rinehart and Winston. 547 pp.

49. Haynes, R. H. 1985. Molecular mechanisms in genetic stability and change: the role of deoxyribonucleotide pool balance. In *Genetic Consequences of Nucleotide Pool Imbalance*, Vol. 31: *Basic Life Sciences*, ed. F. J. deSerres, pp. 1–23. New York/London: Plenum

50. Hsu, T. C. 1961. Chromosome evolution in cell populations. *Int. Rev. Cytol.* 12:69–112

51. Jacky, P. B., Beek, B., Sutherland, G. R. 1983. Fragile sites in chromosomes: possible model for the study of spontaneous chromosome breakage. *Science* 220:69–70

52. Jancey, R. C., Walden, D. B. 1972. Analysis of pattern in distribution of breakage points in the chromosomes of *Zea mays* L. and *D. melanogaster M. Can. J. Genet. Cytol.* 14:429–42

53. Jelaska, S., Papes, D., Pevalek, B., Devide, Z. 1978. *Proc. Fourth Int. Congr. Plant Tiss. Cell Cult.*, p. 101 (Abstr.)

54. Johnson, S. S., Phillips, R. L., Rines, H. W. 1987. Meiotic behavior in progeny of tissue culture regenerated oat plants *(Avena sativa* L.) carrying near-telocentric chromosomes. *Genome* 29:431–38

55. Johnson, S. S., Phillips, R. L., Rines, H. W. 1987. Possible role of heterochromatin in chromosome breakage induced by tissue culture in oats *(Avena sativa* L.). *Genome* 29:439–46

56. Karp, A., Bright, S. W. J. 1985. On the causes and origins of somaclonal variation. In *Oxford Surveys of Plant Molecular and Cell Biology,* ed. B. J. Miflin, 2:199–234. Oxford: Oxford Univ. Press

57. Karp, A., Nelson, R. S., Thomas, E., Bright, S. W. J. 1982. Chromosome variation in protoplast-derived potato plants. *Theor. Appl. Genet.* 63:265–72

58. Kaufman, R. J., Brown, P. C., Schimke, R. T. 1979. Amplified dihydrofolate reductase genes in unstably methotrexate-resistant cells are associated with double-minute chromosomes. *Proc. Natl. Acad. Sci. USA* 76:5669–73

59. Khush, G. S. 1973. *Cytogenetics of Aneuploids.* New York/London: Academic. 301 pp.

60. Khush, G. S., Rick, C. M. 1968. Cytogenetic analysis of the tomato genome by means of induced deficiencies. *Chromosoma* 23:452–84

61. Kunz, B. A. 1982. Genetic effects of deoxyribonucleotide pool imbalance. *Environ. Mutagen.* 4:695–725

62. Kunz, B. A., Haynes, R. H. 1982. DNA repair and genetic effects of thymidylate stress in yeast. *Mutat. Res.* 93:353–75

63. Lapitan, N. L. V., Sears, R. G., Gill, B. S. 1984. Translocations and other karyotypic structural changes in wheat x rye hybrids regenerated from tissue culture. *Theor. Appl. Genet.* 68:547–54

64. Larkin, P. J. 1985. *In vitro* culture and cereal breeding. In *Cereal Tissue and Cell Cultures,* ed. S. W. J. Bright, M. G. K. Jones, pp. 273–96. Boston: Martinus Nijhoff/Dr. W. Junk Publishers

65. Larkin, P. J., Scowcroft, W. R. 1981. Somaclonal variation—a novel source of variability from cell cultures for plant improvement. *Theor. Appl. Genet.* 60: 197–214

66. Larkin, P. J., Ryan, S. A., Brettell, R. I. S., Scowcroft, W. R. 1984. Heritable somaclonal variation in wheat. *Theor. Appl. Genet.* 67:443–55

67. Latunde-Dada, A. O., Lucas, J. A. 1983. Somaclonal variation and reaction to verticillium wilt in *Medicago sativa* L. plants regenerated from protoplasts. *Plant Sci. Lett.* 32:205–11

68. Lee, M., Phillips, R. L. 1987. Genomic rearrangements in maize induced by tissue culture. *Genome* 29:122–28

69. Lee, M., Phillips, R. L. 1987. Genetic variation in progeny of regenerated maize plants. *Genome* 29. In press

70. Lima-De-Faria, A. 1969. DNA replication and gene amplification in heterochromatin. In *Handbook of Molecular Cytology,* ed. A. Lima-De-Faria, pp.

234–82. Amsterdam/London: North Holland

71. Longley, A. E. 1961. Breakage points for four corn translocation series and other corn chromosome aberrations. *Crop Res. US Dept. Agric., Agric. Res. Serv. Tech. Rep. ARS* 34–16

72. Lörz, H., Scowcroft, W. R. 1983. Variability among plants and their progeny from protoplasts of *Su/su* heterozygotes of *Nicotiana tabacum. Theor. Appl. Genet.* 66:67–75

73. Mahfouz, M. N., deBoucand, M. T., Gaultier, J. M. 1983. Caryological analysis of single cell clones of tobacco; relation between the ploidy and the intensity of the callogenesis. *Z. Pflanzenphysiol.* 109:251–57

74. Maliga, P. 1984. Isolation and characterization of mutants in plant cell culture. 1984. *Ann. Rev. Plant Physiol.* 35:519–42

75. McClintock, B. 1950. The origin and behavior of mutable loci in maize. *Proc. Natl. Acad. Sci. USA* 36:344–55

76. McClintock, B. 1951. Chromosome organization and genic expression. *Cold Spring Harbor Symp. Quant. Biol.* 16:13–47

77. McClintock, B. 1978. Mechanisms that rapidly reorganize the genome. *Stadler Genet. Symp.* 10:25–48

78. McClintock, B. 1983. Trauma as a means of initiating change in genome organization and expression. *In Vitro* 19:283–84 (Abstr.)

79. McClintock, B. 1984. The significance of responses of the genome to challenge. *Science* 226:792–800

80. McClintock, B., Kato, T. A., Blumenschein, A. 1981. *Chromosome Constitution of Races of Maize.* Chapingo, Mex.: Dept. Editorial, Colegio de Postgraduados. 517 pp.

81. McCoy, T. J. 1980. *Cytogenetic stability of tissue cultures and regenerated plants* (Avena sativa L.) *and corn* (Zea mays L.). PhD Thesis. Univ. Minn., St. Paul

82. McCoy, T. J., Phillips, R. L. 1982. Chromosome stability in maize *(Zea mays)* tissue cultures and sectoring in some regenerated plants. *Can. J. Genet. Cytol.* 24:559–65

83. McCoy, T. J., Phillips, R. L., Rines, H. W. 1982. Cytogenetic analysis of plants regenerated from oat *(Avena sativa)* tissue cultures; high frequency of partial chromosome loss. *Can. J. Genet. Cytol.* 24:37–50

84. Meins, F. Jr. 1983. Heritable variation in plant cell culture. *Ann. Rev. Plant Physiol.* 34:327–46

85. Mitra, J., Mapes, M. O., Steward, F. C. 1960. Growth and organized development of cultured cells. IV. The behavior of the nucleus. *Am. J. Bot.* 47:357–68

86. Mitra, J., Steward, F. C. 1961. Growth induction in cultures of *Haplopappus gracilis*. II. The behavior of the nucleus. *Am. J. Bot.* 48:358–68

87. Murata, M., Orton, T. J. 1983. Chromosome structural changes in cultured celery cells. *In Vitro* 19:83–89

88. Murata, M., Orton, T. J. 1984. Chromosome fusions in cultured cells of celery. *Can. J. Genet. Cytol.* 26:395–400

89. Murata, M., Orton, T. J. 1984. G-band-like differentiation of mitotic prometaphase chromosomes in celery. *J. Hered.* 75:252–54

90. Murashige, T., Nakano, R. 1967. Chromosome complement as a determinant of the morphogenic potential of tobacco cells. *Am. J. Bot.* 54:963–70

91. Nagl, W. 1979. Differential DNA replication in plants: a critical review. *Z. Pflanzenphysiol. Bd.* 95:283–314

92. Neuffer, M. G. 1966. Stability of the suppressor element in two mutator systems at the A1 locus in maize. *Genetics* 53:541–49

93. Nobécourt, P. 1939. Sur la perennité et l'augmentation de volume des cultures de tissues végétaux. *Compt. Rend. Soc. Biol.* 130:1270–71

94. Orton, T. J. 1980. Chromosomal variability in tissue cultures and regenerated plants of *Hordeum*. *Theor. Appl. Genet.* 56:101–12

95. Orton, T. J. 1984. Genetic variation in somatic tissues: method or madness? *Adv. Plant Pathol.* 2:153–89

96. Orton, T. J. 1984. Somaclonal variation: theoretical and practical considerations. In *Gene Manipulation in Plant Improvement*, ed. J. P. Gustafson, pp. 427–68. New York: Plenum. 668 pp.

97. Papes, D., Jelaska, S., Tomaseo, M., Devide, Z. 1978. Triploidy in callus culture of *Vicia faba* L. investigated by the Giemsa C-banding technique. *Experimentia* 34:1016–17

97a. Peacock, W. J., Dennis, E. S., Gerlach, W. L., Sachs, M. M., Schwartz, D. 1984. Insertion and excision of *Ds* controlling elements in maize. *Cold Spring Harbor Symp. Quant. Biol.* 44:347–54

98. Peschke, V. M., Phillips, R. L., Gengenbach, B. G. 1987. Discovery of transposable element activity among progeny of tissue culture–derived maize plants. *Science* 238:804–7

99. Peterson, P. A. 1986. Mobile elements in maize. *Plant Breed. Rev.* 4:81–122

100. Phillips, R. L. 1978. Molecular cytogenetics of the nucleolus organizer region. In *Maize Breeding and Genetics*, ed. D. B. Walden, pp. 711–41. New York: John Wiley and Sons. 794 pp.

101. Phillips, R. L., Somers, D. A., Hibberd, K. A. 1988. Cell/tissue culture and *in vitro* manipulation. In *Corn and Corn Improvement*, ed. G. F. Sprague. Madison, Wis: Am. Soc. Agron., Inc. In press

102. Pring, D. R., Conde, M. F., Gengenbach, B. G. 1981. Cytoplasmic genome variability in tissue culture-derived plants. *Environ. Exp. Bot.* 21:369–77

103. Pryor, A., Faulkner, K., Rhoades, M. M., Peacock, W. J. 1980. Asynchronous replication of heterochromatin in maize. *Proc. Natl. Acad. Sci. USA* 77:6705–9

104. Rhoades, M. M., Dempsey, E. 1971. On the mechanism of chromatin loss induced by the B chromosome of maize. *Genetics* 71:73–96

105. Rhoades, M. M., Dempsey, E. 1973. Chromatin elimination induced by the B chromosome of maize. *J. Hered.* 64:12–18

106. Rhoades, M. M., Dempsey, E. 1982. Further studies on two-unit mutable systems found in our high-loss studies and on the specificity of interaction on responding and controlling elements. *Maize Genet. Coop. Newslett.* 57:14–17

107. Rhodes, C. A., Phillips, R. L., Green, C. E. 1986. Cytogenetic stability of aneuploid maize tissue cultures. *Can. J. Genet. Cytol.* 28:374–84

108. Rieger, R., Michaelis, A., Green, M. M. 1976. *Glossary of Genetics and Cytogenetics.* New York/Berlin: Springer-Verlag. 647 pp.

109. Rieger, R., Michaelis, A., Schubert, I., Dobel, P., Jank, H. W. 1975. Nonrandom intrachromosomal distribution of chromatid aberrations induced by X-rays, alkylating agents and ethanol in *Vicia faba. Mut. Res.* 27:69–79

110. Rivin, C. J., Cullis, C. A., Walbot, V. 1986. Evaluating quantitative variation in the genome of *Zea mays. Genetics* 113:1009–19

111. Rowley, J. D. 1982. Identification of the constant chromosome regions involved in human hematologic malignant disease. *Science* 216:749–51

112. Rubin, G. M. 1983. Dispersed repetitive DNAs in *Drosophila*. In *Mobile Genetic Elements*, ed. J. A. Shapiro, pp. 329–61. New York: Academic. 688 pp.

113. Ryan, S. A., Scowcroft, W. R. 1987. A somaclonal variant of wheat with addi-

tional β-amylase isozymes. *Theor. Appl. Genet.* 73:459–64

114. Sacristan, M. D. 1971. Karyotypic changes in callus cultures from haploid and diploid plants of *Crepis capillaris* (L.) Wallr. *Chromosoma* 33:273–83

115. Sacristan, M. D., Wendt-Gallitelli, M. F. 1971. Transformation to auxin-autotrophy and its reversibility in a mutant line of *Crepis capillaris* callus culture. *Mol. Gen. Genet.* 110:355–60

116. Saedler, H., Nevers, P. 1985. Transposition in plants: a molecular model. *EMBO J.* 4:585–90

117. Sang, J. H. 1981. *Drosophila* cells and cell lines. In *Advances in Cell Culture,* ed. K. Maramorosch, 1:125–82. London/New York: Academic. 340 pp.

118. Saraiva, L. 1979. *Cytogenetical analysis of the mechanism of chromatin elimination induced by B chromosomes in maize.* PhD thesis. Indiana Univ. Bloomington

119. Schimke, R. T. 1983. Gene amplification in mammalian somatic cells. In *Genetic Rearrangement,* ed. K. F. Chatter, C. A. Cullis, D. A. Hopwood, A. A. W. B. Johnston, and H. W. Woolhouse, pp. 235–52. London/Canberra: Croomhelm. 296 pp.

120. Schimke, R. T., Brown, P. C., Kaufman, R. J., McGrogan, M., Slate, D. L. 1981. Chromosomal and extrachromosomal localization of amplified dihydroreductase genes in cultured mammalian cells. *Cold Spring Harbor Symp. Quant. Biol.* 45:785–97

120a. Schimke, R. T., Sherwood, S. W., Hill, A. B., Johnston, R. N. 1986. Overreplication and recombination of DNA in higher eukaryotes: potential consequences and biological implications. *Proc. Natl. Acad. Sci. USA* 83:2157–61

121. Schwarz-Sommer, Z., Gierl, R., Klosgen, R. B., Wienand, U., Peterson, P. A., Saedler, H. 1985. Plant transposable elements generate the DNA sequence diversity needed in evolution. *EMBO. J.* 4:591–97

122. Scotto, K., Biedler, J. L., Melera, P. W. 1986. Amplification and expression of genes associated with multidrug resistance in mammalian cells. *Science* 232:751–55

123. Shah, D. M., Horsch, R. B., Klee, H. J., Kishore, G. M., Winter, J. A., et al. 1986. Engineering herbicide tolerance in transgenic plants. *Science* 233:478–81

124. Singh, B. D., Harvey, B. L. 1975. Cytogenetic studies on *Haplopappus gracilis* cells cultured on agar and in liquid media. *Cytologia* 40:347–54

125. Singh, R. J. 1986. Chromosomal variation in immature embryo derived calluses of barley (*Hordeum vulgare* L.). *Theor. Appl. Genet.* 72:710–16

126. Silverman, M., Simon, M. 1983. Phase variation and related systems. *Ann. Rev. Genet.* 17:537–57

127. Stalder, G. R., Buhler, E. M., Buchler, U. K. 1965. Possible role of heterochromatin in human aneuploidy. *Humangenetik* 1:307–10

128. Steinrucken, H. C., Schulz, A., Amrhein, N., Porter, C. A., Fraley, R. T. 1986. Overproduction of 5-enolpyruvylshikimate-3-phosphate synthase in a glyphosate-tolerant *Petunia hybrida* cell line. *Arch. Biochem. Biophys.* 244:169–78

129. Sturtevant, A. H. 1925. The effects of unequal crossing-over at the *Bar* locus in *Drosophila. Genetics* 10:117–47

130. Sutherland, G. R. 1983. The fragile X chromosome. *Int. Rev. Cytol.* 81:107–43

131. Sunderland, N. 1973. Nuclear cytology. In *Plant Cell and Tissue Culture,* ed. H. E. Street, pp. 205–39. Blackwell: Oxford Univ. Press

132. Tartof, K. D. 1973. Unequal sister chromatid exchange and disproportionate replication as mechanisms regulating ribosomal gene redundancy. *Cold Spring Harbor Symp. Quant. Biol.* 38:491–500

132a. Taylor, L. P., Walbot, V. 1985. A deletion adjacent to the maize transposable element Mu-1 accompanies loss of *Adh1* expression. *EMBO. J.* 4:869–76

133. Tewari, R., Juyal, R. C., Thelma, B. K., Das, B. C., Rao, S. R. V. 1987. Folate-sensitive fragile sites on the X-chromosome heterochromatin of the Indian mole rat, *Nesokia indica. Cytogenet. Cell Genet.* 44:11–17

134. Therman, E. 1980. *Human Chromosomes.* New York/Berlin: Springer-Verlag. 235 pp.

135. Torrey, J. G. 1961. Kinetin as trigger for mitosis in mature endomitotic plant cells. *Exp. Cell Res.* 23:281–99

136. Tuberosa, R., Phillips, R. L. 1986. Isolation of methotrexate-tolerant cell lines of corn. *Maydica* 31:215–25

137. Vig, B. K. 1982. Sequence of centromere separation: role of centromeric heterochromatin. *Genetics* 102:795–806

138. Walbot, V., Cullis, C. A. 1985. Rapid genomic change in higher plants. *Ann. Rev. Plant Physiol.* 36:367–96

139. Weinberg, G., Ullman, B., Martin, D. W. 1981. Mutator phenotypes in mammalian cell mutants with distinct bio-

chemical defects and abnormal de-oxyribonucleoside triphosphate pools. *Proc. Natl. Acad. Sci. USA* 78:2447–51

140. White, P. R. 1939. Potentially unlimited growth of excised plant callus in an artificial nutrient. *Am. J. Bot.* 26:59–64

141. Woodman, J. C., Kramer, D. A. 1986. *Sixth International Congress of Plant Tissue and Cell Culture,* ed. D. A. Somers, B. G. Gengenbach, D. D. Biesboer,

W. P. Hackett, C. E. Green. Int. Assoc. Plant Tiss. Cult. 215 pp. (Abstr.)

142. Yeoman, M. M. 1981. The mitotic cycle in higher plants. In *The Cell Cycle,* ed. P. C. L. Japess, pp. 161–84. Cambridge: Cambridge Univ. Press

143. Zagorska, N. A., Shamina, Z. B., Butenko, R. G. 1974. The relationship of morphogenetic potency of tobacco tissue culture and its cytogenetic features. *Biol. Plant.* 16:262–74

Ann. Rev. Plant Physiol. Plant Mol. Biol. 1988. 39:439–73

METABOLISM AND PHYSIOLOGY OF ABSCISIC ACID[1]

Jan A. D. Zeevaart

MSU-DOE Plant Research Laboratory, Michigan State University, East Lansing, Michigan 48824

Robert A. Creelman

Department of Biochemistry and Biophysics, Texas A & M University, College Station, Texas 77843

CONTENTS

INTRODUCTION 440
TECHNIQUES 440
    Quantification of ABA 440
    Separation of S- and R-Abscisic Acid 441
    Stable Isotopes and Mass Spectrometry 442
    Extraction and Quantification of Xanthoxin 442
METABOLISM 443
    Biosynthesis in Fungi 443
    Biosynthesis in Higher Plants 444
    Catabolism 453
    Compartmentation 454
EFFECTS OF ABSCISIC ACID 456
    Physiological Responses 456
    Biochemical Responses 461
CONCLUDING REMARKS 463

[1]*Abbreviations used:* ABA, (+)-*S*-abscisic acid; ABA-GE, abscisic acid-β-D-glucosyl ester; DPA, 4'-dihydrophaseic acid; *flc, flacca;* Fluridone, 1-methyl-3-phenyl-5-[3-(trifluoromethyl)-phenyl]-4-(1*H*)-pyridinone; FP, farnesyl phosphate; FPP, farnesyl pyrophosphate; GA, gibberellin; GC-MS, gas chromatography-mass spectrometry; HPLC, high performance liquid chromatography; Me, methyl ester; MVA, mevalonate; m/z, mass/charge; NCI, negative chemical ionization; Norflurazon, 4-chloro-5-(methylamino)-2-(α,α,α-trifluoro-*m*-tolyl)-3(2*H*)-pyridazinone; *not, notabilis;* ODA, 2,7-dimethyl-octa-2,4-dienedioic acid; PA, phaseic acid; SIM, selected ion monitoring; *sit, sitiens; t, trans;* TLC, thin layer chromatography; Xan, xanthoxin

0066-4294/88/0601-0439$02.00

INTRODUCTION

The study of the biochemistry and physiology of ABA has undergone a renaissance in the 1980s. ABA was originally considered a growth inhibitor. We now know that it, like other plant hormones, has multiple roles during the life cycle of a plant. Each of its functions is determined developmentally and environmentally. ABA is ubiquitous in higher plants; it is also produced by certain algae (223, 224) and by several phytopathogenic fungi (46). Outside the plant kingdom, ABA has been found in the brains of several mammals (22, 116), but this ABA may have originated from plants in the animals' diets rather than from synthesis in the brain.

The last review on ABA in this series appeared in 1980 (231), and a comprehensive review of the biosynthesis and catabolism of ABA appeared in the same year (202). Since then, a book on ABA edited by Addicott (4) has been published, as well as a chapter emphasizing chemical aspects of ABA (83). Biosynthesis and catabolism of ABA have been dealt with in several conference and symposium volumes (71, 125, 153, 220, 221), as have the various physiological roles of ABA (42, 98, 181, 190, 191, 207). Excellent reviews describing mutants deficient in ABA have appeared recently (100, 180, 181, 220, 221).

This review highlights developments in the field since the 1980 review. A revised numbering system of ABA, with the methyl groups designated as in Figure 1, has been proposed (14). The finding that several fungi produce ABA led to biosynthetic studies with these organisms in the expectation that the pathway in fungi would be similar to that in higher plants. In fungi, ionylidene derivatives are precursors of ABA, but there is no evidence for this pathway in higher plants. There has been renewed interest in the indirect pathway in which ABA is a breakdown product of xanthophylls with xanthoxin (or a related compound) as an intermediate. We review the extensive evidence that favors this pathway for stress-induced ABA, although definitive proof has not yet been obtained. Mutants deficient in ABA have been used extensively to study the biosynthetic pathway and to clarify the physiological roles of ABA. The changing concept of the role of ABA in plant growth and development is perhaps best illustrated by these mutants: A reduction in the level of ABA, a presumed growth inhibitor, leads to a *reduced* growth rate, in particular of leaves, which can be restored to normal by exogenous ABA (18, 181, 220, 222).

TECHNIQUES

Quantification of ABA

Advantages and disadvantages of various methods used for measurement of ABA have been compared recently (37). For purification of ABA in crude

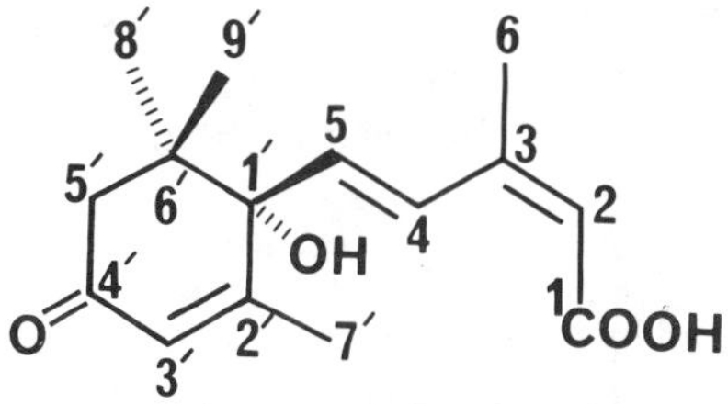

Figure 1 Revised numbering system of (+)-S-abscisic acid (14).

extracts, HPLC has largely replaced TLC. Actual quantification is usually accomplished by GC with an electron capture detector (125), as well as by radio- or enzyme-immunoassays (236). Antibodies of high specificity and affinity against ABA have been produced by a number of laboratories (e.g. 38, 67, 117, 131, 182, 201), and a monoclonal antibody against S-ABA is available commercially (89). Several laboratories (e.g. 230) have developed an indirect enzyme-linked immunoassay in which much less antibody is used than required in the Idetek protocol (89). An enzyme-amplified immunoassay with a sensitivity of 0.05–2.5 pg ABA has been developed to measure the ABA content of mesophyll and guard cells (68). Immunoassays must be validated by physical-chemical methods to ensure that no interfering substances are present in the extracts (234, 236). Immunochemistry has also been used to localize ABA at the subcellular level (209, 210). A serious disadvantage of all immunological methods is that no antibodies against catabolites of ABA are currently available.

Separation of S- and R-*Abscisic Acid*

Commercial sources of ABA, including both ^{14}C- and ^{3}H-labeled analogs, are racemic mixtures of S-ABA (natural) and R-ABA (unnatural). Since the catabolism of the two enantiomers is quite different (see below), it is essential that workers describe which compound(s) they are feeding (S, R, or RS) and that only the natural S-enantiomer be used for metabolic studies.

Two types of methods have been described for the resolution of the enantiomers of ABA: (*a*) HPLC with a chiral stationary phase, and (*b*) immunoaffinity chromatography. In the first method, methyl-RS-1',4'-*cis*-ABA diol was resolved with two different chiral Pirkle columns in series. ABA was regenerated by oxidation with MnO_2 and basic hydrolysis (229). Railton (189) separated the enantiomers of ABA as their methyl esters using a stationary phase of cellulose tris(3,5-dimethylphenylcarbamate) coated on silica gel. In the second method, antibodies selective against S-ABA were bound to Sepharose. The S-enantiomer was retained on the column and subsequently eluted with organic solvent (99, 132). The column is reusable, but because of the low capacity, this technique is more suited for the resolution of radioactive RS-ABA than for the unlabeled racemic mixture.

Pure S-ABA is also available from fungal cultures (6), and the synthesis of S-ABA has been accomplished (M. Soukup, personal communication). The total synthesis of Me-PA has also been reported (213).

Stable Isotopes and Mass Spectrometry

[$^2H_6$]ABA, synthesized by exchange under alkaline conditions, has been used as an internal standard for measuring small amounts of ABA by GC-SIM (196). The use of this compound is limited, however, since the pH must be kept below 8 to prevent exchange. Netting et al (157) introduced three deuterium atoms into the C-6 methyl group of ABA by the same procedures used for synthesis of [^3H]ABA (231). These deuterium atoms are not exchangeable at high pH (157).

$^{18}O_2$ and $H_2^{18}O$ have been used for studying ABA biosynthesis (34, 35, 173) and the conversion of ABA to PA (34, 35). Nonhebel & Milborrow (160, 161) used ^2H_2O to measure the size of the ABA precursor pool. [1,2-^{13}C]Xan has been synthesized for use in investigating differences in the metabolism of Xan in tomato mutants (173).

The original work (64) on the fragmentation pattern of Me-ABA used electron impact MS. Since the molecular ion, M^+, is not abundant (35, 64), this method is not sensitive to small amounts of stable isotope enrichment. Positive chemical ionization gives more intense high-mass ions than electron impact (157), but NCI is the preferred method, since $M^- = 278$ is the base peak in the mass spectrum of Me-ABA (34, 160, 198). The sensitivity to Me-ABA, as measured by NCI, is in the 1–5 ng range (34); greater sensitivity can be obtained by SIM (196, 198).

Takeda et al (215) have reported the chemical ionization mass spectra of free ABA, PA, DPA, and *epi*-DPA, their methyl esters, and some of their conjugates. Mass spectra of the conjugates obtained by desorption chemical ionization at different source temperatures were particularly useful for structural characterization (216).

Extraction and Quantification of Xanthoxin

Xan is readily produced as a degradation product from xanthophylls during extraction and purification. Special precautions must therefore be taken during extraction of Xan to ensure that the amounts measured actually reflect those present in the tissues. Artifactual Xan production was prevented by grinding the tissue in acidic methanol, so that the 5,6-epoxide groups of violaxanthin and neoxanthin were isomerized to 5,8-furanoid groups and thus could no longer produce Xan by oxidative cleavage. The acidic extract was quickly neutralized with sodium bicarbonate to prevent breakdown of Xan, which by itself was relatively stable at 0.1 N HCl at low temperature for 10 min (44). Parry et al (173) purged the extraction solvent with argon. DeVit

(44) separated and measured Xan and *t*-Xan by HPLC, whereas Parry et al (173) used GC-SIM with [1,2-^{13}C]Xan as an internal standard. Xan stored in methanol at room temperature or in a freezer was unstable, undergoing isomerization, breakdown, and conversion to butenone (44). In another laboratory, Xan was stable when stored in ethyl acetate at -20°C (33).

METABOLISM

Biosynthesis in Fungi

Several genera of fungi have been described that can produce ABA as a secondary metabolite (6, 46, 129, 165, 168). Of these fungi, especially *Cercospora rosicola, C. cruenta,* and *Botrytis cinerea* have been used to study the biosynthesis of ABA.

When [^{3}H]MVA was fed to cultures of *C. rosicola* the main products were ABA and 1'-deoxy-ABA (IV, Figure 2); radioactive 1'-deoxy-ABA was converted to ABA with an 11% yield (154). NMR analysis of ABA, obtained from culture filtrates of *C. rosicola* fed [1,2-^{13}C]acetate, demonstrated that biosynthesis occurred via the isoprenoid pathway (12). Incorporation of radioactivity from FP and FPP into ABA was observed (13), but Neill et al (153) reported negligible conversion of FPP to ABA. No incorporation was detected when dephosphorylated compounds were fed to *C. rosicola* (13, 86). FPP was dephosphorylated to FP by the fungal cultures, but the further dephosphorylation of FP to farnesol was not reported (13). Given that this is a likely event and that the incorporation of radioactivity from FPP and FP into ABA was determined by a combination of HPLC and TLC, but not confirmed by MS, it cannot be unequivocally concluded that exogenous FPP and FP are precursors of ABA in fungal cultures.

Most of the work on intermediates in the ABA biosynthetic pathway of fungi has involved C$_{15}$ compounds with carbon skeletons similar to that of ABA. Evidence from labeling studies indicates that ABA is formed from α-ionylidene derivatives. α-Ionylidene ethanol (I), α-ionylidene acetic acid (II), and 4'-hydroxy-α-ionylidene acetic acid (III) were detected in *C. rosicola* (155), and converted to 1'-deoxy-ABA (IV) and ABA (confirmed by MS when ^{2}H-labeled precursors were fed). The 2-*trans* isomers of these compounds, as well as β-ionylidene ethanol and β-ionylidene acetic acid, were not converted (148, 155, 162). Both epimers of 4'-hydroxy-α-ionylidene acetic acid (III) were metabolized to IV and ABA. However, 1'-deoxy-ABA (IV) was stereospecifically hydroxylated at the 1' position to give ABA (155). In cultures of *C. cruenta*, II was also converted to III, IV, and ABA (88), but 4'-hydroxylation was mainly *trans* in relation to the 1' side chain.

Norman et al (162) described 1'-hydroxy-α-ionylidene acetic acid (V) as a minor product of II in *C. rosicola*; the ethyl esters of *cis*- and *t*-V were metabolized to ABA and *t*-ABA, respectively. Neill et al (155) also found that

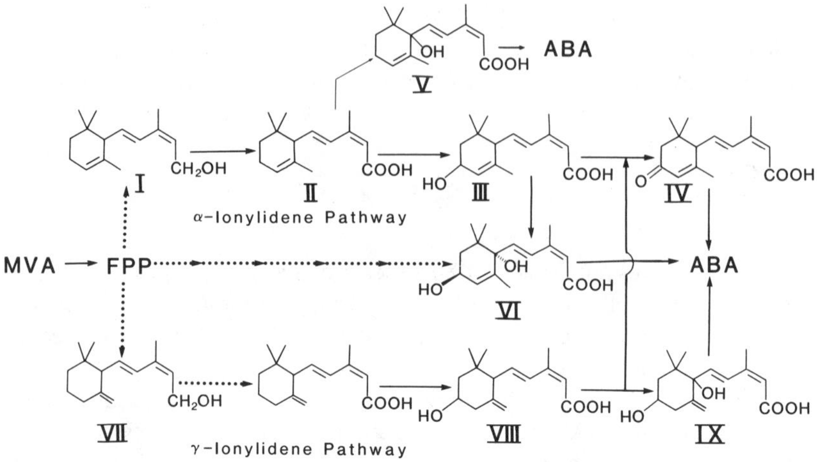

Figure 2 Later stages of the ABA biosynthetic pathway in the fungi *Cercospora rosicola* (α-ionylidene pathway), *C. cruenta* (γ-Ionylidene pathway), *C. pini-densiflorae*, and *Botrytis cinerea*. A single arrow between two compounds does not necessarily indicate that only one enzymatic step is involved. Dotted lines: Conversions not unequivocally demonstrated.

1'-hydroxy-α-ionylidene derivatives were converted to ABA with a low yield. Thus, the conversion of II via V to ABA is presumably a minor pathway in *C. rosicola*.

In cultures of *B. cinerea*, the 1',4'-*t*-diol of ABA (VI) was shown to be a precursor of ABA (85). In *C. pini-densiflorae*, VI was more easily converted to ABA than IV, which led Okamoto et al (165) to propose the conversion of I to ABA via III and VI as the major pathway in this fungus (Figure 2). Thus, the order of hydroxylation and oxidation in the α-ionylidene pathway may differ in different organisms.

Recently, another series of precursors based on the γ-ionylidene structure has been described. It was found that cultures of *C. cruenta* produced γ-ionylidene ethanol (VII) (167), 4'-hydroxy-γ-ionylidene acetic acid (VIII) (169), and 1',4'-dihydroxy-γ-ionylidene acetic acid (IX) (166). Radioactive MVA was incorporated into VIII, which was converted to IV and IX (168). Since IX was converted to ABA more efficiently than IV, it was suggested (168) that in *C. cruenta* the main biosynthetic pathway is via IX rather than via IV, as in *C. rosicola*.

Biosynthesis in Higher Plants

Although the structure of ABA has been known since 1965, the detailed pathway of ABA synthesis in higher plants has remained obscure. Research has focused on two pathways: (*a*) the direct pathway involving a C_{15} precursor derived from FPP, and (*b*) the indirect pathway involving a precursor derived from a carotenoid (Figure 3). The relative importance of each pathway is unknown, and the possibility exists that both operate in higher plants at the same time. In either case, MVA is the ultimate precursor. Since the last review in this series (231), there has been a marked increase in papers presenting data in support of the indirect pathway.

THE DIRECT PATHWAY   Label from MVA has been incorporated into ABA in several tissues (reviewed in 83, 135; 29), but the yield was always very low. Possible reasons for the low incorporation are competition for MVA by other terpenoid pathways (such as sterols) and extensive dilution of the radioactive MVA by a large precursor pool to ABA. Results from experiments with stereospecifically labeled MVA indicate that three residues of the natural R-enantiomer are incorporated into ABA. The stereochemical retention, in ABA, of hydrogens from MVA was similar to that observed with carotenoid biosynthesis (135). Thus, it has not been possible to distinguish between the C_{15} and C_{40} pathway with this approach.

THE INDIRECT PATHWAY   The discovery of Xan in the late 1960s gave rise to the hypothesis that ABA is a breakdown product of carotenoids (218).

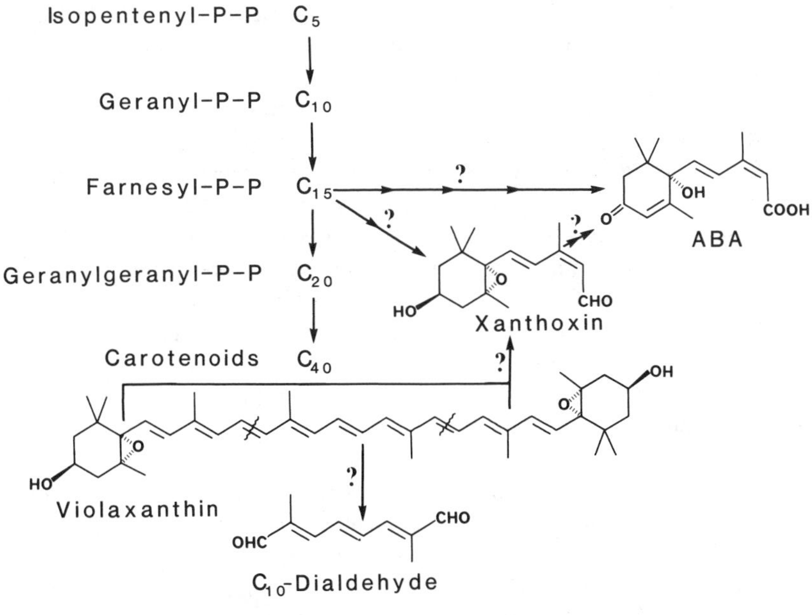

Figure 3   Hypothetical scheme of ABA biosynthesis in higher plants.

Since then, several other lines of evidence have been developed that favor this pathway.

Xanthoxin: a natural precursor of abscisic acid?   Xan, a neutral growth inhibitor, with physiological properties similar to those of ABA, was produced by photochemical, chemical, or enzymatic cleavage of xanthophylls, in particular violaxanthin and neoxanthin (218). In all instances more *t*-Xan was produced than Xan. When applied to tomato and bean shoots, Xan was converted to ABA and PA, whereas *t*-Xan was convered to *t*-ABA (219). If Xan is a natural precursor of ABA, then the immediate precursor to Xan must be 9-*cis*-violaxanthin or violeoxanthin, since cleavage of violaxanthin gives predominantly *t*-Xan. It should be noted, however, that endogenous Xan has never been shown to be derived from violaxanthin, or any other xanthophyll, in vivo. Furthermore, Xan could be an intermediate in the direct pathway (Figure 3).

Because of the artifactual production of Xan from xanthophylls during extraction, the earlier data on Xan levels in plants can no longer be considered reliable. With improved methods, the levels of Xan in bean (44) and tomato leaves (173) were found to be in the range of 1–10 ng g^{-1} fresh weight and did not increase when water stress was imposed. With an increase in ABA in stressed leaves of 300–600 ng hr^{-1} g $^{-}$1 fresh weight, the turnover of Xan would have to be very rapid to produce ABA at such a rate (44, 205). Work with a cell-free system (see below) and results of Xan feeds to bean leaves (205) indicate that the conversion of Xan to ABA is indeed sufficiently rapid to prevent Xan accumulation.

Recently, it has been shown that cell-free extracts from bean, pea, corn, cucumber, and cowpea leaves and bean roots can convert Xan to ABA, while *t*-Xan was converted to *t*-ABA at a rate about 40% of that at which Xan was converted to ABA (205). The enzyme activity requires NAD^+ or $NADP^+$ and appears to be localized in the cytosol. There was no effect of water deficit or cycloheximide on the enzyme activity, and no intermediates between Xan and ABA were detected. The measured rates of ABA production from Xan were so high (usually 3 μg hr^{-1} g^{-1} fresh weight) that this conversion cannot be the rate-limiting step in ABA biosynthesis. Furthermore, the Xan-oxidizing enzymes cannot be the ones whose syntheses are required for stress-induced ABA accumulation (66, 205, 212).

Carotenoid biosynthetic inhibitor and mutant studies   If ABA is derived from a carotenoid, then blocking the carotenoid biosynthetic pathway should prevent ABA accumulation. Inhibitors of carotenoid biosynthesis, such as fluridone and norflurazon, which block the conversion of phytoene to phytofluene, also inhibit accumulation of ABA (59–61, 80, 142, 183, 208). While this suggests a carotenoid origin of ABA, these inhibitors are rather nonspecific, causing photobleaching and chloroplast degradation in weak light, and changes in general metabolism (33, 80). However, in dark-grown barley seedlings neither the size nor the protein composition of plastids was affected by fluridone treatment (61).

Li & Walton (118) performed an experiment in which green bean leaves were pretreated with fluridone, labeled with $^{14}CO_2$ for 24 hr, and then water-stressed for 14 hr. ABA and several xanthophylls were isolated and quantified; specific activities were determined. Fluridone did not inhibit the accumulation of stress-induced ABA in these leaves, but the specific activity was reduced to about the same extent as that of the xanthophylls.

Various corn mutants—viviparous 2, 5, and 9 *(vp2, vp5, vp9)*, pink scutellum *(ps = vp7)*, white seedling *(w3)*, and yellow *(y3, y9)*—have been described that lack the ability to produce certain carotenoids. All maize viviparous mutants are characterized by pale yellow endosperms and white or almost white seedlings. The primary lesions in these mutants are at different steps in the carotenoid pathway (59). These mutants exhibit a pleiotropic phenotype in that they also have a greatly decreased capability to produce ABA (143, 151, 208) and show vivipary. Since application of ABA to these mutants will not cause reversion to a wild-type phenotype, they should not be classified as ABA-deficient mutants. Likewise, since the lesions in these mutants are in the carotenoid biosynthetic pathway, isolation and character-ization of these genes would not contribute to our understanding of ABA biosynthesis.

Mutants deficient in abscisic acid   Green ABA-deficient mutants are found in potato (179), pea (233), tomato (180, 220, 221), and *Arabidopsis* (101). Of these, the only well-characterized mutants are three nonallelic recessive wilty mutants of tomato: *flc, not,* and *sit* (221). ABA levels in the three mutants range from 15% of wild type for *sit,* 26% for *flc,* and 49% for *not* (149, 173) and are closely correlated with the phenotypic expression (220, 221). Of the double mutants, *not flc* and *not sit* exhibited a more severe wilty phenotype than *flc sit* and the single mutants *flc* and *sit.* This suggests that the *not* mutation is somehow distinct from the *flc* and *sit* mutations (220, 222). Since the decreased ABA contents are not the result of increased catabolism (158), the lesions must be in enzymes involved in ABA biosynthesis. The mutants are shorter, produce adventitious roots on stems, and the leaves show epinasty compared with wild type (217). In contrast to an earlier report (217), Neill et al (156) found no differences in auxin and ethylene levels between *flc* and wild-type leaves. Even though ABA is present (albeit in lower amounts than in wild type), application of ABA will cause reversion to the wild phenotype (18, 220, 221).

Stress-induced accumulation of ABA does not occur in *droopy* potato (179), *sit* and *flc* tomato (149), wilty pea (233), and *aba Arabidopsis* (100), but *not* tomato can respond to stress with a slight rise in ABA (149, 173). It is possible that these mutants are already stressed and have reached their maximum accumulation of ABA, or that these mutants may not be able to produce ABA in response to stress. If the latter case is true, this implies the existence of two pathways of ABA biosynthesis in higher plants, one operating in turgid leaves and the other in water-stressed leaves.

The wilty mutants of tomato have a normal complement of carotenoids (158). Thus, if ABA is derived from xanthophylls, the metabolic blocks in these mutants must be in the oxidative steps from carotenoids to ABA. This possibility was investigated by measuring the conversion of [^{13}C]Xan to ABA in the three tomato mutants (173). [^{13}C]Xan was converted to ABA in high yield by wild type (12%) and *not* (16%). However, incorporation into ABA in *flc* and *sit* was only about 1%. These data indicate that *flc* and *sit* are impaired in ABA biosynthesis between Xan and ABA and that the lesion in *not* is at a step prior to Xan. Another possible precursor of ABA, 1',4'-*t*-diol-ABA, is an endogenous compound found in wild-type and mutant tissues; all four genotypes converted this compound to ABA (173). The mutants *flc* and *sit* also accumulate high levels of *t*-ABA alcohol as compared to *not* and wild type (121). It is unlikely that this compound is a precursor of ABA, however, since enzymatic isomerization of the 2-*trans* double bond has never been observed during the later stages of ABA biosynthesis (135).

Double oxidative cleavage of a xanthophyll to two C_{15} compounds (such as xanthoxin) would also yield a C_{10} by-product (Figure 3). Such a putative compound was identified as the C_{10} dicarboxylic acid, ODA (120), which was present in higher levels in *flc* and *sit*, and in lower levels in *not*, in comparison with wild type (122). It was suggested (221) that ODA levels in *flc* and *sit* are abnormally high because their ABA deficiency leads to an increased rate of carotenoid cleavage; the *not* lesion would be at the cleavage step and would thus result in a low rate of both ABA and ODA accumulation. Neill & Horgan (149) pointed out that it is not clear why ODA would accumulate in *flc* and *sit*, since its further metabolism should be independent of that of the C_{15} precursor of ABA. Furthermore, ODA also accumulates in wild-type tomato kept in a chronic wilted state (137). When tomato plants were supplied with ^2H_2O, a fraction of the ABA became labeled, but ODA did not. It was, therefore, concluded that ODA cannot be a by-product of ABA biosynthesis (137). Thus, the relationship of ODA to ABA biosynthesis, if any, remains to be established.

Stable isotope studies   Evidence has been presented that one atom of ^{18}O was incorporated into the carboxyl group of ABA isolated from stressed *Xanthium strumarium* and *Phaseolus vulgaris* leaves that were incubated for 6 hr in the presence of $^{18}O_2$ (35). Using the highly sensitive NCI-MS technique, it was subsequently shown (34) that in stressed *Xanthium* leaves three atoms of ^{18}O from $^{18}O_2$ were incorporated into the ABA molecule. One ^{18}O atom was incorporated rapidly into the carboxyl group of ABA as shown earlier (35), whereas the other two atoms were very slowly incorporated into the ring oxygens. The fourth oxygen atom, in the carboxyl group of ABA, was shown to be derived from water by incubating *Xanthium* leaves in $H_2{}^{18}O$ (34). The low incorporation of ^{18}O into the ring positions implies that there may be other compounds feeding into a large precursor pool; during conversion to this pool, these compounds would incorporate ^{18}O into positions that ultimately form the ring oxygens of ABA. ABA from stressed roots of *Xanthium* incubated in $^{18}O_2$ showed a labeling pattern similar to that of ABA in stressed leaves, but with incorporation of more ^{18}O into the tertiary hydroxyl group at C-1' than found in ABA from stressed leaves (34).

Approximately 20% of ABA-GE in leaves stressed for 24 hr contained one ^{18}O atom, which was located in the carboxyl group (34). Since ABA-GE accumulates at a low rate in *Xanthium* leaves during water stress (244, 245), this result indicates that newly synthesized ABA-GE is derived from stress-induced ABA rather than from unlabeled ABA already present at the onset of stress.

In turgid *Xanthium* leaves, ^{18}O was incorporated into ABA to a much lesser extent than it was in stressed leaves. For example, in the case of ABA, the relative intensity of $M^- + 2$ was 3.1% after 24 hr, and 5.8% after 72 hr. These results suggest that ABA turnover in the turgid leaves under the experimental conditions was very slow. On the other hand, [^{14}C]ABA applied to these leaves was completely catabolized to PA and conjugates after 48 hr. There are two possible explanations for these conflicting observations. First, it is possible that the oxygen atoms in ABA found in turgid leaves do not originate from molecular oxygen. This implies that the biosynthetic pathways for ABA in turgid and water-stressed leaves are different. Second, exogenous ABA may be metabolized differently from endogenous ABA. This implies that in turgid leaves ABA is normally isolated from catabolic enzymes, perhaps in chloroplasts. Exogenous ABA, on the other hand, would encounter these enzymes upon its entrance into the cytoplasm. Excised corn embryos cultured on an agar medium were used as another nonstressed tissue for $^{18}O_2$-labeling experiments. In this material, the labeling pattern of ABA was similar to that found for ABA in water-stressed leaves (F. Fong, D. A. Gage, J. A. D. Zeevaart, unpublished).

In a different approach (118), ^{18}O was introduced into the epoxide oxygen of violaxanthin by means of the xanthophyll cycle (241). Leaves were then stressed and violaxanthin and ABA analyzed. Violaxanthin contained 40–

45% [18]O in the epoxide group; the ABA that accumulated during water stress contained 10–15% [18]O in the ring oxygens, suggesting that a portion of the ABA was derived from violaxanthin. It is possible that part of the ABA was derived from violaxanthin not labeled with ^{18}O, since there are two different pools of violaxanthin (48).

Incorporation of [18]O into Xan has also been detected (173; R. A. Creelman, D. A. Gage, J. A. D. Zeevaart, unpublished), but the enrichment was only 10–15%. This is much lower than the [18]O enrichment found in the carboxyl group of ABA, which implies that Xan is not the natural precursor of ABA. However, the oxygen atom of the aldehyde group rapidly exchanges with water (D. A. Gage, J. A. D. Zeevaart, unpublished). Xan formed in vivo is probably rapidly converted to ABA (44, 205), and thus exchange of [18]O will be minimized. The assumption that Xan is a precursor of ABA remains, therefore, a viable hypothesis.

It is known that the hydroxyl groups of lutein and the epoxide groups of antheraxanthin and violaxanthin are derived from molecular oxygen (241). Furthermore, the turnover of carotenoids in green leaves is low (62). Thus, the low incorporation of [18]O into sites that ultimately form the keto and hydroxyl groups of ABA could represent the biosynthesis of lutein, zeaxanthin, and antheraxanthin and subsequent conversion to violaxanthin. The demonstration of incorporation of one oxygen atom from water into the carboxyl group of ABA is consistent with the idea that xanthophylls are precursors of stress-induced ABA (34). Cleavage of a xanthophyll, such as violaxanthin, by an oxygenase would give rise to Xan where the oxygen atom in the side chain is derived from molecular oxygen. Subsequent oxidation of Xan hydrate by a dehydrogenase would incorporate an oxygen atom from water into the carboxyl group of ABA. On the basis of this evidence, it is reasonable to postulate that ABA is an apo-carotenoid, i.e. a compound derived from a C_{40} carotenoid by oxidative cleavage (34).

Nonhebel & Milborrow (160, 161) measured the incorporation of ^{2}H from [2]H_2O into ABA, Xan, and various carotenoids in tomato shoots. After incubation for 6 days in 70% [2]H_2O, approximately 5% of the total ABA became labeled with one ^{2}H atom and 21% with 3–14 ^{2}H atoms. The precursor pool size was estimated to be approximately 35 times that of ABA (160). Very little or no ^{2}H was detected in either Xan or carotenoids. It was concluded that Xan, lutein epoxide, and violaxanthin are not precursors of ABA (161). Since the carotenoid pool size did not change during the incubation in [2]H_2O (161), and since carotenoids have a very low turnover (62), absence of ^{2}H in the xanthophylls and Xan would be expected. It cannot be ruled out, therefore, that xanthophylls were precursors of the unlabeled ABA which was about 70% of the total ABA (160, 161). Perhaps ABA was synthesized via both the direct and indirect pathway in these tomato shoots. Assuming that only the indirect pathway is stimulated by water stress, then this idea is supported by the observation that the percentage of labeled ABA was less in the wilted than in the turgid shoots (160).

Other inhibitor studies   Stress-induced ABA accumulation was inhibited by cycloheximide, but not by chloramphenicol (184, 212), suggesting that cytosolic protein synthesis is necessary. The transcription inhibitors cordycepin and actinomycin D also prevented the increase in ABA caused by a water deficit (66, 212). Although these inhibitors may have other effects as well, these results suggest that changes in nuclear gene expression are required for stress-induced accumulation of ABA. Several cytokinins inhibited ABA biosynthesis from MVA in avocado mesocarp (31), but stress-induced ABA accumulation in barley leaves was not prevented by a cytokinin (212).

REGULATION OF BIOSYNTHESIS   Initially it was thought that stress-induced ABA accumulation is dependent on the decline of the leaf water potential below a threshold value (reviewed in 231). Subsequent work using the pressure bomb technique (174) and penetrating and nonpenetrating osmotica (36) has led to the conclusion that loss of turgor is the critical parameter of cell water relations that initiates ABA biosynthesis. Accumulation of some ABA prior to the point of zero turgor can be understood by considering that a leaf consists of a heterogeneous population of cells with different osmotic potentials. As the leaf water potential declines, zero turgor will be reached over a certain range and the accumulation of ABA will be gradual rather than sudden (1, 174). Although loss of turgor is at present the most obvious hypothesis to explain stress-induced ABA synthesis, there are associated changes in cell volume (albeit small) and hence in solute concentration. The experimental data do not rule out the possibility that a highly sensitive measurement of solute concentration allows the cell to respond to water withdrawal. There is as yet no explanation for the finding that ABA accumulation depended on the rate at which the stress developed (81). Neither is it clear why flooding caused stomatal closure (90, 247) and a ten-fold increase in ABA without a leaf water deficit (90). It was suggested that flooding inhibited shoot to root translocation and that this caused ABA to accumulate (90).

Water stress can be mimicked by incubating thin leaf slices in hypertonic solutions of osmotica, such as mannitol and sorbitol. Plasmolysis (loss of turgor) caused by such a treatment readily induced ABA production (36, 71, 74, 126, 130, 185, 197, 237). In spinach, an external sorbitol concentration of 0.33 M stimulated ABA biosynthesis, whereas in the xerophytic plants *Nerium oleander* and *Arbutus unedo* 1.0 M sorbitol was required (71). These differences were caused by different osmotic potentials in the leaves of the species used. In xerophytes, the cell sap had a low osmotic potential (74), and thus a high external solute concentration was required to reduce turgor to zero. In this context, results obtained with *Dunaliella parva* are of interest. This salt-tolerant alga had its lowest ABA content at 1.5 M NaCl, the medium that is optimal for growth and photosynthesis. Lower and higher NaCl levels increased the amounts of ABA in cells as well as in the medium (224).

Osmotic stress results in release into the medium of a high proportion of the ABA produced (36, 74, 126, 224, 237). Hartung et al (74) showed that stress-induced efflux of ABA from mesophyll cells was related to the change in cell volume rather than to turgor change. It should be realized, however, that plasmolysis is an unnatural situation for cells of terrestrial plants. During dehydration, continued water loss causes a shrinkage of the elastic cell wall, and the plasmalemma will remain held against the wall (127). This situation (cytorrhysis) is similar to that found in a tissue treated with an osmoticum that does not penetrate the cell walls (36, 163). At zero turgor the relaxation of the plasmalemma and the associated conformational changes may be the signal for ABA synthesis.

Isolated protoplasts are unable to produce ABA in response to lowering the osmotic potential of the incubation medium (71, 106, 126, 237). The reason for this may be that protoplasts behave as osmometers and have no turgor, but it is also possible that production of stress-induced ABA during incubation of the tissue in a sorbitol-containing medium (126) preempted any subsequent ABA synthesis. Barley protoplasts were also unable to catabolize ABA (126). Thus, there is little prospect for studying the biophysical mechanism that triggers ABA synthesis in a system that does not consist of intact cells.

CATABOLISM OF FUNGAL INTERMEDIATES   Regardless of whether ABA is synthesized via the direct or indirect pathway, the immediate precursor of ABA must be a C_{15} compound. A major precursor to ABA in fungi, α-ionylidene acetic acid (II), has greater biological activity than ABA in several bioassays (232). In barley leaves, II was converted to 1'-deoxy-ABA (IV) and conjugates, but not to ABA (113). Recently, conversion of II and IV to ABA with a very low yield (0.01–2.9%) has been demonstrated in several species (170). These low conversions to ABA strongly suggest that II is active by itself (33, 170), and further, that in higher plants, ABA biosynthesis does not occur via 1'-deoxy-ABA (170). Only in *Vicia faba* has a significant conversion of II and IV to ABA been demonstrated (153). Low conversion to ABA in other species could result from failure of II to reach the proper compartment for catabolism, or from the presence of highly active conjugating enzymes. It is also possible that *V. faba* is an exception, possessing an enzyme other species lack.

Catabolism

ISOLATION AND CHARACTERIZATION OF NEW CATABOLITES   This subject has been reviewed recently (37, 125). In addition to the well-known conversion of ABA to PA and DPA, ABA is also conjugated with glucose. ABA-GE appears to be widespread in plants but has been identified conclusively only a few times (16, 39, 150). Its concentration is usually much higher in fruits than in leaves (150). The occurrence of the 1'-O-ABA-β-D-glucoside has been reported in tomato shoots (124). Conjugates of ABA catabolites, such as PA-β-D-glucosyl ester (15) and the 4'-O-DPA-β-D-glucoside (84, 139, 203), have also been identified.

Tietz et al (225) reported 4'-desoxy-ABA as a catabolite of ABA in pea seedlings. However, this compound is an artifact formed during GC-MS measurement by dehydration of 1',4'-t-diol-ABA (136, 214). Both 1',4'-diols of ABA have been identified in plants (40, 164); 1',4'-t-diol can act as a precursor of ABA rather than being a catabolite (164).

DIFFERENCES IN CATABOLISM AND UPTAKE OF S- AND r-ABSCISIC ACID   In leaf discs of V. faba, S-ABA was catabolized much more rapidly than R-ABA, the half-lives being 6–8 hr and 30–32 hr, respectively (132). This may result, at least in part, from differences in uptake of the two enantiomers (see below). The principal catabolites of S-ABA in V. faba (132) and Xanthium (17) were PA, DPA, and their conjugates. In tomato shoots, S-ABA gave as main product 4'-O-DPA-β-D-glucoside via the PA and DPA route (229). In contrast, a large proportion of R-ABA was conjugated to give the glucose ester and 1'-glucoside (17, 132, 229). The formation of a new metabolite, 7'-hydroxy-ABA, in cell suspension cultures was reported (112), but the stereochemistry of this new compound was not specified. Subsequently, evidence has been presented that in Xanthium (17) and Hordeum leaves (30) only R-ABA is hydroxylated in the 7'-position to give 7'-hydroxy-R-ABA. Thus, this new catabolite is an artifact resulting from feeding the unnatural R-enantiomer of ABA. Recently, 7'-hydroxy-ABA (no optical rotation given) has been reported as an endogenous compound in V. faba (115). Evidently, there are differences between species in ABA catabolism (125).

In addition to the passive, diffusive uptake of undissociated ABA (see below), which is determined by pH gradients, carrier-mediated uptake of ABA has been observed in root apices of several species and in cell suspension cultures (7, 138). The carrier is specific for the natural S-enantiomer with a K_m value of approximately 1 μM (7).

REGULATION OF CATABOLISM   In addition to ABA, the catabolites PA and DPA also accumulate in wilted leaves. In contrast, the level of ABA-GE increases only slightly during water stress (25, 69, 175, 244, 245). For bean leaves, it was calculated that the rate of conversion of ABA to PA increased steadily until after approximately 7.5 hr of stress it equalled the rate of synthesis of ABA (175). Thus, stress stimulates ABA synthesis and the high ABA level in stressed leaves is maintained by a balance between synthesis and catabolism.

Upon rehydration (restoration of turgor), the rate of conversion to PA was greatly accelerated, resulting in the rapid disappearance of ABA with a concomitant increase in PA (25, 175, 244, 245). Experiments with $^{18}O_2$ have provided similar results. During the conversion of ABA to PA one ^{18}O atom is incorporated into PA by 8'-hydroxylation, so that isotope enrichment of PA can be taken as a measure of PA synthesis. In stressed *Xanthium* leaves incubated in an atmosphere containing $^{18}O_2$ for up to 24 hr, there was a gradual increase in the ^{18}O content of PA (34). In comparison, when stressed leaves were rehydrated and then incubated with $^{18}O_2$, a large proportion of the PA molecules had incorporated one ^{18}O atom after 5 hr (35). Using different methods and materials, the following half-lives have been estimated for ABA: 3 and 16 hr in stressed bean (69) and *Xanthium* leaves (34), respectively; 7 hr in turgid tomato shoots (160). ABA was degraded more rapidly in leaves in darkness than in light (125, 245). The most likely explanation is that in the light a high proportion of ABA in its anionic form is trapped in the chloroplasts, which would protect ABA from the catabolic enzymes in the cytoplasm. In barley and wheat aleurone layers, the catabolism of ABA to PA was enhanced when the tissue was pretreated with ABA, but this treatment had no effect on the conversion of PA to DPA (227).

When catabolism of radiolabeled *RS*-ABA was compared in turgid and wilted leaves, little difference was observed between the two treatments in wheat (147) and *Xanthium* leaves (25); but in barley leaves, stress inhibited ABA catabolism (30). Continued cytoplasmic protein synthesis is required, as evidenced by the fact that cycloheximide inhibited ABA catabolism in several tissues (29, 30, 227).

Leaf age is a significant factor in determining the rate of ABA catabolism, older leaves having the greater ability to convert ABA (25, 30). Since young leaves have the highest levels of ABA, but the lowest ability to synthesize and catabolize ABA (25, 125, 246), it follows that most of the ABA in young leaves is imported from older leaves via the phloem (246).

Compartmentation

Knowledge of the distribution of ABA within cells and organs is essential to understand its mode of action. ABA is a weak acid (pK 4.7) that dissociates to varying degrees in the different compartments. The protonated form permeates freely across membranes, whereas the dissociated anion is impermeable. Consequently, the distribution between different compartments is determined by the difference in pH between compartments; the greater the difference, the greater the amount of ABA that will accumulate in the more alkaline compartment. Using these rules, along with data on the pH values and volumes of various compartments, Cowan et al (32) developed a model showing that 4.4% (8.2%), 68.4% (40.7%), 17.2% (32.4%), and 10% (18.7%) of the total cell ABA will be found in the apoplast, chloroplasts, cytoplasm, and vacuole, respectively, under conditions of light (dark). These calculated values are in good agreement with experimental data obtained with leaves (71, 76, 94) and roots (11), Thus, the transition from light to darkness and vice versa causes a redistribution of ABA among the cellular compartments. A doubling of ABA in the apoplast in the dark may be significant for stomatal closure, since it is the only pathway from the mesophyll to the guard cells (see below). Although a high percentage of ABA is trapped in the chloroplasts, there is no evidence that these organelles are the sites of ABA synthesis (29, 73) or catabolism (29, 72). Localization of ABA in chloroplasts has also been established by an immunohistochemical technique (210).

The neutral conjugate, ABA-GE, appears to be restricted to the vacuole (20, 111) and does not enter cells without first being hydrolyzed (110, 246). Conjugation of ABA to ABA-GE is irreversible and is a means by which plants sequester ABA from the active pool. There is no evidence to support the idea that conjugated ABA is a source of stress-induced free ABA (39, 114, 133, 150, 244, 245).

At the organ level, attention has focused on the epidermis, since this tissue contains the guard cells that are the targets of ABA action for stomatal closure. Early work indicated that epidermal tissue is unable to produce ABA, so that ABA would have to be transported there from the mesophyll (231). However, recent work has demonstrated that guard cells can both synthesize (28) and catabolize (63, 125) ABA. The ABA content of guard cells has been measured by several workers. Behl & Hartung (10) used compartmental efflux analysis and arrived at 600 fg per guard cell of *Valerianella locusta,* which amounts to a concentration of 0.65 mM in the cells, assuming equal distribution in all compartments. Lahr & Raschke (106) isolated protoplasts from *V. faba* and found 24–150 fg ABA per guard cell protoplast and 20–26 fg per mesophyll protoplast. Cornish & Zeevaart (28) prepared guard cells by sonication of epidermal peels and measured 0.15 and 2.64 fg ABA per guard cell pair from turgid and stressed *V. faba* leaves, respectively. Harris et al (68) dissected out cells from lyophilized turgid and stressed leaves, and measured 0.7 and 17.7 fg ABA per guard cell pair, respectively. ABA concentrations in mesophyll and guard cells were in the range of 7–13 μM (68). It is difficult to reconcile these divergent results. Stressed guard cells obtained by dissection (68) contained 6.7 times more ABA than those isolated by sonication (28). The lower number is presumably due to efflux of ABA from the epidermal strips (68). This leaves unexplained, however, the high values obtained in other experiments (10, 106) in which ABA efflux probably took place as well. Although the presence of ABA in guard cells has now been established, its physiological role in these cells is unclear in view of the finding that ABA acts on guard cells from the outside (70).

EFFECTS OF ABSCISIC ACID

Physiological Responses

STOMATAL CONTROL, WATER RELATIONS, AND PHOTOSYNTHESIS   Following the discovery that applied ABA causes rapid closure of the stomata, investigators have assumed that stress-induced ABA is the signal in the stomatal regulation of water loss. However, a decrease in leaf conductance, as a water deficit develops, often precedes a detectable increase in bulk leaf ABA (26, 41, 47, 79, 190, 191). Nevertheless, ABA could still be the causal agent of stomatal closure if the stress causes a redistribution of ABA already present in turgid leaves. Based on the physicochemical description of ABA distribution within the cell (see above), the amount in the apoplast can double upon transfer to darkness or in response to a pH change in the chloroplast stroma during stress. Since guard cells do not have plasmodesmata (239), ABA originating in the mesophyll can only arrive at the guard cells via the apoplast. Considerable evidence has been obtained recently in support of the stress-mediated ABA redistribution hypothesis. When leaves of *Valerianella* were preloaded with radiolabeled ABA, a rapid movement of radioactive ABA from the mesophyll to the epidermis was observed following dehydration (74). Likewise, an increase in apoplastic ABA was found in *Xanthium* (26) and in cotton leaves (186) prior to an increase in bulk leaf ABA. Pressure-induced dehydration of cotton leaves increased the pH (probably by inhibiting plasma membrane-bound ATPases) and ABA content of exuded sap (75). The amounts of ABA that were released from the symplast into the apoplast were estimated to be adequate for stomatal closure (26, 186). Thus, the early release of ABA into the apoplast, before any changes in bulk leaf ABA levels can be detected, appears sufficient for rapid stomatal closure.

The mesophyll is the most obvious source of ABA released into the apoplastic fluid during stress, but guard cells themselves may release ABA as well. Behl & Hartung (10) applied compartmental efflux analysis to epidermal strips of *Valerianella* in which the guard cells were the only living cells. At moderate stress (0.1 M sorbitol) one guard cell released 0.36 fmol ABA within 20 min. This loss was from the cytoplasmic compartment; the ABA content of the vacuoles did not change. Since the site of ABA action is on the outer surface of the plasmalemma, it is the apoplastic ABA that is of physiological significance, not the ABA content of the guard cells (70). It appears, therefore, that guard cells under stress can release ABA that initiates their own rapid closure (10).

Furthermore, there is evidence that ABA produced by roots in drying soil moves in the transpiration stream and accumulates at or near the guard cells, i.e. at the site of its action (27, 42, 105). In nonwatered *Commelina*, the stomata closed without any change in turgor or ABA content. However, the ABA content of the roots increased several-fold along with that of the lower epidermis of the leaves (249). Movement of ABA from the roots to the shoot was observed, even under conditions of low transpirational flux (248). These results support the hypothesis that ABA produced in stressed roots may act as a chemical signal to the shoot, initiating closure of the stomata before any changes in the water status of the leaves occur, thereby optimizing the plant's water use under conditions of restricted availability (42).

There are several reports (24, 190, 192, 235) that ABA applied via the transpiration stream to leaves affected photosynthesis both via stomatal closure and via a direct effect on carbon fixation. As a possible mechanism, inhibition of the carboxylation of ribulose-1,5-bisphosphate has been suggested (56, 235), but extractable ribulose-1,5-bisphosphate carboxylase activity from ABA-treated leaves was as high as that of control leaves (56). Raschke & Patzke (193) found that the effects of ABA on photosynthesis in *Xanthium* resulted from non-uniform distribution of stomatal pore sizes and that ABA had no direct effect on the photosynthetic machinery in this species.

GROWTH   Reports on the effects of ABA on root growth are contradictory, since exogenous ABA can inhibit root growth (144, 176, 177) as well as promote growth of this organ (144, 177). Mulkey et al (144) observed that the effect of high concentrations of ABA (0.01–1 mM) on root growth in three species was triphasic. A period of promotion, which lasted approximately 12 hr, was followed by a period of inhibition (12 hr). Last, a gradual recovery to about 80% of normal growth rate occurred after 24 hr. With low concentrations of ABA (0.1 μM), only a transient period of stimulation occurred (144). Pilet & Saugy (178) found that the effect of exogenous ABA on root growth in maize depended on the initial elongation rate. Roots with a low growth rate (0.4 mm hr^{-1}) were inhibited at a low ABA concentration (5 nM), whereas fast growing roots (0.8 mm hr^{-1}) were stimulated. At 1 μM ABA, both slow- and fast-growing roots were inhibited. By quantifying ABA in the elongation zones of single maize roots by GC-NCI-MS, a negative correlation between ABA content and growth rate was demonstrated (196, 198). However, below a level of approximately 30 pg ABA per elongation zone, the growth rate was independent of ABA content. Nevertheless, a role of endogenous ABA in root growth is questionable, since ABA content and growth rate, as modified by fluridone and temperature, varied independently (195).

ABA decreased the elongation rate in maize coleoptiles and also inhibited growth induced by auxin, fusicoccin, or acid. In all cases, this effect was due to an inhibition of cell-wall loosening (103, 104). A similar effect of ABA has been observed in light-stimulated expansion of leaves (228).

The opposite effects of ABA on the growth of roots and shoots may be advantageous for survival of plants under stress conditions. In shoots, ABA-induced growth inhibition provides the possibility of maintaining turgor pressure (103). On the other hand, a stimulation of root growth by ABA would enlarge the root system (18) and this, combined with increased hydraulic conductance (18), would increase water uptake.

ROOT GRAVITROPISM   The asymmetric redistribution of a growth-inhibiting substance produced in the root cap is thought to be the basis of gravitropism in roots. Although the nature of the growth-inhibiting compound is unknown, both indole-3-acetic acid and ABA have been proposed as the agent causing gravicurvature (50, 141). Using a maize cultivar that requires light for curvature to occur, Feldman (51) demonstrated a rapid redistribution of ABA from the root cap to the terminal 1.5 mm of the root following a brief exposure to light. A decrease in violaxanthin along with an increase in Xan also occurred during this period. Recent evidence, however, argues against a role for ABA in gravitropism. Treatment with fluridone and norflurazon reduced ABA levels significantly, while allowing curvature to occur (52, 142). The levels of Xan were unaffected by treatment with norflurazon (52), which is unexpected if Xan is derived from a carotenoid (see above). Additional evidence against a role of ABA in root gravitropism comes from experiments with the *w3, vp5,* and *vp7* mutants of maize. The roots of these mutants exhibited normal gravicurvature, while ABA was not detectable (143). The *flc* mutant of tomato also showed normal gravicurvature (238). Other approaches have also led to the conclusion that ABA cannot be the inhibitor that causes gravitropic curvature (144).

HETEROPHYLLY   Aquatic plants often produce two distinct types of leaves on the same plant. Submerged leaves are highly dissected or linear; they have undifferentiated mesophyll and an epidermis with few, if any, stomata. In contrast, aerial leaves are entire and broad; they have differentiated mesophyll and a stomatous epidermis. Interestingly, ABA treatment can induce the formation of aerial leaves on submerged shoots. Such changes in leaf morphology due to ABA have been reported for a fern (123), a monocot (5), and several dicots (43, 95, 96, 140, 243). Leaves produced under ABA treatment were similar to aerial leaves not only in shape but also in stomatal density (5, 95, 123, 243) and anatomy (242). A high external solute concentration also caused aerial-type leaf formation in submerged *Callitriche heterophylla* (43) and *Hippuris vulgaris* (95), suggesting that the effect of low water potential was mediated by stress-induced ABA. Likewise, it is possible that in emerging leaves a water deficiency causes an increase in the ABA content which in turn results in aerial leaf characteristics. In *Proserpinaca palustris* both ABA and GA play a role in the control of leaf morphology. To explain the different leaf forms produced in response to these hormones, it was suggested that GA is produced under long days, while ABA would be produced in aerial leaves that are under short-day conditions (96). Clearly, further work is required to determine the relationships among environmental factors, endogenous hormone levels, and heterophyllous development in aquatic plants.

DORMANCY   Walton (231) concluded in 1980 that "a role for ABA on the induction and maintenance of bud and seed dormancy has been neither unequivocally demonstrated nor disproven." This is still true with respect to bud dormancy in woody species. Studies with seedlings of *Salix* spp. have shown that the day length does not regulate cessation of growth by affecting the ABA levels in the plants (8, 93). It is possible, however, that the sensitivity to the hormone is altered by short-day conditions (8), but factors other than ABA are probably also involved in the control of dormancy.

In the case of immature seeds, several lines of evidence indicate that ABA prevents precocious germination of the developing embryo (vivipary). First, when immature soybean embryos were excised and cultured in vitro, a high percentage germination was observed only when the ABA content was less than 4 μg g^{-1} fresh weight. The ABA content of embryos in situ did not decline to that level until the onset of dehydration (3). The high ABA content of embryos may result partially from in situ ABA synthesis, but an important factor appears to be the pH differential between seed coats and embryos (9, 77). In cotton the seed coat was up to 1.4 pH units more acidic than the developing embryo, which resulted in diffusion of ABA from the seed coat to the embryo (78). Second, exogenous ABA prevented precocious germination of immature embryos of several species when cultured in vitro, e.g. in cotton (77), maize (152), rapeseed (55), soybean (2, 49), and wheat (226). ABA not only prevented embryo germination, but it often caused embryo growth and accumulation of storage proteins as well (see below). However, as the embryos matured and endogenous ABA levels decreased, sensitivity to exogenous ABA declined. Third, ABA-deficient mutants show vivipary under suitable conditions (65, 97, 100). In reciprocal crosses between wild-type and ABA-deficient mutants of *Arabidopsis,* it was demonstrated that the ABA produced in the embryo controls seed dormancy; maternal ABA had only a minor role in this process (97). Precocious germination could also be induced in wild-type maize by treatment of young kernels with fluridone (60). Taken together, these data clearly point to a role of seed-produced ABA in preventing precocious germination during the early stages of embryogenesis.

ADAPTATION TO STRESS   ABA has been called a stress hormone, since it enhances adaptation to various stresses. Freezing tolerance could be increased by ABA both in plants (21, 107) and in cell cultures (23, 171, 194). This increased cold tolerance was not induced in the presence of cycloheximide, which indicates that synthesis of new proteins is involved with stress adaptation (21, 199, 199a). Furthermore, during cold hardening, plants showed a transient increase in ABA content (21), suggesting that ABA may be an endogenous regulator of adaptation to cold stress. ABA also accelerated adaptation of cultured cells to salt (108), at least in part by osmotic adjustment (109). Associated with ABA-mediated salt adaptation was the synthesis of a 26-kD protein (206). It was suggested that this protein plays a role in salt adaptation. ABA accumulation in stressed leaves was not required for accumulation of another stress metabolite, proline (211).

RESPONSE MUTANTS   Several mutants have been isolated that exhibit a decreased response to exogenous ABA. A viviparous mutant of maize, *vp1,* has green seedlings and normal ABA and carotenoid levels. The embryos of this mutant are insensitive to ABA (59, 200). It has, therefore, been classified as a response mutant. The *vp1* mutant shows a pleiotropic phenotype and, in addition to being viviparous, has decreased levels of activities of flavonoid enzymes (45). However, if *vp1* is a response mutant, then expression of this mutation is limited to embryogenesis. Whole plants of *vp1* have a normal stomatal response to ABA, are not wilty (152), and are capable of producing anthocyanins (45). Five *Arabidopsis* mutants selected on 10 μM ABA were 5–20 times less sensitive to exogenous ABA than wild type with regard to seed germination and seedling growth (102). The mutations were at three different loci and termed *abi-1, abi-2,* and *abi-3.* Phenotypically, these mutants resemble the *aba* mutant lines in showing reduced seed dormancy, yet they contain ABA levels similar to wild type. One mutant, *abi-3,* did not exhibit altered rates of transpiration while the others did. It was suggested (102) that the lesions at *abi-1* and *abi-2* are involved in a common step in the perception of ABA (perhaps at, or close to, the binding site). In contrast, *abi-3* would be seed specific, like the *vp1* mutant in maize.

Biochemical Responses

ABA elicits two types of responses in plants. The closure of stomata is an example of a rapid (<5 min) response that can be induced by *S*-ABA, while the *R*-enantiomer is almost without effect (87, 134, 191, 232). In this case, the site of action is the plasmalemma (70, 191). On the other hand, in the slow responses (>30 min), which appear to involve RNA and protein synthesis, *S*- and *R*-ABA are equally effective. Since the fast and slow responses have different molecular requirements, Milborrow (134) concluded that the receptor sites must be different. and that the fast response to ABA cannot be a prerequisite for the slow response. Evidence for ABA-binding proteins has so far only been obtained with protoplasts from guard cells (87). However, in an extensive study with a variety of ABA-sensitive tissues, Smart et al (207) failed to obtain any evidence for ABA-receptor proteins by a number of reversible binding assays.

Not all genetic data are consistent with the hypothesis that, depending on the type of response elicited, there are at least two different ABA receptor sites. Although observations with the mutants *vp1* in maize and *abi-3* in *Arabidopsis* support this view, results with the *abi-1* and *abi-2* mutants of *Arabidopsis* (102) are at variance with this hypothesis.

GUARD CELLS   Guard cells showed equal closure when exposed to ABA solutions over the pH range 5.0–8.0, while no significant uptake of the hormone took place at pH 8.0. This indicates that the sites of ABA action are on the outer surface of the plasmalemma (70). ABA prevents stomatal opening by rapidly blocking H^+ extrusion and K^+ influx (191, 204), and it initiates closure by the rapid release of osmotica, in particular K^+ (128), Cl^-, and malate (191). The latter phenomenon can also be observed in the shrinking of guard cell protoplasts (58).

Hornberg & Weiler (87) used photoaffinity labeling to cross-link radiolabeled ABA to putative binding sites at the plasmalemma of *V. faba* guard cell protoplasts and obtained high-affinity (K_D = 3–4 nM) binding sites specific for guard cell protoplasts. There were ~2 × 10^6 binding sites per protoplast as compared to ~300 K^+ channels as determined by the patch-clamp technique (191). Three proteins that were covalently linked to ABA were separated by gel electrophoresis. One protein had the highest affinity for dissociated ABA, while the other two bound preferentially to the protonated form (87). Could one of these proteins be a modulator of the H^+-ATPase, which is inhibited by ABA? Further work on the nature and function of these proteins in guard cell physiology is anxiously awaited.

BARLEY ALEURONE CELLS   Among the most studied effects of hormones on gene expression is the GA-induced production of hydrolytic enzymes and its inhibition by ABA in barley aleurone layers. When ABA is applied in 25-fold excess of GA, the synthesis of α-amylase, protease, β-glucanase, and ribonuclease is suppressed (reviewed in 91). The best characterized enzyme produced by barley aleurone layers is α-amylase. It consists of two isozyme families (low- and high-pI groups) which have many different characteristics and are encoded by two structural genes on different chromosomes (92). Therefore, using total enzyme activity as a marker for α-amylase synthesis may obscure changes in one isozyme group over the other. ABA inhibited the synthesis of both groups of enzymes when added at the same time as GA, but when added 12 hr after GA the high-pI group was more inhibited than the low-pI group (159). Evidence for a role of ABA in regulating the expression of α-amylase activity at the transcriptional and possibly at the translational level has been reviewed (91).

In addition to its anti-GA action, ABA also causes the production of a large number of the so-called ABA-inducible polypeptides in aleurone layers (82, 119). The accumulation of all ABA-induced polypeptides was inhibited by treatment with GA (82), as was the production of an α-amylase inhibitor (145, 146).

Exogenous PA is as active as ABA in preventing the accumulation of α-amylase (232), which raises the question of whether ABA is active by virtue of its conversion to PA. However, PA is probably not the sole agent preventing α-amylase synthesis; R-ABA, which is not converted to PA in leaves of several species (17, 30, 132), is just as active as S-ABA (232).

There are conflicting reports on induction of ABA-specific proteins by PA. Lin & Ho (119) presented data indicating that PA was unable to induce these proteins. In disagreement with this observation is the work of Ariffin et al (P.M. Chandler, personal communication), showing that the effects of S-ABA, R-ABA, and PA on the ABA-inducible polypeptides are identical at the mRNA and protein levels. Thus, ABA and PA appear to be equally active in their anti-GA actions as well as in their protein synthesis stimulating activities.

DEVELOPING EMBRYOS   ABA not only prevents premature germination during embryogenesis, but it has also been implicated in the regulation of storage-protein synthesis. In embryos of rapeseed (55, 57), soybean (49), and wheat (187, 188, 240) cultured in vitro, exogenous ABA promoted the synthesis of embryo-specific proteins and mRNAs. In rapeseed, a high concentration of sorbitol produced an effect similar to that of ABA, an effect that did not result from increased ABA synthesis in the embryos (53). Treatment with fluridone inhibited accumulation of β-conglycinin in cultured soybean cotyledons (19) and of several storage proteins in embryos of rapeseed (57). The inhibition could be overcome by applied ABA and, in rapeseed, also by a low osmotic potential. In the latter case, it was suggested that ABA and osmotic stress acted independently (57). Thus, it would appear from these results that there may be other factors besides ABA that regulate the synthesis of embryo-specific proteins. Results obtained with ABA-deficient mutants also cast doubt on a pivotal role for ABA in synthesis of embryo-specific proteins. Developing seeds of the *sit* mutant, for example, had a much reduced ABA level as compared to the wild type, yet neither dry weight nor storage proteins were affected (65). In *Arabidopsis,* the *aba* and *abi* mutants showed no differences from the wild type in mRNA levels for the storage protein cruciferin during embryogenesis (54, 172).

CONCLUDING REMARKS

In fungi, it is apparent that ABA is formed from ionylidene derivatives. However, the pathway(s) via which ABA is synthesized in higher plants remains to be elucidated. A considerable amount of evidence indicates that ABA is an apo-carotenoid, but this has not been proven unequivocally, and it remains possible that there is more than one pathway. The step(s) in the pathway stimulated by a water deficit also remains unknown. Turgor sensing presumably resides in the cell wall–plasmalemma interaction, but how loss of turgor is translated into biochemical events is a challenging problem. ABA functions as an endogenous antitranspirant, but the physiological role of the large quantities of ABA in water-stressed leaves is obscure in view of the very small proportion that is needed for stomatal closure. Perhaps the "excess" ABA has a role in ameliorating the deleterious effects of stress. When a stressed leaf is rehydrated, its ABA content returns to a level typical of unstressed leaves with a concomitant accumulation of ABA catabolites. With the exception of PA, none of these appears to have biological activity (125, 232).

Studies with ABA-deficient mutants have been instrumental in determining the role of ABA (or lack thereof) in various physiological processes. Unfortunately, there are currently no ABA mutants available in such plants as cotton, soybean, and wheat in which the effects of ABA on embryo development and protein formation have been studied extensively.

There is a wealth of information on the effects of ABA on growth and development in plants. Since so little is known about many of the basic biochemical processes in plant tissues, it is difficult to assign a specific role to ABA in any process. Originally, ABA was considered a general inhibitor of RNA and protein synthesis, but it is now evident that in certain tissues ABA can induce the formation of its own set of proteins. The physiological responses to ABA range from very rapid to long-term, implying that ABA has different modes of action and different receptors. Of the rapid responses, further progress can be expected on the mechanism by which ABA regulates solute transport in guard cells. Of the slow responses, the study of accumulation of proteins in specialized tissues, as modulated by ABA, is amenable to molecular approaches. It is hoped that the questions raised in this review will inspire investigators to pursue some of the intriguing problems that remain in the areas of the biochemistry and physiology of ABA.

ACKNOWLEDGMENTS

We thank our colleagues for providing us with preprints and unpublished results during the preparation of this manuscript, and others for their reprints. The research in the laboratory of J.A.D.Z. discussed in this review was supported by the National Science Foundation through grant PCM 83-14321 and by the United States Department of Energy under Contract DE-AC02-76ER01338.

Literature Cited

1. Ackerson, R. C. 1982. Synthesis and movement of abscisic acid in water-stressed cotton leaves. *Plant Physiol.* 69:609–13
2. Ackerson, R. C. 1984. Regulation of soybean embryogenesis by abscisic acid. *J. Exp. Bot.* 35:403–13
3. Ackerson, R. C. 1984. Abscisic acid and precocious germination in soybeans. *J. Exp. Bot.* 35:414–21
4. Addicott, F. T., ed. 1983. *Abscisic Acid.* New York: Praeger. 607 pp.
5. Anderson, L. W. J. 1982. Effects of abscisic acid on growth and leaf development in American pondweed (*Potamogeton nodosus* Poir.) *Aquat. Bot.* 13:29–44
6. Assante, G., Merlini, L., Nasini, G. 1977. (+)-Abscisic acid, a metabolite of the fungus *Cercospora rosicola*. *Experientia* 33:1556–57
7. Astle, M. C., Rubery, P. H. 1987. Carrier-mediated ABA uptake by suspension-cultured *Phaseolus coccineus* L. cells: stereospecificity and inhibition by ionones and ABA esters. *J. Exp. Bot.* 38:150–63
8. Barros, R. S., Neill, S. J. 1986. Periodicity of response to abscisic acid in lateral buds of willow (*Salix viminalis* L.). *Planta* 168:530–35
9. Barthe, P., Boulon, B., Gendraud, M., Le Page-Degivry, M.-T. 1986. Intracellular pH and catabolism: two factors determining the level of abscisic acid in embryos of *Phaseolus vulgaris* during maturation. *Physiol. Vég.* 24:453–61
10. Behl, R., Hartung, W. 1986. Movement and compartmentation of abscisic acid in guard cells of *Valerianella locusta:* Effects of osmotic stress, external H^+-concentration and fusicoccin. *Planta* 168:360–68
11. Behl, R., Jeschke, W. D., Hartung, W. 1981. A compartmental analysis of abscisic acid in roots of *Hordeum distichon*. *J. Exp. Bot.* 32:889–97
12. Bennett, R. D., Norman, S. M., Maier, V. P. 1981. Biosynthesis of abscisic acid from [1,2-$^{13}C_2$]acetate in *Cercospora rosicola*. *Phytochemistry* 20:2343–44
13. Bennett, R. D., Norman, S. M., Maier, V. P. 1984. Biosynthesis of abscisic acid from farnesol derivatives in *Cercospora rosicola*. *Phytochemistry* 23:1913–15
14. Boyer, G. L., Milborrow, B. V., Wareing, P. F., Zeevaart, J. A. D. 1985. The nomenclature of abscisic acid and its metabolites. In *Plant Growth Substances 1985*, ed. M. Bopp, pp. 99–100. Berlin: Springer. 420 pp.
15. Boyer, G. L., Zeevaart, J. A. D. 1982. Metabolism of abscisic acid in *Xanthium strumarium*. *Plant Physiol. Suppl.* 69:77
16. Boyer, G. L., Zeevaart, J. A. D. 1982. Isolation and quantitation of β-D-glucopyranosyl abscisate from leaves of

Xanthium and spinach. *Plant Physiol.* 70:227–31

17. Boyer, G. L., Zeevaart, J. A. D. 1986. 7'-Hydroxy (−)-*R*-abscisic acid: a metabolite of feeding (−)-*R*-abscisic acid to *Xanthium strumarium*. *Phytochemistry* 25:1103–5

18. Bradford, K. J. 1983. Water relations and growth of the *flacca* tomato mutant in relation to abscisic acid. *Plant Physiol.* 72:251–55

19. Bray, E. A., Beachy, R. N. 1985. Regulation by ABA of β-conglycinin expression in cultured developing soybean cotyledons. *Plant Physiol.* 79:746–50

20. Bray, E. A., Zeevaart, J. A. D. 1985. The compartmentation of abscisic acid and β-D-glucopyranosyl abscisate in mesophyll cells. *Plant Physiol.* 79:719–22

21. Chen, H. H., Li, P. H., Brenner, M. L. 1983. Involvement of abscisic acid in potato cold acclimation. *Plant Physiol.* 71:362–65

22. Chen, S. C., Wang, L. C. H., Westly, J. C. 1988. Analysis of abscisic acid in the brain of rodents and ruminants. *Agric. Biol. Chem.* In press

23. Chen, T. H. H., Gusta, L. V. 1983. Abscisic acid-induced freezing resistance in cultured plant cells. *Plant Physiol.* 73:71–75

24. Cornic, G., Miginiac, E. 1983. Nonstomatal inhibition of net CO_2 uptake by (±) abscisic acid in *Pharbitis nil*. *Plant Physiol.* 73:529–33

25. Cornish, K., Zeevaart, J. A. D. 1984. Abscisic acid metabolism in relation to water stress and leaf age in *Xanthium strumarium*. *Plant Physiol.* 76:1029–35

26. Cornish, K., Zeevaart, J. A. D. 1985. Movement of abscisic acid into the apoplast in response to water stress in *Xanthium strumarium* L. *Plant Physiol.* 78:623–26

27. Cornish, K., Zeevaart, J. A. D. 1985. Abscisic acid accumulation by roots of *Xanthium strumarium* L. and *Lycopersicon esculentum* Mill. in relation to water stress. *Plant Physiol.* 79:653–58

28. Cornish, K., Zeevaart, J. A. D. 1986. Abscisic acid accumulation by *in situ* and isolated guard cells of *Pisum sativum* L. and *Vicia faba* L. in relation to water stress. *Plant Physiol.* 81:1017–21

29. Cowan, A. K., Railton, I. D. 1986. Chloroplasts and the biosynthesis and catabolism of abscisic acid. *J. Plant Growth Regul.* 4:211–24

30. Cowan, A. K., Railton, I. D. 1987. The catabolism of (±)-abscisic acid by excised leaves of *Hordeum vulgare* L. cv Dyan and its modification by chemical

and environmental factors. *Plant Physiol.* 84:157–63

31. Cowan, A. K., Railton, I. D. 1987. Cytokinins and ancymidol inhibit abscisic acid biosynthesis in *Persea gratissima*. *J. Plant Physiol.* 130:273–77

32. Cowan, I. R., Raven, J. A., Hartung, W., Farquhar, G. D. 1982. A possible role for abscisic acid in coupling stomatal conductance and photosynthetic carbon metabolism in leaves. *Aust. J. Plant Physiol.* 9:489–98

33. Creelman, R. A. 1986. *Stress-induced abscisic acid biosynthesis in higher plants*. PhD thesis. Michigan State Univ., East Lansing. 177 pp.

34. Creelman, R. A., Gage, D. A., Stults, J. T., Zeevaart, J. A. D. 1987. Abscisic acid biosynthesis in leaves and roots of *Xanthium strumarium*. *Plant Physiol.* 85:726–32

35. Creelman, R. A., Zeevaart, J. A. D. 1984. Incorporation of oxygen into abscisic acid and phaseic acid from molecular oxygen. *Plant Physiol.* 75:166–169

36. Creelman, R. A., Zeevaart, J. A. D. 1985. Abscisic acid accumulation in spinach leaf slices in the presence of penetrating and non-penetrating solutes. *Plant Physiol.* 77:25–28

37. Creelman, R. A., Zeevaart, J. A. D. 1987. Separation and purification of abscisic acid and its catabolites by high performance liquid chromatography. In *Modern Methods of Plant Analysis*, ed. H. F. Linskens, J. F. Jackson, 5:39–51. Berlin: Springer. 248 pp.

38. Daie, J., Wyse, R. 1982. Adaptation of the enzyme-linked immunosorbent assay (ELISA) to the quantitative analysis of abscisic acid. *Anal. Biochem.* 119:365–71

39. Dathe, W., Schneider, G., Sembdner, G. 1984. Gradient of abscisic acid and its β-D-glucopyranosyl ester in wood and bark of dormant branches of birch (*Betula pubescens* EHRH.). *Biochem. Physiol. Pflanz.* 179:109–14

40. Dathe, W., Sembdner, G. 1982. Isolation of 4'-dihydroabscisic acid from immature seeds of *Vicia faba*. *Phytochemistry* 21:1798–99

41. Davies, W. J., Metcalfe, J., Lodge, T. A., da Costa, A. R. 1986. Plant growth substances and the regulation of growth under drought. *Aust. J. Plant Physiol.* 13:105–25

42. Davies, W. J., Metcalfe, J. C., Schurr, U., Taylor, G., Zhang, J. 1987. Hormones as chemical signals involved in root to shoot communication of effects of changes in the soil environment. In

Hormone Action in Plant Development. A Critical Appraisal, ed. G. V. Hoad, J. R. Lenton, M. B. Jackson, R. K. Atkin, pp. 201–16. London: Butterworths. 315 pp.

43. Deschamp, P. A., Cooke, T. J. 1984. Causal mechanisms of leaf dimorphism in the aquatic angiosperm *Callitriche heterophylla*. *Am. J. Bot.* 71:319–29

44. DeVit, M. 1986. *Studies with xanthoxin in Phaseolus vulgaris*. MSc thesis. State Univ. New York, Syracuse. 102 pp.

45. Dooner, H. K. 1985. *Viviparous-1* mutation in maize conditions pleiotropic enzyme deficiencies in the aleurone. *Plant Physiol.* 77:486–88

46. Dörffling, K., Petersen, W., Sprecher, E., Urbasch, I., Hanssen, H.-P. 1984. Abscisic acid in phytopathogenic fungi of the genera *Botrytis, Ceratocystis, Fusarium,* and *Rhizoctonia. Z. Naturforsch. Teil C* 39:683–84

47. Dörffling, K., Tietz, D. 1985. Abscisic acid in leaf epidermis of *Commelina communis* L.: Distribution and correlation with stomatal closure. *J. Plant Physiol.* 117:297–305

48. Douce, R., Joyard, J. 1979. Structure and function of the plastid envelope. *Adv. Bot. Res.* 7:1–116

49. Eisenberg, A. J., Mascarenhas, J. P. 1985. Abscisic acid and the regulation of synthesis of specific seed proteins and their messenger RNAs during culture of soybean embryos. *Planta* 166:505–14

50. Feldman, L. J. 1985. Root gravitropism. *Physiol. Plant.* 65:341–44

51. Feldman, L. J., Arroyave, N. J., Sun, P. S. 1985. Abscisic acid, xanthoxin and violaxanthin in the caps of gravistimulated maize roots. *Planta* 166:483–89

52. Feldman, L. J., Sun, P. S. 1986. Effects of norflurazon, an inhibitor of carotenogenesis, on abscisic acid and xanthoxin in the caps of gravistimulated maize roots. *Physiol. Plant.* 67:472–76

53. Finkelstein, R. R., Crouch, M. L. 1986. Rapeseed embryo development in culture on high osmoticum is similar to that in seeds. *Plant Physiol.* 81:907–12

54. Finkelstein, R. R., Somerville, C. 1987. Analysis of the mechanism of abscisic acid action using ABA-sensitive mutants of *Arabidopsis*. In *3rd Int. Meet. Arabidopsis, Abstr. Lectures and Posters,* p. 138. East Lansing: Michigan State Univ. 149 pp.

55. Finkelstein, R. R., Tenbarge, K. M., Shumway, J. E., Crouch, M. L. 1985. Role of ABA in maturation of rapeseed embryos. *Plant Physiol.* 78:630–36

56. Fischer, E., Raschke, K., Stitt, M. 1986. Effects of abscisic acid on photosynthesis in whole leaves: changes in CO_2 assimilation, levels of carbon-reduction-cycle intermediates, and activity of ribulose-1,5-bisphosphate carboxylase. *Planta* 169:536–45

57. Fischer, W., Bergfeld, R., Schopfer, P. 1987. Induction of storage protein synthesis in embryos of mature plant seeds. *Naturwissenschaften* 74:86–88

58. Fitzsimons, P. J., Weyers, J. D. B. 1987. Responses of *Commelina communis* L. guard cell protoplasts to abscisic acid. *J. Exp. Bot.* 38:992–1001

59. Fong, F., Koehler, D. E., Smith, J. D. 1983. Fluridone induction of vivipary during maize seed development. In *3rd Int. Symp. Pre-Harvest Sprouting in Cereals,* ed. J. E. Kruger, D. E. LaBerge, pp. 188–96. Boulder: Westview Press

60. Fong, F., Smith, J. D., Koehler, D. E. 1983. Early events in maize seed development. *Plant Physiol.* 73:899–901

61. Gamble, P. E., Mullet, J. E. 1986. Inhibition of carotenoid accumulation and abscisic acid biosynthesis in fluridone-treated dark-grown barley. *Eur. J. Biochem.* 160:117–21

62. Goodwin, T. W. 1980. *The Biochemistry of the Carotenoids,* 1:126. London: Chapman & Hall. 377 pp. 2nd ed.

63. Grantz, D. A., Ho, T.-H. D., Uknes, S. J., Cheeseman, J. M., Boyer, J. S. 1985. Metabolism of abscisic acid in guard cells, of *Vicia faba* L. and *Commelina communis* L. *Plant Physiol.* 78:51–56

64. Gray, R. T., Mallaby, R., Ryback, G., Williams, V. P. 1974. Mass spectra of methyl abscisate and isotopically labelled analogues. *J. Chem. Soc. Perkins Trans.* 2:919–24

65. Groot, S. P. C. 1987. *Hormonal regulation of seed development and germination in tomato. Studies on abscisic acid- and gibberellin-deficient mutants.* PhD thesis. Agric. Univ. Wageningen. 107 pp.

66. Guerrero, F., Mullet, J. E. 1986. Increased abscisic acid biosynthesis during plant dehydration requires transcription. *Plant Physiol.* 80:588–91

67. Harris, M. J., Dugger, W. M. 1986. The occurrence of abscisic acid and abscisyl-β-D-glucopyranoside in developing and mature citrus fruit as determined by enzyme immunoassay. *Plant Physiol.* 82:339–45

68. Harris, M. J., Outlaw, W. H., Mertens, R., Weiler, E. W. 1988. Water-stress-induced changes in the abscisic acid content of guard cells and other cells of *Vicia faba* L. leaves, as determined by

enzyme-amplified immunoassay. *Proc. Natl. Acad. Sci. USA.* In press
69. Harrison, M. A., Walton, D. C. 1975. Abscisic acid metabolism in water-stressed bean leaves. *Plant Physiol.* 56:250–54
70. Hartung, W. 1983. The site of action of abscisic acid at the guard cell plasmalemma of *Valerianella locusta. Plant Cell Environ.* 6:427–28
71. Hartung, W., Gimmler, H., Heilmann, B. 1982. The compartmentation of abscisic acid (ABA), of ABA-biosynthesis, ABA-metabolism and ABA-conjugation. In *Plant Growth Substances 1982,* ed. P. F. Wareing, pp. 325–33. London: Academic. 683 pp.
72. Hartung, W., Gimmler, H., Heilmann, B., Kaiser, G. 1980. The site of abscisic acid metabolism in mesophyll cells of *Spinacia oleracea. Plant Sci. Lett.* 18: 359–64
73. Hartung, W., Heilmann, B., Gimmler, H. 1981. Do chloroplasts play a role in abscisic acid synthesis? *Plant Sci. Lett.* 22:235–42
74. Hartung, W., Kaiser, W. M., Burschka, C. 1983. Release of abscisic acid from leaf strips under osmotic stress. *Z. Pflanzenphysiol.* 112:131–38
75. Hartung, W., Radin, J. W., Hendrix, D. L. 1988. Abscisic acid movement into the apoplastic solution of water-stressed cotton leaves: role of apoplastic pH. *Plant Physiol.* In press
76. Heilmann, B., Hartung, W., Gimmler, H. 1980. The distribution of abscisic acid between chloroplasts and cytoplasm of leaf cells and the permeability of the chloroplast envelope for abscisic acid. *Z. Pflanzenphysiol.* 97:67–78
77. Hendrix, D. L., Radin, J. W. 1984. Seed development in cotton: feasibility of a hormonal role for abscisic acid in controlling vivipary. *J. Plant Physiol.* 117:211–21
78. Hendrix, D. L., Radin, J. W., Nieman, R. A. 1987. Intracellular pH of cotton embryos and seed coats during fruit development determined by ^{31}P-nuclear magnetic resonance spectroscopy. *Plant Physiol.* 85:588–91
79. Henson, I. E. 1981. Changes in abscisic acid content during stomatal closure in pearl millet (*Pennisetum americanum* [L.] Leeke. *Plant Sci. Lett.* 21:121–27
80. Henson, I. E. 1984. Inhibition of abscisic acid accumulation in seedling shoots of pearl millet (*Pennisetum americanum* L.) following induction of chlorosis by norflurazon. *Z. Pflanzenphysiol.* 114: 35–43
81. Henson, I. E. 1985. Dependence of

abscisic acid accumulation in leaves of pearl millet (*Pennisetum americanum* [L.] Leeke) on rate of development of water stress. *J. Exp. Bot.* 36:1232–39
82. Higgins, T. J. V., Jacobsen, J. V., Zwar, J. A. 1982. Gibberellic acid and abscisic acid modulate protein synthesis and mRNA levels in barley aleurone layers. *Plant Mol. Biol.* 1:191–215
83. Hirai, N. 1986. Abscisic acid. In *Chemistry of Plant Hormones,* ed. N. Takahashi, pp. 201–48. Boca Raton: CRC Press. 277 pp.
84. Hirai, N., Koshimizu, K. 1983. A new conjugate of dihydrophaseic acid from avocado fruit. *Agric. Biol. Chem.* 47: 365–71
85. Hirai, N., Okamoto, M., Koshimizu, K. 1986. The 1',4'-*trans*-diol of abscisic acid, a possible precursor of abscisic acid in *Botrytis cinerea. Phytochemistry.* 25:1865–68
86. Horgan, R., Neill, S. J., Walton, D. C., Griffin, D. 1983. Biosynthesis of abscisic acid. *Biochem. Soc. Trans.* 11:553–57
87. Hornberg, C., Weiler, E. W. 1984. High-affinity binding sites for abscisic acid on the plasmalemma of *Vicia faba* guard cells. *Nature* 310:321–24
88. Ichimura, M., Oritani, T., Yamashita, K. 1983. The metabolism of (2Z, 4E)-α-ionylideneacetic acid in *Cercospora cruenta,* a fungus producing (+)-abscisic acid. *Agric. Biol. Chem.* 47:1895–1900
89. Idetek Inc. 1986. Phytodetek® monoclonal antibody plate preparation and immunoassay procedure. Product Bulletin 3/86 MSD-027. San Bruno: Phytodetek. 2 pp.
90. Jackson, M. B., Hall, K. C. 1987. Early stomatal closure in waterlogged pea plants is mediated by abscisic acid in the absence of foliar water deficits. *Plant Cell Environ.* 10:121–30
91. Jacobsen, J. V. 1983. Regulation of protein synthesis in aleurone cells by gibberellin and abscisic acid. In *The Biochemistry and Physiology of Gibberellins,* ed. A. Crozier, 2:159–87. New York: Praeger. 452 pp.
92. Jacobsen, J. V., Higgins, T. J. V. 1982. Characterization of the α-amylases synthesized by aleurone layers of Himalaya barley in response to gibberellic acid *Plant Physiol.* 70:1647–53
93. Johansen, L. G., Oden, P.-C., Junttila, O. 1986. Abscisic acid and cessation of apical growth in *Salix pentandra. Physiol. Plant.* 66:409–12
94. Kaiser, G., Weiler, E. W., Hartung, W. 1985. The intracellular distribution of

abscisic acid in mesophyll cells—the role of the vacuole. *J. Plant Physiol.* 119:237–45

95. Kane, M. E., Albert, L. S. 1987. Abscisic acid induces aerial leaf morphology and vasculature in submerged *Hippuris vulgaris* L. *Aquat. Bot.* 28:81–88

96. Kane, M. E., Albert, L. S. 1987. Integrative regulation of leaf morphogenesis by gibberellic acid and abscisic acids in the aquatic angiosperm *Proserpinaca palustris* L. *Aquat. Bot.* 28:89–96

97. Karssen, C. M., Brinkhorst-van der Swan, D. L. C., Breekland, A. E., Koornneef, M. 1983. Induction of dormancy during seed development by endogenous abscisic acid: studies on abscisic acid deficient genotypes of *Arabidopsis thaliana* (L.) Heynh. *Planta* 157:158–65

98. Karssen, C. M., Groot, S. P. C., Koornneef, M. 1987. Hormone mutants and seed dormancy in *Arabidopsis* and tomato. In *Developmental Mutants in Higher Plants*, ed. H. Thomas, D. Grierson, Soc. Exp. Biol. Semin. Ser. 32:119–33. Cambridge: Cambridge Univ. Press. 288 pp.

99. Knox, J. P., Galfre, G. 1986. Use of monoclonal antibodies to separate the enantiomers of abscisic acid. *Anal. Biochem.* 155:92–94

100. Koornneef, M. 1986. Genetic aspects of abscisic acid. In *A Genetic Approach to Plant Biochemistry*, ed. A. D. Blonstein, P. J. King, pp. 35–54. Vienna: Springer. 291 pp.

101. Koornneef, M., Jorna, M. L., Brinkhorst-van der Swan, D. L. C., Karssen, C. M. 1982. The isolation of abscisic acid (ABA)-deficient mutants by selection of induced revertants in nongerminating gibberellin-sensitive lines of *Arabidopsis thaliana* (L.) Heynh. *Theor. Appl. Genet.* 61:385–93

102. Koornneef, M., Reuling, G., Karssen, C. M. 1984. The isolation and characterization of abscisic acid-insensitive mutants of *Arabidopsis thaliana*. *Physiol. Plant.* 61:377–83

103. Kutschera, U., Schopfer, P. 1986. Effect of auxin and abscisic acid on cell wall extensibility in maize coleoptiles. *Planta* 167:527–35

104. Kutschera, U., Schopfer, P. 1986. Invivo measurement of cell-wall extensibility in maize coleoptiles: effects of auxin and abscisic acid. *Planta* 169:437–42

105. Lachno, D. R., Baker, D. A. 1986. Stress induction of abscisic acid in maize roots. *Physiol. Plant.* 68:215–21

106. Lahr, W., Raschke, K. 1988. Abscisic-acid contents and concentrations in protoplasts from guard cells and mesophyll cells of *Vicia faba* L. *Planta.* In press

107. Lalk, I., Dörffling, K. 1985. Hardening, abscisic acid, proline and freezing resistance in two winter wheat varieties. *Physiol. Plant.* 63:287–92

108. LaRosa, P. C., Handa, A. K., Hasegawa, P. M., Bressan, R. A. 1985. Abscisic acid accelerates adaptation of cultured tobacco cells to salt. *Plant Physiol.* 79:138–42

109. LaRosa, P. C., Hasegawa, P. M., Rhodes, D., Clithero, J. M., Watad, A.-E. A., Bressan, R. A. 1987. Abscisic acid stimulated osmotic adjustment and its involvement in adaptation of tobacco cells to NaCl. *Plant Physiol.* 85:174–81

110. Lehmann, H. 1983. Time course of the biotransformation of abscisic acid and *trans*-abscisic acid conjugates in cell suspension cultures of *Lycopersicon peruvianum*. *Biochem. Physiol. Pflanz.* 178:21–27

111. Lehmann, H., Glund, K. 1986. Abscisic acid metabolism—vacuolar/extravacuolar distribution of metabolites. *Planta* 168:559–62

112. Lehmann, H., Preiss, A., Schmidt, J. 1983. A novel abscisic acid metabolite from cell suspension cultures of *Nigella damascena*. *Phytochemistry* 22:1277–78

113. Lehmann, H., Schütte, H. R. 1976. Biochemistry of phytoeffectors. 9. The metabolism of α-ionylideneacetic acids in *Hordeum distichon*. *Biochem. Physiol. Pflanz.* 169:55–61

114. Lehmann, H., Schütte, H. R. 1984. Abscisic acid metabolism in intact wheat seedlings under normal and stress conditions. *J. Plant Physiol.* 117:201–9

115. Lehmann, H., Schwenen, L. 1987. Nigellic acid—an endogenous abscisic acid metabolite from *Vicia faba* leaves. *Phytochemistry.* In press

116. Le Page-Degivry, M. Th., Bidard, J. N., Rouvier, E., Bulard, C., Lazdunski, M. 1986. Presence of abscisic acid, a phytohormone, in the mammalian brain. *Proc. Natl. Acad. Sci. USA* 83:1155–58

117. Leroux, B., Maldiney, R., Miginiac, E., Sossountzov, L., Sotta, B. 1985. Comparative quantitation of abscisic acid in plant extracts by gas-liquid chromatography and an enzyme-linked immunosorbent assay using the avidin-biotin system. *Planta* 166:524–29

118. Li, Y., Walton, D. C. 1987. Xantho-

phylls and abscisic acid biosynthesis in water-stressed bean leaves. *Plant Physiol.* 85:910–15

119. Lin. L.-S., Ho, T.-H. D. 1986. Mode of action of abscisic acid in barley aleurone layers. *Plant Physiol.* 82:289–97

120. Linforth, R. S. T., Bowman, W. R., Griffin, D. A., Hedden, P., Marples, B. A., Taylor, I. B. 1987. 2,7-Dimethyl-octa-2,4-dienedioic acid, a possible by-product of abscisic acid biosynthesis in the tomato. *Phytochemistry* 26:1631–34

121. Linforth, R. S. T., Bowman, W. R., Griffin, D. A., Marples, B. A., Taylor, I. B. 1987. 2-*trans*-ABA alcohol accumulation in the wilty tomato mutants *flacca* and *sitiens*. *Plant Cell Environ.* 10:599–606

122. Linforth, R. S. T., Taylor, I. B., Hedden, P. 1987. Abscisic and C_{10} dicarboxylic acids in wilty tomato mutants. *J. Exp. Bot.* 38:1734–40

123. Liu, B.-L. L. 1984. Abscisic acid induces land form characteristics in *Marsilea quadrifolia* L. *Am. J. Bot.* 71:638–44

124. Loveys, B. R., Milborrow, B. V. 1981. Isolation and characterization of 1'-*O*-abscisic acid-β-D-glucopyranoside from vegetative tomato tissue. *Aust. J. Plant Physiol.* 8:571–89

125. Loveys, B. R., Milborrow, B. V. 1984. Metabolism of abscisic acid. In *The Biosynthesis and Metabolism of Plant Hormones,* ed. A. Crozier, J. R. Hillmann, Soc. Exp. Biol. Semin. Ser. 23:71–104. Cambridge: Cambridge Univ. Press. 288 pp.

126. Loveys, B. R., Robinson, S. P. 1987. Abscisic acid synthesis and metabolism in barley leaves and protoplasts. *Plant Sci.* 49:23–30

127. Lucas, W. J., Alexander, J. M. 1981. Influence of turgor pressure manipulation on plasmalemma transport of HCO_3^- and OH^- in *Chara corallina*. *Plant Physiol* 68:553–59

128. MacRobbie, E. A. C. 1981. Effects of ABA in 'isolated' guard cells of *Commelina communis* L. *J. Exp. Bot.* 32:563–72

129. Marumo, S., Katayama, M., Komori, E., Ozaki, Y., Natsume, M., Kondo, S. 1982. Microbial production of abscisic acid by *Botrytis cinerea*. *Agric. Biol. Chem.* 46:1967–68

130. Mawson, B. T., Colman, B., Cummins, W. R. 1981. Abscisic acid and photosynthesis in isolated leaf mesophyll cell. *Plant Physiol.* 67:233–36

131. Mertens, R., Deus-Neumann, B., Weiler, E. W. 1983. Monoclonal antibodies for the detection and quantitation of the endogenous plant growth regulator, abscisic acid. *FEBS Lett.* 160:269–72

132. Mertens, R., Stüning, M., Weiler, E. W. 1982. Metabolism of tritiated enantiomers of abscisic acid prepared by immunoaffinity chromatography. *Naturwissenschaften* 69:595–97

133. Milborrow, B. V. 1978. The stability of conjugated abscisic acid during wilting. *J. Exp. Bot.* 29:1059–66

134. Milborrow, B. V. 1980. A distinction between the fast and slow responses to abscisic acid. *Aust. J. Plant Physiol.* 7:749–54

135. Milborrow, B. V. 1983. Pathways to and from abscisic acid. See Ref. 4, pp. 79–111

136. Milborrow, B. V. 1983. The reduction of $(\pm)$-[2-^{14}C]abscisic acid to the 1',4'-*trans*-diol by pea seedlings and the formation of 4'-desoxy ABA as an artefact. *J. Exp. Bot.* 34:303–8

137. Milborrow, B. V., Nonhebel, H. M., Willows, R. 1988. 2,7-Dimethyl-octa-2,4-dienedioic acid is not a by-product of abscisic acid biosynthesis. *Plant Sci.* In press

138. Milborrow, B. V., Rubery, P. H. 1985. The specificity of the carrier-mediated uptake of ABA by root segments of *Phaseolus coccineus* L. *J. Exp. Bot.* 36:807–22

139. Milborrow, B. V., Vaughan, G. T. 1982. Characterization of dihydrophaseic acid 4'-*O*-β-D-glucopyranoside as a major metabolite of abscisic acid. *Aust. J. Plant Physiol.* 9:361–72

140. Mohan Ram, H. Y., Rao, S. 1982. In-vitro induction of aerial leaves and of precocious flowering in submerged shoots of *Limnophila indica* by abscisic acid. *Planta* 155:521–23

141. Moore, R., Evans, M. L. 1986. How roots perceive and respond to gravity. *Am. J. Bot.* 73:574–87

142. Moore, R., Smith, J. D. 1984. Growth, graviresponsiveness and abscisic-acid content of *Zea mays* seedlings treated with fluridone. *Planta* 162:342–44

143. Moore, R., Smith, J. D. 1984 Gravires-ponsiveness and abscisic-acid content of roots of carotenoid-deficient mutants of *Zea mays*. *Planta* 164:126–28

144. Mulkey, T. J., Evans, M. L., Kuzmanoff, K. M. 1983. The kinetics of abscisic acid action on root growth and gravitropism. *Planta* 157:150–57

145. Mundy, J. 1984. Hormonal regulation of α-amylase inhibitor synthesis in germinating barley. *Carlsberg Res. Commun.* 49:439–44

146. Mundy, J., Hejgaard, J., Hansen, A., Hallgren, L., Jorgensen, K. G., Munck, L. 1986. Differential synthesis *in vitro*

of barley aleurone and starchy endosperm proteins. *Plant Physiol.* 81:630–36

147. Murphy, G. J. P. 1984. Metabolism of R,S-[2-¹⁴C]abscisic acid by non-stressed and water-stressed detached leaves of wheat (*Triticum aestivum* L.). *Planta* 160:250–55

148. Neill, S. J., Horgan, R. 1983. Incorporation of α-ionylidene ethanol and α-ionylidene acetic acid into abscisic acid by *Cercospora rosicola*. *Phytochemistry* 22:2469–72

149. Neill, S. J., Horgan, R. 1985. Abscisic acid production and water relations in wilty tomato mutants subjected to water deficiency. *J. Exp. Bot.* 36:1222–31

150. Neill, S. J., Horgan, R., Heald, J. K. 1983. Determination of the levels of abscisic acid-glucose ester in plants. *Planta* 157:371–75

151. Neill, S. J., Horgan, R., Parry, A. D. 1986. The carotenoid and abscisic acid content of viviparous kernels and seedlings of *Zea mays* L. *Planta* 169:87–96

152. Neill, S. J., Horgan, R., Rees, A. F. 1987. Seed development and vivipary in *Zea mays* L. *Planta* 171:358–64

153. Neill, S. J., Horgan, R., Walton, D. C. 1984. Biosynthesis of abscisic acid. See Ref. 125, pp. 43–70

154. Neill, S. J., Horgan, R., Walton, D. C., Lee, T. S. 1982. The biosynthesis of abscisic acid in *Cercospora rosicola*. *Phytochemistry* 21:61–65

155. Neill, S. J., Horgan, R., Walton, D. C., Mercer, C. A. M. 1987. The metabolism of α-ionylidene compounds by *Cercospora rosicola*. *Phytochemistry* 26:2515–19

156. Neill, S. J., McGaw, B. A., Horgan, R. 1986. Ethylene and 1-aminocyclopropane-1-carboxylic acid production in *flacca*, a wilty mutant of tomato, subjected to water deficiency and pretreatment with abscisic acid. *J. Exp. Bot.* 37:535–41

157. Netting, A. G., Milborrow, B. V., Duffield, A. M. 1982. Determination of abscisic acid in *Eucalyptus haemastoma* leaves using gas chromatography/mass spectrometry and deuterated internal standards. *Phytochemistry* 21:385–89

158. Nevo, Y., Tal, M. 1973. The metabolism of abscisic acid in *flacca*, a wilty mutant of tomato. *Biochem. Genet.* 10:79–90

159. Nolan, R. C., Lin, L.-S., Ho, T.-H. D. 1987. The effect of abscisic acid on the differential expression of α-amylase isozymes in barley aleurone layers. *Plant Mol. Biol.* 8:13–22

160. Nonhebel, H. M., Milborrow, B. V.

1986. Incorporation of ²H from ²H₂O into ABA in tomato shoots: evidence for a large pool of precursors. *J. Exp. Bot.* 37:1533–41

161. Nonhebel, H. M., Milborrow, B. V. 1987. Contrasting incorporation of ²H from ²H₂O into ABA, xanthoxin and carotenoids in tomato shoots. *J. Exp. Bot.* 38:980–91

162. Norman, S. M., Poling, S. M., Maier, V. P., Nelson, M. D. 1985. Ionylidene-acetic acids and abscisic acid biosynthesis by *Cercospora rosicola*. *Agric. Biol. Chem.* 49:2317–24

163. Oertli, J. J. 1985. The response of plant cells to different forms of moisture stress. *J. Plant Physiol.* 121:295–300

164. Okamoto, M., Hirai, N., Koshimizu, K. 1987. Occurrence and metabolism of 1',4'-*trans*-diol of abscisic acid. *Phytochemistry* 26:1269–71

165. Okamoto, M., Hirai, N., Koshimizu, K. 1987. The biosynthesis of abscisic acid in *Cercospora pini-densiflorae*. *Abstr. Bull. Plant Growth Reg. Soc. Am.* 15:4

166. Oritani, T., Ichimura, M., Yamashita, K. 1984. A novel abscisic acid analog, (+)-(2Z,4E)-5-(1',4'-dihydroxy-6',6'-dimethyl-2'-methylene-cyclohexyl)-3-methyl-2,4-pentadienoic acid, from *Cercospora cruenta*. *Agric. Biol. Chem.* 48:1677–78

167. Oritani, T., Niitsu, M., Kato, T., Yamashita, K. 1985. Isolation of (2Z,4E)-γ-ionylideneethanol from *Cercospora cruenta*, a fungus producing (+)-abscisic acid. *Agric. Biol. Chem.* 49:2819–22

168. Oritani, T., Yamashita, K. 1985. Biosynthesis of (+)-abscisic acid in *Cercospora cruenta*. *Agric. Biol. Chem.* 49:245–49

169. Oritani, T., Yamashita, K. 1987. Isolation and structure of 4'-hydroxy-γ-ionylideneacetic acids from *Cercospora cruenta*, a fungus producing (+)-abscisic acid. *Agric. Biol. Chem.* 51:275–78

170. Oritani, T., Yamashita, K., Oritani, T. 1985. Metabolism of (±)-(2Z,4E)-α-ionylideneacetic acid and its 1'-hydroxyl analog in plants. *J. Pestic. Sci.* 10:535–40

171. Orr, W., Keller, W. A., Singh, J. 1986. Induction of freezing tolerance in an embryogenic cell suspension culture of *Brassica napus* by abscisic acid at room temperature. *J. Plant Physiol.* 126:23–32

172. Pang, P. P., Meyerowitz, E. 1987. Seed specific gene expression in *Arabidopsis thaliana*. See Ref. 54, p. 53

173. Parry, A. D., Neill, S. J., Horgan, R. 1988. Xanthoxin levels and metabolism

in wild-type and wilty mutants of tomato. *Planta* 173:397–404

174. Pierce, M., Raschke, K. 1980. Correlation between loss of turgor and accumulation of abscisic acid in detached leaves. *Planta* 148:174–82

175. Pierce, M., Raschke, K. 1981. Synthesis and metabolism of abscisic acid in detached leaves of *Phaseolus vulgaris* L. after loss and recovery of turgor. *Planta* 153:156–65

176. Pilet, P. E., Chanson, A. 1981. Effect of abscisic acid on maize root growth. A critical examination. *Plant Sci. Lett.* 21:99–106

177. Pilet, P. E., Rebeaud, J. E. 1983. Effect of abscisic acid on growth and indolyl-3-acetic acid levels in maize roots. *Plant Sci. Lett.* 31:117–22

178. Pilet, P. E., Saugy, M. 1987. Effect on root growth of endogenous and applied IAA and ABA. A critical reexamination. *Plant Physiol.* 83:33–38

179. Quarrie, S. A. 1982. Droopy: a wilty mutant of potato deficient in abscisic acid. *Plant Cell Environ.* 5:23–26

180. Quarrie, S. A. 1983. Genetic differences in abscisic acid physiology and their potential uses in agriculture. See Ref. 4, pp. 365–19

181. Quarrie, S. A. 1987. Use of genotypes differing in endogenous abscisic acid levels in studies of physiology and development. See Ref. 42, pp. 89–105

182. Quarrie, S. A., Galfre, G. 1985. Use of different hapten-protein conjugates immobilized on nitrocellulose to screen monoclonal antibodies to abscisic acid. *Anal. Biochem.* 151:389–99

183. Quarrie, S. A., Lister, P. G. 1984. Evidence of plastid control of abscisic acid accumulation in barley (*Hordeum vulgare* L.). *Z. Pflanzenphysiol.* 114:295–308

184. Quarrie, S. A., Lister, P. G. 1984. Effects of inhibitors of protein synthesis on abscisic acid accumulation in wheat. *Z. Pflanzenphysiol.* 114:309–14

185. Radin, J. W., Hendrix, D. L. 1986. Accumulation and turnover of abscisic acid in osmotically stressed cotton leaf tissue in relation to temperature. *Plant Sci.* 45:37–42

186. Radin, J. W., Hendrix, D. L. 1987. The apoplastic pool of abscisic acid in cotton leaves in relation to stomatal closure. *Planta*. In press

187. Raikhel, N. V., Quatrano, R. S. 1986. Localization of wheat-germ agglutinin in developing wheat embryos and those cultured in abscisic acid. *Planta* 168:433–40

188. Raihkel, N. V., Wilkins, T. A. 1987. Isolation and characterization of a cDNA clone encoding wheat germ agglutinin. *Proc. Natl. Acad. Sci. USA* 84:6745–49

189. Railton, I. D. 1987. Resolution of the enantiomers of abscisic acid methyl ester by high-performance liquid chromatography using a stationary phase of cellulose tris(3,5-dimethylphenylcarbamate)-coated silica gel. *J. Chromatogr.* 402:371–73

190. Raschke, K. 1982. Involvement of abscisic acid in the regulation of gas exchange: Evidence and inconsistencies. See Ref. 71, pp. 581–90

191. Raschke, K. 1987. Action of abscisic acid on guard cells. In *Stomatal Function*, ed. E. Zeiger, G. D. Farquhar, I. R. Cowan, Ch. 11, pp. 253–79. Stanford: Stanford Univ. Press. 512 pp.

192. Raschke, K., Hedrich, R. 1985. Simultaneous and independent effects of abscisic acid on stomata and the photosynthetic apparatus in whole leaves. *Planta* 163:105–18

193. Raschke, K., Patzke, J. 1988. Abscisic acid can cause apparent depressions of the photosynthetic capacity of leaves through changes in the probability density function of stomatal aperture. *Planta*. In press

194. Reaney, M. J. T., Gusta, L. V. 1987. Factors influencing the induction of freezing tolerance by abscisic acid in cell suspension cultures of *Bromus inermis* Leyss and *Medicago sativa* L. *Plant Physiol.* 83:423–27

195. Reymond, P., Pilet, P. E. 1987. On the importance of ABA in maize root growth. In *Abstracts, General Lectures, Symposia Papers and Posters*, p. 135. 14th Int. Bot. Congr., Berlin. 480 pp.

196. Reymond, P., Saugy, M., Pilet, P. E. 1987. Quantification of abscisic acid in a single maize root. *Plant Physiol.* 85:8–9

197. Rivier, L., Leonard, J. F., Cottier, J. P. 1983. Rapid effect of osmotic stress on the content and exodiffusion of abscisic acid in *Zea mays* roots. *Plant Sci. Lett.* 31:133–37

198. Rivier, L., Saugy, M. 1986. Chemical ionization mass spectrometry of indol-3yl-acetic acid and *cis*-abscisic acid: evaluation of negative ion detection and quantification of *cis*-abscisic acid in growing maize roots. *J. Plant Growth Regul.* 5:1–16

199. Robertson, A. J., Gusta, L. V. 1986. Abscisic acid and low temperature induced polypeptide changes in alfalfa (*Medicago sativa*) cell suspension cultures. *Can. J. Bot.* 64:2758–63

199a. Robertson, A. J., Gusta, L. V., Reaney, M. J. T., Ishikawa, M. 1987. Protein synthesis in bromegrass (*Bromus inermis* Leyss) cultured cells during the induction of frost tolerance by abscisic acid or low temperature. *Plant Physiol.* 84:1331–36

200. Robichaud, C., Sussex, I. M. 1986. The response of viviparous-1 and wild type embryos of *Zea mays* to culture in the presence of abscisic acid. *J. Plant Physiol.* 126:235–42

201. Rosher, P. H., Jones, H. G., Hedden, P. 1985. Validation of a radioimmunoassay for (+)-abscisic acid in extracts of apple and sweet-pepper tissue using high-pressure liquid chromatography and combined gas chromatography-mass spectrometry. *Planta* 165:91–99

202. Sembdner, G., Gross, D., Liebisch, H.-W., Schneider, G. 1980. Biosynthesis and metabolism of plant hormones. In *Encyclopedia of Plant Physiology,* (NS), ed. J. MacMillan, 9:281–444. Berlin: Springer. 681 pp.

203. Setter, T. L., Brenner, M. L., Brun, W. A., Krick, T. P. 1981. Identification of a dihydrophaseic acid aldopyranoside from soybean tissue. *Plant Physiol.* 68:93–95

204. Shimazaki, K., Iino, M., Zeiger, E. 1986. Blue light-dependent proton extrusion by guard-cell protoplasts of *Vicia faba. Nature* 319:324–26

205. Sindhu, R. K., Walton, D. C. 1987. Conversion of xanthoxin to abscisic acid by cell-free preparations from bean leaves. *Plant Physiol.* 85:916–21

206. Singh, N. K., LaRosa, P. C., Handa, A. K., Hasegawa, P. M., Bressan, R. A. 1987. Hormonal regulation of protein synthesis associated with salt tolerance in plant cells. *Proc. Natl. Acad. Sci. USA* 84:739–43

207. Smart, C., Longland, J., Trewavas, A. 1987. The turion: a biological probe for the molecular action of abscisic acid. In *Molecular Biology of Plant Growth Control,* ed. J. E. Fox, M. Jacobs, 44:345–59. New York: Liss. 467 pp.

208. Smith, J. D., Fong, F., Hole, D. J., Magill, C. W., Cobb, B. G. 1988. The origins of abscisic acid found in developing embryos of *Zea mays* L. *Planta.* In press

209. Sossountzov, L., Sotta, B., Maldiney, R., Sabbagh, I, Miginiac, E. 1986. Immunoelectron-microscopy localization of abscisic acid with colloidal gold on Lowicryl-embedded tissues of *Chenopodium polyspermum* L. *Planta* 168:471–81

210. Sotta, B., Sossountzov, L., Maldiney, R., Sabbagh, I., Tachon, P., Miginiac, E. 1985. Abscisic acid localization by light microscopic immunohistochemistry in *Chenopodium polyspermum* L. Effect of water stress. *J. Histochem. Cytochem.* 33:201–8

211. Stewart, C. R., Voetberg, G. 1987. Abscisic acid accumulation is not required for proline accumulation in wilted leaves. *Plant Physiol.* 83:747–49

212. Stewart, C. R., Voetberg, G., Rayapati, P. J. 1986. The effects of benzyladenine, cycloheximide, and cordycepin on wilting-induced abscisic acid and proline accumulations and abscisic acid- and salt-induced proline accumulation in barley leaves. *Plant Physiol.* 82:703–7

213. Takahashi, S., Oritani, T., Yamashita, K. 1986. Total synthesis of (±)-methyl phaseates. *Agric. Biol. Chem.* 50:1589–95

214. Takahashi, S., Oritani, T., Yamashita, K. 1986. Synthesis and biological activities of (±)-deoxy-abscisic acid isomers. *Agric. Biol. Chem.* 50:3205–6

215. Takeda, N., Harada, K., Suzuki, M., Tatematsu, A., Hirai, N., Koshimizu, K. 1984. Structural characterization of abscisic acid and related metabolites by chemical ionization mass spectrometry. *Agric. Biol. Chem.* 48:685–94

216. Takeda, N., Harada, K., Suzuki, M., Tatematsu, A., Hirai, N., Koshimizu, K. 1986. Desorption chemical ionization mass spectra of underivatized abscisic-acid-conjugated metabolites, a useful aid for structural characterization. *Agric. Biol. Chem.* 50:2295–2300

217. Tal, M., Imber, D., Erez, A., Epstein, E. 1979. Abnormal stomatal behavior and hormonal imbalance in *flacca,* a wilty mutant of tomato. V. Effect of abscisic acid and indoleacetic acid metabolism and ethylene evolution. *Plant Physiol.* 63:1044–48

218. Taylor, H. F., Burden, R. S. 1972. Xanthoxin, a recently discovered plant growth inhibitor. *Proc. R. Soc. London Ser. B* 180:317–46

219. Taylor, H. F., Burden, R. S. 1973. Preparation and metabolism of 2-[^{14}C]-*cis,trans*-xanthoxin, *J. Exp. Bot.* 24:873–80

220. Taylor, I. B. 1984. Abnormalities of abscisic acid accumulation in tomato mutants. In *Biochemical Aspects of Synthetic and Naturally Occurring Plant Growth Regulators,* ed. R. Menhenett, D. K. Lawrence, 11:73–90. Wantage: Br. Plant Growth Regul. Group. 121 pp.

221. Taylor, I. B. 1987. ABA-deficient tomato mutants. See Ref. 98, pp. 197–217

222. Taylor, I. B., Tarr, A. R. 1984. Phenotypic interactions between abscisic acid deficient tomato mutants. *Theor. Appl. Genet.* 68:115–19

223. Tietz, A., Kasprik, W. 1986. Identification of abscisic acid in a green alga. *Biochem. Physiol. Pflanz.* 181:269–74

224. Tietz, A., Köhler, R., Ruttkowski, U., Kasprik, W. 1987. Further investigations on the occurrence and the effects of abscisic acid in algae. See Ref. 195, p. 133

225. Tietz, D., Dörffling, K., Wohrle, D., Erxleben, I., Liemann, F. 1979. Identification by combined gas chromatography-mass spectrometry of phaseic acid and dihydrophaseic acid and characterization of further abscisic acid metabolites in pea seedlings. *Planta* 147:168–73

226. Triplett, B. A., Quatrano, R. S. 1982. Timing, localization, and control of wheat germ agglutinin synthesis in developing wheat embryos. *Dev. Biol.* 91:491–96

227. Uknes, S. J., Ho, T.-H. D. 1984. Mode of action of abscisic acid in barley aleurone layers. *Plant Physiol.* 75:1126–32

228. Van Volkenburgh, E., Davies, W. J. 1983. Inhibition of light-stimulated leaf expansion by abscisic acid. *J. Exp. Bot.* 34:835–45

229. Vaughan, G. T., Milborrow, B. V. 1984. The resolution by HPLC of *RS*-[2-14C]Me 1',4'-*cis*-diol of abscisic acid and the metabolism of (−)-*R*- and (+)-*S*-abscisic acid. *J. Exp. Bot.* 35:110–20

230. Walker-Simmons, M. 1987. ABA levels and sensitivity in developing wheat embryos of sprouting resistant and susceptible cultivars. *Plant Physiol.* 84:61–66

231. Walton, D. C. 1980. Biochemistry and physiology of abscisic acid. *Ann. Rev. Plant Physiol.* 31:453–89

232. Walton, D. C. 1983. Structure-activity relationships of abscisic acid analogs and metabolites. See Ref. 4, pp. 113–46

233. Wang, T. L., Donkin, M. E., Martin, E. S. 1984. The physiology of a wilty pea: abscisic acid production under water stress. *J. Exp. Bot.* 35:1222–32

234. Wang, T. L., Griggs, P., Cook, S. 1986. Immunoassays for plant growth regulators—a help or a hindrance? See Ref. 14, pp. 26–34

235. Ward, D. A., Bunce, J. A. 1987. Abscisic acid simultaneously decreases carboxylation efficiency and quantum yield in attached soybean leaves. *J. Exp. Bot.* 38:1182–92

236. Weiler, E. W. 1986. Plant hormone immunoassays based on monoclonal and polyclonal antibodies. In *Modern Methods of Plant Analysis*, ed. H. F. Linskens, J. F. Jackson, 4:1–17. Berlin: Springer. 263 pp.

237. Weiler, E. W., Schnabl, H., Hornberg, C. 1982. Stress-related levels of abscisic acid in guard cell protoplasts of *Vicia faba* L. *Planta* 154:24–28

238. Weyers, J. D. B. 1985. Germination and root gravitropism of *flacca*, the tomato deficient in abscisic acid. *J. Plant Physiol.* 121:475–80

239. Weyers, J. D. B., Hillman, J. R. 1979. Uptake and distribution of abscisic acid in *Commelina* leaf epidermis. *Planta* 144:167–72

240. Williamson, J. D., Quatrano, R. S., Cuming, A. C. 1985. E_m-polypeptide and its messenger RNA levels are modulated by abscisic acid during embryogenesis in wheat. *Eur. J. Biochem.* 152:501–7

241. Yamamoto, H. Y. 1985. Xanthophyll cycles. *Methods Enzymol.* 110:303–12

242. Young, J. P., Dengler, N. G., Horton, R. F. 1987. Heterophylly in *Ranunculus flabellaris:* the effect of abscisic acid on leaf anatomy. *Ann. Bot.* 60:117–25

243. Young, J. P., Horton, R. F. 1985. Heterophylly in *Ranunculus flabellaris* Raf.: The effect of abscisic acid. *Ann. Bot.* 55:899–902

244. Zeevaart, J. A. D. 1980. Changes in the levels of abscisic acid and its metabolites in excised leaf blades of *Xanthium strumarium* during and after water stress. *Plant Physiol.* 66:672–78

245. Zeevaart, J. A. D. 1983. Metabolism of abscisic acid and its regulation in *Xanthium* leaves during and after water stress. *Plant Physiol.* 71:477–81

246. Zeevaart, J. A. D., Boyer, G. L. 1984. Accumulation and transport of abscisic acid and its metabolites in *Ricinus* and *Xanthium*. *Plant Physiol.* 74:934–39

247. Zhang, J., Davies, W. J. 1987. ABA in roots and leaves of flooded pea plants. *J. Exp. Bot.* 38:649–59

248. Zhang, J., Davies, W. J. 1987. Increased synthesis of ABA in partially dehydrated root tips and ABA transport from roots to leaves. *J. Exp. Bot.* 38:2015–23

249. Zhang, J., Schurr, U., Davies, W. J. 1987. Control of stomatal behaviour by abscisic acid which apparently originates in the roots. *J. Exp. Bot.* 38:1174–81

Ann. Rev. Plant Physiol. Plant Mol. Biol. 1988. 39:475–502

CHLOROPLAST DEVELOPMENT AND GENE EXPRESSION

John E. Mullet

Department of Biochemistry and Biophysics, Texas A&M University, College Station, Texas 77843-2128

CONTENTS

INTRODUCTION.. 475
CHLOROPLAST DEVELOPMENT AND PLASTID DIVERSITY........................ 477
   Proplastids 477
   Formation of Chloroplasts from Proplastids 477
   Generation of Plastid Diversity 478
PLASTID CODING CAPACITY AND GENOME ORGANIZATION.................... 479
PHASES OF CHLOROPLAST DEVELOPMENT 481
   Formation of the Shoot and Initiation of Leaf Meristems 481
   Leaf Mesophyll Cell Development 482
REGULATION OF CHLOROPLAST PROTEIN COMPLEX BIOSYNTHESIS........ 483
PLASTID TRANSCRIPTION AND THE DETERMINATION OF RNA LEVELS..... 485
   Regulation of Chloroplast Transcription 485
   Posttranscriptional Mechanisms that Alter RNA Levels 486
REGULATION OF CHLOROPLAST TRANSLATION 488
   The Cytoplasmic Control Principle 490
   Regulation of Chlorophyll-Apoprotein Synthesis 490
CONCLUDING REMARKS 493

INTRODUCTION

Chloroplast development has been studied extensively for many years [see previous reviews for progress to 1967 (80) and 1978 (81)]. Early researchers were attracted to this developmental process because chloroplast-specific

475

0066-4294/88/0601-0475$02.00

pigments were present, such as chlorophylls and carotenoids, that could easily be quantitated. These pigments also provided a way to analyze the biophysical events involved in light-energy absorption, excitation energy transfer, and photochemical trapping (133). A combination of approaches ranging from picosecond (ps) spectroscopy to liquid helium EPR helped define the spatial and energetic constraints of the initial steps in photosynthetic electron transport (47). This data suggested that high-efficiency photon trapping requires quasi-solid-state organization to provide optimal distances and orientation between electron-transfer components. Biochemical and freeze-fracture analyses showed that the structural requirements for high-efficiency photon trapping are met in part by organizing the photosynthetic electron-transfer components into large multiprotein complexes (50, 77). These complexes consist of up to 20 polypeptides and over 200 cofactors. The partial elucidation of the structure and composition of the photosynthetic electron-transport units led to an examination of the mechanisms by which these highly complex photon traps are synthesized and assembled. This process was found to require the coordinated production of polypeptides encoded on nuclear and chloroplast genes as well as the biosynthesis of cofactors such as chlorophyll, carotenoid, heme, and quinone (for reviews see 50, 77).

Chloroplasts are derived from small undifferentiated proplastids which are in most cases inherited maternally by the plant zygote (81). Subsequent development of chloroplasts is coupled to the formation of leaf meristems and the production of mesophyll cells. During development, plastid volume per cell increases dramatically and plastid composition changes in parallel with the acquisition of photosynthetic competence. Plastid composition also varies in response to environmental conditions or to meet specialized needs of the plant. These changes raise numerous questions about the molecular mechanisms that activate and coordinate gene expression during chloroplast biogenesis.

In this review I focus primarily on chloroplast development in vascular plants—in particular, C_3 monocots (barley, wheat). Some examples and data come from studies of other photosynthetic organisms, but comprehensive citations should be sought in other reviews (26, 50, 80, 81, 137, 165). Furthermore, this review focuses on the relationship among leaf development, chloroplast biogenesis, and the expression of chloroplast-encoded genes. This choice is timely in that the chloroplast genome has recently been sequenced in its entirety from two plants (120, 144) and progress on the regulation of chloroplast gene expression has been rapid. Reviews concerning both the expression of nuclear genes encoding chloroplast proteins (44, 89) and the photoregulation of gene expression (44, 62, 136, 162) have been published elsewhere.

CHLOROPLAST DEVELOPMENT AND PLASTID DIVERSITY

In later sections of this review, I trace chloroplast biogenesis and gene expression during leaf development. To lay the groundwork for that discussion, it is useful to consider first the composition of the proplastid and chloroplast, the end points of the developmental process. This allows a description of the major biochemical pathways involved in chloroplast development and a consideration of the origin of plastid diversity.

Proplastids

The progenitors of chloroplasts are small (0.5–1.0 μm diameter), usually spherical organelles termed proplastids (161) (Figure 1). These organelles normally originate maternally during formation of plant zygotes and are maintained in an undifferentiated state in meristematic cells of the developing plant (reviewed in 81). Proplastids may also be formed during seasonal dormancy (46) or during maturation of specialized structures such as pollen and egg cells (68, 122, 175). The prominent inner membrane found in chloroplasts is nearly absent in proplastids. However, the proplastid contains low amounts of plastid DNA, RNA, ribosomes, and soluble proteins (39, 108, 116). This suggests that a basal level of plastid and nuclear gene expression is active in the dividing cells of the meristem, perpetuating the proplastid population. The maintenance of proplastids in meristematic cells may be related to this compartment's role in amino acid (166a) and lipid biosynthesis (30, 50).

Formation of Chloroplasts from Proplastids

Mesophyll cell development is accompanied by an increase in plastid number per cell (Figure 1). In barley, the number of plastids per cell increases from approximately 10 in meristematic cells to 65 in mature mesophyll cells (129, 147). In hexaploid lines of wheat, mature mesophyll cells contain up to 150 plastids (42, 127). The mechanisms regulating plastid division and plastid number per cell are unknown. However, it is interesting to note that a mutant of *Arabidopsis* that forms small plastids shows an increase in plastid number per cell compared to wild-type plants (105). In addition, studies of plastid populations in wheat have led to the conclusion that the number of chloroplasts per cell increases proportionately with increases in cell size. In other words, chloroplasts divide and expand to occupy a constant proportion of the mesophyll cell surface (129).

Plastid size and composition also change during chloroplast development. The increase in plastid volume can be greater than 100-fold, and in mature mesophyll cells, chloroplasts account for a large percentage of the cytoplas-

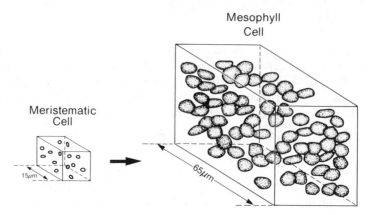

Figure 1 Diagram showing the conversion of a meristematic cell containing 10 1-μm pro-plastids to a leaf mesophyll cell containing 65 disk-shaped chloroplasts 6–8 μm in diameter.

mic volume (108, 129). The increase in plastid volume results in part from the accumulation of proteins, lipids, and cofactors required for photosynthesis. These components are derived from a number of biosynthetic pathways activated during chloroplast biogenesis (Table 1). Many of the biosynthetic processes induced during chloroplast biogenesis require the activation of chloroplast and nuclear genes. In other cases only nuclear genes appear to be involved (Table 1).

Generation of Plastid Diversity

Plastids can be divided into two groups based on photosynthetic competence. Nonphotosynthetic plastids include proplastids, chloroplast precursors in dark-grown plants (etioplasts), and plastids specialized in synthesis or accumulation of carotenoids (chromoplasts), starch (amyloplasts), terpenoids,

Table 1 Biochemical processes induced during chloroplast biogenesis (26, 67, 81)

Biochemical Process	Component	Gene Localization
Terpenoid synthesis	carotenoids, quinone side chain, phytol	nuclear (82)
Tetrapyrrole synthesis	heme, chlorophyll	nuclear (174) & chloroplast[a]
Lipid synthesis	galactolipid, phospholipid, sulfo-lipid	nuclear (30, 50)
Carbon storage/transport	starch	—
Carbon fixation	Calvin cycle/RUBISCO	nuclear & chloroplast (67, 84)
Photosynthetic electron transport	PSI, PSII, ATP synthetase, cytochrome *b/f* complex	nuclear & chloroplast (67, 162)

[a] A chloroplast-encoded RNA is a cofactor in δ-ALA biosynthesis (76, 140, 170, 171).

or lipids. The specialized nonphotosynthetic plastids are normally localized in specific tissues such as petals (chromoplasts), tubers (amyloplasts), or oil glands (161).

Photosynthetic plastids also exhibit variation in composition and function. For example, chloroplasts in mesophyll cells of C_4 plants are deficient in proteins involved in CO_2 fixation, whereas bundle sheath cell chloroplasts are deficient in Photosystem II (PSII) proteins (7, 29, 141). Furthermore, chloroplast composition varies as a function of environmental parameters such as light intensity (18, 100).

The organization and composition of DNA within photosynthetic and nonphotosynthetic plastids are similar (49, 157). This is consistent with data showing interconversion of plastid populations (161). For example, chromoplasts derived from chloroplasts will differentiate back to chloroplasts under appropriate conditions (71). Finally, it is interesting to note that the specialization of nongreen plastids involves induction of a subset of the biosynthetic pathways found in, or which are required for, biogenesis of mesophyll chloroplasts (see Table 1). For example, the terpenoid biosynthetic pathway is selectively activated during the formation of chromoplasts (82). This suggests that the biosynthetic pathways required for chloroplast biogenesis can be induced independently of chloroplast development.

PLASTID CODING CAPACITY AND GENOME ORGANIZATION

Knowledge of the plastid genome is required if we are to understand the involvement of plastid gene expression in the development of chloroplasts and nonphotosynthetic plastids. The composition of the chloroplast genome was reviewed in this series in 1983 (176). Other related reviews have focused on chloroplast transcription (27, 33, 59), genome organization and expression (8, 20, 66, 67, 90, 130), protein synthesis (41), tRNAs (58, 168, 169), and the synthesis of thylakoid membranes (50). In addition, the DNA sequence of tobacco *(Nicotiana tabacum)* (144) and *Marchantia polymorpha* (120) plastid genomes is known and several transcription units (14, 28, 67, 131, 153, 172) have been characterized.

Plastid genomes in vascular plants are circular, and most are 120–180 kbp in size (176). Plastids have been reported to contain 22–900 genome copies (10). The DNA molecules are associated with the inner membrane of the envelope or with internal thylakoid membranes in aggregates of 10–20 DNA molecules (66). In wheat the number of DNA copies per plastid increases during chloroplast development (108; but see 19). In addition, DNA can be found in rings or in a more dispersed arrangement within the plastid (66, 86, 108).

The complete sequence of two plastid genomes and the identification of plastid genes in numerous plants (14, 28, 67, 131, 153, 172; reviewed in 176) makes it possible to estimate that the genome codes for approximately 123 genes. These genes fall into three categories (see Table 2). The first category consists of genes encoding gene products involved in photosynthesis. The second category includes genes required for transcription, translation, or replication of the chloroplast genetic information. The remaining genes include putative sequences for an NADH oxidoreductase complex and 19 open reading frames (>100 codons) of unknown function (120, 144).

In the absence of information concerning the function of the unidentified open reading frames no final conclusions can be made concerning the full coding potential of the genome. It is tempting to speculate, however, that high levels of plastid gene expression may be required only during chloroplast biogenesis and not in nonphotosynthetic plastids such as proplastids, amyloplasts or chromoplasts. This proposal is consistent with low levels of RNA in amyloplasts and chromoplasts (1, 101, 125). Furthermore, the apparent lack of plastid genes involved in terpenoid biosynthesis, lipid biosynthesis, or starch formation (in nonphotosynthetic cells) is consistent with this idea. Enzymes or products from these biosynthetic pathways could be directed to the plastid without concomitant build-up of chloroplast transcription/translation capacity. This appears to be the case for plastid-localized steps in amino acid biosynthesis (166a), which are inferred to occur in plastids lacking plastid ribosomes (i.e. iojap mutant maize; 165a) or plastid DNA (i.e. W_3BUL

Table 2 Genes coded by plastid DNA (120, 144)

I. Genes involved in photosynthesis	
RUBISCO:	*rbcL*
PSI:	*psaA, B*
PSII:	*psbA, B, C, D, E, F, G, H*
ATP synthetase:	*atpA, B, E, F, H, I*
Cytochrome complex:	*petA, B, D*
II. Genes required for chloroplast gene expression	
Transcription:	*rpoA, B, C*
Translation:	*rDNA* (16S, 23S, 4.5S, 5S)
	trn (30 genes)
	rps (12 genes)
	rpl (7 genes)
	infA, secX
Replication:	*ssb*
III. Other genes	
Unidentified:	*ndhA, B, C, D, E, F*
	19 > 100 codons
	18 < 100 codons

mutant of *Euglena;* 39a). In fact, the regulatory advantages obtained by restricting the plastid genome to genes involved in photosynthesis could have provided part of the selective pressure to move genes from the plastid to the nucleus (see 124 for other arguements).

PHASES OF CHLOROPLAST DEVELOPMENT

The development of photosynthetic plastids from proplastids involves activation of a large number of nuclear and chloroplast genes. A central question, therefore, concerns the nature and origin of signals that initiate and control this process. In unicellular photosynthetic organisms, induction of photosynthetic capacity is regulated to a large degree by environmental inputs such as O_2 concentration (77, 178) and carbon, nitrogen, and light availability (70, 109, 137). Chloroplast development in vascular plants can also be influenced by nutrient status and illumination (81, 94, 137, 162). However, seed reserves provide a continuous supply of carbon and nitrogen substrate for seedling germination and early development. Furthermore, in many plants the early phases of leaf and chloroplast development proceed uninhibited in the absence of light. For example, in barley the length of primary leaves and the number of plastids per mesophyll cell are similar in light-grown or dark-grown plants (B. J. Baumgartner, J. E. Mullet, unpublished observation). The volume of plastids in dark-grown barley leaves is 70% that of chloroplasts (83), and nearly all of the soluble proteins found in chloroplasts are present in plastids of barley grown in darkness (61, 83, 84). In contrast to that of monocots such as barley, primary leaf development in dicots is greatly inhibited in the absence of light (31). Inhibition of leaf development appears to lead to a concomitant inhibition of chloroplast development and plastid transcription activity (12, 135, 160). This suggests that development of chloroplasts in vascular plants is regulated in part by internal signals that control differentiation of the leaf.

Formation of the Shoot and Initiation of Leaf Meristems

Leaf primordia are formed in embryos and during vegetative growth from the shoot apex (142). Leaf primordia develop primarily by the action of a basal intercalary meristem. The basal meristem produces files of cells that form the layers of the leaf (i.e. epidermal, mesophyll) and vascular tissue (142). It has been estimated that 44–55% of wheat leaf cells do not contain large numbers of photosynthetic plastids characteristic of mesophyll cells (35). This suggests that cell-specific signals are important determinants of plastid development in the leaf.

Few studies have examined chloroplast gene expression during embryo development and in developing leaf primordia (22, 108). Existing information suggests that plastids remain relatively undifferentiated in the dividing cells of the developing shoot and leaf promordia. Miyamura et al (108) in a recent study of wheat found, however, that plastid size and number per cell were higher in cells of the leaf basal meristem than in plastid populations found in embryonic cells. They also observed a significant increase in plastid DNA content during this early phase of chloroplast development.

Leaf Mesophyll Cell Development

As noted above, cells of the expanding monocot leaf are produced primarily from a meristem located at the leaf base. Therefore, older cells and more developed plastids are located near the leaf tip, and progressively younger cells are located toward the leaf base (129, 142). Chloroplast development along this gradient can be monitored by following the increase of plastid volume per cell as a function of cell age (Figure 2). In barley, plastid volume remains low for approximately one day following cessation of cell division. The plastid compartment then enlarges rapidly until the plastid volume in mature cells is reached. Plastid volume per cell then declines during leaf senescence (84, 167). The kinetics of plastid volume change allow the developmental process to be subdivided into four overlapping periods. I discuss these in sequence, starting with the lag phase (see top of Figure 2).

In a rapidly growing barley leaf, cells continue to enlarge for a period after cell division stops. This creates a zone of cell elongation 2–3 cm long adjacent to the leaf basal meristem (147). In this region, plastid number per cell increases steadily; plastid volume, however, changes little, in part because plastids are dividing. Plastid transcriptional activity increases during the lag period, and carotenoids and chlorophyll begin to accumulate in plastids (B. J. Baumgartner, J. Rapp, J. E. Mullet, unpublished observations). Following the lag phase, plastid volume per cell increases rapidly for 36–48 hr (Figure 2). During this period plastid transcription and translation rates are high, as are mRNA and ribosome levels per plastid (84, 111). This is also the phase in which the bulk of the inner membrane of the chloroplast is synthesized. Once the phase of rapid chloroplast growth is completed, plastid transcription and translation rates, mRNA levels, and ribosome content decline (84, 111). These adjustments occur as the chloroplast population shifts from the rapid synthesis and accumulation of new structures to processes required to maintain chloroplast function in the mature leaf. The final phase of chloroplast development occurs during leaf senescence, in which plastid volume and plastid number per cell decline as the plant recovers materials useful for development of new structures (167).

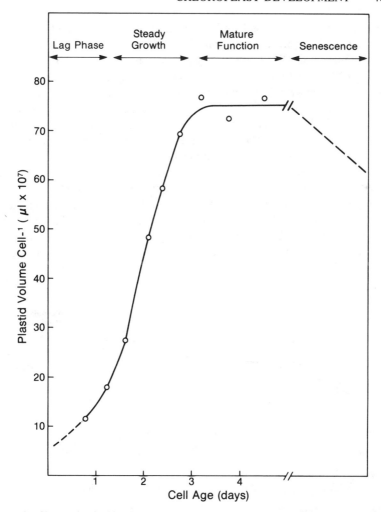

Figure 2 Changes in plastid volume per cell in light-grown barley leaves as a function of the time since cells stop dividing or cell age (B. J. Baumgartner, J. E. Mullet, unpublished data). Phases of chloroplast development are indicated at the top of the figure: a period when plastid volume remains low (lag phase), a phase of rapid chloroplast build-up (growth phase), a period where the chloroplast population is maintained (mature function), and finally leaf and chloroplast senescence (senescence).

REGULATION OF CHLOROPLAST PROTEIN COMPLEX BIOSYNTHESIS

The chloroplast genome contributes only a portion of the subunits found in any given pathway or protein complex present in the plastid. For example, the chloroplast genome encodes 19 of the 52 proteins found in chloroplast ribosomes and 1 of 2 subunits of RUBISCO (Table 2). The complexity of this situation can be appreciated by considering the origin of proteins and cofactors that make up PSII (Figure 3). The production of this complex requires the synthesis of at least 8 different nuclear-encoded proteins. These proteins are translated in the cytoplasm and imported into the chloroplast (3, 102, 138, 143). The imported polypeptides eventually reach the thylakoid membrane and are assembled with at least 8 different polypeptides of chloroplast origin. The chloroplast-encoded PSII proteins are for the most part integral proteins inserted into the thylakoid membrane cotranslationally (9, 63, 64). The production of functional PSII units also requires a large number of cofactors including chlorophyll, heme, carotenoid, quinone, Fe, and Mn.

Coordinated synthesis and assembly of the chloroplast protein complexes is accomplished in part through coactivation of nuclear and chloroplast gene transcription by cellular or environmental signals (75, 151, 177). Chloroplast polypeptide accumulation is also regulated at the levels of translation (16, 17,

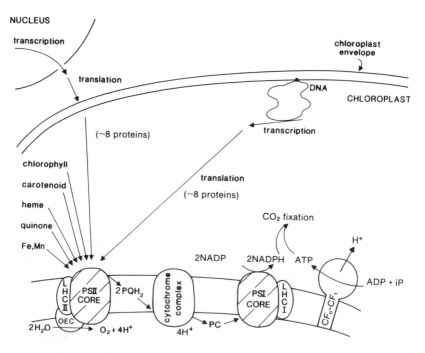

Figure 3 Diagram showing the four major protein complexes of the chloroplast inner membrane, which carry out photosynthetic electron transport. Also shown in the diagram are the proteins and cofactors that are assembled into PSII. LHCI and LHCII refer to chlorophyll *a/b* binding protein units that serve a light-harvesting function. CF_1-CF_0 is the chloroplast ATP synthetase; PC = plastocyanin; PQ = plastoquinone; OEC = proteins of the oxygen evolving complex.

24, 65, 73, 79, 83, 87, 104, 107, 146) and protein turnover (13, 43, 74, 139, 173). Posttranscriptional regulation of plastid gene expression is not surprising for several reasons. First, the assembly of some thylakoid protein complexes occurs in a stepwise fashion, and translation and assembly of polypeptides and cofactors are coupled (43, 74, 83, 154). Second, if plastid transcripts are stable relative to cytoplasmic transcripts or if plastid and nuclear gene transcription does not respond to environmental signals at the same rate, then regulation of protein translation or protein turnover could help coordinate the accumulation of chloroplast and cytoplasmically derived polypeptides.

PLASTID TRANSCRIPTION AND THE DETERMINATION OF RNA LEVELS

In this section two basic questions concerning chloroplast RNA are addressed: (a) What is the basis of changes in overall chloroplast RNA levels during development; and (b) how the relative ratios of chloroplast RNAs established, and what mechanisms cause these ratios to vary during chloroplast biogenesis? I examine first mechanisms that regulate chloroplast transcription and then discuss posttranscriptional events that influence RNA levels.

Regulation of Chloroplast Transcription

Plastid transcription activity and RNA levels are high during rapid chloroplast growth in barley and both decline when this phase of chloroplast development is completed (111). Furthermore, when 4.5-day-old dark-grown barley seedlings are illuminated, rbcL mRNA levels and transcription of rbcL change in parallel during chloroplast maturation (111). In spinach, plastid transcription activity is increased when chloroplast development is induced by light (36). A similar observation has been made in maize (5, 132, 179). These results document the importance of transcription activity in the overall determination of plastid RNA levels.

The induction of plastid transcription during chloroplast biogenesis could result from increased DNA template levels (10). This suggestion is based on the observation that plastid DNA copy number can increase up to 7.5-fold during the transition from proplastids to chloroplasts (108). This view implies that chloroplast transcription activity is template limited. In spinach, however, transcription activity per DNA template varies up to 5-fold during chloroplast biogenesis (36).

Changes in overall plastid transcription could be induced by changes in DNA conformation. Several in vitro studies have shown that chloroplast RNA polymerase is more active and has higher fidelity on supercoiled templates

(92, 123, 150, 158, 159). In addition, compounds that alter plastid DNA conformation in *Chlamydomonas* change plastid transcription (159). However, although these studies are suggestive, no direct link has been made between changes in DNA conformation in vivo and alterations in plastid transcription activity.

Another way to modify plastid transcription activity is to change the level of RNA polymerase in plastids. One RNA polymerase isolated from chloroplasts is a large multisubunit enzyme (23, 95, 96, 148, 156). This enzyme may contain three chloroplast-encoded subunits *(rpoA, B, C)* with homology to *E. coli* RNA polymerase subunits (33, 110, 121). Other work suggests that some of the chloroplast RNA polymerase subunits are encoded on nuclear genes (95). Therefore, the induction of plastid RNA polymerase levels could require the synthesis of a limiting nuclear-encoded subunit. Alternatively, induction could be autocatalytic and directly sensitive to factors that alter RNA polymerase activity.

In addition to the regulation of overall plastid RNA levels, there is also variation in the relative ratios of individual plastid RNAs. These ratios are determined by a number of factors. For example, promoter strength and competition between promoters have been reported to influence the relative transcription rate of chloroplast genes. Promoters of chloroplast tRNA (51, 56, 57), rRNA (78, 97, 152), and protein genes (25, 55, 98) have been examined by comparative sequence analysis (91, 180) and transcription in heterologous and homologous extracts (25, 75, 98). These approaches showed that chloroplast genes, with the possible exception of *trnR1* and *trnS1* (54), have prokaryotic-like promoter elements 10 and 35 bp upstream from the site of transcription initiation and that promoter strength differs from gene to gene. Plastid genes with strong promoters were transcribed at higher rates in vivo relative to genes with weaker promoters (36).

Posttranscriptional Mechanisms that Alter RNA Levels

The stability of plastid RNA in vascular plants has not been determined directly. The influence of RNA stability on RNA levels has been assessed by comparing changes in plastid gene transcription to changes in plastid RNA levels (37, 111). As mentioned above, this type of examination for some genes shows parallel changes between RNA levels and gene transcription (111). However, this is not always the case, and three RNA populations whose stability is influenced by posttranscriptional events are discussed below.

DETERMINATION OF RIBOSOMAL RNA LEVELS   The genes encoding plastid ribosomal RNAs (16S, 23S, 4.5S, 5S) are contiguous and present in one or two copies per genome. The four rRNAs are cotranscribed in some plants

(151); in other plants the 5S RNA may be derived from a separate transcription unit (78; see discussion in 34). The abundance of ribosomal RNA in plastids correlates with a high rate of transcription (37, 111). Such high rates may result from selective transcription of the rDNA transcription unit (52, 93, 114, 128).

In barley, the ratio of transcription of rDNA and protein genes such as *psbA* varies up to 5-fold during chloroplast development (111). The change in relative transcription rate could result from independent regulation of the two RNA polymerases described above. Alternatively, a single RNA polymerase may transcribe rDNA and *psbA*. In this case, the change in relative transcription rate may be explained by a change in RNA polymerase to template ratio or direct alteration of RNA polymerase-promoter interactions.

The high levels of rRNA could also result from stability conferred by association of the RNA with ribosomal proteins. Evidence for this possibility was reported in dark-grown barley where rRNA levels remained high in 4.5–9-day-old plants even though rRNA transcription declined dramatically over the same period (111). In illuminated barley, rRNA transcription declined rapidly ($t_{1/2}$ = 14 hr) whereas rRNA levels decreased more slowly ($t_{1/2}$ = 79 hr), suggesting that rRNA levels are determined in part by RNA stability.

DETERMINATION OF *psbA* mRNA LEVELS   *PsbA* encodes a 32-kDa quinone-binding protein associated with PSII (118, 149). The *psbA* gene product is translated at high rates in mature chloroplast populations of illuminated barley plants (84). Not surprisingly, *psbA* mRNA is also found at high levels in these chloroplasts (84). High levels of *psbA* mRNA are established in part by a strong promoter (55). In addition, two lines of evidence suggest that *psbA* transcript abundance is regulated posttranscriptionally. First, an increase in *psbA* mRNA relative to plastid rRNA or total RNA has been observed when dark-grown spinach (37), maize (9), or *Spirodella* (53) plants are illuminated. In spinach the increase in *psbA* transcript level was not correlated with an increase in *psbA* transcription relative to rDNA transcription (37). Second, illumination of a 4.5-day-old dark-grown barley seedlings caused a 7.5-fold decline in plastid transcription activity (111). *psbA* transcription declined 60% whereas transcription of rDNA, *rbcL*, and *psaA* declined 80–95%. Surprisingly, *psbA* RNA levels decreased little in the plastids of illuminated barley plants, which indicates that an increase in *psbA* transcript stability occurred. The basis of the increase in *psbA* mRNA stability is unknown. It is interesting to note, however, that approximately 80% of the *psbA* transcripts are located in the nonpolysomal fraction of the plastid stroma (85). These transcripts could be stabilized through formation of mRNP particles.

DETERMINATION OF *psb*D-*psb*C TRANSCRIPT ABUNDANCE   *psb*D and *psb*C are plastid genes which encode PSII proteins. *psb*D and *psb*C are contiguous in tobacco (144), barley (14), spinach (2, 69), and pea (15). In barley, *psb*D and *psb*C are part of an approximately 5.7-kbp transcription unit that includes several tRNAs and a 62-amino-acid open reading frame (Figure 4). Six RNAs ranging in size from 1.7 to 5.7 kb plus several RNAs of tRNA size are produced from this transcription unit in dark-grown plants (14). The sizes and approximate location of the transcripts are shown at the top of Figure 4. The transcript 5' ends could be produced by transcription initiation. Alternatively, the transcript 5' ends could arise from RNA processing, as has been documented for several other chloroplast RNAs (60, 113, 172). In a similar way the *psb*D-*psb*C transcript 3' ends could be produced by termination of transcription or RNA processing.

Illumination of a 4.5-day-old dark-grown barley causes a reduction in the level of the six large transcripts produced from the *psb*D-*psb*C transcription unit in dark-grown plants (Figure 4). These six transcripts are replaced by two new transcripts that have a 5' end approximately 150 nucleotides downstream from the 5' end of two transcripts present in dark-grown plants (Figure 4). The shift in transcript population is light induced and occurs concomitantly with a 2-fold decline in the rate of *psb*D-*psb*C transcription (P. E. Gamble, T. Berends, J. E. Mullet, unpublished observation). Therefore, the accumulation of the light-induced transcripts that hybridize to the *psb*D-*psb*C transcription unit results in part from increased stability of these transcripts relative to the RNAs produced in plastids of dark-grown plants.

The three examples described above indicate that plastid RNA levels are determined by regulation of transcription and RNA stability. The rate of transcription of plastid genes is influenced by promoter strength, changes in plastid RNA polymerase activity, and possibily DNA conformation. The molecular basis of the changes in RNA stability may involve protein-RNA associations or RNA processing.

REGULATION OF CHLOROPLAST TRANSLATION

The coordination of subunit and cofactor production in plastids of peas is accomplished in part by coinduction of leaf development and transcription of nuclear and plastid genes (134, 135, 160). In contrast, *Volvox* accumulates RNAs for proteins involved in photosynthesis in dark-grown cells, but translation of the proteins is inhibited (79). Illumination of cultures activates translation of the stored RNA allowing *Volvox* to become photosynthetically competent. Regulation of chloroplast translation by light has also been observed in *Euglena* (24, 104, 107) and *Chlamydomonas* (65). In vascular

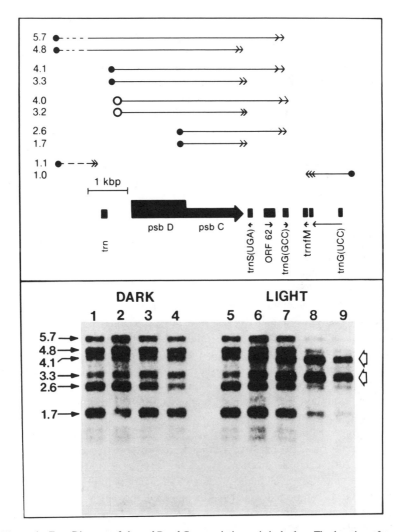

Figure 4   Top: Diagram of the *psbD-psbC* transcription unit in barley. The location of transcripts in plastids of dark-grown plants that hybridize to the transcription unit are indicated by arrows (→). Transcripts that accumulate in plastids of illuminated plants are noted by an open circle (○ →). *Bottom:* Northern hybridization results showing the plastid RNAs from 4.5-, 6-, 7.5- and 9-day-old dark-grown barley plants that hybridize to *psbC* (lanes 1–4). Lanes 5–9 show the changes in plastid RNAs that hybridize to *psbC* when 4.5-day-old dark-grown plants (lane 5) are illuminated for 16, 36, 72, or 108 hr (lanes 6–9). The numbers next to lane 1 of the autoradiogram refer to approximate RNA sizes in kb. The two open arrows at the right indicate the position of the light-induced transcripts.

plants there are several reports of translational control in plastids (16, 17, 73, 83, 87). For example, in dark-grown *Amaranthus* cotyledons, translation of the large subunit of RUBISCO declines in older plants without a parallel change in *rbcL* mRNA level (16, 17). Translational control of the large subunit has also been reported in dark-grown peas (73). In *Spirodella, psbA* gene-product translation is inhibited when plants are placed in darkness (45). The decrease in *psbA* gene-product translation was not accompanied by a decrease in *psbA* mRNA. In the cases described above, it has not been determined if translation is inhibited at the level of translation initiation or in subsequent steps. Furthermore, it is unclear whether the inhibition of translation of the large subunit of RUBISCO or the *psbA* gene product is selective or part of a general inhibition of plastid translation.

The Cytoplasmic Control Principle

In 1977 Ellis (40) proposed the "cytoplasmic control principle" to explain the coordination of chloroplast and nuclear gene expression. In general, this principle states that cytoplasmic translation products control transcription and translation of plastid gene products but not the reverse. This proposal was based on studies of mutants and the effects of protein synthesis inhibitors. These studies showed that cytoplasmic protein synthesis continued when mitochondrial protein synthesis was inhibited but that inhibition of cytoplasmic protein synthesis caused an arrest of organelle subunit production (72, 163). Inhibitors of cytoplasmic translation brought about a similar inhibition of chloroplast protein synthesis (117). In more recent studies, it has become clear that inhibition of chloroplast protein synthesis can also alter production of nuclear gene products (21, 119).

Ellis suggested that an economical way to regulate the synthesis of chloroplast subunits was by negative feedback control by free or unassembled subunits. Although this mechanism has not yet been shown to operate in chloroplasts, it is involved in the production of a variety of proteins in *E. coli*. One of the first-reported and best-documented examples of this type of regulation concerns ribosomal proteins. Unassembled ribosomal proteins inhibit translation of additional ribosomal subunits by binding to mRNA that encodes the proteins (reviewed in 48). This type of regulation may also occur in plastids.

Regulation of Chlorophyll-Apoprotein Synthesis

Chlorophyll is a key chromophore in vascular plants because of its role in photon absorption and primary charge separation. To carry out these functions chlorophyll and other cofactors such as carotenoids, quinones, and hemes are noncovalently bound to proteins. These proteins are in turn assembled into PSI and PSII, as outlined in Figure 3. Chlorophyll can mediate photooxidation of the chloroplast inner membrane in carotenoid-deficient plants (6, 88, 103) or if chlorophyll synthesis exceeds chlorophyll-apoprotein translation. This biological problem underlies much of the complex regulation of chlorophyll and chlorophyll-apoprotein synthesis.

Early studies of dark-grown vascular plants showed that these plants were deficient in PSI and PSII activity (126). The plastid populations of dark-grown plants were also deficient in a number of the chloroplast-encoded chlorophyll-apoproteins of PSI (155, 164; but see 115, 145) and PSII (84, 154). The absence of these proteins did not reflect a general loss of plastid proteins. Most of the soluble and membrane proteins found in chloroplasts were also present in etioplasts at similar levels (73, 84).

In 1985, Herrmann et al (67), noted that mRNA for the *psaA-psaB* genes (which encode the 65–70 kDa PSI chlorophyll-apoproteins) was present in dark-grown spinach even though the PSI chlorophyll-apoproteins were undetectable. A similar observation was made in barley, and it was shown that apoprotein levels increased when plants were illuminated without a concomitant increase in apoprotein mRNA level (83, 84, 87). These results were extended to the PSII chlorophyll-apoproteins of barley (84) and maize (164). Here again, apoproteins were at low levels or undetectable in dark-grown plants whereas RNAs for these genes were readily detectable. As with the PSI chlorophyll-apoproteins, illumination of dark-grown plants caused an accumulation of PSII chlorophyll-apoproteins with little change in mRNA content.

For the increase in chlorophyll-apoprotein levels in illuminated plants there are two straightforward explanations that do not involve a change in RNA: (*a*) The apoproteins are translated in plastids of dark-grown and illuminated plants but they are only stable in plastids of illuminated plants; and (*b*) the translation of chlorophyll-apoprotein mRNA is inhibited in plastids of dark-grown plants relative to light-grown plants.

It is worth noting at this point that nuclear-encoded chlorophyll-*a*/*b*-binding proteins are unstable in chlorophyll-deficient plants (4, 11). The actual turnover rate of the chlorophyll-*a*/*b*-apoproteins in the absence of chlorophyll is not known precisely. However, the proteins can be detected in chlorophyll-deficient plants by in vivo labeling with ^{35}S-methionine (11). Klein & Mullet (83) examined the synthesis of chloroplast-encoded chlorophyll-apoprotein translation in dark-grown barley in vivo and in isolated organelles. Chlorophyll-apoprotein synthesis was very low or undetectable in etioplasts in contrast to other soluble and membrane proteins (Figure 5). Furthermore, illumination of plants caused a selective increase in the incorporation of ^{35}S-methionine into the chlorophyll-apoproteins in the absence of changes in RNA (Figure 5). The rapid increase in amino acid incorporation into the chlorophyll-apoproteins in the absence of changes in mRNA is consistent with activation of translation. Rapid stabilization of the aproproteins, however, cannot be entirely ruled out. Pulse-labeling assays show that if the chlorophyll-apoproteins are being synthesized and rapidly degraded in etioplasts then this is occurring either during translation of the proteins or with a half-time significantly shorter than five minutes (83).

Illumination of dark-grown barley plants caused a rapid increase in ^{35}S-methionine incorporation into the chloroplast-encoded chlorophyll-apoproteins (83). This treatment also causes the photoreduction of protochlorophyllide to chlorophyllide, which is subsequently converted to chlorophyll in 10–30 minutes depending on plant age (reviewed in 32). Therefore, the photoreceptor for chlorophyll-apoprotein translation could be protochlorophyllide and the activator chlorophyll(ide). This idea is consistent with the observations that nearly all of the proteins exhibiting light-induced translation have been shown to bind chlorophyll (83, 84).

The translation of the chlorophyll-apoproteins could be blocked at the level of translation initiation, association of nascent chains with membranes, during elongation, or during the release of the mature apoprotein from the ribosome-mRNA complex. To address these possibilities the distributions of apoprotein mRNA between soluble and membrane phases and within polysome gradients of dark-grown and illuminated plants were compared (85). Surprisingly, the psaA-psaB mRNA was localized almost exclusively in large membrane-bound polysomes in plastids isolated from dark-grown and illuminated plants. The distribution of mRNA indicates that synthesis of the chlorophyll-apoproteins in dark-grown plants is inhibited late in apoprotein translation elongation or at the point of apoprotein-mRNA-ribosome dissociation. Activation of translation by chlorophyll(ide) therefore could occur by interaction between the chromophore and the apoprotein causing chlorophyll-protein dissociation from the ribosome-mRNA complex. The release of the chlorophyll-apoprotein would then allow additional rounds of translation to occur.

Regulation of translation elongation is not unique to chloroplast proteins. The translation of secretory proteins can be inhibited during elongation by the signal recognition particle (166). Catalase mRNA in dark-grown plants also exhibits inhibition of translation elongation (146). Furthermore, analysis of PSII mutants in Chlamydomonas shows that translation of PSII apoproteins is regulated by nuclear gene products and is altered by the absence of translation of other PSII polypeptides (74). This suggests that a coupling exists between translation and assembly of some PSII proteins.

Finally, it should be noted that mutants that fail to synthesize one or more chloroplast-encoded polypeptides of the PSII complex are deficient in other chloroplast-encoded PSII subunits (13, 43, 74). This results in part from increased turnover of unassembled or partially assembled subunits. Protein

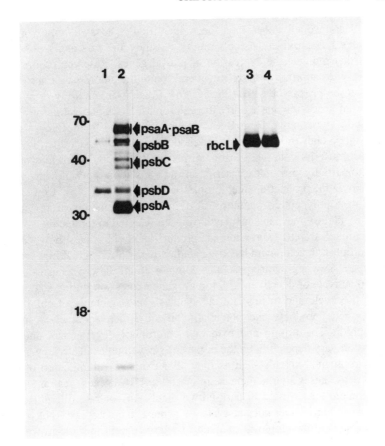

Figure 5   Autoradiogram of membrane (lanes 1, 2) and soluble (lanes 3, 4) proteins radiolabeled with ^{35}S-methionine in plastids isolated from dark-grown plants (lanes 1, 3) or plants illuminated for 16 hr (lanes 2, 4). Molecular masses are indicated at the left of the figure (kDa). Gene products that are the major radiolabeled products in plastids are indicated.

turnover is also increased in the absence of cofactor binding (106) and when abnormal or truncated proteins are synthesized in plastids (99, 112). These activities apparently serve to maintain subunit stoichiometry by removing unassembled or damaged polypeptides.

CONCLUDING REMARKS

In a review in 1977, S. D. Kung noted that three plastid proteins were known to be encoded on chloroplast DNA (90). Over the 10 years since that review, the entire plastid DNA sequence has been obtained for two plants and over 80 chloroplast-encoded genes have been identified. The availability of gene sequences has provided an opportunity to study the *cis-* and *trans*-acting factors that regulate the expression of plastid genes. Research on this topic using in vitro approaches have in general found the chloroplast transcription signals to be prokaryotic-like (25, 55, 98). In addition, an array of posttranscriptional processes that modify the chloroplast RNA population have been observed. These range from RNA splicing (38, 177) to the processing of transcript 5' ends (60, 113, 172) and modulation of RNA stability (36, 111). The posttranscriptional modification of plastid RNA populations may provide a way to partially uncouple transcription and translation, thus allowing greater control over the production of chloroplast-encoded polypeptides. Control over chloroplast polypeptide synthesis also involves regulation of translation. These regulatory mechanisms may have been selected to tighten the coupling between nuclear gene expression and production of chloroplast polypeptides.

Chloroplast development has been studied in a wide range of higher plants and algae. Two regulatory problems common to all these organisms are the need to activate or deactivate chloroplast gene expression during some phase of the life cycle and to regulate the synthesis and assembly of chloroplast/ nuclear gene products and potentially photodestructive cofactors such as chlorophyll. The timing and extent of chloroplast development and gene expression vary depending on the organism's developmental strategy and use of environmental signals such as light. For example, illumination of dark-grown peas causes a large increase in plastid mRNA because leaf and plastid development are arrested in dark-grown plants. In contrast, illumination of dark-grown barley does not cause a large increase in chloroplast RNA because leaf and plastid development proceed to a large extent in dark-grown seedlings.

Finally, there remain questions concerning the cell and tissue specificity of chloroplast development, the molecular basis of light regulation, and mechanisms that determine plastid number, size, and composition of plant cells. The answers to these questions are likely to provide insight into plant development as well as chloroplast biogenesis.

ACKNOWLEDGMENTS

I would like to thank Sharyll Pressley for her skillful secretarial assistance in preparing this review and Bob Klein and Patricia Gamble for their comments on the manuscript. This work was supported in part by the Texas Agricultural Experiment Station, and grants from NIH (GM 37987-01) and NSF (DCB 86-16156). This article was based on literature available prior to August 1, 1987.

Literature Cited

1. Aguettaz, P., Seyer, P., Pesey, H., Lescure, A.-M. 1987. Relations between the plastid gene dosage and levels of 16S rRNA and rbcL gene transcripts during amyloplast to chloroplast change in mixotrophic spinach cell suspensions. *Plant Mol. Biol.* 8:169–77

2. Alt, J., Morris, J., Westhoff, P., Herrmann, R. G. 1984. Nucleotide sequence of the clustered genes for the 44 kd chlorophyll *a* aproprotein and the "32 kd"-like protein of the photosystem II reaction center in the spinach plastid chromosome. *Curr. Genet.* 8:597–606

3. Andersson, B., Jansson, C., Ljungberg, U., Akerlund, H.-E. 1985. Polypeptides in water oxidation. In *Molecular Biology of the Photosynthetic Apparatus,* ed. K. E. Steinback, S. Bonitz, C. J. Arntzen, L. Bogorad, pp. 21–31. Cold Spring Harbor, New York: Cold Spring Harbor Laboratory

4. Apel, K. 1979. Phytochrome-induced appearance of mRNA activity for the apoprotein of the light-harvesting chlorophyll *a/b* protein of barley *(Hordeum vulgare)*. *Eur. J. Biochem.* 97:183–88

5. Apel, K., Bogorad, L. 1976. Light-induced increase in the activity of maize plastid DNA dependent RNA polymerase. *Eur. J. Biochem.* 67:615–20

6. Bartels, P. G., Watson, C. W. 1978. Inhibition of carotenoid synthesis by fluridone and norflurazon. *Weed Sci.* 26:198–203

7. Bassi, R., Peruffo, A. D. B., Barbato, R., Ghisi, R. 1985. Differences in chlorophyll-protein complexes and composition of polypeptides between thylakoids from bundle sheaths and mesophyll cells in maize. *Eur. J. Biochem.* 146:589–95

8. Bedbrook, J. R., Kolodner, R. 1979. The structure of chloroplast DNA. *Ann. Rev. Plant Physiol.* 30:593–620

9. Bedbrook, J. R., Link, G., Coen, D. M., Bogorad, L., Rich, A. 1978. Maize plastid gene expressed during photoregulated development. *Proc. Natl. Acad. Sci. USA* 75:3060–64

10. Bendich, A. J. 1987. Why do chloroplasts and mitochondria contain so many copies of their genome? *BioEssays* 6: 279–82

11. Bennett, J. 1981. Biosynthesis of the light-harvesting chlorophyll *a/b* protein. *Eur. J. Biochem.* 118:61–70

12. Bennett, J., Jenkins, G. I., Hartley, M. R. 1984. Differential regulation of the accumulation of the light-harvesting chlorophyll *a/b* complex and ribulose bisphosphate carboxylase/oxygenase in greening pea leaves. *J. Cell. Biochem.* 25:1–13

13. Bennoun, P., Spierer-Herz, M., Erickson, J., Girard-Bascou, J., Pierre, Y., Delosme, M., Rochaix, J.-D. 1986. Characterization of photosystem II mutants of *Chlamydomonas reinhardtii* lacking the psbA gene. *Plant Mol. Biol.* 6:151–60

14. Berends, T., Gamble, P. E., Mullet, J. E. 1987. Characterization of the barley chloroplast transcription units containing psaA-psaB and psbD-psbC. *Nuc. Acids Res.* 15:5217–40

15. Berends, T., Kubicek, Q., Mullet, J. E. 1986. Localization of the genes coding for the 51 kDa PSII chlorophyll apoprotein, apocytochrome b6, the 65–70 kDa PSI chlorophyll apoproteins and the 44 kDA PSII chlorophyll apoprotein in pea chloroplast DNA. *Plant Mol. Biol.* 6: 125–34

16. Berry, J. O., Nikolau, B. J., Carr, J. P., Klessig, D. F. 1985. Transcriptional and post-transcriptional regulation of ribulose 1,5-bisphosphate carboxylase gene expression in light- and dark-grown Amaranth cotyledons. *Mol. Cell. Biol.* 5:2238–46

17. Berry, J. O., Nikolau, B. J., Carr, J. P., Klessig, D. F. 1986. Translational regulation of light-induced ribulose 1,5-bisphosphate carboxylase gene expression in Amaranth. *Mol. Cell. Biol.* 6:2347–53

18. Bjorkman, O., Boardman, N. K., Anderson, J. M., Thorne, S. W., Goodchild, D. J., Pyliotis, N. A. 1972. Effect of light during growth of *Atriplex patula* on the capacity of photosynthetic reactions, chloroplast components and structures. *Carnegie Inst. Yearb.* 71:115–35

19. Boffey, S. A., Leech, R. M. 1982. Chloroplast DNA levels and the control of chloroplast division in light-grown wheat leaves. *Plant Physiol.* 69:1387–91

20. Bohnert, H. J., Crouse, E. J., Schmitt, J. M. 1982. Organization and expression of plastid genomes. In *Encyclopedia of Plant Physiology, New Series,* ed. B. Parthier, D. Boulter, 14B:475–530. Berlin/Heidelberg/New York: Springer-Verlag. 763 pp.

21. Böner, T., Mendel, R. R., Schiemann, J. 1986. Nitrate reductase is not ac-

cumulated in chloroplast-ribosome-deficient mutants of higher plants. *Planta* 169:202–7

22. Borroto, K. E., Dure, L. III. 1986. The expression of chloroplast genes during cotton embryogenesis. *Plant Mol. Biol.* 7:105–13

23. Bottomley, W., Smith, H. J., Bogorad, L. 1971. RNA polymerases of maize: Partial purification and properties of the chloroplast enzyme. *Proc. Natl. Acad. Sci. USA* 68:2412–16

24. Bouet, C., Schantz, R., Dubertret, G., Pineau, B., Ledoigt, G. 1986. Translational regulation of protein synthesis during light-induced chloroplast development of *Euglena*. *Planta* 167:511–20

25. Bradley, D., Gatenby, A. A. 1985. Mutational analysis of the maize chloroplast ATPase-β subunit gene promoter: the isolation of promoter mutants in *E. coli* and their characterization in a chloroplast *in vitro* transcription system. *EMBO J.* 4:3641–48

26. Bradbeer, J. W. 1980. Development of photosynthetic function during chloroplast development. In *The Biochemistry of Plants*, ed. M. D. Hatch, N. K. Boardman, 8:423–67. New York: Academic. 521 pp.

27. Briat, J. F., Lescure, A. M., Mache, R. 1986. Transcription of the chloroplast DNA: A review. *Biochimie* 68:981–90

28. Briat, J.-F., Bisanz-Seyer, C., Lescure, A.-M. 1987. *In vitro* transcription initiation of the rDNA operon of spinach chloroplast by a highly purified soluble homologous RNA polymerase. *Curr. Genet.* 11:259–63

29. Broglie, R., Coruzzi, G., Keith, B., Chua, N.-H. 1984. Molecular biology of C₄ photosynthesis in *Zea mays:* differential localization of proteins and mRNAs in the two leaf cell types. *Plant Mol. Biol.* 3:431–44

30. Browse, J., McCourt, P., Somerville, C. R. 1985. A mutant of *Arabidopsis* lacking a chloroplast-specific lipid. *Science* 227:763–65

31. Butler, R. D. 1963. The effect of light intensity on stem and leaf growth in broad bean seedlings. *J. Exp. Bot.* 14:142–52

32. Castelfranco, P. A., Beale, S. I. 1983. Chlorophyll biosynthesis: Recent advances and areas of current interest. *Ann. Rev. Plant Physiol.* 34:241–78

33. Cozens, A. L., Walker, J. E. 1986. Pea chloroplast DNA encodes homologues of *Escherichia coli* ribosomal subunit S2 and the β'-subunit of RNA polymerase. *Biochem. J.* 236:453–60

34. Crouse, E. J., Bohnert, H. J., Schmitt, J. M. 1984. Chloroplast RNA synthesis. In *Chloroplast Biogenesis,* ed. R. J. Ellis, 5:83–135. Cambridge: Cambridge Univ. Press. 359 pp.

35. Dean, C., Leech, R. M. 1982. Genome expression during normal leaf development. *Plant Physiol.* 69:904–10

36. Deng, X.-W., Gruissem, W. 1987. Control of plastid gene expression during development: the limited role of transcriptional regulation. *Cell* 49:379–87

37. Deng, X.-W., Stern, D. B., Tonkyn, J. C., Gruissem, W. 1987. Plastid run-on transcription. *J. Biol. Chem.* 262:9641–48

38. Deno, H., Sugiura, M. 1984. Chloroplast tRNA^{Gly} gene contains a long intron in the D stem: nucleotide sequences of two chloroplast genes for tRNA^{Gly}(UCC) and tRNA^{Arg} (UCU). *Prot. Natl. Acad. Sci. USA* 81:405–8

39. Dyer, T. A., Miller, R. H., Greenwood, A. D. 1971. Leaf nucleic acids. I. Characteristics and role in the differentiation of plastids. *J. Exp. Bot.* 22:125–36

39a. Edelman, M., Schiff, J. A., Epstein, H. T. 1965. Studies of chloroplast development in *Euglena*. XII. Two types of satellite DNA. *J. Mol. Biol.* 11:769–74

40. Ellis, R. J. 1977. Protein synthesis by isolated chloroplasts. *Biochim. Biophys. Acta* 463:185–215

41. Ellis, R. J. 1981. Chloroplast proteins: synthesis, transport and assembly. *Ann. Rev. Plant Physiol.* 32:111–37

42. Ellis, J. R., Leech, R. M. 1985. Cell size and chloroplast size in relation to chloroplast replication in light-grown wheat leaves. *Planta* 165:120–25

43. Erickson, J. M., Rahire, M., Malnoë, P., Girard-Bascou, J., Pierre, Y., Bennoun, P., Rochaix, J.-D. 1986. Lack of the D2 protein in a *Chlamydomonas reinhardtii psbD* mutant affects photosystem II stability and D1 expression. *EMBO J.* 5:1754–54

44. Fluhr, R., Kuhlemeir, C., Nagy, F., Chua, N.-H. 1986. Organ specific and light-induced expression of plant genes. *Science* 232:1106–12

45. Fromm, H., Devic, M., Fluhr, R., Edelman, M. 1985. Control of *psbA* gene expression: In mature *Spirodela* chloroplasts light regulation of 32-kd protein synthesis is independent of transcript level. *EMBO J.* 4:291–95

46. Gaff, D. F., Zee, S.-Y., O'Brien, T. P. 1976. The fine structure of dehydrated and reviving leaves of *Borya nitida* Labill. A desiccation tolerant plant. *Aust. J. Bot.* 24:225–36

47. Glazer, A. N., Melis, A. 1987. Photochemical reaction centers: structure, organization and function. *Ann. Rev. Plant Physiol.* 30:11–45
48. Gottesman, S. 1984. Bacterial regulation: global regulatory networks. *Ann. Rev. Genet.* 18:415–41
49. Gounaris, I., Michalowski, C. B., Bohnert, H. J., Price, C. A. 1986. Restriction and gene maps of plastid DNA from *Capsicum annuum. Curr. Genet.* 11:7–16
50. Gounaris, K., Barber, J., Harwood, J. L. 1986. The thylakoid membranes of higher plant chloroplasts. *Biochem. J.* 237:313–26
51. Greenberg, B. M., Hallick, R. B. 1986. Accurate transcription and processing of 19 *Euglena* chloroplast tRNAs in a *Euglena* soluble extract. *Plant Mol. Biol.* 6:89–100
52. Greenberg, B. M., Narita, J. O., De-Luca-Flaherty, C., Gruissem, W., Rushlow, K. A., Hallick, R. B. 1984. Evidence for two RNA polymerase activities in *Euglena gracilis* chloroplasts. *J. Biol. Chem.* 259:14880–87
53. Gressel, J. 1977. Light requirements for the enhanced synthesis of a plastid mRNA during *Spirodela* greening. *Photobiochem. Photobiol.* 27:167–69
54. Gruissem, W., Elsner-Menzel, C., Latshaw, S., Narita, J. O., Schaffer, M. A., Zurawski, G. 1986. A subpopulation of spinach chloropast tRNA genes does not require upstream promoter elements for transcription. *Nuc. Acids Res.* 14:7541–56
55. Gruissem, W., Zurawski, G. 1985. Analysis of promoter regions for the spinach chloroplast *rbc*L, *atp*B and *psb*A genes. *EMBO J.* 4:3375–83
56. Gruissem, W., Zurawski, G. 1985. Identification and mutational analysis of the promoter for a spinach chloroplast transfer RNA gene. *EMBO J.* 4:1637–44
57. Gruissem, W., Greenberg, B. M., Zurawski, G., Prescott, D. M., Hallick, R. B. 1983. Biosynthesis of chloroplast transfer RNA in a spinach chloroplast transcription system. *Cell* 25:815–28
58. Hallick, R. B., Hollingsworth, M. J., Nickoloff, J. A. 1984. Transfer RNA genes of *Euglena gracilis* chloroplast DNA. *Plant Mol. Biol.* 3:169–75
59. Hanley-Bowdoin, L., Chua, N.-H. 1987. Chloroplast promoters. *TIBS* 12:67–70
60. Hanley-Bowdoin, L., Orozco, E. M. Jr., Chua, N.-H. 1985. *In vitro* synthesis and processing of a maize chloroplast transcript encoded by the ribulose 1,5-bisphosphate carboxylase large subunit gene. *Mol. Cell. Biol.* 5:2733–45
61. Hoyer-Hansen, G., Simpson, D. J. 1977. Changes in the polypeptide composition of internal membranes of barley plastids during greening. *Carlsberg Res. Commun.* 42:379–89
62. Harpster, M., Apel, K. 1985. The light-dependent regulation of gene expression during plastid development in higher plants. *Physiol. Plant.* 64:147–52
63. Herrin, D., Michaels, A. 1985. The chloroplast 32 kDa protein is synthesized on thylakoid-bound ribosomes in *Chlamydomonas reinhardtii. FEBS* 184:90–94
64. Herrin, D., Michaels, A., Hickey, E. 1981. Synthesis of a chloroplast membrane polypeptide on thylakoid-bound ribosomes during the cell cycle of *Chlamydomonas reinhardtii* 137. *Biochim. Biophys. Acta* 655:136–45
65. Herrin, D. L., Michaels, A. S., Paul, A.-L. 1986. Regulation of genes encoding the large subunit of ribulose-1,5-bisphosphate carboxylase and the Photosystem II polypeptides D-1 and D-2 during the cell cycle of *Chlamydomonas reinhardtii. J. Cell Biol.* 103:1837–45
66. Herrmann, R. G., Possingham, J. V. 1980. Plastid DNA—the plastome. In *Results and Problems in Cell Differentiation: Chloroplasts,* ed. J. Reinert, 10:45–94. Berlin/New York: Springer-Verlag. 240 pp.
67. Herrmann, R. G., Westhoff, P., Alt, J., Tittgen, J., Nelson, N. 1985. Thylakoid membrane proteins and their genes. In *Molecular Form and Function of the Plant Genome,* ed. L. van Vloten-Doting, G. S. P. Groot, T. C. Hall, pp. 233–56. Amsterdam: Plenum
68. Heslop-Harrison, J. 1972. Sexuality of angiosperms. In *Plant Physiology,* ed. F. C. Steward, 6C:133–89. New York/London: Academic. 450 pp.
69. Holschuh, K., Bottomley, W., Whitfeld, P. R. 1984. Structure of the spinach chloroplast genes for the D2 and 44 kd reaction-centre proteins of photosystem II and for tRNA^{Ser} (UGA). *Nuc. Acids. Res.* 12:8819–34
70. Horrum, M. A., Schwartzbach, S. D. 1980. Nutritional regulation of organelle biogenesis in *Euglena. Plant Physiol.* 65:382–86
71. Huff, A. 1984. Sugar regulation of plastid interconversions in epicarp of citrus fruit. *Plant Physiol.* 76:307–12
72. Ibrahim, N. G., Beattie, D. S. 1976. Regulation of mitochondrial protein synthesis at the polyribosomal level. *J. Biol. Chem.* 251:108–15

73. Inamine, G., Nash, B., Weissbach, H., Brot, N. 1985. Light regulation of the synthesis of the large subunit of ribulose-1,5-bisphosphate carboxylase in peas: evidence for translational control. *Proc. Natl. Acad. Sci. USA* 82:5690–94

74. Jensen, K. H., Herrin, D. L., Plumley, G., Schmidt, G. W. 1986. Biogenesis of photosystem II complexes: transcriptional, translational, and posttranslational regulation. *J. Cell Biol.* 103:1315–25

75. Jolly, S. O., McIntosh, L., Link, G., Bogorad, L. 1981. Differential transcription *in vivo* and *in vitro* of two adjacent maize chloroplast genes: the large subunit of ribulose bisphosphate carboxylase and the 2.2-kilobase gene. *Proc. Natl. Acad. Sci. USA* 78:6821–25

76. Kannangara, C. G., Gough, S. P., Oliver, R. P., Rasmussen, S. K. 1984. Biosynthesis of Δ-aminolevulinate in greening barley leaves. VI. Activation of glutamate by ligation to RNA. *Carlsberg Res. Commun.* 49:417–37

77. Kaplan, S., Arntzen, C. J. 1982. Photosynthetic membrane and structure. In *Photosynthesis I*, ed. Govindjee, 3:65–80. New York/London: Academic. 458 pp.

78. Keus, R. J. A., Bekker, A. F., Kreuk, K. C. J., Groot, G. S. P. 1984. Transcription of ribosomal DNA in chloroplasts of *Spirodela oligorhiza. Curr. Genet.* 9:91–97

79. Kirk, M. M., Kirk, D. L. 1985. Translational regulation of protein synthesis in response to light, at a critical stage of *Volvox* development. *Cell* 41:419–28

80. Kirk, J. T. O., Tilney-Bassett, R. A. E. 1967. *The Plastids.* London: Freeman. 608 pp.

81. Kirk, J. T. O., Tilney-Bassett, R. A. E. 1978. *The Plastids: Their Chemistry, Structure, Growth and Inheritance.* Amsterdam/New York: Elsevier/North-Holland Biomedical Press. 650 pp. 2nd ed.

82. Kirk, J. T. O., Tilney-Bassett, R. A. E. 1978. See Ref. 81, pp. 568–661

83. Klein, R. R., Mullet, J. E. 1986. Regulation of chloroplast-encoded chlorophyll-binding protein translation during higher plant chloroplast biogenesis. *J. Biol. Chem.* 261:11138–45

84. Klein, R. R., Mullet, J. E. 1987. Control of gene expression during higher plant chloroplast biogenesis. *J. Biol. Chem.* 262:4341–48

85. Klein, R. R., Mason, H., Mullet, J. E. 1987. Light-regulated translation of chloroplast proteins. Transcripts of psaA-psaB, psbA and rbcL are associated with polysomes in dark-grown and illuminated barley seedlings. *J. Cell. Biol.* 106:289–302

86. Kowallik, K. V., Herrmann, R. G. 1974. Structural and functional aspects of the plastome DNA regions during plastid development. *Port. Acta Biol. Ser. A* 14:111–26

87. Kreuz, K., Dehesh, K., Apel, K. 1986. The light-dependent accumulation of the P700 chlorophyll *a* protein of the photosystem I reaction center in barley. *Eur. J. Biochem.* 159:459–67

88. Krinsky, N. I. 1979. Carotenoid protection against oxidation. *Pure & Appl. Chem.* 51:649–60

89. Kuhlemeier, C., Green, P. J., Chua, N.-H. 1987. Regulation of gene expression in higher plants. *Ann. Rev. Plant Physiol.* 38:221–57

90. Kung, S. D. 1977. Expression of chloroplast genomes in higher plants. *Ann. Rev. Plant Physiol.* 28:401–37

91. Kung, S. D., Lin, C. M. 1985. Chloroplast promoters from higher plants. *Nuc. Acids Res.* 13:7543–49

92. Lam, E., Chua, N.-H. 1987. Chloroplast DNA gyrase and *in vitro* regulation of transcription by template topology and novobiocin. *Plant Mol. Biol.* 8:415–24

93. Lebrun, M., Briat, J.-F., Laulhere, J.-P. 1986. Characterization and properties of the spinach chloroplast transcriptionally active chromosome isolated at high ionic strength. *Planta* 169:505–12

94. Leech, R. M. 1976. Plastid development in isolated etiochloroplasts and isolated etioplasts. In *Perspective in Experimental Biology*, ed. N. Sunderland, 2:145–62. New York: Pergamon. 515 pp.

95. Lerbs, S., Brautigam, E., Parthier, B. 1985. Polypeptides in DNA-dependent RNA polymerase of spinach chloroplasts: characterization by antibody-linked polymerase assay and determination of sites of synthesis. *EMBO J.* 4:1661–66

96. Lerbs, S., Briat, J.-F., Mache, R. 1983. Chloroplast RNA polymerase from spinach: purification and DNA-binding proteins. *Plant Mol. Biol.* 2:67–74

97. Lescure, A.-M., Bisanz-Seyer, C., Pesey, H., Mache, R. 1985. *In vitro* transcription initiation of the spinach chloroplast 16S rRNA gene at two tandem promoters. *Nuc. Acids Res.* 13:8787–96

98. Link, G. 1984. DNA sequence requirements for the accurate transcription of a protein-encoding plastid gene in a plastid *in vitro* system from mustard (*Sinapis alba* L.). *EMBO J.* 3:1697–1704

99. Liu, X.-Q., Jagendorf, A. T. 1986. A variety of chloroplast-located proteases. In *Regulation of Chloroplast Differentiation*, pp. 597–606. New York: Alan Liss

100. Malkin, S., Fork, D. C. 1981. Photosynthetic units of sun and shade plants. *Plant Physiol.* 67:580–83

101. Mayfield, S. P., Huff, A. 1986. Accumulation of chlorophyll, chloroplastic proteins and thylakoid membranes during reversion of chromoplasts to chloroplasts in *Citrus sinensis* epicarp. *Plant Physiol.* 81:30–35

102. Mayfield, S. P., Rahire, M., Frank, G., Zuber, H., Rochaix, J.-D. 1987. Expression of the nuclear gene encoding oxygen-evolving enhancer protein 2 is required for high levels of photosynthetic oxygen evolution in *Chlamydomonas reinhardtii*. *Proc. Natl. Acad. Sci. USA* 84:749–53

103. Mayfield, S. P., Taylor, W. C. 1984. Carotenoid-deficient maize seedlings fail to accumulate light-harvesting chlorophyll *a/b* binding protein (LHCP) mRNA. *Eur. J. Biochem.* 144:79–84

104. McCarthy, S. A., Schwartzbach, S. D. 1984. Absence of photoregulation of abundant mRNA levels in *Euglena*. *Plant Sci. Lett.* 35:61–66

105. McCourt, P., Kunst, L., Browse, J., Somerville, C. R. 1987. The effects of reduced amounts of lipid unsaturation on chloroplast ultrastructure and photosynthesis in a mutant of *Arabidopsis*. *Plant Physiol.* 84:353–60

106. Merchant, S., Bogorad, L. 1986. Rapid degradation of apoplastocyanin in Cu(II)-deficient cells of *Chlamydomonas reinhardtii*. *J. Biol. Chem.* 261:15850–53

107. Miller, M. E., Jurgenson, J. E., Reardon, E. M., Price, C. A. 1983. Plastid translation in organello and in vitro during light-induced development in *Euglena*. *J. Biol. Chem.* 258:14478–84

108. Miyamura, S., Nagata, T., Kuroiwa, T. 1986. Quantitative fluorescence microscopy on dynamic changes of plastid nucleoids during wheat development. *Protoplasma* 133:66–72

109. Monroy, A. F., Schwartzbach, S. D. 1984. Catabolite repression of chloroplast development in *Euglena*. *Proc. Natl. Acad. Sci. USA* 81:2786–90

110. Sijben-Müller, G., Hallick, R. B., Alt, J., Westhoff, P., Herrmann, R. G. 1986. Spinach plastid genes coding for initiation factor IF-1, ribosomal protein S11 and RNA polymerase α-subunit. *Nuc. Acids Res.* 14:1029–44

111. Mullet, J. E., Klein, R. R. 1987. Transcription and RNA stability are important determinants of higher plant chloroplast RNA levels. *EMBO J.* 6:1571–79

112. Mullet, J. E., Klein, R. R., Grossman, A. R. 1986. Optimization of protein synthesis in isolated higher plant chloroplasts. *Eur. J. Biochem.* 155:331–38

113. Mullet, J. E., Orozco, E. M. Jr., Chua, N.-H. 1985. Multiple transcripts for higher plant *rbcL* and *atpB* genes and localization of the transcription initiation site of the *rbcL* gene. *Plant Mol. Biol.* 4:39–54

114. Narita, J. O., Rushlow, K. E., Hallick, R. B. 1985. Characterization of a *Euglena gracilis* chloroplast RNA polymerase specific for riobosomal RNA genes. *J. Biol. Chem.* 260:11194–99

115. Nechushtai, R., Nelson, N. 1985. Biogenesis of photosystem I reaction center during greening of oat, bean and spinach leaves. *Plant Mol. Biol.* 4:377–84

116. Newcomb, E. H. 1967. Fine structure of protein-storing plastids in bean root tips. *J. Cell Biol.* 33:143–63

117. Nivison, H. T., Stocking, C. R. 1983. Ribulose bisphosphate carboxylase synthesis in barley leaves. *Plant Physiol.* 73:906–11

118. Nixon, P. J., Dyer, T. A., Barber, J., Hunter, C. N. 1986. Immunological evidence for the presence of the D1 and D2 proteins in PSII cores of higher plants. *FEBS* 209:83–86

119. Oelmüller, R., Levitan, I., Bergfeld, R., Rajasekhar, V. K., Mohr, H. 1986. Expression of nuclear genes as affected by treatments acting on the plastids. *Planta* 168:482–92

120. Ohyama, K., Fukuzama, H., Kohchi, T., Shirai, H., Sano, T., et al. 1986. Chloroplast gene organization deduced from complete sequence of liverwort *Marchantia polymorpha* chloroplast DNA. *Nature* 322:572–74

121. Ohme, M., Tanaka, M., Chunwongse, J., Shinozaki, K., Sugiura, M. 1986. A tobacco chloroplast DNA sequence possibly coding for a polypeptide similar to *E. coli* RNA polymerase β-subunit. *FEBS Lett.* 200:87–90

122. Opik, H. 1966. Changes in cell fine structure in the cotyledons of *Phaseolus vulgaris* L. during germination. *J. Exp. Bot.* 17:427–39

123. Orozco, E. M., Mullet, J. E., Chua, N.-H. 1985. An *in vitro* system for accurate transcription initiation of chloroplast protein genes. *Nuc. Acids Res.* 13:1283–1302

124. Palmer, J. D. 1985. Comparative organ-

ization of chloroplast genomes. *Ann. Rev. Genet.* 19:325–54

125. Piechulla, B., Imlay, K. R. C., Gruissem, W. 1985. Plastid gene expression during fruit ripening in tomato. *Plant Mol. Biol.* 5:373–84

126. Plesnicar, M., Bendall, D. S. 1973. The photochemical activities and electron carriers of developing barley leaves. *Biochem J.* 136:803–12

127. Pyke, K. A., Leech, R. M. 1987. The control of chloroplast number in wheat mesophyll cells. *Planta* 170:416–20

128. Reiss, T., Link, G. 1985. Characterization of transcriptionally active DNA-protein complex from chloroplasts of mustard (*Sinapsis alba* L.). *Eur. J. Biochem.* 148:207–12

129. Robertson, D., Laetsch, W. M. 1974. Structure and function of developing barley plastids. *Plant Physiol.* 54:148–59

130. Rochaix, J. D. 1985. Genetic organization of the chloroplast. *Int. Rev. Cytol.* 93:57–91

131. Rock, C. D., Barkan, A., Taylor, W. C. 1987. The maize plastid *psbB-psbF-petB-petD* gene cluster: Spliced and unspliced *petB* and *petD* RNAs encode alternative products. *Curr. Genet.* 12:69–77

132. Rodermel, S. R., Bogorad, L. 1985. Maize plastid photogenes: mapping and photoregulation of transcript levels during light-induced development. *J. Cell. Biol.* 100:463–76

133. Sauer, K. 1975. Primary events and the trapping of energy. In *Bioenergetics of Photosynthesis*, ed. Govindjee, 3:116–175. New York: Academic. 683 pp.

134. Sasaki, Y., Tomoda, Y., Kamikubo, T. 1984. Light regulates the gene expression of ribulose bisphosphate carboxylase at the levels of transcription and gene dosage in greening pea leaves. *FEBS Lett.* 173:31–34

135. Sasaki, Y., Tomoda, Y., Tomi, H., Kamikubo, T., Shinozaki, K. 1985. Synthesis of ribulose bisphosphate carboxylase in greening pea leaves. *Eur. J. Biochem.* 152:179–86

136. Schafer, E., Briggs, W. R. 1986. Photomorphogenesis from signal perception to gene expression. *Photobiochem. Photobiophys.* 12:305–20

137. Schiff, J. A. 1980. Development, inheritance and evolution of plastids and mitochondria. In *The Biochemistry of Plants*, ed. N. E. Tolbert, 1:209–72. New York: Academic

138. Schmidt, G. W., Bartlett, S. G., Grossman, A. R., Cashmore, A. R., Chua, N.-H. 1981. Biosynthetic pathways of two polypeptide subunits of the light-harvesting chlorophyll *a/b* protein complex. *J. Cell Biol.* 91:468–78

139. Schmidt, G. W., Mishkind, M. L. 1983. Rapid degradation of unassembled ribulose 1,5-bisphosphate carboxylase small subunits in chloroplasts. *Proc. Natl. Acad. Sci. USA* 80:2632–36

140. Schön, A., Krupp, G., Gough, S., Berry-Lowe, S., Kannangara, C. G., Söll, D. 1986. The RNA required in the first step of the chlorophyll biosynthesis is a chloroplast glutamate tRNA. *Nature* 322:281–84

141. Schuster, G., Ohad, I., Martineau, B., Taylor, W. C. 1985. Differentiation and development of bundle sheath and mesophyll thylakoids in maize. *J. Biol. Chem.* 260:11866–73

142. Sharman, B. C. 1942. Developmental anatomy of the shoot of *Zea mays* L. *Ann. Bot.* 1:245–82

143. Sheen, J.-Y., Bogorad, L. 1986. Differential expression of six light-harvesting chlorophyll *a/b* binding protein genes in maize leaf cell types. *Proc. Natl. Acad. Sci. USA* 83:7811–15

144. Shinozaki, K., Ohme, M., Tanaka, M., Wakasugi, T., Hayashida, N., et al. 1986. The complete nucleotide sequence of the tobacco chloroplast genome: its gene organization and expression. *EMBO J.* 5:2043–49

145. Shlyk, A. A., Chaika, M. T., Fradkin, L. I., Rudoi, A. B., Averina, N. G., Savchenko, G. E. 1986. Biogenesis of photosynthetic apparatus in etiolated leaves during greening. *Photobiochem. Photobiophys.* 12:87–96

146. Skadsen, R. W., Scandalios, J. G. 1987. Translational control of photo-induced expression of the *Cat2* catalase gene during leaf development in maize. *Proc. Natl. Acad. Sci. USA* 84:2785–89

147. Smith, H. 1970. Changes in plastid ribosomal-RNA and enzymes during the growth of barley leaves in darkness. *Phytochemistry* 9:965–75

148. Smith, H. J., Bogorad, L. 1974. The polypeptide subunit structure of the DNA-dependent RNA polymerase of *Zea mays* chlorophasts. *Proc. Natl. Acad. Sci. USA* 71:4839–42

149. Steinback, K. E., McIntosh, L., Bogorad, L., Arntzen, C. J. 1981. Identification of the triazine receptor protein as a chloroplast gene product. *Proc. Natl. Acad. Sci. USA* 78:7463–67

150. Stirdivant, S. M., Crossland, L. D., Bogorad, L. 1985. DNA supercoiling affects in vitro transcription of two maize chloroplast genes differently.

Proc. Natl. Acad. Sci. USA 82:4886–90

151. Strittmatter, G., Kössel, H. 1984. Cotranscription and processing of 23S, 4.5S and 5S rRNA in chloroplasts from *Zea mays*. *Nuc. Acids. Res.* 12:7633–47

152. Strittmatter, G., Gozdzicka-Jozefiak, A., Kössel, H. 1985. Identification of an rRNA operon promoter from *Zea mays* chloroplasts which excludes the proximal tRNA $_{GAC}^{Val}$ from the primary transcript. *EMBO J.* 4:599–604

153. Sun, E., Shapiro, D. R., Wu, B. W., Tewari, K. K. 1986. Specific *in vivo* transcription of 16S rRNA gene by pea chloroplast RNA polymerase. *Plant Mol. Biol.* 6:429–39

154. Sutton, A., Sieburth, L. E., Bennett, J. 1987. Light-dependent accumulation and localization of photosystem II proteins in maize. *Eur. J. Biochem.* 164:571–78

155. Takabe, T., Takabe, T., Akazawa, T. 1986. Biosynthesis of P700-chlorophyll *a* protein complex, plastocyanin, and cytochrome b_6/f complex. *Plant Physiol* 81:60–66

156. Tewari, K. K., Goel, A. 1983. Solubilization and partial purification of RNA polymerase from pea chloroplasts. *Biochem.* 22:2142–48

157. Thompson, J. A. 1980. Apparent identity of chromoplast and chloroplast DNA in the daffodil. *Narcissus pseudonarcissis. Z. Naturforch.* 356:1101–3

158. Thompson, R. J., Mosig, G. 1985. An ATP-dependent supercoiling topoisomerase of *Chlamydomonas reinhardtii* affects accumulation of specific chloroplast transcripts. *Nuc. Acids Res.* 13:873–91

159. Thompson, R. J., Mosig, G. 1987. Stimulation of a *Chlamydomonas* chloroplast promoter by novobiocin *in situ* and in *E. coli* implies regulation by torsional stress in the chloroplast DNA. *Cell* 48:281–87

160. Thompson, W. F., Everett, M., Polans, N. O., Jorgensen, R. A., Palmer, J. D. 1985. Phytochrome control of RNA levels in developing pea and mung-bean leaves. *Planta* 158:487–500

161. Thomson, W. W., Whatley, J. M. 1980. Development of non-green plastids. *Ann. Rev. Plant Physiol.* 31:375–94

162. Tobin, E. M., Silverthorne, J. 1985. Light regulation of gene expression in higher plants. *Ann. Rev. Plant Physiol.* 36:569–93

163. Tzagoloff, A. 1971. Assembly of the mitochondrial membrane system. *J. Biol. Chem.* 246:3050–56

164. Vierling, E., Alberte, R. S. 1983. Regu-

lation of synthesis of the photosystem I reaction center. *J. Cell Biol.* 97:1806–14

165. Virgin, H. I., Egneus, H. S. 1983. Control of plastid development in higher plants. In *Encyclopedia of Plant Physiology,* ed. W. Shropshire, H. Mohr, 16A:289–311. Berlin/New York: Springer-Verlag. 832 pp.

165a. Walbot, V., Coe, E. H. Jr. 1979. Nuclear gene *iojap* conditions a programmed change to ribosome-less plastids in *Zea mays. Proc. Natl. Acad. Sci. USA* 76:2760–64

166. Walter, P., Blobel, G. 1981. Translocation of proteins across the endoplasmic reticulum. III. Signal recognition protein (SRP) causes signal sequence-dependent and site-specific arrest of chain elongation that is released by microsomal membranes. *J. Cell Biol.* 91:557–61

166a. Wallsgrove, R. M., Lea, P. J., Miflin, B. J. 1983. Intracellular localization of aspartate kinase and the enzymes of threonine and methionine biosynthesis in green leaves. *Plant Physiol.* 71:780–84

167. Wardley, T. M., Bhalla, P. L., Dalling, M. J. 1984. Changes in the number and composition of chloroplasts during senescence of mesophyll cells of attached and detached primary leaves of wheat (*Triticum aestivum* L.). *Plant Physiol.* 75:421–24

168. Weil, J. H. 1979. Cytoplasmic and organellar tRNAs in plants. In *Nucleic Acids in Plants,* ed. T. C. Hall, J. W. Davies, 2:143–92. Boca Raton: CRC Press. 229 pp.

169. Weil, J. H., Parthier, B. 1982. Transfer RNA and aminoacyl-tRNA synthetase in plants. In *Encyclopedia of Plant Physiology, New Series,* ed. D. Boulter, B. Parthier, 14A:65–112. Berlin/New York: Springer-Verlag. 743 pp.

170. Weinstein, J. D., Beale, S. I. 1985. RNA is required for enzymatic conversion of glutamate to δ-aminolevulinate by extracts of *Chlorella vulgaris. Arch. Biochem. Biophys.* 239:87–93

171. Weinstein, J. D., Mayer, S. M., Beale, S. I. 1986. Stimulation of δ-aminolevulinic acid formation in algal extracts by heterologous RNA. *Plant Physiol.* 82:1096–1101

172. Westhoff, P. 1985. Transcription of the gene encoding the 51 kd chlorophyll a-apoprotein of the photosystem II reaction centre from spinach. *Mol. Gen. Genet.* 201:115–23

173. Wettern, M., Galling, G. 1985. Degradation of the 32-kilodalton thylakoid-membrane polypeptide of *Chlamydomonas reinhardtii* Y-1. *Planta* 166:474–82

174. Wettstein, D. V., Henningsen, K. W., Boynton, J. E., Kannangara, C. G., Nielsen, O. F. 1971. The genetic control of chloroplast development of barley. In *Autonomy and Biogenesis of Mitochondria and Chloroplasts,* ed. N. K. Boardman, A. W. Linnane, R. M. Smillie, pp. 205–23. Amsterdam/London: North-Holland

175. Whatley, J. M. 1979. Plastid development in the primary leaf of *Phaseolus vulgaris:* variation between different types of cells. *New Phytol.* 82:1–10

176. Whitfeld, P. R., Bottomley, W. 1983. Organization and structure of chloroplast genes. *Ann. Rev. Plant Physiol.* 34:279–310

177. Zaita, N., Torazawa, K., Shinozaki, K., Sugiura, M. 1987. *Trans* splicing *in vivo:* joining of transcripts from the "divided" gene for ribosomal protein S12 in the chloroplasts of tobacco. *FEBS Lett.* 210:153–56

178. Zhu, Y. S., Hearst, J. E. 1986. Regulation of expression of genes for light-harvesting antenna proteins LH-I and LH-II; reaction center polypeptides RC-L, RC-M, and RC-H; and enzymes of bacteriochlorophyll and carotenoid biosynthesis in *Rhodobacter capsulatus* by light and oxygen. *Proc. Natl. Acad. Sci. USA* 83:7613–17

179. Zhu, Y. S., Kung, S. D., Bogorad, L. 1985. Phytochrome control of levels of mRNA complementary to plastid and nuclear genes of maize. *Plant Physiol.* 79:371–76

180. Zurawski, G., Clegg, M. T. 1987. Evolution of higher-plant chloroplast DNA-encoded genes: implications for structure-function and phylogenetic studies. *Ann. Rev. Plant Physiol.* 38:391–418

Ann. Rev. Plant Physiol. Plant Mol. Biol. 1988. 39:503–32

PLANT MITOCHONDRIAL GENOMES: ORGANIZATION, EXPRESSION AND VARIATION

Kathleen J. Newton

Division of Biological Sciences, University of Missouri, Columbia, Missouri 65211

CONTENTS

INTRODUCTION... 503
PLANT MITOCHONDRIAL GENOMES.. 504
   Organization.. 504
   Chloroplast DNA Sequences.. 506
   Plasmids.. 506
GENES.. 507
   Ribosomal RNAs, Transfer RNAs, and Ribosomal Proteins............................... 507
   Protein-Coding Genes.. 509
   Transcription.. 510
   Codon Usage.. 512
   Identification of Additional Genes.. 512
CYTOPLASMIC MALE STERILITY.. 515
   Maize cms-T.. 516
   Maize cms-C.. 518
   Maize cms-S.. 518
   RNA Plasmids in cms-S *Mitochondria*.. 520
   CMS in Petunia.. 521
   Sublimons.. 521
ABNORMAL GROWTH MUTANTS.. 522
DIFFERENTIAL EXPRESSION OF PLANT MITOCHONDRIAL GENES.............. 523
PROSPECTS.. 523

INTRODUCTION

Our understanding of plant mitochondrial genomes has advanced considerably in the last five years. The present review serves to update aspects of the

503

previous review in this series by Leaver & Gray (93). I emphasize several studies that have significantly contributed to our current understanding of the structure, expression, and variation of plant mitochondrial genomes. Douce (43) has recently described the biology and biochemistry of plant mitochondria. Techniques for mitochondrial DNA (mtDNA) and RNA (mtRNA) isolations, gene cloning, and genome mapping were detailed in a recent volume of *Methods of Enzymology* (35, 67, 106, 167). The interested reader is referred to other reviews for more comprehensive coverage of mitochondrial genome organization (8, 140), evolution (132), cytoplasmic male sterility (68, 90, 101), and analysis of mtDNA in somatic hybrids (66).

PLANT MITOCHONDRIAL GENOMES

Organization

Recent studies have elucidated several general features of higher plant mitochondrial genomes: 1. The genomes are larger than those from mammals and fungi. 2. Restriction endonuclease mapping indicates that plant mitochondrial genomes are usually organized as multiple circular molecules, with conversion of circle types mediated by recombination between repeated sequences. 3. Mitochondrial genomes from closely related species are highly conserved in primary sequence, but vary greatly in linear gene order. 4. Chloroplast DNA sequences are found in mitochondrial DNA. 5. In addition to high molecular weight mtDNA, plasmid-like molecules are present in mitochondria.

Higher-plant mitochondrial genomes, which vary in size from 200 kb to approximately 2500 kb, are much larger than their animal or fungal counterparts (1, 2, 10, 24, 37, 62, 135, 180). Analyses of cucurbit mtDNAs demonstrated a seven-fold range in genome size within this family, from approximately 330 kb in watermelon to approximately 2500 kb in muskmelon (180). Despite the large size differences, there appears to be no correlation of mitochondrial genome size with repeated sequences (180), mitochondrial volume (9), or the number of detectable translation products (168).

The relatively small sizes of mitochondrial genomes from *Brassica* species have made them attractive for studying mitochondrial genome organization and evolution. Summation of restriction fragment lengths suggested that the mitochondrial genomes of *Brassica* species are approximately 200 kb in size (95). Subsequent restriction mapping of mtDNA from *Brassica campestris* indicated that its genome is organized as a tripartite structure: A "master" circle of 218 kb recombines through a directly repeated 2-kb sequence to form two subgenomic circles of 135 and 83 kb (135).

Other mitochondrial genomes from crucifers have been characterized (133; J. Palmer, personal communication). Most were also found to have tripartite structures: Recombination between a set of direct repeats on a master circle results in two smaller circles, each with one copy of the repeated sequence. The recombination events apparently occur frequently (133). A tripartite genome structure has also been reported for spinach mitochondria (169). Recombination between 6-kb directly repeated elements on a 327-kb master circle leads to subgenomic circles of 93 kb and 134 kb.

In maize and wheat mitochondria, genomic organization is much more complex. The 430-kb wheat mitochondrial genome contains a minimum of 10 repeats (144). In the fertile WF9N strain of maize, a 570-kb master circle contains 6 major repeats, 5 direct and 1 inverted, which can give rise to multiple circular size classes (104).

Brassica hirta has the only characterized plant mitochondrial genome in which no recombination repeats have been found (134). Restriction endonuclease mapping of this 208-kb genome showed that it is organized as a single circular molecule. Because the *Brassica hirta* mitochondrial genome lacks recombination repeats, a multipartite genome structure, while common, is not an obligatory feature of higher plant mitochondria (134).

An exception to the circular organization of higher-plant mitochondrial genomes has been reported for one type of maternally inherited male-sterile cytoplasm of maize, *cms-S*. The mtDNA of *cms-S* is characterized by the presence of relatively large amounts of two linear plasmids, S1 and S2 (143). Schardl and co-workers (148) presented evidence that the *cms-S* genome becomes linearized through recombination between the ends of the linear S1 and S2 plasmids and homologous sequences in the main mitochondrial genome. Thus, linear molecules predominate over circular ones in *cms-S* mitochondria (148).

Despite the possibilities for generating diversity in genomic organization by recombination among repeats, mitochondrial genomes appear to be relatively stable. In general, individuals of a specific line have the same mtDNA restriction fragment patterns, and these patterns are conserved during maternal transmission to progeny plants (130). Palmer & Shields (135) hypothesized that the 218-kb master circle is the only replicating molecule in the mitochondria of *Brassica campestris*. The observed stability of plant mitochondrial genomes could be explained if only molecules containing the total complexity of the genome replicate, even if subsequent recombination among repeats occurs randomly.

Under certain circumstances, mtDNA structural variation does arise and is inherited. This can be surmised both from studies of mitochondrial mutants (see below) and from variation detected among related species. Mitochondrial gene localization studies in crucifers have demonstrated rapid structural evolution of their mitochondrial genomes. The linear sequence arrangements among closely related species can differ dramatically, although there is a high degree of sequence homology (133, 134). Indeed, the nucleotide substitution rate for mtDNAs of related *Brassica* species is even slower than that for the corresponding chloroplast DNAs (134).

When plant tissues are subjected to conditions of prolonged culture, variation in mtDNA restriction fragment patterns may be observed (60, 117). In tobacco suspension cultures, the majority of the mtDNA can exist as amplified smaller circular DNAs derived from the main mitochondrial genomes (33). *Oenothera berteriana* mtDNA from tissue culture cells is also arranged as smaller circular molecules (19, 20). In *O. berteriana,* both long and short (ten nucleotide) repeats participate in rearrangement events (109, 155). However, tissue culture does not invariably lead to detectably altered mitochondrial genomes (reviewed by Hanson; 66).

Chloroplast DNA Sequences

Sequences highly homologous to chloroplast DNA are present in the mitochondrial genomes of many species. The phenomenon was first reported in maize. A 12-kb portion of the chloroplast inverted repeat, containing the chloroplast 16S ribosomal RNA gene and two chloroplast transfer RNA genes, was identified in the mitochondrial genome (166). Subsequently, a sequence homologous to the chloroplast ribulose-1,5-bisphosphate carboxylase large subunit gene was also found in mtDNA (105). Because the shared chloroplast and mitochondrial sequences are more than 90% homologous to one another, recent transfer events are suggested, presumably from the chloroplast to the mitochondrion. It is unlikely, however, that the chloroplast genes are expressed correctly in mitochondria (105).

A wide range of other higher-plant mitochondrial genomes have been surveyed for the presence of chloroplast DNA sequences (170, 171). Different chloroplast sequences are present at various locations in the mitochondrial genomes of different species. These studies demonstrate that DNA transfer from chloroplasts to mitochondria is common in higher plants and that most of the events are recent.

Interorganellar DNA transfer appears to be a general phenomenon in higher plants. In addition to DNA transfer from chloroplast to mitochondrion, chloroplast and mitochondrial DNA sequence homologies have been detected in nuclear DNA (84, 156, 173). A few sequences have been reported to be present in all three genomes (183).

Plasmids

Linear and circular DNA plasmids and double-stranded RNAs (dsRNA) have been reported in many higher-plant mitochondria (reviewed by Pring & Lonsdale; 140). The array of plasmids found within a species can be quite variable. Characteristic, high-abundance mitochondrial plasmids have been reported in some cytoplasmic male-sterile cytoplasms, although their exact relationships to the male-sterile phenotypes remain to be clarified. These include, for example, the 6.4- and 5.4-kb S plasmids of *cms-S* maize (143) and the 11.3-kb plasmid of *Brassica* species (136).

The only type of plasmid found in the mitochondria from every maize variety is either a 2.3-kb linear plasmid or a shorter 2.1-kb derivative (7, 80, 83, 118, 125). The plasmid has one or more proteins tightly associated at the 5' termini (81) and shows homology with the linear S2 plasmid of *cms-S* maize (88). DNA sequencing showed that the 2.3-kb plasmid ends in identical 170-bp terminal inverted repeats and that it has a transcribed open reading frame (ORF), potentially coding for a 33-kD polypeptide (97). The predicted amino acid sequence shares significant homology with a 130-kD protein potentially coded for by an open reading frame (ORF1) of the S2 plasmid (97). Although no full-length integrated forms of this plasmid were found in the main mitochondrial genome of maize, homologies to nuclear and chloroplast DNA were detected (7).

The chloroplast-homologous region of the maize 2.3-kb plasmid contains two transfer RNA genes, $tRNA_{CCA}^{Pro}$ and $tRNA_{UGG}^{Trp}$, that are present in the main mitochondrial genome of maize relatives (teosintes) and other higher plants (97, 115). One of these, the $tRNA_{UGG}^{Trp}$ gene, appears to be the only functional copy present in the maize mitochondrion, and it has been hypothesized that the universal maintenance of the 2.3-kb plasmid in maize lines may result from the fact that it carries an essential gene (97).

GENES

Biogenesis of mitochondria is controlled by both nuclear and mitochondrial genes. In yeast, it is estimated that there are several hundred mitochondrial polypeptides (123), very few of which are coded for by the mitochondrial genome. Most mitochondrial proteins are nuclearly encoded, synthesized on cytoplasmic ribosomes and transported posttranslationally into the mitochondria (177). Mitochondria do have their own protein synthesis apparatus, and the mtDNA of all eukaryotes codes for a small number of polypeptides as well as for mitochondrial ribosomal RNAs and transfer RNAs (2, 62).

Ribosomal RNAs, Transfer RNAs, and Ribosomal Proteins

Mitochondrial ribosomes in higher plants differ from those in animals and fungi. They contain a 5S ribosomal RNA (92) and they have 18S (27, 28) and 26S rRNAs (32), which are larger than those found in human (12S and 16S) or yeast (15S and 21S) mitochondria (2, 177). The maize mitochondrial 18S rRNA (1968 nucleotides; 28) is larger than the maize cytosolic 17S rRNA

(1905 nucleotides; 119). Sequence analysis of other higher plant mitochondrial rRNA genes shows that they are very similar to the maize genes (21, 108, 162, 163). Plant mitochondrial rRNAs are more similar to bacterial and chloroplast rRNAs in primary and secondary structure than they are to mammalian or fungal mitochondrial rRNAs (21, 28, 32, 162, 163). The ribosomal 16S, 23S, and 5S RNA genes are closely linked in bacteria and chloroplasts. In higher-plant mitochondria, the 18S and 5S genes are closely linked: 108 bp apart in maize (27) and 582 bp apart in *Oenothera* (21). However, the 26S gene is located several kilobases distant (16 kb in maize; 73, 165).

Several transfer RNA genes have been identified in higher-plant mtDNA (57, 111–115, 137, 138). To date, ten potential tRNA coding sequences have been located on the restriction map of the maize mitochondrial genome (102). Mapping in maize (102) and also wheat (14) suggests that the tRNA genes are present at several locations in the mitochondrial genomes. An initiator methionine tRNA gene terminates one nucleotide 5' to the coding sequence for the 18S rRNA gene of wheat (163). In the maize WF9N mitochondrial genome, these two genes are more than 70 kb apart (102). In soybean mtDNA, an initiator tRNA gene is located near the cytochrome oxidase subunit II coding sequence (58). As in yeast mitochondria, separate initiator and elongator methionine tRNAs have been characterized in higher-plant mitochondria (57, 59, 114, 137). This contrasts with mammalian mitochondria, in which only one methionine tRNA gene is found (2). Thus, it would not be surprising if the number of plant mitochondrial tRNA genes exceeds the 22 encoded by mammalian mtDNA (2, 5, 129). Twenty-five tRNA genes are coded for by yeast mtDNA (177).

In mammals, none of the approximately 85 mitochondrial ribosomal proteins appears to be encoded by mtDNA (150). Yeast and *Neurospora* mtDNAs each encode one small subunit ribosomal protein: the var1 protein in yeast (25, 172) and the S5 protein in *Neurospora* (89). Mitochondria from maize may synthesize a 44-kD var1-like protein (93); however, until the corresponding gene is isolated and sequenced, this identification remains tentative. Sequence analysis has identified a potential ribosomal protein gene in tobacco mtDNA. The predicted amino acid sequence of the open reading frame, *rps13,* shares partial homology with the *E. coli* small subunit ribosomal protein S13 (11). Bland and co-workers (11) suggested that *rps13* may function as a ribosomal protein in higher plants because the predicted protein is hydrophilic, characteristic of ribosomal proteins, and highly basic, characteristic of proteins associated with nucleic acids. The tobacco *rps13* sequence hybridizes to maize and wheat mtDNAs, but not to pea and bean mtDNAs (11). Northern analyses suggest that this gene is transcribed in tobacco and maize (11).

Protein-Coding Genes

Mitochondria synthesize a few of the subunits of the enzyme complexes located in the inner membrane (176). Other polypeptides of these enzyme complexes are nuclear gene products (177). In addition, many of the proteins responsible for the maintenance and expression of mitochondrial genomes (e.g. mtDNA replication, transcription, RNA processing, and translation) are nuclearly encoded.

In general, mitochondrial genomes encode subunits I, II, and III of the cytochrome *c* oxidase complex, subunits 6 and 8 of the ATPase complex, and the apocytochrome *b* of the bc1 complex (2, 177). However, the location of genes coding for a few other mitochondrial polypeptides may vary between nucleus and mitochondria, depending on the organism. For example, the ATPase subunit 9 *(atp9)* is a nuclear gene product in animals (2) but a mitochondrial gene product in yeast and higher plants (64, 177). Although there is an *atp9* sequence in *Neurospora* mtDNA, the functional *atp9* gene is thought to be nuclear (179). In higher plants, the alpha subunit of ATPase *(atpA)* is mitochondrially synthesized (16, 65), but it is a nuclear gene product in yeast and animals (2, 177).

Previously unassigned reading frames (URFs) in mammalian mtDNA were recently shown to encode seven subunits of the NADH-ubiquinone oxidoreductase (complex I; NADH dehydrogenase) (29, 30). In yeast, these genes are absent from mtDNA (177). In *Neurospora crassa,* six subunits of this complex are mitochondrially synthesized (76). Open reading frames homologous to the mammalian complex I genes have also been identified in several organisms, including *Aspergillus* (23, 24, 147) and *Chlamydomonas* (139). Homology with *ndh1* (URF1) has been detected in maize mtDNA (147). A potential *ndh1* coding sequence, with regions of homology to the corresponding animal sequences, has been cloned from watermelon mtDNA (164).

Several approaches have been used to isolate plant mitochondrial genes. Because ribosomal RNA represents the majority of mtRNA, the use of end-labeled mtRNA as a hybridization probe to a cosmid library of mtDNA led to the identification of rRNA coding sequences (165). In a similar manner, end-labeled mtRNA can also be used to identify other highly expressed regions. DNA clones that hybridize to abundant RNAs, other than rRNAs, have been isolated and sequenced. Subsequent gene assignments have been made on the basis of predicted amino acid sequence homologies to known genes from other organisms. This method has led to the identification of potential genes for subunits 9 *(atp9)*, 6 *(atp6)*, and alpha *(atpA)* of the ATPase (18, 38, 40) in maize mitochondria. An alternative approach to using end-labeled RNA to probe for actively transcribed regions is to use cloned cDNAs (71). Because plant mtRNAs are not polyadenylated, it is necessary to add poly-A tails prior to the cDNA synthesis reaction. Analysis of a cDNA clone from *Oenothera* led to the isolation of the cytochrome oxidase subunit III *(coxIII)* gene (71).

Heterologous gene probes can help to identify potential genes in mtDNA of plants. For example, cloned yeast mitochondrial genes can be used as hybridization probes to plant mtDNA, in order to identify those genes conserved between plants and yeast (35). The maize cytochrome oxidase subunit II gene *(coxII*, initially called *moxI)* was cloned using this strategy (54). Sequence homologies between the maize gene and *coxII* genes from other organisms confirmed its identity. A similar approach was used to isolate apocytochrome *b (cob)*, cytochrome oxidase subunit I *(coxI)*, and *atpA* genes from maize (36, 74, 75).

In common with other organisms, the mitochondrial genomes of plants may also contain a gene for subunit 8 of the ATPase complex. DNA sequence analysis of regions that flanked the *Oenothera coxII* gene led to the identification of another open reading frame that shared some homology to the ATPase subunit 8 *(atp8)* gene of yeast mitochondria (70). Interestingly, the *Oenothera coxII* and *atp8* reading frames are overlapping and cotranscribed (70).

Evidence that these gene sequences are functional in plant mitochondria can be obtained by analyzing proteins synthesized in this organelle. Proteins synthesized by purified mitochondria, in the presence of radiolabeled amino acids, are separated by gel electrophoresis and the labeled polypeptides are detected by autoradiography (94). Antibody probes can be used to ascertain whether any of the translation products correspond to particular proteins (51, 65).

The first evidence that the alpha subunit of the F_1-ATPase is a mitochondrial gene in higher plants was obtained from such studies (65). The alpha subunit of purified maize F_1-ATPase co-migrated with a mitochondrially synthesized polypeptide. This polypeptide was also precipitated with antiserum against the yeast alpha subunit (65). It was concluded that the alpha subunit was a product of mitochondrial protein synthesis in maize. The alpha subunit is also synthesized in *Vicia faba* mitochondria (16).

Subunits I and II of cytochrome oxidase have also been identified as mitochondrial translation products in maize because specific, labeled polypeptides are precipitated by the corresponding yeast antibodies (51). In addition, a DCCD-binding protein, subunit 9 of ATPase, is synthesized by plant mitochondria (64).

Transcription

Transcripts homologous to specific plant mitochondrial genes identify RNAs that are generally larger than necessary to encode the corresponding polypeptides. In maize, individual probes for protein-coding genes identify more than one major RNA species when hybridized to Northern blots of electro-phoretically separated mtRNAs (18, 36, 38, 40, 45, 54, 74, 75). Cucurbit mitochondria, exhibiting a seven-fold variation in genome size, also have multiple, large transcripts homologous to protein-coding genes (168). In *Brassica campestris* mitochondria, only 3 of 9 examined protein-coding genes hybridized to more than one major transcript (107). However, the transcripts, which range in size from approximately 500 to 10,000 nucleotides (107), are also usually larger than the corresponding coding regions.

At present, the basis for the complex transcript patterns in maize mitochon-dria is unknown. Intron processing alone does not account for them because most of the characterized plant mitochondrial genes, including *coxI, coxIII, cob, atpA, atp6, atp9,* and the ribosomal RNA genes, do not contain introns (18, 28, 32, 36, 38, 40, 71, 74, 75). The *coxII* genes of maize and other monocots are interrupted by a single intron (13, 54, 79), but the dicot *coxII* genes that have been sequenced have no introns (69, 121). If the *ndh1*-homologous region in watermelon mtDNA is functional, it would include as many as four introns (164). The *rps13* gene of tobacco is located between the *atp9* gene and an open reading frame that has homology with an internal region of *ndh1* (11). If this ORF corresponds to a region of the functional NADH:Q1 gene in tobacco, then *rps13* and *atp9* would be located within an intron of the *ndh1* gene. Similarly, the maize *rps13* gene would also be located within an intron of the *ndh1* gene (11). Neither the *rps13* gene nor the putative *ndh1* genes have been shown to be translated in plant mitochondria, and they could represent pseudogenes. Nonetheless, complex transcript pat-terns would be detected when sequences homologous to one gene are located within the transcriptional unit of another gene, or if two or more genes are co-transcribed.

Large transcripts would be observed if long untranslated segments are present on either side of the coding region. Multiple transcripts would be observed if the genes have more than one RNA initiation or termination site, or if extensive posttranscriptional processing occurs. The 5' ends of primary unprocessed transcripts can be specifically labeled by in vitro capping reac-tions. These probes can then be used to identify sites of transcription initiation in DNA. Such experiments demonstrated that the maize *cob* and *atp9* genes have multiple transcription initiation sites (122). In contrast, the maize 26S rRNA and the co-transcribed 18S and 5S rRNA genes each have a single major initiation site. The rRNA primary transcripts are processed to yield the mature rRNAs (122). The identification of primary and processed transcripts for additional genes should help to elucidate the role of RNA processing in the expression of plant mitochondrial genes.

Very little is known about the nucleotide signals that regulate transcription and translation in plant mitochondria. In order to identify potential regulatory sequences, the DNA sequences that flank several genes have been examined. A sequence located near the 5' transcriptional start site of the maize *coxl* gene is homologous at 7 positions to a 9-nucleotide consensus sequence that marks the initiation of transcription in yeast mitochondrial genes (75). However, a different sequence is found near the 5' starts for the *coxII* transcripts from pea (121). Other gene sequences have been examined for potential promoter sequences (58, 70, 71). No clear consensus has emerged, and none of the sequences has yet been demonstrated to function in transcription.

An octanucleotide sequence found 5' to several plant mitochondrial genes was proposed to function as a ribosome binding site (35, 36), but it does not appear to be highly conserved (153). Additional experiments will be necessary to demonstrate whether this sequence can actually function as a ribosome binding site.

The 3' termini of the *Oenothera coxII* and *atpA* transcripts (determined by S1 nuclease protection studies) were found to include an identical 50-nucleotide sequence, which also showed homologies to the predicted 3' ends of three maize transcripts (154). Computer-assisted sequence analysis of these regions predicted stem-loop secondary structures similar to bacterial terminators (154).

Codon Usage

Exceptions to the universal code have been found for mitochondrial genes from animals and fungi (2, 24, 37, 177). Although genes from higher-plant mitochondria have been less intensively investigated, only one exception to the universal code has been proposed. Coding regions for particular plant mitochondrial genes have been deduced by aligning predicted amino acid sequences with those corresponding to genes from other mitochondria, chloroplasts, and bacteria. Direct evidence for specific codon usages in plant mitochondrial genes is lacking. No amino acid sequences for plant mitochondrial polypeptides have yet been reported. Thus, codon usage in plant mitochondria must be deduced from highly conserved nucleotide and amino acid sequences from other organisms.

The universal translational start codon, AUG, is used in plant mitochondria. Plant mitochondria apparently use UGA as a translational stop codon (18, 54, 152), in accordance with the universal code. UGA specifies tryptophan in mitochondria from other organisms (2, 177). However, the plant mitochondrial genetic code differs from the universal code in apparently using CGG to specify tryptophan instead of arginine (13, 36, 54, 69, 75, 79, 152).

Identification of Additional Genes

Although much of the "extra" DNA in higher-plant mitochondria is probably noncoding, plant genomes do include genes that are not present in the mitochondrial genomes of animals and fungi. Only plant mitochondrial genomes encode a 5S ribosomal RNA and the alpha subunit of ATPase. It is certainly possible that additional functional genes are also present. Evidence for this proposition is provided by the number of plant mitochondrial translation products.

Following incorporation of labeled amino acids into proteins synthesized by isolated maize mitochondria, between 30 and 50 labeled polypeptides can be resolved by 2-dimensional electrophoresis (65). Although certain of these polypeptides may result from various modifications of a smaller number of primary protein synthesis products, it is probable that plant mitochondria code for at least 20 polypeptides (65). Similar numbers of polypeptides are synthesized by mitochondria from different plants, even when their genomes vary greatly in size (93, 168).

All plant mitochondrial genomes probably encode apocytochrome *b*, subunits I, II, and III of cytochrome oxidase, and the alpha, 6, and 9 subunits of ATPase. Limited evidence suggests that the *atp8*, *rps13*, and *ndh1* genes are also present in the mitochondrial genomes of some plants. Thus, ten potential protein-coding genes with homologies to genes from other organisms have been identified. If plant mitochondria encode the same additional components of the NADH dehydrogenase complex as do animal mitochondria, six additional polypeptides of known function should be identified as products of plant mitochondrial genes. Further studies are necessary to ascertain whether all these genes are present or expressed in mitochondria from all plant species. For example, the putative *rps13* gene does not appear to be present in bean and pea mtDNA (11) and the *ndh1* gene may not be expressed in all plant mitochondria. In *Brassica campestris*, mitochondrial transcripts homologous to the watermelon *ndh1* gene were not detected (107), although such transcripts have been detected in both watermelon (164) and maize (48).

Are additional polypeptides encoded by plant mitochondrial genomes? The extra coding capacity is present, but is it used? Different approaches are being employed to address this question. One approach, most feasible with a small plant mitochondrial genome, is to construct a transcriptional map. Makaroff & Palmer (107) used clones spanning the *Brassica campestris* mitochondrial genome as hybridization probes to Northern blots of mtRNAs separated by gel electrophoresis; 24 abundant and non-overlapping transcripts larger than 500 nucleotides were identified. Approximately 30% (61 kb) of the 218-kb genome is highly transcribed, but low abundance, overlapping transcripts were also detected (107). Nine of the 24 abundant transcripts were accounted for by known mitochondrial genes (*coxI, coxII, coxIII, atpA, atp6, atp9, rps13,* 26S rRNA, and the co-transcribed 18S + 5S rRNAs). Transcripts corresponding to *cob* were less abundant, and transcripts corresponding to *ndh1* were not detected (107).

The 15 unidentified abundant transcripts are in the size range expected for protein-coding genes (107). If the abundant independent transcripts identified in the *B. campestris* mitochondrial genome do specify protein products, there would be a minimum of 22 polypeptides coded for by plant mitochondria (107). This is in good agreement with the estimates of 20 or more polypeptides synthesized by isolated maize mitochondria (65).

Different results were obtained by Carlson et al (26). RNA hybridization probes were prepared by allowing isolated *Brassica napus* mitochondria to synthesize RNA in the presence of radiolabeled UTP. Hybridization to mtDNA blots suggested that most of the genome was transcribed, including the chloroplast-homologous sequences present in this mitochondrial genome. This result contrasts with that of Makaroff & Palmer; however, different methodologies were used in the two studies. Perhaps large transcripts are initially synthesized by plant mitochondria and then rapidly processed to stable, translatable transcripts. The possibility has not been excluded that the RNA hybridization to chloroplast-homologous sequences resulted from plastid contamination of the mitochondrial preparations used by Carlson et al (26). Makaroff & Palmer (107) did not find detectable transcription of the chloroplast-homologous sequences.

The DNA sequences of the clones that hybridize to the abundant, unknown *Brassica* transcripts can be examined for homologies to mitochondrial and nuclear genes encoding mitochondrial proteins in other organisms. This should help to determine if additional mitochondrial proteins of known function are coded for by plant mtDNA. Once the sequences of the open reading frames defining actual transcripts are identified, correlations with mitochondrially synthesized proteins can be made using either of two methods. DNA sequences comprising open reading frames can be cloned into expression vectors in order to obtain fusion proteins in *E. coli,* which can then be purified and used to elicit antibodies (35). Alternatively, antibodies can be raised to synthetic oligopeptides corresponding to predicted amino acid sequences (41, 184). Immunoprecipitation of mitochondrially synthesized proteins with antibodies would constitute strong evidence that the open reading frames correspond to particular mitochondrial proteins. Both of these approaches have been successfully used in confirming that open reading frames associated with cytoplasmic male sterility in maize are translated into specific proteins (41, 110, 184, 188).

Genes can also be defined using mutations. This genetic approach to identifying functional plant mitochondrial genes is currently being used in maize. Analysis of certain abnormal growth mutants, known as nonchromosomal stripe (NCS), suggest that the mutations interrupt essential mitochondrial genes, two of which do not correspond to previously characterized plant mitochondrial genes (48, 124). Characterization of the mtDNA sequences altered in the NCS mutants may lead to the identification of additional functional genes in plant mitochondria.

CYTOPLASMIC MALE STERILITY

Cytoplasmic male-sterile (CMS) plants are normal in appearance but fail to shed functional pollen (44, 68, 90). The sterility is maternally inherited. CMS has been investigated at the molecular level most intensively in maize, sorghum, and petunia (3, 4, 17, 39, 41, 53, 86, 99, 101, 142, 148, 160, 184, 186), although the phenomenon is widespread in the plant kingdom. The evidence is now overwhelming that the genetic determinants of CMS reside in the mitochondrial, rather than the chloroplast genome. However, it is also clear that the nuclear genes control the expression of the CMS phenotype.

In general, CMS results either from interspecific crosses (alloplasmic CMS) or from intraspecific crosses (68). Backcrossing the hybrid plant by the pollen parent, and repeating this process for several generations, yields plants containing the cytoplasm of the original maternal parent and the nuclear genotype of the original male parent. In such plants, the interaction between nuclearly encoded and mitochondrially encoded gene products may be disturbed, resulting in alterations of mitochondrial function. Nuclear-cytoplasmic incompatibilities could reflect nuclear gene effects on mitochondrial gene expression (transcription, processing of RNAs, or translation). Alternatively, the enzymes of the inner mitochondrial membrane, which contain subunits coded for by both the nuclear and the mitochondrial genomes, may be partially dysfunctional as a result of aberrant subunit interactions.

CMS is only a subset of phenotypic abnormalities that can be observed when a "foreign" nucleus is introduced into the cytoplasm of another species. Other nuclear-cytoplasmic incompatibilities include floral abnormalities, vegetative disturbances, and lethal and subnormal seedlings (63). However, CMS is the most frequently observed type of nuclear-cytoplasmic incompatibility in angiosperms (63). It is interesting that this defective phenotype is only manifest during microsporogenesis.

CMS has been used in breeding programs to generate hybrids, because it eliminates the expense of hand emasculation procedures (44). Nuclear "restorer of fertility" *(Rf)* genes are known that can counteract, or compensate for, the cytoplasmically determined pollen sterility. *Rf* genes are most useful when they are dominant because the F1 hybrid plants (with CMS cytoplasm and heterozygous for nuclear restorer genes) are then fertile. CMS plants in species such as maize have been classified according to which inbred lines restore fertility (6, 44, 90). In maize, there are three different CMS types, known as S, T, and C, each of which is restored to fertility by the action of different nuclear *Rf* genes.

The three CMS types can also be distinguished from one another and from the fertile N mitochondria by molecular criteria. Their restriction endonuclease digest patterns (15, 99, 142) and protein synthesis profiles (50, 52, 53) differ. Within N and each CMS type, subgroups have been identified according to various criteria, including mtDNA restriction enzyme and hybridization analyses (99, 118, 126, 141, 160, 161). Variation within N, C, or S mtDNAs is much smaller than variation between cytoplasmic types.

Maize cms-T

Maize plants carrying *cms-T* cytoplasm are susceptible to race T of the fungal pathogen *Cochliobolus heterostrophus,* also referred to in the literature as *Helminthosporium maydis, Drechslera maydis,* and *Bipolaris maydis* (61, 120). They are also sensitive to methomyl, a carbamate insecticide (87). The race T pathogen produces a toxin that affects the permeability of the mitochondrial membrane, resulting in uncoupling of oxidative phosphorylation and leakage of NAD^+ (116).

Cms-T plants are restored to fertility by two genes, *Rf1* and *Rf2* (44). The dominant alleles of both genes must be provided by the nucleus in order to restore fertility to *cms-T* plants. *Cms-T* mitochondria are distinguished by characteristic mtDNA restriction fragment profiles (15, 99, 142), and they synthesize a novel 13-kD polypeptide (52). Synthesis of this "T polypeptide" is greatly reduced in mitochondria of plants restored to male fertility by the nuclear restorer genes (50).

The coding sequence for the T polypeptide has been identified. Dewey and co-workers (39) identified a 9-kb *Bam*HI fragment in a mtDNA library from *cms-T* that hybridized more intensely to end-labeled *cms-T* mtRNA than it did to N mtRNA. A region (3547 bp) of this cloned fragment was sequenced. It contained two long open reading frames that could encode polypeptides of 12,961 (ORF13) and 24,675 (ORF 25) daltons. ORF25 hybridized to transcripts in both N and T mitochondria, but a region of ORF13 hybridized to transcripts only in T mitochondria (39). Furthermore, ORF13 transcripts were specifically altered in restored *cms-T* plants (39, 41). Interestingly, only *Rf1* appears to be necessary for the modification of ORF13 transcripts in *cms-T* mitochondria (41).

The region of the mitochondrial genome including ORF13 and ORF25 in *cms-T* appears to result from at least seven recombination events (39). It shares sequence homologies with several other maize genes, including coding and/or flanking sequences from the 26S ribosomal RNA gene, *atp6,* and a chloroplast tRNA[Arg] gene. Sequences normally located 5' to the *atp6* gene precede the open reading frame for ORF13, and much of the coding region of ORF13 consists of sequences with homology to the 26S ribosomal RNA coding and 3' flanking regions (39, 41).

Direct evidence that the ORF13 sequence (renamed *urf13-T*) does encode the 13-kD T polypeptide was established by studies with antibodies elicited to chemically synthesized oligopeptides, corresponding to portions of the ORF13 predicted amino acid sequence (41, 184). A labeled 13-kD protein specifically synthesized by isolated T mitochondria was immunoprecipitated by the antisera. Dewey et al (41) showed that the antiserum also bound to a 13-kD membrane polypeptide on Western blots of T, but not N, mitochondrial proteins. The presence of *Rf1* was sufficient to reduce the amount of the 13-kD polypeptide in *cms-T* mitochondria (41), which is consistent with the effect of this nuclear gene on transcript processing. In addition to precipitating the T polypeptide, an antiserum used by Wise et al (184) also precipitated a 6-kD polypeptide that co-migrates with the DCCD-binding protein (subunit 9 of ATPase). This was observed for *cms-T,* but not N mitochondria, suggesting that the T polypeptide may associate with the ATPase complex (184).

CMS-T REVERSIONS   Several plants regenerated from tissue cultures of *cms-T* were found to have simultaneously mutated to male fertility and fungal toxin insensitivity (22, 56, 178). Multiple restriction fragment differences were detected between the mutant and *cms-T* progenitor mtDNAs (55, 82, 178). The most consistent difference (observed in all but one of several revertants) was the loss of a 6.6-kb *Xho*I restriction fragment (178). Conversely, in each of the male-sterile, toxin-sensitive regenerated plants, this restriction fragment was retained. Additionally, mitochondria from fertile revertants do not synthesize the 13-kD polypeptide characteristic of *cms-T* mitochondria (42).

The revertants that have lost the 6.6-kb *Xho*I fragment have deletions that remove the *urf13-T* open reading frame (47, 146, 185). The events that ultimately result in the deletions include recombination between homologous sequences, which are not recombinogenic in *cms-T* plants. For example, analysis of mtDNA cosmid clones from *cms-T* and one revertant, V3, suggested that recombination occurred between imperfectly repeated 127-bp sequences, one of which is located at the 3' end of the *urf13-T* coding region (146). A subsequent deletion removed the *urf13-T* coding sequences and at least 5 kb of upstream DNA. Other deletion mutants have been found to have a 3-kb deletion removing *urf13-T,* (185). ORF25, the open reading frame downstream of *urf13-T* remains unaltered in the *cms-T* revertants.

In the exceptional case of the tissue culture–induced revertant, T-4, which retained the 6.6-kb *Xho*I fragment, a mutation has altered the *urf13-T* coding region (185). A 5-bp insertion within this open reading frame (adjacent to a guanine-to-adenine transition) generates a frameshift mutation, resulting in a premature stop codon. Instead of a 13-kD polypeptide, T-4 mitochondria would synthesize an 8.3-kD protein. *urf13-T*-homologous transcripts are not altered in the T-4 mutant (185).

Analysis of the tissue culture–induced revertants of *cms-T* provides further evidence for the involvement of the 13-kD T polypeptide in T-type cytoplasmic male sterility and fungal-toxin sensitivity. Although the presence of the *Rf1* gene is sufficient to reduce synthesis of this polypeptide, the action of both *Rf1* and *Rf2* is required to restore fertility, suggesting we need further studies to help us discern the molecular basis of T-type male-sterility. It is significant that none of the molecular analyses of *cms-T* has been performed on the affected tissue, tapetal cells during microsporogenesis (96, 181). Perhaps modulation of *urf13-T* expression during this developmental stage requires the action of both nuclear restorer genes.

Maize cms-C

C-type male sterile plants are restored to fertility by the action of a single dominant nuclear gene, *Rf4* (85). There is limited evidence for other restorers of *cms-C* (90). Isolated *cms-C* mitochondria synthesize a novel, soluble, 17.5-kD polypeptide that appears to replace a 15.5-kD membrane-bound polypeptide, sythesized by N, T, and S mitochondria (50, 93). Unlike *cms-T*, the presence of *Rf* genes does not detectably reduce the synthesis of the novel protein in *cms-C* mitochondria (50). Recently, it was found that the *cms-C* mitochondrial genome has three rearrangements, relative to N, that result in the formation of chimeric genes (R. Dewey, personal communication). It is not yet known if any of the fused genes correspond to the 17.5-kD aberrant polypeptide.

Maize cms-S

One dominant nuclear gene, *Rf3,* restores fertility to plants carrying *cms-S* cytoplasms (90). Mitochondria from *cms-S* plants synthesize several minor high molecular weight polypeptides which are not synthesized by mitochondria from N, T, or C plants (50). The synthesis of these proteins is not suppressed when *Rf3* is present.

Spontaneous reversions to fertility occur in field-grown *cms-S* plants with certain nuclear genotypes, while such reversions have not been observed in *cms-T* or *cms-C* (90, 91). Reversions can result either from nuclear mutations, which are transmitted in a Mendelian fashion, or from cytoplasmic changes, inherited only through the female parent (91). Analysis of the cytoplasmic revertants should ultimately lead to the identification of the genes responsible for the *cms-S* phenotype. However, multiple rearrangements of the mitochondrial genome are observed when *cms-S* plants revert to fertility under the influence of nuclear genes (see below).

S-type mitochondrial genomes are characterized by the presence of autonomously replicating, linear episomal DNAs, S1 and S2; these molecules are not present in maize N, T, or C mitochondria (143). S1 is 6397 bp and S2 is 5453 bp long (100, 131). Both episomes have 208-bp terminal inverted repeats, which associate with proteins (81), and they share 1254 bp of homology near one end. The relationship between the presence of the S episomes and the expression of sterility is unclear. Mitochondria of plants restored to fertility by *Rf3* retain S1 and S2 episomes. Plasmid-like molecules, R1 (7.4 kb) and R2 (5.4 kb), have been reported in 12 Latin American races of maize with male-fertile cytoplasms designated RU (182). A relative of maize, *Zea diploperennis,* also has similar plasmids, D1 (7.5 kb) and D2 (5.4 kb) (174). The S2, R2, and D2 molecules appear to be identical. The R1, D1, and S1 molecules are related, homologous for much of their lengths. Both N and *cms-S* cytoplasms contain sequences in their main mitochondrial genomes that show homology to S1 and S2 sequences (88, 118).

In *cms-S,* the free S1 and S2 episomes cause the linearization of the main mitochondrial genome (148). The genome contains a pair of 186-bp repeats with homologies to the 208-bp terminal inverted repeats (TIR) of the S molecules. Recombination between the TIRs and these integrated sequences leads to linearized chromosomes terminated by S1 and S2 sequences (148). Further recombination events can place S1 and S2 sequences at internal positions within the linearized chromosomes (148). In N cytoplasm, the sequences with homologies to S1 and S2 are found adjacent to recombinationally active 5.27-kb repeat regions on the 570-kb master circle (72, 104).

In all fertile cytoplasmic revertants from *cms-S* in one nuclear background, M825, the free S1 and S2 episomes disappear and, simultaneously, rearrangements involving integrated copies of S2 sequences are detected in regions of the main mitochondrial genome (91, 98). The genome, instead of being largely linear, is now circular (149). The integrated S1 and S2 sequences have undergone rearrangements, and the S2 sequences have lost most of one terminal inverted repeat (149).

In contrast, cytoplasmic revertants of *cms-S* in the WF9 inbred line retain the S episomes (46). Fertile revertant WF9 strains are similar to the fertile RU strains of maize, which contain the R1 and R2 molecules. Therefore, *cms-S* sterility is not uniquely associated with the presence of autonomously replicating S1 and S2 episomes, nor is cytoplasmic reversion necessarily correlated with their disappearance. It is also apparent that the nuclear genotype has a major influence not only on the molecular changes associated with cytoplasmic reversion of *cms-S* to male fertility, but also on episome maintenance.

The S molecules have several open reading frames (100, 131). One open reading frame, ORF2 (1017 bp), lies within the 1254-bp region of homology between S1 and S2. S2 contains an additional reading frame, ORF1 (3513 bp), and S1 has two additional reading frames, ORF3 (2782 bp) and ORF4 (768 bp). The ORFs are transcribed (175). Recall that the S-type mitochondria synthesize minor, additional high molecular weight polypeptides (50); recent studies provide evidence that at least two of the ORFs code for large polypeptides.

In one study, a portion of S2 ORF1 was cloned into an expression vector and the β-galactosidase fusion protein was used to elicit antiserum (110). By Western blot and immunoprecipitation analyses, the resulting antibodies were shown to bind specifically to a set of proteins present in, and synthesized by, S but not N mitochondria. The major polypeptide precipitated by the antiserum in S mitochondria was approximately 125 kD (110). In a second study, antiserum was raised to an S polypeptide of approximately 130 kD (excised from gels). This antibody also reacted with ORF1 fusion proteins (187). On Western blots, the antibodies detected similar amounts of the polypeptide in mitochondria from *cms-S*, restored *cms-S*, and WF9 S revertant plants that retained the S episomes (187). In *cms-S* revertants that lost the S molecules, the polypeptide was not detected. In non-male-sterile, N-type mitochondria in the B37 nuclear background (B37N), only low levels of the polypeptide could be detected (187).

The S1 ORF3 coding sequence was also shown to correspond to a high molecular weight polypeptide in *cms-S* mitochondria (188). Antibodies raised to an ORF3 fusion protein reacted with a 103 kD polypeptide from mitochondria of *cms-S*, restored *cms-S* plants, and in *cms-S* WF9-induced fertile revertants, which retain S episomes. The antibody did not identify this polypeptide in N-type mitochondria (188).

The relationship between the synthesis of the large polypeptides and the *cms-S* phenotype is as yet unclear. Unlike *cms-T,* in which nuclear restorer genes reduce specifically the synthesis of the T polypeptide, *Rf3* apparently has no effect on the synthesis of the high molecular weight S polypeptides. The expression of the ORF1 and ORF3 open reading frames is independent of the fertility status of the plant (188). The open reading frames in the S episomes may code for proteins involved in plasmid replication or maintenance (131, 175), and they may have no role in determining sterility. However, all of the studies have been performed on mitochondria isolated from young seedlings. It is not known whether these genes are regulated differently in floral tissues during microsporogenesis (188).

RNA Plasmids in cms-S *Mitochondria*

In one line of maize, LBN, carrying the L-subgroup of *cms-S* cytoplasm in the W182BN nuclear background, two double-stranded RNAs, LBN1 (2.9 kb) and LBN2 (0.84 kb), were found in the mitochondria (159). Some S-cytoplasm lines, as well as RU and one S revertant, were found to have abundant single-stranded mtRNAs (S/RU-RNA-a and S/RU-RNA-b) in their mitochondria (151). The single-stranded mtRNAs hybridized strongly with the dsRNAs from LBN cytoplasm. None of these RNA plasmids hybridized to mitochondrial DNA plasmids or to the main mitochondrial genome (151). The synthesis of double-stranded and single-stranded RNAs in isolated S-type mitochondria was recently demonstrated (49). In the presence of actinomycin D, which eliminates DNA-directed RNA synthesis, no RNA synthesis was observed in N mitochondria, but four RNA species were synthesized by S mitochondria. Two of these RNAs (2850 bp and 900 bp) appeared to be the double-stranded forms of two single-stranded RNAs (49).

Although these RNAs are apparently not causally related to the CMS phenotype, the origin and possible function of the RNA plasmids are of interest. It has been suggested that the source of these RNA species could be of viral origin (49, 103). In this regard, it is intriguing that a double-stranded RNA is located in mitochondria of the fungus that causes Dutch elm disease (145).

CMS in Petunia

Only one type of CMS has been defined in *Petunia,* with one dominant nuclear *Rf* gene (68, 77). The mitochondrial genomes of CMS and fertile lines of *Petunia* differ in several restriction fragments, reflecting divergence between the genomes. Analysis of recombinant mitochondrial genomes in somatic hybrid plants resulting from fusion of protoplasts from sterile and fertile lines (78) led to the identification of a DNA arrangement associated with CMS (12).

By DNA sequence analysis of this region, Young & Hanson (186) identified an open reading frame, *Pcf,* that contained the 5' flanking and amino-terminal segment of *atp9,* parts of the *coxII* coding region, and the carboxyl terminus and 3' flanking region of an unidentified reading frame, *urfS.* Normal *atp9* and *coxII* genes were also present and expressed in the sterile cytoplasm. The *Pcf* gene was only transcribed in CMS lines, and the highest levels of transcripts were found in anthers (186).

Unlike *cms-T* in maize, where *urf13-T* transcripts differed between sterile and restored lines (39), no significant differences in *Pcf* transcripts were detected between sterile and restored *Petunia* lines (186). Therefore, it was suggested that the restorer gene in *Petunia* functions on a protein level, possibly affecting translation, degradation, or assembly (186).

Sublimons

Restriction enzyme analyses suggested that the genomic environments in which functional genes are found differ among mitochondrial types in maize (161). In the fertile WF9N mitochondrial genome, two copies of the *atpA* gene have been located near the borders of a 12-kb repeat on the 570-kb "master circle" (74, 34). The two copies of the genes can be distinguished by the 3' flanking sequences. In N mitochondria of most maize lines, the two *atpA* genes are present in equal amounts (161). Similarly, in *cms-S* cytoplasms, there are often two distinguishable *atpA* genes (161). In a few N and

S cytoplasms, one type of atpA gene was found to predominate. However, low levels of the other *atpA* arrangement could be detected following prolonged exposures of mtDNA blots hybridized with the *atpA* probe (161). Additional arrangements, some of which were characteristic of CMS lines, were also present in low amounts in N-type mitochondrial genomes. The faint hybridization signals indicated that the extra *atpA* genes are present on low-abundance molecules (161). Such substoichiometric molecules have been called "sublimons."

The origin of sublimons is at present unknown. They may result from infrequent recombination events between very short regions of homology (161). They could be generated continuously or they may be generated rarely and would be maintained as sublimons by differential replication. Small et al (161) proposed that selective amplification of preexisting substoichiometric molecules could result in sudden genome reorganization. They further suggested that events of this kind could lead to the evolution of different cytoplasmic types, including *cms*.

ABNORMAL GROWTH MUTANTS

While most studies of plant mitochondrial genome variability have concentrated on cytoplasmic male sterility, certain maternally inherited, abnormal-growth mutants in maize also result from mtDNA alterations (124). Nonchromosomal stripe (NCS) plants are characterized by poor growth, decreased yields, and variable leaf striping (31, 157). The observed mtDNA alterations appear to result from specific rearrangements (48, 124). Different regions of the mtDNA are altered in each of two mutants, NCS2 and NCS3. Restriction enzyme analysis and hybridization with cloned probes indicated that the affected plants contain a mixture of normal and mutant mitochondrial genomes (48, 124). Quantitative variation in defective phenotypes among sibling plants presumably results from somatic segregation of different numbers of mutant and normal mitochondria (124).

Analysis of gene expression in the NCS2 mutant suggests that this mutation defines a gene that encodes a mitochondrial protein of as yet unknown function (48). When mitochondrial translation products from NCS2 plants were compared with those from related plants with normal growth, greatly reduced synthesis of a single, approximately 24-kD polypeptide was detected. Probes corresponding to the mitochondrial DNA region altered in NCS2 hybridized to an aberrant set of transcripts, while transcripts homologous to previously characterized plant mitochondrial genes were similar in mutant and normal plants (48). The gene defined by the NCS2 mutation maps to a region of the maize mitochondrial genome, where no other genes have previously been located (K. Newton, D. Lonsdale, unpublished data).

Although the NCS2 and NCS3 mutations arose in the *cms-T* mitochondrial genotype, they are genetically independent of the CMS trait (124). The mitochondrial gene interrupted by the NCS2 mutation is conserved and expressed in C, S, and non-male-sterile N maize, as well as in other higher plants (D. Roussell, K. Newton, unpublished observations).

Several NCS mutants have been identified. Abnormal plants have also been discovered in families of the WF9 line, carrying N and S mitochondrial genotypes, as well as T (44; S. Gabay-Laughnan, J. Laughnan, personal communication). Alterations in the mtDNA of these mutants have been detected (K. Newton, unpublished results). NCS plants arise spontaneously in WF9 or WF9-related lines (31, 44, 124, 157); therefore, the nuclear genome controls the generation of this class of mitochondrial mutations. Nuclear control of spontaneous reversions to fertility is also observed for *cms-S*. The mechanism by which the WF9 nuclear constitution leads to mutations in mtDNA remains to be established.

DIFFERENTIAL EXPRESSION OF PLANT MITOCHONDRIAL GENES

Very few studies have been directed towards understanding potential differences in mitochondrial gene expression during development, or in response to environmental changes (93). In one study, quantitative and apparent qualitative differences in the synthesis of a few mitochondrial polypeptides were found when mitochondria from different maize organs and developmental stages were compared (125). The relatively higher levels of the *Petunia* CMS *Pcf* transcripts in anthers also argues that some mitochondrial genes may be differentially expressed (186). Although it has been reported that one polypeptide is induced by heat shock in maize mitochondria (127, 158), more recent results suggest that the synthesis of this protein resulted from bacterial contamination of the mitochondrial preparations and that maize mitochondria do not synthesize heat shock proteins (128).

PROSPECTS

Although much has been learned about higher-plant mitochondrial genomes, many questions remain. It is apparent that plant mitochondrial genomes change primarily by rearranging. These rearrangements may be mediated by recombination between relatively long or very short repeats. However, not every repeated sequence recombines. What controls the recombinational activity of mitochondrial repeats? The nucleus controls the replication and maintenance of mitochondrial episomes and, presumably, mitochondrial subgenomes. Studies to identify the enzymes responsible for recombination and replication in plant mitochondria are needed. Genetic and molecular analyses of the nuclearly encoded components should help to clarify the roles of nuclear genes in plant mitochondrial genome organization.

Several mitochondrial genes have now been identified in higher plants. However, evidence from transcript analysis and in organelle protein synthesis studies indicate that additional protein-coding genes remain to be identified. It is also possible that a few genes are variously represented or expressed in mitochondria from different plant species.

We understand little about the processes of gene regulation in plant mitochondria. The functions of the proposed regulatory signals will need to be tested in in vitro transcription and/or translation systems. Alternatively, the development of a successful transformation system for plant mitochondria will allow researchers to assay for transient gene expression.

While considerable progress has been made in understanding a few types of cytoplasmic male sterility, the molecular events that restrict the phenotypic consequences of these defects to specific developmental stages remain to be elucidated. The susceptibility of *cms-T* plants to fungal toxins has limited the use of CMS in maize breeding programs. It would be useful to understand how toxin susceptibility is correlated with the CMS phenotype. The availability of a second class of plant mitochondrial defects with abnormal growth phenotypes should aid in the identification of additional mitochondrial genes and help us understand the roles of mitochondria in plant cells. The molecular characterization of nuclear genes affecting mitochondrial functions, such genes that restore fertility and those that control other types of nuclear-mitochondrial incompatibilities, should yield insights into the interactions of nuclear and mitochondrial genes in controlling mitochondrial biogenesis and function at different developmental stages.

ACKNOWLEDGMENTS

I thank D. Roussell, M. Hunt, and C. Knudsen for their helpful comments and J. Hack for assistance in preparing the manuscript. Work in the author's laboratory was supported by a grant from the National Science Foundation and a McKnight individual investigator award.

Literature Cited

1. Anderson, S., Bankier, A. T., Barrell, B. G., de Bruijn, M. H. L., Coulson, A. R., et al. 1981. Sequence and organization of the human mitochondrial genome. *Nature* 290:457–65
2. Attardi, G. 1985. Animal mitochondrial DNA: an extreme example of genetic economy. *Int. Rev. Cytol.* 93:93–145
3. Bailey-Serres, J., Hanson, D. K., Fox, T. D., Leaver, C. J. 1986. Mitochon-drial genome rearrangement leads to extension and relocation of the cytochrome c oxidase subunit I gene in *Sorghum. Cell* 47:567–76
4. Bailey-Serres, J., Dixon, L. K., Liddell, A. D., Leaver, C. J. 1986. Nuclear-mitochondrial interactions in cytoplasmic male-sterile *Sorghum. Theor. Appl. Genet.* 73:252–60
5. Barrell, B. G., Anderson, S., Bankier,

A. T., et al. 1980. Different pattern of codon recognition by mammalian mitochondrial tRNAs. *Proc. Natl. Acad. Sci. USA* 77:3164–66

6. Beckett, J. B. 1971. Classification of male-sterile cytoplasms in maize (*Zea mays* L.) *Crop Sci.* 11:724–27

7. Bedinger, P., de Hostos, E. L., Leon, P., Walbot, V. 1986. Cloning and characterization of a linear 2.3 kb mitochondrial plasmid of maize. *Mol. Gen. Genet.* 205:206–12

8. Bendich, A. J. 1985. Plant mitochondrial DNA: unusual variation on a common theme. In *Genetic Flux in Plants,* ed. B. Hohn, E. S. Dennis, pp. 111–38. Wien: Springer-Verlag

9. Bendich, A. J., Gauriloff, L. P. 1984. Morphometric analysis of cucurbit mitochondria: the relationship between chondriome volume and DNA content. *Protoplasma* 119:1–7

10. Bibb, M. J., Van Etten, R. A., Wright, C. T., et al. 1981. Sequence and gene organization of mouse mitochondrial DNA. *Cell* 26:167–180

11. Bland, M. M., Levings, C. S. III, Matzinger, D. F. 1986. The tobacco mitochondrial ATPase subunit 9 gene is closely linked to an open reading frame for a ribosomal protein. *Mol. Gen. Genet.* 204:8–16

12. Boeshore, M. L., Hanson, M. R., Izhar, S. 1985. A variant mitochondrial DNA arrangement specific to *Petunia* stable sterile somatic hybrids. *Plant Mol. Biol.* 4:125–32

13. Bonen, L., Boer, P. H., Gray, M. W. 1984. The wheat cytochrome oxidase subunit II gene has an intron insert and three radical amino acid changes relative to maize. *EMBO J.* 3:2531–36

14. Bonen, L., Gray, M. W. 1980. Organization and expression of the mitochondrial genome of plants. I. The genes for wheat mitochondrial ribosomal and transfer RNA: evidence for an unusual arrangement. *Nucleic Acids Res.* 8:319–35

15. Borck, K. S., Walbot, V. 1982. Comparison of the restriction endonuclease digestion patterns of mitochondrial DNA from normal and male sterile cytoplasms of *Zea mays* L. *Genetics* 102:109–28

16. Boutry, M., Briquet, M., Goffeau, A. 1983. The alpha subunit of a plant mitochondrial F_1-ATPase is translated in mitochondria. *J. Biol. Chem.* 258:8524–26

17. Boutry, M., Faber, A.-M., Charbonnier, M., et al. 1984. Microanalysis of plant mitochondrial protein synthesis products: detection of variant polypeptides associated with cytoplasmic male sterility. *Plant Mol. Biol.* 3:445–52

18. Braun, C. J., Levings, C. S. III. 1985. Nucleotide sequence of the F_1-ATPase alpha subunit gene from maize mitochondria. *Plant Physiol.* 79:571–77

19. Brennicke, A. 1980. Mitochondrial DNA from *Oenothera berteriana.* Purification and properties. *Plant Physiol.* 65:1207–10

20. Brennicke, A., Blanz, P. 1982. Circular mitochondrial DNA species from *Oenothera* with unique sequences. *Mol. Gen. Genet.* 187:461–67

21. Brennicke, A., Möller, S., Blanz, P. A. 1985. The 18S and 5S ribosomal genes in *Oenothera* mitochondria: sequence rearrangements in the 18S and 5S RNA genes of higher plants. *Mol. Gen. Genet.* 198:404–10

22. Brettell, R. I. S., Thomas, E., Ingram, D. S. 1980. Reversion of Texas male-sterile cytoplasm maize in culture to give fertile, T-toxin resistant plants. *Theor. Appl. Genet.* 58:55–58

23. Brown, T. A., Davies, R. W., Ray, J. A., et al. 1983. The mitochondrial genome of *Aspergillus nidulans* contains reading frames homologous to the human URFs 1 and 4. *EMBO J.* 2:427–35

24. Brown, T. A., Waring, R. B., Scazzocchio, C., Davies, R. W. 1985. The *Aspergillus nidulans* mitochondrial genome. *Curr. Genet.* 9:113–17

25. Butow, R. A., Perlman, P. S., Grossman, L. I. 1985. The unusual *var1* gene of yeast mitochondrial DNA. *Science* 228:1496–1501

26. Carlson, J. E., Erickson, L. R., Kemble, R. J. 1986. Cross hybridization between organelle RNAs and mitochondrial and chloroplast genomes in *Brassica.* *Curr. Genet.* 11:161–63

27. Chao, S., Sederoff, R. R., Levings, C. S. III. 1983. Partial sequence analysis of the 5S to 18S rRNA gene region of the maize mitochondrial genome. *Plant Physiol.* 71:190–93

28. Chao, S., Sederoff, R., Levings, C. S. III. 1984. Nucleotide sequence and evolution of the 18S ribosomal RNA gene in maize mitochondria. *Nucleic Acids Res.* 12:6629–44

29. Chomyn, A., Cleeter, M. W. J., Ragan, C. I., Riley, M., Doolittle, R. F., Attardi, G. 1986. URF6, last unidentified reading frame of human mtDNA, codes for an NADH dehydrogenase subunit. *Science* 234:614–18

30. Chomyn, A., Mariottini, P., Cleeter, M. W. J., Ragan, C. I., Matsuno-Yagi, A., et al. 1985. Six unidentified reading frames of human mitochondrial DNA

encode components of the respiratory-chain NADH dehydrogenase. *Nature* 314:592–97

31. Coe, E. H. Jr. 1983. Maternally inherited abnormal plants in maize. *Maydica* 28:151–67

32. Dale, R. M. K., Mendu, N., Ginsburg, H., Kridl, J. C. 1984. Sequence analysis of the maize mitochondrial 26S rRNA gene and flanking regions. *Plasmid* 11:141–50

33. Dale, R. M. K., Wu, M., Kiernan, M. C. C. 1983. Analysis of four tobacco mitochondrial DNA size classes. *Nucleic Acids Res.* 11:1673–85

34. Dawson, A. J., Hodge, T. P., Isaac, P. G., Leaver, C. J., Lonsdale, D. M. 1986. Location of the genes for cytochrome oxidase subunits I and II, apocytochrome b, α-subunit of the F_1 ATPase and the ribosomal genes on the mitochondrial genome of maize. *Curr. Genet.* 10:561–64

35. Dawson, A. J., Jones, V. P., Leaver, C. J. 1986. Strategies for the identification and analysis of higher plant mitochondrial genes. *Methods Enzymol.* 118:470–85

36. Dawson, A. J., Jones, V. P., Leaver, C. J. 1984. The apocytochrome b gene in maize mitochondria does not contain introns and is preceded by a potential ribosome binding site. *EMBO J.* 3:2107–13

37. de Bruijn, M. H. L. 1983. *Drosophila melanogaster* mitochondrial DNA, a novel organization and genetic code. *Nature* 304:234–41

38. Dewey, R. E., Levings, C. S. III, Timothy, D. H. 1985. Nucleotide sequence of the ATPase subunit 6 gene of maize mitochondria. *Plant Physiol.* 79:914–19

39. Dewey, R. E., Levings, C. S. III, Timothy, D. H. 1986. Novel recombinations in the maize mitochondrial genome produce a unique transcriptional unit in the Texas male-sterile cytoplasm. *Cell* 44:439–49

40. Dewey, R. E., Schuster, A. M., Levings, C. S. III, Timothy, D. H. 1985. Nucleotide sequence of F_0-ATPase proteolipid (subunit 9) gene of maize mitochondria. *Proc. Natl. Acad. Sci. USA* 82:1015–19

41. Dewey, R. E., Timothy, D. H., Levings, C. S. III. 1987. A mitochondrial protein associated with cytoplasmic male sterility in the T cytoplasm of maize. *Proc. Natl. Acad. Sci. USA* 84:5374–78

42. Dixon, L. K., Leaver, C. J., Bretell, R. I. S., Gengenbach, B. G. 1982. Mitochondrial sensitivity to *Drechslera*

maydis T-toxin and the synthesis of a variant mitochondrial polypeptide in plants derived from maize tissue cultures with Texas male-sterile cytoplasm. *Theor. Appl. Genet.* 63:75–80

43. Douce, R. 1985. *Mitochondria in Higher Plants: Structure, Function, and Biogenesis.* Orlando, Fla: Academic. 327 pp.

44. Duvick, D. N. 1965. Cytoplasmic pollen sterility in corn. *Adv. Genet.* 13:1–56

45. Eckenrode, V. K., Levings, C. S. III. 1986. Maize mitochondrial genes. *In Vitro Cell & Dev. Biol.* 22:169–76

46. Escote, L. J., Gabay-Laughnan, S. J., Laughnan, J. R. 1985. Cytoplasmic reversion to fertility in cms-S maize need not involve loss of linear mitochondrial plasmids. *Plasmid* 14:264–67

47. Fauron, C. M.-R., Abbott, A. G., Brettell, R. I. S., Gesteland, R. F. 1987. Maize mitochondrial DNA rearrangements between the normal type, the Texas male sterile cytoplasm, and a fertile revertant cms-T regenerated plant. *Curr. Genet.* 11:339–46

48. Feiler, H. S., Newton, K. J. 1987. Altered mitochondrial gene expression in the nonchromosomal stripe 2 mutant of maize. *EMBO J.* 6:1535–39

49. Finnegan, P. M., Brown, G. G. 1986. Autonomously replicating RNA in mitochondria of maize plants with S-type cytoplasm. *Proc. Natl. Acad. Sci. USA* 83:5175–79

50. Forde, B. G., Leaver, C. J. 1980. Nuclear and cytoplasmic genes controlling synthesis of variant mitochondrial polypeptides in male-sterile maize. *Proc. Natl. Acad. Sci. USA* 77:418–22

51. Forde, B. G., Leaver, C. J. 1979. Mitochondrial genome expression in maize: possible involvement of variant mitochondrial polypeptides in cytoplasmic male-sterility. In *The Plant Genome*, ed. D. R. Davies, D. A. Hopwood, pp. 131–46. Norwich: John Innes Charity. 273 pp.

52. Forde, B. G., Oliver, R. J. C., Leaver, C. J. 1978. Variation in mitochondrial translation products associated with male-sterile cytoplasms in maize. *Proc. Natl. Acad. Sci. USA* 75:3841–45

53. Forde, B. G., Oliver, R. J. C., Leaver, C. J., et al. 1980. Classification of normal and male-sterile cytoplasms in maize. I. Electrophoretic analysis of variation in mitochondrially synthesized proteins. *Genetics* 95:443–50

54. Fox, T. D., Leaver, C. J. 1981. The *Zea mays* mitochondrial gene coding

cytochrome oxidase subunit II has an intervening sequence and does not contain TGA codons. *Cell* 26:315–23

55. Gengenbach, B. G., Connelly, J. A., Pring, D. R., Conde, M. F. 1981. Mitochondrial DNA variation in maize plants regenerated during tissue culture selection. *Theor. Appl. Genet.* 59:161–67

56. Gegenbach, B. G., Green, C. E., Donovan, C. M. 1977. Inheritance of selected pathotoxin resistance in maize plants regenerated from cell cultures. *Proc. Natl. Acad. Sci. USA* 74:5113–17

57. Gottschalk, M., Brennicke, A. 1985. Initiator methionine tRNA gene in *Oenothera* mitochondria. *Curr. Genet.* 9:165–68

58. Grabau, E. A. 1987. Cytochrome oxidase subunit II gene is adjacent to an initiator methionine tRNA gene in soybean mitochondria. *Curr. Genet.* 11:287–93

59. Gray, M. W., Spencer, D. F. 1983. Wheat mitochondrial DNA encodes a eubacteria-like initiator methionine transfer RNA. *FEBS Lett.* 161:323–27

60. Grayburn, W. S., Bendich, A. J. 1987. Variable abundance of a mitochondrial DNA fragment in cultured tobacco cells. *Curr. Genet.* 12:257–61

61. Gregory, P., Earle, E. D., Gracen, V. E. 1977. Biochemical and ultrastructural aspects of southern leaf blight disease. In *Host Plant Resistance to Pests*, ed. P. A. Hedin, pp. 90–114. ACS Symp. Ser. No. 62. Washington, DC: Am. Chem. Soc.

62. Grivell, L. 1983. Mitochondrial DNA. *Sci. Am.* 248:78–89

63. Grun, P. 1976. *Cytoplasmic Genetics and Evolution*. New York: Columbia Univ. Press. 435 pp.

64. Hack, E., Leaver, C. J. 1984. Synthesis of a dicyclohexylcarbodiimide-binding proteolipid by cucumber (*Cucumis sativus* L.). *Curr. Genet.* 8:537–42

65. Hack, E., Leaver, C. J. 1983. The alpha-subunit of the maize F_1-ATPase is synthesized in the mitochondrion. *EMBO J.* 2:1783–89

66. Hanson, M. R. 1984. Stability, variation and recombination in plant mitochondrial genomes via tissue culture and somatic hybridization. *Oxford Surv. Plant Mol. Cell Biol.* 1:33–52

67. Hanson, M. R., Boeshore, M. L., McClean, P. E., O'Connell, M. A., Nivison, H. T. 1986. The isolation of mitochondria and mitochondrial DNA. *Methods Enzymol.* 118:437–53

68. Hanson, M. R., Conde, M. F. 1985. Functioning and variation of cytoplas-

mic genomes: lessons from cytoplasmic-nuclear interactions affecting male fertility in plants. *Int. Rev. Cytol.* 94:213–67

69. Heisel, R., Brennicke, A. 1983. Cytochrome oxidase subunit II gene in mitochondria of *Oenothera* has no intron. *EMBO J.* 2:2173–78

70. Heisel, R., Brennicke, A. 1985. Overlapping reading frames in *Oenothera* mitochondria. *FEBS Lett.* 193:164–68

71. Heisel, R., Schobel, W., Schuster, W., Brennicke, A. 1987. The cytochrome oxidase subunit I and subunit III genes in *Oenothera* mitochondria are transcribed from identical promoter sequences. *EMBO J.* 6:29–34

72. Houchins, J. P., Ginsburg, H., Rohrbaugh, M., Dale, R. M. K., Schardl, C. L., et al. 1986. DNA sequence analysis of a 5.27-kb direct repeat occurring adjacent to the regions of S-episome homology in maize mitochondria. *EMBO J.* 5:2781–88

73. Iams, K. P., Sinclair, J. H. 1982. Mapping the mitochondrial DNA of *Zea mays*: ribosomal gene localization. *Proc. Natl. Acad. Sci. USA* 79:5926–29

74. Isaac, P. G., Brennicke, A., Dunbar, S. M., Leaver, C. J. 1985. The mitochondrial genome of fertile maize (*Zea mays* L.) contains two copies of the gene encoding the α-subunit of the F_1-ATPase. *Curr. Genet.* 10:321–28

75. Isaac, P. G., Jones, V. P., Leaver, C. J. 1985. The maize cytochrome c oxidase subunit I gene: sequence, expression and rearrangement in cytoplasmic male sterile plants. *EMBO J.* 4:1617–23

76. Ise, W., Haiker, H., Weiss, H. 1985. Mitochondrial translation of subunits of the rotenone-sensitive NADH: ubiquinone reductase in *Neurospora crassa*. *EMBO J.* 4:2075–80

77. Izhar, S. 1978. Cytoplasmic male sterility in *Petunia*. III. Genetic control of microsporogenesis and male fertility restoration. *J. Hered.* 69:22–26

78. Izhar, S., Schlicter, M., Swartzberg, D. 1983. Sorting out of cytoplasmic elements in somatic hybrids of *Petunia* and the prevalence of the heteroplasmon through several meiotic cycles. *Mol. Gen. Genet.* 190:468–74

79. Kao, T.-H., Moon, E., Wu, R. 1984. Cytochrome oxidase subunit II gene of rice has an insertion sequence within the intron. *Nucleic Acids Res.* 12:7305–15

80. Kemble, R. J., Bedbrook, J. R. 1980. Low molecular weight circular and linear DNA in mitochondria from normal and male sterile *Zea mays* cytoplasm. *Nature* 284:565–66

81. Kemble, R. J., Thompson, R. D. 1982. S1 and S2, the linear mitochondrial DNAs present in a male sterile line of maize, possess terminally attached proteins. *Nucleic Acids Res.* 10:8181–90

82. Kemble, R. J., Flavell, R. B., Bretell, R. I. S. 1982. Mitochondrial DNA analyses of fertile and sterile maize plants derived from tissue culture with the Texas male sterile cytoplasm. *Theor. Appl. Genet.* 62:213–17

83. Kemble, R. J., Gunn, R. E., Flavell, R. B. 1983. Mitochondrial DNA variation in races of maize indigenous to Mexico. *Theor. Appl. Genet.* 65:129–44

84. Kemble, R. J., Mans, R. J., Gabay-Laughnan, S., Laughnan, J. R. 1983. Sequences homologous to episomal mitochondrial DNAs in the maize nuclear genome. *Nature* 304:744–47

85. Kheyr-Pour, A., Gracen, V. E., Everett, H. L. 1981. Genetics of fertility restoration in the C-group of cytoplasmic male sterility in maize. *Genetics* 98:379–88

86. Kim, B. D., Mans, R. J., Conde, M. F., Pring, D. R., Levings, C. S. III. 1982. Physical mapping of homologous segments of mitochondrial episomes from S-male-sterile maize. *Plasmid* 7:1–14

87. Koeppe, D. E., Cox, J. K., Malone, C. P. 1978. Mitochondrial heredity: a determinant in the toxic response of maize to the insecticide methomyl. *Science* 201:1227–29

88. Koncz, C., Sumegi, J., Udvardy, A., Racsmany, M., Dudits, D. 1981. Cloning of mtDNA fragments homologous to mitochondrial S2 plasmid-like DNA in maize. *Mol. Gen. Genet.* 183:449–58

89. Lambowitz, A. M., LaPolla, R. J., Collins, R. A. 1979. Mitochondrial ribosome assembly in *Neurospora,* two-dimensional gel electrophoretic analysis of mitochondrial ribosomal proteins. *J. Cell. Biol.* 82:17–31

90. Laughnan, J. R., Gabay-Laughnan, S. J. 1983. Cytoplasmic male sterility in maize. *Ann. Rev. Genet.* 17:27–48

91. Laughnan, J. R., Gabay-Laughnan, S. J., Carlson, J. E. 1981. Characteristics of cms-S reversion to male fertility in maize. *Stadler Symp.* 13:93–114

92. Leaver, C. J., Harmey, M. A. 1976. Higher plant mitochondrial ribosomes contain a 5S ribosomal RNA component. *Biochem. J.* 157:275–77

93. Leaver, C. J., Gray, M. W. 1982. Mitochondrial genome organization and expression in higher plants. *Ann. Rev. Plant Physiol.* 33:373–402

94. Leaver, C. J., Hack, E., Forde, B. G.

1983. Protein synthesis by isolated plant mitochondria. *Methods Enzymol.* 97:476–84

95. Lebacq, P., Vedel, F. 1981. SalI restriction enzyme analysis of chloroplast and mitochondria DNAs in the genus *Brassica. Plant Sci. Lett.* 23:1–9

96. Lee, S. L. J., Warmke, H. E. 1979. Organelle size and number in fertile and T-cytoplasmic male-sterile corn. *Am. J. Bot.* 66:141–48

97. Leon, P., Walbot, V., Bedinger, P. 1988. Molecular analysis of the linear 2.3 kb plasmid of maize mitochondria: apparent capture of tRNA genes. *Proc. Natl. Acad. Sci.* In press

98. Levings, C. S. III, Kim, B. D., Pring, D. R., Conde, M. F., Mans, R. J., et al. 1980. Cytoplasmic reversion of cms-S in maize: association with a transpositional event. *Science* 209:1021–23

99. Levings, C. S. III, Pring, D. R. 1976. Restriction endonuclease analysis of mitochondrial DNA from normal and Texas cytoplasmic male-sterile maize. *Science* 193:158–60

100. Levings, C. S. III, Sederoff, R. R. 1983. Nucleotide sequence of the S-2 mitochondrial DNA from the S cytoplasm of maize. *Proc. Natl. Acad. Sci. USA* 80:4055–59

101. Lonsdale, D. M. 1987. Cytoplasmic male sterility: a molecular perspective. *Plant Physiol. Biochem.* 25:265–71

102. Lonsdale, D. M. 1987. The physical and genetic map of the mitochondrial genome from the WF9-N fertile cytoplasm. *Maize Genet. Coop. Newsl.* 61:148–49

103. Lonsdale, D. M. 1986. Viral RNA in mitochondria? *Nature* 323:299

104. Lonsdale, D. M., Hodge, T. P., Fauron, C. M.-R. 1984. The physical map and organization of the mitochondrial genome from the fertile cytoplasm of maize. *Nucleic Acids Res.* 12:9249–61

105. Lonsdale, D. M., Hodge, T. P., Howe, C. J., Stern, D. B. 1983. Maize mitochondrial DNA contains a sequence homologous to the ribulose-1,5-bisphosphate carboxylase large subunit gene of chloroplast DNA. *Cell* 34:1007–14

106. Lonsdale, D. M., Hodge, T. P., Stoehr, P. J. 1986. Analysis of the genome structure of plant mitochondria. *Methods Enzymol.* 118:453–70

107. Makaroff, C. A., Palmer, J. D. 1987. Extensive mitochondrial specific transcription of the *Brassica campestris* mitochondrial genome. *Nucleic Acids Res.* 15:5141–56

108. Manna, E., Brennicke, A. 1985. Primary and secondary structure of 26S ribosomal RNA of *Oenothera* mitochondria. *Curr. Genet.* 9:505–15

109. Manna, E., Brennicke, A. 1986. Site-specific circularisation at an intragenic sequence in *Oenothera* mitochondria. *Mol. Gen. Genet.* 203:377–81

110. Manson, J. C., Liddell, A. D., Leaver, C. J., Murray, K. 1986. A protein specific to mitochondria from S-type male-sterile cytoplasm of maize is encoded by an episomal DNA. *EMBO J.* 5:2275–80

111. Marechal, L., Guillemaut, P., Weil, J.-H. 1985. Sequences of two bean mitochondria tRNA^Tyr which differ in the level of post-transcriptional modification and have a prokaryotic-like large extra-loop. *Plant Mol. Biol.* 5: 347–51

112. Marechal, L., Guillemaut, P., Grienenberger, J.-M., Jeannin, G., Weil, J.-H. 1985. Structure of bean mitochondrial tRNA^Phe and localization of the tRNA^Phe gene on the mitochondrial genomes of maize and wheat. *FEBS Lett.* 184:289–93

113. Marechal, L., Guillemaut, P., Grienenberger, J.-M., Jeannin, G., Weil, J.-H. 1985. Sequence and codon recognition of bean mitochondria and chloroplast tRNAs^Trp: evidence for a high degree of homology. *Nucl. Acids Res.* 13:4411–16

114. Marechal, L., Guillemaut, P., Grienenberger, J.-M., Jeannin, G., Weil, J.-H. 1986. Sequences of initiator and elongator methionine tRNAs in bean mitochondria. *Plant Mol. Biol.* 7:245–53

115. Marechal, L., Runeberg-Roos, P., Grienenberger, J. M., Colin, J., Weil, J. 1987. Homology in the region containing a tRNA^Trp gene and a complete or partial tRNA^Pro gene in wheat mitochondrial and chloroplast genomes. *Curr. Genet.* 12:91–98

116. Matthews, D. E. P., Gregory, P., Gracen, V. E. 1979. *Helminthosporium maydis* race T toxin induces leakage of NAD^+ from T cytoplasm corn mitochondria. *Plant Physiol.* 63: 1149–53

117. McNay, J. W., Chourey, P. S., Pring, D. R. 1984. Molecular analysis of genomic stability of mitochondrial DNA in tissue cultured cells. *Theor. Appl. Genet.* 67:433–37

118. McNay, J. W., Pring, D. R., Lonsdale, D. M. 1983. Polymorphism of mitochondrial DNA 'S' regions among

normal cytoplasms of maize. *Plant Mol. Biol.* 12:177–89

119. Messing, J., Carlson, J., Hagen, G., et al. 1984. Cloning and sequencing of the ribosomal RNA genes in maize: the 17S region. *DNA* 3:31–40

120. Miller, R. J., Koeppe, D. E. 1971. Southern corn leaf blight: susceptible and resistant mitochondria. *Science* 173: 67–69

121. Moon, E., Kao, T.-H., Wu, R. 1985. Pea cytochrome oxidase subunit II gene has no intron and generates two mRNA transcripts with different 5'-termini. *Nucleic Acids Res.* 13:3195–212

122. Mulligan, R. M., Maloney, A. P., Walbot, V. 1988. RNA processing and multiple transcription initiation sites result in transcript size heterogeneity in maize mitochondria. *Mol. Gen. Genet.* In press

123. Neuport, W., Schatz, G. 1981. How proteins are transported into mitochondria. *Trends Biochem. Sci.* 6:1–4

124. Newton, K. J., Coe, E. H. Jr. 1986. Mitochondrial DNA changes in abnormal growth mutants of maize. *Proc. Natl. Acad. Sci. USA* 83:7363–66

125. Newton, K. J., Walbot, V. 1985. Maize mitochondria synthesize organ-specific polypeptides. *Proc. Natl. Acad. Sci. USA* 82:6879–83

126. Newton, K. J., Walbot, V. 1985. Molecular analysis of mitochondria from a fertility restorer line of maize. *Plant Mol. Biol.* 4:247–52

127. Niebiolo, C. M., White, E. M. 1985. Corn mitochondrial protein synthesis in response to heat shock. *Plant Physiol.* 79:1129–32

128. Nieto-Sotelo, J., Ho, T.-H. D. 1987. Absence of heat shock protein synthesis in isolated mitochondria and plastids from maize. *J. Biol. Chem.* 262:12288–92

129. Ojala, D., Montoya, J., Attardi, G. 1981. tRNA punctuation model of RNA processing in human mitochondria. *Nature* 290:470–74

130. Oro, A. E., Newton, K. J., Walbot, V. 1985. Molecular analysis of the inheritance and stability of the mitochondrial genome of B37N maize. *Theor. Appl. Genet.* 70:287–93

131. Paillard, M., Sederoff, R. R., Levings, C. S. III. 1985. Nucleotide sequence of the S-1 mitochondrial DNA from the S cytoplasm of maize. *EMBO J.* 4:1125–28

132. Palmer, J. D. 1985. Evolution of chloroplast and mitochondrial DNA in plants and algae. In *Molecular Evolutionary*

Genetics, ed. R. J. MacIntyre, pp. 131–240. New York: Plenum

133. Palmer, J. D., Herbon, L. A. 1986. Tricircular mitochondrial genomes of *Brassica* and *Raphanus:* reversal of repeat configurations by inversion. *Nucleic Acids Res.* 14:9755–65

134. Palmer, J. D., Herbon, L. A. 1987. Unicircular structure of the *Brassica hirta* mitochondrial genome. *Curr. Genet.* 11: 565–70

135. Palmer, J. D., Shields, C. R. 1984. Tripartite structure of the *Brassica campestris* mitochondrial genome. *Nature* 307: 437–40

136. Palmer, J. D., Shields, C. R., Cohen, D. B., Orton, T. J. 1983. An unusual mitochondrial DNA plasmid in the genus *Brassica. Nature* 301:725–28

137. Parks, T. D., Dougherty, W. G., Levings, C. S. III, Timothy, D. H. 1984. Identification of two methionine transfer RNA genes in the maize mitochondrial genome. *Plant Physiol.* 76:1079–82

138. Parks, T. D., Dougherty, W. G., Levings, C. S. III, Timothy, D. H. 1985. Identification of an aspartate transfer RNA gene in maize mitochondrial DNA. *Curr. Genet.* 9:517–19

139. Pratje, E., Schnierer, S., Dujon, B. 1984. Mitochondrial DNA of *Chlamydomonas reinhardtii:* the DNA sequence of a region showing homology with mammalian URF2. *Curr. Genet.* 9:75–82

140. Pring, D. R., Lonsdale, D. M. 1985. Molecular biology of higher plant mitochondrial DNA. *Int. Rev. Cytol.* 97:1–46

141. Pring, D. R., Lonsdale, D. M., Gracen, V. E., Smith, A. G. 1987. Mitochondrial DNA duplication/deletion events and polymorphism of the C group of male sterile maize cytoplasms. *Theor. Appl. Genet.* 73:646–53

142. Pring, D. R., Levings, C. S. III. 1978. Heterogeneity of maize cytoplasmic genomes among male-sterile cytoplasms. *Genetics* 89:121–36

143. Pring, D. R., Levings, C. S. III, Hu, W. W. L., Timothy, D. H. 1977. Unique DNA associated with mitochondria in the "S"-type cytoplasm of male-sterile maize. *Proc. Natl. Acad. Sci. USA* 74:2904–8

144. Quetier, F., Lejeune, B., Delorme, S., Falconet, D. 1985. Molecular organization and expression of the mitochondrial genome of higher plants. In *Encyclopedia of Plant Physiology 18,* ed. R. Douce, D. A. Day, pp. 25–36. Berlin: Springer-Verlag

145. Rogers, H. J., Buck, K. W., Brasier, C.

M. 1987. A mitochondrial target for double-stranded RNA in diseased isolates of the fungus that causes Dutch elm disease. *Nature* 329:558–60

146. Rottman, W. H., Brears, T., Hodge, T. P., Lonsdale, D. M. 1987. A mitochondrial gene is lost via homologous recombination during reversion of CMS T maize to fertility. *EMBO J.* 6:1541–46

147. Scazzocchio, C., Brown, T. A., Waring, R. B., et al. 1983. Organization of the *Aspergillus nidulans* mitochondrial genome. In *Mitochondria 1983: Nucleomitochondrial Interactions,* ed. R. J. Schweyen, K. Wolf, F. Kaudewitz, pp. 303–12. New York: de Gruyter

148. Schardl, C. L., Lonsdale, D. M., Pring, D. R., Rose, K. R. 1984. Linearization of maize mitochondrial chromosomes by recombination with linear episomes. *Nature* 310:292–96

149. Schardl, C. L., Pring, D. R., Lonsdale, D. M. 1985. Mitochondrial DNA rearrangements associated with fertile revertants of S-type male-sterile maize. *Cell* 43:361–68

150. Schieber, G., O'Brien, T. W. 1985. Site of synthesis of the proteins of mammalian ribosomes. *J. Biol. Chem.* 260: 6367–72

151. Schuster, A. M., Sisco, P. H., Levings, C. S. III. 1983. Two unique RNAs in cms-S and RU maize mitochondria. In *Plant Molecular Biology,* ed. R. B. Goldberg, 12:437–44. UCLA Symp. Mol. Cell. Biol.

152. Schuster, W., Brennicke, A. 1985. TGA-termination codon in the apocytochrome b gene from *Oenothera* mitochondria. *Curr. Genet.* 9:157–63

153. Schuster, W., Brennicke, A. 1986. Pseudocopies of the ATPase α-subunit gene in *Oenothera* are present on different circular molecules. *Mol. Gen. Genet.* 204:29–35

154. Schuster, W., Heisel, R., Isaac, P. G., Leaver, C. J., Brennicke, A. 1986. Transcript termini of messenger RNAs in higher plant mitochondria. *Nucl. Acids. Res.* 14:5943–54

155. Schuster, W., Heisel, R., Manna, E., Schobel, W., Wissinger, B., Brennicke, A. 1987. Molecular analysis of the mitochondrial genome in *Oenothera berteriana. Plant Physiol. Biochem.* 25: 259–64

156. Scott, N. S., Timmis, J. N. 1984. Homologies between nuclear and plastid DNA in spinach. *Theor. Appl. Genet.* 67:279–88

157. Shumway, L. K., Bauman, L. F. 1967.

Nonchromosomal stripe of maize. *Genetics* 55:33–38

158. Sinibaldi, R. M., Turpen, T. 1985. A heat shock protein is encoded within mitochondria of higher plants. *J. Biol. Chem.* 260:15382–85

159. Sisco, P. H., Garcia-Arenal, F., Zaitlin, M., Earle, E. D., Gracen, V. 1984. LBN, a male-sterile cytoplasm of maize, contains two double-stranded RNAs. *Plant Sci. Lett.* 34:127–34

160. Sisco, P. H., Gracen, V. E., Everett, H. L., Earle, E. D., Pring, D. R., et al. 1985. Fertility restoration and mitochondrial nucleic acids distinguish at least five subgroups among cms-S cytoplasms of maize (*Zea mays* L.). *Theor. Appl. Genet.* 71:5–15

161. Small, I. D., Isaac, P. G., Leaver, C. J. 1987. Stoichiometric differences in DNA molecules containing the *atpA* gene suggest mechanisms for the generation of mitochondrial genome diversity in maize. *EMBO J.* 6:865–69

162. Spencer, D. F., Bonen, L., Gray, M. W. 1981. Primary sequence of wheat mitochondrial 5S ribosomal ribonucleic acid: functional and evolutionary implications. *Biochemistry* 20:4022–29

163. Spencer, D. F., Schnare, M. N., Gray, M. W. 1984. Pronounced structural similarities between the small subunit ribosomal RNA genes of wheat mitochondria and *Escherichia coli*. *Proc. Natl. Acad. Sci. USA* 81:493–97

164. Stern, D. B., Bang, A. G., Thompson, W. F. 1986. The watermelon mitochondrial URF-1 gene: evidence for a complex structure. *Curr. Genet.* 10:857–69

165. Stern, D. B., Dyer, T. A., Lonsdale, D. M. 1982. Organization of the mitochondrial ribosomal RNA genes of maize. *Nucleic Acids Res.* 10:3333–40

166. Stern, D. B., Lonsdale, D. M. 1982. Mitochondrial and chloroplast genomes of maize have a 12-kilobase DNA sequence in common. *Nature* 299:698–702

167. Stern, D. B., Newton, K. J. 1986. Isolation of plant mitochondrial RNA. *Methods Enzymol.* 118:488–96

168. Stern, D. B., Newton, K. J. 1985. Mitochondrial gene expression in Cucurbitaceae: conserved and variable features. *Curr. Genet.* 9:395–405

169. Stern, D. B., Palmer, J. D. 1986. Tripartite mitochondrial genome of spinach: physical structure, mitochondrial gene mapping, and locations of transposed chloroplast DNA sequences. *Nucleic Acids Res.* 14:5651–66

170. Stern, D. B., Palmer, J. D. 1984. Extensive and widespread homologies between mitochondrial DNA and chloroplast DNA in plants. *Proc. Natl. Acad. Sci. USA* 81:1946–50

171. Stern, D. B., Palmer, J. D., Thompson, W. F., Lonsdale, D. M. 1983. Mitochondrial DNA sequence evolution and homology to chloroplast DNA in angiosperms. In *Plant Molecular Biology,* pp. 467–77. New York: Alan R. Liss

172. Terpstra, P., Zanders, E., Butow, R. A. 1979. The association of var1 with the 38S mitochondrial ribosomal subunit in yeast. *J. Biol. Chem.* 254:12653–61

173. Timmis, J. N., Scott, N. S. 1983. Sequence homology between spinach nuclear and chloroplast genomes. *Nature* 305:65–67

174. Timothy, D. H., Levings, C. S. III, Hu, W. W. L., Goodman, M. M. 1983. Plasmid-like mitochondrial DNAs in diploperennial teosinte. *Maydica* 28:139–49

175. Traynor, P. L., Levings, C. S. III. 1986. Transcription of the S-2 maize mitochondrial plasmid. *Plant Molecular Biol.* 7:255–63

176. Tzagoloff, A. 1982. *Mitochondria.* New York: Plenum. 342 pp.

177. Tzagoloff, A., Myers, A. M. 1986. Genetics of mitochondrial biogenesis. *Ann. Rev. Biochem.* 55:249–85

178. Umbeck, P. F., Gengenbach, B. G. 1983. Reversion of male-sterile T-cytoplasm maize to male fertility in tissue culture. *Crop Sci.* 23:584–88

179. van den Boogaart, P., Samallo, J., Agsteribbe, E. 1982. Similar genes for a mitochondrial ATPase subunit in the nuclear and mitochondrial genomes of *Neurospora crassa. Nature* 298:187–89

180. Ward, B. L., Anderson, R. S., Bendich, A. J. 1981. The size of the mitochondrial genome is large and variable in a family of plants *(Cucurbitaceae) Cell* 25:793–803

181. Warmke, H. E., Lee, S.-L. J. 1977. Mitochondrial degeneration in Texas cytoplasmic male-sterile corn anthers. *J. Hered.* 68:213–22

182. Weissinger, A. D., Timothy, D. H., Levings, C. S. III, Hu, W. W. L., Goodman, M. M. 1982. Unique plasmid-like DNAs from indigenous maize races of Latin America. *Proc. Natl. Acad. Sci. USA* 79:1–5

183. Whisson, D. L., Scott, N. S. 1985. Nuclear and mitochondrial DNA have sequence homology with a chloroplast gene. *Plant Mol. Biol.* 4:267–73

184. Wise, R. P., Fliss, A. E., Pring, D. R., Gengenbach, B. G. 1987. *Urf13-T* of T cytoplasm maize mitochondria encodes

a 13 kD polypeptide. *Plant Mol. Biol.* 9:121–26

185. Wise, R. P., Pring, D. R., Gengenbach, B. G. 1987. Mutation to fertility and toxin insensitivity in Texas (T) cytoplasm maize is associated with a frameshift in a mitochondrial open reading frame. *Proc. Natl. Acad. Sci. USA* 84:2858–62

186. Young, E. G., Hanson, M. R. 1987. A fused mitochondrial gene associated with cytoplasmic male sterility is developmentally regulated. *Cell* 50:41–49

187. Zabala, G., O'Brien-Vedder, C., Walbot, V. 1987. S2 episome of maize mitochondria encodes a 130-kilodalton protein found in male sterile and fertile plants. *Proc. Natl. Acad. Sci. USA* 84: 7861–65

188. Zabala, G., Walbot, V. 1988. An S1 episomal gene of maize mitochondria is expressed in male sterile and fertile plants of the S-type cytoplasm. *Mol. Gen. Genet.* In press

Ann. Rev. Plant Physiol. Plant Mol. Biol. 1988. 39:533–94

ENZYMATIC REGULATION OF PHOTOSYNTHETIC CO_2 FIXATION IN C_3 PLANTS*

Ian E. Woodrow

C.S.I.R.O. Division of Plant Industry, P.O. Box 1600, Canberra City, ACT 2601, Australia

Joseph A. Berry[1]

Carnegie Institution of Washington, 290 Panama Road, Stanford, California 94305

> *"Reality is only an approximation to our models."*
> –Henry Miziorko, at a workshop on Rubisco, the University of Arizona, April 1987.

CONTENTS

INTRODUCTION 534
BIOCHEMICAL INTERPRETATION OF NET CO_2 UPTAKE 534
    The CO_2 Compensation Point, a Special Case 535
    Experimental Approaches 537
    Kinetic Analysis of Rubisco in vivo 542

*Abbreviations: CA-1-P—2-carboxyarabinitol 1-phosphate; CABP—2-carboxyarabinitol 1,5-bisphosphate; DHAP—dihydroxacetone phosphate; G3P—glyceraldehyde 3-phosphate; G6P—glucose 6-phosphate; F26P—fructose 2,6-bisphosphate; F6P—fructose 6-phosphate; FBP—fructose 1,6-bisphosphate; FBPase—FBP phosphatase; PGA—3-phosphoglyceric acid; PCR-cycle—photosynthetic carbon reduction cycle; Pi—orthophosphate; PSI—photosystem 1; PSII—photosystem 2; Q_A—the primary acceptor of PSII; Ru5P—ribulose 5-phosphate; Rubisco—RuBP-carboxylase/oxygenase; RuBP—ribulose 1,5-bisphosphate; SBPase—sedoheptulose 1,7-bisphosphate phosphatase; SPS—sucrose phosphate synthase; UDPG—UDP-glucose

[1]C.I.W.–D.P.B. Publication No. 1009.

533

0066-4294/88/0601-0533$02.00

REGULATORY SEQUENCES... 552
    Regulatory Properties of the PCR Cycle.. 553
    Feedback Regulation of Photosynthesis 556
    Operation of the Feedback Sequence... 561
    Feedback on Energy Input from Rubisco 566
    Feedforward Regulation of Enzyme Activity.. 568
CONTROL OF THE RATE OF PHOTOSYNTHESIS 571
    Definition of Limitation... 573
    Control of the Rate of CO_2 Uptake... 574
CONCLUDING REMARKS 580

INTRODUCTION

The physiological process of photosynthetic CO_2 uptake by plant leaves is a manifestation of reactions catalyzed by ribulose-1,5-bisphosphate carboxylase/oxygenase (Rubisco). In this review we consider the metabolic and biochemical factors that exert "local control" over these reactions in vivo. Mechanisms of regulation (particularly of Rubisco) play a prominent role here, but the phenomenon of central interest is the overall regulation of the biochemical process of photosynthesis. Our goal is to show how some of these mechanisms, which effect "local" control of this key enzyme, are linked to other steps in the photosynthetic process, thus forming "regulatory sequences" that connect the velocity of the carboxylation and oxygenation reactions to independent factors of the environment (e.g. the intensity of light or the concentration of CO_2). These sequences bring us, inescapably, to consideration of limiting factors in photosynthesis—a topic that has been considered mostly in qualitative terms. To facilitate a natural progression toward application of quantitative methods we pay particular attention in this part of the review to quantitative procedures that have been used to examine control of complex metabolic systems by co-limiting factors. Finally, we note that our approach to this problem is conditioned in part by the view that, through evolutionary processes, the photosynthetic system has been selected to achieve efficient use of limited resources for primary production. We conclude our review with consideration of some general ways in which the regulation of CO_2 fixation conforms to this view.

BIOCHEMICAL INTERPRETATION OF NET CO_2 UPTAKE

Studies of the biochemical regulation of net CO_2 uptake in photosynthesis must focus first upon the operation of Rubisco in vivo. This enzyme is the gatekeeper standing between the internal metabolic systems of the mesophyll cells and the observer, who may monitor the rate of net gas exchange with the environment. Obviously, the more known about the rules governing passage through the gate, the more we may infer concerning the internal operation of the system. In recent years a number of reviews of Rubisco have appeared (5, 121, 137, 158, 166, 174, 261). Here, we consider some aspects of the enzyme from C_3 plants that are relevant to understanding its function in vivo.

The CO_2 Compensation Point, a Special Case

To presume that physiological responses of intact leaves can be interpreted from biochemical studies of Rubisco involves a leap of faith. Our confidence for making this leap may be bolstered by reviewing studies that have established the biochemical basis of a single, well-defined physiological property—the CO_2 compensation point (Γ). When a leaf of a C_3 plant is illuminated in a closed space it will either take up or evolve CO_2 (depending upon the initial concentration of CO_2) until the CO_2 concentration reaches a stable steady-state value ($\Gamma \approx 45$ μbar CO_2 at 25°C and 210 mbar O_2) at which the rate of net CO_2 assimilation (A) is zero. The value of Γ is independent of the intensity of illumination (above a threshold intensity), strongly dependent upon $[O_2]$ and temperature, and under any given condition, similar for leaves of all C_3 plants. Tregunna and coworkers (252) postulated that the responses of Γ to O_2 and temperature indicate that the rates of photorespiration and gross CO_2 uptake are somehow linked. The basis for this linkage became clear with the discovery that Rubisco catalyzes the first step in both metabolic sequences (19, 20).

Laing et al (147) derived an equation relating the ratio of the rates of carboxylation (v_c) and oxygenation (v_o) of RuBP to the concentration of the substrates, CO_2 and O_2,

$$\frac{v_c}{v_o} = \frac{V_c/K_c}{V_o/K_o} \cdot \frac{[CO_2]}{[O_2]} = \tau \cdot \frac{[CO_2]}{[O_2]},$$

1.

where v_c and v_o are the rates of carboxylation and oxygenation, K_c, K_o, V_c, and V_o are the corresponding K_m and V_{max} terms for the carboxylase and oxygenase functions of Rubisco, respectively, and τ is a constant used to abbreviate the ratio of kinetic constants. From the equation for net CO_2 exchange (64)

$$A = v_c \left(1 - 0.5 \frac{v_o}{v_c}\right) - R_d$$

2.

it may be deduced that $v_o/v_c = 2$ at a $[CO_2]$ and $[O_2]$ where $A = -R_d$, the rate of normal respiration continuing in the light. Ignoring the respiration term for the moment we may write that $\Gamma \approx 0.5[O_2]/\tau$. Laing et al (147) evaluated τ from measurements of the kinetic constants conducted over a range of temperatures and showed that the observed dependence of Γ values on $[O_2]$ and temperature could be approximated using this model and kinetic constants derived from in vitro studies with purified Rubisco. Jordan & Ogren (126) developed a more accurate procedure for determining τ based upon a simultaneous assay of the v_c/v_o ratio. These workers demonstrated much better agreement between Γ values predicted from their measurements of τ and Γ values for C_3 plants (18, 127, 128, 134). Furthermore, they showed that τ is not significantly affected by the concentration of RuBP, the pH of the medium, or the activation state of Rubisco, all of which may vary in vivo (128).

The next step in the development of this quantitative test fell to the physiologists. As first indicated by Laisk (149), the "true" compensation point of photosynthesis is not at Γ where $A = 0$ but at a slightly lower CO_2 concentration where $A = -R_d$. He defined this point as Γ^* and showed that $\tau = 0.5[O_2]/\Gamma^*$. The difference between Γ and Γ^*, although only a few μbar, is significant given the precision of the measurements of τ by Jordan & Ogren (128). Brooks & Farquhar (22) developed gas exchange procedures to measure Γ^* for leaves of spinach as a function of temperature, and the corresponding values of τ obtained by this procedure are compared with those obtained in vitro by Jordan & Ogren (128) in Figure 1. The close agreement

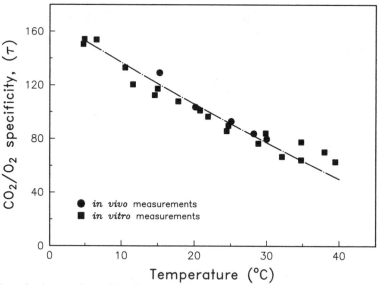

Figure 1 A comparison of the temperature dependence of the CO_2/O_2 specificity of spinach Rubisco. Values were obtained from measurements of the CO_2-compensation point (Γ^*) of intact leaves by Brooks & Farquhar (22) and from simultaneous determination of RuBP-carboxylase and RuBP-oxygenase activities with purified spinach Rubisco in vitro by Jordan & Ogren (128). The dashed line is drawn using the $Q_{10}=0.74$ (Table 1).

between the two sets of data provides some confidence that kinetic measurements with isolated Rubisco conducted in vitro are relevant to interpreting physiological responses of intact leaves.

Experimental Approaches

The experimental approaches that provide the basis for mechanistic analyses of the rates of CO_2 uptake during steady-state photosynthesis include: (a) biochemical studies of the components of the system—particularly Rubisco; (b) physiological measurements of the systemic properties of the intact tissue at defined steady states; and (c) predictive or interpretive models of the relevant processes. It is not practical to provide a complete review of these approaches here. Nevertheless, a brief introduction to these topics may be useful.

KINETIC PROPERTIES OF RUBISCO   There are many reports of measurements of the kinetic constants of Rubisco from C_3 species [for a compilation see Keys (137)]. Most of these measurements have been made with preparations having somewhat lower activity than the native enzyme as a result of damage during preparation or aging. Of course what one wishes to know are the values of the kinetic constants that apply in vivo, and there is no completely sound basis for selecting such constants. The values listed in Table 1 are calculated from measurements of the enzyme from spinach. There may, however, be small significant differences in the kinetic properties of Rubisco from different species of C_3 plants (59, 217). The set of constants provided here—the $K_m(CO_2)(K_c)$, the inhibitory constant for oxygen with

Table 1   Suggested values of kinetic constants to approximate the response of net CO_2 uptake by Rubisco of C_3 plants in vivo.

Constant	Symbol	Y_{25}, value at 25°C	Q_{10}	Reference
CO_2/O_2 specificity	τ	88 M M^{-1}	0.74[b]	(22, 128)
		2360 bar bar^{-1}	0.67[b]	
$K_m(CO_2)$	K_c	9 μM	1.8[c]	(128, 212)
		270 μbar	2.1[c]	
$K_i(O_2)$ on K_c	K_o^i	535 μM	1.0	(128)
		400 mbar	1.2	
$K_m(RuBP)$	K_R^i	28μM	1.9	(9)
Activity	V_c	3.6 μmol mg^{-1}min^{-1}	2.4[c]	(9, 212)
	k_{cat}	3.3 s^{-1}site^{-1}		

[a]These values are based upon measurements conducted in vitro with Rubisco and have been selected assuming that the activity of the native enzyme is greater than or equal to that observed in vitro. The Q_{10} values approximating the temperature dependence of these constants may be used to calculate the value (Y_T) at any temperature (T) between 5 and 35°C according to $Y_T = Y_{25} \cdot Q_{10}^{(T-25)/10}$, where Y_{25} is the value at 25°C.
[b]See also a polynomial given in Ref. 22.
[c]To extrapolate below 15°C use: $Y_T = Y_{25} \cdot Q_{10}^{-1} (1.8 Q_{10})^{(T-15)/10}$.

respect to CO_2 on the carboxylase (K_o^i), the K_m(RuBP) (K_R), and the catalytic constant for the carboxylase function (k_{cat})—are generally all that is required to calculate v_c and v_o. Direct measurements of the kinetic constants for the oxygenase function are not as accurate as those for the carboxylase function, and v_o can be defined from v_c by an expression—$v_o = v_c[O_2]/\tau[CO_2]$—which is based upon the more precise measurements of the specificity factor.

The constants for the gaseous substrates are given (Table 1) both in terms of the dissolved concentration and on an equivalent partial pressure basis. The latter are convenient for interpretation of gas exchange measurements. Some controversy has arisen concerning the use of partial pressure as a basis for expressing these constants (90). Either basis is valid, and Henry's law can be used to convert between them. Note that the apparent temperature coefficients (Q_{10}) for the constants are slightly different depending on the convention used. Difficulties in controlling the concentration of the gaseous substrates dissolved in the aqueous phase during in vitro assays (and in the case of CO_2, errors in equating experimental measurements obtained in terms of total dissolved inorganic carbon with an equivalent level of CO_2) severely limit the reliability of the K_c and K_o^i values. Jordan & Ogren (128) used precision gas mixing equipment to obtain the $[CO_2]/[O_2]$ ratios in their assays of τ. The K_R value in Table 1 is the apparent K_m (RuBP) and does not take into account chelation of the active form RuBP^{4-} with Mg^{2+} present in the assay system (30). Inhibitory constants for various phosphorylated compounds with respect to RuBP summarized by Ashton (6) and Jordan et al (125) are also apparent constants valid only at 5–10 mM Mg^{2+}. Provided $[Mg^{2+}]$ does not vary significantly during steady-state photosynthesis, these apparent constants are adequate approximations of the operative values. Von Caemmerer and coworkers (254, 256) present kinetic expressions that take into account chelation by Mg^{2+}, and the true constants can be calculated knowing the dissociation constants for the respective Mg^{2+} complexes. The k_{cat} of Rubisco reported here (Table 1) was obtained from assays of whole-leaf extracts and an assay for Rubisco protein (see below). Since the apparent k_{cat} values decline rapidly after extraction of Rubisco from the leaf, the values of purified Rubisco seldom approach this activity, and the adequacy of these constants must be judged, among other things, on how well they account for in vivo responses.

GAS EXCHANGE STUDIES   Obtaining accurate measurements of the net flux of CO_2 uptake is now routine. For the present purposes, it is important to obtain estimates of the velocity of specific reactions in the photosynthetic process. For example, the rate of carboxylation (v_c) can be obtained from measurements of A and R_d, if the $[CO_2]/[O_2]$ ratio under that condition is known. By combining Equations 1 and 2, we may write that

$$v_c = \frac{A + R_d}{1 - 0.5[O_2]/\tau[CO_2]}. \qquad\qquad 3.$$

The rate of other biochemical steps of photosynthetic carbon metabolism that are linked to CO_2 uptake or photorespiration can be calculated from v_c and the v_o/v_c ratio using stoichiometric relationships summarized by Farquhar & von Caemmerer (255). This requires that one know the activity of CO_2 and O_2 in the chloroplast stroma during steady-state photosynthesis. (Unless otherwise noted, terms in square brackets denote the chemical activity of that species in the chloroplast.)

The partial pressures of CO_2 and O_2 can be measured only for the bulk air surrounding the leaf. Whenever there is a net flux, the mole fraction CO_2 must be lower (and that of O_2 must be higher) in the stroma than in the bulk air. Note that in the special case where A is zero (at Γ) the diffusion gradients are negligible. The CO_2-diffusion gradient can be calculated from the net flux, knowing the conductance of the diffusion pathway. Simultaneous measurements of the rates of assimilation (A) and transpiration (E) permit measurement of the conductance of the stomata and boundary layer (12, 255) and may be used to obtain reliable estimates of the concentration of CO_2 in the intercellular air spaces (224).

There must, however, be an additional gradient across the cell wall and aqueous phase pathway that separates the intercellular air spaces and the stroma. Theoretical arguments have led to conflicting opinions as to the magnitude of this gradient (172, 193). Direct estimates of this gradient have been obtained recently by Evans et al (60), who conducted simultaneous measurements of A and E (which permit calculation of the gradient of CO_2 between the bulk air and the intercellular spaces), and carbon isotope discrimination (which permits calculation of the gradient of CO_2 between the bulk air and the stroma) by gas exchange techniques (60). They found that the concentration of CO_2 in the stroma could be as much as 60 μbar less than that in the intercellular air spaces during active CO_2 uptake (e.g. $A \approx 30$ μmol $m^{-2}s^{-1}$) and calculated that the aqueous-phase conductance of wheat leaves (g_{aq}) is 0.5–1 mol $m^{-2}s^{-1}$. This conductance should be a constant for a given leaf. There have not been sufficient measurements of this parameter yet to permit any general conclusions concerning the value of g_{aq} for leaves of different species or as a function of developmental factors such as light intensity or nutrient status during growth. Uncertainty concerning the value to use for g_{aq} is a significant problem, but provided this conductance is similar to that of wheat, the chloroplast $[CO_2]$ will not be particularly sensitive to this parameter. Corrections of the O_2 partial pressure are only significant at very low $[O_2]$ and are generally ignored in experiments at normal atmospheric levels.

FREEZE-CLAMP STUDIES   Development of the freeze-clamp apparatus (11) opened the way for studies of the metabolic basis of steady-state photosynthesis. This device is designed to obtain metabolite flux measurements from gas exchange studies, and while the leaf is under steady-state conditions in the gas exchange cuvette, to take a sample of the leaf by clamping it between two cold metal pistons that can be brought into contact with both surfaces of the leaf. Metabolic processes of the tissue are arrested in <1 s, and the leaf sample can be divided and extracted to assay for enzyme activities and metabolite pool sizes (11, 168, 213, 226, 254). The measured pools include material from all compartments of the leaf. Studies of the subcellular compartmentation of metabolites in protoplasts (83) and non-aqueous fractionation of the frozen leaf samples (46, 47, 83) have been important in determining the distribution of metabolites among the stroma, cytosol, and vacuole.

THE RUBISCO CONTENT   The concentration of active sites is widely assumed to be ~4 mM (121), but as pointed out by Walker et al (261), this may be too low. Crystals have been observed in electron microscopy of chloroplasts (235), and the concentration of crystalline Rubisco has been estimated to be ~8–10 mM (184). The Rubisco content of leaf extracts may be quantified in terms of activity or protein. Activity measurements are adequate to interpret measurements of CO_2 response kinetics, but an estimate of the concentration of Rubisco catalytic sites is needed to interpret the significance of RuBP pool size measurements and to calculate the apparent k_{cat} of Rubisco. The molar concentration of Rubisco in an extract can be obtained by incubating with ^{14}C-carboxyarabinitol-1,5-bisphosphate (CABP), which binds very tightly and specifically to Rubisco sites. Determination of protein-bound ^{14}C by immunoprecipitation, gel filtration, or precipitation with polyethylene glycol (35, 91, 285) permits a direct estimate of the molar concentration of Rubisco. Immunoelectrophoresis (129) has also been used to quantify Rubisco in crude extracts. Procedures based upon in vivo assays (267) or separation and quantification of denatured protein may eventually supplant present methods.

RUBISCO ACTIVATION STATE   The activation level of Rubisco may vary over a wide range according the prevailing environmental conditions. Very rapid extraction of quick-frozen leaf tissue and processing the extract at ice temperature in buffer containing a low [CO_2] seems to preserve the in vivo level of activation for at least 1 min (160, 218). Assays for Rubisco activity conducted immediately after extraction and again after incubation of an aliquot of the extract with saturating concentrations of Mg^{2+} and CO_2 at room temperature give an estimate of the steady-state activation level. An alternative approach takes advantage of the observation that CABP forms a much tighter complex with the active than with the inactive form of Rubisco (91). Thus, ^{14}C-CABP bound to an inactive Rubisco catalytic site can be displaced by exchange with unlabeled CABP, while that bound to an activated Rubisco catalytic site cannot be displaced (91). The latter method, since it is not based on catalytic activity, can be used when the inhibitor CA-1-P is present (214).

CA-1-P CONTENT   Extracts of some species may contain 2-carboxyarabini-tol-1-phosphate, a strong inhibitor of Rubisco activity (16, 89, 214). General-ly, when this compound is present, one observes diurnal variations in the extractable activity of Rubisco (164, 257). The concentration of the com-pound can be measured by titrating the activity of a known quantity of Rubisco with the solution of CA-1-P (16). Provided the CA-1-P concentration is less than that of Rubisco, analysis of the quantity of Rubisco and its catalytic activity in a leaf extract can be used to estimate the CA-1-P concen-tration (140, 214).

MODELS OF CO_2 FIXATION   Knowledge of physiological and biochemical mechanisms of photosynthetic metabolism have been drawn together in the form of conceptual or mathematical models of the photosynthetic process. The usefulness of models stems primarily from the complexity of the photosynthetic system and from the strong emphasis on quantitative analysis of photosynthetic responses. The two essential components of any such model are: (*a*) the structure and stoichiometric relationships of the component reac-tions of the pathway, and (*b*) kinetic expressions for the controlling steps. The review of Farquhar & von Caemmerer (63) provides a comprehensive in-troduction to these essentials. Models differ primarily in their complexity and in the details of the kinetic expressions used. The model of Farquhar et al (64), for example, assumes that kinetic control of the rate of CO_2 uptake resides either with Rubisco or with the light-dependent reactions that regener-ate RuBP, and the intervening reactions are not considered. The models of Woodrow (273) and Laisk & Walker (151) include reactions of the PCR cycle and the synthesis of sucrose and starch. However, while these models are useful for analyzing regulatory responses within the photosynthetic system there is not yet enough information regarding the activities and kinetic constants of the constituent enzymes to predict accurately the responses of the pathway to changes in external variables. In this regard a simpler model (93) that focuses upon the main points where environmental factors interact with the system may be more useful. In the following sections we develop a general kinetic model for Rubisco, then consider regulatory sequences that link this enzyme to the remainder of the photosynthetic process.

Kinetic Analysis of Rubisco in vivo

A GENERAL KINETIC MODEL FOR RUBISCO   In order to analyze the multiple
levels of kinetic interactions known for this enzyme, we may write a general
expression of the form,

$$v_c \approx (E_t \cdot k_{cat}) \times f_1([CO_2],[O_2]) \times f_2(R_t,E_t) \times f_3(C_t) \times f_4(X), \qquad 4.$$

where the functions $f_1 \ldots f_4$ are factors ($0 \leq f \leq 1$) giving the fractional
activity as influenced by: 1) the concentration of gaseous substrates; 2) the
total concentration of RuBP (R_t) and competitive inhibitors (with respect to
RuBP) in relation to the Rubisco site concentration (E_t); 3) the stromal
concentration of the tight binding inhibitor, 2-carboxyarabinitol-1-phosphate
(C_t); and 4) the effectors (X) that influence the activation of Rubisco. The rate
(v_c) can be approximated by substituting the measured "local" concentration
values into Equation 4 and is related to the rate of net CO_2 uptake according to
Equation 3. We emphasize, however, that Equation 4 serves only as a simple
and clear way to approximate the Rubisco rate equation and that the multi-
plicative properties of these terms has not been rigorously established.

RESPONSES TO CO_2 AND O_2 The influence of the gaseous substrates has
generally been assumed to follow the classical Michaelis-Menten expression
for an enzyme with a single substrate (CO_2) in the presence of a competitive
inhibitor (O_2):

$$f_1 = \frac{[CO_2]}{[CO_2] + K_c(1 + [O_2]/K_o^i)}. \qquad 5.$$

This equation is based upon in vitro studies in the presence of a saturating
concentration of RuBP, but there has been some debate as to whether these
kinetics are appropriate when RuBP is not saturating or the enzyme is not
fully activated. More complex kinetics could apply, depending upon the order
of substrate binding. Farquhar (61) developed equations to describe these
kinetics assuming that the reaction mechanism is ordered with: (*a*) CO_2
binding first, (*b*) RuBP binding first, or (*c*) either binding first (i.e. random).
From analysis of kinetic experiments in vitro, Badger & Collatz (9) were able
to eliminate the first possibility but were not able to choose between the
second and third alternatives. In his general kinetic model, Farquhar (61)
assumes the second possibility, in part because the random mechanism does
not lend itself to an analytical solution.

Recent experiments have established that the reaction mechanism is strictly
ordered, with RuBP binding first, followed by a slow step—enolization of

RuBP on the enzyme (185). This apparently creates the binding site for CO_2 or O_2 (186), making it improbable that the apparent $K_m(CO_2)$ is dependent on the level of RuBP or the activation level of the enzyme. The latter should affect the number of functional sites available for reaction with the gaseous substrates, but the apparent K_m for CO_2 or O_2 of these sites should be independent of the steady-state population of functional sites. Therefore, the model of Farquhar (61) should provide a good approximation for the kinetics of Rubisco.

In order to test whether the Rubisco kinetic constants with respect to CO_2 and O_2 calculated from in vitro studies are valid for the in vivo enzyme, the modulation of enzyme activity by factors 2–4 must be eliminated (i.e. f_2, f_3, f_4 = 1). If this is not done, the physiological response may become saturated more abruptly and at a lower $[CO_2]$ than would be expected from in vitro measurements. Laisk & Oya (150) used a creative experimental design to avoid this problem. CO_2 was provided to a leaf in short (1 s) pulses interspersed with CO_2-free periods of variable duration selected to keep the time-average rate of CO_2 uptake constant. In this way the $[CO_2]$ during the pulse could be increased to saturating levels while the time-average rate of net CO_2 uptake remained essentially constant and well below the maximum that could be attained under steady-state conditions. When the rate of net CO_2 uptake was plotted against $[CO_2]$ in the pulses, Laisk & Oya (150) obtained a classical rectangular hyperbola with an apparent $K_c = 28$ μmol and V_{max} = 2.8 μmol mg^{-1}protein.

These kinetic constants for Rubisco in the complete photosynthetic system agree fairly well with those presently measured in vitro. This is remarkable, in part because these measurements were made in the early 1970s, before the activation process was understood. These experiments also provided a clear demonstration that the abrupt (nonrectangular hyperbola) saturation curve for the steady-state CO_2 uptake vs $[CO_2]$ in vivo is a reflection of regulation of Rubisco by factors 2–4. Laisk (149) suggested that the rate of RuBP synthesis becomes limiting at high $[CO_2]$. Walker and coworkers (157) independently came to the same conclusion by comparing the $[CO_2]$ response of CO_2 uptake by illuminated intact chloroplasts to that obtained upon lysing the chloroplasts into a medium containing a rate-saturating concentration of RuBP.

Most comparisons of the dependence of CO_2 uptake on CO_2 and O_2 have been restricted to low $[CO_2]$ and high light, conditions where $f_2, f_3, f_4 \approx 1$. The use of such low substrate concentrations precludes calculation of the apparent K_c, K_o^i and k_{cat} values. It is possible, however, to approximate the rate of CO_2 uptake using the Rubisco rate equation, measurements of the concentration of substrates, and kinetic constants estimated from in vitro studies. Seemann et al (213, 217) confirmed the validity of this approximation in experiments with spinach leaves at a low $[O_2]$ and a $[CO_2]$ of 100 μbar. In this study no allowance was made for the gradient in [CO$_2$] due to the finite aqueous phase conductance, but the uncertainty introduced by this omission is not larger than that introduced by uncertainty in the magnitude of the kinetic constants. In a conceptually similar a series of studies, Evans and coworkers (54–56, 58) analyzed the slope of the [CO$_2$] response curve at the CO$_2$-compensation point ((dA/dc)$_\Gamma$) of *Triticium aestivum* (wheat) using the derivative of the rate expression for CO$_2$ uptake (see 255). The observed values of (dA/dc)$_\Gamma$ are highly correlated with the Rubisco content of the corresponding leaves, and these values are quantitatively consistent with the response predicted from in vitro kinetic measurements. Evan's studies, however, did not resolve differences between genotypes in g_{aq} (estimated by ^{13}C discrimination) and in the k_{cat} of Rubisco by analyzing (dA/dc)$_\Gamma$ in vivo. A significant concern in experiments of this type is the assumption that Rubisco is fully activated. Von Caemerer & Edmondson (254) show that the activation state of well-nourished leaves of *Raphanus sativa* decline at [CO$_2$]<100μbar. Furthermore, phosphate-deficient leaves do not fully activate their Rubisco—even at normal ambient CO$_2$ and saturating light (21, 161). Variation in the activation level may underlie the rather large variation in (dA/dc)$_\Gamma$.

DEPENDENCE ON RuBP   In assessing the degree to which regulatory mechanisms influence the rate of CO$_2$ assimilation it is necessary, among other things, to have a clear picture of the range over which changes in the concentration of RuBP significantly affect the activity of Rubisco. The kinetics of the Rubisco-catalyzed reaction do not obey a simple rate equation with respect to RuBP, partly because the concentrations of the enzyme and substrate are similar (34, 61, 121). Peisker (181) derived a rate equation for conditions where the concentration of RuBP (R$_t$) is less than that of the enzyme (E$_t$). Farquhar (61) used similar conditions in a general kinetic model for Rubisco according to which the response of the catalytic velocity to R$_t$ at a constant CO$_2$ concentration has the form of a nonrectangular hyberbola. This relationship defines the second term in Equation 4 and is given by:

$$f_2 = \frac{(E_t + K_R^1 + R_t) - [(E_t + K_R^1 + R_t)^2 - 4R_t \times E_t]^{1/2}}{2E_t}, \qquad 6.$$

where R$_t$ is the total concentration (free and bound) of RuBP, and K_R^1 the apparent Michaelis constant for RuBP. Farquhar et al (64) also noted that, since K_R^1 is small in comparison to E$_t$ in vivo, the solution to this equation can be approximated by

$$f_2 = \begin{cases} R_t/E_t, & \text{for } R_t < E_t \\ 1, & \text{for } R_t \geq E_t \end{cases}. \qquad 7.$$

Initial results supporting this interpretation (34, 85, 104, 272) have been contradicted by studies with intact leaves which demonstrate that R_t does not apparently control the rate of CO_2 uptake when that rate (at constant $[CO_2]$) declines from the "light-saturated" level in response to a decrease in light intensity. It is generally accepted that regulation of Rubisco activity as described by f_3 and f_4 is the primary reason for this observation. Nevertheless, it will be important to establish the role of other factors, such as competitive inhibitors with respect to RuBP.

Since the activation state of Rubisco changes relatively slowly ($t_{1/2} \approx 5$ min) in response to an alteration in external conditions, it is possible to conduct short-timeframe experiments on Rubisco in vivo during which the rate of CO_2 assimilation may change but the activation state remains approximately constant. Mott et al (169) conducted a series of such experiments and assumed that the rapid (≈ 30 s) adjustments to the rate of photosynthesis in response to a change in light intensity results entirely from modulation of the [RuBP]. They used freeze-clamp experiments at a constant $[CO_2]$ to demonstrate that the relationship between the rate of carboxylation and [RuBP] approximates a nonrectangular hyperbola (Figure 2). Freeze-clamp samples taken following a transition to a low light intensity, however, showed that after the rapid drop in R_t there was a rise in the level of this substrate until, in the steady state, R_t is approximately equal to that occurring prior to the decrease in light. Prinsley et al (192) reported similar results, but these workers suggested that increases in [PGA] and [Pi] (competitive inhibitors with respect to RuBP) also play a significant role in determining the rate of photosynthesis during the transient.

Effects of phosphorylated compounds on K_R' can be described by an equation of the general form

$$K_R' = K_R \left(1 + \frac{P_1}{K_{P1}} + \frac{P_2}{K_{P1}} ... \right), \qquad\qquad 8.$$

where P_n is the activity of any phosphorylated compound present in the stroma and K_{P_n} is its corresponding K_i with respect to RuBP. Note that the concentrations used here should be the activity of the free compound, which may differ significantly from the total concentration if the compound binds tightly to Rubisco or another ligand (6, 254). The expected relationship between v_c and R_t according to Equation 6 are plotted (Figure 2) for the case where only RuBP is present, and where the total phosphate moieties in RuBP and PGA are assumed constant (i.e. $[PGA] + 2 \times [RuBP] = 60$ mM). The latter simulation, which should approximate the situation in vivo, indicates that an RuBP concentration approximately twice that of Rubisco active-sites is sufficient to sustain a v_c of 80% of the RuBP-saturated velocity. It should be noted that the presence of inactive Rubisco or of the inhibitor CA-1-P must be considered when using Equation 6 to interpret in vivo measurements because (*a*) inactive Rubisco binds RuBP (125, 146), thus rendering a portion of the measured RuBP unavailable for catalysis; and (*b*) the operative value for E_t should reflect the activated sites that are free of inhibitor.

Most measurements of the relationship between RuBP and the rate of CO_2 uptake under normal steady-state conditions indicate that the total concentra-

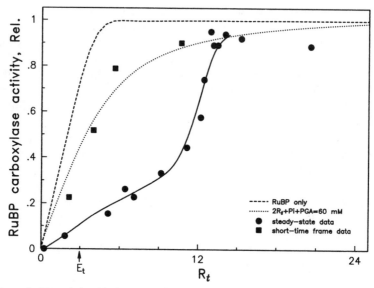

Figure 2 The relationship between the relative activity of RuBP-carboxylase and the total concentration of RuBP (R_t) at constant [CO_2]. The dotted and dashed lines are simulations showing the expected influence of physiological levels of PGA and Pi, which are competitive inhibitors with respect to RuBP. The data points are from freeze-clamp studies with intact leaves that show the response in short–time frame experiments where the activation state of Rubisco remains essentially constant (■) or in experiments where time is sufficient for the activation state of Rubisco to adjust to a new steady-level at each rate (●). The solid line (not a simulation) is drawn to illustrate the sigmoid relationship observed under steady-state conditions. The simulations are drawn according to Equation 6, where: (dashed line) there are no inhibitors; or (dotted line) the concentration of inhibitors (Pi+PGA) varies with R_t such that the total of phosphate moieties in the two pools is constant at 60 mM (estimated from Ref. 254). In the latter case the value of K'_R in Equation 6 was adjusted for each value of R_t using K_i(PGA+Pi) = 0.9 mM and [PGA+Pi] = 60 − 2R_t in Equation 8. The steady-state data points are taken from von Caemmerer & Edmondson (254), and the short–time frame experiments are from Mott et al (169). The rates of CO_2 uptake were scaled assuming that v_c was 95% saturated with respect to RuBP at "rate saturating" light, and the concentration values were scaled to give E_t = 3 mM assuming 21.5 μmol Rubisco m^{-2} and 60 nmole Rubisco mg^{-1} chl for the two sets of data, respectively.

tion of RuBP is only less than the concentration of Rubisco active sites when light intensities are very low (i.e. well below the light intensity required for half maximum assimilation rates) or when CO_2 partial pressures are very high. Furthermore, when light intensity changes are used to perturb the steady-state assimilation rate, there is a significant range over which assimilation changes and RuBP either remains constant (11, 183, 254) or declines (46, 47, 182). Low levels of RuBP have been measured in isolated cells and chloroplasts (34, 85, 104, 272), and these probably reflect a transient rather than the true steady-state condition. Von Caemmerer & Edmondson (254) estimated that, at constant [CO_2], the relationship between v_c and [RuBP] has a sigmoid form (Figure 2; see also Figure 4 in Ref. 11). Such sigmoid relationships are normally associated with cooperative binding of a substrate to the enzyme, but in vitro studies of Rubisco indicate no evidence for such kinetics. We discuss this phenomenon in greater detail in a later section.

REGULATION BY CA-1-P   An inhibitor of Rubisco, carboxyarabinitol-1-phosphate (CA-1-P), has been detected in extracts of *Solanum tuberosum* and *Phaseolus vulgaris,* and the same (or a similar) compound has been detected in many (but not all) species (214, 220, 257). The substance binds tightly to the activated form of Rubisco [K_D = 32 nM (16)] inhibiting the activity of Rubisco, and the third term of Equation 4 is given by, $f_3 = 1 - EI/E_t$ where EI is the concentration of enzyme-inhibitor complex. Substituting the equation for the concentration of EI (16), we may write that

$$f_3 = 1 - \frac{(C_t + E_t + K_D) - [(C_t + E_t + K_D)^2 - 4C_tE_t]^{1/2}}{2E_t}, \qquad 9.$$

where C_t is the level of CA-1-P, E_t the concentration of Rubisco sites, and K'_D the apparent dissociation constant. For values of $C_t < 0.8E_t$ the concentration of EI can probably be approximated by assuming EI = C_t. Effects of compounds that may compete with CA-1-P for binding to Rubisco have not been investigated.

Evidence obtained with *Phaseolus vulgaris* implicates CA-1-P as an important regulator of Rubisco activity (214). In leaves held overnight in darkness, the concentration of this compound may approach or exceed that of Rubisco sites, and the activity of Rubisco extracted from these leaves is only 10% of the normal value. When a leaf is illuminated, however the level of CA-1-P declines to a steady-state value that depends on the intensity of illumination (213, 214). After prolonged exposure to saturating light, the compound is not detectable (213, 214). It is proposed that the concentration of CA-1-P is regulated by a futile cycle (16) involving simultaneous synthesis and degradation. Little is known about the synthetic reaction, but it appears that the rate of the degradative reaction is dependent upon the turnover of PSII (214). This would be sufficient to regulate the steady-state concentration of CA-1-P.

Kobza & Seemann (140) examined the combined roles of CA-1-P, the activation state, and the concentration of RuBP in controlling the rate of CO_2 uptake in species that, under a given set of conditions, have different levels of CA-1-P. In *Phaseolus vulgaris,* most of the regulation of Rubisco seems to be via CA-1-P. The uninhibited Rubisco sites generally remain in the activated form, although at high [CO_2] the CO_2-Mg^{2+}-dependent level of activation may fall (225). *Spinacea oleracea* apparently does not produce CA-1-P under any circumstance, but this compound is found in the related species, *Beta vulgaris.* In contrast to *P. vulgaris,* regulation of Rubisco in *B. vulgaris* also involves the CO_2-Mg^{2+}-dependent control of the activation state. In all three species, however, the net result was similar: The activity of Rubisco (as controlled by CA-1-P and/or the activation state) changes with light intensity, and the concentration of RuBP is apparently saturating, except at strongly limiting light. The level of RuBP in *P. vulgaris* responds more with light than that in *S. oleracea* which may reflect some interaction between the RuBP and CA-1-P pools.

CONTROL OF ACTIVATION STATE   A specific lysine residue of Rubisco (ly 201 of the spinach enzyme) participates in a reversible reaction in which CO_2 and Mg^{2+} add to the inactive Rubisco (E) to form a catalytically active ternary complex (ECM):

$$E + CO_2 + Mg^{2+} \rightleftharpoons ECM + H^+ \qquad 10.$$

The mass action equilibrium expression for this reaction solved for the fraction of Rubisco in the activated form provides a good approximation to experimental data for activation as a function of pH at air levels of CO_2 in vitro (Figure 3). When RuBP is present, the activation level is more sensitive to pH (Figure 3). Mott & Berry (167) attribute this additional pH dependence to the involvement of a protonated group on the enzyme in forming the binary enzyme–RuBP(ER) complex ($E + H^+ + R \rightleftharpoons ER$). An equilibrium expression assuming the overall release of two H^+ in the activation of the ER complex approximates the pH dependence of the activation level in the presence of RuBP (Figure 3).

Several other phosphorylated compounds (particularly 6-phosphogluconate and NADPH) affect the activation process, apparently by binding preferentially to the E or the ECM forms of Rubisco (10, 31, 95, 146, 163). Kinetic studies indicate that these compounds interact with the catalytic site (10, 125) and are competitive with respect to RuBP. High concentrations of RuBP typically present in vivo would therefore tend to suppress any effect of

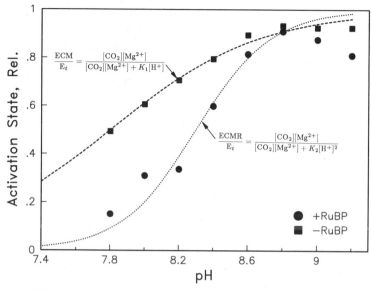

Figure 3   The fraction of Rubisco in the activated state as a function of pH at air level CO_2 in the presence or absence of RuBP. These data are taken as the initial and steady-state rates observed in time-course experiments when the 0.5 mM RuBP was added to Rubisco preincubated at air level CO_2 in the assay mixture at the indicated pH [from Mott et al (167, 168)]. Theoretical expressions for the equilibrium activation level assuming that the activation reaction involves a stoichiometry of one H^+ (in the absence of RuBP) or 2 H^+ (in the presence of RuBP) approximate the observed responses where $K_1 = 170$ M and $K_2 = 1.31 \times 10^{11}$. The mechanism leading to a lower steady-state Rubisco activity in the presence of RuBP has not been established and may involve another inactive form (i.e. not the ER complex as assumed here). Nevertheless, the activity of Rubisco at physiological pH appears to be lower in the presence than in the absence of RuBP.

these substances. At very low light intensities, when $[RuBP] < E_t$, these compounds could affect activation, but this effect must be offset against their effect as competitive inhibitors of catalysis (166, 249). Parry et al (179) propose that the effect of Pi is allosteric (as distinct from the above mechanism), but there is no evidence that there is a separate effector binding site for Pi on the enzyme (see 125). A particularly interesting and unresolved matter concerns the possible role of light-dependent changes in $[Ca^{2+}]$ in the stroma in regulating of Rubisco activity. There are light-dependent changes in the stromal $[Ca^{2+}]$ (145), but studies on the effects of this ion on the activation state or activity of Rubisco are contradictory (cf. 14, 30, 180).

The equilibrium expressions provide a basis for calculating the activation level as a function of $[CO_2]$, $[Mg^{2+}]$, $[RuBP]$, and the pH of the medium (e.g. 256). Studies attempting to relate the responses of activation state to the levels of Rubisco effectors in situ, however, have highlighted several phenomena that are difficult to reconcile with our current knowledge of the equilibrium process. First, it is well established that $[Mg^{2+}]$ should increase and $[H^+]$ decrease in the stroma upon illumination (101). These changes taken in combination should increase the activation level of Rubisco. Nevertheless, the strong light-dependent activation of Rubisco occurs over a range of light intensities (>100 μmol m^{-2}s^{-1}) where changes in these factors are assumed to be "saturated" (53, 175, 187, 189). Second, it is difficult to explain how the enzyme can be fully activated at normal atmospheric levels of CO_2, because calculations based on the equilibrium (Equation 10) indicate that it should be strongly $[CO_2]$-dependent and less than 50% activated [unless, as pointed out by Mott et al (169), the stromal pH $\geq$ 8.6]. Moreover, recent evidence (254) indicates that the activation level is not significantly affected by CO_2 until $[CO_2] < 100$ μbar. Third, it is well established than an **increase** in $[CO_2]$ can result in a **decrease** in the in vivo activation level— directly contradicting the results of in vitro studies. Such a decrease was observed, for example, by Sharkey et al (225) when $[CO_2]$ was increased from near 300 to 500 μbar under optimal conditions for photosynthesis. Fourth, Heldt et al (102) showed that light-dependent activation of Rubisco in intact chloroplasts requires the presence of an optimal concentration of Pi in the suspending medium. This result is not explicable by any known direct effect of Pi on the activation equilibrium. Finally, the rate of catalysis in vitro decays with time (146). An unpublished experiment (G. Lorimer and J. Pierce) shows that this decline is not explained by a loss of the ECM form. The possibility that Rubisco is regulated by yet another mechanism cannot be ignored.

Despite uncertainty about the mechanisms, it is possible to consider the level of Rubisco activation on an ad hoc basis. The recent study of von Caemmerer & Edmondson (254) gives complete response curves for CO_2 uptake by leaves of *Raphanus sativa* to light at constant $[CO_2]$ and to $[CO_2]$ at saturating light. They also present information on the activation state of Rubisco, concentrations of RuBP and other PCR cycle intermediates, and total enzyme levels. We have analyzed these measurements according to Equation 4 and find (data not shown) that these measurements are quantitatively sufficient to account for $>95\%$ of the variation of net CO_2 uptake with light or $[CO_2]$. These studies demonstrate that gas exchange analysis of steady-state metabolic flux together with freeze-clamp studies of metabolite concentrations and Rubisco activation levels provide an adequate basis upon which to interpret the local control of CO_2 uptake by Rubisco in vivo.

RUBISCO-ACTIVASE   Ogren and coworkers propose that a chloroplast protein—"Rubisco-activase"—functions to maintain Rubisco in the activated form in vivo (174). According to their hypothesis, this protein catalyzes an energy-dependent activation of the Rubisco-RuBP complex (190, 202) driven by ATP (248):

$$ER + CO_2 + Mg^{2+} + ATP \xrightarrow{\text{activase}} ECMR + ADP + Pi. \qquad 11.$$

While both the mechanism of the activase reaction and its kinetics remain to be established, there is substantial evidence for this activity. A mutant of *Arabidopsis thaliana* that lacks peptides identified as Rubisco activase has a low level of Rubisco activation in the light and requires high $[CO_2]$ for growth (191, 202, 233). Consistent with the proposed model, the level of activation of the mutant is normal in the dark (in the absence of RuBP) and is much lower than that of the wild type in the light (when RuBP is present) (203). Antibodies to Rubisco activase from spinach have been used to demonstrate the existence of a cross-reacting protein in other photosynthetic organisms, including C_4 plants and green algae (204).

If one accepts the activation reaction given in Equation 11, and that both activation and inactivation occur by the equilibrium shown in Equation 10, then the steady-state proportion of active Rubisco will be a function of the rate constants for the three partial reactions. Assume, for example, that the rates of these reactions are first order with respect to the active $[E_a]$ and inactive $[E_i]$ forms of the enzyme present in the system (i.e. ER and ECMR when RuBP is saturating). The rates of the partial reactions are: $v_{act} = k_{act}[E_i]$; $v_{on} = k_{on}[E_i]$; $v_{off} = k_{off}[E_a]$, where the subscripts *act, on,* and *off* designate the activase reaction and the forward and reverse partial reactions of Equation 10, respectively. The steady-state proportion of active enzyme is thus given by

$$f_4 = \frac{E_a}{E_a + E_i} = \frac{k_{act} + k_{on}}{k_{act} + k_{on} + k_{off}}. \qquad 12.$$

Note that when k_{act} is small compared to $k_{on} + k_{off}$ the activation level should be determined by effects of Mg^{2+}, CO_2, and pH on the values of k_{on} and k_{off} (146, 159), but when $k_{act} >> k_{on} + k_{off}$ the activation level can approach 1.0. Recent experiments by Streusand & Portis (248) indicate that k_{act} (all else being equal) may increase as a sensitive function of the ATP/ADP ratio. It may also be significant that RuBP has been shown to be a strong inhibitor of both k_{on} and k_{off} in the equilibrium activation reaction (125, 146). Thus, for a given value of k_{act}, the steady-state level of activation should be higher in the presence of a high concentration of RuBP, which is consistent with the apparent RuBP requirement for the "activase effect" (190).

Activase may function to link the activation of Rubisco to the light intensity (174, 190, 248). In this regard, it should be noted that the ATP/ADP ratio in vivo is, among other things, affected by the activity of Rubisco, since ATP use is linked to the production of PGA. For example, an immediate effect of a rise in light intensity would be an increase in the ATP/ADP ratio and the rate of PGA reduction. Activation of Rubisco and the other "light-activated" enzymes would follow, and this concomitant rise in activation and flux would tend to stabilize the concentration of PCR cycle intermediates (see below). Since the reaction catalyzed by PGA kinase is close to equilibrium, the steady-state change in the ATP/ADP ratio would be relatively small. The effectiveness of this homeostatic regulatory mechanism will depend upon the degree of change of other effectors such as CA-1-P, Mg^{2+}, and H^+ and their effects on Rubisco and the "light-activated" enzymes. An increase in the $[CO_2]$ may, however, be expected to have the opposite effect on the activation state of Rubisco because the increased flux will tend to reduce the ATP/ADP ratio. This concept seems to explain the observation noted above that increasing CO_2 in vivo can result in a decline of Rubisco activation state. Sharkey et al (226) measured a decrease in the ATP/ADP ratio of intact leaves during the transient following an increase in $[CO_2]$, and this change in ATP/ADP ratio correlates with deactivation of Rubisco during the transient (199). These results support the notion that Rubisco-activase participates in regulation of the activation state of Rubisco in vivo.

SUMMARY   Measurements of the rate of CO_2 fixation by intact leaves can now be accounted for fairly well by measurements of several elements known to exert "local" control over Rubisco. Experimental approaches to quantifing the extent of "local" control by these different elements have been developed and demonstrated, but many details concerning the mechanisms and kinetics of reactions that affect the V_{max} of Rubisco (by controlling the concentration of CA-1-P or by controlling the activation state) remain to be elaborated. The studies discussed in this section provide some indication of how the "local" control of Rubisco may be linked to other elements of the photosynthetic system. These links are developed below.

REGULATORY SEQUENCES

We have given Rubisco a central position in the foregoing discussion because it is strategically placed to regulate the rate of photosynthesis. As the catalyst of the first largely irreversible reaction in the CO_2-fixation sequence, Rubisco alone (if we ignore diffusion of CO_2 to the site of reaction) will determine the flux unless subject to feedback regulation. There are now clear indications that under some conditions (e.g. high light and limiting CO_2) the levels of Rubisco effectors are such that feedback inhibition of enzyme activity is negligible and the rate of photosynthesis is determined solely by the amount of Rubisco. Under other conditions (e.g. low light and high CO_2) Rubisco activity is held well below its maximum. What remains now is to describe our current understanding of the sequences of regulatory events that lead to changes in the levels of Rubisco effectors and to define the conditions under which these "regulatory sequences" are most operative.

The most fundamental regulatory sequences influencing Rubisco activity can be defined by considering the structure of the photosynthetic system (Figure 4). This system can be described broadly as a convergent metabolic pathway (266). One branch mediates the input of CO_2 from the atmosphere and the other the input of quantum energy for the ultimate reduction of CO_2 to sucrose, the principal product of photosynthesis. A fundamental requirement for a steady state in such a system is that the two input fluxes conform to a specific stoichiometry and that the input flux does not exceed the capacity of the distal reactions to sustain that flux. If the converging sequences contain largely irreversible reactions, there must be regulatory mechanisms that can coordinate the rate of these reactions such that the soichiometric requirements are fulfilled. In the photosynthetic system, the input of CO_2 is mediated by a series of reversible diffusion processes (62) and the largely irreversible reactions catalyzed by Rubisco. The absorption of quanta involves processes in which quantum energy is either dissipated as heat and fluorescence or used for photochemistry within a few picoseconds (50). From that point on, the input flux is essentially irreversible, although there are other distal mechanisms by which energy can be dissipated (8). There are, therefore, three fundamental regulatory sequences that must be in place to ensure that steady-state photosynthesis can be approached. First, at low light intensities, Rubisco activity must be modulated such that it balances the input of quantum energy. Second, at high light intensity when energy input exceeds the capacity of Rubisco to fix CO_2, a regulatory sequence must ensure that excess energy is dissipated. Finally, if under certain circumstances the maximum activity of reactions involved in product synthesis is approached, then a sequence that can reduce both the rate of energy input and Rubisco activity must be activated. In the previous section, we have discussed several mechanisms whereby the reactions of CO_2 fixation are regulated in response to a change in the rate of energy input. We elaborate upon the details of this sequence in the following discussion and describe some of the recent advances in our understanding of regulatory sequences linking the activity of Rubisco to the input of quantum energy and the reactions of the sucrose synthetic pathway to both input processes. Since the PCR cycle plays a crucial role in mediating all three regulatory sequences, we first examine some of the salient features of this cycle and then link these to events occurring in the cytosol and the thylakoids.

Regulatory Properties of the PCR Cycle

Many of the important regulatory properties of the PCR cycle have been discussed in detail in reviews by Walker (258), Edwards & Walker (52), and

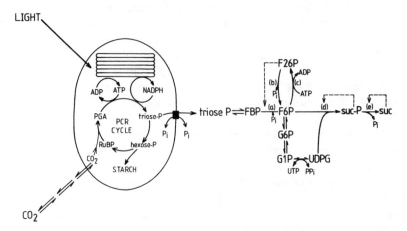

Figure 4   A diagram showing the structure of the photosynthetic system. Inputs of fixed carbon, mediated by Rubisco, and reducing equivalents, mediated by PSII, converge in the photosynthetic carbon reduction (PCR) cycle. Two major branch points of the cycle lead to the production of starch in the chloroplast and the export of triose-P to the cytosol via the phosphate translocator. Synthesis of sucrose (suc) in the cytosol is linked to the release of Pi which is returned to the stroma via the phosphate translocator in exchange for triose-P. Enzymes of the cytosolic pathway are: (a) fructose 1,6-bisphosphatase; (b) fructose 2,6-bisphosphatase; (c) fructose 6-phosphate, 2-kinase; (d) sucrose phosphate synthase, and (e) sucrose phosphate phosphatase—the dashed lines indicate possible feedback mechanisms.

Leegood et al (155), and in a paper by Woodrow (273). We do not elaborate upon these discussions, but draw attention to two basic properties of the PCR cycle that are critical for understanding how the cycle mediates the regulatory interactions among the cytosol, the electron transport system, and Rubisco.

The first property involves the constancy of total stromal phosphate (esterified plus inorganic), which is a consequence of the compartmentation of the PCR cycle and the fact that the main protein catalyzing metabolite export from the chloroplast—the phosphate translocator—does not sustain a net phosphate flux (68, 69). This conservation of phosphate is of great importance because it requires that a change in the level of any phosphorylated intermediate be compensated by an equal and opposite change (in terms of phosphate) elsewhere in the cycle (66, 273). Therefore, a change in the activity of any enzyme in the PCR cycle can affect both the substrate concentrations and the activities of other enzymes in the chloroplast regardless of whether they are adjacent on a metabolic scheme or whether they are connected by a classical allosteric (or other) feedback mechanism. For example, a change in the concentration of RuBP would most likely be associated with balancing changes in PGA and Pi, since these compounds account for about 80% of the stromal phosphate (83). This association was discussed previously with respect to the relationship between the apparent K_m(RuBP) of Rubisco and the RuBP concentration (see Figure 2). But it may also be of great significance in connecting the activity of Rubisco (which influences the steady-state [RuBP]) to the level of stromal Pi and thus to the rate of electron transport. We discuss this interaction and the importance of phosphate conservation in cytosol-stroma interactions in the following sections. It is important to note that the compartmentation of the cytosol by membranes that are only slowly permeable to Pi may also result in the conservation of total cytosolic phosphate in the minutes-to-hours range. The same rules of balancing esterified and inorganic phosphate concentrations would therefore be relevant to consideration of the regulation of sucrose synthesis.

The second property of the PCR cycle involves regulatory responses at the two major branch points. At the first of these, triose-P is withdrawn from the chloroplast for the ultimate synthesis of sucrose; and at the second, hexosephosphate is used for the synthesis of starch (Figure 4). In contrast to a linear divergent pathway, the competing reactions must carry a specific range of fluxes for the inputs and outputs to balance and the system to approach a steady state [see reviews by Walker (258), Edwards & Walker (52)]. Laisk (149) and Woodrow et al (278) proposed that a corollary to this need for restricting the steady-state fluxes at the branch points is that enzymes catalyzing the reactions adjacent to these branch points must adopt a restricted range of kinetic parameters in order for the system to adopt a stable steady state. With the exception of the phosphate translocator, the kinetic properties of these competing reactions (namely, stromal aldolase, transketolase, and hexose phosphate isomerase) are subject to regulation only through modulation of their product pool sizes. It is therefore more relevant to consider the required balance at the branch points in terms of the kinetic parameters of enzymes catalyzing the adjacent largely irreversible reactions. This means that in the stroma the FBPase, SBPase, Ru5P kinase, and ADP-glucose pyrophosphorylase will be important in determining metabolite partition at both branch points; and in the cytosol the FBPase and the phosphate translocator (which is also subject to regulation by cytosolic Pi) can influence the competition for triose phosphate (see 273).

Should a regulatory mechanism perturb the balance between the kinetic constants of these latter enzymes, a transient state would occur during which either metabolite export or recycling is favored. The effect of this perturbation would be to change the relative levels of esterified and inorganic phosphate in the stroma until a new steady state is attained. For example, a rise in the cytosolic Pi concentration (cytosolic triose-phosphate and PGA levels remaining constant) would affect both the apparent K_m and V_{max} of the phosphate translocator with respect to stromal triose-phosphate (84, 188) such that this enzyme, at least temporarily, could out-compete the enzymes of the stroma for triose-phosphate, and proportionally more Pi would be imported from the cytosol than is incorporated into the esterified phosphate pool via the Ru5P kinase reaction (Figure 4). As the concentration of stromal Pi increases and that of esterified phosphate decreases, a new steady state would be approached in which the ratio of Pi to esterified phosphate in both compartments is altered and the rate of starch synthesis reduced. Changes in the levels of cytosolic triose-phosphate and PGA can similarly affect the kinetic characteristics of the phosphate translocator, the partition to starch and sucrose, and the ratio of Pi to esterified phosphate in the stroma.

This ability of regulation of the branch-point reactions to modulate the ratio of esterified to inorganic phosphate (within the confines of phosphate conservation), and the steady-state rates of starch synthesis and photosynthetic CO_2 assimilation has been clearly demonstrated in a series of experiments with isolated chloroplasts where the properties of the phosphate translocator were altered by adding Pi, PGA, and triose-phosphate to the external medium [see reviews by Walker & Crofts (259), Walker (258), Edwards & Walker (52)]. These fundamental stoichiometric relationships and regulatory responses of the PCR cycle will be important in describing regulatory sequences that permit reactions in the sucrose synthetic pathway to influence both the partition of carbon and the initial events of photosynthesis. Details of these mechanisms are elaborated in the following discussions.

Feedback Regulation of Photosynthesis

Many biosynthetic pathways contain product-sensitive feedback sequences that can inhibit early irreversible reactions in the pathway and thus stabilize the supply of product (253, 284). A natural consequence of such feedback loops is that, over a wide range of conditions, the flux is sensitive to the activities of some or all the enzymes distal to the first irreversible reaction (209). This sensitivity, as mentioned in the introduction to this section, becomes significant when the flux approaches the maximum activity of these distal enzymes. In the following discussion, we first examine the biochemical evidence for a sequence linking sucrose—the primary end product exported from leaf mesophyll cells—to the photoacts and to Rubisco. We then look at evidence of the operation of this sequence, and of whether, under normal conditions, the activities of cytosolic enzymes contribute significantly to the determination of the rate of photosynthesis.

THE STRUCTURE OF THE FEEDBACK SEQUENCE   The enzymological evidence for a metabolic feedback loop originating from sucrose (see Figure 4) is somewhat variable. Both sucrose phosphatase (96) and sucrose phosphate synthase (SPS) (115, 119, 133, 201) from a variety of species are inhibited by sucrose, but there are some exceptions to this rule (3, 72, 92, 115, 116, 118,

271). The effect of sucrose-phosphate on SPS is also variable. There is evidence that in spinach sucrose-phosphate is an inhibitor of SPS activity (3), but in wheat germ this compound exerts little influence over SPS (200). The enzymes involved in the synthesis and degradation of fructose 2,6-bisphosphate in spinach also appear not to be affected by sucrose (243).

The structure of the sequence linking the activity of SPS to chloroplast metabolism is, however, relatively well defined and involves the recently identified inhibitor of cytosolic FBPase, fructose-2,6-bisphosphate (F26P) (43, 44, 111, 241, 246). The concentration of F26P is regulated by a futile cycle involving F6P,2-kinase and a specific F26P phosphatase. The former enzyme synthesizes F26P from F6P and ATP (41) and the latter catalyzes the production of F6P and Pi from F26P (43). A fall in the activity of SPS (relative to that of the cytosolic FBPase) should lead to accumulation of F6P, UDPG, and metabolites close to equilibrium with these compounds (e.g. G6P). This rise in F6P concentration would probably effect an amplified (i.e. greater than proportional) rise in F26P because the former compound stimulates F6P,2-kinase and inhibits the phosphatase (43). This amplified response was verified by Stitt et al (244), who showed that, under conditions that reduce the activity of SPS, a 50% rise in F6P was accompanied by a doubling of F26P. These authors also measured a decrease in the metabolic flux through the cytosolic FBPase and a build-up of cytosolic DHAP [which is presumably in equilibrium with FBP (83)]. An increase in substrate concentration together with a reduction in metabolic flux is unequivocal evidence that FBPase is regulated under these conditions (144).

When assessing the sensitivity of this mechanism to either a build-up of products or a reduction in SPS activity, one must consider two other regulatory properties. First, it is probable that the accumulation of esterified phosphates in the cytosol is compensated by a reduction in the level of Pi. This opposite movement of the G6P-F6P and Pi pools together with the influence of all three metabolites on SPS activity (49) probably results in a relatively sensitive relationship between the F6P concentration and the catalytic velocity of SPS. In other words, a large change in SPS activity may only result in a relatively small increase in substrate concentration and thus a relatively weak feedback effect. Second, the rise in triose phosphate and fall in Pi that accompany a build-up of F6P would tend to affect F26P phosphatase and F6P,2-kinase in a manner opposite to that of F6P (41, 242). The sensitivity of the F26P level to F6P may thus be somewhat reduced.

The notion that changes in cytosolic Pi and esterified phosphates balance each other such that there is no net change in the level of cytosolic phosphate (i.e. that cytosolic phosphate is a conserved moiety) depends upon there being slow exchange between the cytosol and adjacent compartments. Significant exchange with the chloroplast is unlikely because the phosphate translocator does not sustain a net phosphate flux (69), and other forms of exchange are thought to be slow (97). Although it is thought that the large vacuolar Pi pool serves to buffer the cytosolic Pi pool from changes in phosphate supply from the environment (73, 74, 194) and perhaps to maintain cytosolic Pi at a level optimal for photosynthesis, most evidence indicates that exchange between these compartments requires several hours to equilibrate (194, 277). It appears, therefore, that cytosolic Pi can fluctuate over the short term and act as a regulatory intermediate. In this regard the short-term variation in Pi considered here differs in a fundamental way from a nutritional deficiency of phosphorus (21, 161).

The decrease in the cytosolic Pi/triose phosphate ratio and the inactivation of FBPase exert their influence over chloroplast metabolism by altering the kinetic characteristics of the phosphate translocator, as discussed earlier. Under these conditions the reactions of the PCR cycle are more able to compete for triose phosphate, and recycling of carbon is favored over export to the cytosol. The ultimate effect of this change appears to be an increase in the stromal PGA/Pi and ADP/ATP ratios, activation of several stromal enzymes, and an increase in the rate of starch synthesis (265, 273, 278). In addition, if the feedback is strong enough, there are mechanisms in place by which both the rate of photosynthetic electron transport and the activity of Rubisco can be attenuated. In the next sections, we focus especially on the role of Pi in regulatory mechanisms that can affect the activities of both the electron transport system and Rubisco.

REGULATION OF ELECTRON TRANSPORT   The initial step in the electron transport chain is the abstraction of an electron from water by a reaction linked to the photoact of PSII. As noted in the introduction to this section, these electrons must ultimately be used to reduce products of carbon metabolism, and to a lesser extent, alternative electron acceptors (8). If, under conditions where the PCR cycle or output pathway activity is limiting, the input of quantum energy increases, then the initial steps in the electron transport system must be regulated such that the excess energy is dissipated and the required rate of electron transport is adhered to. Duysens & Sweers (51) suggested that the primary acceptor of PSII (Q_A) could accumulate in the reduced form under conditions of strong illumination or in the absence of suitable electron acceptors and lead to an increase in fluorescence and a corresponding decrease in the yield of electrons. These workers were careful to state, however, that another mechanism not involving redox feedback or increased fluorescence might also be important in vivo. Evidence for this view comes from studies showing that, at high light when the ΔpH and the phosphorylation potential (i.e. [ATP]/[ADP][Pi]) are relatively large, the NADPH/NADP ratio is actually less than that occurring in low light (99,

250). A feedback mechanism involving the redox state of the adenine nucleo-tide pool is difficult to reconcile with these measurements, and studies in which the ΔpH was modified during steady-state photosynthesis in isolated chloroplasts demonstrated that an excessive proton gradient could reduce the rate of electron transport (231, 232). It was also shown that ΔpH could be modified by altering the rate of ATP turnover (230–232).

In analyzing the regulatory sequence involving ATP turnover and the ΔpH, it is important to recognize that comparisons of phosphorylation potentials in isolated thylakoid systems (e.g. 142) and intact chloroplasts undergoing even limited photosynthesis (85) suggest that the reaction catalyzed by the ATP synthase is significantly displaced from equilibrium during steady-state photosynthesis [for a more detailed discussion, see Horton (114)]. If this is true, then ΔpH will be strongly influenced by modification of the substrate levels of this reaction. Attention has focused on the influence of Pi because the ADP concentration apparently remains quite constant over a broad range of conditions (46, 98, 99).

The theory that a reduction in the stromal Pi concentration restricts the rate of photophosphorylation and ultimately the rate of CO_2 fixation has been questioned on the basis that the ATP synthase is apparently, under most conditions, saturated with Pi (229). The apparent K_m(Pi) measured in several studies (cf 77, 130, 219, 251) is less than 1mM (60–600 μM), and metabolite measurements indicate a stroma Pi concentration of 4–35 mM (cf 46, 77, 156, 205, 272). Changes in the Pi level are, on this basis, unlikely to affect ATP synthase activity, but two points must be considered when evaluating this proposal.

First, there is evidence that the concentration of free Pi influencing the activity of the ATP synthase is significantly less than that estimated from the Pi content of non-aqueously fractionated leaf tissue (46, 205). Part of this discrepancy is thought to result from contamination of the chloroplast extracts by Pi from the vacuolar pool (46, 247). Other evidence indicates that a considerable proportion of stromal Pi could be bound to both the protein (6, 27, 76, 179) and membrane (195) components of the chloroplast. In addition to this presumably freely exchangeable component, Furbank et al (78) have shown that there may be a pool of up to 2 mM Pi that is unavailable—at least for several minutes—to the ATP synthase. These authors suggest that "shield-ing" of the latter enzyme by Rubisco may be the mechanism underlying the exclusion of Pi (see also 195).

Second, in assessing whether a change in the stromal Pi concentration could significantly affect the level of the other substrates of the reaction (ADP and H^+) we must have a clear understanding of whether the calculated K_m(Pi) values for the ATP synthase (1, 130, 219) accurately represent the operative in vivo K_m(Pi). At present this value is difficult to assess because it is a function of both the intra-thylakoid and stromal pH (1, 251) and of the concentration of other stromal elements. Experiments with sollen "leaky" chloroplasts showed that both the apparent K_m(ADP) and K_m(Pi) values are significantly higher than those calculated using isolated thylakoids (77). The effect may be caused by binding of certain substrates to Rubisco and by a process related to Rubisco binding to the thylakoid membrane (76). If the free stromal Pi concentration is in the apparent K_m(Pi) range and there is a decline in its concentration, the immediate effect would be a rise in the ADP concentration and ΔpH in order to maintain the flux through the reaction. Similarly, a decrease in ADP concentration in the range of the K_m(ADP) may also cause acidification of the thylakoid space. In either case, the acidification itself would not reduce the steady-state rate of photosynthesis unless it is linked to a process that can alter the input of quantum energy. Clearly, more work is required to clarify whether changes in the Pi and ADP levels can elicit a feedback response through effects on ΔpH.

Recent experiments on functional photosynthetic systems have yielded new information regarding the mode by which PSII is regulated by feedback sequences. Weis et al (268) used chlorophyll fluorescence to show that the primary acceptor of PSII, Q_A, can remain mostly oxidized—even at light intensities sufficient to "saturate" the rate of photosynthesis (see also 48, 143, 269). To explain how the steady state is maintained under these conditions, Weis et al (268) postulate an alternative feedback mechanism that would involve reversible conversion of PSII centers to a form with reduced photochemical and fluorescence yields, and they suggest that acidification of the thylakoid space is the "local" condition that controls this mechanism. Studies using fluorescence procedures to quantify both the redox feedback and this "pH-dependent" feedback on PSII were able to account quantitatively for the regulation of PSII during steady-state photosynthesis by intact leaves when light, [CO_2] or [O_2] were varied (269). More work is required to assess this proposal and the significance of possibly contradictory observations (78, 79, 178). Other regulatory mechanisms that balance the ratio of production of NADPH and ATP to their consumption, the distribution of quanta to PSI and PSII, and regulation of other quenching mechanisms need to be considered. Nevertheless, the sequence of events linking changes in the cytosolic and stromal [Pi] to the thylakoid pH and in turn to the photochemical efficiency of PSII seems to provide a plausible explanation for regulation of the irreversible input of quantum energy (Figure 4).

REGULATION OF RUBISCO   We do not have a clear mechanistic view of the regulation of Rubisco activity in response to the stromal changes effected by feedback from the sucrose synthetic pathway. In experiments with isolated chloroplasts, both Heldt et al (102) and Furbank et al (78) showed that a drop in stromal Pi is accompanied by a rise in RuBP and a decline in flux. This result indicates that Rubisco is subject to regulation under these conditions, but only Heldt's group detected a decrease in the apparent activation state of the enzyme. It is conceivable that the lower ATP/ADP ratio under feedback conditions brings about a change in Rubisco activity through the "activase" system (174), or that there is an allosteric effect of the change in Pi on enzyme activity (179). As outlined in the earlier section on Rubisco, the clarification of this part of the feedback regulatory sequence awaits further information on the basic regulatory properties of Rubisco and associated chloroplast proteins.

Operation of the Feedback Sequence

MANIPULATION OF CYTOSOLIC Pi   The potential importance of cytosolic Pi in regulating photosynthesis was first demonstrated in experiments with isolated chloroplasts in which the external Pi concentration was directly manipulated (33). Since that time, similar studies have been made of photosynthesis in intact leaves by modulating the concentration of Pi in the cytosol using a variety of techniques. Infiltration of photosynthetic tissues with mannose, 2-deoxyglucose, or glucosamine—all of which sequester cytosolic Pi as phosphorylated compounds that are not readily metabolized—causes an increase in the rate of starch synthesis and a depression of the rate of photosynthesis (29, 94, 108, 109). This inhibition can be overcome by feeding Pi to mannose-treated leaf discs (229, 263, 264); and at CO_2 partial pressure well above ambient, Pi feeding can actually stimulate the rate of photosynthesis (262).

The importance of cytosolic Pi in regulating photosynthesis has also been reinforced by studies of oscillations in the rate of CO_2 fixation. These complex transients generally appear after a perturbation to the photosynthetic system under conditions of high light and CO_2 (e.g. 173, 228, 263). Oscillation can be restricted by feeding Pi to leaves and enhanced by feeding mannose (229, 263), a result suggesting that modification of the cytosolic Pi concentration can influence the distribution of flux control among elements of the photosynthetic system (see also 175, 279) and that, under these conditions, feedback from the cytosol is significant in determining the rate of photosynthesis. Stitt (237) conducted an experiment along similar lines where photosynthesis was interrupted by a brief period of low light. During this period it is thought that a depletion of cytosolic metabolites results in a temporary increase in the cytosolic Pi concentration. Upon re-illumination, there was a transient enhancement of the rate of photosynthesis which was interpreted to indicate that stromal Pi can restrict photosynthesis under steady-state conditions of high light and CO_2.

The cytosolic Pi status may also be involved in the lack of stimulation of photosynthesis observed, under some conditions, when the O_2 concentration is lowered and photorespiration restricted. This effect was first noticed in experiments with leaves from plants grown at "normal" temperatures that were illuminated at a lower temperature (26, 36, 122, 123, 165). More recent studies have involved a combination of high CO_2 partial pressure and depressed temperature. This "O_2 insensitivity" is thought to occur when the activities of enzymes in the sucrose synthetic pathway partially limit the rate of photosynthesis—i.e. when there is a significant feedback from the cytosolic reactions. Upon transfer from normal to low O_2 the feedback mechanism affects Rubisco activity such that both v_c and v_o decline and the net rate of CO_2 assimilation remains unchanged. Feedback inhibition of the rate of electron transport must also occur in order to balance the changes in v_c and v_o. The hypothesis that the cytosolic Pi pool mediates the feedback sequence under O_2-insensitive conditions is supported by experiments showing that feeding of Pi can restore some O_2 sensitivity to previously O_2-insensitive tissues, whereas the reverse can be effected by feeding the tissues with mannose (94). Studies of metabolite pool sizes under O_2-insensitive conditions show that, compared to control plants, there is a rise in the PGA/triose phosphate and ADP/ATP ratios (153, 226). Assuming that the reactions catalyzed by PGA kinase and G3P dehydrogenase are close to equilibrium and that the stromal pH is constant (46), Leegood & Furbank (153) calculated that there is a kinetic restriction placed upon the rate of electron transport and photophosphorylation.

Sharkey and coworkers (286) employed changes in CO_2 under "O_2- insensitive" conditions to examine the engagement of this feedback mechanism. Upon increasing the [CO_2], the electron transport rate decreased by 25–30%. An accompanying and immediate decline in the level of photochemical quenching of fluorescence indicates that the degree of redox-linked feedback regulation of PSII increases under these conditions. Over the next few minutes, however, they observed compensating changes in the levels of fluorescence quenching (photochemical and nonphotochemical) indicating an acidification of the thylakoid space and a relaxation of the redox feedback on Q_A. As there was little change in the rate of electron transport, it appears that the feedback regulation of PSII was maintained, and the feedback sequence became increasingly mediated by ΔpH. Dietz et al (48) also observed a rise in nonphotochemical quenching on increasing CO_2 from near ambient levels to ~3000 μbar. These results are consistent with the general interpretation that stromal Pi may reach a suboptimal steady-state level in low [O_2] or high [CO_2] or under a combination of both conditions, but there is apparently an initial period when the primary feedback may occur by another mechanism. Furbank et al (79) suggested that an alternate feedback sequence may involve the redox state of the pyridine nucleotide pool. In experiments with isolated chloroplasts, these authors showed that an increase in ΔpH is not necessarily associated with a reduction in stromal Pi.

Sharkey et al (225) and Leegood & Furbank (153) monitored the effect of "O_2-insensitive conditions" on Rubisco activity and the RuBP concentration in leaves. The former group demonstrated that, upon transfer to low $[O_2]$, Rubisco is inactivated to a similar extent to the drop in electron transport. They speculated that this may be the result of the lower ATP/ADP ratio reducing "Rubisco-activase" activity (216). They also noted that the change in activation state occurs relatively slowly and that the initial decline in v_o and v_c probably results from a transient drop in the RuBP concentration (199). Leegood & Furbank (153), however, measured a relatively small decline in RuBP, which is probably insufficient to account for the change in flux through the Rubisco reactions.

These studies of whole tissues verify the hypothesis that feedback inhibition of both the rate of electron transport and Rubisco activity can occur and that cytosolic Pi plays a central role in this regulatory sequence (107, 109, 223, 260). Nevertheless, a clearer view of whether this sequence affects the rate of CO_2 fixation (and not just assimilate partition) under more natural conditions comes from studies in which sucrose is allowed to accumulate or is externally supplemented.

THE SENSITIVITY OF CO_2 FIXATION TO CYTOSOLIC Pi   When considering the effectiveness of the cytosolic feedback sequence in regulating the rate of CO_2 fixation, one must recognize that it would be theoretically possible for cytosolic feedback to alter the fluxes of the photosynthetic system such that the PCR cycle pool sizes and rate of CO_2 fixation remain unchanged. This would involve parallel and compensating changes in the activities of the cytosolic and stromal FBPases and the ADP-glucose pyrophosphorylase, assuming that the reaction catalyzed by aldolase, the phosphate translocator, hexose phosphate isomerase, and phosphoglucomutase are at equilibrium (273). Nevertheless, there are two factors that complicate this simple means of altering the relative rates of assimilate export. First, there is no evidence for an effector that can initiate such a sequence without affecting several other important reactions in the chloroplast (e.g. Pi also affects the phosphate translocator, G3P dehydrogenase, SBPase, Ru5P kinase, Rubisco, and the ATP synthase). Second, the starch and sucrose synthetic pathways do not carry true export fluxes because they are coupled to the release of Pi which, in turn, can affect chloroplast metabolism.

If the regulation of assimilate partition and CO_2 fixation are inextricably associated through the effect of Pi on both processes, it is important to distinguish between conditions that affect metabolite partition and those that affect electron transport and Rubisco activity and thus the rate of CO_2 fixation. Experiments with isolated chloroplasts by Steup et al (236) and Heldt et al (103) addressed this problem. In both studies, the properties of the phosphate translocator were modulated by changing the external Pi concentration. Figure 5 shows the sensitivity of both CO_2 assimilation and starch synthesis to the concentration of external Pi [data from Steup et al (236)]. It is important to note that there is a significant range of Pi levels over which the sensitivity of CO_2 assimilation is close to zero, whereas that of starch synthesis is at its maximum. Although the presence of relatively high levels of PGA and triose phosphate in the cytosol may complicate this relationship in intact tissues, there is probably also a range of Pi/triose phosphate ratios over

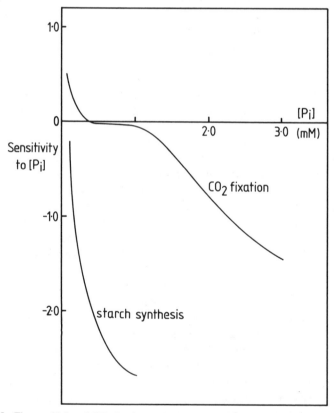

Figure 5   The sensitivity of CO_2 fixation and starch synthesis to the concentration of Pi in a suspension of isolated chloroplasts. The data of Steup et al (236) were used in the calculations. The sensitivity is defined as: $\partial v/\partial[Pi] \times [Pi]/v$ where v is the rate of either CO_2 fixation or starch synthesis. There is a range of Pi levels over which the rate of CO_2 fixation is insensitive to Pi but the rate of starch sythesis is extremely sensitive.

which assimilate partition is altered but CO_2 fixation remains relatively unaffected.

FEEDBACK FROM SUCROSE   Plants vary considerably in the degree to which they favor starch and sucrose as their major storage compound. In barley, for example, sucrose is the major storage product (88, 227) and appears to be located in the vacuole (65, 81, 132, 177). Sugarbeet, on the other hand, only accumulates sucrose for a short period after the onset of illumination (70, 71) and stores most assimilate as starch (80, 82). The relative sizes of the starch and sucrose pools may also vary during the course of a day and night (28, 71, 83, 106) which, to some degree, may reflect the changes in the level of extractable SPS (106, 115, 136, 197) or F6P,2-kinase activity (136). In some species, these diurnal changes appear to be the result of a light-related regulatory mechanism (227, 238), while in others there is evidence for an endogenous rhythm mechanism that is independent of light-intensity changes (117, 118, 136, 197, 238). Nevertheless, it is by no means clear whether, superimposed upon these "coarse" control mechanisms, there is a classical metabolic feedback sequence that can moderate the rate of CO_2 assimilation in response to a change in the level of sucrose [see reviews by King et al (138), Neales & Incoll (170), Herold (107), Gifford & Evans (86)].

THE EFFECT OF SUCROSE   Experiments with leaves, leaf discs, and pro-toplasts have shown a correlation between high levels of sucrose and a change in the partition of assimilate between starch and sucrose, but the evidence for a depression of the rate of assimilation is not conclusive. Herold et al (110) showed that the rate of CO_2 assimilation is not altered but starch synthesis is stimulated when sugarbeet leaf discs are floated on a sucrose solution. Similar changes in partition were demonstrated by Stitt et al (244, 245) in experiments with spinach leaf discs where sucrose was allowed to build up in the tissue. There was, however, some depression of the rate of photosynthesis, but it should be noted that the experiments were done at levels of CO_2 well in excess of ambient. Because they detected a decline in the rate of sucrose synthesis and an accompanying build-up of UDP-glucose and fructose 6-phosphate, these authors concluded that the mechanism altering metabolite partitioning involved, in part, the inhibition of SPS.

Further evidence for inhibition of SPS activity during a build-up of sucrose was found by Clausen & Lenz (32) and Rufty & Huber (197) using leaves of eggplant and soybean, respectively. But similar experiments with spinach leaves showed no change in the extractable activities of either SPS or cytosol-ic FBPase (197). The rates of photosynthesis did not change during the sucrose build-up in the spinach leaf experiments, nor was there a marked temporal correlation between the decline in SPS activity and photosynthetic rate in soybean leaves (197). The effect of sucrose on photosynthesis in protoplasts also varies between species. Foyer et al (72) found that sucrose pretreatment of spinach protoplasts depressed the rate of CO_2 fixation but did not affect the extractable activities of SPS or cytosolic FBPase. Wheat and barley protoplasts, on the other hand, showed little metabolic response to the sucrose treatment. Hills (113) examined the metabolism of freshly isolated asparagus cells and concluded that the inhibition of sucrose synthesis immediately upon cell isolation was unlikely to result from feedback inhibition by sucrose.

There is clearly a need for more work to resolve whether a build-up of sucrose can initiate a feedback sequence by inhibiting the latter reactions of the sucrose synthetic pathway. But even if this can be shown, the evidence from most of the whole-tissue studies indicates that the resulting feedback may exert its greatest effect on the partition of assimilates and have little effect on the rate of CO_2 fixation. The important matter in this context, therefore, concerns the nature of the regulatory mechanisms that affect partition and CO_2 assimilation and the degree to which they can operate independently.

SUMMARY   There is a considerable amount of evidence from both the biochemical and whole-tissue studies that classical product feedback, or feedback originating from diurnal changes in the activities of enzymes such as SPS (198), exerts its greatest effect on metabolite partition and has little effect on the rate of CO_2 fixation. Nevertheless, there is evidence that under some relatively extreme conditions, such as shortly after a transition to a low temperature (153) or under conditions of water stress (cf 223), there may be a period during which the feedback regulatory sequence affects the rate of CO_2 fixation. It will be important to establish whether feedback under such conditions can be sustained in the long term, or whether processes such as adjustments of the total cytosolic phosphate pool by exchange with the vacuole (e.g. 73, 194) can minimize the cytosolic limitation and reinstate energetically efficient photosynthesis (see 226, 229).

Feedback on Energy Input from Rubisco

The quantity and kinetic properties of Rubisco largely determine the maximum capacity to provide electron acceptors at any given $[CO_2]$, $[O_2]$, and temperature. It is important that, when the absorbed flux of quantum energy is sufficient to "saturate" this capacity, a feedback mechanism comes into play to allow the dissipation of excess energy. We have noted that the "local" conditions involved in feedback regulation of the photochemical reactions are associated with acidification of the thylakoid space and redox regulation of the photochemical traps. Engagement of these regulatory mechanisms has been examined in intact leaves by varying the intensity of illumination over the range encompassing both "rate-limiting" and "rate-saturating" levels or by varying the concentration of CO_2 over a corresponding range. Weis et al (268) showed that large changes in the scattering of light at 540 nm by intact leaves begin at a light intensity or $[CO_2]$ that marks the transition between these light-limited and -saturated states and continue to increase as the intensity of illumination is increased or the $[CO_2]$ is decreased. This indicates that, as the absorbed flux of quantum energy approaches the maximum capacity of Rubisco to sustain that flux, there is a marked rise in ΔpH which appears to be associated with regulation of the ATP synthase. They also used light-induced absorbance changes to show that the reaction centers of PSI, which are mostly in the active, reduced form in limiting light, accumulate in the oxidized form as the intensity of illumination exceeds saturation. This accumulation correlates with the apparent acidification of the thylakoid space, and may be the result of an effect of the pH gradient on the transport of electrons from PSII by the cytochrome-b/f-complex (cf 40, 270). Nonphotochemical quenching of fluorescence also increases with the light scattering changes, indicating an effect of the ΔpH (or some associated change) on the reaction centers of PSII. Significantly, the primary acceptor of PSII remained mostly in the active, oxidized configuration until the intensity of illumination was 3–4-fold saturating [see also Dietz et al (48)]. This indicates that regulation of PSII follows a hierarchical pattern with redox mechanisms becoming important after the pH-dependent mechanism becomes fully engaged.

Dietz & Heber (46, 47) examined the reactions catalyzed by PGA kinase and G3P dehydrogenase during changes in light intensity and CO_2 concentration. They showed that the overall mass action ratio for these reactions does not increase substantially as light becomes "rate-saturating," and that the reactions are not greatly displaced from equilibrium. Changes in the levels of PGA, DHAP, and Pi can therefore affect the steady-state ATP/ADP and NADPH/NADP ratios, which along with Pi could affect the electron transport system by a feedback mechanism. As discussed above, the conservation of total chloroplast phosphate provides a mechanism by which any PCR-cycle enzyme can affect the Pi, PGA, and DHAP concentrations. Von Caemmerer & Edmondson (254), for example, note that the total level of phosphate in PGA and RuBP in intact leaves increases with light intensity and reaches the apparent upper limit at a light intensity slightly lower than that required to saturate CO_2 assimilation. It is conceivable that a concomitant decline in Pi could initiate feedback inhibition of the photoacts by the mechanisms alluded to above. This hypothesis is consistent with the optical studies of the state of the thylakoid described in the preceding paragraph.

Feedforward Regulation of Enzyme Activity

It is necessary, under certain circumstances, to regulate the input of energy into the photosynthetic system to keep pace with limiting steps in the PCR cycle and carbon export pathways. A more common occurrence, however, is that the input of quantum energy limits the rate of photosynthesis and that Rubisco activity (v_c and v_o) must be regulated such that the stoichiometric relationship between CO_2 fixation and energy transduction is fulfilled. This requirement could in theory be met by simply reducing the size of intermediate pools in parallel with the flux and letting the RuBP concentration regulate Rubisco activity. However, in a previous section we indicated that the RuBP concentration does not vary over a wide range of irradiances and that a mechanism controlling the degree of activation of Rubisco appears to be an important mode by which v_c and v_o are modulated to match the rate of energy input. Feedforward activation of other PCR-cycle enzymes (namely, SBPase, FBPase, Ru5P kinase, and G3P dehydrogenase) to match the light-dependent flux may also play a role in the sequence regulating Rubisco. In this section, we do not expand upon the mechanistic details of the regulatory sequence connecting the input of quantum energy to Rubisco outlined in previous sections and other reviews (e.g. 4, 23, 42, 114, 152) but briefly examine the implications and possible advantages of "light regulation" of enzyme activity as a means of feedforward control of the PCR cycle and CO_2 fixation.

The current view of the importance of light regulation of the activity of PCR cycle enzymes (other than Rubisco) has been summarized by Leegood et al (155). (a) It may serve as a mechanism preventing enzymes from becoming oxidized (and thereby inactivated) by certain products of chloroplast metabolism. (b) Inactivation of enzymes in darkness prevents the operation of several energy-wasteful futile cycles. (c) Modulation of enzyme activities may serve to maintain a balance between carbon export and the regeneration of RuBP, and to regulate PCR-cycle metabolite levels and the flux. There is little doubt regarding the value of inactivating enzymes in darkness to prevent the waste of energy stored during periods of illumination, but what remains puzzling is that changes in enzyme activity occur over a wide spectrum of light intensities (e.g. 135, 148, 152, 154, 278) and that steady-state activation states are approached relatively slowly (e.g. 135, 148, 199, 281, 283).

Regulation of the activation state of the PCR-cycle enzymes in parallel with changes in the rate of photosynthesis has a significant effect on the steady-state relationship between substrate concentration and reaction velocity. Woodrow (274) used a model of the regulatory mechanism of SBPase (280, 282) to predict that concomitant changes in the enzyme activation state and the CO_2 fixation flux would result in a sigmoidal relationship between SBP concentration and the flux through the SBPase reaction, even though the in vitro substrate kinetics are hyperbolic. He also predicted that a more pronounced sigmoidal relationship would result if parallel changes in pH and the Mg^{2+} concentration were considered. Similar predictions can also be made regarding the relationship between the flux and the substrate concentration of all the light-regulated enzymes. The degree of sigmoidicity of the substrate-velocity curves will depend upon the steady-state kinetic properties of the enzymes and the range of photosynthetic fluxes over which variations in activation state and effector concentrations occur. In Figure 2, we show how parallel changes in the Rubisco activation state and the rate of photosynthesis can result in a highly sigmoid substrate response. This curve is quite different from that expected from the steady-state rate equation of active Rubisco (also shown in Figure 2).

One of the most remarkable characteristics of the PCR cycle is the constancy of the intermediate pool sizes, despite large changes in the rate of photosynthesis [see review by Leegood et al (155)]. This observation supports the notion that multivariate interactions in vivo lead to apparent sigmoidal substrate kinetics where there is a large range of fluxes over which changes in substrate concentrations are small. The discrepancy between the pool sizes anticipated by the rate equations for active enzyme species and those occurring in vivo has been confirmed in experiments that take advantage of slow changes in enzyme activation state. For example, the activation state of Rubisco requires several minutes to assume a new level after a change in light intensity. Mott et al (169) showed that when the light intensity is reduced the RuBP concentration drops quickly in response to the lower flux, but soon rises again to the "normal" level as the enzyme activation state declines.

We may speculate that this "normal" state offers advantages in terms of the steady-state efficiency of photosynthesis or the ability of the system to respond to a change in environmental conditions. For example, we have developed arguments that Pi mediates at least part of the feedback sequence influencing the input of quantum energy. Given an unregulated PCR cycle, changes in esterified phosphates would occur in parallel with changes in the flux, and since total stromal phosphate is conserved, the concentration of Pi would change over the full range of fluxes. These changes in Pi could well effect proportional changes in the strength of feedback inhibition of energy input. Mechanisms that stabilize the levels of esterified and inorganic phosphate may therefore "poise" the PCR cycle such that feedback effects on the electron transport system are minimized and energetic efficiency is maximized over the widest range of fluxes. Regulation of Rubisco is especially important in this regard because (a) the RuBP pool contains a relatively large proportion of stromal phosphate, and (b) this pool is effectively isolated from the other intermediates of the PCR cycle by the largely irreversible reactions that regulate its synthesis and consumption (273). Therefore, phosphate can be "stored" in RuBP without greatly affecting the partition of assimilate between starch and sucrose.

Modulation of the activation state of PCR-cycle enzymes may also serve to prevent the sequestrations of important chloroplast metabolites. Rubisco, for example, can bind several metabolites (such as NADPH) (e.g. 6), and should the RuBP or CA-1-P concentration drop significantly, a proportion of the compounds would become unavailable for normal function.

The slow imposition of changes to the activation state of the "light-regulated" PCR-cycle enzymes is also puzzling because, if this mechanism serves to aid energetically efficient photosynthesis, it is logical that it should operate as quickly as possible. Certain suggestions regarding the significance of slow "hysteretic" responses have been made (e.g. 75), but these have not stood up to critical examination (e.g. 209). One possibility is that the slow steps in the activation/inactivation mechanisms enhance the sigmoidal substrate responses by enabling a degree of "kinetic cooperativity" (2, 282). Some insight may also be gained by examining the "cost" versus the "benefit" of these enzyme activation mechanisms. If modulation of enzyme activity in response to irradiance serves to increase the overall efficiency of photosynthesis, then this "benefit" must be offset against the energetic and nutritional "cost" of the regulatory mechanisms. Consider, for example, the basic reaction describing the activation/inactivation of the enzymes regulated by thioredoxin (Th) (234):

$$E_i + Th_{red} \overset{k_a}{\underset{k_i}{\rightleftharpoons}} E_a + Th_{ox}, \qquad\qquad 13.$$

where E_i and E_a are inactive and active enzyme forms, respectively. The relaxation time (r) of the system is approximated by

$$r = (k_a[Th_{red}] + k_i[Th_{ox}])^{-1}. \qquad\qquad 14.$$

A rapid response thus requires a relatively high concentration of thioredoxin. But as the size of the reduced thioredoxin pool grows, the energetic cost of maintaining it also grows, because oxidation of thioredoxin can occur either directly or indirectly through, for example, interaction with O_2 (154). Rapid activation/inactivation of Rubisco may similarly require high levels of "activase" and rates of ATP turnover. The relatively slow changes in activity of the light-regulated enzymes may therefore simply reflect an optimal balance between energetic and nutritional "costs" and "benefits" and may not have an important regulatory significance in the non-steady state. The energy cost of such regulatory mechanisms has been analyzed in animal systems, and it has been shown that up to 20% of the basal respiratory rate of the cell may be used by regulatory, energy-requiring futile cycles (87, 141).

CONCLUSIONS   One of the most intriguing aspects of the regulatory sequences linking the sucrose synthetic pathway, the PCR cycle, and the electron transport chain is the existence of several feedforward mechanisms. Regulation of the "light-activated" enzymes, cytosolic FBPase (see also 239, 240), ADP-glucose pyrophosphorylase, and SPS by feedforward processes ensures that these enzymes can transmit a large range of metabolic fluxes with relatively small changes in their substrate pool sizes. We have suggested that one of the main functions of these mechanisms is to reduce the overall "strength" of the feedback sequences and thus to maintain the energetic efficiency of photosynthesis over a wide range of irradiances. If this proposition is true, it means that, over this range of irradiances, the rate of photosynthesis will not be sensitive to the activities of enzymes of the PCR cycle and starch and sucrose synthetic pathways since changes in these activities would produce minimal changes in intermediate pool sizes. In other words, the flux will be determined by the rate of quantum energy input and the efficiency of energy transduction rather than the activities of any distal enzymes. Assessments of the degree to which enzymes of the photosynthetic system determine the flux therefore not only provide fundamental information about flux control (and how the flux can be altered) but also provide a measure of the strength of the feedback regulatory sequences. In the following section we examine studies of the determination of the photosynthetic flux by enzymes of the photosynthetic system.

CONTROL OF THE RATE OF PHOTOSYNTHESIS

A great many studies have sought to define the factors that limit the rate of photosynthesis. If one ignores for the moment any limitation imposed by light, temperature, the availability of water, and stress factors, the answer seems to be protein. Nitrogen, a major constituent of proteins, is the mineral nutrient that plants require in greatest quantity, and nitrogen is the nutrient that most often limits plant growth. Field & Mooney (67) reported a strong correlation between the light-saturated rate of net CO_2 assimilation (A_{max}) measured in the natural environment and the total protein nitrogen content of the leaves of C_3 plants with a wide variety of growth forms. This correlation is not surprising because key reactions of the photosynthetic process are associated with specific assemblages of polypeptides, and it is estimated that 75% of the total leaf protein is associated with the chloroplast (54). This sort of correlative evidence does not, however, provide an insight into the degree to which individual proteins might limit the rate of photosynthesis, because A_{max}

is also highly correlated with changes in the levels of Rubisco, Ru5P-kinase, NADP-G3P-dehydrogenase, cytochrome f, coupling factor, and carbonic anhydrase (7, 17, 54, 57, 162). As we have discussed here, such an insight can only be gained from experiments involving direct manipulation of the levels or activities of individual enzymes in the context of the whole system or from studies of the regulation of the photosynthetic system and the kinetic properties of the constituent enzymes.

Taking an evolutionary view, one may reasonably propose that selective mechanisms have tended to allow the expression of genes for proteins such that optimal rates of photosynthesis can occur and no single step is limiting— at least under an assimilation-weighted average condition for the natural environment. A theoretical treatment by Cowan (37) illustrates this point. He considered the optimal partitioning of nitrogen between two enzymes— carbonic anhydrase (CA) and Rubisco—that are directly involved in CO_2 fixation. The presumed role of CA in higher-plant leaves is to facilitate transport of CO_2 to the sites of its fixation by Rubisco, and previous studies had estimated that the high activity of CA present in chloroplasts increases the rate of CO_2 fixation of a given amount of Rubisco in the chloroplast by about 10% (120). Cowan derived a kinetic expression for CO_2 uptake of a leaf mesophyll cell as a function of nitrogen partitioning, assuming that a given total amount of nitrogen could be partitioned to the two functions (i.e. N_{CA} + N_R = a constant). The theoretical optimum $N_{CA}:N_R = 1:22$) is very close to the observed content of CA in several species. According to this logic the most "expensive" components (i.e. those with the lowest specific activities and highest molecular weights) should be the most limiting with respect to the rate of CO_2 fixation. But it must be emphasized that the optimal distribution of protein, and therefore the degree to which various enzymes limit the flux, is also strongly dependent upon the characteristics of the various regulatory sequences.

We have discussed evidence indicating that feedback inhibition of Rubisco and the electron transport system may be quite weak under most "natural" conditions. In view of this finding, the relatively low specific activity of Rubisco, and the large amount of protein devoted to Rubisco and the light-harvesting function [Evans (54) estimated that leaves devote similar amounts of protein to these components], it is not surprising that both protein complexes have emerged as important points of flux control in this and other analyses of the regulation of photosynthesis. Nevertheless, the argument of Cowan illustrates that control is likely to be shared among all proteins that participate in the process, albeit to different extents. Furthermore, given the large variation in the intensity of environmental factors that affect the rate of photosynthesis in a typical natural habitat, the distribution of flux control is likely to change dramatically over the course of a day. It is therefore necessary to develop a quantitative approach to assess how this control is shared, and how this may change with environmental conditions.

Definition of Limitation

It was recognized some time ago that certain enzymes are particularly important in determining the metabolic flux. Krebs (144) called these enzymes "pacemakers" and specified that they must catalyze reactions the velocities of which are restricted by the activities or concentrations of enzymes and not by substrate concentrations. A similar definition was elaborated by Bücher & Rüssmann (24), who used the term "rate limiting" and recognized that several such enzymes could coexist in one pathway. It therefore became necessary to develop a quantitative means of expressing limitation in order to be able to compare the relative importance of the enzymes whose activities are important in determining the flux. Higgins (112) proposed that the "control strength" of an enzyme could be defined as the increase in flux brought about by an infinitesimal increase in enzyme activity. As a differential, this expression is relevant to a particular steady state and does not define the effect of large changes in enzyme activity on the flux.

Differential coefficients of this kind, which express the sensitivity of the steady-state flux to "independent" variables such as enzyme concentration and kinetic constants, have been applied widely in analyzing and characterizing metabolic systems (e.g. 38, 100, 131, 171, 206–208, 210). Using the nomenclature of Burns et al (25), the "flux control coefficient" for an enzyme (E) is given by

$$C_E^J = \frac{\partial J/J}{\partial P/P},$$ 15.

where J is the metabolic flux of interest and P the independent variable whose change brings about a change in the flux. The latter variable is typically enzyme concentration or V_{mzx}. The control coefficient is a dimensionless quantity because the increments are expressed as fractions, and can be positive, negative, or zero. A coefficient of unity, for example, indicates that small changes in the parameter P effect proportional changes in the flux. Kacser & Burns (131) have demonstrated that the control coefficients for each enzyme in a system with respect to a single flux sum to unity as long as P is chosen such that it is proportional to the velocity of the respective reactions.

There has recently been debate about the use of the control analysis of Kacser & Burns (131) for determining control coefficients from the kinetic properties of individual enzymes (e.g. 39, 211). These criticisms, however, have not questioned the validity or usefulness of the control coefficient for characterizing flux control in the steady state, but rather have expressed the need for caution when making the assumptions required to predict systemic properties from those of individual components. Crabtree & Newsholme (38) have expressed some reservations about the use of the control coefficient for describing metabolic systems because of the possibility that a high control coefficient may be incorrectly interpreted as indicating that the relevant enzyme must be important during a change in flux resulting from a change in the level of an external effector. They also stress that, as a differential and a property of a given steady state, the control coefficient does not yield information about the effect on the flux of relatively large changes in enzyme activity or, conversely, the degree to which an enzyme is limiting during a large change in flux (see 124).

There are no strict rules that dictate how and with respect to which parameters the determination or limitation of a metabolic flux is defined, but the most useful definitions are made with respect to the independent variables of the delimited system because system variables (e.g. intermediate pool sizes) can only be altered by changing the independent variables.

Control of the Rate of CO_2 Uptake

Before examining some of the studies of enzymes of the PCR cycle and sucrose synthesis, it is worth noting that, unlike "downstream" reactions, the processes mediating CO_2 diffusion to the stroma can limit the rate of photosynthesis simply by restricting the supply of this substrate. Many of the quantitative studies of the limitation to the rate of photosynthesis have in fact been directed at these elements, and reviews by Farquhar & Sharkey (62) and Jones (124) provide an assessment of many of them. Several more recent studies have successfully developed expressions, involving both empirical and imposed equations, for the control coefficients of the stomata and aerodynamic boundary layer (13, 275). An alternative approach advocated by Sharkey (223) defines the limitation due to CO_2 diffusion to the carboxylation site as the difference between the steady-state rate of photosynthesis and that which would occur if the diffusive conductance to CO_2 were infinite. This expression defines the maximum value of the limitation imposed by stomatal conductance and has the disadvantage that it is a complex function of several independent variables and is therefore not useful in evaluating the impact on an existing system of changes to these variables.

FLUX CONTROL BY RUBISCO   The most widely accepted appraisal of the role of Rubisco in determining the rate of CO_2 assimilation was put forward by Farquhar et al (64). These authors proposed that the rate of steady-state CO_2 fixation is determined either by the maximum capacity of Rubisco at the prevailing $[CO_2]$ and $[O_2]$ or the light-dependent capacity to regenerate RuBP. This model has been of great value because not only does it define the criteria for high-gain photosynthesis with respect to CO_2, but it also provides a basis for linking separate kinetic models of Rubisco and carbon metabolism with the input of photochemical energy from the light reactions. It was originally proposed that it would be possible simply to determine the state of the system by measuring the RuBP pool size (64). However, it is now clear that this measurement, taken by itself, is meaningless because v_c and v_o are also controlled in vivo by additional mechanisms that regulate both the amount of active Rubisco and the apparent kinetic properties of the active enzyme species.

In the light of these new findings regarding the regulation of Rubisco activity, it is necessary to redefine the criteria used to assess the degree to which Rubisco determines the rate of photosynthesis. Assuming that we now have an accurate picture of the mechanisms by which Rubisco can be regulated, unequivocal evidence that the quantity of Rubisco is limiting can be obtained by demonstrating that the enzyme is fully active, and the RuBP closely approaches a rate-saturating level (from our earlier discussion, this would appear to be close to twice the active-site concentration). Such a demonstration is technically possible, and several published experiments (213, 215, 254) can be analyzed to show that Rubisco is at least the primary limitation to the photosynthetic rate when the irradiance is saturating and CO_2 is at or below normal ambient levels. However, it becomes more difficult to determine the degree to which Rubisco limits the rate of photosynthesis at subsaturating light intensities or high levels of CO_2 because attenuation of Rubisco activity by regulatory mechanisms becomes more significant.

We have stated a case suggesting that feedforward regulation of Rubisco and other light-activated enzymes is in part a means of "poising" the PCR cycle such that feedback inhibition of the electron transport system is minimized and the efficiency of photosynthetic electron transport maximized. If one were to propose that the progressive inactivation of Rubisco observed with decreasing irradiance actually detracts from the rate of photosynthesis, then it follows that (a) the feedforward mechanism is not "tuned" for maximum high-gain photosynthesis, and (b) a quantitative assessment of the degree to which Rubisco activity limits the rate of photosynthesis will show a significant degree of limitation at low irradiances.

The complexity of the Rubisco regulatory system makes it very difficult to use kinetic models to draw quantitative conclusions concerning the limitation to the rate of photosynthesis imposed by Rubisco. Woodrow et al (276; see also 279) have taken advantage of the fact that both the internal CO_2 and O_2 levels can be manipulated as fixed (independent) variables. By changing the levels of these gases, it is possible to simulate the effect on the photosynthetic system of changing the maximum activity of Rubisco. The advantage of using this approach is that, from what we understand of the kinetics of Rubisco, the

$K_m(CO_2)$ and $K_m(O_2)$ values do not vary greatly with changes in the level of RuBP or the activation state, and therefore it is possible to predict relatively accurately the fractional effects of CO_2 and O_2 changes on both the carboxylase and oxygenase activities of Rubisco over a range of conditions. The authors derived expressions for control coefficients which reflect the effects of small changes in both the Rubisco activities on the net rate of CO_2 fixation and the RuBP flux (J_R).

Variation in the rate of CO_2 uptake (proportional to J_R and the control coefficient with respect to the V_{max} for Rubisco ($C^{JR}_{Rubisco}$) to light intensity is shown in Figure 6. At a rate-saturating light intensity the control coefficient is over 0.75, indicating that Rubisco is the major controller of the flux. However, as the light intensity is decreased, control by Rubisco decreases until, at less than 400 μmol quanta $m^{-2}s^{-1}$, this step exerts no flux control. One may conclude that small changes in Rubisco activation or the level of CA-1-P (both of which modulate V_{max}) would not affect the flux at low light intensities. These changes would have a much larger effect at

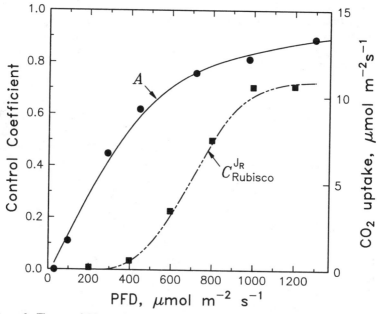

Figure 6   The rate of CO_2 uptake for an intact soybean leaf as a function of the incident photon flux density (*PFD*) and the corresponding control coefficient of the RuBP-flux ($v_c + v_o$) with respect to Rubisco. Values of the control coefficient were calculated from gas-exchange measurements of the response coefficients of CO_2 uptake with respect to [CO_2] (e.g. $\partial A/\partial[CO_2] \times [CO_2]/A$) and [$O_2$] at the indicated *PFD* levels and ambient levels of CO_2 and O_2. For details of the measurements and calculations see Woodrow et al (276).

saturating light, but as most studies indicate that Rubisco is already fully active at high light (160, 182, 254), regulatory mechanisms that control the activation state of Rubisco may also not be significantly limiting under these conditions. Systematic studies of control coefficients and the level of Rubisco activation over a wide variety of conditions are required to develop a picture of whether, at intermediate light intensities, for example, regulatory mechanisms can actually detract from the rate of photosynthesis. This example, which is typical of healthy C_3 plants, would argue that feedback inhibition of energy input and transduction due to suboptimal poising of enzyme activation states is minimal. Nevertheless, from published responses, one could be fairly confident that mechanisms regulating Rubisco activity will come much more into play with leaves at higher temperatures (139) and under conditions of phosphorus deficiency (21).

FLUX CONTROL BY OTHER ENZYMES   From the foregoing discussion of regulatory sequences, it is evident that all of the enzymes of the PCR cycle, the starch and sucrose synthetic pathways, and probably the electron transport system have the potential to affect Rubisco activity and therefore the rate of CO_2 fixation. At low light intensities, for example, Rubisco activity is clearly highly regulated by feedback processes and, in view of the almost linear relationship between irradiance and CO_2 uptake (Figure 6) under these conditions, the rate of photosynthesis is most probably highly limited by the quantum efficiencies of PSI and PSII. Under conditions of intermediate light intensities, however, both Rubisco and the photosystems may be subject to feedback inhibition, some of which may originate from the PCR cycle and export pathways. It remains a challenge to quantify the degree to which these enzymes limit the rate of photosynthesis, because the analysis of Woodrow et al (276) suggests that, in a healthy C_3 plant, the influence of this component is relatively small.

Numerous conclusions concerning the importance of enzyme activities in limiting the rate of photosynthesis have been drawn from relationships between substrate pool sizes and the flux. A strategy of comparing equilibrium constants with mass action ratios was employed by Bassham & Krause (15) in a study of steady-state photosynthesis in *Chlorella*. From the calculated free-energy changes, it was concluded that Rubisco, FBPase, SBPase, and Ru5P kinase are the potentially regulatory enzymes of the PCR cycle, and, except for Rubisco, these enzymes participate in the regulatory sequence affecting the branch points of the PCR cycle. Heldt & Stitt (105) and Gerhardt et al (83) have made similar calculations for enzymes of the sucrose synthetic pathway and have estimated that the cytosolic FBPase, SPS, and sucrose phosphatase have relatively high, negative free-energy changes and therefore are potentially regulatory enzymes. Comparison of phosphorylation potentials and ΔpH values in thylakoid suspensions where ATP is not consumed and in isolated chloroplasts undergoing photosynthesis revealed that the reaction catalyzed by ATP synthase is also generally displaced from equilibrium (85, 142). Dietz & Heber (46) showed that, over a range of fluxes, there was little change in the mass action ratio for the combined PGA kinase-G3P dehydrogenase reaction and the stromal aldolase reaction, and concluded that these enzymes probably do not limit the rate of photosynthesis. However, the reactions catalyzed by stromal hexose phosphate isomerase and phosphoglucomutase both show some deviation from their calculated equilibrium constants and therefore have the potential to regulate the rate of starch synthesis and presumbly, to some degree, the rate of photosynthesis (45). There have been numerous other studies of the relationships among PCR pool sizes [see review by Leegood et al (155)], the ATP/ADP and NADPH/NADP ratios, ΔpH [see review by Horton (114)], and the flux. While these have greatly expanded our understanding of the regulation of photosynthesis, their interpretation in terms of flux limitation has been complicated by two fundamental problems. 1. Regulation of the activation state of the PCR cycle enzymes and the ATP synthase results in a very complex relationship between metabolite pool sizes and the flux that is not easily interpreted. 2. There is still no clear definition of the nature and strength of feedback processes influencing the initial irreversible steps in photosynthesis (i.e. the quantum yield of the photosystems and the activity of Rubisco). This information is essential if we are to make estimates of the degree to which enzymes limit the flux from information regarding the kinetic and regulatory properties of the constituent enzymes.

The implicit logic of most of these studies of pool sizes is that "nonregulatory" and "potentially regulatory" enzymes can be distinguished simply by comparing the equilibrium constant and mass action ratio for a given reaction under a defined set of conditions (196). It has not always been appreciated, however, that regulatory enzymes identified by this and other means are not necessarily important in determining the flux (196). In fact, Kacser & Burns (131) showed that there is **no** general correlation between the mass action ratio/equilibrium constant quotient and the flux control coefficient (see below). This conclusion is not only relevant to control coefficients but will hold for any quantitative or qualitative expression relating the activity of an enzyme or process to the determination of a flux. These authors also showed that if a reaction is at equilibrium, the relevant enzyme does not limit the flux. Nevertheless, the problem with any arbitrary division between equilibrium and nonequilibrium reactions is that the relationship between the displacement from equilibrium and the control coefficient (or any similarly based quantitative expression) is dependent upon the nature and strength of the feedback sequence linking the substrate pool with the initial, largely irreversible steps in the system. Thus, if the substrate of a reaction causes a severe and sensitive inhibition of the first largely irreversible reaction in a sequence, even the smallest movement of the mass action ratio (which could occur when the reaction is close to equilibrium) would have a profound impact on the flux. Quantitative and qualitative measurements of the sensitivity both of pool sizes to changes in the activity of an enzyme and of the feedback sequence to alterations in pool sizes are required to draw valid conclusions regarding flux limitation.

Computer models of photosynthesis have also been used to examine the sensitivity of the rate of CO_2 assimilation to various parameters (151, 273). There are of course many assumptions underlying the construction of such models and there are not enough data concerning the regulation of the system for such models to predict accurately the impact on the flux of changes in the activities of the enzymes. Nevertheless, both models do yield valuable insights into the way the rate of photosynthesis is limited under a variety of conditions. The Laisk & Walker (151) model was designed to simulate photosynthesis under high light and CO_2 conditions. Their analysis and that of Woodrow (273) show that under these conditions the limitation of the photosynthetic rate may be shared among several elements of the system, and Laisk & Walker (151) conclude that the activity of the sucrose synthetic enzymes and therefore the rate of cytosolic Pi production may be of great importance in this regard. Although the model of Woodrow (273) does not include all of the sucrose synthetic pathway and cytosolic Pi is not a variable, the accompanying analysis does show an interesting relationship between the flux control coefficients of the cytosolic (C_1) the stromal FBPases (C_2) and the ADP-glucose pyrophosphorylase (C_3):

$$\frac{C_2}{v_2} + \frac{C_3}{v_3} = \frac{C_1}{v_1} \qquad\qquad 16.$$

where v_1, v_2, and v_3 are the fluxes through the cytosolic and stromal FBPase and the ADP-glucose pyrophosphorylase catalyzed reactions, respectively. The control coefficients of at least these three enzymes therefore cannot change independently of each other, and limitation of the rate of photosynthesis by the enzymes of the sucrose synthetic pathway may by necessity be accompanied by a significant degree of limitation by the other enzymes at the triose phosphate and hexose phosphate branch points of the PCR cycle.

Part of the difficulty in interpreting experiments aimed at examining limitation by elements other than Rubisco and the photosystems is that the answer is going to be quantitative (i.e. relatively small) over the full range of light intensities (276). In order to overcome this difficulty, many studies have been undertaken using elevated CO_2 concentrations or depressed temperatures. The Walker group had considerable success, using a step change in conditions and the ensuing transient rate, in showing that the limitation to the rate of photosynthesis under these conditions is both widely distributed and in part involves the reactions of the sucrose synthetic pathway. The most direct evidence comes from their experiments involving the feeding of mannose and Pi to intact tissues (229, 262, 263). The stimulation of steady-state photosynthesis and alteration of the condition under which oscillations in the rate of photosynthesis occur, which can be elicited by feeding Pi, demonstrates that, at high CO_2, the cytosolic reactions that liberate Pi (this will include all of the reactions of the cytosol to a greater or lesser degree because of the conservation of cytosolic Pi and presumably any processes involved in the movement of Pi from the vacuole to the cytosol) partially limit the rate of photosynthesis. Similar conclusions have been drawn by workers examining the sensitivity of the rate of photosynthesis to O_2 either at high CO_2 levels and slightly depressed temperatures (221, 222) or at ambient CO_2 and a low temperature (153).

CONCLUDING REMARKS

The major function of the mature leaf is to supply carbon to the rest of the plant for a range of metabolic and storage functions. In the long term, the requirements of these "sink" tissues for carbon may vary considerably, but over the shorter term a more or less uniform supply of carbon to the developing tissues may contribute to greater metabolic efficiency and therefore may be of adaptive advantage. It may also be of adaptive advantage for the photosynthetic system to be able to make best use of the available light and CO_2 and therefore to produce the greatest amount of end product for growth. Since the conditions in most natural habitats tend to fluctuate frequently over the course of a full day, there is ostensibly a fundamental contradiction between the need of the photosynthetic system to have a high gain (i.e. to be sensitive to variations in external input parameters) and the need to produce a relatively continuous supply of sucrose for export to developing tissues. The mechanisms by which linear metabolic systems deal with this conflict are discussed by Savageau (209) and broadly involve the operation of feedback loops that can partially desensitize the rate of product supply from changes in the substrate levels (253, 284). In other words, there is a trade-off between efficient use of substrates and stable substrate supply.

   The photosynthetic system appears, however, to have adapted efficiently to the conflicting demands placed upon it. It is responsive to fluctuations in the range of light intensities and CO_2 partial pressures likely to be encountered in the natural habitat, and the supply of sucrose over the course of a full day generally remains uniform (e.g. 70). The main reason for these properties is that rather than being a simple linear sequence, the photosynthetic system has two major branch points which terminate in the relatively large transient storage pools of starch and sucrose. By channeling assimilated carbon into these pools, the plant can insulate the rate of sucrose supply to developing tissues from the changes in the rate of photosynthesis that occur during the course of a day and night.

Nevertheless, the system cannot operate with maximum gain by simply eliminating feedback sequences because, when the rate of energy input exceeds the maximum capacity of the PCR-cycle enzymes to sustain the corresponding flux, a feedback mechanism must operate to facilitate the dissipation of excess energy and thus to maintain the condition for a steady state and to protect against photoinhibition (cf 17, 143). The most energetically efficient feedback mechanisms would therefore be largely inoperative over a large spectrum of light intensities and only switch on when the maximum capacity of the enzymes of the PCR cycle and adjoining pathways has been closely approached.

Biological responses are rarely discontinuous, but a "metabolic switch" can be approximated by strongly sigmoidal kinetics. Such enzyme responses have been termed "ultra-sensitive" because there is an appreciable range of substrate or effector concentrations over which there is little or no change in the catalytic velocity, but once the "threshold" has been reached, there is a rapid increase to almost the maximum velocity (141). There is no evidence of cooperative kinetics in the PCR cycle extreme enough to constitute a metabolic switch. However, we have discussed how parallel variation in the levels of several effectors of a single enzyme in response to irradiance, for example, can result in highly sigmoid substrate response characteristics. Such responses appear to have the capacity to reduce significantly the feedback strength of regulatory sequences originating from the PCR cycle and the sucrose synthetic pathway.

The subcellular compartmentation of reactions and the resulting conservation of stromal and cytosolic phosphate play an important role in these and other regulatory responses. For example, we have developed the argument that Pi mediates at least part of the feedback sequence influencing the input of quantum energy, and that these initial steps of photosynthesis may be attenuated when the stromal concentration of Pi declines from its optimal range. While the kinetics of this mechanism are not yet clear, it is apparent that the "ultra-sensitive" responses of certain stromal enzymes must play a major role in this mechanism through their influence on the concentration of esterified intermediates and, consequently, Pi. For example, the sigmoid response of Rubisco to RuBP levels in vivo is such that the photosynthetic flux may change over a wide range without an appreciable change in RuBP (see Figure 2). Among other things, this would tend to maintain a stable—and perhaps near optimal—level of stromal Pi. But as the flux approaches the maximum capacity of Rubisco, the level of RuBP must increase by a large amount to attain an additional increment of flux. The level of stromal Pi would thus become correspondingly sensitive to changes in flux in this range and initiate feedback inhibition of the photoacts. The regulatory mechanisms underlying the sigmoid responses of Rubisco and other enzymes of the photosynthetic system are therefore important in affecting the extent to which feedback mechanisms may detract from the efficient use of absorbed light energy when light is a factor limiting the rate of photosynthesis.

Amplified responses of enzyme activity may also result from the subcellular compartmentation of reactions and the resulting conservation of stromal and cytosolic phosphate since several photosynthetic enzymes are both activated by phosphate esters and inhibited by inorganic phosphate. The best example of this is the highly sensitive regulatory sequence governing assimilate partition. The three critical enzymes involved in this process (namely, ADP-glucose pyrophosphorylase and cytosolic and stromal FBPase) are all, either directly or indirectly, activated by esterified phosphates and inhibited by Pi. Opposite changes in these effectors permit assimilate partition to be altered—for example, by feedback from the sucrose pathway—almost independently of CO_2 assimilation, despite the capacity of this sequence to affect the electron transport system and possibly the activation state of Rubisco.

The participation of futile cycles in these amplified control mechanisms is also noteworthy. Regulation of cytosolic FBPase through the futile F6P-F26P cycle is the best-documented of these mechanisms (238), but it is clearly not the only such energy consuming cycle. The thioredoxin system operates against a constant decay to O_2, the synthesis and degradation of CA-1-P most probably consume energy, and it is proposed that the activation of Rubisco by "Rubisco activase" involves a direct ATP-dependent reaction acting in opposition to the passive decay of activated Rubisco. These cycles will have metabolic "costs" that presumably must be offset by improved energetic efficiency of the photosynthetic process. The superimposition of these complex regulatory processes upon existing allosteric mechanisms is one of the remarkable features of the photosynthetic system. Studies of the characteristics of these mechanisms, which may serve principally to modify the feedback processes in photosynthesis, seem important to understanding the high energetic efficiency of the system.

Studies of the regulation of photosynthesis have tended to focus at the enzyme level, and this has helped us enormously to understand the responses of enzymes to their "local" conditions. Recent progress in understanding the factors controlling Rubisco under conditions of steady-state photosynthesis in vivo have been given considerable attention in this review. Nevertheless, these studies by themselves provide no understanding of the rationale, in terms of the overall performance of the system, for the observed responses. This can only come from viewing these responses in the larger context of the system of photosynthetic reactions.

The structure of the photosynthetic system is remarkable because the two major input pathways—one involving the transduction of quantum energy and the other the supply of electron acceptors principally by carboxylation or oxygenation of RuBP—are essentially irreversible. These pathways must therefore be subject to a feedback regulation that maintains the stoichiometric balance required for steady-state photosynthesis. We have examined two such regulatory sequences as well as a third that permits the distal reactions involved in sucrose synthesis to feed back on both of these initial steps of photosynthesis. Our knowledge of these sequences is still at a rudimentary level, and much work is still required—particularly regarding the molecular details of the mechanisms that regulate Rubisco and the electron transport system. Other key regulatory problems have not been addressed in detail here. One of the most important involves the mechanisms that maintain the pools of PCR cycle intermediates in the face of rapid fluctuations in the rate of CO_2 fixations.

Finally, we have only touched on the problems associated with understanding the basis for differences in photosynthetic rate. This must be dealt with if we are to develop rational strategies for improving photosynthetic performance by plants or to examine questions of optimization of the photosynthetic system to different environments. It is our opinion that this requires a quantitative approach, whereas qualitative or comparative approaches have been most frequently used. The methods of control analysis (131) and metabolic systems analysis (210) have been widely used in other areas of biology and appear to be well suited for application to quantitative analysis of the marginal control of photosynthesis by co-limiting factors.

Literature Cited

1. Aflalo, C., Shavit, N. 1983. Steady-state kinetics of photophosphorylation: limited access of nucleotides to the active site on the ATP synthetase. *FEBS Lett.* 154:175–79
2. Ainslie, G. R. Jr., Shill, J. P., Neet, K. E. 1972. Transients and cooperativity. A slow transition model for relating transients and cooperative kinetics of enzymes. *J. Biol. Chem.* 247:7058–96
3. Amir, J., Preiss, J. 1982. Kinetic characterization of spinach leaf sucrose phosphate synthase. *Plant Physiol.* 69:1027–30
4. Anderson, L. E. 1986. Light/dark modulation of enzyme activity in plants. *Adv. Bot. Res.* 12:1–46

5. Andrews, T. J., Lorimer, G. H. 1987. Rubisco: structure, mechanisms, and prospects for improvement. In *The Biochemistry of Plants,* ed. M. D. Hatch, 10:131–218. New York: Academic
6. Ashton, A. R. 1982. A role for ribulose-1,5-bisphosphate carboxylase as a metabolite buffer. *FEBS Lett.* 145:1–7
7. Atkins, C. A., Patterson, B. D., Graham, D. 1972. Plant carbonic anhydrases. *Plant Physiol.* 50:214–23
8. Badger, M. R. 1985. Photosynthetic oxygen exchange. *Ann. Rev. Plant Physiol.* 36:27–53
9. Badger, M. R., Collatz, G. J. 1977. Studies on the kinetic mechanism of ribulose 1,5-bisphosphate carboxylase

and oxygenase reactions, with particular reference to the effect of temperature on kinetic parameters. *Carnegie Inst. Wash. Yearb.* 76:355–61

10. Badger, M. R., Lorimer, G. H. 1981. Interaction of sugar phosphates with the catalytic site of ribulose 1,5-bisphosphate carboxylase. *Biochemistry* 20:2219–25

11. Badger, M. R., Sharkey, T. D., von Caemmerer, S. 1984. The relationship between steady state gas exchange of bean leaves and the levels of carbon-reduction-cycle intermediates. *Planta* 160:305–13

12. Ball, J. T. 1987. Calculations related to gas exchange. In *Stomatal Function*, ed. E. Zeiger, G. D. Farquhar, I. R. Cowan, pp. 445–76. Stanford: Stanford Univ. Press

13. Ball, J. T., Woodrow, I. E., Berry, J. A. 1986. A model predicting stomatal conductance and its contribution to the control of photosynthesis under different environmental conditions. In *Progress in Photosynthesis Research,* ed. J. Biggins, 4:221–24. Dordrecht: Martinus Nijhoff

14. Barcena, J. A. 1983. Differential effect of Ca^{2+} and Mn^{2+} on activation and catalysis of spinach ribulose-1,5-bisphosphate carboxylase/oxygenase. *Biochem. Int.* 7:755–60

15. Bassham, J. A., Krause, G. H. 1969. Free energy changes and metabolic regulation in steady state photosynthetic carbon reduction. *Biochim. Biophys. Acta* 189:207–21

16. Berry, J. A., Lorimer, G. H., Pierce, J., Seemann, J. R., Meek, J., Freas, S. 1987. Isolation, identification, and synthesis of 2-carboxyarabinitol-1-phosphate, a diurnal regulator of ribulose-bisphosphate carboxylase activity. *Proc. Natl. Acad. Sci. USA* 84:734–38

17. Björkman, O. E. 1981. Responses of higher plants to quantum flux. In *Encylopedia of Plant Physiology (NS),* ed. O. L. Lange, P. S. Nobel, C. B. Osmond, H. Zeigler, 12A:57–107. New York: Springer-Verlag

18. Björkman, O. E., Gauhl, E., Nobs, M. A. 1970. Comparative studies of *Atriplex* species with and without β-carboxylation photosynthesis and their first generation hybrids. *Carnegie Inst. Wash. Yearb.* 68:620–33

19. Bowes, G., Ogren, W. L. 1972. Oxygen inhibition and other properties of soybean ribulose 1,5-diphosphate carboxylase. *J. Biol. Chem.* 247:2171–76

20. Bowes, G., Ogren, W. L., Hageman, R.

H. 1971. Phosphoglycolate production catalyzed by ribulose diphosphate carboxylase. *Biochem. Biophys. Res. Commun.* 45:716–22

21. Brooks, A. 1986. Effects of phosphorus nutrition on ribulose-1,5-bisphosphate carboxylase activation, photosynthetic quantum yield and amounts of some Calvin-cycle metabolites in spinach leaves. *Aust. J. Plant Physiol.* 13:221–37

22. Brooks, A., Farquhar, G. D. 1985. Effects of temperature on the CO_2/O_2 specificity of ribulose-1,5-bisphosphate carboxylase/oxygenase and the rate of respiration in the light. *Planta* 165:397–406

23. Buchanan, B. B. 1980. Role of light in the regulation of chloroplast enzymes. *Ann. Rev. Plant Physiol.* 31:341–74

24. Bücher, T., Rüssmann, W. 1963. Gleichgewicht und Ungleichgewicht im System der Glykolyse. *Angew. Chem.* 19:881–93

25. Burns, J. A. et al. 1985. Control analysis of metabolic systems. *Trends Biochem. Sci.* 10:16

26. Canvin, D. T. 1978. Photorespiration and the effect of oxygen on photosynthesis. In *Photosynthetic Carbon Assimilation,* ed. H. Siegelman, G. Hind, pp. 61–76. New York: Plenum

27. Charles, S. A., Halliwell, B. 1980. Properties of freshly purified and thiol treated spinach chloroplast fructose bisphosphatase. *Biochem. J.* 185:689–93

28. Chatterton, N. J., Silvius, J. E. 1979. Photosynthate partitioning into starch in soybean leaves. *Plant Physiol.* 64:749–53

29. Chen-She, S. H., Lewis, D. H., Walker, D. A. 1975. Stimulation of photosynthetic starch formation by sequestration of cytoplasmic orthophosphate. *New Phytol.* 74:383–92

30. Christeller, J. T. 1981. The effects of bivalent cations on ribulose bisphosphate carboxylase/oxygenase. *Biochem. J.* 193:839–44

31. Chu, D. K., Bassham, J. A. 1975. Regulation of ribulose 1,5-diphosphate carboxylase by substrates and other metabolites. *Plant Physiol.* 55:720–26

32. Clausen, W., Lenz, F. 1983. Investigations over the relation between the activity of sucrose phosphate synthase and the net photosynthesis rates as well as the sucrose and starch contents of leaves of *Solanum melongena* L. *Z. Pflanzenphysiol.* 109:459–68

33. Cockburn, W., Baldry, C. W., Walker,

D. A. 1967. Some effects of inorganic phosphate on O_2 evolution by isolated chloroplasts. *Biochim. Biophys. Acta* 143:614–24

34. Collatz, G. J. 1978. The interaction between photosynthesis and ribulose-P_2 concentration—effects of light, CO_2 and O_2. *Carnegie Inst. Wash. Yearb.* 77:248–51

35. Collatz, G. J., Badger, M. R., Smith, C., Berry, J. A. 1979. A radioimmune assay for RuP_2 carboxylase protein. *Carnegie Inst. Wash. Yearb.* 78:171–75

36. Cornic, G., Louason, G. 1980. The effects of O_2 on net photosynthesis at low temperature (5°C). *Plant, Cell Environ.* 3:147–57

37. Cowan, I. R. 1986. Economics of carbon fixation in higher plants. In *On the Economy of Plant Form and Function*, ed. T. J. Givnish. Cambridge: Cambridge Univ. Press

38. Crabtree, B., Newsholme, E. A. 1985. A quantitative approach to metabolic control. *Curr. Top. Cell Regul.* 25:21–76

39. Crabtree, B., Newsholme, E. A. 1987. A systematic approach to describing and analysing metabolic control systems. *Trends Biochem. Sci.* 12:4–12

40. Crofts, A. R., Wraight, C. A. 1983. The electrochemical domain of photosynthesis. *Biochim. Biophys. Acta* 726:149–85

41. Cséke, C., Buchanan, B. B. 1983. An enzyme synthesizing fructose 2,6-bisphosphate occurs in leaves and is regulated by metabolic effectors. *FEBS Lett.* 155:139–42

42. Cséke, C., Buchanan, B. B. 1986. Regulation of the formation and utilization of photosynthate in leaves. *Biochim. Biophys. Acta* 853:43–63

43. Cséke, C., Stitt, M., Balogh, A., Buchanan, B. B. 1983. A product regulated fructose 2, 6-bisphosphatase occurs in green leaves. *FEBS Lett.* 162:103–6

44. Cséke, C., Weeden, N. F., Buchanan, B. B., Uyeda, K. 1982. A special fructose bisphosphate functions as a cytoplasmic regulatory metabolite in green leaves. *Proc. Natl. Acad. Sci. USA* 79:4322–26

45. Dietz, K.-J. 1987. Control functions of hexosemonophosphate isomerase and phosphoglucomutase in starch synthesis of leaves. In *Progress in Photosynthesis Research*, ed. J. Biggins, 3:329–32. Dordrecht: Martinus-Nijhoff

46. Dietz, K.-J., Heber, U. 1984. Rate limiting factors in leaf photosynthesis.

1. Carbon fluxes in the Calvin Cycle. *Biochim. Biophys. Acta* 767:432–43

47. Dietz, K.-J., Heber, U. 1986. Light and CO_2 limitation of photosynthesis and states of the reactions regenerating ribulose 1,5-bisphosphate or reducing 3-phosphoglycerate. *Biochim. Biophys. Acta* 848:392–401

48. Dietz, K.-J., Schreiber, U., Heber, U. 1985. The relationship between the redox state of Q_A and photosynthesis in leaves at various carbon-dioxide, oxygen and light regimes. *Planta* 166:219–26

49. Doehlert, D. C., Huber, S. C. 1984. Phosphate inhibition of spinach leaf sucrose phosphate synthase as affected by glucose 6-phosphate and phosphoglucose isomerase. *Plant Physiol.* 76:250–53

50. Duysens, L. N. M. 1979. Transfer and trapping of excitation energy in photosystem II. In *Chlorophyll Organization and Energy Transfer in Photosynthesis: Ciba Foundation Symposium 61 (NS)*, ed. G. Wolstenholme, D. Fitzsimons, pp. 323–340. New York: Exerpta Medica

51. Duysens, L. N. M., Sweers, H. E. 1963. Mechanism of two photochemical reactions in algae as studied by means of fluorescence. In *Microalgae and Photosynthetic Bacteria*, ed. Japanese Soc. Plant Physiol., pp. 353–72. Tokyo: Univ. Tokyo Press

52. Edwards, G. E., Walker, D. A. 1983. C_3, C_4: *Mechanisms and Cellular and Environmental Regulation of Photosynthesis*. Berkeley/Los Angeles: Univ. Calif. Press

53. Enser, U., Heber, U. 1980. Metabolic regulation by pH gradients. *Biochim. Biophys. Acta* 592:577–91

54. Evans, J. R. 1983. *Photosynthesis and nitrogen partitioning in leaves of* Triticum aestivum *and related species*. PhD thesis. Australian National University, Canberra

55. Evans, J. R. 1983. Nitrogen and photosynthesis in the flag leaf of wheat (*Triticum aestivum* L.). *Plant Physiol.* 72:297–302

56. Evans, J. R. 1986. The relationship between carbon-dioxide-limited photosynthetic rate and ribulose-1,5-bisphosphate-carboxylase content in two nuclear-cytoplasm, substitution lines of wheat, and the coordination of ribulose-bisphosphate-carboxylation and electron-transport capacities. *Planta* 167:351–58

57. Evans, J. R. 1987. The relationship between electron transport components and

photosynthetic capacity in pea leaves grown at different irradiances. *Aust. J. Plant Physiol.* 14:157–70

58. Evans, J. R., Austin, R. B. 1986. The specific activity of ribulose-1,5-bisphosphate carboxylase in relation to genotype in wheat. *Planta* 167:344–50

59. Evans, J. R., Seemann, J. R. 1984. Differences between wheat genotyoes in specific activity of ribulose-1,5-bisphosphate carboxylase and the relationship to photosynthesis. *Plant Physiol.* 74:759–65

60. Evans, J. R., Sharkey, T. D., Berry, J. A., Farquhar, G. D. 1986. Carbon isotope discrimination measured concurrently with gas exchange to investigate CO_2 diffusion in leaves of higher plants. *Aust. J. Plant Physiol.* 13:281–92

61. Farquhar, G. D. 1979. Models describing the kinetics of ribulose bisphosphate carboxylase/oxygenase. *Arch. Biochem. Biophys.* 193:456–68

62. Farquhar, G. D., Sharkey, T. D. 1982. Stomatal conductance and photosynthesis. *Ann. Rev. Plant Physiol.* 33:317–45

63. Farquhar, G. D., von Caemmerer, S. 1982. Modelling of photosynthetic response to environmental conditions. In *Encylopedia of Plant Physiology (NS)*, ed. O. L. Lange, P. S. Nobel, C. B. Osmond, H. Ziegler. 12B:549–87. New York: Springer-Verlag

64. Farquhar, G. D., von Caemmerer, S., Berry, J. A. 1980. A biochemical model of photosynthetic CO_2 fixation in leaves of C_3 species. *Planta* 149:78–90

65. Farrar, S. C., Farrar, J. F. 1986. Compartmentation and fluxes of sucrose in intact leaf blades of barley. *New Phytol.* 103:645–57

66. Fell, D. A., Sauro, H. M. 1985. Metabolic control and its analysis: additional relationships between elasticities and control coefficients. *Eur. J. Biochem.* 148:555–61

67. Field, C., Mooney, H. A. 1986. The photosynthesis-nitrogen relationship in wild plants. See Ref. 37, pp. 25–55

68. Fliege, R., Flügge, U. I., Werdan, K., Heldt, H. W. 1978. Specific transport of inorganic phosphate, 3-phosphoglycerate and triose phosphates across the inner membrane of the envelope of spinach chloroplasts. *Biochim. Biophys. Acta* 502:232–47

69. Flügge, U. I., Heldt, H. W. 1984. The phosphate—triose phosphate—phosphoglycerate translocator of the chloroplast. *Trends Biochem. Sci.* 9:530–33

70. Fondy, B. R., Geiger, D. R. 1980. Effect of rapid changes in sink source ratio on export and distribution of products of photosynthesis in leaves of *Beta vulgaris* L. and *Phaseolus vulgaris* L. *Plant Physiol.* 66:945–49

71. Fondy, B. R., Geiger, D. R. 1982. Diurnal pattern on translocation and carbohydrate metabolism in source leaves of *Beta vulgaris* L. *Plant Physiol.* 70:671–76

72. Foyer, C. H., Rowell, J., Walker, D. A. 1983. The effect of sucrose on the rate of *de novo* sucrose biosynthesis in leaf protoplasts from spinach, wheat and barley. *Arch. Biochem. Biophys.* 220:232–38

73. Foyer, C., Spencer, C. 1986. The relationship between phosphate status and photosynthesis in leaves. *Planta* 167:369–75

74. Foyer, C. H., Walker, D. A., Spencer, C., Mann, B. 1981. Observations on the phosphate status and intracellular pH of intact cells, protoplasts and chloroplasts from photosynthetic tissue using phosphorus-31 nuclear magnetic resonance. *Biochem. J.* 202:429–34

75. Frieden, C. 1970. Kinetic aspects of regulation of metabolic processes. The hysteretic enzyme concept. *J. Biol. Chem.* 245:5788–99

76. Furbank, R. T., Foyer, C. H., Walker, D. A. 1986. Inhibition of photophosphorylation by ribulose-1,5-bisphosphate carboxylase. *Biochim. Biophys. Acta* 852:46–54

77. Furbank, R. T., Foyer, C. H., Walker, D. A. 1987. Interactions between ribulose-1,5-bisphosphate carboxylase and stromal metabolites. II. Corroboration of the role of this enzyme as a metabolite buffer. *Biochim. Biophys. Acta* 894:165–73

78. Furbank, R. T., Foyer, C., Walker, D. A. 1988. Regulation of photosynthesis in isolated chloroplasts during orthophosphate limitation. *Biochim. Biophys. Acta.* In press

79. Furbank, R. T., Horton, P. 1987. Regulation of photosynthesis in isolated barley protoplasts: the contribution of cyclic photophosphorylation. *Biochim. Biophys. Acta* 894:332–38

80. Geiger, D. R. 1975. Phloem loading in source leaves. In *Transport and Transfer Processes in Plants*, ed. I. F. Wardlaw, J. B. Passioura, pp. 167–83. New York: Academic

81. Geiger, D. R., Ploeger, B. J., Fox, T. C., Fondy, B. R. 1983. Sources of sucrose translocated from illuminated sugar beet source leaves. *Plant Physiol.* 72:964–70

82. Geiger, D. R., Swanson, C. A. 1965. Evaluation of selected parameters in a sugar beet translocation system. *Plant Physiol.* 40:942–47

83. Gerhardt, R., Stitt, M., Heldt, H. W. 1987. Subcellular metabolite levels in spinach leaves. *Plant Physiol.* 83:399–407

84. Giersch, C. 1977. A kinetic model for translocators in the chloroplast envelope as an element of computer simulation of the dark reactions of photosynthesis. *Z. Naturforsch. Teil C* 32:263–70

85. Giersch, C., Heber, U., Kaiser, G., Walker, D. A., Robinson, S. P. 1980. Intercellular metabolite gradients and flow of carbon during photosynthesis of leaf protoplasts. *Arch. Bioch. Biophys.* 205:246–59

86. Gifford, R. M., Evans, L. T. 1981. Photosynthesis, carbon partitioning and yield. *Ann. Rev. Plant Physiol.* 32:485–509

87. Goldbeter, A., Koshland, D. E. 1987. Energy expenditure in the control of biochemical systems by covalent modification. *J. Biol. Chem.* 262:4460–71

88. Gordon, A. J., Ryle, G. J. A., Mitchell, D. F., Powell, C. E. 1980. The dynamics of carbon supply from leaves of barley plants grown in long and short days. *J. Exp. Bot.* 33:241–50

89. Gutteridge, S., Parry, M. A. J., Keys, A. J., Burton, S., Mudd, A., et al. 1986. A nocturnal inhibitor of carboxylation in leaves. *Nature* 324:274–76

90. Hall, N. P., Keys, A. J. 1983. Temperature dependence of enzymic carboxylation and oxygenation of ribulose 1,5-bisphosphate in relation to effects of temperature on photosynthesis. *Plant Physiol.* 72:945–48

91. Hall, N. P., Pierce, J., Tolbert, N. E. 1981. Formation of a carboxyarabinitol bisphosphate complex with ribulose bisphosphate carboxylase/oxygenase and the theoretical specific activity of the enzyme. *Arch. Biochem. Biophys.* 212:115–19

92. Harbron, S., Foyer, C. H., Walker, D. A. 1981. the purification and properties of sucrose phosphate synthase from spinach leaves: the involvement of this enzyme and fructose bisphosphatase in the regulation of sucrose biosynthesis. *Arch. Biochem. Biophys.* 212:237–46

93. Harley, P. C., Weber, J. A., Gates, D. M. 1985. Interactive effects of light, leaf temperature, CO_2 and O_2 on photosynthesis in soybeans. *Planta* 165:249–63

94. Harris, G. C., Cheesbrough, J. K., Walker, D. A. 1983. Effects of mannose on photosynthetic gas exchange in spinach leaf discs. *Plant Physiol.* 71:108–11

95. Hatch, A. L., Jensen, R. G. 1980. Regulation of ribulose 1,5-bisphosphate carboxylase from tobacco: changes in pH response and affinity for CO_2 and Mg^{2+} induced by chloroplast intermediates. *Arch. Biochem. Biophys.* 205:587–91

96. Hawker, J. S., Smith, G. M. 1984. The occurrence of sucrose phosphatase (EC3.1.3.24) in vascular and nonvascular plants. *Phytochemistry* 23:245–49

97. Heber, U., Heldt, H. W. 1981. The chloroplast envelope: structure function and role in leaf metabolism. *Ann. Rev. Plant Physiol.* 32:139–68

98. Heber, U., Neimanis, S., Dietz, K.-.J., Viil, J. 1986. Assimilatory power as a driving force in photosynthesis. *Biochim. Biophys. Acta* 852:144–55

99. Heber, U., Takahama, U., Neimanis, S., Shimizu-Takahama, M. 1982. Transport as the basis of the Kok effect. Levels of some photosynthetic intermediates and activation of light-regulated enzymes during photosynthesis of chloroplasts and green leaf protoplasts. *Biochim. Biophys. Acta* 679:287–99

100. Heinrich, R., Rappoport, T. A. 1974. A linear steady state treatment of enzymatic chains. General properties, control and effector strength. *Eur. J. Biochem.* 42:89–95

101. Heldt, H. W. 1979. Light-dependent changes in stromal H^+ and Mg^{2+} concentrations controlling CO_2 fixation. In *Encyclopedia of Plant Physiology (NS)*, ed. M. Gibbs, E. Latzko, 6:202–8. New York: Springer-Verlag

102. Heldt, H. W., Chon, C. J., Lorimer, G. H. 1978. Phosphate requirement for the light activation of ribulose 1,5-bisphosphate carboxylase in intact spinach chloroplasts. *FEBS Lett.* 92:234–38

103. Heldt, H. W., Chon, C. J., Maronde, D., Herold, A., Stankovic, Z. S., et al. 1977. The role of orthophosphate and other factors in the regulation of starch formation in leaves and isolated chloroplasts. *Plant Physiol.* 59:1146–55

104. Heldt, H. W., Laing, W., Lorimer, G. H., Stitt, M., Wirtz, W. 1981. On the regulation of CO_2 fixation by light. In *Photosynthesis IV: Regulation of Carbon Metabolism (Proc. 5th Int. Cong. Photosynth.)*, ed. G. Akoyunoglou, pp. 213–26. Philadelphia: Balaban

105. Heldt, H. W., Stitt, M. 1986. The regu-

lation of sucrose synthesis in leaves. In *Progress in Photosynthesis Research,* ed. J. Biggins, 3:675–84. Dordrecht: Martinus Nijhoff

106. Hendrix, D. L., Huber, S. C. 1986. Diurnal fluctuations in cotton leaf carbon export, carbohydrate content and sucrose synthesizing enzymes. *Plant Physiol.* 81:584–86

107. Herold, A. 1980. Regulation of photosynthesis by sink activity—the missing link. *New Phytol.* 86:131–44

108. Herold, A., Lewis, D. H. 1977. Mannose and green plants. Occurrence, physiology and metabolism and use as a tool to study orthophosphate. *New Phytol.* 79:1–40

109. Herold, A., Lewis, D. H., Walker, D. A. 1976. Sequestration of cytoplasmic orthophosphate by mannose and its differential effect on photosynthetic starch synthesis in C_3 and C_4 species. *New Phytol.* 76:397–407

110. Herold, A., McGee, E. E. M., Lewis, D. H. 1980. The effect of orthophosphate concentration and exogenously supplied sugars on the distribution of newly fixed carbon in sugar beet leaf discs. *New Phytol.* 85:1–13

111. Herzog, B., Stitt, M., Heldt, H. W. 1984. Control of photosynthetic sucrose synthesis by fructose 2,6-bisphosphate. III. Properties of the cytosolic fructose 1,6-bisphosphatase. *Plant Physiol.* 75:561–65

112. Higgins, J. 1965. Dynamics and control in cellular reactions. In *Control of Energy Metabolism,* ed. B. Chance, R. K. Estabrook, J. R. Williamson, pp. 13–46. New York: Academic

113. Hills, M. J. 1986. Photosynthetic characteristics of mesophyll cells isolated from cladophylls of *Asparagus officinalis* L. *Planta* 169:38–45

114. Horton, P. 1985. Interactions between electron transfer and carbon assimilation. In *Photosynthetic Mechanisms and the Environment,* ed. J. Barber, N. R. Baker, pp. 137–87. New York: Elsevier

115. Huber, S. C. 1981. Interspecific variation in activities and regulation of leaf sucrose phosphate synthase. *Z. Pflanzenphysiol.* 102:443–50

116. Huber, S. C. 1983. Role of sucrose phosphate synthase in partitioning of carbon in leaves. *Plant Physiol.* 71:818–21

117. Huber, S. C., Bickett, D. M. 1984. Evidence for control of carbon partitioning by fructose 2,6-bisphosphate in spinach leaves. *Plant Physiol.* 74:445–47

118. Huber, S. C., Kerr, P. S., Kalt-Torres, W. 1985. Regulation of sucrose forma-

tion and movement. In *Regulation of Carbon Partitioning in Photosynthetic Tissue, 8th Symp. Plant Physiology, Riverside, Calif.,* ed. R. L. Heath, J. Preiss, pp. 199–214. Rockville, Md: Am. Soc. Plant Physiol.

119. Huber, S. C., Kerr, P. S., Rufty, T. W. 1985. Diurnal changes in sucrose phosphate synthase in leaves. *Physiol. Plant.* 64:81–87

120. Jacobson, B. S., Fong, F., Heath, R. L. 1975. Carbonic anhydrase of spinach. *Plant Physiol.* 55:474–86

121. Jensen, R. G., Bahr, J. T. 1977. Ribulose 1,5- bisphosphate carboxylase/oxygenase. *Ann. Rev. Plant Physiol.* 28:379–408

122. Jolliffe, P. A., Tregunna, E. B. 1968. Effects of temperature, CO_2 concentration, and light intensity on oxygen inhibition of photosynthesis in wheat leaves. *Plant Physiol.* 43:902–6

123. Jolliffe, P. A., Tregunna, E. B. 1973. Environmental regulation of the oxygen effect on apparent photosynthesis in wheat. *Can. J. Bot.* 51:841–53

124. Jones, H. G. 1985. Partitioning of stomatal and non-stomatal limitations to photosynthesis. *Plant, Cell Environ.* 8:95–104

125. Jordan, D. B., Chollet, R., Ogren, W. L. 1983. Binding of phosphorylated effectors by active and inactive forms of ribulose 1,5-bisphosphate carboxylase. *Biochemistry* 22:3410–18

126. Jordan, D. B., Ogren, W. L. 1984. A sensitive assay procedure for simultaneous determination of ribulose-1,5-bisphosphate carboxylase and oxygenase activities. *Plant Physiol.* 54:237–45

127. Jordan, D. B., Ogren, W. L. 1983. Species variation in kinetics properties of ribulose 1,5-bisphosphte carboxylase/oxygenase. *Arch. Biochem. Biophys.* 227:425–33

128. Jordan, D. B., Ogren, W. L. 1984. The CO_2/O_2 specificity of ribulose-1,5-bisphosphate carboxylase/oxygenase. *Planta* 161:308–13

129. Joseph, M. C., Randall, D. D. 1981. Photosynthesis in polyploid tall fescue. *Plant Physiol.* 68:894–98

130. Junge, W., Schoenknecht, G., Lill, H. 1987. Complete tracking of proton flow mediated by CF_0-CF_1 and by CF_0. See Ref. 45, pp. 133–40

131. Kacser, H., Burns, J. A. 1973. The control of flux. *Symp. Soc. Exp. Biol.* 27:65–104

132. Kaiser, G., Martinoia, E., Wiemken, A. 1982. Rapid appearance of photosynthetic products in the vacuoles isolated from barley mesophyll protoplasts by a new

fast method. *Z. Pflanzenphysiol.* 107: 103–13

133. Kalt-Torres, W., Kerr, P. S., Usuda, H., Huber, S. C. 1987. Diurnal changes in maize leaf photosynthesis. I. Carbon exchange rate, assimilate export rate and enzyme activities. *Plant Physiol.* 83: 283–88

134. Keck, K. W., Ogren, W. L. 1976. Differential oxygen response of photosynthesis in soybean and *Panicum miliodes*. *Plant Physiol.* 48:552–55

135. Kelly, G. J., Zimmermann, G., Latzko, E. 1976. Light induced activation of fructose 1,6-bisphosphatase in isolated intact chloroplasts. *Biochem. Biophys. Res. Commun.* 70:193–99

136. Kerr, P. S., Huber, S. C. 1987. Coordinate control of sucrose formation in soybean leaves by sucrose-phosphate synthase and fructose 2,6-bisphosphate. *Planta* 170:197–204

137. Keys, A. J. 1986. Rubisco: its role in photorespiration. *Proc. R. Soc. London B* 313:325–36

138. King, R. W., Wardlaw, I. F., Evans, L. T. 1967. Effect of assimilate utilization on photosynthetic rate in wheat. *Planta* 77:261–76

139. Kobza, J., Edwards, G. E. 1987. Influences of leaf temperature on photosynthetic carbon metabolism in wheat. *Plant Physiol.* 83:69–74

140. Kobza, J., Seemann, J. R. 1988. Mechanism for the light regulation of ribulose 1,5-bisphosphate carboxylase activity and photosynthesis in intact leaves. *Proc. Natl. Acad. Sci. USA.* In press

141. Koshland, D. E. 1987. Switches, thresholds and ultrasensitivity. *Trends Biochem. Sci.* 12:225–29

142. Kraayenhof, R. 1969. 'State 3–State 4 transition' and phosphate potential in 'Class 1' spinach chloroplasts. *Biochim Biophys. Acta* 180:213–15

143. Krause, G. H., Laasch, H. 1987. Energy-dependent chlorophyll fluorescence quenching in chloroplasts correlated with quantum yield of photosynthesis. *Z. Naturforsch. Teil C* 42:581–84

144. Krebs, H. A. 1957. Control of metabolic processes. *Endeavour* 16:125–32

145. Kreimer, G., Surek, B., Heimann, K., Burchert, M., Lukow, L., et al. 1987. Calcium metabolism in chloroplasts and protoplasts. See ref. 45, pp. 345–57

146. Laing, W. A., Christeller, J. T. 1976. A model for the kinetics of activation and catalysis of ribulose 1,5-bisphosphate carboxylase. *Biochem. J.* 159:563–70

147. Laing, W. A., Ogren, W. L.; Hageman, R. H. 1974. Regulation of soybean net photosynthetic CO_2 fixation by the interaction of CO_2, O_2, and ribulose-1,5-bisphosphate carboxylase. *Plant Physiol.* 54:678–85

148. Laing, W. A., Stitt, M., Heldt, H. W. 1981. Changes in the activity of ribulose-phosphate kinase and fructose- and sedoheptulose-bisphosphatase in chloroplasts. *Biochim. Biophys. Acta* 637:348–59

149. Laisk, A. 1977. *Kinetics of Photosynthesis and Photorespiration in* C_3 *Plants*. Moscow: Nauka

150. Laisk, A., Oya, V. M. 1974. Photosynthesis of leaves subjected to brief impulses of CO_2. *Soviet Plant Physiol.* 21:928–35

151. Laisk, A., Walker, D. A. 1986. Control of phosphate turnover as a rate limiting factor and possible cause of oscillations in photosynthesis: a mathematical model. *Proc. Roy. Soc. London* Ser. B 227:281–302

152. Leegood, R. C. 1985. Regulation of photosynthetic CO_2-pathway enzymes by light and other factors. *Photosynth. Res.* 6:247–59

153. Leegood, R. C., Furbank, R. T. 1986. Stimulation of photosynthesis by 2% oxygen at low temperatures is restored by phosphate. *Planta* 168:84–93

154. Leegood, R. C., Walker, D. A. 1982. Regulation of fructose-1,6-bisphosphatase activity in leaves. *Planta* 156:449–56

155. Leegood, R. C., Foyer, C. H., Walker, D. A. 1985. Regulation of the Benson-Calvin cycle. In *Photosynthetic Mechanisms and the Environment*, ed. J. Barber, and N. R. Baker, pp. 191–258. New York: Elsevier

156. Lilley, R. McC., Chon, C. J., Mosbach, A., Heldt, H. W. 1977. The distribution of metabolites between spinach chloroplasts and medium during photosynthesis in vivo. *Biochim. Biophys. Acta* 460:259–72

157. Lilley, R. McC., Walker, D. A. 1975. Carbon dioxide assimilation by leaves, isolated chloroplasts and ribulose-1,5-bisphosphate carboxylase from spinach. *Plant Physiol.* 55:1087–92

158. Lorimer, G. H. 1981. The carboxylation and oxygenation of ribulose 1,5-bisphosphate: the primary events in photosynthesis and photorespiration. *Ann. Rev. Plant Physiol.* 32:349–83

159. Lorimer, G. H., Badger, M. R., Andrews, T. J. 1976. The activation of ribulose-1,5-bisphosphate carboxylase by carbon dioxide and magnesium ions. Equilibria, kinetics, a suggested mechanism, and physiological implications. *Biochemistry* 15:529–36

160. Mächler, F., Nösberger, J. 1980. Regulation of ribulose bisphosphatase carboxylase activity in intact wheat leaves by light, CO_2, and temperature. *J. Exp. Bot.* 31:1485–91

161. Madhusudana, R., Abadia, J., Terry, N. 1987. The role of orthophosphate in the regulation of photosynthesis in vivo. See Ref. 45, pp. 325–28

162. Makino, A., Mae, T., Ohira, K. 1983. Photosynthesis and ribulose 1,5-bisphosphate carboxylase in rice leaves. *Plant Physiol.* 73:1002–7

163. McCurry, S. D., Pierce, J., Tolbert, N. E., Orme-Johnson, W. H. 1981. On the mechanism of effector-mediated activation of ribulose bisphosphate carboxylase/oxygenase. *J. Biol. Chem.* 256:6623–28

164. McDermott, D. K., Zeiher, C. A., Porter, C. A. 1983. Physiological activity of RuBP carboxylase in soybeans. In *Current Topics in Plant Biochemistry and Physiology*, ed. D. D. Randall, D. G. Blevins, R. Larson, p. 230. Columbia, Mo: Univ. Missouri

165. McVetty, P. B. E., Canvin, D. T. 1981. Inhibition of photosynthesis by low oxygen concentrations. *Can. J. Bot.* 59:721–25

166. Miziorko, H. M., Lorimer, G. H. 1983. Ribulose-1,5-bisphosphate carboxylase/oxygenase. *Ann. Rev. Biochem.* 5:507–35

167. Mott, K. A., Berry, J. A. 1986. Effects of pH on activity and activation of ribulose 1,5-bisphosphate carboxylase at air level CO_2. *Plant Physiol.* 82:77–82

168. Mott, K. A., Jensen, R. G., Berry, J. A. 1986. Limitation of photosynthesis by RuBP regeneration rate. In *Biological Control of Photosynthesis*, ed. R. Marcelle, H. Clijsters, M. Van Pouche, pp. 33–43. Dordrecht: Martinus Nijhoff

169. Mott, K. A., Jensen, R. G., O'Leary, J. W., Berry, J. A. 1984. Photosynthesis and ribulose 1,5-bisphosphate concentrations in intact leaves of *Xanthium strumarium* L. *Plant Physiol.* 76:968–71

170. Neales, T. F., Incoll, L. D. 1968. The control of leaf photosynthetic rate by the level of assimilate concentration in the leaf. A review of the hypothesis. *Bot. Rev.* 34:107–25

171. Newsholme, E. A., Crabtree, B. 1976. Substrate cycles in metabolic regulation and in heat generation. *Biochem. Soc. Symp.* 41:61–110

172. Nobel, P. S. 1974. *Introduction to Biophysical Plant Physiology.* San Francisco: W. H. Freeman

173. Ogawa, T. 1982. Simple oscillations in photosynthesis of higher plants. *Biochim. Biophys. Acta* 681:103–9

174. Ogren, W. L., Salvucci, M. E., Portis, A. R. 1986. The regulation of Rubisco activity. *Philos. Trans. Soc. London Ser. B* 313:337–46

175. Oja, V., Laisk, A., Heber, U. 1986. Light-induced alkalization of the chloroplast stroma in vivo as estimated from the CO_2 capacity of intact sunflower leaves. *Biochim. Biophys. Acta* 849:355–65

176. Osmond, C. B., Oja, V., Laisk, A. 1988. Regulation of carboxylation during sun shade acclimation in *Helianthus annuus* using a rapid response gas exchange system. *Aust. J. Plant Physiol.* 15:In press

177. Outlaw, W. H. Jr., Fisher, D. B., Christy, A. L. 1975. Compartmentation in *Vicia faba* L. leaves. II. Kinetics of ^{14}C-sucrose in tissues following pulse-labeling. *Plant Physiol.* 55:704–11

178. Oxborough, K., Horton, P. 1986. Inhibition of high energy state quenching in spinach chloroplasts by low concentrations of antimycin A. In *Progress in Photosynthesis Research*, ed. J. Biggins, 2:489–92. The Hague: Martinus Nijhoff

179. Parry, M. A. J., Schmidt, C. N. G., Cornelius, M. J., Keys, A. J., Millard, B. N., Gutteridge, S. 1985. Stimulation of ribulose bisphosphate carboxylase activity by inorganic orthophosphate without an increase in bound activating CO_2: Cooperativity between the subunits of the enzyme. *J. Exp. Bot.* 36:1396–1404

180. Parry, M. A. J., Schmidt, C. N. G., Keys, A. J., Gutteridge, S. 1983. Activation of ribulose 1,5-bisphosphate carboxylase by Ca^{2+}. *FEBS Lett.* 159:107–11

181. Peisker, M. 1974. A model describing the influence of oxygen on photosynthetic carboxylation. *Photosynthetica* 8:47–50

182. Perchorowicz, J. T., Jensen, R. G. 1983. Photosynthesis and activation of ribulose bisphosphate carboxylase in wheat seedlings. *Plant Physiol.* 71:955–60

183. Perchorowicz, J. T., Raynes, D. A., Jensen, R. G. 1981. Light limitation of photosynthesis and activation of ribulose bisphosphate carboxylase in wheat seedlings. *Proc. Natl. Acad. Sci. USA* 78:2985–89

184. Pickersgil, R. W. 1986. An upper limit to the active site concentration of ribulose bisphosphate carboxylase in chloroplasts. *Biochem. J.* 236:311

185. Pierce, J. 1986. Determinants of substrate specificity and the role of metal in the reactions of ribulosebisphosphate carboxylase/oxygenase. *Plant Physiol.* 81:943–45

186. Pierce, J., Lorimer, G. H., Reddy, G. S. 1986. Kinetic mechanism of ribulosebisphosphate carboxylase: evidence for an ordered, sequential reaction. *Biochemistry* 25:1636–44

187. Portis, A. R. 1981. Evidence of a low stromal Mg^{2+} concentration in intact chloroplasts in the dark. *Plant Physiol.* 67:985–89

188. Portis, A. R. 1983. Analysis of the role of the phosphate translocator and external metabolites in steady-state chloroplast photosynthesis. *Plant Physiol.* 71:936–43

189. Portis, A. R., Heldt, H. W. 1976. Light-dependent changes of the Mg^{2+} concentration in the stroma in relation to the Mg^{2+} dependency of CO_2 fixation in intact chloroplasts. *Biochim. Biophys. Acta* 449:434–44

190. Portis, A. R. Jr., Salvucci, M. E., Ogren, W. L. 1986. Activation of ribulosebisphosphate carboxylase/oxygenase at physiological CO_2 and ribulosebisphosphate concentrations by Rubisco activase. *Plant Physiol.* 82:967–71

191. Portis, A. R. Jr., Salvucci, M. E., Ogren, W. L., Werneke, J. 1987. Rubisco activase: a new enzyme in the regulation of photosynthesis. See Ref. 45, pp. 371–78

192. Prinsley, R. T., Dietz, K-.J., Leegood, R. C. 1986. Regulation of photosynthetic carbon assimilation in spinach leaves after a decrease in irradiance. *Biochim. Biophys. Acta* 849:254–63

193. Raven, J. A., Glidewell, S. M. 1981. Processes limiting photosynthetic conductance. In *Physiological Processes Limiting Plant Productivity*, ed. C. B. Johnson, pp. 109–36. London-Butterworths

194. Rebeille, F., Bilingy, R., Martin, J.-B., Douce, R. 1983. Relationship between the vacuolar phosphate pool in *Acer psuedoplatanus* cells. *Arch. Biochem. Biophys.* 225:143–48

195. Robinson, S. P., Giersch, C. 1987. Inorganic phosphate concentration in the stroma of isolated chloroplasts and its influence on photosynthesis. *Aust. J. Plant Physiol.* 14:451–62

196. Rolleston, S. S. 1972. A theoretical background in the use of measured concentrations of intermediates in study of the control of intermediary metabolism. *Curr. Top. Cell Regul.* 5:47–75

197. Rufty, T. W., Huber, S. C. 1983. Changes in starch formation and activities of sucrose phosphate synthase and cytoplasmic fructose 1,6-bisphosphatase in response to source-sink alterations. *Plant Physiol.* 72:474–80

198. Rufty, T. W., Kerr, P. S., Huber, S. C. 1983. Characterisation of diurnal changes in activities of enzymes involved in sucrose biosynthesis. *Plant Physiol.* 73:428–33

199. Sage, R. F., Seemann, J. R., Sharkey, T. D. 1987. The time course for deactivation and reactivation of ribulose-1,5-bisphosphate carboxylase following changes in CO_2 and O_2. See Ref. 45, pp. 285–88

200. Salerno, G. L., Pontis, H. G. 1977. Studies on the sucrose phosphate synthetase kinetic mechanism. *Arch. Biochem. Biophys.* 180:298–302

201. Salerno, G. L., Pontis, H. G. 1978. Studies on sucrose phosphate synthetase: the inhibitory action of sucrose. *FEBS Lett.* 86:263–67

202. Salvucci, M. E., Portis, A. R. Jr., Ogren, W. L. 1986. A soluble chloroplast protein catalyzes ribulosebisphosphate carboxylase/oxygenase activation in vivo. *Photosynth. Res.* 7:193–201

203. Salvucci, M. E., Portis, A. R. Jr., Ogren, W. L. 1986. Light and CO_2 response of ribulose-1,5-bisphosphate carboxylase/oxygenase activation in *Arabidopsis* leaves. *Plant Physiol.* 80:655–59

204. Salvucci, M. E., Werneke, J. M., Ogren, W. L., Portis, A. R. Jr. 1987. Rubisco activase; purification, subunit composition and species distribution. See Ref. 45, pp. 379–82

205. Santarius, K. A., Heber, U. 1965. Changes in the intracellular levels of ATP, ADP, AMP and Pi and regulatory function of the adenylate system in leaf cells during photosynthesis. *Biochim. Biophys. Acta* 102:39–54

206. Savageau, M. A. 1969. Biochemical systems analysis. II. The steady-state solutions for an *n*-pool system using a power-law approximation. *J. Theor. Biol.* 25:370–79

207. Savageau, M. A. 1970. Biochemical systems analysis. III. Dynamic solutions using a power-law approximation. *J. Theor. Biol.* 26:215–26

208. Savageau, M. A. 1971. Parameter sensitivity as a criterion for evaluating and comparing the performance of biochemical systems. *Nature* 229:542–44

209. Savageau, M. A. 1972. The behaviour of intact biochemical control systems. *Curr. Top. Cell Regul.* 6:63–130

210. Savageau, M. A. 1976. *Biochemical*

Systems Analysis. Menlo Park, Calif: Addison-Wesley

211. Savageau, M. A. 1987. Control of metabolism: where is the theory? *Trends Biochem. Sci.* 12:219–20

212. Seemann, J. R. 1982. *Environmental regulation of photosynthesis in desert annuals*. PhD thesis. Stanford University

213. Seemann, J. R. 1986. Mechanisms for the regulation of CO_2 fixation by ribulose 1,5-bisphosphate. See Ref. 168, pp. 71–82

214. Seemann, J. R., Berry, J. A., Freas, S. M., Krump, M. A. 1985. Regulation of ribulose bisphosphate carboxylase activity in vivo by a light-modulated inhibitor of catalysis. *Proceed. Natl. Acad. Sci. USA* 82:8024–28

215. Seemann, J. R., Sharkey, T. D. 1986. Salinity and nitrogen effects on photosynthesis, ribulose 1,5-bisphosphate carboxylase and metabolite pool size in *Phaseolus vulgaris* L.. *Plant Physiol.* 82:555–60

216. Seemann, J. R., Sharkey, T. D. 1987. The effect of abscisic acid and other inhibitors on photosynthetic capacity and the biochemistry of CO_2 assimilation. *Plant Physiol.* 84:696–700

217. Seemann, J. R., Tepperman, J. M., Berry, J. A. 1981. The relationship between photosynthetic performance and the levels and kinetic properties of RuP_2 carboxylase/oxygenase from desert winter annuals. *Carnegie Inst. Wash. Yearb.* 80:67–72

218. Seftor, R. E., Bahr, J. T., Jensen, R. G. 1986. Measurement of the enzyme-CO_2-Mg^{2+} form of spinach ribulose 1,5 bisphosphate carboxylase/oxygenase. *Plant Physiol.* 80:599–600

219. Selman, B. R., Selman-Preiner, S. 1981. The steady state kinetics of photophosphorylation. *J. Biol. Chem.* 256:1722–26

220. Servaites, J. C., Parry, M. A. J., Gutteridge, S., Keys, A. J. 1986. Species variation in the predawn inhibition of ribulose 1,5-bisphosphate carboxylase/oxygenase. *Plant Physiol.* 82:1161–63

221. Sharkey, T. D. 1986. Theoretical and experimental observations on O_2-insensitivity of C_3 photosynthesis. See Ref. 168 pp. 115–25

222. Sharkey, T. D. 1985. O_2-insensitive photosynthesis in C_3 plants. Its occurrence and a possible explanation. *Plant Physiol.* 78:71–75

223. Sharkey, T. D. 1985. Photosynthesis in intact leaves of C_3 plants: physics, physiology and rate limitations. *Bot. Rev.* 51:53–106

224. Sharkey, T. D., Imai, K., Farquhar, G. D., Cowan, I. R. 1982. A direct con-

formation of the standard method of estimating intercellular partial pressure of CO_2. *Plant Physiol.* 69:657–59

225. Sharkey, T. D., Seemann, J. R., Berry, J. A. 1986. Regulation of ribulose-1,5-bisphosphate carboxylase activity in response to changing partial pressure of O_2 and light in *Phaseolus vulgaris*. *Plant Physiol.* 81:788–91

226. Sharkey, T. D., Stitt, M., Heineke, D., Gerhardt, R., Raaschke, K., Heldt, H. W. 1986. Limitation of photosynthesis by carbon metabolism. O_2-insensitive CO_2 uptake results from limitation of triose phosphate utilization. *Plant Physiol.* 81:1123–29

227. Sicher, R. C., Kremer, D. F., William, G. H. 1984. Diurnal carbohydrate metabolism of barley primary leaves. *Plant Physiol.* 76:165–69

228. Sivak, M. N., Walker, D. A. 1984. New perspectives in the understanding of the regulation of photosynthesis and its relation to chlorophyll fluorescence kinetics through the study of oscillations. *Proc. Symp. Oscillations in Physiological Systems: Dynamics and Control*, 11–13 September, Oxford, pp. 91–96

229. Sivak, M. N., Walker, D. A. 1986. Photosynthesis in vivo can be limited by phosphate supply. *New Phytol.* 102:499–512

230. Slovacek, R. E., Hind, G. 1977. Influence of antimycin A and uncouplers on anaerobic photosynthesis in isolated chloroplasts. *Plant Physiol.* 60:538–42

231. Slovacek, R. E., Hind, G. 1980. Energetic factors affecting carbon dioxide fixation in isolated chloroplasts. *Plant Physiol.* 65:526–32

232. Slovacek, R. E., Hind, G. 1981. Correlation between photosynthesis and the transthylakoid proton gradient. *Biochim. Biophys. Acta* 635:393–404

233. Somerville, C. R., Portis, A. R. Jr., Ogren, W. L. 1982. A mutant of *Arabidopsis thaliana* which lacks activation of RuBP carboxylase in vivo. *Plant Physiol.* 70:381–87

234. Soulé, J. M., Buc, J., Meunier, J.-.C., Pradel, J., Ricard, J. 1981. Molecular properties of chloroplastic thioredoxin *f* and the regulation of the activity of fructose 1,6-bisphosphatase. *Eur. J. Biochem.* 119:497–502

235. Sprey, B. 1977. Lamellae-bound inclusions in isolated spinach chloroplasts. *Z. Pflanzenphysiol.* 83:159–79

236. Steup, M., Peavey, D. G., Gibbs, M. 1976. The regulation of starch metabolism by inorganic phosphate. *Biochem. Biophys. Res. Commun.* 72:1554–61

237. Stitt, M. 1986. Limitation of photo-

synthesis by carbon metabolism. *Plant Physiol.* 81:1115–22

238. Stitt, M. 1987. Fructose 2,6-bisphosphate and plant carbohydrate metabolism. *Plant Physiol.* 84:201–4

239. Stitt, M., Heldt, H. W. 1985. Control of photosynthetic sucrose synthesis by fructose 2,6-bisphosphate. VI. Regulation of the cytosolic fructose 1,6-bisphosphatase in spinach leaves by an interaction between metabolic intermediates and fructose 2,6-bisphosphate. *Plant Physiol.* 79:599–608

240. Stitt, M., Heldt, H. W. 1986. Control of photosynthetic sucrose synthesis by fructose 2,6-bisphosphate. VI. Regulation of the cytosolic fructose 1,6-bisphosphatase in spinach leaves by an interaction between metabolic intermediates and fructose 2,6-bisphosphate. *Plant Physiol.* 79:599–608

241. Stitt, M., Cséke, C., Buchanan, B. B. 1984. Regulation of fructose 2,6-bisphophate concentration in spinach leaves. *Eur. J. Biochem.* 143:89–93

242. Stitt, M., Gerhardt, R., Wilke, I., Heldt, H. W. 1987. The contribution of fructose 2,6-bisphosphate to the regulation of photosynthetic sucrose synthesis. *Physiol. Plant.* 69:377–86

243. Stitt, M., Herzog, B., Gerhardt, R., Kurzel, B., Heldt, H. W., et al. 1984. Regulation of photosynthetic sucrose synthesis by fructose 2,6-bisphosphate. In *Proceedings of the 6th International Congress on Photosynthesis,* ed. C. Sybesma. The Hague: Junk

244. Stitt, M., Herzog, B., Heldt, H. W. 1984. Control of photosynthetic sucrose synthesis by fructose 2,6-bisphosphate. I. Coordination of CO_2 fixation and sucrose synthesis. *Plant Physiol.* 75:548–53

245. Stitt, M., Kurzel, B., Heldt, H. W. 1984. Control of photosynthetic sucrose synthesis by fructose 2,6-bisphosphate. II. Partitioning between sucrose and starch. *Plant Physiol.* 75:554–60

246. Stitt, M., Mieskes, G., Soling, H. D., Heldt, H. W. 1982. On a possible role of fructose 2,6-bisphosphate in regulating photosynthetic metabolism in leaves. *FEBS Lett.* 145:217–22

247. Stitt, M., Wirtz, W., Heldt, H. W. 1980. Metabolite levels during induction in the chloroplast and extrachloroplast compartments of spinach protoplasts. *Biochim. Biophys. Acta* 593:85–102

248. Streusand, V. J., Portis, A. R. Jr. 1987. Rubisco activase mediates ATP-dependent activation of ribulose bisphosphate carboxylase. *Plant Physiol.* 85:152–54

249. Streusand, V. J., Portis, A. R. Jr. 1987.

Effects of 6-phosphogluconate and RuBP on Rubisco activation state and activity. See Ref. 45, pp. 383–86

250. Takahama, U., Shimizu-Takahama, M., Heber, U. 1981. The redox state of the NADP system in illuminated chloroplasts. *Biochim. Biophys. Acta* 637:530–39

251. Tran-Anh, T., Rumberg, B. 1987. Coupling mechanism between proton transport and ATP synthesis in chloroplasts. See Ref. 45, pp. 185–88

252. Tregunna, E. B., Krotkov, G., Nelson, C. D. 1966. Effect of oxygen on the rate of photorespiration in detached tobacco leaves. *Physiol. Plant* 19:723–33

253. Umbarger, H. E. 1956. Evidence for a negative feedback in the biosynthesis of isoleucine. *Science* 123:848

254. von Caemmerer, S., Edmondson, D. L. 1986. Relationship between steady state gas exchange, in vivo ribulose bisphosphate carboxylase activity and some carbon reduction cycle intermediates in *Raphanus sativus. Aust. J. Plant Physiol.* 13:669–88

255. von Caemmerer, S., Farquhar, G. D. 1981. Some relationships between the biochemistry of photosynthesis and the gas exchange of leaves. *Planta* 153:376–87

256. von Caemmerer, S., Farquhar, G. D. 1985. Kinetics and activation of Rubisco and some preliminary modelling of RuP_2 pool sizes. In *Kinetics of Photosynthesis and Photorespiration in C_3 Plants,* ed. J. Viil, 1:46–48. Tallinn: Valgus

257. Vu, C. V., Allen, L. H. Jr., Bowes, G. 1983. Effects of light and elevated atmospheric CO_2 on the ribulose bisphosphate carboxylase activity and ribulose bisphosphatate level of soybean leaves. *Plant Physiol.* 73:729–34

258. Walker, D. A. 1976. Regulatory mechanisms in photosynthetic carbon metabolism. *Curr. Top. Cell. Regul.* 2:204–41

259. Walker, D. A., Crofts, A. R. 1970. Photosynthesis. *Ann. Rev. Biochem.* 39:389–428

260. Walker, D. A., Herold, A. 1977. Can the chloroplast support photosynthesis unaided? In *Photosynthetic Organelles: Structure and Function,* ed. Y. Fujita, S. Katoh, K. Shinata, pp. 295–310. Spec. Iss. *Plant and Cell Physiol.*

261. Walker, D. A., Leegood, R. C., Sivak, M. N. 1986. Ribulose bisphosphate carboxylase- oxygenase: its role in photosynthesis. *Philos. Trans. R. Soc. London Ser. B* 313:305–24

262. Walker, D. A., Osmond, C. B. 1986. Measurement of photosynthesis in vivo with a leaf disc electrode: correlations between light dependence of steady-state

photosynthetic O_2 evolution and chlorophyll a fluorescence transients. *Proc. R. Soc. London Ser. B* 227:267–80

263. Walker, D. A., Sivak, M. N. 1985 In vivo chlorophyll a fluorescence transients associated with changes in the CO_2 content of the gas phase. In *Regulation of Carbon Partitioning in Photosynthetic Tissues,* ed. R. L. Heath, J. Preiss, pp. 93–108. Washington DC: Am. Soc. Plant Physiol.

264. Walker, D. A., Sivak, M. N. 1985. Can phosphate limit photosynthetic carbon assimilation in vivo? *Physiol. Veg.* 23:829–41

265. Walker, D. A., Sivak, M. N. 1986. Photosynthesis and phosphate: a cellular affair? *Trends Biochem. Sci.* 11:176–79

266. Webb, J. L. 1963. *Enzyme and Metabolic Inhibitors,* 1:348–68. New York: Academic

267. Weber, J. A., Tenhunen, J. D., Gates, D. M., Lange, O. L. 1987. Effect of photosynthetic photon flux density on carboxylation efficiency. *Plant Physiol.* 85:109–14

268. Weis, E., Ball, J. T., Berry, J. A. 1987. Photosynthetic control of electron transport in leaves of *Phaseolus vulgaris:* evidence for regulation of photosystem 2 by the proton gradient. See Ref. 178, pp. 553–56

269. Weis, E., Berry, J. A. 1987. Quantum efficiency of photosystem II in relation to 'energy'-dependent quenching of chlorophyll fluorescence. *Biochim. Biophys. Acta* 894:198–208

270. West, K. R., Wiskich, J. T. 1968. Photosynthetic control by isolated pea chloroplasts. *Biochem. J.* 109:527–32

271. Whitaker, D. P. 1984. Purification and properties of sucrose-6-phosphatase from *Pisum sativa* shoots. *Phytochemistry* 23:2429–30

272. Wirtz, W., Stitt, M., Heldt, H. W. 1980. Enzymic determination of metabolites in the subcellular compartments of spinach protoplasts. *Plant Physiol.* 66:187–93

273. Woodrow, I. E. 1986. Control of the rate of photosynthetic carbon dioxide fixation. *Biochim. Biophys. Acta* 851:181–92

274. Woodrow, I. E. 1986. Control of transient and steady-state photosynthesis by light-regulated enzymes. In *Kinetics of Photosynthesis and Photorespiration in C_3 Plants,* ed. J. Viil, pp. 106–15. Tallinn: Valgus

275. Woodrow, I. E., Ball, J. T., Berry, J. A. 1986. A general expression for the control of the rate of photosynthetic CO_2 fixation by stomata, the boundary layer

and radiation exchange. See Ref. 13, pp. 225–28

276. Woodrow, I. E., Ball, J. T., Berry, J. A. 1988. A quantitative assessment of the degree to which RuBP carboxylase/oxygenase determines the steady state rate of photosynthesis in C_3 plants. In *Rubisco 87. Proceedings of an International Workshop on Rubisco.* Unpublished manuscript. Submitted to *Nature*

277. Woodrow, I. E., Ellis, J. R., Jellings, A., Foyer, C. H. 1984. Compartmentation and fluxes of inorganic phosphate in photosynthetic cells. *Planta* 161:525–30

278. Woodrow, I. E., Furbank, R. T., Brooks, A., Murphy, D. J. 1985. The requirements for steady state in the C_3 reductive pentose phosphate pathway of photosynthesis. *Biochim. Biophys. Acta* 807:23–71

279. Woodrow, I. E., Mott, K. A. 1988. A quantitative assessment of the degree to which RuBP carboxylase/oxygenase determines the steady-state rate of photosynthesis during sun-shade acclimation in *Helianthus annuus. Aust. J. Plant Physiol.* In press

280. Woodrow, I. E., Murphy, D. J., Latzko, E. 1984. Regulation of stromal sedoheptulose 1,7-bisphosphatase activity by pH and Mg^{2+} concentration. *J. Biol. Chem.* 259:3791–95

281. Woodrow, I. E., Murphy, D. J., Walker, D. A. 1983. Regulation of photosynthetic carbon metabolism. The effect of inorganic phosphate on stromal sedoheptulose 1,7-bisphosphatase. *Eur. J. Biochem.* 132:121–23

282. Woodrow, I. E., Walker, D. A. 1983. Regulation of stromal sedoheptulose-1,7-bisphosphatase activity and its role in controlling the reductive pentose phosphate pathway. *Biochim. Biophys. Acta* 722:508–16

283. Woodrow, I. E., Walker, D. A. Light-mediated activation of stromal sedoheptulose bisphosphatase. *Biochem. J.* 191:845–49

284. Yates, R. A., Pardee, A. B. 1956. Control of pyrimidine biosynthesis in *Escherichia coli* by a feedback mechanism. *J. Biol. Chem.* 221:757–70

285. Yokota, A., Canvin, D. T. 1985. Ribulose bisphosphate carboxylase/oxygenase content determined with [^{14}C] carboxy pentitol bisphosphate in plants and algae. *Plant Physiol.* 77:735–39

286. Sharkey, T. D., Berry, J. A., Sage, R. F. 1988. Regulation of photosynthetic electron transport rates as determined by room temperature chlorophyll a fluorescence in *Phaseolus vulgaris.* Submitted to *Planta*

AUTHOR INDEX

A

Aasa, R., 392
Abadia, J., 544, 558
Abbott, A. G., 517
Abbott, L. K., 227-29
Abe, H., 27-29, 33, 37, 43
Abe, M., 306, 307
Abe, T., 32
Abita, J. P., 106
Abramowicz, D. A., 391, 392, 396
Acevedo, E., 286
Ackerson, R. C., 288
Adair, W. S., 324, 325
Addicott, F. T., 440
Aggeler, J., 55, 57
Agrawal, V. P., 117
Agsteribbe, E., 509
Ahle, S., 62, 67, 72
Ahn, A., 65
Ailhaud, G. P., 106
Ainslie, G. R. Jr., 570
Ajoika, R. S., 81
Akabori, K., 395
Akazawa, T., 491
Akerlund, H.-E., 395, 484
Akhrem, A. A., 32
Akhtar, M., 342
Aksenova, N. P., 186, 187, 200
Al-Awqati, Q., 77, 85
Albersheim, P., 87, 150, 232, 312, 321, 322, 325, 338, 345
Albert, L. S., 459
Alberte, R. S., 148, 386, 491
Alberti, M., 389
Albertini, D. F., 81
Al-Doori, A. H., 183
Aldridge, E. F., 168
Alexander, D. C., 346
Alexander, J. M., 452
Alfsen, A., 71, 75, 88
Allakhverdiev, S. I., 389, 391
Allaway, W. G., 256, 258, 259
Allen, L. H. Jr., 541, 547
Allen, M. F., 232, 233
Allevi, P., 38, 40, 42
Allred, D. R., 147, 148
Alt, J., 390, 478-80, 486, 488, 491
Altstiel, L., 59, 73, 75-77
Ambronn, H., 334
Ames, R. N., 230, 232
Amesz, J., 380
Amherdt, M., 57, 83, 84
Amir, J., 556, 557
Ammerlaan, A., 364
Amrhein, N., 429

Ananyev, G. M., 391
Anastasia, M., 38, 40, 42
Anderson, D. M., 164, 165, 169
Anderson, E., 7, 8
Anderson, J. M., 147, 148, 479
Anderson, L. E., 568
Anderson, L. W. J., 288, 458, 459
Anderson, M. A., 150
Anderson, R. G. W., 54, 59, 68, 73, 79-82
Anderson, R. S., 504
Anderson, S., 508
Anderson, W. P., 24
Andersson, B., 147, 148, 387, 389, 395, 484
Andreae, M., 78, 77
ANDRÉASSON, L.-E., 379-411; 387, 392-95
Andres, A., 59, 60, 80
Andrews, J., 119, 123, 124
Andrews, T. J., 535, 551
Androlewicz, M. J., 65
Aoki, S., 388
Apel, K., 144, 147, 148, 476, 485, 490, 491
Appelbaum, E., 311
Appels, R., 418, 420, 422, 429
Appleqvist, L.-A., 120
Apps, D. K., 77
ap Rees, T., 102, 127, 360, 362, 368
Arakai, S., 388
Arima, M., 27-29, 44
Armbrust, E. V., 165
Armstrong, K. C., 416, 417, 419, 420, 422, 426
Arnaud, Y., 185
Arnold, D. J., 228
Arnon, D. I., 390
Arntzen, C. J., 380, 389, 476, 481, 487
Arroyave, N. J., 458
Arteca, R. N., 34, 37, 42, 44
Arthur, E. D., 281
Asada, K., 391
Ashford, A. E., 144, 147, 229
Ashmore, C. J., 282
Ashmore, S. E., 417, 422
Ashton, A. R., 538, 545, 559, 570
Ashwell, G., 79, 80
Asimi, S., 229
Assante, G., 442, 443
Astié, M., 196, 197
Astier, C., 387, 389
Astle, M. C., 453
Atherton, J. G., 176

Atkins, C. A., 364, 551, 572
Attardi, G., 504, 507-9, 512
Auderset, G., 189, 179
Austin, R. B., 544
Ausubel, F. M., 305, 313
Averina, N. G., 491
Avery, G. S. Jr., 5, 13
Avery, J. S., 269-72, 274, 275
Averyhart-Fullard, V., 325, 331
Ayers, A. R., 306
Ayers, S. B., 306
Aylmore, L. A. G., 228, 247
Azcon, R., 232
Azcon-Aguilar, C., 223, 228, 232

B

Baba, J., 27
Baba, S., 325
Babcock, G. T., 382, 387, 390, 391, 393, 395, 396
Baboszewski, B., 339
Bachmaa, J. M., 34, 37, 42, 44
Bacic, A., 338
Bacic, T., 150
Badger, M. R., 537, 540, 542, 547, 548, 551, 553, 558
Badila, P., 181
Bagyaraj, D. J., 234
Bahr, J. T., 535, 540, 544
Baianu, I. C., 396
Bailey, J. A., 234, 312, 313, 329, 336, 337
Bailey-Serres, J., 515
Bainton, D. F., 55, 82
Baker, D. A., 359, 360, 457
Baker, J. H., 228
Baker, N. R., 268
Bakhuizen, R., 306
Baldry, C. W., 561
Ball, J. T., 539, 560, 567, 574-77, 579
Balogh, A., 557
Bambridge, H. E., 103, 105
Banfalvi, Z., 298, 304
Banfield, J., 368
Bang, A. G., 509, 511, 513
Bangerth, F., 358, 359, 371
Bankier, A. T., 508
Barbato, R., 479
Barber, C. E., 306
Barber, J., 380, 387, 389, 390, 395, 476-79, 487
Barbier, M., 424, 430
Barcena, J. A., 549
Barclay, G. F., 363
Barea, J. M., 223, 232, 233

Barendse, G. W. M., 180, 204, 207
Barigozzi, C., 422
Barkan, A., 479, 480
Barker, A. V., 228
Barley, K. P., 260
Barlow, E. W. R., 276, 368
Barrell, B. G., 504, 508
Barros, R. S., 459
Barrow, N. J., 228
Barry, B. A., 391
Bartels, P. G., 491
Barthe, P., 459
Bartlett, S. G., 484
Bartnicki-Garcia, S., 78, 322
Bar-Zvi, D., 59, 61, 76
Bassham, J. A., 548, 577
Bassi, R., 479
Bateman, J., 310
Batley, M., 298, 301, 305, 306, 312
Battey, N. H., 193, 195, 196
Bauer, W. D., 298, 299, 301, 306-8, 312, 314, 325, 334
Bauerle, R., 102
Baulcombe, D. C., 86
Bauman, L. F., 522, 523
Baumgartner, B., 143, 145
Bavrina, T. V., 187
Bayley, S. T., 322
Baylis, G. T. S., 228
Bayliss, M. W., 414-16, 418
Bayne, H. C., 235
Baysdorfer, C., 202
Beachy, R. N., 463
Beale, S. I., 478, 492
Beattie, D. S., 490
Beck, K. A., 76
Beck, W. F., 392, 394
Becker, D. W., 391
Becker, W. M., 144, 148, 150
Beckett, J. B., 515
Bedbrook, J. R., 479, 484, 487, 507
Bedinger, P., 507
Beeckmans, E., 61
Beek, B., 423
Beever, R. E., 229, 231
Beevers, H., 127, 128
Beevers, L., 232
Begg, J. E., 273
Begusch, H., 389
Behl, R., 455-57
Beiby, J. P., 223
Beisiegel, U., 80, 81
Bekker, A.F., 486, 487
Bell, J. N., 312, 313, 329, 336, 337
Bendall, D. S., 491
Bendayan, M., 141-43, 145
Bender, G. L., 298, 312
Bendich, A. J., 479, 485, 504, 506

Bennett, A. B., 346, 364, 365, 368
Bennett, H. W., 41
Bennett, J., 481, 485, 491
Bennett, R. D., 443
Bennoun, P., 395, 485, 492
Bensink, J., 16
Benson, R. J., 59, 62, 83
Benzion, G., 418, 419, 422, 426, 428, 430
Bereman, P. D., 109
Berends, T., 479, 480, 488
Berg, S. P., 395
Bergfeld, R., 274, 462, 463, 490
Bergman, J. F., 81
Bergmann, H., 150
Bergström, J., 390
Beringer, J. E., 223
Berjak, P., 274, 275
Berne, M., 76, 77
BERNIER, G., 175-219; 176, 177, 179-83, 186-97, 199, 200, 202-9, 358, 361
Bernier, I., 76
BERRY, J. A., 533-94; 535, 537, 539-41, 543-50, 560, 567, 569, 574-77, 579
Berry, J. O., 484, 490
Berry-Lowe, S., 478
Berzborn, R. J., 389
Besnard-Wibaut, C., 191, 193, 204, 206
Bessoule, J.-J., 117
Bethlenfalvay, G. J., 228, 229, 235
Bevege, D. I., 223, 234
Bhalla, P. L., 482
Bharuca, B., 62, 65
Bhullar, S. S., 361, 371
Bhuvaneswari, T. V., 298, 299, 306, 307, 312
Bibb, M. J., 504
Bickett, D. M., 565
Bidard, J. N., 440
Biedler, J. L., 429, 430
Bielesky, R. L., 231
Biermann, S. J., 62, 65
Biggins, J., 380
Bignais, P. M., 75
Bilingy, R., 558, 566
Billot, J., 180, 182
Binder, B. J., 164, 165, 169
Bingham, E. T., 425-27
Birecka, H., 342
Birnberg, P. R., 206
Bisanz-Seyer, C., 479, 480, 486
Bisher, M., 75
Bishop, D. G., 121
Bishop, P. D., 312
Bismuth, F., 180, 181, 204
Bisseling, T., 149, 313

Björkman, O. E., 536, 572, 581
Björn, G. S., 168
Björn, L. O., 167, 168
Bjorkman, O., 479
Black, M. T., 380, 390
Black, R. L., 227-30
Blackett, N. M., 58
Bland, M. M., 508, 511, 513
Blank, G. S., 65
Blanken, H. J. den, 382
Blankenship, R. E., 393
Blanz, P., 506
Blanz, P. A., 508
Blaschek, W., 78
Bliel, J., 81
Bligny, R., 118
Blitz, A. L., 59, 70, 76
Blizzard, W. E., 248, 249, 260
Blobel, G., 492
Bloch, K., 118
Bloom, W., 70
Blough, N. V., 395, 396
Blubaugh, D. J., 388
Blumenschein, A., 422
Blumenthal, R., 68, 69, 75
Boardman, N. K., 280, 282, 479
Bodner, U., 389
Bodson, M., 177, 188, 189, 191, 192, 202-5, 207, 358, 361
Boer, P. H., 511, 512
Boeshore, M. L., 504, 521
Boffey, S. A., 479
Bogorad, L., 162, 163, 168, 381, 389, 484-87, 493
Bohlool, B. B., 307-10
Böhm, W., 247
Bohner, J., 358, 359
Bohnert, H. J., 389, 479, 487
Boisard, J., 284
Bolan, N. S., 228
Bolanos, J. A., 275
Bold, H. C., 161
Boller, T., 230, 343-45
Bollmann, M. P., 203
Bolton, P., 111-13, 116, 117
Bolwell, G. P., 301, 312
Bomsel, M., 75, 88
Bonen, L., 508, 511, 512
Böner, T., 490
Bonetti, A., 24, 38, 40, 42
Bonfante-Fasolo, P., 222, 224, 226, 229, 234
Bonner, B. A., 166
Bonnerjea, J., 384, 385
Bonnerot, C., 121, 122
Bonnet-Masimbert, M., 205
Bonnett, H. T., 57
Bonzon, M., 179
Bookjans, G. B., 381
Bookland, R., 147, 151

Boote, K. J., 247-49
Booth, A. G., 80
Bopp, M., 87, 342
Borck, K. S., 516
Borgman, E., 344
Borner, T., 108
Borrel-Flood, C., 122
Borroto, K. E., 482
Borthakur, D., 306
Bos, S., 159
Boska, M., 391, 395, 396
Botha, C. E. J., 86
Bottin, H., 386, 387
Bottomley, W., 389, 479, 480, 486, 488
Bouet, C., 485, 488
Boulon, B., 459
Boundy, J. A., 325
Boussac, A., 391, 395, 396
Boutry, M., 509, 510, 515
Bowen, G. D., 223, 234
Bowen, J. C., 365
Bowes, G., 535, 541, 547
Bowles, D. J., 145
Bowling, D. J. F., 256
Bowman, W. R., 448
Boxer, S. G., 386
Boyer, G. L., 183, 200, 440, 441, 453-55, 462
Boyer, J. S., 245, 248, 249, 260, 273, 277, 284-87, 325, 336, 337, 455
Boylan, K. L. M., 298
Boynton, J. E., 478
Bozarth, C. S., 325, 336, 337
Bracciano, D. M., 36, 47
Bracker, C. E., 78
Bradbeer, J. W., 476, 478
Bradford, K. J., 246, 440, 447, 458
Bradley, D., 486, 494
Braell, W. A., 71, 74
Branton, D., 59, 61-71, 73, 75-77, 326
Branton, E., 59, 69-71
Brasier, C. M., 521
Braun, C. J., 509, 511, 512
Braun, P., 35, 36, 40, 41
Braune, W., 164
Brautigam, E., 486
Bray, E. A., 455, 463
Brearley, T. H., 390
Brears, T., 517
Brede, J. M., 201
Breekland, A. E., 460
Breeman, A. M., 158-60, 169
Bremner, P. M., 247
Brenner, M. L., 206, 453, 460
Brennicke, A., 506, 508-12, 521
Bressan, R. A., 460
Bretell, R. I. S., 517
Breton, J., 382

Bretscher, M., 81
Bretscher, M. S., 75, 79, 81, 82
Brettel, K., 383, 386, 388-91
Brettell, R. I. S., 418, 420, 422, 424, 426, 428, 429, 517
Breuer, J., 120, 122
Brewin, N. J., 144, 148, 237
Briantais, J.-M., 390
Briat, J.-F., 479, 480, 486, 487
Bridgen, M. P., 186, 197
Bridle, K. A., 200
Brierley, H. L., 298, 305
Briggs, W. R., 476
Bright, S. W. J., 414, 427
Brightman, A., 179
Brinkhorst-van der Swan, D. L. C., 447, 460
Briquet, M., 509, 510
Briskin, D. P., 369
Bristow, K. L., 248, 249
Britt, R. D., 384, 385, 392-94
Britz, S. J., 202
Broadhurst, R. W., 383
Brodie, J., 158-60
Brodsky, F. M., 62, 65, 66
Broglie, K. E., 306
Broglie, R. M., 306, 479
Brok, M., 391, 393
Bromley, J.-L., 227, 230, 232, 235
Broniatowski, M., 271
Brooking, I. R., 360, 371
Brooks, A., 536, 537, 544, 555, 558, 568, 577
Brooks, C. J. W., 32
Brosius, J., 65
Brot, N., 485, 490, 491
Brouwer, R., 260
Browers, M. A., 420
Brown, G. G., 521
Brown, J. M., 259
Brown, K. M., 272
Brown, M. F., 226, 231, 232, 234
Brown, M. R., 229
Brown, M. S., 54, 73, 79-82, 228, 229
Brown, P. C., 429, 430
Brown, P. T. H., 431
Brown, R., 269
Brown, S. C., 367
Brown, T. A., 504, 509, 512
Brown, T. J., 163
Brown, W. E., 312
Brown, W. J., 83
Browse, J., 103, 123-25, 477, 478
Browse, J. A., 125
Browse, J. R., 120
Bruder, G., 62, 74

Brudvig, G. W., 387, 392, 394, 395
Brulfert, J., 183, 185
Brumell, D. A., 274
Brun, W. A., 453
Brusztyn-Pettegrew, H., 60, 83
Bryant, J. A., 421
Brysk, M. M., 323
Buban, T., 182, 184
Buc, J., 570
Buchanan, B. B., 557, 568
Bücher, T., 573
Buchler, U. K., 422
Buck, K. W., 521
Buckley, K., 80
Buckner, J. S., 106, 116
Bugnon, F., 269
Buhler, E. M., 422
Bulard, C., 440
Bunce, J. A., 288, 457
Bunger-Kibler, S., 358, 371
Buoniconti, P., 83
Burchert, M., 549
Burden, R. S., 445, 446
Burgess, J., 145
Burgess, N., 128
Burggraaf, A. J. P., 223
Bürgi, H.-B., 393
Burgoyne, R. D., 75
Burkhart, W., 343
Burkholder, P. R., 5
Burnham, C. R., 418, 427
Burns, D. J. W., 229, 231
Burns, J. A., 573, 578
Burr, B., 425, 426
Burr, F. A., 425, 426
Burschka, C., 451, 452, 456
Bursztajn, S., 82
Burton, S., 541
Burtt, B. L., 273
Bush, M. G., 205, 206
Buta, J. G., 27, 42
Butenko, R. G., 415
Butler, R. D., 481
Butler, W. L., 143
Butler, W. M., 394
Butow, R. A., 508
Buttner, W., 391
Buvat, R., 184
Buwalda, J. G., 224, 233
Bylina, E. J., 389

C

Caldwell, M. M., 250, 251
Calissendorff, C., 248, 249
Callaham, D. A., 306
Callahan, F. E., 391, 392
Callow, J. A., 229
Calvert, H. E., 298, 299, 301, 306, 307, 312, 334
Cambardella, C., 230, 232
Cameron, I. L., 270

Cammarata, K., 395
Camp, P. J., 102
Campbell, C. E., 422
Campbell, C. H., 59, 76, 80, 83
Campbell, G. S., 248, 249
Campillo, E. D., 343-45
Cangelosi, G. A., 306
Canny, M. J., 261
Canter-Cremers, H., 298, 301, 305
Canter Cremers, H. C. J., 306
Cantley, L., 76, 77
Cantrell, M. A., 324, 326
Canvin, D. T., 104, 540, 562
Capaccio, L. C. M., 229
Carlemalm, E., 143
Carling, D. E., 232
Carlson, J. E., 508, 514, 518, 519
Carlson, P. S., 424
Carmi, A., 183
Carpentier, J.-L., 58, 82
Carpentier, R., 396
Carpita, N., 86
Carr, G. R., 223
Carr, J. P., 484, 490
Carraro, L., 341, 342
Carroll, B. J., 308, 310, 311
Carroll, E. J. Jr., 60
Carroll, J. W., 162, 168
Carter, A. S., 310
Carvey, D. H., 81
Casey, J. L., 392
Cashmore, A. R., 484
CASSAB, G. I., 321-53; 322, 324-27, 329-32, 334, 335
Cassagne, C., 116, 117
Castelfranco, P. A., 492
Catcheside, D. G., 428
Catt, J. W., 322, 345
Caughey, I., 109-12
Cavalieri, A. J., 337
Cerana, R., 24, 38, 40, 42
Cesaro, A., 340
Cetas, C. B., 201
Chabot, B. F., 281
Chabot, J. R., 281
Chadwick, B., 382
Chaika, M. T., 491
Chailakhyan, M. Kh., 176, 183, 186-88, 199, 200, 205
Champagnat, P., 176, 201
Chan, H. W.-S., 102, 126, 128
Chandler, M. R., 298
Chandler, V. L., 425
Chandra Sekhar, K. N., 198
Chaney, W. R., 272
Chang, C., 198
Chang, L. O., 323
Chang, W. C., 366
Chanson, A., 457
Chao, S., 507, 508, 511

Chapin, F. S., 227, 235
Chapman, D. J., 390
Chappell, T. G., 74
Charbonnier, M., 515
Charles, D. J., 103, 105
Charles, S. A., 559
Charles-Edwards, D. A., 268
Charnofsky, K., 164
Chatterton, N. J., 565
Cheesbrough, J. K., 561, 562
Cheeseman, J. M., 455
Chen, C.-M., 183
Chen, H., 306
Chen, H. H., 460
Chen, J. A., 313, 322, 324, 325, 328, 329, 331, 336
Chen, S. C., 440
Chen, T. H. H., 460
Chen-She, S. H., 561
Cheniae, G. M., 391, 392, 394, 395
Cherry, J. H., 103, 105
Chestnut, M. H., 76
Chiba, N., 346
Childs, J., 75
Chilvers, G. A., 229
Cho, S. H., 125
Cho, Y. P., 323, 328, 335
Chollet, R., 538, 546, 548, 549, 551
Chomyn, A., 509
Chon, C. J., 550, 559, 561, 564
Chouard, P., 185
Chourey, P. S., 371, 506
Chow, E. P., 65
Chrispeels, M. J., 142, 143, 145, 151, 323, 326, 328, 335, 336, 344
Christ, R. A., 288
Christeller, J. T., 538, 546, 548-51
Christensen, M., 232, 233
Christoffersen, R. E., 344
Christy, A. L., 565
Chu, D. K., 548
Chua, N.-H., 283, 476, 479, 484, 486, 488, 494
Chunwongse, J., 486
Church, D. L., 341
Chuzel, M., 122
Chvatchko, Y., 85
Chwirot, W. B., 181
Ciana, A., 340
Citharel, B., 120
Ciuffreda, P., 38, 40, 42
Clarke, A. E., 150, 325, 343
Clarke, J. A., 340
Clarkson, D. T., 227, 230
Clausen, W., 565
Clauss, H., 163, 164, 166
Cleary, W., 429
Cleeter, M. W. J., 509

Clegg, M. T., 486
Cleland, C. F., 179, 205
Cleland, R. E., 276-79
Clermont, Y., 58
Clithero, J. M., 460
Clough, B. F., 256, 258, 259, 272, 285
Cobb, B. G., 446, 447
Cochet, T., 191, 206
Cockburn, W., 561
Cockshull, K. E., 176, 179, 185, 193
Coe, E. H. Jr., 184, 480, 514, 522, 523
Coen, D. M., 484, 487
Cohen, D. B., 507
Cohen, J. D., 37, 273
Cohen, S., 82
Cohn, Z. A., 54, 87
Colbeau, A., 75
Colclasure, G. C., 35, 37
Cole, J., 393-96
Cole, L., 62, 66
Coleman, J. O. D., 62, 66
Coleman, W. J., 387, 396
Colin, J., 507, 508
Collatz, G. J., 537, 540, 542, 544, 545, 547
Collins, G. B., 125
Collins, J., 299
Collins, R. A., 508
Colman, B., 451
Colombo, R., 38, 40, 42
Côme, D., 180, 182
Comfurius, P., 75
Conde, M. F., 414, 504, 515, 517, 519, 521
Condit, C. M., 325, 329, 337
Conjeaud, H., 391, 396
Connelly, J. A., 517
Connolly, J. A., 55, 58
Connolly, J. L., 82
Considine, J. A., 197
Constabel, F., 60, 62
Cook, H., 361, 362
Cook, M. G., 357
Cook, S., 441
Cooke, T. J., 459
Coombe, B. G., 367
Cooper, D. C., 5
Cooper, J. B., 313, 322, 324, 326, 327, 329, 331, 335, 340
Cooper, K. M., 227, 230, 232-34
Cooper, T. G., 127, 128
Corde, J. P., 108
Cordero, R. E., 192-95
Cordonnier, M.-M., 166, 167
Cornelius, J. M., 385, 386
Cornelius, M. J., 549, 559, 561
Cornic, G., 457, 562
Cornish, K., 453-57

Cornish, P. S., 251
Cornu, A., 198, 198
Coruzzi, G., 479
Cosgrove, D. J., 268, 276-79, 287
Cottier, J. P., 451
Cottrell, J., 184, 193, 194
Coulson, A. R., 504
Cousson, A., 187, 203, 209
Cowan, A. K., 445, 451, 453-55, 462
Cowan, I. R., 279, 455, 539, 572
Cox, G., 226, 229, 231, 234
Cox, J. K., 516
Cozens, A. L., 479, 486
Crabtree, B., 573, 574
Craig, S., 141, 143-45
Cram, W. J., 86
Cramer, C. L., 301, 312, 313, 329, 336, 337
Cramer, W. A., 380, 390
Crasnier, M., 339
CREELMAN, R. A., 439-73; 440, 442, 443, 447, 449-54
Creighton, H. B., 5
Creissen, C. P., 427
Cremer, F., 180
Cress, W. A., 228
Creutz, C. E., 65
Crèvecoeur, M., 208
Criddle, R. S., 105, 106, 109
Critchley, C., 395, 396
Croen, K., 77
Croes, A. F., 204, 207
Crofts, A. R., 396, 556, 567
Cronan, J. E., 106
Crook, E. M., 322
Crossland, L. D., 486
Crouch, M. L., 459, 462
Crouse, E. J., 479, 487
Crowther, R. A., 59, 61, 63, 64, 68, 69, 73
Croxdale, J. G., 191
Croy, R. R. D., 149
Croze, E. M., 58
Crozier, A., 27, 28, 44
Csatorday, K., 389
Cséke, C., 557, 568
Cseplo, A., 122
Cuculis, J. J., 142
Culafić, L., 177
Cullis, C. A., 429
Cuming, A. C., 462
Cumming, B. G., 203
Cummins, W. R., 451
Cunningham, E. M., 159
Cusset, G., 269
Cutler, J. M., 276, 284, 286, 287
Czernuszewicz, R. S., 393

D

da Costa, A. R., 456
Daft, M. J., 234, 235
Daie, J., 441
Dainty, J., 254, 256, 286
Daiss, J. L., 62, 67
DALE, J. E., 267-95; 184, 191, 193, 194, 268, 270-72, 274, 275, 280, 281, 283, 288
Dale, R. M. K., 506-8, 511, 519
Dalhuizen, R., 344
Dalling, M. J., 358, 361, 370, 371, 482
Dalton, F. N., 253
D'Amato, F., 414, 415, 429
Damoder, R., 392, 394, 396
Damon, S., 365
Danielius, R. V., 388, 390
Daniels, M. J., 306
Darbyshire, B., 367
Darlington, C. D., 10
Darvill, A. G., 87, 150, 187, 209, 274, 312, 321, 322, 325, 338, 345
Darville, A. G., 232
Das, B. C., 423
Date, R. A., 298
Dathe, W., 453, 455
Datta, K., 325, 331
Davenport, D., 167
Davenport, T. L., 177
Davidson, J. L., 273
Davies, A. O., 123
Davies, J. P., 162, 164
Davies, J. T., 249, 250
Davies, P. A., 428, 431
Davies, P. J., 206
Davies, R. W., 504, 509, 512
Davies, T. M., 310
Davies, W. J., 248, 268, 279, 282, 285, 288, 440, 451, 456-58
Davis, R. W., 328, 331, 336, 337
Davis, W., 200, 205
Dawidowicz, E. A., 75, 83
Dawson, A. J., 504, 510-12, 514, 521
Dawson, A. L., 189
Day, D. A., 298, 308, 310
Dazzo, F. B., 306, 307
Dean, C., 481
Deaven, L. L., 422
Debelle, F., 298
De Boer, S., 228
deBoucand, M. T., 415
de Bruijn, M. H. L., 504, 512
De Chalain, T. M. B., 274, 275
de Faria, S. M., 299

Degli Agosti, R., 179
De Greef, J. A., 208, 283
de Groot, A., 393
Dehesh, K., 144, 147, 148, 485, 490, 491
Dehne, H.-W., 231, 234
de Hostos, E. L., 507
Deichgräber, G., 55
Deisenhofer, J., 387
Deitzer, G. F., 179
Dejaegere, R., 61, 144
Dekker, J. P., 380, 387, 391, 393, 395
Delarue, Y., 271
Delben, F., 340
Deleens, E., 202
Della Penna, D., 346
Delmer, D. P., 77, 86, 321, 322, 345
Delorme, S., 505
Delosme, M., 485, 492
De Luca, V., 104
DeLuca-Flaherty, C., 487
Delves, A. C., 298, 299, 310
Demandre, C., 120-22
Demeter, S., 389
Demetriou, D. M., 387
Dempsey, E., 421, 425
De Munk, W. J., 208
Deng, X.-W., 485-87, 494
Dengler, N. G., 271, 274-76, 280, 459
Denne, M. P., 271
Dennenberg, R. J., 389
Dennin, K. A., 183, 186, 187, 201
Dennis, D. T., 104
Dennis, E. S., 188, 426
Dennis, M. J., 55
Deno, H., 494
DePaepe, R., 429
de Paula, J. C., 392, 394, 395
Depka, B., 389
De Proft, M., 208
DEPTA, H., 53-99; 57, 60-63, 66, 67, 69, 77, 78, 85, 86
Derrenbacker, E. C., 322
Desbiez, M.-O., 201
Deschamp, P. A., 459
de Silva, J. V., 122
De Silva, N. S., 189
Deurs, B. van, 58
Deus-Neumann, B., 441
Devedzhyan, A. G., 199
Devic, M., 490
Devide, Z., 416, 422
DeVit, M., 442, 443, 446, 450
de Vitry, C., 391
De Vlaming, P., 198
de Vos, G. F., 306
Devreux, M., 429
Dewey, R. E., 509, 511, 514-17, 521

de Wit, C. T., 16
Dexheimer, J., 226, 229, 230, 232, 234
Dexheimer, S. L., 384, 385, 392-94
Dey, P. M., 343-45
Dhillon, S. S., 429
Diakoff, S., 164, 165
Diamantidis, G., 339
Diaz, C. L., 306
Dick, P. S., 360, 362
Dickinson, C. H., 304, 312
Dickson, D. P. E., 385
Dickson, L. G., 158, 159
Dieckert, J. W., 145
Dieckert, M. C., 145
Dietrich-Glaubitz, R., 394
Dietz, K.-J., 232, 540, 545, 547, 559, 560, 562, 567, 578
Digby, J., 274
Dimler, R. J., 325
Dinar, M., 368
Diner, B. A., 387-91
Dingkuhn, M., 372
Diomaiuto-Bonnand, J., 180, 189
Dirksen, C., 250, 251
Dismukes, G. C., 387, 391-94, 396
Dixon, L. K., 515, 517
Dixon, R. A., 301, 312
Djordjevic, M. A., 298, 301, 304, 305, 307, 309, 312, 313
Djordjevic, S. P., 306
Dobel, P., 418, 421
Doehlert, D. C., 557
Doerschung, E. B., 425
Doll, S., 366, 369
Dommergues, Y. R., 298, 299
Dommes, J., 180
Donaldson, D. D., 147, 151
Donaubauer, J. R., 32
Doney, D. L., 370
Donkin, M. E., 447, 448
Donn, G., 429
Donovan, C. M., 517
Donovan, G. R., 368
Doohan, M. E., 55
Doolittle, R. F., 509
Dooner, H. K., 460
Dörffling, K., 440, 443, 453, 456, 460
Doriaux, M., 84
Dorne, A. J., 108
Dörnemann, D., 382
Douce, R., 103, 108, 118, 122, 124, 450, 504, 558, 566
Dougall, D. K., 322
Dougherty, W. G., 508
Douglas, C., 301, 306, 312
Douglas, C. J., 306

Downie, J. A., 237, 298, 305, 306, 313
Downs, R. J., 283
Draber, W., 389, 390
Drapier, D., 122, 124, 125
Drew, M. C., 257
Dreyfus, B., 298, 299
Drickamer, K., 62, 65
Driessche, E. van, 61
DRING, M. J., 157-74; 158-61, 163-69
Dron, M., 389
Drubin, D. G., 62, 67, 76
Dubacq, J.-P., 122, 124, 125
Dubertret, G., 485, 488
Dudits, D., 507, 519
Dudko, N. D., 200
Duffield, A. M., 442
Dugger, W. M., 441
Duineveld, Th. L. J., 208
Dujon, B., 509
Duke, S. O., 148
Dulieu, H., 269, 271, 424, 430
Dunaway, C., 162, 168
Dunbar, S. M., 510, 511, 521
Duncan, M. J., 164, 165, 167
Dungy, L. J., 36, 47
Dunlap, J. R., 208, 371
Dunn, W. A., 81
Dunstone, R. L., 356, 370
Durant, J. P., 166
Duranton, J., 381, 384, 385
Dure, L. III, 482
Durr, M., 230
Dutky, S. R., 29, 32, 44, 45
Dutton, A. H., 143
Duvick, D. N., 515, 516, 523
Duysens, L. N. M., 383, 386, 388, 553, 558
Dyer, T. A., 389, 477, 487, 508, 509
Dygdala, R. S., 181
Dyke, R. W. van, 76, 77
Dylan, T., 306

E

Eaglesham, A. R. J., 311
Earle, E. D., 515, 516, 520
Earle, F. R., 325
Eastwell, K. C., 103
Eaton-Rye, J. J., 388
Ebel, J., 104, 110, 111
Eberly, S. L., 162, 169
Eckenrode, V. K., 511
Ecker, J. R., 328, 331, 336, 337
Eckert, H.-J., 395
Edallo, S., 419, 428
Edelhoch, H., 59, 62, 64, 70-72
Edelman, J., 344, 367
Edelman, M., 481, 490

Edmondson, D. L., 538, 540, 544-47, 550, 567, 575, 577
Edreva, A. M., 344
Edwards, E. E., 165
Edwards, G. E., 553, 555, 556, 577
Edwards, G. R., 209
Edwards, K., 301, 312
Egelhoff, T. T., 298, 304, 305
Egin-Buhler, B., 104
Egneus, H. S., 476
Ehlers, W., 247
Ehret, D. L., 368, 370
Ehrlich-Rogozinski, S., 87
Ehwald, R., 256, 363
Eidenbock, M. P., 257, 258, 277
Eisenberg, A. J., 188, 459, 462
Ekelund, N. G. A., 167
Ekhato, I. V., 32
Elgersma, A., 207
El Giahmi, A. A., 235
Ellis, J. R., 477, 558
Ellis, R. J., 479, 490
Elmes, R. P., 236
Elsner-Menzel, C., 486
Elston, J. F., 287
Emanuel, G. B., 256
Emons, A. M. C., 55
Engelke, A. L., 194
England, R. R., 396
English, P. D., 345
Enser, U., 550
Epel, B., 162
Epp, O., 387
Epstein, E., 288, 447
Epstein, H. T., 481
Epstein, L., 327, 328
Erez, A., 288, 447
Erickson, J. M., 389, 485, 492
Erickson, L. R., 514
ERICKSON, R. O., 1-22; 7, 8, 10-17, 19, 20, 192-95, 272, 275
Erxleben, I., 453
Esau, K., 322, 334
Eschrich, W., 363, 365
Escote, L. J., 519
Espelie, K. E., 342
Esquerré-Tugayé, M. T., 336
Etienne, A.-L., 387, 389, 391, 395
Evans, D. A., 414, 424, 426
Evans, D. E., 62, 66
Evans, E. H., 385
Evans, G. C., 280
Evans, I. J., 237, 305
Evans, J. R., 537, 539, 544, 571, 572
Evans, L., 55
Evans, L. T., 176, 177, 184, 185, 187, 191, 193, 194, 205-7, 356, 357, 370, 565

Evans, M. C. W., 383-85, 388
Evans, M. L., 457, 458
Evelo, R., 393
Everard, J. D., 257
Everdeen, D., 341
Everett, H. L., 515, 516, 518
Everett, M., 481, 488
Evert, F., 363
Evert, R. F., 86
Evola, S. V., 425, 426
Ewijk, W. van, 141

F

Faber, A.-M., 515
Faiz, S. M. A., 247-49
Fajer, J., 382
Falconet, D., 505
Falkowski, P. G., 390
Fan, J. Y., 58
Fardeau, J.-C., 228, 229
Farnden, K. J. F., 149
Farquhar, G. D., 279, 455,
    535-39, 541-44, 549, 553,
    574, 575
Farquhar, M. G., 57-59, 73,
    83, 84
Farrar, J. F., 357, 360, 565
Farrar, S. C., 565
Faulkner, K., 421
Fauron, C. M.-R., 505, 517,
    519
Faust, M. A., 162, 163, 169,
    182, 184
Fazekas de St. Groth, C., 247
Fearey, B. L., 382, 386
Fedoroff, N., 425
Feenstra, W. J., 310
Feher, G., 390
Feiler, H. S., 513, 514, 522
Feldman, L. J., 458
Felker, F. C., 366, 371
Fell, D. A., 554
Fensom, D. S., 363
Fenton, J. M., 386
Fereres, E., 286
Ferguson, I. B., 231
Ferrante, G., 122, 125
Field, C., 85, 571
Fieuw, S., 364, 367
Figura, K. von, 80, 83
Filser, M., 309
Finan, T. M., 306
Finch, J. T., 59, 61, 68, 73
Fincher, G. B., 325, 338
Findlay, J. B. C., 145
Fine, R. E., 59, 60, 62, 70, 75,
    76, 79, 80, 82-84
Finkelstein, R. R., 459, 462,
    463
Finlayson, S. A., 103, 104
Finnegan, P. M., 521

Firmin, J. L., 237, 298, 301,
    312
Firn, R. D., 274
Fischbach, G., 82
Fischer, E., 457
Fischer, W., 462, 463
Fiscus, E. L., 253, 255, 257,
    260
Fish, L. E., 381
Fisher, D. B., 359, 565
Fisher, D. G., 363
Fisher, R. A., 7
Fisher, R. F., 298, 305
Fishman, J. B., 84
Fitter, A. H., 233, 282
Fitzsimons, P. J., 461
Flavell, R. B., 507, 517
Fleming, G. R., 386
Fliege, R., 554
Flippen-Anderson, J. L., 29,
    32, 44, 45
Fliss, A. E., 514, 515, 517
Flood, R. G., 178, 197
Flügge, U. I., 554, 558
Fluhr, R., 476, 490
Foard, D. E., 269, 271
Fondy, B. R., 360, 565, 580
Fong, F., 446, 447, 460, 572
Fontaine, D., 183, 185
Ford, M. A., 361
Ford, R. C., 380, 388
Ford, V. S., 197
Forde, B. G., 510, 515, 516,
    518, 519
Foreman, R. E., 164, 165, 167
Forgac, M., 76, 77
Fork, D. C., 479
Förster, V., 394, 395
Fortin, M. G., 144, 148
Forward, R. B., 167
Foster, K. W., 169, 310
Foster, V., 150
Fourcy, A., 229
Fowke, L. C., 55, 57, 58, 60-
    62, 66, 67, 77, 85, 87
Fox, T. C., 565
Fox, T. D., 510-12, 515
Foyer, C., 232, 561, 562
Foyer, C. H., 556, 558-60, 566
Frachisse, J.-M., 201
Fradkin, L. I., 491
Fraley, R. T., 429
Franceschi, V. R., 145, 147,
    342, 360, 362
Francis, D., 189-91
Frank, G., 395, 484
Frank, H. A., 382
Franke, J. E., 384
Franke, W. W., 58, 73, 85
Frankel, O. H., 197
Franssen, H. J., 313, 330, 334
Franzén, L.-G., 387, 390, 392,
    395

Frasch, W. D., 395
Freas, S., 541, 547
Freas, S. M., 541, 547, 548
Freeling, M., 271, 425
Freeman, H. C., 386, 387
Frens, G., 142
Frentzen, M., 108, 109, 119
Freudenberg, K., 344
Freundt, H., 60-62, 66, 67, 69
Frieden, C., 570
Friend, D. J. C., 202, 203
Friend, D. S., 58
Fromm, H., 490
Frost, J. W., 298, 301, 312
Frova, C., 371
Fry, J. C., 274, 322
Fry, S. C., 232, 274, 275, 322,
    327, 340, 341
Fuchs, C., 271
Fujimura, Y., 395
Fujioka, S., 204
Fujita, E., 382
Fujita, F., 47
Fujita, T., 179
Fujita, Y., 162, 163, 168, 388,
    390
Fukui, K., 426
Fukuzama, H., 476, 479, 480
Fukuzawa, H., 380, 381, 384,
    387
Fuller, F. F., 313
Fuller, W. A., 368
Fung, S., 32
Furbank, R. T., 555, 558-63,
    566, 568, 580
Furrer, R., 383

G

Gabay-Laughnan, S. J., 504,
    506, 515, 518, 519
Gabriel, D. W., 298, 305, 312
Gaff, D. F., 477
Gage, D. A., 442, 449, 450,
    454
Gahan, P. B., 189
Galey, F. R., 59
Galfre, G., 144, 148, 441
Gallagher, S. R., 60
Galle, A.-M., 121
Galliard, T., 102, 126-28
Galling, G., 485
Galloway, C. T., 65
Galston, A. W., 41, 43, 47,
    187, 209, 341
Gamble, P. E., 446, 447, 479,
    480, 488
Ganago, I. B., 391
Garavito, M., 143
Garavito, R. M., 143, 380
Garcia-Arenal, F., 520
Gardiner, S. E., 125
Gardner, C. O., 360

Gardner, G., 389
Gardner, W. R., 246-49, 253
Garland, W., 104
Garraway, M. O., 342
Garrec, J.-P., 229
Gaspar, T., 208, 342
Gasser, C. S., 188
Gast, P., 383
Gatenby, A. A., 86, 486, 494
Gates, D. M., 540, 541
Gauhl, E., 536
Gaultier, J. M., 415
Gauriloff, L. P., 504
Gautheret, R. J., 414
Gay, A. P., 359
Gay, J. L., 227
Gay, M. R., 324
Gayler, K. R., 344, 365
Gaynor, J. J., 306
Gebhart, B., 324, 330
Geen, G. H., 163
Gegenbach, B. G., 517
Geiger, D. R., 358, 360, 363, 565, 580
Geiger, R., 389
Geisow, M. J., 75
Geissmann, T., 340
Gendel, S., 168
Gendraud, M., 459
Gengenbach, B. G., 414, 425, 514, 515, 517
Gennity, J. M., 122
Georgieva, I. D., 344
Georgieva-Hanson, V., 62, 65
Gerard, M. J., 230
Gerats, A. G. M., 198, 209
Gerdemann, J. W., 230, 232
Gerhardt, B., 127, 128
Gerhardt, P., 86
Gerhardt, R., 540, 552, 554, 557, 562, 563, 565, 566, 577
Gerken, S., 380, 396
Gerlach, W. L., 426
Germann, I., 158, 159
Gerola, F. M., 341, 342
Gesteland, R. F., 517
Getz, H. P., 363, 365-67, 369
Geuns, J. M. C., 39, 43
Geuze, H. J., 80, 83
Ghanotakis, D. F., 387, 390, 395, 396
Ghisi, R., 479
Giafynazzi-Pearson, V., 230
Gianinazzi, S., 222, 224, 226-30, 232, 234
GIANINAZZI-PEARSON, V., 221-44; 222-24, 226-30, 232, 234
Giaquinta, R. T., 231, 355, 357, 360, 362-64
Gibbs, M., 564
Gibson, A. H., 308, 310

Gierl, R., 426
Giersch, C., 545, 547, 555, 559, 578
Gifford, R. M., 203, 356, 359, 360, 565
Giles, K. L., 166, 167
Gill, B. S., 418-20, 422
Gillie, J. K., 382, 386
Gilman, R., 1
Gimmler, H., 440, 451, 452, 455
Ginaninazzi, S., 223, 228
Ginsburg, H., 507, 508, 511, 519
Girard-Bascou, J., 485, 492
Girvin, M. E., 380
Givan, C. V., 102
Glad, G., 120, 121
Glasziou, K. T., 344, 365
Glazer, A. N., 380, 389, 476
Gleadow, R. M., 358, 361, 370, 371
Gleiter, H., 392, 394, 396
Glick, B. S., 57, 83
Glickman, J., 77
Glidewell, S. M., 539
Gloudemans, T., 313, 330, 334
Gluckin, D. S., 345
Glund, K., 455
Goddard, D. R., 11, 12
Goddard, R. D., 275
Goel, A., 486
Goffeau, A., 509, 510
Golbeck, J. H., 384-86, 389
Goldberg, R. B., 188
Goldbeter, A., 571
Goldenthal, K. L., 58
Goldschmidt, E. E., 111, 184
Goldstein, J. L., 54, 73, 79-82
Goldstein, M. S., 59, 68, 80
Goldstein, R. A., 386
Gollin, D. J., 187, 209
Gonthier, R., 189
Goodchild, D. J., 141, 143-45, 148, 479
Goodenough, U. W., 324, 325, 330
Goodin, D. B., 392
Goodman, H. M., 429
Goodman, M. M., 519
Goodman, R. N., 306, 311
Goodwin, R. H., 11
Goodwin, T. W., 450
Gorden, P., 58, 82
Gordon, A. J., 360, 565
Gostan, J., 161
Gottesman, S., 490
Gottlieb, L. D., 197
Gottschalk, M., 508
Goud, B., 73
Goudrian, J., 16
Gough, S. P., 478
Gould, A. R., 417, 422

Gounaris, I., 479
Gounaris, K., 387, 390, 395, 476-79
Gourret, J. P., 229
Govers, F., 313
Govindjee, 386-89, 396
Goydych, W., 298, 312
Gozdzicka-Jozefiak, A., 486
Grabau, E. A., 508, 512
Gräber, P., 394
Gracen, V. E., 515, 516, 518, 520
Graebe, J. E., 205, 206
Grafstrom, R. H., 425
Graham, D., 551, 572
Graham, J. H., 228, 233
Graham, J. S., 312
Graham, T. L., 306
Grambow, H. J., 342
Grange, R. I., 370
Grantz, D. A., 455
Gray, M. W., 504, 508, 511-13, 518, 523
Gray, R. T., 442
Grayburn, W. S., 506
Greacen, E. L., 248, 260
Greaves, A. W., 32
Green, C. E., 414, 416, 419, 517
Green, G. H., 163
Green, M. M., 415, 421, 428
Green, P. B., 195, 196, 209, 268, 274, 276
Green, P. J., 283, 476
Green, S. A., 82
Greenberg, B. M., 486, 487
Greene, L. A., 82
Greenland, A. J., 281
Greenwood, A. D., 477
Greenwood, J. S., 142, 143, 145, 151
Gregorini, G., 205
Gregory, L. E., 25, 27, 34-36, 38, 41-44
Gregory, L. M., 271, 275, 276
Gregory, P., 516
Gregory, R. A., 182, 209
Grellert, E. A., 59
Greppin, H., 166, 167, 179, 189, 201, 208, 284, 342
Gressel, J., 179, 181, 182, 186-88, 191, 199, 487
GRESSHOFF, P. M., 297-319; 298, 299, 307-11
Grief, C., 327
Grienenberger, J.-M., 507, 508
Griffin, D., 443
Griffin, D. A., 448
Griffin, F. M., 82
Griffing, L. R., 57, 58, 60-62, 66, 67, 77, 85, 87
Griffiths, G., 57, 83, 84, 141
Griggs, P., 441

Grigor'eva, N. Y., 200
Grime, J. P., 282
Grimmett, M. M., 182
Grippiolo, R., 226
Grisebach, H., 284
Grivell, L., 504, 507
Groose, R. W., 425, 426
Groot, G. S. P., 486, 487
Groot, S. P. C., 440, 460, 463
Gross, D., 440
Gross, G. G., 340
Grossman, A. R., 484, 493
Grossman, L. I., 508
Grove, M. D., 24, 27, 28, 44
Gruber, P. J., 150
Gruissem, W., 480, 485-87, 494
Grun, P., 515
Grund, C., 62, 67, 68, 71
Grunfeld, C., 58
Gubler, F., 144, 147
Guerra, D. J., 108-11, 119
Guerrero, F., 446, 451
Gueze, H. J., 142
Guha-Mukherjee, S., 284
Guiking, P., 162, 163, 169
Guiles, R. D., 384, 385, 392-94
Guillemaut, P., 508
Guiry, M. D., 158-60
Gunn, R. E., 507
Gunstone, F. D., 114, 122, 126
Guo, H. Y., 229
Gupta, L. M., 117
Gurney-Smith, M., 324
Gurr, M. I., 118, 120, 123
Guss, J. M., 386, 387
Gussin, A. E. S., 345
Gusta, L. V., 460
Gustafson, J. R., 418, 420, 422, 429
Gutowsky, H. S., 396
Gutteridge, S., 541, 547, 549, 559, 561
Guy, L. W., 179
Gwynne, H., 112, 114, 115

H

Haber, A. H., 269, 271
Haberlandt, G., 322, 332, 333
Hack, E., 509, 510, 513, 514
Hackett, W. P., 201, 202
Häder, D.-P., 158
Hadwiger, L. A., 343
Haehnel, W., 386
Hageman, R. H., 535
Hagemann, R., 395
Hagen, G., 508
Hagerman, R. H., 232
Hahlbrock, K., 110, 111, 284, 301, 312
Haiker, H., 509

Hainfield, J. F., 61
Hainsworth, J. M., 247
Hake, S., 271
Halban, P., 57, 84
Hale, K. A., 227, 230
Halevy, A. H., 176, 178, 185, 202
Halfer, C., 422
Hall, J. L., 274
Hall, K. C., 451
Hall, M. A., 274
Hall, N. P., 538, 540, 541
Hallgren, L., 462
Hallick, R. B., 479, 486, 487
Halliwell, B., 559
Halloran, G. M., 178, 197, 358, 361, 370
Halperin, W., 194, 268
Halverson, J. L., 306, 307, 312
Hamada, K., 41-44
Hamilton, D. L., 425
Hammerschmid, R., 336
Hammond, L. C., 247-49
Hamzi, H. Q., 194
Handa, A. K., 460
Hanhart, C. J., 207
Hanley-Bowdoin, L., 479, 488, 494
Hannapel, D. J., 109
Hansen, A., 462
Hanson, D. D., 188
Hanson, D. K., 515
Hanson, M. R., 504, 506, 515, 521, 523
Hanson, P. J., 256
Hanspal, M., 59, 69-71
Hanssen, H.-P., 440, 443
Hansson, Ö., 387, 388, 390, 392-95
Hanssum, B., 394
Harada, K., 442
Harbron, S., 556
Hardham, A. R., 196
Hardie, K., 233
Harding, J., 112, 114, 115
Hardman, J. K., 168
Hardy, P. H. Jr., 142
Harikumar, P., 77
Harkin, J. M., 344
Harley, J. L., 222, 224, 227, 230, 231, 233-35
Harley, P. C., 541
Harmey, M. A., 507
Harold, F. M., 229
Harpster, M., 476
Harris, G. C., 561, 562
Harris, M., 420
Harris, M. J., 441, 455, 456
Harris, N., 144, 149
Harris, P., 118
Harris, P. J., 150, 338
Harris, R. V., 122
Harris, W. F., 15

Harris, W. M., 332
Harrison, J. R., 65
Harrison, M. A., 453, 454
Harrison, S. C., 62-66, 68, 70, 72
Harrowell, P. R., 386, 387
Hartley, M. R., 481
Hartmann, D., 60-62, 66, 67, 69, 77
Hartmann-Bouillon, M.-A., 117
Hartung, W., 440, 451, 452, 455-57, 461
Harvey, B. L., 415
HARWOOD, J. L., 101-38; 102-4, 111-18, 122-27
Hasegawa, P. M., 460
Hashimoto, T., 128
Hasilik, A., 80, 83
Hatch, A. L., 548
Hatch, M. D., 365
Hattori, A., 162, 163
Hattori, H., 24
Haupt, W., 158, 166
Haury, J. F., 162, 163
Hausmann, K., 58
Havelange, A., 189, 192, 203, 205
Hawes, C. R., 55, 62, 66
Hawke, J. C., 103, 104, 123
Hawker, J. S., 367, 368, 556
Hayami, H., 32
Hayashi, T., 337
Hayashida, N., 380, 381, 384, 476, 479, 480, 488
Hayes, J. M., 382, 386
Hayes, P. M., 362, 363
Hayman, D. S., 227, 228, 235
Haynes, R. H., 422
Heald, J. K., 453, 455
Hearst, J. E., 389, 481
Heath, R. L., 387, 572
Heathcote, P., 380, 383-86
Heber, U., 540, 545, 547, 550, 558-62, 567, 578
Hedden, P., 441, 448
Hedrich, R., 457
Heide, O. M., 205, 206
Heilmann, B., 440, 451, 452, 455
Heimann, K., 549
Heineke, D., 540, 552, 562, 563, 566
Heinrich, R., 573
Heinz, E., 109, 119, 122-25
Heinz, K.-P., 102
Heise, K., 103
Heise, K.-P., 102
Heisel, R., 506, 509-12
Hejgaard, J., 462
Hejnowicz, Z., 12
Heldt, H. W., 540, 545, 547, 550, 552, 554, 557-59, 561-66, 568, 571, 577

Helgeson, J. P., 324, 326
Hellyer, A., 103-5, 112, 114, 115
Helmy, S., 75, 83
Hemleben, V., 188
Hemrika-Wagner, A. M., 167
Henderson, D. W., 286
Hendricks, R. H., 323
Hendrix, D. L., 451, 456, 459, 565
Henningsen, K. W., 478
Henry, E. W., 36, 47
Henry, R. J., 367
Henson, I. E., 200, 205, 446, 447, 451, 456
Henwood, J. A., 144, 147, 148
Hepler, P. K., 179
Hepper, C. M., 223, 224, 228
Herbon, L. A., 504-6
Herkelrath, W. N., 247-49
HERMAN, E. M., 139-55; 141, 144, 145, 147, 149, 151
Hermans, J., 382
Hermo, L., 58
Herold, A., 561, 563-65
Herrin, D., 484
Herrin, D. L., 485, 488, 492
Herrmann, R. G., 381, 390, 478-80, 486, 488, 491
Herth, W., 85
Herzog, B., 557, 565
Herzog, V., 83
Hesketh, J. D., 247
Heslop-Harrison, J., 477
Hesse, F., 247
Heukart, M., 68
Heuser, J., 63, 64, 68
Heuser, J. E., 55, 80, 324-26, 330
Hewett, J., 365
Hewitt, F. R., 24
Hewitt, J., 368
Hibberd, K. A., 418
Hickey, E., 484
Hideg, É., 389
Higashi, S., 307
Higgins, A., 310
Higgins, J., 573
Higgins, T. J. V., 462
Hikichi, M., 181, 209
Hildebrand, D. F., 125
Hill, A. B., 429
Hill, D. F., 59, 61, 80
Hillman, J. R., 456
Hillman, W. S., 160
Hillmer, S., 57, 77, 85
Hills, G. J., 322, 324, 327
Hills, M. J., 566
Hinchee, M. A. W., 188
Hind, G., 559
Hinkley, M. A., 223
Hinson, K., 308
Hirai, N., 440, 442-45, 453

Hiroi, T., 280
Hirosawa, T., 169
Hirsch, A. M., 306
Hirsch, P. R., 237
Hitchcock, C., 124, 126
Hitz, W. D., 356, 360
Hiyama, T., 382, 384
Ho, I., 223, 234
HO, L. C., 355-78; 357, 359-61, 368, 370, 371
Ho, T.-H. D., 454, 455, 462, 523
Hoad, G. V., 359, 361, 368, 371
Hobé, J. H., 389
Hodge, T. P., 504-6, 517, 519, 521
Hodges, A., 18
Hodgson, J. M., 102
Hoff, A. J., 382, 383, 393
Hoffman, B. M., 385
Hoffman, H., 301, 312
Hoffman, L. M., 147, 151
Hofstra, G., 177
Høj, P. B., 105-11, 115, 381, 384, 385
Holden, C., 10, 11
Hole, D. J., 446, 447
Hollingsworth, M. J., 479
Hollingsworth, R. I., 306, 307
Holmes, N., 62, 65
Holmes, N. J., 65
Holschuh, K., 488
Homann, P. H., 387, 396
Hombrecher, G., 298, 305
Homna, M. A., 305
Hong, J. C., 325, 330, 331
Hooykaas, P., 313
Hopkins, C. R., 80, 81
Hore, P. J., 383
Horgan, P. A., 28, 45
Horgan, R., 200, 205, 440, 442, 443, 446-48, 450, 452, 453, 455, 459, 460
Horie, T., 16
Horio, T., 395
Horisberger, M., 141, 142, 144
Hornberg, C., 451, 452, 461
Horridge, J. S., 185, 193
Horrum, M. A., 481
Horsch, R. B., 429
Horsley, D., 62, 66
Horton, P., 390, 559, 560, 568, 578
Horton, R. F., 288, 459
Hortsch, M., 141
Horvath, B., 298, 305
Hösel, W., 344
Hoskyns, P., 209
Hotta, Y., 10
Houchins, J. P., 519
Hough, T., 24
Howald, I., 85

Howe, C. J., 506
Howland, G. P., 165
Hoyer-Hansen, G., 481
Hrabak, E. M., 306
Hsiao, T. C., 246, 277, 278, 284, 286, 287
Hsu, B.-D., 389
Hsu, F. C., 364
Hsu, T. C., 420
Hu, W. W. L., 505, 507, 518, 519
Hua, E. L., 65, 76
Huang, A. H. C., 150
Huang, K. P., 111, 114, 115
Huang, S. Z., 305, 307, 309
Hubbard, A. L., 81
Hubbard, J. A. M., 384
Huber, D. J., 86, 346
Huber, R., 387
Huber, S. C., 556, 557, 565, 566
Hübner, R., 57, 86
Huck, M. G., 247, 248
Huet, C., 73
Huff, A., 479, 480
Hughes, A. P., 280
Hughes-Jones, N. C., 142
Huguet, T., 429
Humbeck, K., 162, 163
Humphrey, G. H., 163
Humphreys, G. B., 256
Humphries, E. C., 268
Hung, L., 306
Hungerford, W. E., 202
Hunter, C. N., 389, 487
Hunter, J. E., 365
Hunziker, D., 391, 392, 396
Hurd, R. G., 268, 359
Husken, D., 285
Huth, K., 159
Hynes, M. F., 306
Hynes, R. O., 329

I

Iams, K. P., 508
Ibrahim, N. G., 490
Ichikawa, Y., 32
Ichimura, M., 443, 444
Ideguchi, T., 395
Idziak, E.-M., 164, 165
Ielpi, L., 306
Iino, M., 461
Ikeda, M., 27
Ikegami, I., 383, 384
Ikekawa, M., 28, 32
Ikekawa, N., 27-29, 32-34, 37, 42-44
Ikeuchi, M., 387, 391
Imahori, K., 164, 167
Imai, K., 539
Imaoka, A., 395
Imber, D., 288, 447

Imhoff, C., 183, 185
Imlay, K. R. C., 480
Inamine, G., 485, 490, 491
Incoll, L. D., 565
Ingram, D. S., 517
Innes, R. W., 298, 301, 305, 312
Inoue, H., 356
Inoue, T., 396
Inoue, Y., 387, 389, 391, 392, 394-96
Irace, G., 59
Irvin, R. T., 28, 45
Isaac, P. G., 510-12, 516, 521, 522
Isaacson, R. A., 390
Isavalev, L., 80
Ise, W., 509
Ishiguro, M., 32
Ishikawa, M., 460
Isogai, Y., 387
Israel, H. W., 323
Itoh, S., 383, 384, 387, 389, 391, 396
Iwaki, M., 384, 396
Izhar, S., 521

J

Jaarsveld, P. P. van, 59, 72
Jackson, A. P., 62, 65
Jackson, M. B., 451
Jacky, P. B., 423
Jacobs, T. W., 305
Jacobs, W. P., 177, 207, 208
Jacobsen, E., 310
Jacobsen, J. V., 462
Jacobson, B. S., 572
Jacobson, J. V., 144, 147
Jacqmard, A., 177, 188, 189, 191, 192, 205, 358
Jacques, R., 176
Jaffe, L., 166
Jaffe, M. J., 200
Jagendorf, A. T., 493
Jakobsen, I., 223
James, A. T., 118, 120, 122, 123
Jan, L., 55
Jan, Y., 55
Jancey, R. C., 418
Janick, J., 204
Jank, H. W., 418, 421
Jansen, A. M. A. C., 141
Jansen, M. A. K., 389
Jansen, T., 382
Jansson, C., 382, 395, 484
Jarvis, P. G., 276, 283, 287
Jasani, B., 144
Jaworski, J. G., 110, 111, 118
Jeanmaire, C., 226, 232
Jeannin, G., 508
Jeffcoat, B., 184, 193, 194

Jefford, T. G., 367
Jeffree, C. E., 274, 275
Jegla, D. E., 184
Jelaska, S., 416, 422
Jellings, A., 558
Jenkins, G. I., 481
Jenner, C. F., 361, 366, 368, 371
Jennings, A. C., 361
Jensen, C. J., 103
Jensen, K. H., 485, 492
Jensen, R. G., 535, 540, 544-50, 569, 577
Jepsen, B. R., 381
Jeschke, W. D., 258, 259, 455
Jeune, B., 271
Jiménez-Martinez, A., 322
Jing, J., 277, 278, 284, 286, 287
Joachim, S., 55, 85
Johanningmeier, U., 389
Johansen, E., 306
Johansen, L. G., 459
John, M., 298, 304, 305
Johnson, C. B., 283
Johnson, C. E., 385
Johnson, C. R., 234
Johnson, D. R., 232
Johnson, G. A., 259
Johnson, M. L., 62, 64
Johnson, M. P., 283
Johnson, S. S., 420-23
Johnson, V. A., 360
Johnston, A. W. B., 149, 237, 298, 301, 305, 306, 312
Johnston, I. R., 322
Johnston, R. N., 429
Johri, M. M., 184
Jolles, P., 76
Jolliffe, P. A., 562
Jolliot, A., 121, 122
Jolly, S. O., 484, 486
Jones, A. L., 125, 126
Jones, A. V. M., 123
Jones, E. V., 71, 73
Jones, H., 258, 259
Jones, H. G., 441, 574
Jones, J. W., 247-49
Jones, M. G. K., 223
Jones, M. M., 286, 287
Jones, Q., 325
Jones, R. L., 86
Jones, V. P., 504, 510-12, 514
Jones, W. K., 384, 385
Jong, K., 273
Joo, F., 122
Jordan, B. R., 111, 113, 114, 116, 117
Jordan, D. B., 536-38, 546, 548, 549, 551
Jordan, D. C., 297, 299
Jordan, W. R., 246
Jorgensen, K. G., 462

Jorgensen, R. A., 481, 488
Jorna, M. L., 447
Jörnvall, H., 346
Joseph, C., 180, 182, 205
Joseph, M. C., 540
Journet, E.-P., 103
Joyard, J., 108, 119, 122, 124, 450
Juang, T. C., 229
Juguelin, H., 117
Julienne, M., 125
Junge, W., 394, 395, 559
Junttila, O., 459
Jurgenson, J. E., 485, 488
Jurik, T. W., 281
Jursinic, P. A., 389, 396
Justin, A. M., 120-22
Juyal, R. C., 423

K

Kacser, H., 573, 578
Kader, J.-C., 121, 125
Kadota, K., 59, 71, 76
Kadota, T., 76
Kahlem, G., 196
Kaihara, S., 200, 204
Kaiser, G., 455, 545, 547, 559, 565, 578
Kaiser, W. M., 451, 452, 456
Kajiura, H., 395
Kakuno, T., 395
Kalinich, J. F., 40, 43, 47
Kallarackal, J., 360
Kalt-Torres, W., 556, 565
Kamalay, J. C., 188
Kamate, K., 187
Kambara, T., 387
Kamikubo, T., 481, 488
Kamiya, A., 162, 163, 169
Kanarek, L., 61, 144
Kanaseki, T., 59, 71
Kanchanapoom, M. L., 189, 191
Kandeler, R., 177, 179, 208, 209
Kane, M. E., 459
Kanemoto, S., 32
Kannangara, C. G., 103, 478
Kao, T.-H., 511, 512
Kaplan, J., 79, 81
Kaplan, S., 476, 481
Karege, F., 179, 201
Karp, A., 414, 427
Karssen, C. M., 440, 447, 460, 461
Kartenbeck, J., 58, 70, 73, 82
Karunaratne, R. S., 228
Kashyap, L., 306
Kasperbauer, M. J., 280
Kasprik, W., 440, 452
Kassam, A. H., 287
Kastelija, D., 75

Kataoka, H., 164, 165
Katayama, M., 443
Kates, M., 122, 125
Kato, N., 27, 43
Kato, T. A., 422, 444
Katoh, S., 391
Katona, L., 323, 326
Katsumi, M., 35-38, 40
Kaufman, L. S., 162
Kaufman, P. B., 273
Kaufman, R. J., 429, 430
Kaufmann, M. R., 255, 257
Kaur-Sawhney, R., 187, 209
Kavanagh, A. J., 13, 20, 272
Kavon, D. L., 199
Kawaguchi, A., 110
Kawamori, A., 392, 394
Kawano, K., 356
Kawase, M., 275
Kay, J., 113, 114, 117
Keck, K. W., 536
Kedersha, N. L., 59, 61, 80
Keegstra, K., 119, 325
Keen, J. H., 58, 59, 62-64, 66, 70-73, 76
Keen, N. T., 343
Keith, B., 479
Kekwick, R. G. O., 103, 109-13
Keller, G. A., 142, 143, 145
Keller, W. A., 416, 417, 419, 420, 422, 426, 460
Kelly, G. J., 568
Kelly, R. B., 59-62, 67, 76, 80, 81, 84
Kelly, S., 77
Kelly, W. G., 62, 67, 122
Kemble, R. J., 506, 507, 514, 517, 519
Kemp, A., 204, 207
Kemp, D. R., 273, 281, 282
Kerr, P. S., 556, 565
Kessel, R. M. J., 112
Kessell, R., 112, 114, 115
Kessissoglou, D. P., 394
Keus, R. J. A., 486, 487
Key, J. L., 39, 42, 75, 325, 330, 331
Keys, A. J., 535, 537, 538, 541, 547, 549, 559, 561
Khan, M., 125
Kheyr-Pour, A., 518
Khripach, V. A., 32
Khush, G. S., 418, 427
Kidby, D. K., 223
Kiefer, S., 341
Kieliszewski, M., 325
Kiernan, M. C. C., 506
Kijne, J. W., 306
Killham, K., 232
Kilmartin, J. V., 65
Kim, B. D., 515, 519
Kim, J. H., 276, 287

Kim, S. K., 27
Kimpel, J. A., 75
Kinden, D. A., 226, 231, 232, 234
Kindl, H., 102
Kinet, J.-M., 176, 177, 179-83, 186-97, 199, 200, 202-9, 358, 359, 361, 372
King, N. J., 322
King, R. W., 179, 185, 193, 194, 201, 205-7, 565
Kinraide, T. B., 366
Kirby, E. J. M., 193, 197, 360, 371
Kirchhausen, T., 62-66, 68, 70, 72
Kirk, D. L., 162, 169, 485, 488
Kirk, J. T. O., 475-79, 481
Kirk, M. M., 162, 169, 485, 488
Kirsch, W., 381
Kishore, G. M., 429
Kitsuwa, T., 27, 28, 43
Klämbt, H. D., 86, 87
Klausner, R. D., 68, 69
Klee, H. J., 429
Klein, M. P., 392, 394
Klein, R. R., 478, 481, 482, 485-87, 490-94
Kleinig, H., 103
Klement, Z., 311
Klepper, B., 247, 248, 260
Klessig, D. F., 484, 490
Klimov, V. V., 389, 391, 394, 396
Klis, F. M., 336, 344
Klosgen, R. B., 426
Klute, A., 255
Knapp, P. H., 182
Knauer, D., 363, 365, 367
Kneen, B. E., 310
Knievel, D. P., 364, 366
Knight, C. D., 298, 305
Knox, J. P., 441
Knox, R. B., 197
Koba, S., 27
Kobayashi, M., 382
Kobza, J., 541, 548, 577
Koch, K. E., 234
Kochankov, V. G., 183
Kochwa, S., 70
Koehler, D. E., 446, 447, 460
Koeppe, D. E., 516
Kogel, K. H., 87
Koguchi, M., 32, 34, 35, 38, 39, 41, 42, 44
Kohchi, T., 380, 381, 384, 387, 476, 479, 480
Köhler, R., 440, 452
Kohtz, D. S., 62, 65
Kohtz, J. D., 62, 65
Koide, R., 227, 235, 258

Koike, H., 389, 395, 396
Kolattukudy, P. E., 106, 116, 342
Kolloffel, C., 275
Kolodner, R., 479
Komor, E., 365, 366, 370
Komori, E., 443
Koncz, C., 507, 519
Kondo, H., 34, 37-39, 42
Kondo, S., 443
Kondorosi, A., 298, 304, 305
Kondorosi, E., 298, 304, 305
Konforti, B. B., 74
Konstantinova, T. N., 186, 187, 200
Koornneef, M., 207, 440, 447, 448, 460, 461
Kopcewicz, J., 200
Köpke, U., 247
Koshimizu, K., 442-44, 453
Koshioka, M., 205
Koshland, D. E., 571, 581
Koske, R. E., 223
Kössel, H., 484, 486, 487
Kosslak, R. M., 308, 309, 311
Kough, J., 223, 228
Kough, J. K. L., 223, 228
Kovaleva, L. V., 188
Kovganko, N., 32
Kowallik, K. V., 479
Kowallik, W., 162, 163
Kozempel, M., 25
Kozinka, V., 256
Kraayenhof, R., 559, 578
Kramer, D. A., 426
Kramer, P. J., 246, 257, 259, 260
Krause, G. H., 560, 577, 581
Krauss, A., 368
Krauss, R. W., 162
Krebs, H. A., 557, 573
Kreimer, G., 549
Krekule, J., 182, 183, 193, 204, 205, 207, 208
Kremer, D. F., 565
Kreuk, K. C. J., 486, 487
Kreutz-Jeanmaire, C., 230
Kreuz, K., 485, 490, 491
Krick, T. P., 453
Kridl, J. C., 507, 508, 511
Kriechbaum, D. G., 206
Krikun, J., 232, 233
Krinsky, N. I., 491
Krishna, K. R., 234
Krishnan, H. B., 145, 147
Krishtalik, L. I., 387
Krizek, D. T., 36, 38, 42
Kronquist, K. E., 59, 61, 80
Krotkov, G., 535
Krotzky, A., 310
Krump, M. A., 541, 547, 548
Krupp, G., 478
Kubicek, Q., 488

Kucey, R. M. N., 228, 232, 234
Kück, U., 381
Kuempel, P. L., 298, 301, 305, 312
Kuhlemeier, C., 283, 476
Kuhn, D. N., 108, 109, 112, 116
Kuismanen, E., 83
Kuldau, G. A., 306
Kumar, N., 68
Kumke, J., 166, 168
Kung, S. D., 479, 485, 486, 493
Kunkel, B., 306
Kunst, L., 477
Kunz, B. A., 422
Kunze, U., 388
Kuo, T. M., 105-8
Kuroiwa, T., 477-79, 482, 485
Kursanov, A. L., 364
Kurtz, D. M., 384, 385
Kurzel, B., 557, 565
Kutschera, U., 274, 458
Kuwabara, T., 387, 388, 395
Kuzmanoff, K. M., 457, 458

L

Laasch, H., 560, 581
Labella, F., 340
Labes, M., 306
Lachno, D. R., 457
Lado, P., 24, 38, 40, 42
Laetsch, W. M., 339, 477, 478, 482
LaFavre, J. S., 311
Lafitte, C., 336
Lagenfelt, G., 387, 392
Lagoutte, B., 381, 384, 385
Lagrimini, L. M., 342, 343
Lahr, W., 452, 455, 456
Laing, W., 545, 547
Laing, W. A., 535, 546, 548, 550, 551, 568
Lainson, R. A., 268
Laisk, A., 536, 541, 543, 550, 555, 561, 579
Laites, G. G., 344
Lakhvich, F. A., 32
Lalk, I., 460
Lam, E., 486
Lamb, C. J., 283, 301, 312, 313
Lamb, J. W., 306, 309
Lambers, H., 310
Lambowitz, A. M., 508
Lamoreaux, R. J., 272
Lamport, D. T. A., 322-28, 336, 340, 341, 343, 345
Landre, P., 270

Lang, A., 16, 158, 176, 180-83, 185, 187, 199, 201, 205, 206
Lange, O. L., 540
Langenbeck-Schwich, B., 342
Lang-Unasch, N., 313
Lantz, S., 395
Lapitan, N. L. V., 419, 420, 422
LaPolla, R. J., 508
Larkin, P. J., 414, 424, 425, 428, 431
Larkins, A. P., 144, 148
LaRosa, P. C., 460
Larrieu, C., 180, 181
Larsson, C., 395
LaRue, T. A., 310
Latshaw, S., 486
Latunde-Dada, A. O., 427
Latzko, E., 121, 568
Lauenroth, W. K., 250
Laughnan, J. R., 504, 506, 515, 518, 519
Laulhere, J.-P., 487
Laurière, C., 344
Lauzac, M., 181
Lavergne, J., 393
Lavorel, J., 395
Law, C. N., 178, 197
Law, I. J., 298, 299, 306, 307, 312
Lawley, K., 62, 67, 68, 71
Lawlor, D. W., 232, 235
Lawton, M. A., 283
Lay-Yee, M., 180
Layzell, D. B., 308
Lazaroff, N., 164, 165
Lazarus, C. M., 86
Lazdunski, M., 106, 440
Lea, P. J., 477, 480
Leach, J. E., 324, 326
Leaver, C. J., 504, 507-23
Lebacq, P., 504
Lebrun, M., 487
Lechuga-Deveze, C., 161
Ledoigt, G., 485, 488
Lee, D. R., 202
Lee, J. W., 368
Lee, J.-Y., 389
Lee, K. A., 123, 124
LEE, M., 413-37; 417-19, 421-23, 426, 428
Lee, S.-L. J., 518
Lee, T. S., 443
Leech, R. M., 477, 479, 481
Leegood, R. C., 535, 540, 545, 554, 562, 563, 566, 568-70, 578, 580
Leenen, P. J. M., 141
Lee-Stadelman, O. Y., 276, 287
Lehmann, H., 452, 453, 455
Lehmbeck, J., 381
Leigh, J. A., 306

Leigh, R. A., 258, 259, 368
Lejeune, B., 505
Lejeune, P., 200, 205
Lem, N. W., 123
Lemischka, I. R., 329
Lemmon, S. K., 71, 73
Lemmon, V. T., 71, 73
Lenk, R., 179
Lenton, J. R., 280
Lenz, F., 565
Leon, P., 507
Leonard, J. F., 451
Leonard, M., 358, 361, 372
Leonard, R. T., 60, 231
Le Page-Degivry, M.-T., 440, 459
Lerbs, S., 486
Leroux, B., 441
Le Saint, A.-M., 180, 189
Lescure, A.-M., 479, 480, 486
Lesemann, C., 208
Lessire, R., 110, 111, 116, 117
Lester, G. E., 371
Le Tacon, F., 223
Leunissen, J. L. M., 143
Levings, C. S. III, 505, 507-9, 511-21
Levitan, I., 490
Levy, Y., 232, 233
Lewis, D. H., 231, 232, 281, 561, 563, 565
Lewis, M., 223
Ley, A., 390
Leyden, R. F., 232
Leyton, L., 233
Li, P. H., 460
Li, P. M., 395
Li, X., 394
Li, Y., 447, 449
Lichtenthaler, H. K., 280
Liddell, A. D., 514, 515, 520
Liebisch, H.-W., 440
Liedvogel, B., 102, 103
Liemann, F., 453
Lill, H., 559
Lilley, R. McC., 543, 559
Lima-De-Faria, A., 420, 421
Lin, C. M., 486
Lin, C.-P., 382
Lin, L.-S., 462
Lin, R. I. S., 59
Lin, W., 360, 362
Linden, C. D., 65
Lindenmayer, A., 15
Linderman, R. G., 228
Lindsey, D. L., 228
Linforth, R. S. T., 448
Ling-Lee, M., 229
Link, G., 484, 486, 487, 494
Linotte, A., 358
Linssen, P. W. T., 275
Lippoldt, R. E., 59, 62, 64, 70-72

Lipps, M. J., 167
Lisanti, M. P., 65, 76
Lister, P. G., 446, 451
Little, G., 344
Liu, B.-L. L., 458, 459
Liu, X.-Q., 493
Liverman, J. L., 283
Livingston, G. A., 33
Ljungberg, U., 395, 484
Lloyd, E. J., 281
Loach, K., 280
Lobban, C. S., 160
Löbler, M., 86, 87
LoBrutto, R., 385
Lockau, W., 383, 384
Lockhart, J. A., 276
Lockhart, J. C., 164, 166
Lodge, T. A., 456
Lodish, H. F., 80
Logman, G. J. J., 306
Lombardo, G., 341, 342
Long, S. R., 298, 301, 304, 305, 312
Longland, J., 440, 461
Longley, A. E., 418
Longstreth, D. J., 275
Lonsdale, D. M., 504-9, 515-17, 519, 521
Looney, N. E., 206
Lorimer, G. H., 535, 541, 543, 545, 547-51, 561
Lörz, H., 424, 430, 431
Lösel, D., 234
Lotova, G. N., 199
Lottspeich, F., 324
Lou, C. H., 366, 367
Louason, G., 562
Louden, L., 86
Louvard, D., 57, 62, 73, 83
Louwerse, J., 313
Loveys, B. R., 440, 441, 451-55, 463
Loyal, R., 104
Lozhnikova, V. N., 176, 200, 205
Lucas, J. A., 304, 312, 427
Lucas, W. J., 57, 58, 452
Lucocq, J. M., 58, 84
Ludford, P. M., 323
Ludi, A., 393
Ludlow, M. M., 245
Lugtenberg, B. J. J., 301, 305
Lugtenberg, E. J. J., 306
Lukow, L., 549
Lumsden, P. J., 179
Luna, E., 59, 69-71
Lüning, K., 158-62, 164-66, 168
Lusby, W. R., 29, 32, 44
Luskey, B. D., 59
Luxova, M., 256
Lyman, H., 162
Lynch, D. V., 113, 118, 125

Lyndon, R. F., 188-96, 268
Lyttleton, P., 58, 298

M

Ma, Q. S., 298, 305
Macdonald, R. M., 223
MacDougal, D. T., 282
Macey, M. J. K., 113, 116, 127, 128
Macháčková, I., 207, 208
Mache, R., 479, 486
Mächler, F., 540, 577
Mäder, M., 87, 342
Mackay, L. B., 271, 274-76
MacLachlan, G., 337
MacRobbie, E. A. C., 461
Madhusudana, R., 544, 558
Mae, T., 572
Magill, C. W., 446, 447
Mahadevan, S., 345
Maheshwari, R., 345
Mahfouz, M. N., 415
Maier, V. P., 443
Maier-Maercker, U., 288
Maison-Peteri, B., 387, 395
Maizonnier, D., 198
Makaroff, C. A., 511, 513, 514
Makarow, M., 85, 86
Maki, S. L., 206
Makino, A., 572
Maksymowych, R., 15, 192-95, 268, 271, 272, 275
Maldiney, R., 441, 455
Maliga, P., 430
Malik, N. S. A., 298, 299, 301, 306, 307, 312, 334
Malkin, R., 380-84
Malkin, S., 479
Mallaby, R., 442
Malmberg, R. L., 198, 209
Malnoë, P., 389, 485, 492
Malone, C. P., 516
Maloney, A. P., 511
Mamiya, G., 125
Mancha, M., 117, 119
MANDAVA, N. B., 23-52; 24, 25, 27, 28, 32-38, 40-44, 47
Mander, L. N., 193, 194, 206, 207
Manen, J. F., 144
Manichon, H., 247
Mann, B., 558
Manna, E., 506, 508
Manners, J. M., 227
Manolson, M. F., 77
Mans, R. J., 506, 515, 519
Mansfield, R. W., 383, 384
Manson, J. C., 514, 520
Mäntele, W., 382
Manteuffel, R., 75, 144, 145
Manzine, G., 340

Mapelli, S., 371
Mapes, M. O., 416
Marc, J., 189, 191, 193, 203
Marcus, A., 325, 331
Marcus, S. E., 145
Marechal, L., 507, 508
Mares, D. J., 368
Maretzki, A., 365-67, 370
Mariottini, P., 509
Markhart, A. H., 256
Markovic, O., 346
Markwalder, H. U., 341
Maronde, D., 564
Marples, B. A., 448
Marre, E., 24, 38, 40, 42, 47
Marre, M. T., 24, 38, 40, 42
Marschner, H., 368
Martens, P., 275
Martin, B., 55
Martin, B. A., 123
Martin, D. W., 422
Martin, E. S., 447, 448
Martin, G. C., 184
Martin, J.-.B., 558, 566
Martineau, B., 479
Martinoia, E., 565
Martin-Tanguy, J., 209
Marumo, S., 24, 27-29, 32-34, 37-39, 42, 44, 443
Marx, C., 226, 232
Marx, G. A., 197
Mascarenhas, J. P., 188, 459, 462
Maskell, C. S., 223, 224
Mason, H., 487, 492
Masoner, M., 284
Masuda, Y., 274
Matheson, N. K., 344
Mathew, W. D., 80
Mathews, A., 298, 301, 307, 308, 310, 311
Mathis, P., 380, 381, 383, 386-88, 390, 391, 393, 396
Matile, P., 229, 343, 345
Matlin, K., 57, 83
Matsubara, H., 384
Matsuda, K., 273, 284, 286
Matsumoto, A. K., 64
Matsumura, S., 105, 106
Matsunami, H., 387
Matsuno-Yagi, A., 509
Mattaliano, R. J., 65
Matthees, D., 25
Matthews, D. E. P., 516
Matthews, M. A., 245, 277, 287
Mattson, D. M., 324, 325
Matzinger, D. F., 508, 511, 513
Mau, S. L., 150
Mauch, F., 343
Maupin, P., 55
Mauro, V. P., 148

Mauzerall, D., 390
Mawson, B. T., 451
Maxfield, F. R., 77, 81
Maximov, N. A., 268
Maxwell, J. O., 284, 286
Maycox, P. R., 145
Mayer, S. M., 478
Mayfield, S. P., 395, 480, 484, 491
Mazau, D., 336
Mazliak, K. P., 122
Mazliak, P., 120-22, 124
McCarthy, S. A., 485, 488
McClean, P. E., 504
McClintock, B., 422, 425, 430
McCormack, J. H., 345
McCormick, S., 188
McCourt, P., 125, 477, 478
McCoy, T. J., 415, 417, 419-22, 424, 426-28
McCully, M. E., 261
McCurdy, D. W., 142, 149
McCurry, S. D., 548
McDaniel, C. N., 177, 180, 183, 186, 187, 201
McDermott, A. E., 384, 385, 392-94
McDermott, D. K., 541
McGaw, B. A., 447
McGee, E. E. M., 565
McGrogan, M., 429
McHughen, A., 197
McIndoo, J., 198, 209
McInroy, S. G., 299
McIntosh, L., 484, 486, 487
McKanna, J. A., 58
McKean, M. L., 27, 28
McKeon, T. M., 117-19
McLaren, J. S., 283
McManus, T. T., 122, 123
McMorris, T. C., 32
McNay, J. W., 506, 507, 516, 519
McNeil, D. L., 308, 310
McNeil, M., 150, 232, 321, 322, 325, 345
McVetty, P. B. E., 562
McWilliam, J. R., 251
Mead, J. F., 59
Meagher, R. B., 325, 329, 337
Mecham, R. P., 324, 330
Medina-Filho, H. P., 414
Meek, J., 541, 547
Meeks-Wagner, D. R., 188
Meeson, B. W., 162, 163, 169
Mehdy, M. C., 301, 312
Mei, R., 395
Meiburg, R. F., 390
Meicenheimer, R. D., 15, 193, 195
Meins, F. Jr., 414
Melera, P. W., 429
Melis, A., 380, 389, 390, 476

Mellman, I. S., 54, 87
Mello, R. J., 59, 68, 73, 80, 82
Mellon, J. E., 324, 326
Mendel, R. R., 490
Mendu, N., 507, 508, 511
Menge, J. A., 228, 231
Mentze, J., 273
Mercer, C. A. M., 443
Merchant, S., 493
Meredith, F. I., 233
Merisko, E. M., 57, 59, 73, 74, 88
Merlini, L., 442, 443
Mersey, B. G., 57, 58, 60-62, 66, 67, 77, 85
Mertens, R., 441, 453, 455, 456, 462
Messing, J., 508
Metcalfe, J., 456
Metcalfe, J. C., 440, 457
Mets, L., 386, 389
Metz, J. G., 389, 392
Metzger, J. D., 182, 206
Meudt, W. J., 27, 29, 32, 34, 37, 41, 42, 44
Meunier, J.-.C., 570
Meyer, B., 391, 396
Meyer, D. I., 141
Meyer, F. H., 221
Meyer, H. G., 142
Meyer, I., 389
Meyer, L., 309
Meyer, S., 366, 369
Meyer, Y., 87
Meyerowitz, E. M., 177, 197, 198, 463
Michaelis, A., 415, 418, 421, 428
Michaels, A., 484
Michaels, A. S., 485, 488
Michalowski, C. B., 479
Michel, B. E., 253
Michel, H., 387
Michelena, V. A., 284, 286, 287
Michelini, F. J., 13, 272
Michelis, M. I., 38
Michiel, D. F., 65
Michl, J., 82
Miernyk, J. A., 103, 128
Mierzwa, R. J., 86, 363
Mieskes, G., 557
Miflin, B. J., 477, 480
Miginiac, E., 183, 185, 202, 204, 205, 441, 455, 457
Miki, K., 387
Mikkelsen, J. D., 109-11, 115
Miksche, J. P., 429
Milborrow, B. V., 24, 440-42, 445, 448, 450, 451, 453-55, 461
Milburn, J. A., 360

Milford, G. F. J., 280, 361, 370
Millar, F. K., 323
Millard, B. N., 549, 559, 561
Miller, A., 342
Miller, A.-F., 392, 395
Miller, C., 144
Miller, D. H., 326
Miller, D. M., 253-56
Miller, E. E., 247-50
Miller, J. H., 164, 165
Miller, K. F., 273
Miller, K. W., 257
Miller, M. B., 190, 192
Miller, M. E., 485, 488
Miller, R. H., 477
Miller, R. J., 516
Miller, R. M., 226, 234
Miller, R. W., 325
Millerd, A., 144
Milstein, C., 65
Milthorpe, F. L., 268, 270-73, 281, 285
Milyaeva, E. L., 188
Mims, C. T., 55, 67
Minchin, P. E. H., 363
Mishkind, M. L., 142, 485
Mitchell, D. F., 360, 565
Mitchell, J. W., 24, 25, 27, 33, 36, 41-43
Mitchell, K., 125
Mitra, J., 416
Miyachi, S., 162, 163, 169
Miyamura, S., 477-79, 482, 485
Miyao, M., 387, 390, 392, 395, 396
Miziorko, H. M., 535, 549
Moens, P. B., 10
Moerman, M., 313, 313
Mohan, S. B., 103
Mohan Ram, H. Y., 459
Mohapatra, S. S., 298, 299
Mohr, H., 167, 490
Møller, B. L., 381, 384, 385
Möller, S., 508
Monroe, J. J., 65
Monroy, A. F., 481
Monselise, S. P., 184
Monsi, M., 280
Montavon, M., 179, 201
Montesano, R., 82
Montezinos, D., 86
Montoya, J., 508
Mooibroek, M. J., 65
Moon, E., 511, 512
Moon, G. J., 256, 258, 259
Mooney, H. A., 571
Moore, A. T., 322
Moore, P. J., 150, 338
Moore, R., 446, 447, 458
Moore, T. S. Jr., 150, 232, 233
Moran, K. J., 229

Moran, L. B., 429
Morandi, D., 226, 232, 234
Moreau, P., 117
Moreau, R. A., 121, 126, 127
Morgan, C. L., 356
Morgan, D. C., 279, 282, 283
Morgan, J. M., 286
Morgan, P. W., 179, 189, 194, 206, 208
Mori, K., 27, 32, 33, 37, 44, 45
Mori, S., 86
Morisaki, M., 28, 32, 33, 37, 44
Morishita, T., 27-29, 32, 33, 37, 43
Morita, M., 27
Moriyama, Y., 77
Morizawa, Y., 32
Morré, D. J., 58
Morré, D. M., 58
Morré, J., 179
Morris, C., 62, 73
Morris, D. A., 281
Morris, J., 488
Morris, L. J., 122
Morrison, J. C., 245
Morrison, N., 148
Morrow, D. L., 86
Morton, R. K., 361
Mosbach, A., 559
Mosig, G., 162, 164, 486
Moskowitz, N., 65, 76
Moss, B. L., 159, 169
Mossar, A., 82
Mosse, B., 222, 223, 228, 236
Mott, K. A., 540, 545, 546, 548-50, 561, 569, 575
Mott, R. L., 323
Mountifield, A. C., 359
Moustacas, A. M., 339
Moyer, M., 343
Mudd, A., 541
Mudd, J. B., 102, 122, 123, 126, 129
Mueller, S. C., 65-67
Mueller, W. A., 54, 87
Mukherjee, K. D., 120, 121
Muldoon, E. P., 324, 326-28, 336
Mulkey, T. J., 457, 458
Müller, H., 73
Muller, M., 124
Muller, P., 306
Müller, S., 163, 164, 166
MULLET, J. E., 475-502; 382, 446, 447, 451, 478-82, 485-88, 490-94
Mulligan, J. T., 298, 301, 305
Mulligan, R. M., 511
Mullins, M. G., 194, 206
Munakata, K., 24, 27, 29
Munck, L., 462

Mundy, J., 462
Munns, R., 253-57, 286, 287
Murashige, T., 415
Murata, M., 386, 387, 417, 418, 420, 422
Murata, N., 125, 387, 388, 390, 392, 395, 396
Murata, T., 395
Murfet, I. C., 176-78, 181, 182, 185, 194, 199, 207, 209
Murofushi, N., 204
Murphy, D. J., 102, 120, 121, 123, 124, 555, 558, 568
Murphy, G. J. P., 454
Murray, D., 283
Murray, D. R., 364
Murray, K., 514, 520
Muscatine, L., 231
Myers, A. M., 507-9, 512

N

Nabedryk, E., 382
Nachbaur, J., 75
Nachtwey, D. S., 270
Nagahashi, G., 339, 343
Nagai, J., 118
Nagao, R. T., 325, 330, 331
Nagata, T., 477-79, 482, 485
Nagata, Y., 164, 165, 167
Nagl, W., 429
Nagy, F., 476
Nagy, M., 230
Nakagawa, C. K., 28, 45
Nakamura, C., 416, 417, 419, 420, 422, 426
Nakamura, K., 27, 33, 37
Nakanishi, K., 27, 43
Nakano, R., 415
Nakao, P., 307-10
Nakatani, H. Y., 396
Nakayama, H., 392, 394, 395
Nakazato, M., 382
Nanba, O., 387, 390
Nandi, P. K., 59, 62, 64, 70-72
Nanoyama, Y., 24
Nap, J.-P., 313, 330, 334
Napp-Zinn, K., 177, 178
Nari, J., 339
Narita, J. O., 486, 487
Nash, B., 485, 490, 491
Nasini, G., 442, 443
Natsume, M., 443
Nayudu, M., 298, 312
Neales, T. F., 565
Nechushtai, R., 387, 491
Nee, T., 115
Neet, K. E., 570
Negbi, M., 182
Neill, S. J., 440, 442, 443, 446-48, 450, 452, 453, 455, 459, 460

Neimanis, S., 558, 559
Nelson, C. D., 535
Nelson, C. E., 233
Nelson, M. D., 443
Nelson, N., 381, 385, 387, 478-80, 491
Nelson, O. E., 366, 371
Nelson, R. S., 427
Nemec, S., 233
Nemson, J. A., 288
Nes, W. D., 102, 129
Nes, W. R., 27, 28
Nessel, A., 342
Nester, E. W., 306
Netting, A. G., 442
Neuffer, M. G., 425
Neukom, H., 340, 341
Neumann, D., 75, 144, 145
Neuport, W., 507
Neushul, M., 166
Nevalainen, L. T., 85
Nevers, P., 426
Nevins, D. J., 86
Nevo, Y., 447, 448
Newcomb, E. H., 54, 57, 58, 60, 62, 65, 144, 150, 477
Newcomb, W., 298, 299
Newman, E. I., 246, 253, 254, 258
Newman, G. R., 144
Newsholme, E. A., 573, 574
NEWTON, K. J., 503-32; 504, 505, 507, 511, 513, 514, 516, 522, 523
Newton, P., 270, 271, 280, 281
Nguyen, T., 150
Nicholas, D. J. D., 232, 234
Nichols, B. W., 120, 122, 124, 126
Nickoloff, J. A., 479
Nicolas, M. E., 371
Niebiolo, C. M., 523
Nieden, U.-z., 144, 145
Nieden, U. zur, 75
Niehaus, K., 306
Nielsen, O. F., 478
Nieman, R. A., 459
Nieto-Sotelo, J., 324, 326, 327, 329, 331, 335, 523
Niitsu, M., 444
Nikolau, B. J., 103-5, 484, 490
Nilson, E. B., 360
Nimmo, J. R., 250
Nishida, I., 110, 125
Nishimura, M., 387, 391
Nishioka, H., 200
Nishizawa, N., 86
Nitschke, W., 384
Nivison, H. T., 490, 504
Nixon, P. J., 389, 487
Noat, G., 339
Nobécourt, P., 414

Nobel, P. S., 250, 251, 275, 281, 539
Nobs, M. A., 536
Nockolds, C., 229
Noin, M., 191, 206
Nolan, R. C., 462
Noma, M., 206
Nonhebel, H. M., 442, 448, 450, 451, 454
Norman, H. A., 113, 125
Norman, S. M., 443
Norris, J. R., 382
Norris, V. A., 386, 387
Northcote, D. H., 104, 322
Nösberger, J., 540, 577
Nougarède, A., 186, 189, 191, 193
Novaes-Ledieu, M., 322
Nover, L., 75
Novick, P., 85
Nozaki, K., 32
Nugent, J. H. A., 384
Nuijs, A. M., 380, 383, 386, 388, 390
Nukina, M., 27
Nultsch, W., 169
Nutman, P. S., 307
Nye, P., 227

O

Oades, J. M., 228, 229
O'Brien, T., 283
O'Brien, T. P., 477
O'Brien, T. W., 508
O'Brien-Vedder, C., 520
Ockleford, C. D., 62, 79, 82
O'Connell, M. A., 504
Oden, P.-C., 459
Oelmüller, R., 490
Oertli, J. J., 452
Oettmeier, W., 389
Offler, C. E., 361-64
Ogawa, T., 561
Ogawa, Y., 179, 201
Ogren, W. L., 535-38, 546, 548-51, 561
Ogur, M., 10, 11
Oh, J. S., 248
Ohad, I., 168, 387, 479
Ohira, K., 572
Ohki, K., 168
Ohkoushi, N., 395
Ohlrogge, J. B., 105-12, 116, 119
Ohme, M., 380, 381, 384, 476, 479, 480, 486, 488
Ohnishi, J.-I., 123, 124, 126
Ohnishi, T., 385
Ohno, T., 391
Oh-oka, H., 384, 395
Ohyama, K., 380, 381, 384, 387, 476, 479, 480

Oja, V., 550, 561
Ojala, D., 508
Okabe, M., 169
Okada, K., 27, 32
Okamoto, M., 443, 444, 453
Okamura, M. Y., 390
Okayama, S., 387, 391
O'Kelley, J. C., 162, 166, 168
Okhuma, S., 77
Okita, T. W., 145, 147
Okker, R. J. H., 301, 305
Okusanya, B. O., 234
O'Leary, J. W., 545, 546, 550, 569
Oliver, A. J., 232, 234
Oliver, R. J. C., 515, 516
Oliver, R. P., 478
Ollinger, O., 394
Olson, A. C., 322
Olsson, J. E., 298, 307-10
O'Malley, P. J., 382, 391
Omata, T., 387
O'Neill, P., 24
Ong, C. K., 268
Ono, T., 389, 392, 394-96
Oo, K. C., 114, 116
Oparka, K. J., 361, 362
Op den Kamp, J. A. F., 75
Opik, H., 477
Orci, L., 57-59, 61, 82-84, 142, 143
Oren, R., 256
Ores, C., 65, 70
Oritani, T., 442-44, 452, 453
Orme-Johnson, W. H., 548
Ormrod, J. C., 190, 191
Oro, A. E., 505
Orozco, E. M. Jr., 486, 488, 494
Orr, A. R., 196
Orr, W., 460
Orton, T. J., 417, 418, 420, 422, 507
Oshima, K., 32
Osmesher, J., 115
Osmond, C. B., 561, 580
Osteryoung, K., 365
Osumi, T., 128
Otto, V., 149
Oursel, A., 120
Outlaw, W. H. Jr., 188, 191, 202, 203, 441, 455, 456, 565
Ouwehand, L., 391, 393, 395
Overath, P., 105, 109
Owens, T. G., 386
Owens, V., 185
Owens, V. A., 191
Owusu-Bennoah, E., 227
Oxborough, K., 560
Oya, V. M., 543
Ozaki, Y., 443
Ozga, S. D., 306

P

Pace, R. J., 396
Pacovsky, R. S., 228, 229, 235
Paddock, E. F., 424
Padley, F. B., 114, 122, 126
Paillard, M., 518-20
Paillerets, C. de, 75, 88
Painter, T. J., 338
Pairunan, A. K., 228
Pakrasi, H. B., 389
Palace, G. P., 384
Palade, G. E., 57, 59, 73
Paleg, L. G., 206
Palevitz, B. A., 55, 142
Pallotta, M. A., 418, 420, 422, 428, 429, 431
Palmer, J. D., 481, 488, 504-7, 511, 513, 514
Palmer, J. H., 189, 191, 193
Palmerley, S. M., 227-30
Palter, K. B., 74
Pan, R.-L., 389
Pang, P. C., 234
Pang, P. P.-Y., 198, 463
Pao, C.-I., 179, 189, 194, 206
Paoletti, S., 340
Paolillo, D. J. Jr., 185, 191
Papes, D., 416, 422
Pappin, D. J. C., 145
Pardee, A. B., 556, 580
Parham, P., 62, 65, 66
Parish, G., 229
Parks, T. D., 508
Parr, E. L., 142
Parrish, D. J., 288
Parry, A. D., 442, 443, 446-48, 450
Parry, M. A. J., 541, 547, 549, 559, 561
Parsons, L. R., 257
Parthier, B., 479, 486
Passaniti, A., 62, 67
PASSIOURA, J. B., 245-65; 246-48, 252-58, 260, 261
Pastan, I., 58, 62-64, 81
Pastan, I. H., 58, 59, 61, 62, 66, 70-73, 81
Pate, J. S., 364
Patel, N., 169
Patrick, J. W., 358, 361-64
Patterson, B. D., 551, 572
Patterson, D. J., 58
Patzer, E. J., 74
Patzke, J., 457
Paul, A.-L., 485, 488
Paul, E. A., 228, 232, 234
Paulet, P., 181
Pauloin, A., 76
Paulson, J. C., 58, 84
Pavlová, L., 207, 208
Payne, G. S., 62, 66, 71, 85
Pe, M. E., 188

Peacock, W. J., 188, 421, 426
Pearman, I., 361
Pearse, B. M. F., 54, 59-65, 67-70, 72, 73, 75, 79-82
Pearson, V., 230
Peavey, D. G., 564
Pecoraro, V. L., 394
Pees, E., 301, 305, 306
Pegg, G. F., 343
Peisker, M., 544
Pence, M. K., 298, 301, 307, 334
Penel, C., 179, 201, 208, 284, 342
Penny, D., 273
Penny, P., 273
Peoples, M. B., 364
Perchorowicz, J. T., 110, 547, 577
Percy, J. M., 77
Perenzin, M., 419, 428
Perlman, P. S., 508
Perrelet, A., 57, 83, 84
Peruffo, A. D. B., 479
Perumalla, C. J., 258, 259
Pesacreta, T. C., 57, 58
Peschke, V. M., 425
Pesey, H., 486
Peters, K. N., 298, 301, 312
Petersen, J., 383
Petersen, O. W., 58
Petersen, W., 440, 443
Peterson, C. A., 256, 258, 259
Peterson, D. F., 422
Peterson, D. M., 366, 371
Peterson, P. A., 425, 426
Petrouleas, V., 389, 393
Petschow, B., 183
Pettitt, T. P., 125, 126
Pevalek, B., 416, 422
Pfeffer, P. E., 24
Pfeffer, S. R., 59-62, 67, 76, 80, 81
Pfeiffer, S., 57, 83
Pfister, K., 389
Pham Thi, A. T., 122
Phan, C. T., 306
Pharis, R. P., 33, 35, 37, 42, 43, 183, 193, 194, 203, 205-7
Philip, J. R., 246, 247
Philip, S., 306
Philippe, L., 208
Phillips, D. A., 205, 299, 307, 310
Phillips, J. M., 324
PHILLIPS, R. L., 413-37; 415-23, 425, 426, 428-30
Picaud, M., 390
Pick, U., 395
Picken, A. J., 370
Pickersgil, R. W., 540
Picot, D., 380

Piechulla, B., 480
Pieczonka, M. M., 82
Pierard, D., 188
Pierce, J., 540, 541, 543, 547, 548
Pierce, M., 298, 301, 307, 308, 334, 451, 453, 454
Pierre, Y., 485, 492
Pieters, G. A., 280
Pilate, G., 205
Pilch, P. F., 59, 62, 75, 76, 83
Pilet, P. E., 345, 442, 457, 458
Pineau, B., 485, 488
Pinkerton, A., 251
Pitman, M. G., 256
Plazinski, J., 298, 301, 305
Plesnicar, M., 491
Plijter, J. J., 383, 388, 390, 393, 395
Plimmer, J. R., 24
Ploeger, B. J., 565
Plumley, G., 485, 492
Poethig, R. S., 269-71
Polak, J. M., 142, 143
Polans, N. O., 481, 488
Poling, S. M., 443
Pollard, J. K., 323
Pollard, M., 126
Pollard, M. R., 114, 115, 117
Pollard, T. D., 55
Pollock, C. J., 281, 367
Pons, F., 223
Ponsana, P., 260
Pontis, H. G., 556, 557
Poole, R. J., 77
Pooley, A., 268
Porter, C. A., 429, 541
Porter, G. A., 364
Porter, K. R., 55, 68, 149
Porter, L., 230, 232
Porter-Jordan, K., 75, 83
Portis, A. R., 535, 550, 551, 555, 561
Portis, A. R. Jr., 549, 551
Possingham, J. V., 273, 479
Potts, W. C., 207
Poulhe, R., 185
Powell, C. E., 360, 565
Powell, C. L., 228, 232
Powell, J., 159, 169
Pradel, J., 570
Prasad, K., 70-72
Prat, D., 429
Pratje, E., 509
Pratt, L. H., 142, 143, 149, 166, 167
Preiss, A., 453
Preiss, J., 556, 557
Prescott, D. M., 486
Pressman, E., 182
Preston, C., 395, 396
Preston, G. K., 247
Preston, R. D., 322, 323

Pretorius, H. T., 62, 64
Price, C. A., 479, 485, 488
Price, G. D., 298, 299, 310
Pring, D. R., 414, 504-7, 514-19
Prinsley, R. T., 545
Privitera, E., 422
Pruitt, R. E., 177, 197, 198
Pryce, R. J., 24
Pryor, A., 421
Pugh, E. L., 122
Pühler, A., 306
Punnett, T., 322
Purohit, S., 148
Purse, J. G., 181, 200, 208
Puszkin, S., 59, 60, 62, 65, 70, 76, 80
Pusztai, A., 144
Puvanesarajah, V., 306
Pyke, K. A., 477
Pyliotis, N. A., 479

Q

Quader, H., 55, 77
Quail, P. H., 77, 78
Quarrie, S. A., 440, 441, 446-48, 451
Quatrano, R. S., 459, 462
Queen, G., 340
Quetier, F., 505

R

Raaschke, K., 540, 552, 562, 563, 566
Raats, P. A. C., 250, 251, 253
Racker, E., 76, 77
Racsmany, M., 507, 519
Radin, J. W., 257, 258, 273, 277, 288, 451, 456, 459
Radmer, R., 394, 395
Ragan, C. I., 509
Rahire, M., 389, 395, 484, 485, 492
Raikhel, A. S., 73
Raikhel, N. V., 142, 462
Railton, I. D., 441, 445, 451, 453-55, 462
Rainbird, R. M., 364
Rains, D. W., 287
Rajasekhar, V. K., 490
Rajeevan, M. S., 187
Raju, M. V. S., 177, 191
Rall, S., 188
Ramachandran, K. L., 65
Rambourg, A., 58
Randall, D. D., 102, 540
Rao, M. M., 27, 28, 40, 44
Rao, S. R. V., 423, 459
Rappoport, T. A., 573
Raschke, K., 283, 286, 288, 440, 451-57, 461

Rashka, K., 147, 151
Rasmussen, O. F., 381
Rasmussen, S. K., 478
Ratnayake, R. T., 231
Rau, W., 177, 206
Ravazzola, M., 57, 83, 84
Raven, J. A., 233, 455, 539
Ray, J. A., 509
Ray, P. M., 276, 277
Rayapati, P. J., 446, 451
Rayle, D. L., 273
Raymond, B., 273
Raynes, D. A., 547
Read, N. D., 275
Reaney, M. J. T., 460
Reardon, E. M., 485, 488
Rebeaud, J. E., 457
Rebeille, F., 118, 558, 566
Reddy, G. S., 543
Reddy, P. M., 164, 167
Rédei, G. P., 178
Redman, R. E., 284, 286
Redmond, J. W., 298, 301,
  305, 306, 312
Rees, A. F., 459, 460
Rees-Jones, R., 77, 85
Reese, T. S., 55, 80
Reeves, J. P., 77
Regensburger, B., 309
Reggio, H., 62, 73
Reichardt, L. F., 80
Reicosky, D. C., 246, 248
Reid, C. P. P., 230, 232
Reid, J. B., 194, 207
Reid, M. S., 180
Reingold, A. L., 393
Reisener, H. J., 87
Reiss, T., 487
Remacle, B., 202, 203
Rembur, J., 189, 191
Renger, G., 387, 389-95
Rennie, P. J., 55, 57, 58, 60-
  62, 66, 67
Rentschler, H.-G., 159, 160,
  167, 169
Rethy, R., 164, 165, 167
Reuling, G., 460, 461
Reymond, P., 442, 457, 458
Reys, E., 78
Rhoades, M. M., 421, 425
Rhodes, C. A., 419
Rhodes, D., 460
Rhodes, J. A., 81
Rhodes, L. H., 230, 232
Riazi, A., 273, 284, 286
Ricard, J., 339, 570
Rich, A., 484, 487
Rich, P. R., 380
Richards, J. H., 250, 251
Richards, O. W., 13, 20, 272
Richardson, N., 167
Richter, E., 363
Rick, C. M., 198, 418

Rideal, E. K., 249, 250
Ridge, R. W., 305-7
Ridley, S. M., 117
Rieger, R., 415, 418, 421, 428
Riehle, W. G., 232
Rietema, H., 159, 160, 169
Riezman, H., 85
Riley, M., 509
Rinaudo, M., 340
Rines, H. W., 415, 417-23,
  426, 428, 430
Rinne, R. W., 123
Ritchie, J. T., 246, 248
Rivett, A. J., 338
Rivier, L., 442, 451, 457
Rivin, C. J., 429
Robards, A. W., 86
Robb, M. E., 86
Robbins, W. E., 29, 32, 44, 45
Roberto, F., 123, 124
Roberts, K., 322, 324, 327
Roberts, L. W., 325
Roberts, P., 115
Robertson, A. J., 460
Robertson, D., 477, 478, 482
Robertson, J. G., 58, 144, 148,
  149, 237, 298
Robichaud, C., 460
Robinson, B. L., 164, 165
ROBINSON, D. G., 53-99; 55,
  57, 58, 60-63, 66, 67, 69,
  77, 78, 85, 86
Robinson, G., 10, 18
Robinson, H. H., 396
Robinson, J. M., 198, 202
Robinson, M. S., 62, 67, 70,
  72, 81
Robinson, N. L., 368
Robinson, P., 120, 123
Robinson, S. P., 451, 452, 463,
  545, 547, 559, 578
Robson, A. D., 227-29
Roby, D., 336
Rochaix, J.-D., 389, 395, 479,
  484
Rochester, C. P., 121
Rock, C. D., 479, 480
Rock, C. O., 106
Rodermel, S. R., 485
Rodier, F., 366, 369
Roelfsen, B., 75
Roering, S., 323, 326
Rogers, D. C., 84
Rogers, H. J., 521
Rohozinski, J., 209
Rohrbaugh, M., 519
Rohwedder, W. K., 27, 28, 44
Roland, J. C., 275
ROLFE, B. G., 297-319; 298,
  299, 301, 304-7, 309, 312,
  313
Rolleston, S. S., 578
Rollins, M. L., 322

Romani, G., 24, 38, 40, 42
Romano, E. L., 142
Romano, M., 142
Romberger, J. A., 182, 209
Rome, L. H., 59, 61, 80,
  83
Roscher, E., 162
Rose, K. R., 505, 515, 519
Rosen, G. U., 10, 11
Rosher, P. H., 441
Ross, G. J. S., 224
Ross, J. J., 185, 199
Ross, S. D., 183, 203, 206,
  207
Rossen, L., 237, 298, 301,
  305, 312
Rosset, J., 142
Rostas, K., 305
Roth, J., 58, 84, 140, 142,
  143
Roth, T. F., 55, 61, 62, 67,
  68, 70, 72
Rothman, J. E., 57, 59, 60, 64,
  71, 74, 83
Rothstein, S. J., 86, 342, 343
Rottman, W. H., 517
Roughan, P. G., 101, 109, 119,
  120, 122-25
Roughley, R. J., 298
Rouvier, E., 440
Rowan, K. S., 24
Rowell, J., 556, 566
Rowley, J. D., 418
Rózsa, Zs., 389
Rubery, P. H., 453
Rubin, G. M., 427
Rubinstein, J. L. R., 59
Rudoi, A. B., 491
Rufty, T. W., 556, 565, 566
Ruiz-Herrera, J., 78
Rumberg, B., 559, 560
Rumeau, D., 336
Runeberg-Roos, P., 507, 508
Rünger, W., 177, 180
Rush, J. D., 385
Rushing, J. W., 346
Rushlow, K. A., 487
Rushlow, K. E., 487
Russell, D. W., 54, 79, 80, 82
Rüssmann, W., 573
Rutherford, A. W., 380, 382-
  93, 395, 396
Ruttkowski, U., 440, 452
Ruyters, G., 162, 163, 169
Ryan, C. A., 312
Ryan, K. G., 363
Ryan, S. A., 424, 428, 431
Ryback, G., 442
Ryberg, M., 144, 147, 148
Ryder, T. B., 312
Ryle, G. J. A., 177, 360, 565
Ryrie, I. J., 395
Ryser, U., 55, 57

S

Saase, J. M., 24
Sabbagh, I., 441, 455
Sabularse, D., 86
Sacher, J. A., 365
Sachs, J., 273
Sachs, K. B., 188
Sachs, M. M., 426
Sachs, R. M., 176, 177, 179-83, 186-97, 199-209
Sachs, T., 271
Sacristan, M. D., 416, 418, 421, 422, 427
Sadava, D., 323, 328, 335, 336
Sadler, N. L., 360, 362
Sàdliková, H., 192, 194
Saedler, H., 426
Saenger, W., 380
Safir, G. R., 233
Sage, R. F., 552, 563, 568
Sager, J. C., 162, 163, 169
Sahagian, G. G., 80
Saimi, H. S., 344
Saint-Côme, R., 189, 191
Saito, H., 27, 28, 43
Saito, T., 199
Sakakibara, M., 32
Sakurai, H., 385
Sala, C., 39
Sala, F., 39
Sala, O. E., 250
Salamini, F., 419, 428
Salerno, G. L., 556, 557
Salim, M., 256
Sallai, A., 389
Salmon, J., 188
Salnikow, J., 395
Salpeter, M. M., 323
Salvucci, M. E., 535, 550, 551, 561
Samallo, J., 509
Sanchez, J., 113, 114, 116, 117, 122
Sandelius, A. S., 179
Sanders, F. E., 222, 224, 227-30, 234
Sanderson, J., 250, 251, 260
Sandusky, P. O., 394
Sane, P. V., 389, 395
Sang, J. H., 422
Sano, T., 380, 381, 384, 387, 476, 479, 480
San Pietro, A., 385
Santarius, K. A., 559
Saraiva, L., 421
Saranak, J., 169
Saraste, J., 83
Sargent, L., 307, 309
Sarkar, S. K., 306
Saroff, H., 72
Sasaki, Y., 481, 488
Sassa, T., 27

Sasse, J. M., 35-40, 42, 43
Sato, M., 32
Sato, N., 125
Satoh, K., 387-91, 395
Sauer, A., 103
Sauer, K., 391-96, 476
Saugy, M., 442, 457
Saunders, J. W., 427
Saurer, W., 273
Sauro, H. M., 554
Sautter, C., 150
Savageau, M. A., 556, 570, 573, 580
Savchenko, G. E., 491
Sawhney, S., 182
Sawhney, V. K., 197, 198, 358
Sawyer, D. T., 387
Sax, K. B., 10, 11
Saygin, Ö., 393-96
Sayre, R. T., 389
Scandalios, J. G., 485, 492
Scannerini, S., 222, 224, 226, 229
Scarascia, G. T., 429
Scarmats, P., 65
Scarth, R., 197
Scazzocchio, C., 504, 509, 512
Schafer, E., 149, 476
Schaffer, M. A., 486
Schantz, R., 485, 488
Schardl, C. L., 505, 515, 519
Scharf, K.-D., 75
Scharfetter, E., 208
Scharschmidt, B. F., 76, 77
Schatz, G. H., 387, 388, 390, 507
Scheibe, J., 162-65, 168
Schekman, R., 62, 66, 71, 85
Schell, J., 304, 305
Schell, J. M., 306
Scheller, H. V., 381, 384, 385
Scherrer, R., 86
Schieber, G., 508
Schiemann, J., 490
Schiff, J. A., 162, 476, 481
Schiller, B., 390
Schimke, R. T., 429, 430
Schjeide, O. A., 59
Schlagnhaufer, C., 34, 37, 42
Schlesinger, M. J., 74
Schlessinger, J., 81
Schlicter, M., 521
Schlipfenbacher, R., 324
Schlodder, E., 388-91
Schloss, P., 342
Schlossmann, D. M., 71, 74
Schmalstig, J., 360, 363
Schmid, E., 73
Schmid, R., 162-65
Schmid, S. L., 64, 71, 74
Schmidt, C. N. G., 549, 559, 561
Schmidt, G. W., 484, 485, 492

Schmidt, J., 298, 304, 305, 453
Schmidt, M. G., 279
Schmitt, J. M., 479, 487
Schnabl, H., 451, 452
Schnare, M. N., 508
Schneider, G., 440, 453, 455
Schneider, H.-J. A. W., 168
Schneider, J. A., 27, 43
Schneider, W. T., 54, 79, 80, 82
Schnepf, E., 55
Schnierer, S., 509
Schobel, W., 506, 509-12
Schoeder, H.-U., 383
Schoenknecht, G., 559
Schofield, P. R., 298, 304, 305, 312, 313
Schön, A., 478
Schook, W. I., 59, 60, 62, 65, 70, 76, 80
Schopfer, P., 274, 283, 458, 462, 463
Schoser, G., 162
Schregardus, D., 8
Schreiber, U., 560, 562, 567
Schröder, W. P., 395
Schubert, I., 418, 421
Schuch, W., 301, 312
Schuller, K. A., 308, 310
Schulte-Altedorneburg, M., 365
Schulz, A., 429
Schulz, W., 301, 312
Schulze, A., 390
Schulze, E.-D., 245, 256
Schulze-Lohoff, E., 83
Schumann, R., 162, 163
Schürmann, R., 162, 163
Schurr, U., 440, 457
Schuster, A. M., 509, 511, 520
Schuster, G., 387, 479
Schuster, W., 506, 509-12
Schütte, H. R., 452, 455
Schuz, R., 110, 111
Schwabe, W. W., 179, 183, 196, 199-201
Schwartz, A. L., 75, 80, 83
Schwartz, D., 426
Schwartzbach, S. D., 481, 485, 488
Schwarzbauer, J. E., 329
Schwarz-Sommer, Z., 426
Schwebel-Dugue, N., 202
Schwencke, J., 230
Schwenen, L., 453
Scorza, R., 204
Scott, N. S., 506
Scotto, K., 429
Scowcroft, W. R., 414, 424-26, 428, 430, 431
Seabrook, J. E. A., 203
Sears, R. G., 419, 420, 422
Sederoff, R. R., 507, 508, 511, 518-20

Seemann, J. R., 537, 540, 541, 543, 547, 548, 550, 552, 562, 563, 568, 575
Seftor, R. E., 540
Seibert, M., 389, 392, 395
Seibles, T., 339, 343
Seidlová, F., 177, 192-94, 204
Sellen, D. C., 276
Selman, B. R., 559
Selman-Preiner, S., 559
Sembdner, G., 440, 453, 455
Sen, R., 223, 224
Senger, H., 162, 163, 168, 382
Sengupta-Gopalan, C., 311
Seow, H.-F., 62, 65
Sequeira, L., 306, 311, 312, 324, 326, 342
Servaites, J. C., 547
Sétif, P., 380, 381, 383-86
Setter, T. L., 453
Seyama, Y., 125
Seyer, P., 381
Shackel, K. A., 245
Shafiev, M. A., 389, 391
Shah, D. M., 429
Shahan, K. W., 276, 284, 286, 287
Shamina, Z. B., 415
Shannon, J. C., 364, 366
Shannon, L. M., 144, 145, 147, 340
Shao, L. M., 366
Shapiro, D. R., 479, 480
Shapiro, L. S., 65
Sharifi, E., 298, 312
Sharkey, T. D., 279, 539, 540, 547, 548, 550, 552, 553, 562, 563, 566, 568, 574, 575, 580
Sharma, R., 284
Sharma, S. R., 298
Sharman, B. C., 194, 195, 273, 481, 482
Sharon, N., 87
Sharp, J. K., 150
Sharp, R. E., 248
Sharp, R. R., 393
Sharp, W. R., 414, 426
Shaw, P. J., 144, 147, 148, 327
Shearman, C. A., 237, 305
Sheats, J. E., 393
Shecter, J., 81
Sheen, J.-Y., 484
Sheikh, N. A., 224
Sherwood, J. E., 306, 307
Sherwood, S. W., 429
Shevlin, D. E., 166
Shia, M., 59, 62, 76
Shields, C. R., 504, 505, 507
Shihara, I., 166
Shill, J. P., 570
Shillo, R., 177, 179, 184

Shimakata, T., 110-12, 115
Shimazaki, K., 461
Shimbayashi, K., 322
Shimizu-Takahama, M., 558, 559
Shine, J., 298, 299, 305, 312
Shine, W. E., 119
Shinozaki, K., 380, 381, 384, 476, 479-81, 484, 486, 488
Shinozaki, M., 179, 181, 183, 209
Shipman, L. L., 382
Shirai, H., 380, 381, 384, 387, 476, 479, 480
Shlyk, A. A., 491
Show, D. R., 82
Showalter, A. M., 312, 313, 329, 336, 337
Shumway, J. E., 459, 462
Shumway, L. K., 522, 523
Shurkin, J., 18, 19
Shuvalov, V. A., 383, 386
Sicher, R. C., 565
Sidall, J. B., 32
Sidebottom, C., 112, 114, 115
Sidebottom, C. M., 112
Siderer, Y., 389
Sidwell, B. A., 24
Siebertz, H. P., 122, 124
Sieburth, L. E., 485, 491
Signer, E. R., 306
Sijben-Müller, G., 486
Silk, W. K., 14, 20, 272, 277, 278, 284, 286, 287
Silva, W. I., 59, 60, 80
Silverman, M., 418
Silverstein, S. C., 82
Silverthorne, J., 476, 478, 481
Silvius, J. E., 565
Simcox, P. D., 103
Simmonds, J., 204
Simon, M., 418
Simon, R., 306
Simoncsits, A., 305
Simoni, R. D., 105, 106, 109
Simons, K., 57, 83, 84
Simpson, D., 382
Simpson, D. J., 481
Simpson, J. R., 251
Sinclair, J., 396
Sinclair, J. H., 508
Sinclair, T. R., 245, 246
Sindhu, R. K., 446, 450
Sing, V. O., 78
Singer, S. J., 141-43
Singer, S. R., 186, 187
Singh, B. D., 415
Singh, J., 460
Singh, N. K., 460
Singh, R. J., 422
Singh, S. P., 206
Singh, S. S., 114, 115
Singleton, P. W., 308, 309

Sinibaldi, R. M., 523
Sisco, P. H., 515, 516, 520
Sivak, M. N., 232, 535, 540, 558, 561, 566, 580
Sjut, V., 359, 361, 368, 371
Skadsen, R. W., 485, 492
Skaggs, D. P., 24
Skinner, M. F., 234
Skoog, F., 194
Skriver, L., 122
Skurai, A., 204
Slabas, A. R., 103-5, 112, 114, 115
Slack, C. R., 101, 103, 109, 119, 120, 122-25
Slate, D. L., 429
Slot, J. W., 80, 83, 142
Slovacek, R. E., 559
Small, G. J., 382, 386
Small, I. D., 516, 521, 522
Smart, C., 440, 461
Smart, J. E., 65
Smets, G., 144
Smit, H. W. J., 383, 386
Smith, A. G., 188, 516
Smith, C. J., 274, 540
Smith, D. C., 231
Smith, F. A., 227, 230-35
Smith, G. M., 556
Smith, H., 279, 280, 282, 283, 477, 482
Smith, H. J., 486
Smith, J. A. C., 275, 429
Smith, J. D., 446, 447, 458, 460
Smith, J. J., 324, 326-28
Smith, K. B., 306
Smith, L. A., 113, 125
Smith, M. V., 24
SMITH, S. E., 221-44; 222, 224, 227, 230-35
Smith, W. E., 86
Smith, W. K., 232, 233, 275
Snellgrove, R. C., 232, 234, 235
Snider, M. D., 84
So, H. B., 251
Sokol, R. C., 164
Sol, K., 344
Soling, H. D., 557
Söll, D., 478
Somers, D. A., 418
Somerville, C., 125, 463
Somerville, C. R., 477, 478, 551
Sommerfeld, M. R., 164
Son, C. L., 230
Sopory, S. K., 284
Soressi, G. P., 371
Sorger, P. K., 73
Sossountzov, L., 441, 455
Sotta, B., 183, 441, 455
Součková, D., 207, 208

Soudain, P., 180, 182
Soulé, J. M., 570
Southwick, S. M., 177
Sovonick, S. A., 358
Spätjens,, E. E. M., 390
Spanswick, R. M., 364
Sparace, S. A., 126
Sparrow, R. W., 396
Spaink, H. P., 301, 305
Spaulding, D. W., 27-29, 32, 40, 44
Spelta, M., 38, 40, 42
Spencer, A. K., 340
Spencer, C., 558
Spencer, D. F., 508
Spencer, F. G., 27, 28, 44
Spencer, G. F., 24
Spencer, L., 387
Speth, V., 149
Spierer-Herz, M., 485, 492
Splittstosser, W. E., 234
Spratling, L., 166
Sprecher, E., 440, 443
Spremulli, G. H., 162, 169
Spremulli, L. L., 162, 169
Sprent, J. I., 299
Sprey, B., 540
Squicciarini, J., 59, 76, 80, 83
Srinivasan, A. N., 393
Stabenau, H., 162
Stacey, G., 306, 307, 312
Stade, S., 298
Staehelin, L. A., 147, 148, 150, 326, 327, 335, 338, 380
Stafstrom, J. P., 326, 327, 335, 338
Stahmann, M. A., 340
St. Aubin, G., 261
St. John, B. J., 227, 230, 232, 234, 235
St. John, J. B., 113, 123, 124
St. John, T. V., 228, 232
Stalder, G. R., 422
Stanfield, S., 306
Stankovic, Z. S., 564
Stanley, J., 148
Stanley, K., 62, 73
Stanley, K. K., 63, 65, 72
Stapleton, S. R., 110, 111
Starr, L., 283
Stebbins, G. L., 197
Stebler, M., 393
Steer, C. J., 61, 69, 75-77, 79, 80
Steffens, G. L., 27, 42
Stehlik, D., 383
Steinback, K. E., 389, 487
Steinman, R. M., 54, 87
Steinmetz, S. A., 324, 325
Steinmüller, K., 162, 163
Steinrucken, H. C., 429
Stenberg, P. E., 55, 82

Stepka, W., 11
Steponkus, P. L., 276, 284, 286, 287
Steppuhn, J., 382
Stern, D. B., 486, 487, 504-6, 508, 509, 511, 513
Stern, H., 10
Sternberger, L. A., 142
Steudle, E., 256, 258, 259, 276, 285
Steup, M., 564
Steven, A. C., 61, 75
Stevens, M. A., 368
Stevenson, T. T., 325
Steward, F. C., 323, 416
Stewart, A. C., 395
Stewart, C. R., 446, 451, 460
Stiekema, W., 313, 330, 334
Stinson, J. R., 188
Stirdivant, S. M., 486
Stitt, M., 102, 457, 540, 545, 547, 552, 554, 557, 559, 561-63, 565, 566, 568, 571, 577, 582
Stobart, A. K., 120, 126
Stock, M., 208
Stocking, C. R., 490
Stoehr, P. J., 504
Stolinski, C., 142
Stone, B. A., 325, 338, 343
Stone, D. K., 76, 77
Stout, J. T., 418
Stoutjesdijk, P., 279
Streusand, V. J., 549, 551
Stribley, D. P., 224, 232-35
Stringam, G. R., 418
Strittmatter, G., 484, 486, 487
Strobel, G. A., 345
Stross, R. G., 164
Strous, G. J. A. M., 80, 83
Strullu, D. G., 229
Stuart, D. A., 323, 326, 328, 335
Studer, D., 309
Stults, J. T., 442, 449, 450, 454
Stummann, B. M., 381
Stumpf, P. K., 102-6, 108, 110, 111, 113-19, 121-24, 126-28
Stüning, M., 441, 453, 462
Sturm, A., 344
Sturtevant, A. H., 428
Stutte, G. W., 184
Stymne, S., 120, 121, 126
Styring, S., 387, 391, 393, 395, 396
Sucoff, E. I., 256
Suge, H., 199
Sugiura, M., 484, 486, 494
Sukakibara, M., 27
Sumegi, J., 507, 519
Sumper, M., 324

Sun, E., 479, 480
Sun, P. S., 458
Sunderland, N., 269, 271, 414, 416
Surek, B., 549
Suresh, H. M., 234
Surholt, E., 344
Sussex, I. M., 184, 269, 270, 460
Sutherland, G. R., 423
Sutherland, J. M., 299
Suttle, J. C., 194, 207
Sutton, A., 485, 491
Suziki, Y., 27, 29
Suzuki, K., 388
Suzuki, M., 442
Svec, W. A., 382
Svendsen, I. B., 105-8, 381, 384, 385
Swaminathan, V., 229
Swanson, C. A., 565
Sward, R. J., 223
Swartzberg, D., 521
Swe, K. L., 179, 181, 183
Sweers, H. E., 558
Sweet, G. B., 203, 205
Syamsunder, J., 234
Sybesma, C., 380
Syvertsen, J. P., 228, 233
Szabo, A. G., 122

T

Taatjes, D. J., 58, 84
Tabata, K., 387, 391
Tachon, P., 441, 455
Taiz, L., 337
Takabe, T., 491
Takahama, U., 558, 559
Takahashi, A., 76
Takahashi, H., 199
Takahashi, M., 391
Takahashi, N., 27-29, 38, 41-44, 204
Takahashi, S., 442, 453
Takahashi, Y., 384, 388, 390, 391
Takano, T., 77
Takatori, S., 164, 167
Takatsumo, S., 27
Takatsuto, S., 27-29, 32-34, 37, 42-44
Takeda, N., 442
Takematsu, T., 27, 32, 34, 35, 38, 39, 41, 42, 44
Takemura, R., 55, 57, 82
Takenchi, Y., 34, 35, 38, 39, 41
Takeno, K., 33, 35, 37, 42, 43
Takeuchi, Y., 32, 34, 42, 44
Takimoto, A., 179, 181, 183, 200, 204, 209
Takishima, K., 125

Tal, M., 288, 447, 448
Talmadge, K. W., 325
Talpasyi, E. R. S., 164, 167
Tamkun, J. W., 329
Tamura, N., 391, 394, 395
Tanaka, A., 356
Tanaka, M., 380, 381, 384, 476, 479, 480, 486, 488
Tanaka, S., 395
Tanchak, M. A., 85
Tang, X.-S., 387, 389, 395
Tanimoto, E., 274, 345
Tanner, C. B., 253, 258
Tardieu, F., 247
Tarr, A. R., 440, 447
Tartakoff, A. M., 145
Tartof, K. D., 429
Tatematsu, A., 442
Tavitian, B. A., 382
Taylor, A. O., 166
Taylor, C. D., 165
Taylor, G., 279, 282, 285, 440, 457
Taylor, H. F., 445, 446
Taylor, H. M., 246-48
Taylor, I. B., 440, 447, 448
Taylor, J. S., 205
Taylor, L. P., 426
Taylor, W. C., 479, 480, 491
Teede, H.-J., 102
Telser, J., 385
Tenbarge, K. M., 459, 462
ten Hoopen, A., 158-60, 169
Tenhunen, J. D., 540
Tepperman, J. M., 537, 543
Terborgh, J., 166
Termaat, A., 257
Terpstra, P., 508
Terry, L. A., 159, 169
Terry, N., 285, 288, 544, 558
Tester, M., 224, 227, 235
Tewari, K. K., 479, 480, 486
Tewari, R., 423
Theg, S. M., 390, 396
Thellier, M., 201
Thelma, B. K., 423
Therman, E., 427
Theurer, J. C., 370
Thibault, J. F., 340
Thiele, R., 166
Thigpen, S. P., 189
Thom, M., 365-67, 370
Thomas, B., 176, 179
Thomas, D. R., 128
Thomas, E., 223, 427, 517
Thomas, J., 162, 168
Thomas, J. F., 189, 191
Thomas, J. P., 168
Thomas, M. D., 323
Thomas, P. A., 366
Thomasson, M., 271
Thompson, E. W., 322, 323

Thompson, G. A., 113, 118, 122, 125
Thompson, J. A., 479
Thompson, J. F., 323
Thompson, M. J., 27-29, 32-37, 39-45
Thompson, R. D., 507, 519
Thompson, R. J., 162, 164, 486
Thompson, W. F., 481, 488, 506, 509, 511, 513
Thomson, L. W., 103
Thomson, W. W., 477, 479
Thornber, J. P., 381
Thorne, G. N., 361, 370
Thorne, J. H., 356, 357, 360, 362, 364
Thorne, S. W., 479
Thornley, J. H. M., 268, 359
Thornley, W. R., 369
Thorpe, M. R., 363
Thorpe, T., 342
Throneberry, G. O., 228
Thurnauer, M. C., 383
Tiburcio, A. F., 187, 209
Tierney, M. L., 322, 325, 329, 336
Tietz, A., 372, 440, 452
Tietz, D., 453, 456
Tilney-Bassett, R. A. E., 475-79, 481
Timmer, L. W., 232
Timmis, J. N., 506
Timothy, D. H., 505, 507-9, 511, 514-19, 521
Tinker, P. B., 222, 224, 226-35, 246, 247
Tischer, E., 429
Tisdall, J. M., 228, 229
Tittgen, J., 382, 478-80, 491
Titus, D. E., 144, 148, 150
Tobin, E. M., 476, 478, 481
Todhunter, J. A., 40, 43, 47
Tokuyasu, K. T., 141-43, 145
Tolbert, N. E., 540, 541, 548
Tomaseo, M., 416, 422
Tombs, M. P., 112
Tomi, H., 481, 488
Tomoda, Y., 481, 488
Tomos, A. D., 258, 259, 277, 365
Tonkyn, J. C., 486, 487
Tooze, J., 57, 84
Tooze, S. A., 57, 84
Toppan, A., 336
Topper, J. N., 395, 396
Torazawa, K., 484
Torok, I., 298, 305
Torrey, J. G., 298, 301, 306, 415
Torti, G., 371
Toselli, P. A., 59, 70, 76
Toth, R., 226, 234

Toubart, P., 187, 209
Touzé, A., 336
Towill, L. R., 164
Toyoshima, Y., 395
Traas, J. A., 55
Tran-Anh, T., 559, 560
Tran Thanh Van, K., 187, 203, 204, 207, 209
Tran Thanh Van, M., 185
Trappe, J. M., 223, 234
Traynor, P. L., 519, 520
Trebst, A., 389, 390
Treede, H.-J., 102
Tregunna, E. B., 535, 562
Trelease, R. N., 128, 150
Trémolières, A., 120-22, 124, 125
Trewavas, A. J., 37, 204, 440, 461
Trinh, T. H., 187
Trinick, M. J., 298
Triplett, B. A., 459
Tripp, V. W., 322
Trissl, H.-W., 388
Trouvelot, A., 222
Trowbridge, I. S., 80
Tsai, A.-L., 385
Tsai, C. Y., 371
Tsai, D. S., 34, 37, 42
Tso, J., 391
Tsujimoto, H. Y., 390
Tuberosa, R., 430
Tucker, M. L., 344
Tünnermann, M., 164, 165
Turgeon, B. G., 298, 299, 306, 307
Turgeon, R., 360, 363
Turner, J. E., 325
Turner, J. F., 368
Turner, N. C., 245, 246, 261, 286, 287
Turnham, E., 103
Turpen, T., 523
Turrill, F. M., 275
Turvey, P. M., 363
Tuttle, A., 425
Tyler, J. M., 326
Tyree, M. T., 276, 283, 287
Tzagoloff, A., 490, 507-9, 512

U

Uchiyama, M., 27, 29, 43
Udvardy, A., 507, 519
Ueda, H., 32
Ui, N., 128
Uknes, S. J., 454, 455
Ullman, B., 422
Ullmann, P., 208
Ulrich, A., 285
Umbarger, H. E., 556, 580
Umbeck, P. F., 517
Unanue, E. R., 63, 65, 69

Ungemach, J., 342
Ungewickell, E., 62-67, 69-72, 74
Unkeless, J. C., 82
Unser, G., 284
Urbasch, I., 440, 443
Urech, K., 230
Usami, M., 76
Usuda, H., 556
Uyeda, K., 557

V

Vaadia, Y., 257
Valade, J., 198
Valk, P. van der, 55, 57, 58
van Bel, A. J. E., 364
van Brussel, A. A. N., 301, 305, 306
Vance, C. P., 298, 312
van Cleve, B., 144, 147, 148
van Dam, H., 313, 330, 334
van Deenen, L. L. M., 75
van den Boogaart, P., 509
VandenBosch, K. A., 142, 144, 150
Van den Ende, G., 204, 207
van der Velde, H. H., 162, 163, 167, 169
van der Wulp, D., 162, 163, 169
Van de Walle, C., 180
Van Dijck, R., 208
van Dijk, M. A., 393
van Dorssen, R. J., 390
Van Driessche, E., 144
Van Etten, C. H., 325
Van Etten, R. A., 504
van Gorkom, H. J., 380, 383, 386-88, 390, 391, 393, 395
Van Holst, G. J., 313, 322, 324, 326, 327, 329, 331, 335
van Kammen, A., 313
van Kan, P. J. M., 388, 390
van Kessel, C., 308, 309
Van Loenen-Martinet, E. P., 207
VÄNNGÅRD, T., 379-411; 386, 392-94
van Obberghen, E. V., 58
van Rensen, J. J. S., 389, 390
Van Rijn, L., 207
Van Spronsen, P. C., 306
Van Staden, J., 183
Van Volkenburgh, E., 273, 277-79, 287, 288, 458
Varndell, I. M., 142, 143
VARNER, J. E., 321-53; 312, 313, 322-29, 331, 332, 335-37, 340
Vartanian, N., 250
Vasile, E., 73, 82

Vass, I., 389, 395
Vassalli, J.-D., 57, 84
Vassalli, P., 82
Vater, J., 395
Vaughan, G. T., 441, 453
Vaughn, K. C., 148
Vedel, F., 504
Veen, A. H., 15
Veilleux, R. E., 186, 197
Veldink, G. A., 143
Velthuys, B. R., 387
Veluthambi, K., 345
Verbeke, J. A., 197
Verbelen, J.-P., 143, 283
Veres, J. S., 275
Vergnalle, C., 125
Verkleij, A. J., 75
Verma, D. P. S., 144, 148, 150, 313
Vermaas, W. F. J., 388
Vernooy-Gerritsen, M., 143
Vernotte, C., 387, 389, 390, 395
Vesper, S. J., 298, 299, 306, 307, 312
Vessey, J. K., 308
Vick, B. A., 128, 129
Vidaver, W., 257
Vierling, E., 148, 491
Vig, B. K., 422
Vigers, G. P. A., 64, 68, 69
Vigh, L., 122
Viil, J., 559
Villanueva, J. R., 322
Villiger, W., 143
Vince-Prue, D., 160, 161, 176, 177, 179, 181, 182, 186-88, 191, 199
Virgin, H. I., 164, 165, 167, 476
Vishniac, W., 164, 165
Vliegenthart, J. F. G., 143
Voetberg, G., 446, 451, 460
Vogelmann, T. C., 162, 163
Völker, M., 392, 394, 395
von Caemmerer, S., 535, 538-41, 544-47, 549, 550, 567, 574, 575, 577
Vonlanthen, M., 144
Vonlanthen, M. T., 141
von Wettstein-Knowles, P., 117
Vorob'eva, L. V., 205
Vreeland, V., 339
Vu, C. V., 541, 547

W

Waaland, J. R., 158, 159
Wada, K., 32-34, 37-39, 42, 44, 384, 395
Wagner, E., 179
Wakasugi, T., 380, 381, 384, 476, 479, 480, 488

Walbot, V., 425, 426, 429, 480, 505, 507, 511, 514, 516, 520, 523
Walden, D. B., 418
Waldron, L. J., 285, 288
Walker, A. J., 357, 359
Walker, D. A., 232, 535, 540, 541, 543, 545, 547, 553-56, 558-63, 566, 568-70, 578-80
Walker, D. B., 197
Walker, F., 336
Walker, G. C., 306
Walker, J. E., 479, 486
Walker, K. A., 112-14, 117
Walker, N. A., 224, 235
Walker, W. S., 335
Walker-Simmons, M., 312, 441
Wall, D. A., 79-81
Wall, J. S., 61, 325
Wallace, L. L., 234
Wallace, W., 232, 234
Wallen, D. G., 163
Wallsgrove, R. M., 477, 480
Walter, C., 87
Walter, G., 75
Walter, P., 492
Walton, D. C., 440, 442, 443, 445-47, 449-55, 459, 461-63
Wang, J. H., 65, 247
Wang, L. C. H., 440
Wang, T. L., 441, 447, 448
Wang, X., 125
Ward, B. L., 504
Ward, D. A., 457
Wardell, W. L., 189
Warden, J. T., 384, 385, 389
Wardlaw, I. F., 358, 565
Wardley, T. M., 482
Wareing, P. F., 184, 200, 205, 358, 440, 441
Waring, R. B., 504, 509, 512
Warm, E., 177, 180
Warmke, H. E., 518
Warren, G., 62, 73
Warren Wilson, J., 299, 358
Warren Wilson, P. M., 299
Warthen, J. D., 24, 25
Wasielewski, M. R., 382, 386
Watad, A.-E. A., 460
Watanabe, K., 179, 181, 209
Watanabe, S., 125
Watanabe, T., 382
Watson, B. T., 256
Watson, C. W., 491
Watson, J. M., 298, 304, 305
Waung, L. Y., 345
Waykole, P., 340
Wayne, R. O., 179
Ways, J. P., 65
Weaire, P. J., 112, 113
Weatherley, P. E., 247-49

Webb, J. A., 360
Webb, J. L., 553
Webb, S. P., 386
Webber, A. N., 387
Weber, E., 144, 145
Weber, J., 309
Weber, J. A., 540, 541
Weeden, N. F., 557
Weete, J. D., 223
Weidner, M., 160
Weil, J. H., 479, 507, 508
Weiler, E. W., 441, 451-53, 455, 456, 461, 462
Weimken, A., 230
Weinberg, G., 422
Weineke, U., 298, 304, 305
Weinheimer, W. H., 418
Weinstein, J. D., 478
Weinstein, J. V., 58, 68, 84
Weintraub, H., 75, 88
Weir, R., 286, 287
Weis, E., 560, 567
Weiss, H., 509
Weiss, W., 391, 393, 395
Weissbach, H., 485, 490, 491
Weissinger, A. D., 519
Welch, W. J., 74
Wellensiek, S. J., 179, 181, 182, 206
Wells, B., 144, 148, 149, 237, 298, 305
Welsch, M. A., 306
Wendt-Gallitelli, M. F., 416, 427
Wensink, J., 393, 395
Wenzl, S., 324
Werb, Z., 55, 57, 82
Werdan, K., 554
Werneke, J., 551
Wernsman, E. A., 429
Wessel, D., 58
Wesselius, J. C., 389
West, J. A., 159, 160, 166, 169
West, K. R., 567
Westhoff, P., 478-80, 486, 488, 491, 494
Westly, J. C., 440
Wettern, M., 485
Wettstein, D. V., 478
Weyers, J. D. B., 456, 458, 461
Wharfe, J., 122, 124
Whatley, J. M., 477, 479
Wheeler, A. W., 268
Whisson, D. L., 506
Whitaker, D. P., 557
White, E. M., 523
White, J. A., 229
White, P. R., 414
Whitelam, G. C., 283
Whitfeld, P. R., 389, 479, 480, 488

Whitmore, F. W., 340
Whittaker, E., 10, 18
Whyte, A., 62
Widger, W. R., 380, 390
Wiedenmann, B., 55, 62, 67, 68, 71, 74, 76, 77
Wiemken, A., 229, 345, 565
Wiemken-Gehrig, V., 345
Wienand, U., 426
Wiering, H., 198
Wijffelman, C. A., 298, 301, 305, 306, 312
Wijsman, H. J. W., 198
Wild, A., 35, 36, 40, 41, 227
Wild, A. E., 54
Wild, J. A., 234
Wildner, G. F., 389
Wilke, I., 557
Wilkins, T. A., 462
Willard, J. J., 327, 328
Willatt, S. T., 248
Willemot, C., 117, 122-24
Willenbrink, J., 363-67, 369
William, G. H., 565
Williams, E. D., 144
Williams, E. G., 24
Williams, G. J. III, 275
Williams, J. P., 123, 125
Williams, M. A., 237
Williams, R. F., 177, 270
Williams, V. P., 442
Williamson, J. D., 462
Willing, R. P., 188
Willingham, M. C., 53, 58, 59, 61, 62, 66, 70-73, 81
Willows, R., 448
Wilson, G., 81
Wilson, G. L., 280
Wilson, K. E., 298, 301, 312
Wilson, L. G., 274, 322
Wilson, M. J., 80
Wilson, R. B., 392
Wilson, R. G., 191, 193
Winkler, D., 62, 73
Winkler, F. K., 63, 65, 72
Winter, J. A., 429
Wirtz, W., 545, 547, 559
Wise, R. P., 514, 515, 517
Wiskich, J. T., 567
Wissinger, B., 506
Witt, H. T., 380, 387-91, 393-96
Witt, I., 380
Wohrle, D., 453
Wolf, A.-M. A., 110
Wolf, C., 122
Wolf, D. D., 288
Wolff, I. A., 325
Wollman, F.-A., 380, 381
Wolswinkel, P., 364
Wood, C., 128
Wood, E. A., 144, 148
Woodman, J. C., 426

WOODROW, I. E., 533-94; 120, 121, 541, 554, 555, 558, 561, 563, 568-70, 574-77, 579
Woods, J. M., 84
Woods, J. W., 61, 62, 67
Woodward, M. P., 61, 70, 72
Woolhouse, H. W., 230
Woolley, J. T., 247, 259
Worley, J. F., 24, 25, 27-29, 32, 36, 38, 41, 42, 44, 45
Worton, R. G., 422
Woude, W. J. van der, 78
Woudstra, M., 339
Woychik, J. H., 325
Wraight, C. A., 567
Wright, B., 65
Wright, C. T., 504
Wright, M. J., 273
Wu, B. W., 395, 479, 480
Wu, M., 506
Wu, R., 511, 512
Wu, S. H., 366
Wurtels, E. S., 104, 105
Wydrzynski, T., 389
Wyn Jones, R. G., 258, 259
Wynne, M. J., 161
Wyse, R. E., 363-66, 369, 370, 441

X

Xie, X.-S., 76, 77

Y

Yachandra, V. K., 384, 385, 392-94
Yagil, E., 197
Yamada, M., 110, 123-26
Yamada, Y., 387
Yamaguchi, I., 27, 29, 204
Yamaguchi, J., 356
Yamamoto, H. Y., 449, 450
Yamamoto, Y., 387, 391
Yamaoka, T., 346
Yamashiro, D. J., 77
Yamashita, J., 395
Yamashita, K., 442-44, 452, 453
Yan, W. M., 366, 367
Yanagi, M., 395
Yanina, L. I., 199
Yates, R. A., 556, 580
Yazawa, N., 32, 34, 42, 44
Yeoman, M. M., 421
Yerkes, C. T., 396
Ying, B., 28, 32
Yocum, C. F., 387, 390, 394-96
Yokota, A., 540
Yokota, T., 27-29, 38, 41-44
Yopp, J. H., 34-40, 42-44

Yoshida, K., 181, 209
Yoshihara, K., 27, 43
Yoshikawa, M., 343
Young, C. C., 229
Young, D. H., 343
Young, E. G., 515, 521, 523
Young, J. P., 288, 459
Youvan, D. C., 389
Yuan, R., 425
Yuasa, M., 387

Z

Zaat, S. A. J., 301, 305, 306
Zabala, G., 514, 520
Zablackis, E., 339
Zagorska, N. A., 415
Zaita, N., 484

Zaitlin, M., 520
Zalik, S., 103
Zamski, E., 365
Zanders, E., 508
Zaragosa, L. J., 275
Zaremba, S., 62, 66, 70, 72
Zarilla, G., 169
Zee, S.-Y., 477
ZEEVAART, J. A. D., 439-73;
    176, 179, 181-83, 187,
    188, 194, 198-201, 204,
    205, 207, 208, 283, 288,
    440-42, 449, 450-57, 462
Zeiger, E., 461
Zeiher, C. A., 541
Zelechowska, M., 144, 148,
    150
Zetsche, K., 162, 163
Zhang, J., 440, 451, 457

Zhang, W. C., 366, 367
Zhang, Y.-H., 57
Zhu, Y. S., 481, 485
Ziegler, K., 384
Zime, V., 358
Zimmerman, D. C., 128,
    129
Zimmermann, G., 568
Zimmermann, J. L., 387-89,
    392, 396
Zimmermann, U., 276, 285
Zollinger, M., 141, 145
Zuber, H., 395, 484
Zucchinali, C., 419, 428
Zur, B., 247-49
Zurawski, G., 389, 486, 487,
    494
Zwaal, R. F. A., 75
Zwar, J. A., 462

SUBJECT INDEX

A

Abscisic acid, 35, 43, 233
  numbering system, 441
  sink organs, 371-72
Abscisic acid, metabolism and
    physiology, 439-73
  introduction, 440
  biochemical responses, 461
  barley aleurone cells, 462
  developing embryos, 462-
    63
  guard cells, 461
  biosynthesis in fungi, 443-44
  biosynthesis in higher plants,
    444-45
  carotenoid biosynthetic in-
    hibitor and mutant
    studies, 446-47
  catabolism of fungal in-
    termediates, 452-53
  deficient mutants, 447-48
  direct pathway, 445
  hypothetical scheme, 445
  indirect pathway, 445-51
  inhibitor studies, 451
  regulation, 451-52
  stable isotope studies, 448-
    51
  xanthoxin precursor, 446
  catabolism
  differences between S- and
    R-abscisic acid, 453
  isolation and characteriza-
    tion of new catabo-
    lites, 453
  regulation, 453-54
  compartmentation, 454-56
  conclusion, 463-64
  physiological responses
  adaptation to stress, 460
  dormancy, 459-60
  floral morphogenesis, 208-9
  growth, 457-58
  heterophylly, 458-59
  leaf growth, 288
  response mutants, 460
  root gravitropism, 458
  stomatal control, water re-
    lations, and photo-
    synthesis, 456-57
  stress mediated redistribu-
    tion hypothesis, 456
  techniques
  extraction and quantifica-
    tion of xanthoxin, 442-
    43
  quantification, 440-441
  separation of S- and R-
    abscisic acid, 441-42

stable isotopes and mass
    spectrometry, 442
Acceptors A_0 and A_1, 383-86
Acetylcholine esterase, 82-83
Acetylcholine receptors, 82-83
Acetyl-CoA, 115, 128
  source for fatty acid synthe-
    sis, 102-3
Acetyl-CoA:ACP transacylase,
    109
  rate-limiting nature, 115
Acetyl-CoA carboxylase
  as multifunctional protein,
    103-5
  regulatory factors, 104
Acetyl-CoA hydrolase, 102
Acid phosphatase, 345
Actinomycin D, 40
Acyl-ACP acyltransferase, 115
Acyl-ACP hydrolases, 119
Acyl-ACP thioesterase, 114
Acyl carrier protein (ACP), 105-
    9, 115
  elongation of fatty acids, 116
  isoforms, 106-9
    role, 108
    N-terminal amino acid
    sequences, 107
Acyl-CoA, 114, 117
Acyl-CoA oxidase, 127
Acyl transferase, 119
Agalactolipid, 87
Albumin storage protein, 149
Algal development, photocon-
    trol, 157-74
  action spectra for responses
    and photoreceptors, 166
  blue and red light re-
    sponses, 169
  blue light responses, 168-
    69
  green light responses, 169-
    70
  photoreversible pigment
    systems other than
    phytochrome, 167-
    68
  phytochrome, 166-67
  red/far-red reversible re-
    sponses, 166-67
  introduction, 157-58
  nonperiodic control, 161
  metabolic development,
    161-64
  metabolic development
    types, 162
  reproductive development
    types, 166
  reproductive structures,
    165-66

types of algae influenced,
    161
  vegetative morphology,
    164-65
  vegetative morphology de-
    velopment types, 164
  photoperiodic control, 158
  contrasts with photoperiod-
    ism in higher plants,
    160-61
  photoperiodic algal species,
    158-59
  types of development, 159-
    60
  summary, 170
Alginate, 339
Allophycocyanin, 168
1-Amino-cyclopropane-1-
    carboxylic acid, 176
Amino-levulinic acid, 162
Ammonia, 394
α-Amylase, 84-86, 462
Anaerobiosis
  floral morphogenesis control,
    180-81
  preception sites, 182
Anderson, Dorothy, 7
Anderson, Edgar, 1, 6-9
Antheraxanthin, 450
Anthesin, 200
Anthocyanin, 426
Antiflorigen, 176, 182, 198-201
Apigenin-7-O-glucoside, 301
Arabinosidase, 344
Arachidonate, 126
Aspartic acid, 62
Assembly polypeptides, 62, 66-
    67
ATP, 74, 103
ATPase, 40, 74, 231
  coated vesicles, 76-77
  localization in VA mycorrhi-
    zae, 225-26
  mitochondrial gene, 510, 513
ATP synthase, 559, 578
Avery, George, 5
Auxin, 37-39, 43, 207-8, 223

B

Barley aleurone cells
  abscisic acid, 462
Benzoic acid, 204
Bishop, Sherman, 9
Borgstedt, Elinor, 10
Botanist growth and develop-
    ment, 1-22
  California, 16-18
  computers and calculators,
    18-19

foreword, 1
Gustavus Adolphus College, 3-5
looking forward, 19-20
kinematics, 20
Penn, 10-14
phyllotaxis and other things, 14
Rochester, 9-10
roots, 2
Shaw's garden, 6-8
si quaeris peninsulam amoenam, 2-3
summer school, 5-6
Western Cartridge Company, 8
Brassinolide, 25-26, 28, 41, 45
structure, 30
synthesis, 32
Brassinone, 27-28
Brassinosteroids, 23-52
background, 24-25
biological activity, 33-39
characteristics, 42-43
effect on tissue, 36
light effect, 38
order of activity, 34-35
synergistic interaction with auxin, 37-38
tissue sensitivity, 36
summarization, 39
brassinolide isolation, 25-26
chemistry, 28-31
distribution in plants, 29
structures, 30-31
structures of analogues, 34
effects in nucleic acid and protein metabolism, 39-41
introduction, 23-24
pollen
brassin activity sources, 25
practical applications in agriculture, 41-42
problems and prospects, 45, 47-48
proposed biosynthesis, 43-45
pathway diagram, 46
proposed functions, 42-43
synthesis and structure-activity relationship, 32-33
structural requirements for brassin activity, 33
terminology, 26-28
trivial names, 27
Brassins, 24
activity
discription, 33
sources, 25
structural requirements for, 33
practical applications, 41

α-Bungarotoxin, 82
Butenone, 443

C

Calcium 65, 78, 395-96
Calmodulin, 65
Capsular polysaccharides, 306
Carbon
VA mycorrhizae, 234-35
Carbon dioxide
see Photosynthetic CO₂ fixation in C₃ plants
2-Carboxyarabinitol 1-phosphate, 542, 570, 576, 582
concentration estimation, 541
Rubisco activity regulation, 547-48
Carbonic anhydrase, 572
Caspari, Ernst, 9
Casparian strip, 258
Castasterone, 27-28
Cellulase, 343-44
Cell wall proteins, 321-53
cell wall enzymes, 340-46
acid phosphatase, 345
arabinosidase, 344
cellulase, 343-44
β-fructofuranosidase, 344
α-galactosidase, 345
β-1,3-glucanase, 343
β-glucosidase, 344
β-glucuronidase, 345
hydrolases, 343
α-mannosidase, 344
pectinesterase, 345-46
peroxidase, 340
polygalacturonase, 346
conclusions, 346
extensin, 323-25
characterization, 326-28
localization, 332, 335-36
plant defense role, 336
plant growth role, 335-36
fiber tensile strength, 333
introduction, 321-22
structural proteins
biochemical characterization, 326-28
cellular localization and function, 331-37
history, 322-26
interaction with other cell wall components, 337—340
molecular biology studies, 328-31
Charles, Donald R., 9-10
Chitosomes, 78
2-Chloroethyltrimethylammonium chloride, 178-79
p-Chloromercuribenzene sulfonic acid, 362-63

Chlorophyll
see Photosystem I, electron transport; Photosystem II, electron transport
Chlorophyll-apoprotein
synthesis regulation, 490-93
Chloroplast
abscisic acid localization, 455
DNA sequences, 506
genome sequencing, 380
Chloroplast development and gene expression, 475-502
biochemical processes induced during chloroplast biogenesis, 478
chloroplast translation regulation, 488, 490
cytoplasmic control principle, 490
chlorophyll-apoprotein synthesis regulation, 490-93
conclusions, 493-95
development phases, 481
leaf mesophyll cell development, 482-83
plastid volume changes, 483
shoot formation and leaf meristem initiation, 481-82
introduction, 475-76
plastid coding capacity and genome organization, 479-81
genes coded by plastid DNA, 480
plastid diversity, 477
chloroplast formation from proplastids, 477-78
meristematic cell conversion to mesophyll cell, 478
plastid diversity generation, 478-79
proplastids, 477
plastid transcription and RNA level determination, 485
chloroplast transcription regulation, 485-86
determination of psbd-psbc transcript abundance, 488-89
posttranscriptional mechanisms that alter levels, 486-88
psbA mRNA level determination, 487
ribosomal RNA level determination, 486-87
regulation of chloroplast protein complex biosynthesis, 483-85

major protein complexes, 484
Cholesterol, 75
Chromosomal variation
  see Somaclonal variation, chromosomal basis
Clathrin, 63-67, 71, 74-75, 79
  characteristics, 62
  coated vesicles assembly dynamics, 73
  forms in cells, 74
  polymerization, 72
  structure and organization, 64-65
  term, 61
Coated vesicles, 53-99
  associated enzyme activities
    ATPase, 76-77
    glucan synthases, 77-78
    kinases, 76
  coat architecture and coat-vesicle interactions, 68
  coat proteins and their properties
    assembly polypeptides, 66-67
    clathrin, 61-65
    light chains, 62-66
    list, 62
    terminology, 61-62
    triskelions, 63-66, 68
    tubulin, 67-68
  coated membranes in the cell
    coated pits, 55-57, 81-82
    Golgi apparatus, 57
    other membranes, 57-58
    plasma membrane, 54-57, 77-78, 82-83
  coated pits, 85-86
    formation, 82
  dissociation and reassembly of coat proteins, 69-73
    methods releasing coat proteins, 70
    triskelions reassembly, 72
    triskelions release, 71
  dynamics of coat proteins, 73-75
  existence in cell, 58
  functions in animal cells
    exocytosis, 82-83
    interactions between receptor molecules and coat proteins, 80-81
    receptor, cargo and adaptor molecules, 80
    receptor-mediated endocytosis, 81-82
    recognition properties, 81
    sorting at trans Golgi compartment, 83-84
    specific transport vesicles, 79-81
  functions in plants, 85-86
  functions in yeasts, 84-85
  introduction, 54
  isolation
    from plants, 60-61
    purification methods, 59-60
    sizes and yields, 61
    sources, 59
  lipids, 75
  perspectives for the future, 86-88
  receptosome, 58
Collenchyma, 334-35
Concanavalin A, 145
β-Conglycinin, 462
Cordycepin, 40
Coumestrol, 301
Cryptochrome, 163, 165
  algal development control, 168-69
Cytochromes
  immunocytochemical localization, 147-48
Cytochrome b_{559}, 390
Cytochrome oxidase, 510, 513
Cytokinins, 38-40, 43, 183, 233, 371
  floral morphogenesis, 192, 194, 197, 201
  role in control, 204-5
Cytoplasmic male sterility, 515-15
  maize cms-C, 518
  maize cms-S, 505, 518-20
  maize cms-T, 516-17
    cms-T reversions, 517-18
  Petunia, 521
  RNA plasmids in cms-S mitochondria, 520-21
  sublimons, 521-22
Cytosol, 74

D

2,4-D, 194, 197
Daidzein, 301
Deoxyrobonucleotide pools
  chromosome variation, 422-23
Dihydrofolate reductase, 429-30
4'-Dihydrophaseic acid, 454
Dihydroxyacetone phosphate, 557, 567
2,4-Dinitrophenol, 188
2,7-Dimethyl-octa-2,4-dienedioic acid, 448
DNA, 423, 429
  activation of cryptic transposable elements, 425
  interorganellar transfer, 506
  late replication and chromosome rearrangement, 421-22
  sequences in chloroplast, 506

DNA polymerase, 40
Dyhydroxyflavone, 301

E

Eicosanoic acid, 117, 126
Eicosapentaenoate, 126
Elastase, 64, 72
Electron microscope
  see Macromolecules, immunocytochemical localization
Elson, J. A., 4
Endoplasmic reticulum, 117
  fatty acid synthesis pathways, 113
Engler, Adolph, 6
Enolpyruvylshikimate-3-phosphate synthase, 429
Enoyl-ACP reductase, 110
  forms, 112
Erickson, Charles, 1
Eriodictyl, 301
Escherichia coli, 109
  fatty acid synthetase, 105
    acyl carrier protein, 105-6
Ethylene, 43, 180-81, , 207, 234, 336
  extensin gene expression, 328
  floral morphogenesis, 208
Evans, Lloyd T., 16
Exopolysaccharides, 306-7
Extensin, 323, 331, 338, 340-41
  cell wall tensile strength, 334
  characterization, 326-28
  genomic clone, 328
  localization, 332, 335-36
  role, 324
    disease resistance, 336
    plant growth, 335-36

F

Farnesyl pyrophosphate, 443
Fatty acid metabolism, 101-38
  acetyl-CoA
    source for fatty acid synthesis, 102-3
  acetyl-CoA carboxylase as multifunctional protein, 103-5
    regulatory factors, 104
  conclusions and future directions, 129
  desaturation reactions
    oleic acid biosynthesis, 118-19
    other desaturation reactions, 126
    polyunsaturated fatty acids formation, 120-25
  elongase, 117

elongation of fatty acids, 116-17
introduction, 101-2
oxidation, 126
  lipoxygenase, 128-28
  β-oxidation, 127-28
  ricinoleic acid formation, 126, 27
plant fatty acid synthetase, 105
  acyl carrier protein, 105-9
  chain termination mechanisms, 114-16
  fatty acid synthesis site, 112-14
  molecular nature, 109-12
synthesis pathways, 113
Fatty acid synthetase, 106, 114
  enzyme type, 105
  chain termination mechanisms, 114-16
  fatty acid synthesis site, 112-14
  molecular nature, 109-12
Ferrodoxin, 118
Floral evocation and morphogenesis control, 175-219
  conclusions, 209
  correlative influences, 182-84
  environmental control
    autonomous induction, 180-81
    day length, 178-80
    efficiency of inductive factors, 181
    genetics of sensitivity to environmental factors, 178
    perception sites, 181-82
    photoinduction, 179-80
  experimental systems and model plants, 176-77
  floral evocation, 184
    cellular changes, 189-91
    component processes and nature of evocation, 187-92
    direct light effects, 191
    gene expression and evocation, 187-88
    meristem competence, 184-85
    partial evocation, 192
    response of different apex components, 191-92
    start and end of evocation, 186-87
  floral morphogenesis, 192-93
    appendage production rate, 193
    axillary meristems initiation, 194

genetics, 197-98
gibberellins role, 205-7
internode growth, 194-95
leaf growth, 193-94
meristem shape and size, 193
organ number and fusion, 196-97
phyllotactic changes, 195-96
primordium size decrease, 195-96
introduction, 176
theories of endogenous control
  assimilate import in the apex, 202-3
  auxins role, 207-8
  cytokinins role, 204-5
  electrical signal, 201
  florigen/antiflorigen concept, 198-201
  florigen and antiflorigen identification, 200-1
  multifactorial control model, 203-9
  nutrient diversion hypothesis, 201-3
  photosynthetic or carbohydrate input timing, 203
  supporting evidence for florigen/antiflorigen concept, 198-200
  supporting evidence for nutrient diversion hypothesis, 202
Florigen/antiflorigen concept, 176, 182
  florigen and antiflorigen identification, 200-1
  supporting evidence, 198-200
Fluridone, 446-47, 458-59, 462
Folic acid, 423
β-Fructofuranosidase, 344
Fructosans, 367
Fructose 1,6-bisphosphate phosphatase, 555, 557, 577, 582
Fructose 2,6-bisphosphate, 557
Fructose 6-phosphate, 557, 565
Fusicoccin, 278

G

α-Galactosidase, 145, 345
Gene amplification, 429-30
Genistein, 301
Gerber, Bernard, 15
Gernald, M. L., 7
Gibberellic acid, 181, 192, 195, 206

Gibberellin, 24, 33, 35-36, 43, 223, 233, 368, 371, 459, 462
  floral evocation, 183, 191
  floral morphogenesis, 193-94, 202
  role, 205-7
  fruit size increase, 358
1,3-β-Glucan, 77-78
β-1,3-Glucanase, 343
Glucan synthases
  coated vesicles, 77-78
Glucose, 363, 365-66, 370
β-Glucosidase, 344
β-Glucuronidase, 345
Glutamate oxaloacetate transaminase, 223
Glutamic acid, 62
Glutamine synthase, 429
Glutamine synthetase, 232
Glyoxysomes, 150
Glycogen, 234
Goddard, David R., 1, 9-11
Golgi apparatus, 73, 82-83, 85
  coated vesicles and sorting, 83-84
  coating, 57
  immunocytochemical observations
    protein transport, 144-47
Green, Paul B., 13-14, 17
Green, Margaret, 17
Greenman, Jesse, 6
Guard cells
  abscisic acid, 455, 461

H

Harris, William F., 15
Hejnowicz, Zygmunt, 12-13
Heterochromatin, 418, 420, 429
  late replication
    chromosome bridges, 423
    chromosome rearrangement, 420-22
Hexadecadienoate, 125
Hexadecenoic acid, 126
Highkin, Harry R., 16
Hillman, William S., 16
Holden, Connie, 11
Hotta, Yasuo, 17
β-Hydroxyacyl-ACP dehydrase, 110-12
Hydroxylamine, 394
Hydroxylysine, 323
Hydroxyproline, 322-31, 334, 336

I

IAA, 194-95, 207-8, 204, 274, 458

Immunocytochemical localization of macromolecules see Macromolecules, immunocytochemical localization
Invertase, 188, 231
α-Ionylidene acetic acid, 443-44, 452
Iron-sulfur centers, 384-86
Isodityrosine, 327, 340
Isoperoxidase, 341

J

Jasmonic acid, 129

K

β-Ketoacyl-ACP reductase, 109-10
forms, 111
β-Ketoacyl-ACP synthetase, 109-10, 114-15
forms, 110-111
Kinases
coated vesicles, 76
Kinetin, 38-39
Küchler, A. W., 9

L

LaRue, C. D., 5
Leaf expansion control, 267-95
cell division and cell enlargement, 269
division, 269-72
expansion, 272
grass leaf, 273
conclusions, 289
enlargement mechanics
epidermis role, 273-75
intercellular spaces and their significance, 275-76
introduction, 267-68
mechanisms of leaf cell growth
biophysical considerations, 276-78
primary leaves of Phaseolus, 278-79
regulation by light, 279
light quality, 282-83
light quantity, 280-82
phytochrome role, 283-84
the light environment, 279-80
water and leaf expansion
abscisic acid and leaf growth, 288
cell division effects, 285
cell enlargement and turgor control, 285-88

methodological difficulties, 284-85
osmotic adjustment, 286-87
water stress, 287-88
Leghemaglobin, 149
Legume nodule initiation, genetic analysis, 297-319
compatibility of infecting rhizobia, 306-7
exopolysaccharides, 306-7
conclusions, 313-14
early nodulin gene expression, 313
genetic organization of Rhizobium nodulation genes, 304-6
host factors controlling nodulation, 307-10
autoregulation, 307-9
nitrogen fixation, 309
pseudoinfections, 307
inducible plant defense systems, 311-13
phytoalexins, 304, 312
pathogen recognition, 312
introduction, 297-98
isolation of nodulation mutants, 310-11
nonnodulating mutants, 311
supernodulation mutants, 310
regulation of Rhizobium nodulation genes, 301
interaction steps during nodule formation, 303
Rhizobium infection and nodule ontogeny, 298-301
in soybean, 302
variation in ontogenies summary, 300
shikimic acid and phytoalexins
biosynthetic pathways, 304
Legumin, 145
Leucine, 62
Light
leaf expansion control, 278-79
regulation, 279-84
see also Algal development, photocontrol; Floral evocation and morphogenesis control
Light-chain polypeptides, 65, 74
Lignin, 234
Lindenmayer, Aristid, 14
α-Linolenic acid, 122-23
linolenate, 125
desaturation inhibition, 124
formation, 123-24
Linoleoyl glycerolipids, 123
Lipids
coated vesicles, 75

Lipopolysaccharides, 306-7
Lipoxygenase, 128-29
Lowicryl, 143-45, 147-50
LR White, 143-44, 147-51
Luteolin, 301
Lysosomal enzymes, 83
O-Lysophospholipid acyltransferase, 121-22

M

Macromolecules, immunocytochemical localization, 139-55
general principles, 140
introduction, 139-40
methods, 140-41
fixation and antigenicity retention, 141
post-embedding immunocytochemistry, 143-44
pre-embedding immunocytochemistry, 142-43
visualization of bound antibodies, 141-42
observations, 144
cell walls, 150
cytoplasmic proteins, 149
Golgi apparatus-mediated protein transport, 144-47
heterologous expression of proteins, 151
membranes, 147-49
microbodies, 150
protein targeting, 151
Magnesium, 538, 549-50
Malonyl-CoA, 114-17
Malonyl-CoA:ACP transacylase, 109-10
comparative aspects, 111
Manganese, 395
oxygen evolution in PSII binding, 391-92
center models, 393-94
spectroscopy, 392-93
α-Mannosidase, 344
Mannosyl-6-phosphate receptor, 83
Membrane proteins
immonocytochemical observations, 148
Mes, Margaretta G., 16
Mevalonate, 445
Michelini, Mike, 13
Microbodies, 150
Mitochondria
β-oxidation of fatty acids, 127-28

Mitochondrial genomes organization, expression, and variation, 503-32
abnormal growth mutants, 522-23
chloroplast DNA sequences, 506
cytoplasmic male sterility, 515-16, 523
maize *cms-C*, 518
maize *cms-S*, 505, 518-20
maize *cms-T*, 516-17
maize *cms-T* reversions, 517-18
Petunia, 521, 523
RNA plasmids in *cms-S* mitochondria, 520-21
sublimons, 521-22
genes, 507
codon usage, 512
differential expression, 523
identification of additional genes, 512-15
protein-coding genes, 509-10
rRNA, tRNA, and ribosomal proteins, 507-8
transcription, 510-12
introduction, 503-4
organization, 504-6
stability, 505
plasmids, 506-7
prospects, 523-24
Monensin, 145
Monogalactosyldiacylglycerol, 122-25
Mycorrhizae
see Vesicular-arbuscular mycorrhizae, symbionts interactions

N

1-N-Naphthylphthalamic acid, 88
Nitrate reductase, 223, 232
Nitrogenase, 298

O

Octadecatetraenoate, 126
Ogur, Maurice, 11
Oleic acid122
biosynthesis, 118-19
desaturation, 121-24
incorporation into phosphatidylcholine species, 121
Oleoyl-ACP, 119
Oleoyl-CoA, 112, 119-20, 122, 126

β-Oleoyl phosphatidylcholine, 120
Oleoyl phosphatidylcholine desaturase, 120-21
Orthophosphate, 554-58, 567, 569, 579-82
CO_2 fixation sensitivity, 563-65
chloroplasts, 564
manipulation, 561-62
photophosphorylation regulation, 559-60
photosynthesis regulation, 561
Rubisco activity regulation, 561
12-Oxo-phytodienoic acid, 128
Oxygen
evolution in PSII
manganese binding, 391-92
manganese spectroscopy, 392-93
models of the manganese center, 393-94
protein and cofactors role, 395-96
reactions of system, 394-95
fatty acid metabolism
desaturation reactions, 118

P

P680, 387, 395
reduction kinetics, 390-91
P700, 382-87
Palmitoyl-ACP, 112, 114, 118-19
Palmitoyl-CoA, 116-17, 127-28
Partially coated reticulum, 57, 85
Pectinesterase, 345-46
Peptidase, 223
Peridinin, 163
Peroxidase, 87, 284, 340-43
Peroxisomes
formation, 150
β-oxidation of fatty acids, 127-28
Phaseic acid, 449, 453-54, 462
Phaseolin, 145, 151
Phaseolus vulgaris
primary leaves
expansion control, 278-81
Phosphate, 554-56
Phosphatidylcholine, 75, 120-24
Phosphatidylethanolamine, 75, 122
Phosphatidylinositol, 75
Phosphatidylserine, 75
Phosphoenolpyruvate carboxylase, 40
3-Phosphoglyceric acid kinase, 567, 578

Phosphorus
VA mycorrhizae, 234
bidirectional transport, 230-31
effect on host-plant metabolism, 232-33
nutrition , 227-28
uptake and metabolism by the fungi, 228-30
Phosphatase, 339
Photosynthetic carbon reduction cycle, 567, 577, 581
regulation of enzymes, 568-71, 578
regulatory enzymes, 577
regulatory properties, 553-56
Photosynthetic CO_2 fixation in C_3 plants, 533-94
conclusions, 580-83
introduction, 534
net CO_2 uptake, 534-35
activation state of Rubisco, 540-41
2-carboxyarabinitol 1-phosphate concentration, 541
CO_2 compensation point, 535-37
CO_2 fixation models, 541
control of Rubisco activation state, 548-50
dependence on RuBP, 544-47
experimental approaches, 537
freeze-clamp studies, 540
gas exchange studies, 538-39
kinetic constants for Rubisco, 543
kinetic model for Rubisco, 542
kinetic properties of Rubisco, 537-38
regulation by 2-carboxyarabinitol 1-phosphate, 547-48
responses to CO_2 and O_2, 542-44
Rubisco-activase, 550-52
Rubisco content of leaves, 540
summary, 552
photosynthesis rate control, 571-73
CO_2 uptake rate control, 574
definition of limitation, 573-74
flux control by other enzymes, 577

flux control by Rubisco, 574-77
regulatory sequences, 552-53
CO_2 fixation sensitivity to cytosolic Pi, 563-65
conclusions of feedforward regulation, 571
cytosolic Pi manipulation, 561-63
electron transport regulation, 558-60
feedback from sucrose, 565
feedback on energy input from Rubisco, 566-67
feedback regulation of photosynthesis, 556-61
feedback sequence operation, 561-66
feedback sequence structure, 556-58
feedforward regulation of enzyme activity, 568-71
regulatory properties of the PCR cycle, 553
Rubisco regulation, 560-61
sucrose effect, 565-66
summary of feedback sequence operation, 566
Photosynthetic system structure, 554
Photosystem I, 484, 491, 560
Photosystem I, electron transport, 380-81
acceptors A_0 and A_1, 383-84
electron transfer reactions, 381
introduction, 379-80
path and rates of electron transfer, 385-87
iron-sulfur centers, 384-85
P700, 382
polypeptide composition, 381-82
summary, 396-97
Photosystem II, 484, 491-92, 554, 558, 560, 562, 567
Photosystem II, electron transport, 387
acceptor side
cytochrome b_{559}, 390
electron transfer reactions, 387-89
PSII heterogeneity, 390
Q_{400}, 389
role of D1 and D2 proteins, 389-90
summary of electron transfer reactions, 388
donor side
nature of Z and D, 391

P680 reduction kinetics, 390-91
introduction, 379-80
oxygen evolution
manganese binding, 391-92
manganese center models, 393-94
manganese spectroscopy, 392-93
system reactions, 394-95
summary, 396-97
Phycobiliproteins, 168
Phycochromes, 168
Phycocyanin, 163-64
Phycoerythrin, 162-63
Phytoalexins, 234, 312
biosynthetic pathways, 304
Phytochrome, 161
algae, 166-67
immunocytochemical observations, 149
leaf expansion control, 283-84
photoinduction, 179
Phytoene, 446
Phytofluene, 446
Phytohemmagglutinin, 145
Pilet, Paul E., 16
Plasma membrane, 77-78, 82-86, 117
coats, 54-57
Plasmids
mitochondria, 505-7, 520-21
Plasmodesmata, 362-63
Plastid
see Chloroplast development and gene expression
Plastocyanin, 386-7
Plastoquinone, 388, 391
Platt, Robert B., 8
Pollen
brassin activity sources, 25
brassinosteroids content, 28-29
Polygalacturonase, 346
Polyphosphate kinase, 229
Polyproline II, 326
Potassium, 232-33, 256, 370
assimilate unloading, 364
Preer, John, 11
Proline, 233
Protochlorophyllide, 162
Pyruvate dehydrogenase, 102

Q

Q_{400}, 389

R

Receptosome, 58
Rhodopsin, 169

Ribosomes
coated vesicle isolation, 60
Ribulose 1,5-bisphosphate, 548-51, 569-70, 575, 581-82
CO_2 uptake, 544-47
Rubisco activation state, 549
Ribulose 1,5-bisphosphate carboxylase, 40
activity
relation to RuPB concentration, 546
coated vesicle isolation, 60
Ribulose 1,5-bisphosphate carboxylase/oxygenase, 534, 545-47, 554-55, 563
activation state, 540-41, 569
control, 548-50
activity regulation, 560-61, 577
2-carboxyarabinitol 1-phosphate
concentration, 541
regulation of activity, 547-48
CO_2
assimilation rate control, 574
compensation point, 535-37
content of leaves, 540
feedback regulation of photochemical reactions, 566-57
kinetic constants, 543
kinetic models, 542
kinetic properties, 537-38
metabolite binding, 570
photosynthesis rate control, 572, 574-77
photo flux density, 576
regulatory sequences influencing activity, 553
response to RuPB changes, 546, 581-82
temperature
CO_2/O_2 specificity, 536
Ribulose 1,5-bisphosphate carboxylase/oxygenase activase, 550-52, 570
Richards, F. J., 14
Ricinoleic acid
formation, 126-27
Rhizobia
see Legume nodule initiation, genetic analysis
RNA
mitochondria, 507-8, 520-21
levels
chloroplast transcription regulation, 485-86
determination of $psbd$-$psbc$ transcript abundance, 488

psbA mRNA, 487-88
ribosomal RNA, 486-87
RNA polymerase, 40
Roots
see Water transport in and to
roots
Rosen, Gloria, 11
Routledge, Lewis, 15

S

Sachs, Roy, 16
Salicylic acid
floral morphogenesis, 179
Sax, Kathie, 11
Sialyl transferase, 84
Silk, Wendy K., 14, 17
Sjostrom, Stella, 1
Sink organs, metabolism and
compartmentation of
sugars, 355-78
characteristics of sink organs
reversible and irreversible
sink organs, 361-62
utilization and storage sink
organs, 360
conclusions, 372
determination of sink strength
genetic determination, 370-
71
genetic expression, 371-72
import regulation
apoplastic unloading, 363-
65
physical compartmentation,
368-70
symplastic unloading, 362-
63
sugar conversion inside
sink cells, 367-68
sugar unloading from
phloem conducting tis-
sues, 362-65
sugar uptake by sink cells,
365-67
introduction, 356-57
sink strength and assimilate
import
definitions of sink strength,
357-59
sink strength and sink competition, 359-60
Soil
water uptake by roots
cylindrical flow model,
246-50
environmental influences,
257
roots in dry soil, 250-51
Somaclonal variation, chromosomal basis, 413-37
activation of cryptic transposable elements, 425

chromosome breakage, 425
qualitative variants origin,
426-27
transposable-element activity in tissue culture,
425-26
chromosome aberrations
cultured cells, 417-18
nonrandom chromosome
rearrangement, 418
regenerated plants, 417-20
variation generation, 418
chromosome rearrangements,
427
changes in sequence copy
number, 429
gene amplification, 429-30
morphology of regenerated
plants, 427-28
position effects, 428
qualitative variation, 428
conclusions, 430-31
generalities from past work,
415-16
introduction, 414-15
observing chromosome aberrations, 416-17
origin of chromosome
rearrangement in tissue
culture, 420
late-replicating heterochromatin, 420-22
mitotic recombination, 424-
25
nucleotide pool imbalance,
422-23
Sorbitol, 451-52
Spencer, Warren, 9
Spermine, 78
Starch, 231, 367-68, 371, 565
storage sinks, 361
synthesis
sensitivity to Pi concentration, 564
Stearoyl-ACP, 116, 118
Stearoyl-ACP desaturase, 118
Stearoyl-CoA, 117-18
Stern, Curt, 9
Stern, Herbert, 17
Sterols
brassinosteroid synthesis, 45-
45
Steward, F. C., 10
Subtilisin, 64
Sucrose, 556, 579-81
floral evocation, 188
floral morphogenesis, 202
level
CO_2 assimilation, 565
metabolite partitioning, 565
metabolic response, 566
sink organs, 365
aplastic unloading, 363

conversion, 367-68
import regulation, 359
resynthesis, 367
storage sinks, 361
symplastic unloading, 362-
63
transport, 368-69
uptake, 365-66, 369
Sucrose phosphatase, 556, 577
Sucrose phosphate synthetase,
367
Sucrose phosphate synthase,
556-57, 565-66, 577
Sucrose synthetase, 364

T

Temperature
floral morphogenesis control,
178, 181
gibberellins, 206
low temperature, 180
perception sites, 182
Rubisco
CO_2/O_2 specificity, 536
Thermolysin, 64
Thylakoids
immunocytochemical observations, 147-48
Thymidine, 423
Tilney, Lewis, 15
Transferrin, 84
Trehalase, 345
Trehalose, 234
Triskelions, 63-66, 68-69, 73,
75
reassembly, 72
release from coated vesicles,
71
Trypsin, 64
Tubulin, 62, 67-68
Tyrosine, 391

U

Uricase II, 150

V

Vacuoles
sucrose uptake, 369-70
Veen, Arthur, 15
Vesicles
see Coated vesicles
Vesicular-arbuscular mycorrhizae, symbionts interactions,
221-41
conclusions, 236-37
endophyte physiology in culture, 222-24
spore biochemistry, 223
spore germination, 222-23

infection and host-fungus in-
    terface development,
    224-27
arbuscule formation, 226
ATPase localization, 225-
    26
infection spread, 224
introduction, 221-22
physiological compatibility
    and mycorrhizal efficien-
    cy, 235-36
efficiency determinants,
    236
symbiont physiology and the
    mycorrhizal condition
bidirectional nutrient trans-
    fer between symbionts,
    230-32
carbon distribution, 234-35
effects on host-plant
    metabolism, 232-33
endophyte contribution,
    228-230
extramatrical hyphae de-
    velopment in soil, 229
hyphal growth, 228
Nonnutritional mod-
    ifications of physiolo-
    gy, 233-34
plant growth and phosphor-
    us nutrition, 227-28
phosphorus uptake and
    metabolism by fungi,
    228-30
translocation to roots, 230
water relations, 233

Vesicular stomatitis virus, 83-
    84
Violaxanthin, 446, 449-50, 458
Vitamin K$_1$, 383-84

W

Water
leaf expansion control
    abscisic acid and leaf
        growth, 288
    cell division effects, 285
    cell enlargement and turgor
        control, 285-88
    methodological difficulties,
        284-85
Water transport in and to roots,
    245-65
concluding remarks, 261-62
introduction, 245-46
pathways
    axial, 260-61
    radial, 258-60
    variation of uptake along
        roots, 260
transport across whole root
    systems
    basic hydraulic and osmotic
        properties, 252-56
    circadian rhythms, 257
    endogenous and environ-
        mental influences,
        257-58
    possible experimental arte-
        facts, 256

pressure axis offset in-
    terpretation, 255
three-compartment model,
    254-55
two-compartment model
uptake from soil
    cylindrical flow model,
        246-50
    root clumping effect,
        247
    interfacial resistance be-
        tween root and soil,
        247-49
    modification of relation be-
        tween suction and
        water content, 249-50
    reverse flow of water to
        dry soils, 251
    sands and sandy soils,
        249
    uptake in dry soils, 250
    uptake recovery after water-
        ing dry soils, 250
Went, Frits W., 7, 16
Willier, Benjamin, 9

X

Xanthoxin, 445, 458
abscisic acid precursor, 446,
    450
extraction and quantification,
    442-43
Xyloglucan, 337-38

CUMULATIVE INDEXES

CONTRIBUTING AUTHORS, VOLUMES 31–39

A

Abeles, F. B., 37:49–72
Akazawa, T., 36:441–72
Albersheim, P., 35:243–75
Aloni, R., 38:179–204
Amasino, R. M., 35:387–413
Anderson, J. M., 37:93–136
Andréasson, L., 39:379–411
Appleby, C. A., 35:443–78

B

Badger, M. R., 36:27–53
Bandurski, R. S., 33:403–30
Barber, J., 33:261–95
Bauer, W. D., 32:407–49
Beale, S. I., 34:241–78
Beard, W. A., 38:347–89
Bell, A. A., 32:21–81
Benson, D. R., 37:209–32
Benveniste, P., 37:275–308
Bernier, G., 39:175–219
Berry, J. A., 39:533–94
Berry, J., 31:491–543
Bickel-Sandkötter, S., 35:97–120
Björkman, O., 31:491–543
Boller, T., 37:137–64
Bottomley, W., 34:279–310
Boudet, A. M., 38:73–93
Boyer, J. S., 36:473–516
Brady, C. J., 38:155–78
Brenner, M. L., 32:511–38
Buchanan, B. B., 31:341–74
Burnell, J. N., 36:255–86

C

Cassab, G. I., 39:321–53
Castelfranco, P. A., 34:241–78
Chapman, D. J., 31:639–78
Chua, N., 38:221–57
Clarke, A. E., 34:47–70
Clarkson, D. T., 31:239–98; 36:77–115
Clegg, M. T., 38:391–418
Cogdell, R. J., 34:21–45
Cohen, J. D., 33:403–30
Conn, E. E., 31:433–51
Cosgrove, D., 37:377–405
Creelman, R. A., 39:439–73
Cronshaw, J., 32:465–84
Cullis, C. A., 36:367–96

D

Dale, J. E., 39:267–95
Darvill, A. G., 35:243–75
Delmer, D. P., 38:259–90
Dennis, D. T., 33:27–50
Depta, H., 39:53–99
Diener, T. O., 32:313–25
Digby, J., 31:131–48
Dilley, R. A., 38:347–89
Dring, M. J., 39:157–74
Dutcher, F. R., 38:317–45

E

Edwards, G. E., 36:255–86
Eisbrenner, G., 34:105–36
Eisinger, W., 34:225–40
Ellis, R. J., 32:111–37
Elstner, E. F., 33:73–96
Erickson, R. O., 39:1–22
Etzler, M. E., 36:209–34
Evans, H. J., 34:105–36
Evans, L. T., 32:485–509
Evenari, M., 36:1–25

F

Farquhar, G. D., 33:317–45
Feldman, J. F., 33:583–608
Feldman, L. J., 35:223–42
Fincher, G. B., 34:47–70
Firn, R. D., 31:131–48
Flavell, R., 31:569–96
Fork, D. C., 37:335–61
Freeling, M., 35:277–98
Fry, S. C., 37:165–86
Furuya, M., 35:349–73

G

Galston, A. W., 32:83–110
Galun, E., 32:237–66
Gantt, E., 32:327–47
Gianinazzi-Pearson, V., 39:221–44
Giaquinta, R. T., 34:347–87
Gifford, E. M. Jr., 34:419–40
Gifford, R. M., 32:485–509
Glass, A. D. M., 34:311–26
Glazer, A. N., 38:11–45
Good, N. E., 37:1–22
Gordon, M. P., 35:387–413
Graebe, J. E., 38:419–65

Graham, D., 33:347–72
Gray, M. W., 33:373–402
Green, P. B., 31:51–82
Green, P. J., 38:221–57
Greenway, H., 31:149–90
Gresshoff, P. M., 39:297–319
Guerrero, M. G., 32:169–204
Gunning, B. E. S., 33:651–98

H

Haehnel, W., 35:659–93
Halstead, T. W., 38:317–45
Hanson, A. D., 33:163–203
Hanson, J. B., 31:239–98
Hara-Nishimura, I., 36:441–72
Hardham, A. R., 33:651–98
Harding, R. W., 31:217–38
Harris, N., 37:73–92
Harwood, J. L., 39:101–38
Hatch, M. D., 36:255–86
Haupt, W., 33:205–33
Heath, R. L., 31:395–431
Heber, U., 32:139–68
Heidecker, G., 37:439–66
Heldt, H. W., 32:139–68
Hepler, P. K., 36:397–439
Herman, E. M., 39:139–55
Higgins, T. J. V., 35:191–221
Hirel, B., 36:345–65
Hitz, W. D., 33:163–203
Ho, L. C., 39:355–78
Ho, T.-H. D., 37:363–76
Hoffman, N. E., 35:55–89
Horsch, R., 38:467–86
Howell, S. H., 33:609–50
Huber, S. C., 37:233–46
Hull, R., 38:291–315

J

Jackson, M. B., 36:145–74

K

Kamiya, N., 32:205–36
Kaplan, A., 35:45–83
Kauss, H., 38:47–72
King, R. W., 36:517–68
Klee, H., 38:467–86
Kolattukudy, P. E., 32:539–67
Kowallik, W., 33:51–72
Kuhlemeier, C., 38:221–57

L

Labavitch, J. M., 32:385–406
Lang, A., 31:1–28
Laties, G. G., 33:519–55
Leaver, C. J., 33:373–402
Lee, M., 39:413–37
Leong, S. A., 37:187–208
Letham, D. S., 34:163–97
Lin, W., 37:309–34
Lloyd, C. W., 38:119–39
Loewus, F. A., 34:137–61
Loewus, M. W., 34:137–61
Lorimer, G. H., 32:349–83
Losada, M., 32:169–204
Lucas, W. J., 34:71–104

M

Møller, I. M., 37:309–34
Maliga, P., 35:519–42
Malkin, R., 33:455–79
Mandava, N. B., 39:23–52
Marx, G. A., 34:389–417
Meins, F. Jr., 34:327–46
Melis, A., 38:11–45
Messing, J., 37:439–66
Miernyk, J. A., 33:27–50
Mimura, T., 38:95–117
Minchin, P. E. H., 31:191–215
Moore, T. S. Jr., 33:235–59
Moreland, D. E., 31:597–638
Morgan, J. M., 35:299–319
Morris, R. O., 37:509–38
Mullet, J. E., 39:475–502
Munns, R., 31:149–90

N

Nakamoto, H., 36:255–86
Neilands, J. B., 37:187–208
Nester, E. W., 35:387–413
Newton, K. J., 39:503–32

O

O'Leary, M. H., 33:297–315
Oaks, A., 36:345–65
Ogren, W. L., 35:415–42
Outlaw, W. H. Jr., 31:299–311

P

Palni, L. M. S., 34:163–97
Passioura, J. B., 39:245–65

Pate, J. S., 31:313–40
Patterson, B. D., 33:347–72
Payne, P. I., 38:141–53
Pharis, R. P., 36:517–68
Phillips, D. A., 31:29–49
Phillips, R. L., 39:413–37
Pickard, B. G., 36:55–75
Possingham, J. V., 31:113–29
Powles, S. B., 35:15–44
Pradet, A., 34:199–224
Pratt, L. H., 33:557–82
Preiss, J., 33:431–54

R

Ragan, M. A., 31:639–78
Ranjeva, R., 38:73–93
Raymond, P., 34:199–224
Reinhold, L., 35:45–83
Rennenberg, H., 35:121–53
Roberts, J. A., 33:133–62
Roberts, J. K. M., 35:375–86
Robinson, D., 39:53–99
Rogers, S., 38:467–86
Rolfe, B. G., 39:297–319
Roughan, P. G., 33:97–132
Rubery, P. H., 32:569–96

S

Sachs, M. M., 37:363–76
Satoh, K., 37:335–61
Satter, R. L., 32:83–110
Schnepf, E., 37:23–47
Schubert, K. R., 37:539–74
Schulze, E.-D., 37:247–74
Schwintzer, C. R., 37:209–32
Sexton, R., 33:133–62
Sharkey, T. D., 33:317–45
Shimmen, T., 38:95–117
Shropshire, W. Jr., 31:217–38
Silk, W. K., 35:479–518
Silverthorne, J., 36:569–93
Slack, C. R., 33:97–132
Smith, H., 33:481–518
Smith, S. E., 39:221–44
Smith, T. A., 36:117–43
Snell, W. J., 36:287–315
Somerville, C. R., 37:467–507
Spanswick, R. M., 32:267–89
Spiker, S., 36:235–53
Steponkus, P. L., 35:543–84
Stocking, C. R., 35:1–14
Stoddart, J. L., 31:83–111
Stone, B. A., 34:47–70

Strotmann, H., 35:97–120
Sweeney, B. M., 38:1–9
Sze, H., 36:175–208

T

Taiz, L., 35:585–657
Tang, P.-S., 34:1–19
Tazawa, M., 38:95–117
Theg, S. M., 38:347–89
Theologis, A., 37:407–38
Thomas, H., 31:83–111
Thomson, W. W., 31:375–94
Thorne, J. H., 36:317–43
Ting, I. P., 36:595–622
Tjepkema, J. D., 37:209–32
Tobin, E. M., 36:569–93
Tran Thanh Van, K. M., 32:291–311
Trelease, R. N., 35:321–47
Troughton, J. H., 31:191–215

V

van Huystee, R. B., 38:205–19
Vänngård, T., 39:379–411
Varner, J., 39:321–53
Vega, J. M., 32:169–204
Velthuys, B. R., 31:545–67
Vennesland, B., 32:1–20
Virgin, H. I., 32:451–63

W

Walbot, V., 36:367–96
Walton, D. C., 31:453–89
Wareing, P. F., 33:1–26
Wayne, R. O., 36:397–439
Weiler, E. W., 35:85–95
Whatley, J. M., 31:375–94
Whitfeld, P. R., 34:279–310
Wiemken, A., 37:137–64
Woodrow, I. E., 39:533–94

Y

Yang, S. F., 35:155–89
Yanofsky, M. F., 35:387–413

Z

Zaitlin, M., 38:291–315
Zeevaart, J. A. D., 39:439–73
Zeiger, E., 34:441–75
Zurawski, G., 38:391–418

CHAPTER TITLES, VOLUMES 31–39

PREFATORY CHAPTERS

Some Recollections and Reflections	A. Lang	31:1–28
Recollections and Small Confessions	B. Vennesland	32:1–20
A Plant Physiological Odyssey	P. F. Wareing	33:1–26
Aspirations, Reality, and Circumstances: The Devious Trail of a Roaming Plant Physiologist	P.-S. Tang	34:1–19
Reminiscences and Reflections	C. R. Stocking	35:1–14
A Cat Has Nine Lives	M. Evenari	36:1–25
Confessions of a Habitual Skeptic	N. E. Good	37:1–22
Living in the Golden Age of Biology	B. M. Sweeney	38:1–9
Growth and Development of a Botanist	R. O. Erickson	39:1–22

MOLECULES AND METABOLISM

Bioenergetics

Efficiency of Symbiotic Nitrogen Fixation in Legumes	D. A. Phillips	31:29–49
Role of Light in the Regulation of Chloroplast Enzymes	B. B. Buchanan	31:341–74
Mechanisms of Electron Flow in Photosystem II and Toward Photosystem I	B. R. Velthuys	31:545–67
The Carboxylation and Oxygenation of Ribulose 1,5-Bisphosphate: The Primary Events in Photosynthesis and Photorespiration	G. H. Lorimer	32:349–83
The Physical State of Protochlorophyll(ide) in Plants	H. I. Virgin	32:451–63
Blue Light Effects on Respiration	W. Kowallik	33:51–72
Oxygen Activation and Oxygen Toxicity	E. F. Elstner	33:73–96
Photosystem I	R. Malkin	33:455–79
The Cyanide-Resistant, Alternative Path in Higher Plant Respiration	G. G. Laties	33:519–55
Photosynthetic Reaction Centers	R. J. Cogdell	34:21–45
Adenine Nucleotide Ratios and Adenylate Energy Charge in Energy Metabolism	A. Pradet, P. Raymond	34:199–224
Structure, Function, and Regulation of Chloroplast ATPase	H. Strotmann, S. Bickel-Sandkötter	35:97–120
Photorespiration: Pathways, Regulation, and Modification	W.L. Ogren	35:415–42
Photosynthetic Electron Transport in Higher Plants	W. Haehnel	35:659–93
Photosynthetic Oxygen Exchange	M. R. Badger	36:27–53
Pyruvate,P_i Dikinase and NADP-Malate Dehydrogenase in C_4 Photosynthesis: Properties and Mechanism of Light/Dark Regulation	G. E. Edwards, H. Nakamoto, J. N. Burnell, M. D. Slack	36:255–86
Crassulacean Acid Metabolism	I. P. Ting	36:595–622
Photoregulation of the Composition, Function, and Structure of Thylakoid Membranes	J. M. Anderson	37:93–136

Physiology of Actinorhizal Nodules — J. D. Tjepkema, C. R. Schwintzer, D. R. Benson — 37:209–32

Membrane-Bound NAD(P)H Dehydrogenases in Higher Plant Cells — I. M. Møller, W. Lin — 37:309–34

The Control by State Transitions of the Distribution of Excitation Energy in Photosynthesis — D. C. Fork, K. Satoh — 37:335–61

Products of Biological Nitrogen Fixation in Higher Plants: Synthesis, Transport, and Metabolism — K. R. Schubert — 37:539–74

Membrane-Proton Interactions in Chloroplast Bioenergetics: Localized Proton Domains — R. A. Dilley, S. M. Theg, W. A. Beard — 38:347–89

Photochemical Reaction Centers: Structure, Organization, and Function — A. N. Glazer, A. Melis — 38:11–45

Genetic Analysis of Legume Nodule Initiation — B. G. Rolfe, P. M. Gresshoff — 39:297–319

Photosynthetic Electron Transport in Higher Plants — T. Vänngård, L. Andréasson — 39:379–411

Enzymatic Regulation of Photosynthetic CO2 Fixation in C3 Plants — I. E. Woodrow, J. A. Berry — 39:533–94

Small Molecules

Photocontrol of Carotenoid Biosynthesis — R. W. Harding, W. Shropshire, Jr. — 31:217–38

A Descriptive Evaluation of Quantitative Histochemical Methods Based on Pyridine Nucleotides — W. H. Outlaw, Jr. — 31:299–311

Cyanogenic Compounds — E. E. Conn — 31:433–51

Biochemistry and Physiology of Abscisic Acid — D. C. Walton — 31:453–89

Mechanisms of Action of Herbicides — D. E. Moreland — 31:597–638

Modern Methods for Plant Growth Substance Analysis — M. L. Brenner — 32:511–38

Structure, Biosynthesis, and Biodegradation of Cutin and Suberin — P. E. Kolattukudy — 32:539–67

Compartmentation of Nonphotosynthetic Carbohydrate Metabolism — D. T. Dennis, J. A. Miernyk — 33:27–50

Cellular Organization of Glycerolipid Metabolism — P. G. Roughan, C. R. Slack — 33:97–132

Phospholipid Biosynthesis — T. S. Moore, Jr. — 33:235–59

Chemistry and Physiology of the Bound Auxins — J. D. Cohen, R. S. Bandurski — 33:403–30

myo-Inositol: Its Biosynthesis and Metabolism — F. A. Loewus, M. W. Loewus — 34:137–61

The Biosynthesis and Metabolism of Cytokinins — D. S. Letham, L. M. S. Palni — 34:163–97

Chlorophyll Biosynthesis: Recent Advances and Areas of Current Interest — P. A. Castelfranco, S. I. Beale — 34:241–78

Immunoassay of Plant Growth Regulators — E. W. Weiler — 35:85–95

The Fate of Excess Sulfur in Higher Plants — H. Rennenberg — 35:121–53

Ethylene Biosynthesis and its Regulation in Higher Plants — S. F. Yang, N. E. Hoffman — 35:155–89

Polyamines — T. A. Smith — 36:117–43

Nitrogen Metabolism in Roots — A. Oaks, B. Hirel — 36:345–65

Siderophores in Relation to Plant Growth and Disease — J. B. Neilands, S. A. Leong — 37:187–208

Sterol Biosynthesis — P. Benveniste — 37:275–308

Gibberellin Biosynthesis and Control — J. E. Graebe — 38:419–65

Plant Growth-Promoting Brassinosteroids — N. B. Mandava — 39:23–52

Fatty Acid Metabolism — J. L. Harwood — 39:101–38

Metabolism and Physiology of Abscisic Acid — J. A. D. Zeevaart, R. A. Creelman — 39:439–73

Macromolecules

The Molecular Characterization and Organization of Plant Chromosomal DNA Sequences — R. Flavell — 31:569–96

Chloroplast Proteins: Synthesis, Transport,
  and Assembly   R. J. Ellis   32:111–37
The Assimilatory Nitrate-Reducing System
  and Its Regulation   M. G. Guerrero, J. M. Vega,
                                                  M. Losada   32:169–204
Cell Wall Turnover in Plant Development   J. M. Labavitch   32:385–406
Phosphoenolpyruvate Carboxylase: An
  Enzymologist's View   M. H. O'Leary   33:297–315
Regulation of the Biosynthesis and
  Degradation of Starch   J. Preiss   33:431–54
Phytochrome: The Protein Moiety   L. H. Pratt   33:557–82
Arabinogalactan-Proteins: Structure,
  Biosynthesis, and Function   G. B. Fincher, B. A. Stone,
                                                  A. E. Clarke   34:47–70
Synthesis and Regulation of Major Proteins in
  Seeds   T. J. V Higgins   35:191–221
Leghemoglobin and Rhizobium Respiration   C. A. Appleby   35:443–78
Plant Lectins: Molecular and Biological
  Aspects   M. E. Etzler   36:209–34
Plant Chromatin Structure   S. Spiker   36:235–53
Rapid Genomic Change in Higher Plants   V. Walbot, C. A. Cullis   36:367–96
Topographic Aspects of Biosynthesis,
  Extracellular Secretion, and Intracellular
  Storage of Proteins in Plant Cells   T. Akazawa, I. Hara-Nishimura   36:441–72
Alteration of Gene Expression During
  Environmental Stress in Plants   M. M. Sachs, T.-H. D. Ho   37:363–76
Cellulose Biosynthesis   D. P. Delmer   38:259–90
Some Aspects of Calcium-Dependent
  Regulation in Plant Metabolism   H. Kauss   38:47–72
Some Molecular Aspects of Plant Peroxidase
  Biosynthetic Studies   R. B. van Huystee   38:205–19

ORGANELLES AND CELLS

Function
The Chloroplast Envelope: Structure,
  Function, and Role in Leaf Metabolism   U. Heber, H. W. Heldt   32:139–68
Physical and Chemical Basis of Cytoplasmic
  Streaming   N. Kamiya   32:205–36
Plant Protoplasts as Physiological Tools   E. Galun   32:237–66
Electrogenic Ion Pumps   R. M. Spanswick   32:267–89
Viroids: Abnormal Products of Plant
  Metabolism   T. O. Diener   32:313–25
Light-Mediated Movement of Chloroplasts   W. Haupt   33:205–33
Influence of Surface Charges on Thylakoid
  Structure and Function   J. Barber   33:261–95
Mitochondrial Genome Organization and
  Expression in Higher Plants   C. J. Leaver, M. W. Gray   33:373–402
Plant Molecular Vehicles: Potential Vectors
  for Introducing Foreign DNA into Plants   S. H. Howell   33:609–50
Photosynthetic Assimilation of Exogenous
  HCO_3^- by Aquatic Plants   W. J. Lucas   34:71–104
Aspects of Hydrogen Metabolism in
  Nitrogen-Fixing Legumes and Other
  Plant-Microbe Associations   G. Eisbrenner, H. J. Evans   34:105–36
Organization and Structure of Chloroplast
  Genes   P. W. Whitfeld, W. Bottomley   34:279–310
The Biology of Stomatal Guard Cells   E. Zeiger   34:441–75
Membrane Transport of Sugars and Amino
  Acids   L. Reinhold, A. Kaplan   35:45–83
Plant Transposable Elements and Insertion
  Sequences   M. Freeling   35:277–98

Study of Plant Metabolism in vivo Using
NMR Spectroscopy — J. K. M. Roberts — 35:375–86
H+-Translocating ATPases: Advances Using
Membrane Vesicles — H. Sze — 36:175–208
Light Regulation of Gene Expression in
Higher Plants — E. M. Tobin, J. Silverthorne — 36:569–93
Dynamics of Vacuolar Compartmentation — T. Boller, A. Wiemken — 37:137–64
Fructose 2,6-Bisphosphate as a Regulatory
Metabolite in Plants — S. C. Huber — 37:233–46
Structural Analysis of Plant Genes — G. Heidecker, J. Messing — 37:439–66
Phosphorylation of Proteins in Plants:
Regulatory Effects and Potential
Involvement in Stimulus Response
Coupling — R. Ranjeva, A. M. Boudet — 38:73–93
Membrane Control in the Characeae — M. Tazawa, T. Shimmen,
T. Mimura — 38:95–117
Agrobacterium-Mediated Plant Transformation
and Its Further Applications to Plant
Biology — H. Klee, R. Horsch, S. Rogers — 38:467–86
Regulation of Gene Expression in Higher
Plants — C. Kuhlemeier, P. J. Green,
N. Chua — 38:221–57
Cell Wall Proteins — J. Varner, G. I. Cassab — 39:321–53
Coated Vesicles — D. Robinson, H. Depta — 39:53–99
Plant Mitochondrial Genomes: Organization,
Expression, and Variation — K. J. Newton — 39:503–32

Organization
Phycobilisomes — E. Gantt — 32:327–47
Microtubules — B. E. S. Gunning, A. R. Hardham — 33:651–98
Biogenesis of Glyoxysomes — R. N. Trelease — 35:321–47
Plant Cell Expansion: Regulation of Cell Wall
Mechanical Properties — L. Taiz — 35:585–657
Organization of the Endomembrane System — N. Harris — 37:73–92
Cross-Linking of Matrix Polymers in the
Growing Cell Walls of Angiosperms — S. C. Fry — 37:165–86
The Plant Cytoskeleton: The Impact of
Fluorescence Microscopy — C. W. Lloyd — 38:119–39
Immunocytochemical Localization of
Macromolecules with the Electron
Microscope — E. M. Herman — 39:139–55

Development
Plastid Replication and Development in the
Life Cycle of Higher Plants — J. V. Possingham — 31:113–29
Development of Nongreen Plastids — W. W. Thomson, J. M. Whatley — 31:375–94
Auxin Receptors — P. H. Rubery — 32:569–96
Heritable Variation in Plant Cell Culture — F. Meins, Jr. — 34:327–46
Calcium and Plant Development — P. K. Hepler, R. O. Wayne — 36:397–439
Cellular Polarity — E. Schnepf — 37:23–47
Biophysical Control of Plant Cell Growth — D. Cosgrove — 37:377–405
Genes Specifying Auxin and Cytokinin
Biosynthesis in Phytopathogens — R. O. Morris — 37:509–38
Chloroplast Development and Gene
Expression — J. E. Mullet — 39:475–502

TISSUES, ORGANS, AND WHOLE PLANTS

Function
Quantitative Interpretation of Phloem
Translocation Data — P. E. H. Minchin, J. H. Troughton — 31:191–215
The Mineral Nutrition of Higher Plants — D. T. Clarkson, J. B. Hanson — 31:239–98
Transport and Partitioning of Nitrogenous
Solutes — J. S. Pate — 31:313–40

Infection of Legumes by*Rhizobia*   W. D. Bauer   32:407–49
Phloem Structure and Function   J. Cronshaw   32:465–84
Photosynthesis, Carbon Partitioning, and
    Yield   R. M. Gifford, L. T. Evans   32:485–509
Regulation of Pea Internode Expansion by
    Ethylene   W. Eisinger   34:225–40
Regulation of Ion Transport   A. D. M. Glass   34:311–26
Phloem Loading of Sucrose   R. T. Giaquinta   34:347–87
Phytoalexins and Their Elicitors: A Defense
    Against Microbial Infection in Plants   A. G. Darvill, P. Albersheim   35:243–75
Factors Affecting Mineral Nutrient
    Acquisition by Plants   D. T. Clarkson   36:77–115
Ethylene and Responses of Plants to Soil
    Waterlogging and Submergence   M. B. Jackson   36:145–74
Cell-Cell Interactions in *Chlamydomonas*   W. J. Snell   36:287–315
Phloem Unloading of C and N Assimilates in
    Developing Seeds   J. H. Thorne   36:317–43
Water Transport   J. S. Boyer   36:473–516
Plant Chemiluminescence
F. B. Abeles   37:49–72
Fruit Ripening   C. J. Brady   38:155–78
Physiological Interactions Between Symbionts
    in Vesicular- Arbuscular Mycorrhizal Plants   S. E. Smith, V. Gianinazzi-Pearson 39:221–44
Metabolism and Compartmentation of
    Imported Sugars in Sink Organs in Relation
    to Sink Strength   L. C. Ho   39:355–78
Water Transport in and to Roots   J. B. Passioura   39:245–65

Development
Organogenesis—A Biophysical View   P. B. Green   31:51–82
Leaf Senescence   H. Thomas, J. L. Stoddart   31:83–111
The Establishment of Tropic Curvatures in
    Plants   R. D. Firn, J. Digby   31:131–48
Mechanisms of Control of Leaf Movements   R. L. Satter, A. W. Galston   32:83–110
Control of Morphogenesis in In Vitro
    Cultures   K. M. Tran Thanh Van   32:291–311
Cell Biology of Abscission   R. Sexton, J. A. Roberts   33:133–62
Genetic Approaches to Circadian Clocks   J. F. Feldman   33:583–608
Developmental Mutants in Some Annual Seed
    Plants   G. A. Marx   34:389–417
Concept of Apical Cells in Bryophytes and
    Pteridophytes   E. M. Gifford, Jr.   34:419–40
Regulation of Root Development   L. J. Feldman   35:223–42
Osmoregulation and Water Stress in Higher
    Plants   J. M. Morgan   35:299–319
Cell Division Patterns in Multicellular Plants   M.Furuya   35:349–73
Quantitative Descriptions of Development   W. K. Silk   35:479–518
Early Events in Geotropism of Seedling
    Shoots   B. G. Pickard   36:55–75
Gibberellins and Reproductive Development
    in Seed Plants   R. P. Pharis, R. W. King   36:517–68
Rapid Gene Regulation by Auxin   A. Theologis   37:407–38
Plants in Space   T. W. Halstead, F. R. Dutcher   38:317–45
Differentiation of Vascular Tissues   R. Aloni   38:179–204
The Control of Floral Evocation and
    Morphogenesis   G. Bernier   39:175–219
Photocontrol of Development in Algae   M. J. Dring   39:157–74
The Control of Leaf Expansion   J. E. Dale   39:267–95

POPULATION AND ENVIRONMENT

Physiological Ecology
Mechanisms of Salt Tolerance in
    Nonhalophytes   H. Greenway, R. Munns   31:149–90

Photosynthetic Response and Adaptation to
  Temperature in Higher Plants   J. Berry, O. Björkman   31:491–543
Metabolic Responses of Mesophytes to Plant
  Water Deficits   A. D. Hanson, W. D. Hitz   33:163–203
Stomatal Conductance and Photosynthesis   G. D. Farquhar, T. D. Sharkey   33:317–45
Responses of Plants to Low, Nonfreezing
  Temperatures: Proteins, Metabolism, and
  Acclimation   D. Graham, B. D. Patterson   33:347–72
Light Quality, Photoperception, and Plant
  Strategy   H. Smith   33:481–518
Photoinhibition of Photosynthesis Induced by
  Visible Light   S. B. Powles   35:15–44
Carbon Dioxide and Water Vapor Exchange
  in Response to Drought in the Atmosphere
  and in the Soil   E.-D. Schulze   37:247–74

Genetics and Plant Breeding
— Isolation and Characterization of Mutants in
  Plant Cell Culture   P. Maliga   35:519–42
Analysis of Photosynthesis with Mutants of
  Higher Plants and Algae   C. R. Somerville   37:467–507

Genetics of Wheat Storage Proteins and the
  Effect of Allelic Variation on
  Bread-Making Quality   P. I. Payne   38:141–53

Pathology and Injury
Initial Events in Injury to Plants by Air
  Pollutants   R. L. Heath   31:395–431
Biochemical Mechanisms of Disease
  Resistance   A. A. Bell   32:21–81
Crown Gall: A Molecular and Physiological
  Analysis   E.W. Nester, M. P. Gordon,
                                                  R. M. Amasino, M. F. Yanofsky   35:387–413
Role of the Plasma Membrane in Freezing
  Injury and Cold Acclimation   P. L. Steponkus   35:543–84
Plant Virus-Host Interactions   M. Zaitlin, R. Hull   38:291–315

Evolution
Evolution of Biochemical Pathways: Evidence
  from Comparative Biochemistry   D. J. Chapman, M. A. Ragan   31:639–78
Evolution of Higher Plant Chloroplast
  DNA-Encoded Genes: Implications for
  Structure-Function and Phylogenetic Studies   G. Zurawski, M. T. Clegg   38:391–418
The Chromosomal Basis of Somaclonal
  Variation   M. Lee, R. L. Phillips   39:413–37

Annual Reviews Inc.

A NONPROFIT SCIENTIFIC PUBLISHER

4139 El Camino Way
P.O. Box 10139
Palo Alto, CA 94303-0897 • USA

ORDER FORM

Now you can order
TOLL FREE
1-800-523-8635
(except California)

Annual Reviews Inc. publications may be ordered directly from our office by mail or use our Toll Free Telephone line (for orders paid by credit card or purchase order, and customer service calls only); through booksellers and subscription agents, worldwide; and through participating professional societies. Prices subject to change without notice.   ARI Federal I.D. #94-1156476

- **Individuals:** Prepayment required on new accounts by check or money order (in U.S. dollars, check drawn on U.S. bank) or charge to credit card — American Express, VISA, MasterCard.
- **Institutional buyers:** Please include purchase order number.
- **Students:** $10.00 discount from retail price, per volume. Prepayment required. Proof of student status must be provided (photocopy of student I.D. or signature of department secretary is acceptable). Students must send orders direct to Annual Reviews. Orders received through bookstores and institutions requesting student rates will be returned. You may order at the Student Rate for a maximum of 3 years.
- **Professional Society Members:** Members of professional societies that have a contractual arrangement with Annual Reviews may order books through their society at a reduced rate. Check with your society for information.
- **Toll Free Telephone orders:** Call 1-800-523-8635 (except from California) for orders paid by credit card or purchase order and customer service calls only. California customers and all other business calls use 415-493-4400 (not toll free). Hours: 8:00 AM to 4:00 PM, Monday-Friday, Pacific Time.

Regular orders: Please list the volumes you wish to order by volume number.
Standing orders: New volume in the series will be sent to you automatically each year upon publication. Cancellation may be made at any time. Please indicate volume number to begin standing order.
Prepublication orders: Volumes not yet published will be shipped in month and year indicated.
California orders: Add applicable sales tax.
Postage paid (4th class bookrate/surface mail) **by Annual Reviews Inc.** Airmail postage or UPS, extra.

ANNUAL REVIEWS SERIES		Prices Postpaid per volume USA & Canada/elsewhere	Regular Order Please send:	Standing Order Begin with:
			Vol. number	Vol. number
Annual Review of ANTHROPOLOGY				
Vols. 1-14	(1972-1985)	$27.00/$30.00		
Vols. 15-16	(1986-1987)	$31.00/$34.00		
Vol. 17	(avail. Oct. 1988)	$35.00/$39.00	Vol(s). _____	Vol. _____
Annual Review of ASTRONOMY AND ASTROPHYSICS				
Vols. 1-2, 4-20	(1963-1964; 1966-1982)	$27.00/$30.00		
Vols. 21-25	(1983-1987)	$44.00/$47.00		
Vol. 26	(avail. Sept. 1988)	$47.00/$51.00	Vol(s). _____	Vol. _____
Annual Review of BIOCHEMISTRY				
Vols. 30-34, 36-54	(1961-1965; 1967-1985)	$29.00/$32.00		
Vols. 55-56	(1986-1987)	$33.00/$36.00		
Vol. 57	(avail. July 1988)	$35.00/$39.00	Vol(s). _____	Vol. _____
Annual Review of BIOPHYSICS AND BIOPHYSICAL CHEMISTRY				
Vols. 1-11	(1972-1982)	$27.00/$30.00		
Vols. 12-16	(1983-1987)	$47.00/$50.00		
Vol. 17	(avail. June 1988)	$49.00/$53.00	Vol(s). _____	Vol. _____
Annual Review of CELL BIOLOGY				
Vol. 1	(1985)	$27.00/$30.00		
Vols. 2-3	(1986-1987)	$31.00/$34.00		
Vol. 4	(avail. Nov. 1988)	$35.00/$39.00	Vol(s). _____	Vol. _____

ANNUAL REVIEWS SERIES	Prices Postpaid per volume USA & Canada/elsewhere	Regular Order Please send:	Standing Order Begin with:
		Vol. number	Vol. number

Annual Review of COMPUTER SCIENCE

Vols. 1-2	(1986-1987)................$39.00/$42.00		
Vol. 3	(avail. Nov. 1988)............$45.00/$49.00	Vol(s). _____	Vol. _____

Annual Review of EARTH AND PLANETARY SCIENCES

Vols. 1-10	(1973-1982)................$27.00/$30.00		
Vols. 11-15	(1983-1987)................$44.00/$47.00		
Vol. 16	(avail. May 1988)............$49.00/$53.00	Vol(s). _____	Vol. _____

Annual Review of ECOLOGY AND SYSTEMATICS

Vols. 2-16	(1971-1985)................$27.00/$30.00		
Vols. 17-18	(1986-1987)................$31.00/$34.00		
Vol. 19	(avail. Nov. 1988)............$34.00/$38.00	Vol(s). _____	Vol. _____

Annual Review of ENERGY

Vols. 1-7	(1976-1982)................$27.00/$30.00		
Vols. 8-12	(1983-1987)................$56.00/$59.00		
Vol. 13	(avail. Oct. 1988)............$58.00/$62.00	Vol(s). _____	Vol. _____

Annual Review of ENTOMOLOGY

Vols. 10-16, 18-30	(1965-1971; 1973-1985)........$27.00/$30.00		
Vols. 31-32	(1986-1987)................$31.00/$34.00		
Vol. 33	(avail. Jan. 1988)............$34.00/$38.00	Vol(s). _____	Vol. _____

Annual Review of FLUID MECHANICS

Vols. 1-4, 7-17	(1969-1972, 1975-1985)........$28.00/$31.00		
Vols. 18-19	(1986-1987)................$32.00/$35.00		
Vol. 20	(avail. Jan. 1988)............$34.00/$38.00	Vol(s). _____	Vol. _____

Annual Review of GENETICS

Vols. 1-19	(1967-1985)................$27.00/$30.00		
Vols. 20-21	(1986-1987)................$31.00/$34.00		
Vol. 22	(avail. Dec. 1988)............$34.00/$38.00	Vol(s). _____	Vol. _____

Annual Review of IMMUNOLOGY

Vols. 1-3	(1983-1985)................$27.00/$30.00		
Vols. 4-5	(1986-1987)................$31.00/$34.00		
Vol. 6	(avail. April 1988)............$34.00/$38.00	Vol(s). _____	Vol. _____

Annual Review of MATERIALS SCIENCE

Vols. 1, 3-12	(1971, 1973-1982)............$27.00/$30.00		
Vols. 13-17	(1983-1987)................$64.00/$67.00		
Vol. 18	(avail. August 1988)...........$66.00/$70.00	Vol(s). _____	Vol. _____

Annual Review of MEDICINE

Vols. 1-3, 6, 8-9	(1950-1952, 1955, 1957-1958)		
11-15, 17-36	(1960-1964, 1966-1985)........$27.00/$30.00		
Vols. 37-38	(1986-1987)................$31.00/$34.00		
Vol. 39	(avail. April 1988)............$34.00/$38.00	Vol(s). _____	Vol. _____

Annual Review of MICROBIOLOGY

Vols. 18-39	(1964-1985)................$27.00/$30.00		
Vols. 40-41	(1986-1987)................$31.00/$34.00		
Vol. 42	(avail. Oct. 1988)............$34.00/$38.00	Vol(s). _____	Vol. _____